Geo-Economy of the Future

Elena G. Popkova · Bruno S. Sergi
Editors

Geo-Economy of the Future

Sustainable Agriculture and Alternative Energy

Volume I. Green Growth, Environmental Protection and Sustainable Agriculture in the Geo-economy of the Future

Springer

Editors
Elena G. Popkova
MGIMO University
Moscow, Russia

Bruno S. Sergi
University of Messina
Messina, Italy

ISBN 978-3-030-92302-0 ISBN 978-3-030-92303-7 (eBook)
https://doi.org/10.1007/978-3-030-92303-7

This Springer imprint is published by the registered company Springer Nature Switzerland AG
The registered company address is: Gewerbestrasse 11, 6330 Cham, Switzerland

Introduction to Volume I: Green Growth, Environmental Protection and Sustainable Agriculture in the Geo-economy of the Future

Sustainable Development of the Geo-economy and Environmental Management Best Practices

This book opens a new perspective on geo-economy, which for the first time means not economic geography and globalization (differences in the sociocultural and economic situation of countries—from the standpoint of the humanities), but land economics (from the standpoint of earth and planetary sciences). This makes it possible to systematize, in-depth study, and critically rethink current and future economic practices from the standpoint of consequences for the environment and sustainable development.

The uniqueness of this book lies also in the fact that, in contrast to the established environmental economy in science, which focuses on the problems of climate change and the reduction of production and consumption waste, the spotlight of this book is a geo-economy with a focus on responsible environmental management, sustainable agriculture, and alternative energy. Thus, this book fills a gap in the existing scientific knowledge at the intersection of economic geography and environmental economy, opening a new and wide field for scientific research on land economics.

The relevance of this book is explained by the fact that in the area of the new perspective of the geo-economy that has opened thanks to it, there is a wide range of acute, poorly studied, and unsolved issues. These issues include uncertainty about how to make the transition to a green economy and how to manage environmental management. The theoretical concept of the green economy is well developed and widely represented in the available scientific literature. However, it is not clear how to implement this concept in practice. The core of this concept is presented in the Sustainable Development Goals: SDG 12 (responsible consumption and production) and partly SDG 3 (in terms of the contribution of the green economy to health care).

This concept covers a wide range of areas of economic activity, because of which it has blurred lines, is not understood, and is not supported by many potentially interested economic entities in its implementation. For example, it is unclear whether the disposition of green investments is sufficient to qualify an enterprise as a subject of a green economy. The concept of a green economy does not give a clear and unambiguous answer to this question—the answer must be flexible and requires further in-depth scientific study from the standpoint of the geo-economy. For example, if an enterprise creates a large carbon footprint and causes serious damage to the environment, and its green investments are aimed at eliminating only superficial negative consequences of this damage, it is obvious that it cannot and should not belong to a green economy.

It also raises the question of what is meant by green investment. If the selection criteria are set too high, green investments will become unavailable for small and medium-sized enterprises (i.e., for mass implementation), and if the criteria are set too low, large businesses will formally approach green investments without actually contributing to the sustainable development of geo-economy. Consequently, the criteria should be flexible and take into account the specifics of different business practices and different business entities.

Among the problems under consideration is the ambiguity in the interpretation of the concept of "environmental quality," as well as the lack of a scientific and methodological approach to assessing and ensuring the environmental efficiency of economic practices. This concept is based on SDGs 13 (combating climate change), and 14 and 15 (biodiversity conservation). These concepts are based on SDG 12 (responsible consumption and production) and partially SDG 3 (in terms of the contribution of the "green" economy to health care). In particular, questions about whether and how to separate or combine environmental quality from the overall quality of goods and services need to be considered.

Likewise, it is worth considering whether environmental efficiency is viewed in isolation (and how to measure it in isolation) or in unity with the overall performance of an enterprise, country, and region. Although there are separate criteria for measuring and categorizing the given characteristics, coherent methodological support has not been developed yet. For practical application, the criteria and categories need clarification and should be supplemented with specified and detailed recommendations.

Separately, it is worth noting such an issue as the incompleteness of the concept of sustainable agriculture and the development of rural areas. This concept is based on SDG 2. In some cases, agricultural sustainability refers to the stability of the operation (normal, smooth functioning) of agricultural enterprises; in other cases, it refers to their contribution to the fight against hunger and ensuring food security.

Much less studied, but no less significant is the third area of sustainable development of agriculture and rural areas, namely the responsible land use. Although the responsible environmental management in the context of studying a new perspective of geo-economy applies to all sectors of the economy, it is most significant for agriculture, since, firstly, its conduct presupposes the closest naturality, and, secondly, agriculture has the greatest potential not only to reduce environmental damage but also to improve its condition based on restorative land use.

The described problems are comprehensively considered and studied in detail in the first volume of the book, and their promising solutions are also proposed. The purpose of the first volume of the book is to develop the scientific concept of the geo-economy of the future and to develop applied recommendations for its practical implementation while systematically considering the experience and prospects of sustainable green growth, environmental protection, and sustainable agriculture. The problems noted have determined the logic and structure of this volume of the book—they are sequentially explored in the first three parts (first, second, and third).

The first part is devoted to the green economy and environmental management with a detailed consideration of international experience (e.g., countries such as Russia and China) and industry practices. They also consider the health benefits of sustainable development of geo-economy and take into account the experience of sustainable development of geo-economy in the context of the COVID-19 pandemic and crisis. The second part examines the environmental quality and environmental efficiency of business practices. The third part explores sustainable agriculture and rural development.

The original perspective of the study of geo-economy provides a multidisciplinary approach to this book, due to which we hope it might be of interest and useful for representatives of various fields of science, including economic geography, environmental economics, agricultural sciences, business economics, regional economics (in terms of rural tourism and rural development), as well as the earth and planetary sciences.

Introduction to Volume II: Transition to the Geo-economy of the Future based on Responsible Nature Use, Sustainable Communities and Alternative Energy

A Scientific View of the Transition to the Geo-economy of the Future

The geo-economy of the future develops quickly but faces many problems on its path. One of the important problems is the lack of a clear idea of how the state regulation of sustainable development of geo-economy should be carried out, as well as what technologies should be used for responsible environmental management. This indicates a lack of scientific knowledge on how to build a system of state and corporate governance for the sustainable development of the geo-economy. This concept is based on SDG 16 (rule of law, justice, and strong institutions). As the practical experience of several countries and regions shows, often the state and corporate governance of geo-economy are in an imbalance—they are carried out separately and contradict each other, reducing overall efficiency.

We should also pay attention to such a problem as distancing from the practice of the existing concept of responsible industries, sustainable regions, cities, and communities. The basis of this concept is SDG 11. Issues such as empirical and engineering support—the development of guidelines for action on the organization of responsible environmental management by business entities and territories—deserve attention. Also, the accumulated experience of establishing responsible industries, sustainable regions, cities, and communities needs to be collected, generalized, and broadcasted among the academic community and beyond.

And finally, the last (but no less significant) of the considered issues is related to the incompleteness of the concept of sustainable and alternative energy. The core of this concept is SDG 7. In relation to this concept, the question of how to establish and institutionalize selected practices of using alternative energy and how to achieve a full-scale transition to sustainable energy deserves attention.

The existing concept offers only particular recommendations, most of which are academic. We need practice-oriented and universal recommendations, as well as ready-made frameworks. Also, the question of how to combine clean energy with other priorities for sustainable development of geo-economy, for example, environmentally responsible agriculture, needs to be considered.

The above problems are studied comprehensively and solved in Volume II of this book. The goal of Volume II is to form a systemic scientific view at the transition to the geo-economy of the future based on responsible nature use, sustainable communities, and alternative energy. This goal predetermined the structure of Volume II, which consists of three parts (fourth, fifth, and sixth). The fourth part focuses on responsible industries, sustainable regions, cities, and communities. The fifth part explores sustainable agriculture and rural development. The sixth (final) part of the book studies sustainable and alternative energetics.

Volume II (similarly to Volume I) is multidisciplinary—it contains studies that would be of high interest for representatives of economic geography, corporate economics, regional economics (in the aspect of sustainable development of cities and regions), sociology and social sciences (in the aspect of sustainable communities), management (in particular state and corporate governance), and earth and planetary sciences.

Contents

Environmental Quality and Ecological Efficiency of Economic Practices

Environmental Quality and Ecological Efficiency of Economic Practices

Green Economy and Ecological Management: International Experience and Sectorial Practices

Scenario Approach to the Emergence of Green Economy in Russia

Ruslan H. Ilyasov, Gilyan V. Fedotova, and Irina S. Larionova

Abstract *Purpose/objectives*: the purpose of this article is to study the processes of development of science and technology, their implementation in the agro-industrial complex based on forecasting and modeling the future state of the green economy in the context of worldwide loss of biodiversity. The widespread transition to the principles of a green economy carried out by the most advanced countries of the world dictates the need to search and modernize the existing technological structure towards expanding the usage of innovative technologies and production mechanisms. *Methodology*: in this investigation general scientific methods of analysis, synthesis, genesis of new knowledge, methods of modeling and forecasting the development of future events, methods of statistical and economic data analysis, methods of regulatory analysis of strategic documents and methods of generalization, analogy and comparison of initial data were used. *Results*: the study made it possible to present all ongoing innovation trends in the main sectors of the agro-industrial complex such as crop production, livestock, processing and fishery, systematically and comprehensively. Moreover, this study allowed assessing the significance of the main factors affecting the change in the state of industries' development and analyzing the main target indicators of the elaborated development scenarios of agriculture in the context of innovative search and expansion of the technological component in the production process. In addition, the investigation allowed forming recommendations for the application of development scenarios in modern planning and management conditions. *Conclusions/Relevance*: scenarios are an effective tool for long-term strategic planning of innovative development of agriculture, therefore, it becomes necessary to study scenarios and assess the factors affecting their development. In addition,

R. H. Ilyasov
Chechen State University, Grozny, Russia

G. V. Fedotova (✉)
Volga Region Research Institute of Production and Processing of Meat and Dairy Products, Volgograd, Russia

Plekhanov Russian University of Economics, Moscow, Russia

I. S. Larionova
Moscow State Academy of Internal Medicine and Biotechnology Named After K.I. Scriabin, Moscow, Russia

E. G. Popkova and B. S. Sergi (eds.), *Geo-Economy of the Future*,
https://doi.org/10.1007/978-3-030-92303-7_1

the article examines the main global challenges (economic, environmental, social, and technological) of the formation of a green economy of the future. Development scenarios of the agricultural industry have been developed basing on these challenges.

Keywords Scenarios · Innovations · Biodiversity · Green economy · Agro-industries

JEL Code P42 · Q01 · Q13 · Q18

1 Introduction

The modern economy is functioning in the context of recurring crises, the growth of global challenges and threats, which entails the need to search for new approaches and concepts to form a future model of sustainable development. The practiced today model of the consumer attitude towards the environment, the constant consumption of natural and biological resources up to their complete disappearance can no longer provide all growing needs of mankind to support life on the planet. In fact, today we can observe the depletion of the ecosystem, that support functioning of all life-forms on our planet caused by anthropogenic impact and predatory attitude of man to nature.

Many international organizations such as UN, UNIDO, WMO, World Bank and others are concerned about the deterioration of the state of the environment and depletion of natural resources around the world, which is confirmed by many international conferences and intergovernmental agreements signed during the 20-21st centuries. As part of these activities, the main global threats to the world community were identified (Fig. 1).

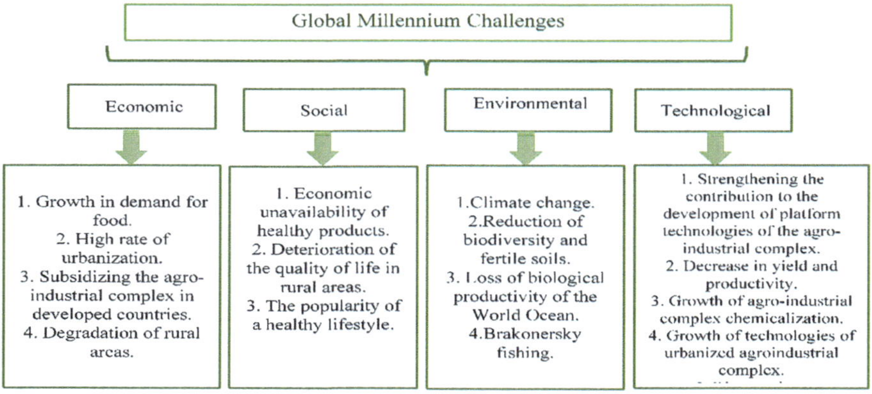

Fig. 1 Global threats to the development of the world agro-industrial complex in the twenty-first century. *Source* Developed and compiled by the authors

Modern humanity is constantly facing crises of the most diverse nature, from political to economic. Some crises are caused by the growth of social instability, but a number of crises are directly caused by the deterioration of the climatic conditions of life on the planet and the reduction of biological and natural resources. Along with the growth of the Earth's population and the demand for agricultural raw materials, especially in the countries of Africa, Asia, Latin America, which do not have sufficient potential for food production and the provision of social conditions for life, there is a reduction in agricultural areas suitable for land use and grazing animals (Sukhdev et al., 2010). If serious measures will not be taken to stabilize the situation, a world social crisis caused by the struggle for survival between developed and poor countries may occur. Therefore, it is necessary to search for innovative mechanisms for the future development of the world agro-industrial complex and the transition to a model of a green economy.

2 Materials and Methods

The innovative development of the agro-industrial complex is aimed at preserving and restoring the existing biodiversity of the planet. The latest advances in science and technology make it possible to maximize harvest and productivity, reduce production losses and losses during storage of agricultural products, thereby reducing the pressure on the environment. The growing demand for food raw materials from a number of developing countries requires the search and development of new fundamentally new technological approaches to production processes. The emerging green economy in a number of EU countries, such as Scandinavia and the United States, makes it possible to achieve high yields without burdening ecosystems with additional production costs. Other countries of the East and Asia with limited agricultural resources have already adopted this approach. In the future, there will be an absolute change in the existing paradigm of management; therefore, a timely transition to a green platform will help to maintain its competitive position in the global market and to ensure the safety of the environment.

For several years, scientists from a number of countries have been working on the phenomenon of the green economy and the preservation of the planet biodiversity at the existing level of anthropogenic impact on the Earth's ecosystem. The existing scientific debate about the need to reduce the area of human impact and restore the damaged ecosystem is reflected in the works of the following economists (Chapple, 2008; Fedotova & Slozhenkina, 2020; Fedotova et al., 2019; Plotnikov et al., 2015; Qu et al., 2001; Terentyev, 2011). Nevertheless, these issues need to be considered, since the world community has recognized them as priorities for sustainable development at the leadership level of many countries.

3 Results

The concept of green economy appeared in 2011 within the framework of the UNEP report, when the international document (UNEP, 2011) was presented to the world. Since that moment, this term has firmly entered the international usage and has become a new conceptual approach that will allow many countries of the world to overcome poverty and reduce the pressure on the environment. The peculiarity of the green economy is manifested in such features as low-carbon, energy efficiency, environmental friendliness, which allow achieving economic growth without wasting natural resources.

The most advanced countries of the European Union, such as Scandinavia and the United States, have been implementing the transition to a green platform (reducing gas emissions by 20%, increasing energy efficiency by 20%, increasing the level of renewable energy sources up to 20%) for several years. In fact, these goals are aimed at reducing the consumption of carbon fuels and replacing them with biofuels. According to economists, the cost of this transition should be at the level of 2% of world GDP, which will significantly change the approaches to the development of the world economy (Table 1).

Table 1 shows the effect of the development of traditional "brown" and "green" economies from investing in the development of additional funds in the amount of 2% of GDP. We see that, unfortunately, until 2020 the green economy does not show positive growth, but after 2020 a certain progress is visible in comparison with the "brown" one. To ensure this growth, a number of conditions must be met:

- creation of an effective regulatory mechanism and management system;
- growth of public investment in the industry of the green economy,
- organization of state financing in the extractive industries,
- use of market instruments to stimulate the green economy,
- investments in the formation of human resources for the green economy,
- strengthening of the international environmental cooperation.

First of all, it should be noted that the green economy will be based on the latest achievements of science and technology, fundamentally new technologies and approaches to production, including in the agro-industrial complex. Therefore, high innovative activity and readiness of the traditional economy will be an important factor for a quick and high-quality transition to a green platform. If we consider the rating of countries by the share of organizations in the processing industry that are introducing technological innovations into the production process, we see that Russia lags significantly behind many countries of the world (Fig. 2).

The level of innovation activity of Russian companies 10.2 proves that there is a big problem in the commercialization of innovative developments in the country. The low level of this indicator is a serious obstacle to maintaining the competitiveness of Russian goods in the world food market. The main problem

Table 1 The results of allocating 2% of GDP to turn the world economy green versus the results of investing 2% of GDP into a regular one

	2011	2015		2020		2030		2050	
Index		Brown	Green (%)	Brown	Green (%)	Brown	Green (%)	Brown	Green (%)
GDP in US dollars	69.344	79.306	−0.8	92.583	−0.4	119.307	2.7	172.049	15.7
GDP per capita	9.992	10.959	−0.8	12.205	−0.4	14.577	2.4	19.476	13.9
Employment (million people)	3.187	3.419	0.6	3.722	−0.6	4.204	−1.5	4.836	0.6
Calories Per Capita	2.787	2.857	0.3	2.946	0.3	3.050	1.4	3.273	3.4
Absorb forests (billion ha)	3.94	3.92	1.4	3.89	3.2	3.83	7.9	3.71	21
Water requirement (km^3/ year	4.864	5.275	−3.7	5.792	−7.2	6.784	−13.2	8.434	−21.6
Landfill (billion tons)	7.88	8.4	−4.9	9.02	−15.1	10.23	−38.3	12.29	−87.2
Emissions to biological capacity	1.51	1.6	−7.5	1.68	−12.5	1.84	−21.5	2.23	−47.9
Energy requirement (million tons / year)	12.549	13.674	−3.1	15.086	−9.1	17.755	−19.6	21.687	−39.8
Share of renewable energy in demand (%)	13	13	15	13	17	12	19	12	27

Source Compiled by the authors based on (https://geoline-tech.com/smartfarm/, https://toeplitz.ru/hydro/prognoz-razvitija-rynka-gidroponiki.html)

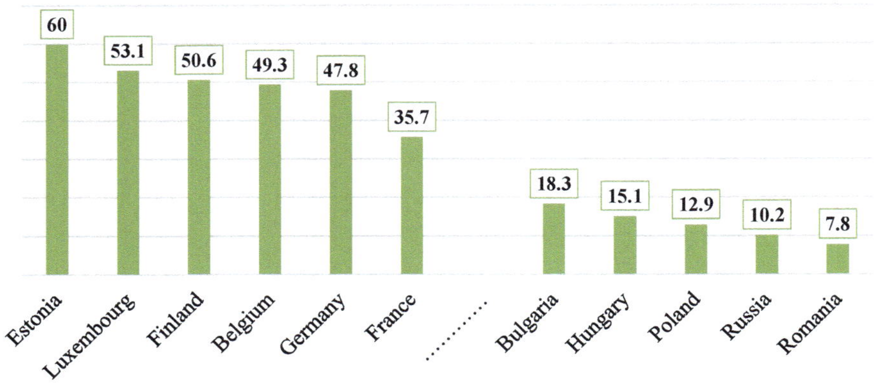

Fig. 2 The share of innovatively active companies in the processing industry by countries of the world, in 2018. *Source* Compiled by the authors based on (https://habr.com/ru/company/dca/blog/267361/)

lies in the absence of close and effective interaction between the state, business and the sphere of science and education. In Russia, innovations in agriculture are introduced inertially and partially, as pilot projects at the level of individual large agricultural holdings. To intensify the innovative development of the agro-industrial complex, the report (https://fb.ru/article/261672/blokcheyn---eto-kak-rabotaetblokcheyn-preimuschestva-primenenie-perspektivyi, http://mcx.ru/upload/iblock/264/264dfabe7e526b6a79ffe5697c34ed4f.pdf) was developed and approved. Within the framework of this forsight two development scenarios were adopted: Local growth and Global breakthrough. The listed development scenarios as their ultimate goal set an increase in the share of Russian exports of agricultural products to a level from 1.5% (Local growth) to 3–4% (Global breakthrough). After 2020, these scenarios will differ significantly in the dynamics of the development of industries. The scenarios will ensure the high competitiveness of Russian products through the use of the latest scientific achievements and innovative developments. Due to the growth of the competitiveness of the agro-industrial complex, the following tasks will be solved:

- expansion of external product markets for Russian products,
- reduction of the country's import dependence on foreign food,
- increasing the level of food security of the country,
- creation of new jobs in the agro-industrial complex, which entails an increase in the quality of life in rural areas,
- growth of investment attractiveness of the Russian agro-industrial complex, including for foreign investors,
- additional financing to the state budget.

Under the scenario of Local growth, a sustained increase actually in all crops and livestock production will be achieved. The growth rates will not be high, but the

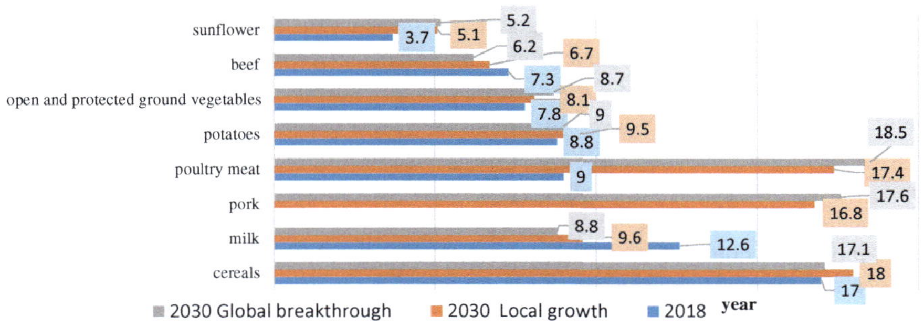

Fig. 3 Structure of the Russian agro-industrial complex by 2030 according to development scenarios in comparison with 2018, shares. *Source* Compiled by the authors based on (Fedotova & Slozhenkina, 2020; https://habr.com/ru/post/149593/)

overall positive dynamics of development will be observed. Cereals, in particular wheat and corn, will play a key role.

According to the scenario of Global breakthrough, an explosive growth of grain production is initially ensured, and then the growth rate slows down, but the overall positive dynamics remains. In parallel, the dynamics of livestock production will increase, with the exception of milk production. In both scenarios, this indicator does not demonstrate a positive growth rate, which requires a more detailed study of the problems of dairy farming in the country.

The presented scenarios are aimed at intensifying agricultural production; finally they will shape the structure of the future agro-industrial complex of Russia in different ways (Fig. 3).

Thus, under the scenario of Local growth, by 2030 the volume of grain crops should amount to 18% of the total production volume, the share of such livestock industries as pig and poultry farming will be high, while livestock breeding will decline. Under the scenario of Global breakthrough, the share of cereals will be 17.1%, while the share of livestock products will significantly increase (pork production—17.6%, poultry—18.5%). Therefore, it can be noted that the scenarios are focused on various branches of the agro-industrial complex, scenario of Local growth—on crop production, scenario of Global Breakthrough—on animal husbandry.

To achieve these goals, first of all, significant state support of agricultural producers is obvious, as well as the formation of infrastructure and the development of an institutional environment for efficient production and processing of agricultural raw materials. In addition, it is necessary to eliminate the contradictions in legislative acts and to create conditions for transparency and an effective mechanism of state regulation (http://threshold21.com/integrated_planning/tools/T21/timeline.html). Of course, these transformations require investment resources that can be provided by introducing a system of preferences and benefits for investors.

Today, the branches of the domestic agro-industrial complex should be export-oriented, since access to world markets will allow the Russian economy to leave the energy-dependent economic model of functioning. Selling your own energy resources does not seem to be a promising and effective policy, since resources are depleted and export potential disappears. For this purpose it is necessary to supply final, processed products to world markets. The presence of huge land resources allows Russia to increase the volume of agricultural land, develop them effectively and provide the world market with agricultural raw materials.

The implementation of the established goals of the innovative development of the agro-industrial complex should be ensured considering the interests of all parties. The future state of the agro-industrial complex can be described with the following parameters:

- increase in the share of Russian agricultural products on world markets,
- increase in the share of finished food products of domestic production on the internal food market,
- increase in the share of domestic goods on the capital goods market,
- increase in the share of innovative products in the total volume of agricultural products,
- increasing the level of implementation of the advances in science and technology in the agricultural sector,
- reduction of the unemployment rate in rural areas.

4 Conclusion

At the end of the study, it should be noted that in the modern conditions of modernization and innovative development of the agro-industrial complex, the main trends in industries can be:

- smart agriculture,
- search for new technologies for the production of agricultural products,
- transition to a new paradigm of production and attitude to the environment,
- formation of an agricultural export-oriented infrastructure,
- development of digital platforms in order to ensure interactions between participants of the process of agricultural production,
- training specialists with relevant competencies in the field of agro-industrial complex and innovative development,
- maximum greening of the economy and the formation of prerequisites for a sustainable transition to a green economy.

The presented scenarios for the innovative development of the agro-industrial complex allow looking at the possible options for the development of industries by 2030 in a comprehensive manner. Undoubtedly, the growth of the technological component in the production process of traditional agricultural production is aimed at increasing productivity while minimizing the pressure on the environment. Basing on the scenario approach it is possible to form an industry-specific system for modeling the state of the biosphere and its biodiversity under various anthropogenic loads on the environment in the future. The paradigm shift and the adoption of the green economy concept make it possible to search for and develop new solutions to complete routine tasks and maximize efforts at the current level of costs.

The scenario approach will allow the participants of this process to monitor regularly existing global trends in the development of the world market, to evaluate the latest achievements of science and technology and their applicability in Russian conditions, regularly make adjustments to the development of the industry taking into account environmental changes, and to strengthen the country's agricultural potential by world market, to ensure the integration of the science and education system with the agro-industries, namely, to form horizontal and vertical links of the information-analytical system for forecasting.

References

"Smart Farming": An overview of leading producers and technologies. Available at: https://geoline-tech.com/smartfarm/ (in Russ).

Big Data A to I. Part 1: Principles of the work with Big Data, paradigm MapReduce DCA Blog (Data-CentricAlliance), BigData. Available at: https://habr.com/ru/company/dca/blog/267361/ (in Russ).

Blockchain is... How blockchain works, benefits, application, prospects. Available at: https://fb.ru/article/261672/blokcheyn---eto-kak-rabotaetblokcheyn-preimuschestva-primenenie-perspektivyi (in Russ).

Chapple, K. (2008). *Defining the green economy: A primer on green economic development*. University of California Center for Community Innovation, November. Available at: http://communityinnovation.berkeley.edu/reports/Chapple-DefiningtheGreenEconomy.pdf

Fedotova, G., Gorlov, I., Oschenkina, M., & Glushchenko, A. (2019). Trends in scientific and technical development and increase of competitiveness of Russian agriculture. *Bulletin of the Academy of Knowledge, 3*(32), 251–255. (in Russ).

Fedotova, G., & Slozhenkina, M. (2020). Influence of climatic changes on the structure of the world agro-industrial complex. News of the South-West State University. Seri*es: Economics. Sociology. Management, 10*(3), 23–35.

Forecast of scientific and technological development of the agro-industrial complex of the Russian Federation for the period up to 2030. Available at: http://mcx.ru/upload/iblock/264/264dfabe7e526b6a79ffe5697c34ed4f.pdf. Accessed: 11 Nov 2020.

Global forecast for the development of the hydroponic market up to 2025. Available at: https://toeplitz.ru/hydro/prognoz-razvitija-rynka-gidroponiki.html (in Russ).

Historical Development and Applications of the T21 Model.: Millennium Institute (2012). Available at: http://threshold21.com/integrated_planning/tools/T21/timeline.html. Accessed 11 Nov 2020.

Internet of Things - what is it? Available at: https://habr.com/ru/post/149593/ (in Russ).

Plotnikov, V., Fedotova, G., Popkova, E., & Kastyurin, A. (2015). Harmonization of strategic planning indicators of territories' socioeconomic growth. *Regional and Sectoral Economic Studies, 15-2*(July–December), 105–114.

Qu, W., Barney, G., Symalla, D., & Martin, L. (2001). The Threshold-21: National sustainable development model. Integrated global models of sustainable development: Encyclopedia technology, information, and systems management resources: Part of Encyclopedia of Life Support Systems (EOLSS). In Akira Onishi (ed). *Oxford (UK): Developed under the Auspices of the UNESCO*, EOLSS Publ, II (pp. 78–87).

Sukhdev, P., Stone, S., & Nuttall, N. (2010). *Green economy developing countries success stories*. United Nations Environment Programme, 26 p.

Terentyev, N. (2011). *Climatic risks and "green" technologies: New factors in the development of companies*. Scientific works. Institute for Economic Forecasting RAS (pp. 115–135).

UNEP. (2011). *Towards a green economy: Pathways to sustainable development and poverty eradication—A synthesis report for government officials*. Available at: http://old.ecocongress.info/5_congr/docs/doklad.pdf. Accessed 11 Nov 2020.

On the Improvement of the Environmental Management System in China

Viktor N. Ovchinnikov and Jia Hao

Abstract Recently, Chinese science and technology have developed dramatically. The socio-economic standard of living improved significantly. Yet, the ecological situation is still severe. To ensure the sustainable development of the economy and social sphere, one must pay due attention to the improvement and protection of the environment. In this paper, we present the current state of the ecological situation management system in China. We analyzed the strategic measures of China on environmental management and its development trends.

Keywords Environmental management · Environment · Strategic measures · Trends in environmental management

JEL Codes Q57 · F52 · E65

1 Introduction

The environment is a general term denoting the organisms, land, water, and climatic resources closely related to human life and its development. Ecological problems caused by inefficient environmental management negatively impact human activities. There are two causes of environmental problems.

The first is the irrational development and use of natural resources. The second is the violation of ecological imperatives in industrial and economic activity.

In China, the current problems in environmental management are: insufficient investment, shallow awareness of environmental protection, unsubstantiated research and development, and inefficient use of resources. However, taking strong measures contributes to rapid progress in environmental protection.

Today, China has chosen the path of sustainable development, improving the ecological environment, and implementing social economy principles. Faced with global environmental problems, China addresses them by improving the legal system

V. N. Ovchinnikov (✉) · J. Hao
Southern Federal University, Rostov-on-Don, Russia
e-mail: vovchinnikov@sfedu.ru

E. G. Popkova and B. S. Sergi (eds.), *Geo-Economy of the Future*,
https://doi.org/10.1007/978-3-030-92303-7_2

and the management system and actively promoting international cooperation in the field of environmental protection.

Environmental protection has become the focus of the global community and expert discussion. This is related to the development of long-term monitoring of global environmental changes.

2 Materials and Methods

In this study, we used the heuristic-cognitive potential of the evolutionary-genetic approach and the system-functional analysis of economic phenomena and processes. This allowed us to trace the evolution of the environmental management system and the state of the environment in China, as well as to identify the directions of its improvement in the new circumstances.

3 Results

3.1 Condition of the Existing Environmental Management System in China

Today, China's environmental protection system has several shortcomings.

3.1.1 Poor Public Awareness of Environmental Protection

Recently, China's economic development has accelerated, which resulted in significant deterioration of the environmental situation. One of the reasons for this is the lack of public awareness (including the business community) on the state of environmental protection. This hinders the adoption of effective measures for its improvement. Society has no clear and complete understanding of the close connection between economic interests and environmental protection. This fact affects the sustainability of the overall societal development (Su, 2011).

3.1.2 Poorly Substantiated Ways of Using Natural Resources

Rapid societal development and increased use of natural resources entail deterioration of the environment, worsened further by an insufficient justification of environmental safety measures. Today, advanced enterprises have realized the importance of environmental protection and begun to take measures to protect the environment, control and monitor its pollution. However, this format cannot meet all the needs of environmental protection.

3.1.3 Low Investment into Environmental Protection

Lack of financial resources in environmental protection funds is a key factor limiting environmental rehabilitation scope (Su, 2011). In the long term, significant financial resources are required to protect the environment (Yang, 2014). However, the financial needs of environmental protection are currently unmet, which hinders the implementation of environmental projects.

3.2 The Reaction of China to Global Challenges in Ecology

3.2.1 Ecological Orientation of the Sustainable Development Strategy in China

Environmental protection is one of the most important tasks of China's national policy. Environmental policy is focused on ensuring the simultaneous planning, implementation, and development of ecological construction with the aim of achieving the unity of economic, social, and ecological goals. The 13th Five-Year Plan emphasizes the concepts of development based on innovation, coordination, environmental friendliness, openness, and sharing of natural resources. This strategy aims to ensure the well-being of people and promote their all-around advancement, with the ultimate goal of intensifying environmental policy measures.

3.2.2 Improvement of the Institutional-Legal Framework for the Environmental-Economic Management

China is currently developing a legal system for promoting green economy; its institutional platform is based on the "Constitution of the People's Republic of China," the "Environmental Protection Law of the PRC," the "Water Pollution Prevention Law," the "Air Pollution Prevention Law," the "Solid Waste Pollution Prevention Law," the "Marine Environmental Protection Law," and the "Forest Law" (Su, 2011).

China has prioritized environmental law enforcement, opened information channels on environmental problems, and strengthened public disclosure of environmental violations. Moreover, it has established environmental management systems and legislative oversight of the environmental situation by the National People's Congress. The government and local authorities are responsible for implementing environmental policy measures; their environmental protection departments exercise oversight and management functions.

3.2.3 Strategic Transformation of the System for Pollution Prevention and Comprehensive Improvement of the Urban Environment

By creating a socialist market-economic system, China changed its development strategy, preferring clean production models and moving towards sustainable development. A "triad of transformation" was defined. In industrial production, there was a shift from an emphasis on preventing pollution at the final stage of manufacturing to controlling the entire production process. In controlling emissions—a shift from controlling the high concentration of one ingredient to full control of all pollutants. In pollution control—a shift from the decentralized control of specific sources to a combination of centralized and decentralized control (Yang, 2014).

The proposed algorithm for implementing the strategy includes preparing a general city plan and adjusting the planning functions of the city hall, as well as including its environmental protection and urban environmental improvement activities into the overall city planning (Guo, 2015).

In compliance with the general plan, many cities are adjusting the industrial layout according to the functional zoning during the transformation of old areas and the development of new areas. Moreover, they intensify control over industrial pollution, separate the areas of industrial zones and areas of residence, control industrial and domestic pollution of the urban environment, and build new residential areas with a reasonable layout and abundant social facilities.

3.2.4 Land Development and Environmental Protection in Rural Areas

The development and improvement of large freshwater basins have yielded tangible positive results. There are plans to continue cleaning, strengthening, and improving the shorelines and diverting floods of major rivers and lakes: the middle and lower reaches of the Yangtze River, the Huanghe, Huai, Haihe, Songhua, and Liaohe Rivers, as well as Taihu Lake. Projects for trans-basin water transfer, "south-to-north water supply," and agricultural land reclamation and reconstruction have been scheduled. Greater attention has been paid to the planning and management of land allocated for residential and industrial development, and the overall extent and structure of these areas have been better controlled. The Chinese government places great importance on ensuring the ecological purity of the hydrosphere and soil. The government promotes nature-oriented agriculture. Recently, the area of forest cover has been greatly increased, effectively preventing the development of soil erosion and improving the ecological environment and farming conditions. The development and promotion of coal-saving technologies have accelerated the spread of natural gas, solar, wind, geothermal, and small hydropower technologies in rural areas. Measures to prevent pollution in rural enterprises have been strengthened (Yang, 2014).

3.2.5 Improving Environment and Biodiversity

In this field, there are plans to organize large-scale planting of trees and intensify afforestation of rural areas to improve protection and rational use of forest resources. There are also plans to improve the protection and management of pasture resources, prohibit indiscriminate overgrazing and excessive cultivation of pastures, introduce a combination of public, collective, and individual forms of their use, and control processes of soil degradation and desertification. For that purpose, ten major forest ecological projects have been developed to "protect and improve the natural ecological environment and ensure sustainable use of resources." China has enacted a series of laws to protect the marine environment, established a national marine environment monitoring and surveillance network, and completed the division of ecological function zones. Moreover, it improved control over continental shelf construction, offshore oil exploration, and marine waste dumping activities; organized work to prevent "red tides" and protect marine fishery resources; and established a national marine reserve.

Chinese government applies a combination of technologies to protect ecological sites and establishes nature reserves, plant breeding facilities, and gene banks. The "Nature Conservation Action Plan of China" and "Biodiversity Conservation Action Plan of China" set the strategy course, important areas and priority projects for the conservation of the country's biodiversity.

3.2.6 Ecology, Technology, Education, and Marketing

Environmental protection research institutes intensified their activities by organizing the selection, evaluation, and promotion of the best environmental protection technologies. There are plans to implement an eco-labeling plan for products, intensify environmental advertising and education activities, promote environmental protection knowledge, raise awareness, and gradually shape the environmental-friendly mentality of the population (Guo, 2015).

These measures are intended to mobilize society for broad participation in environmental outreach and education, and improve the system of professional environmental education to train the necessary number of technological and managerial personnel (Chen, 2020).

3.2.7 Develop International Cooperation in the Field of Environmental Protection

China actively participates in environmental activities conducted by the United Nations, maintains cooperative relationships with environmental organizations worldwide, conducts exchanges and implements cooperation programs on several environmental projects, and is a party to international environmental conventions.

3.3 New Policy of Government Regulation over Environmental Protection

The core idea of the new environmental policy is to take preventive measures, strengthen the system of environmental rehabilitation, and avoid or reduce pollution and damage caused by it (Su, 2011).

3.3.1 Key Environmental Management Measures of the Government

1. Incorporating environmental protection measures in the long-term national economic and social development plan and the annual plan, including measures to prevent cases of environmental pollution and minimize environmental damage.
2. Consistent introduction of environmental impact assessment into construction projects (when the basic facility and environmental infrastructure facilities are designed, built, and put into production) (Zhang, 2016). Today, environmental impact indicators and the implementation of "three simultaneous" national construction projects have reached 95%.
3. Improvement of laws, regulations, and rules on environmental protection, which allows for legalization and standardization of environmental management.
4. Structural units of environmental management shall be established at all levels of government. Specialized departments of the State Council, sectoral agencies, and departments of large and medium-sized enterprises should carry out environmental management activities.

3.3.2 Financial Tools of Ecological-Economic Policy

1. Enterprises have started to allocate funds to prevent pollution, taking them from the plan of fixed capital investment.
2. State environmental protection authorities provide financial support to enterprises; the city administration allocates funds for the construction of environmental protection facilities; the state charges fees for the dumping of pollutants.
3. The state has introduced a preferential income tax on the enterprises with a closed resource cycle: using wastewater, waste gases, residues, and other waste as raw materials (alternatively, such enterprises can be exempt from paying tax for five years).

Tax incentives have been established for companies investing in the construction of wastewater treatment plants; the "greening" investments in fixed assets are exempt from taxes.

Value-added tax exemptions are introduced for construction materials produced from coal rock, fly ash, and other waste, as well as for the production that uses liquid waste and waste used for gold and silver extraction.

Projects that apply for preferential credit for pollution source treatment facilities are exempt from building tax.

Another financial tool is tariff preference: temporary introduction of preferential tariff rates on imported equipment for the treatment of municipal wastewater and wastewater from paper production.

Establishes a 30% tax reduction on the production and sale of cars, SUVs, and small buses that meet global emissions standards (Yang, 2014).

Special tax incentives are introduced for the production of agricultural products. Income from products produced by converting farmland to forests and meadows (hunting and gathering wild fruits, mushrooms, berries) in the western region is exempt from tax for ten years.

3.3.3 Technological Focus on Improving the Efficiency of Natural Resource Use and Reducing Emissions

1. Today, industrial enterprises are actively adopting advanced technologies to improve resource and energy efficiency.
2. Following the adopted laws, enterprises that use outdated technology, use resources inefficiently, or pollute the environment must be retrofitted within a specified time frame.
3. Enterprises are required to use harmless, non-toxic, or low-toxic raw materials and power sources in production.
4. In adjusting the national economic structure, it is important to optimize environmental protection and expand its technological functionality. Following the principles of international cooperation, China strives to ensure national sovereignty, preferring to solve environmental problems by taking into account the real interests of all countries and the long-term global interests (Zhang, 2016).

3.4 Formation of a System of Environmental and Nature Conservation Measures in the New Conditions

3.4.1 Improving the Environmental Protection Awareness System

With the dynamic development of the world economy, the resources of the Earth are used more and more intensively. This leads to depletion of finite resources and deterioration of the ecological environment. In this situation, China should actively pursue a policy of environmental protection, organize events to raise environmental awareness, and start environmental education as early as kindergarten (Li, 2016).

3.4.2 Creation of the Urban Environment Protection System

The rapid development of China's social economy has increased the demand for natural resources, which has intensified the problem of resource conservation and pollution. China faces such environmental problems as car exhausts pollution and industrial waste gases. It is important to organize a comprehensive environmental cleanup, accelerate the development and application of new energy vehicles. Industrial enterprises must adopt resource-saving technologies, minimize environmental pollution, and improve their environmental efficiency. Settlements must modernize the wastewater treatment system (especially industrial wastewater with a high concentration of pollutants), regulate the functioning of sewer networks, and ensure centralized treatment of household wastewater.

3.4.3 Introduction of Innovations in Environmental Management Technology

Environmental protection requires applying advanced technologies. It is important to create green technology research laboratories in the basic sectors of the economy, provide them with financial and political support, expand research on environmental protection technologies, and promote the use of new energy types.

3.4.4 Increasing Attention to Environmental Conservation in Natural Resource Exploitation

In order to achieve the effective use of natural resources, it is necessary to implement strategic sustainability, to develop measures for environmental protection and environmental monitoring, to carry out restoration activities (Peng, 2020). First of all, these are measures to protect biological resources and endangered species, to create rescue stations for endangered animal and plant species, and to establish wildlife protection associations. Second, it is necessary to actively plant forest belts in the habitats of the Northern provinces and at the same time "revert" some of the farmland to forests and lakeshore areas. The state must provide targeted subsidies for this purpose. Another way is to eliminate sources of pollution and restore the ecological environment.

3.4.5 Active Development of a Waste-Free Economy and Sustainable Development

Today China is implementing the circular economy concept (with a closed production cycle), which focuses on saving and reuse of resources, allowing to reduce waste emissions and environmental pollution. It is important to rely on the experience of other countries (Germany, Japan, and the USA), which had started to fight pollution earlier than China. The circular economy concept constitutes an ideology of environmental protection, the creation of "green" industries, and sustainable economic development.

3.4.6 Development of Ecotourism and Strengthening Environmental Protection

At present, the tourism industry in China is developing rapidly, and the environmental problems associated with it are becoming more and more relevant. It is important to develop tourism resources without destroying the environment. To this end, China has introduced fines for destroying the scenic environment. In such areas, it is necessary to organize a special service to broadcast instructions to tourists and install billboards urging them to protect the environment. Along with this, it is necessary to create waste disposal stations and improve sanitary conditions.

These are the main directions for establishing the new environmental policy measures in China (Guo, 2015).

4 Conclusion

Environmental protection is now a global concern, and environmental problems have a negative impact on economic and social development. China actively protects the environment and promotes coordination and unity of economic, social, and ecological goals by implementing sustainable development strategy. Insufficient public awareness of environmental protection, inefficient use of resources, and insufficient investment in environmental protection seriously limit the effect of environmental protection. In the current situation, China should strengthen attention to environmental protection issues, raise public awareness about environmental protection, ensure the rational use of resources, and promote innovation in environmental protection technology.

Management of the ecological environment must encompass two aspects: the first is to solve current ecological problems, and the second is to develop new energy- and resource-saving technologies on the path to sustainable development. The government plays a key role in the creation of the ecological environment (Guo, 2015).

First of all, the government allocates resources between human activity areas according to the needs and possibilities of social, economic, and cultural development of society. Second, it draws up strategic plans for national development and directs human, industrial, natural, and financial capital to improve the environment. Third, the government determines the rules and regulations, incentives, and guarantees in economic environmental management. Finally, the government can use the available funds to encourage public organizations to play an active role in improving the environmental ethics of citizens, raising their awareness, and promoting their environmental culture. In ensuring the economic development of the country, the government must assume environmental responsibility and take effective measures to implement policies that guarantee the well-being of the environment.

The current period is an important stage in the realization of China's strategic opportunities for economic development, as well as the formation of a scientific view of the national development and the harmony between people and nature. Authorities at all levels must take proactive steps to educate on the environmental issues, promote systemic innovation, enforce environmental obligations, and improve the legal protection of the environment. In the course of industrialization and urbanization, there must be a new impetus to intensify environmental protection activities and promote comprehensive progress in all areas of society: politics, economy, ecology, and culture.

References

Chen, A. (2020). Research on the management of ecological civilization education for follege students. *Journal of Taiyuan City Vocational and Technical College*. Retrieved from https://kns.cnki.net/kcms/detail/detail.aspx?dbcode=CJFD&dbname=CJFDAUTO&filename=CSZY202012020&v=JbJmFBz%25mmd2BMf6EYqQ8epORA1ANH3fCLGJ4%25mmd2Fi5jXzDP0yPRoTyy%25mmd2FdrbINyRkK5Gv7KX.

Guo, Y. (2015, April 8). *Research on the construction of ecological civilization in China in the context of synergistic development*. Retrieved from https://kns.cnki.net/kcms/detail/detail.aspx?dbcode=CDFD&dbname=CDFDLAST2020&filename=1016254350.nh&v=th2d6s5LojwMHtnDr3Zhd7bZOOWahRMR2pfsvUBV%25mmd2F6HPMVsvxMoBH3tJWVhfCKDr.

Li, L. (2016). *Research of ecological crisis in the realm of modernity*. Retrieved from https://kns.cnki.net/kcms/detail/detail.aspx?dbcode=CDFD&dbname=CDFDLAST2016&filename=1016126136.nh&v=xn0%25mmd2F5AtetuUmwjd9JyWHrm3%25mmd2FNvlm0mAzJY%25mmd2BDeOeAu%25mmd2BQ7nisksSxvna0vK0GbCaKu.

Peng, J. (2020). Improving management system of ecological environmental protection at the grassroots level. *Environmental Economy*. Retrieved from https://kns.cnki.net/kcms/detail/detail.aspx?dbcode=CJFD&dbname=CJFDAUTODAY&filename=HJJI202024022&v=BdkUe5vJ8FdP5hcCQ8maBlxEZKx4utS2%25mmd2F6oqxk88kj%25mmd2FhVxULQWhlSvdaKiUVsBAR.

Su, L. (2011, June 13). *On the role of government in the development strategy of China's low carbon economy*. Retrieved from https://kns.cnki.net/kcms/detail/detail.aspx?dbcode=CDFD&dbname=CDFD1214&filename=1011211081.nh&v=2q3yV1A0oF1LZTiivmpcisgr8kognjIbZmXBT4iOK%25mmd2F3kLfHKF05ckVMdyuQD87py.

Yang, Q. (2014). *Research on eco-environmental governance in the construction of Ecological civilization in contemporary China*. Retrieved from https://kns.cnki.net/kcms/detail/detail.aspx?dbcode=CDFD&dbname=CDFD1214&filename=1014318357.nh&v=kGza9YO5Chiibrvw%25mmd2B%25mmd2BU3yzYVHzFlb0LY1QyMIKtSlnXCw0I0dL%25mmd2BFLyd2N1oweNfo.

Zhang, Z. (2016, June 4). *On the practice of ecological civilization construction with Chinese characteristics*. Retrieved from https://kns.cnki.net/kcms/detail/detail.aspx?dbcode=CDFD&dbname=CDFDLAST2016&filename=1016077956.nh&v=iaJQR0q35O93DFjS0KcVFZ59pUstwQuvSPr4xM5LkuWaO5o67frrTm%25mmd2BvYZ6De6N5.

Formation of the Green Economy Model in the Region

Andrey V. Glotko, Zhanna V. Mirzoyan, and Zalina L. Kantsalieva

Abstract The research relevance lies in the need to elaborate on the mechanisms of transition of the regional economy to green production. This research is of particular relevance for the agricultural territories of Russia due to the strategic guidelines of the regions until 2035, including the transition to a green economy. The study is conducted on the example of the Republic of Altay. The novelty of the research objectives lies in the development of theoretical and methodological provisions for the implementation of a green economy model at the regional level through the development of specific economic and institutional mechanisms. The scientific novelty lies in the fact that the region becomes a territory of advanced development when the mechanisms of transition to a green economy at the national level have not yet been fully formed.

Keywords Green economy model · Sectoral approach · Tourism · Indicator system

JEL Codes Q5 · Q51 · Q56

1 Introduction

In current conditions, the socio-economic development of countries and their regions is facing the problem of pollution, environmental degradation, excessive consumption of natural resources, and increasing anthropogenic pressure on the biosphere, which ultimately affects climate change and the quality of life. Hence, it becomes necessary to introduce qualitatively new models of economic growth involving the sustainable development of economic activity while increasing the efficiency of consumption of natural resources, ensuring the energy efficiency of production, solving the problems of environmental pollution, and increasing the welfare of the

A. V. Glotko (✉) · Z. V. Mirzoyan
Novosibirsk State University of Economics and Management, Novosibirsk, Russian Federation

Z. L. Kantsalieva
Kabardino-Balkarian State Agrarian University Named After V.M. Kokova, Nalchik, Russia
e-mail: kan-307-zal@list.ru

E. G. Popkova and B. S. Sergi (eds.), *Geo-Economy of the Future*,
https://doi.org/10.1007/978-3-030-92303-7_3

population. Nowadays, one of such models of economic growth is the model of a green economy. This model involves a qualitative change in the production and consumption patterns of the economy, the introduction of green principles in the system of strategic management, and the formation of environmental sustainability of production and consumption (Ivanova & Levchenko, 2017; Kalner, 2013; Karnyshev, 2011; Knyazeva & Kirusheva, 2016; Lipina et al., 2016; Lyaskovskaya & Grigoryeva, 2018; UNEP, 2011; Zakharova, 2015).

In the second half of 2017, the Republic of Altay adopted a draft of the "Strategy for the socio-economic development of the region until 2035," which was formed taking into account the principles of a green economy.

Nowadays, it is no coincidence that the region has returned to the idea of greening the economy from the point of expanding its production capabilities in the areas of organic agriculture, clean energy, tourism, and recreation. At this stage, especially relevant are scientific developments allowing to see the strategic guidelines and implement a model of a green economy, taking into account the current legal, financial, and economic conditions and the achieved level of development of the Republic of Altay.

2 Materials and Methods

The problems of forming a model of a green economy are sufficiently studied by foreign and Russian scholars (Adarina et al., 2019; Frolov et al., 2017; Grand View Research, 2017; Kuttubaeva et al., 2017; Kuzminov et al., 2018; Medvedeva, 2018; Popov & Semyachkov, 2014; World Bank, 2001; Yashalova, 2014).

The main components of the model of a green economy are economic growth, environmental responsibility, and social development, which ensure sustainable development in general. At the same time, the ecological component of a green economy is a priority and stabilizing element. Therefore, we will consider several main components of a green economy. We divide these elements into three groups:

- Components of the model of a green economy involving the formation of environmental sustainability of production and consumption processes environmentally favorable for human activity (green technology, green sectors of the economy, etc.);
- Components of the model of a green economy involving the formation of economic sustainability, favorable economic conditions for production and economic activity, and not worsening the environmental situation (green economic growth, green economic sectors, green investments, green economic mechanisms and tools, etc.);
- Components of the green economy model aimed at ensuring social sustainability and improving the quality of human life, taking into account environmental and economic sustainability (green jobs, green sectors of the economy, green food standards, etc.).

In the transition to a green economy, there is an experience in forming an environmental rating among Russian cities during transition to a green economy. This rating was developed by the consulting company *Ernst & Young* for the Ministry of Natural Resources of the Russian Federation, taking into account the recommendations of the Organization for Economic Cooperation and Development (OECD) (Arustamov, 2017). According to this rating, the Republic of Altay is in the lead. In the rating of cities developed by the consulting company *Ernst & Young*, in 2013, the city of Gorno-Altaysk (the administrative center of the Republic of Altay) took the second place, giving primacy to Moscow.

The published environmental rating of the all-Russian public organization *Green Patrol* ranks the Republic of Altay from 2nd to 4th during 2008–2018. The Ecological and Economic Index, formed in 2012, shows the leadership of the Republic of Altay in the ranking.

3 Results

One of the main directions of developing the model of a green economy in the Republic of Altay is tourism.

Currently, the Republic of Altay puts tourism in the first place, proclaiming it a strategic direction of regional economic development (Table 1).

Table 1 Dynamics of indicators of tourism development of the Republic of Altay in 2014–2018

Criterion	2014	2015	2016	2017	2018
Tourist flow, thousand visits	1500.1	1750.1	1986.3	2049.9	2115.0
of which foreign citizens, thousand people	11.2	12,5	12.0	10.6	13.7
Number of organizations involved in the provision of tourism services, units	793	800	804	907	902
including:					
rural tourism	460	434	434	434	434
accommodation (hotels, tourist camps, etc.)	230	240	279	377	377
Travel agencies	103	126	73	73	73
Tour operators	–	–	18	23	18
Number of accommodations at the tourist facilities (without rural houses), units	12,350	13,222	15,383	17,517	17,517
including year-round facilities	5057	6710	9611	9716	9716
Volume of sold tourist products, million rubles	2750	3200	3600	3900	4000
Tax revenues to the consolidated budget from tourism development, million rubles	70	122.7	98.9	147.2	182.1

Source Compiled by the authors based on Gazukina et al., (2015), Kuttubaeva et al., (2017)

It should be noted that the distribution of tourist flow by municipal districts of the republic is uneven. The most visited areas are the Chemalsky, Maiminsky, and Turochaksky districts. These districts have the most attractive tourist sites, a considerable number of tourist bases, and great transport accessibility. The number of tourist accommodations (excluding rural houses) is gradually increasing. In 2014, there were only 1350 beds. In 2018, this figure increased to 17,517 seats. During the summer tourist season, the accommodation loading reaches 80% on weekdays and 100% on weekends.

Environmental consequences of the development of tourism include the destruction of the local natural environment, reducing the recreational attractiveness of tourist sites, and reducing the attractiveness of the regional tourist product. Moreover, they narrow the opportunities for developing the local economy and generate conflicts within the local community.

There are several other negative consequences in addition to the environmental problems of the development of tourism. The development of tourism in places considered undesirable for tourism by indigenous or traditional communities, as well as the lack of systematic protection and restoration of cultural and historical monuments, reduces the recreational attractiveness of tourist sites in the republic.

Thus, the republic faces a need to reduce the negative effects of tourism and actively develop the potential of green tourism in the socio-economic development of the region.

The potential of green tourism lies in its positive economic and social consequences. Let us note some of the possible consequences:

- The sphere of tourism can ensure the promotion of economic growth in the region through the effective use of local resources (natural, land, labor, etc.), the development of entrepreneurship in the region, the development of related sectors of the economy, and increased tax revenues to the local budget;
- The sphere of tourism can ensure the development of infrastructure in the region, including by attracting investment resources and tax revenues to the local budget;
- The sphere of tourism can provide growth of personal incomes and welfare of the local population, ensuring employment of the local population by creating jobs at tourist enterprises and the implementation of opportunities for self-employment among the population;
- The sphere of tourism can improve the educational and cultural level of tourists and the local population.

Green tourism should be seen not as a separate type of tourism but as a generalized characteristic of all types of tourism meeting the following requirements:

- Competent use of environmental potential, absence of the impact on ecological processes, and contribution to the conservation of nature and biodiversity;
- Optimal use of cultural and historical resources, consideration of the specific socio-cultural features of local communities, and the preservation of the cultural heritage, traditions, and customs of the region;

- Promotion of tourist satisfaction with the quality of tourism products, employment, growth of personal income and tax revenues to the budget, etc.

Consequently, the formation and development of the model of a green economy in the tourism sector of the region should provide conditions for ensuring the environmental safety of the territory and conservating natural landscapes, biodiversity, and water resources. The development of tourism on the model of the green economy provides opportunities for the involvement of the local population in tourism activities to generate income while preserving culture, customs, and traditional way of life. Since the Republic of Altay is marked with a low level of socio-economic development, the introduction of the model of a green economy in the sphere of tourism should act as a factor of economic development.

The main directions for the development of the model of a green economy in the field of tourism at the state level can include the following:

- Providing the necessary information and methodological assistance on the greening of tourism for local executive bodies, tourism enterprises and organizations, and local communities;
- Involving all stakeholders (tourism enterprises, public organizations, and local communities) in the implementation of the model of a green economy in the field of tourism;
- Developing of regulations on the greening of tourism and the development of certain types of tourism in the context of a green economy;
- Developing support systems for local communities, tourist enterprises, and protected areas of Russia implementing programs on green tourism;
- Stimulating the inflow of funds into the sphere of tourism, which, in addition to economic efficiency, would also solve environmental and social problems. Of particular importance is the mobilization of local investment for the implementation of projects on the greening of tourist activities;
- Organizing a training system on the greening of tourism for representatives of the tourism business, local communities, etc.

The Republic of Altay has several protected areas of federal importance—two biosphere reserves (Altay and Katun), as well as the National Park *Sailugemsky*. There are also 50 specially protected natural areas of national importance: four natural parks, two sanctuaries, and 44 natural monuments. The total area of the protected areas in the Republic of Altay is 1223.06 thousand hectares.

The main directions of development of ecological tourism in the Republic of Altay in the context of a green economy should include:

1. Identification of priorities for developing ecological tourism in the strategic documents for tourism development in the region. The main legal act regulating tourist activity in the Republic of Altay is the Law of the Republic of Altay "On Tourism" (December 5, 2008 No. 121-RZ). Nevertheless, this law does not define the concept of green tourism or ecotourism;
2. Development of ecological tourism through the active involvement of representatives of the local community in the formation and implementation of ecological

tourism products while conducting training of the local population and developing a system of state support. This measure will lead to the creation of new jobs, the revival of customs and crafts, the formation of a favorable image of the territory, increasing its investment attractiveness, etc. (Atlas of Emerging Jobs, 2019);
3. Creation and certification of environmental trails and green routes;
4. Organization of a system of training in the field of ecological tourism. Training programs should provide basic skills in organizing tourism activities following the requirements of environmental sustainability. Moreover, these programs should be aimed at creating economic incentives for the local population to preserve the unique objects of nature and develop related industries and tourist infrastructure.

The effectiveness of implementing the principles of green tourism in the region also depends on the development of a system for its promotion. The important issues in this direction include the following:

- Creating conditions (information materials, Internet resources, and information centers) to inform tourists about the rules of conduct when visiting natural sites;
- Organizing cooperation between tour operators, protected areas, nonprofit organizations, and government agencies to promote green tourism;
- Supporting information work with tourists within the framework of the relevant national and regional program on managing protected areas;
- Implementing ecological certification and ecological labeling schemes to provide tourists with more information on the choice of tour operators promoting green tourism and eco-friendly products;
- Development of networks with the participation of protected areas, tourist organizations, and nonprofit organizations for promoting and marketing green tourism in the region.

The development of ecological tourism in the region allows solving two problems:

- To preserve the natural resource potential of the Republic of Altay;
- To use this potential in economic activity to obtain a stable source of income.

4 Conclusion

In forming the model of a green economy at the regional level, it is necessary to theoretically reconsider this concept to create socio-economic and legal mechanisms that would lead to the goal posed. A green economy is not an end in itself; it is a desirable state of economic relations in which addressing environmental security becomes the norm for all economic agents rather than a reaction to another environmental catastrophe (Pavlov, 2020). The formation of norms must be conditioned by economic, socio-psychological, administrative, and legal measures.

The introduction and development of a green economy model in the region should be based on the following basic principles:

- Principle of comprehensiveness implies the interconnection of all components of the model of a green economy in the process of regional development;
- Principle of efficiency and effectiveness determines that the implementation and development of a green economy should ensure the achievement of goals and objectives by ensuring regional socio-economic development while minimizing environmental risks and implementing resource conservation;
- Principle of unity involves the consistency of the actions of all subjects of the regional economy participating in the implementation of the model of a green economy;
- Principle of informativeness determines the openness of information in the social, environmental, and economic spheres of the region, which is necessary for managerial decision-making in implementing the model of a green economy.

Effective mechanisms and tools must be identified for the development of priority green sectors of the economy. In the framework of the model of a green economy at the regional level, all mechanisms can be grouped according to the intended purpose and the priority areas of the green economy. The main types of green economy mechanisms include legal regulation, economic, and institutional mechanisms. In the priority areas of a green economy, there are such mechanisms as rational land use in agriculture and forestry, recycling and disposal of waste, promotion of alternative energy, rational consumption of resources, and resource efficiency. The main sectors of the economy of the Republic of Altay, contributing to the greening of the entire economy, are green agriculture (organic farming).

For regions similar to the Republic of Altay, namely, the regions with a significant environmental potential for the national economy but a low level of economic development, the transition to a green economy must become a strategic priority. For this purpose, it is necessary to develop and implement the mechanisms mentioned above. However, this does not mean that it is necessary to transform the existing state regulation system completely. At the current stage, the primary methods and tools of state regulation have been created. The program-targeted approaches, namely, state programs, are considered the most promising methods of forming a green economy model.

References

Adarina, R. T., Glotko, A. V., Kuttubaeva, T. A., Shvakova, O. N., & Yankovskaya, X. G. (2019). "Green economy" indicators as a foundation for the development of the regional economy. *International Journal of Innovative Technology and Exploring Engineering, 8*(8), 3220–3227.

Arustamov, E. A. (2017). Ratings and criteria for assessing the environmental condition of Russian cities and regions. *Internet Magazine Naukovedenie, 9*(4), 3–10.

Atlas of Emerging Jobs. (n.d.). *About the project*. Retrieved from https://atlas100.ru/en/about/ Accessed June 25, 2019.

Frolov, D. P., Popkova, E. G., & Strekalova, A. S. (2017). Strategic planning of regional ecological development: Adaptation of elinor ostrom's principles in Russia. In E. G. Popkova (Ed.), *Russia and the European Union: Development and perspectives* (pp. 39–45). Springer.

Gazukina, Y. G., Zyablitskaya, T. S., Kuttubaeva, T. A., Shvakova, O. N., Davaasuren, N., Jamgan, N., … Sukhbat, N. (2015). Tourist and recreational activities as a factor in the sustainable development of mountainous areas. In T. A. Kuttubaeva (Ed.)., *Gorno-Altaysk*. Russian Historical Society of Gorno-Altaisk State University.

Grand View Research. (2017). *Organic food & beverage market size worth $320.5 billion by 2025*. Retrieved from http://www.grandviewresearch.com/press-release/global-organic-food-beverages-market. Accessed July 29, 2018.

Ivanova, N. I., & Levchenko, L. V. (2017). "Green" economy: The essence, principles and prospects. *Herald of Omsk University. Series "Economics", 2*(58), 19–28.

Kalner, V. D. (2013). Rating industry and the realities of the "green economy" in Russia. Ecological factor in the system of socio-economic indicators of the regions. *Ecology and Industry in Russia, 9*, 57–61.

Karnyshev, A. D. (2011). Institutional and socio-psychological factors of environmentally friendly economy. *Bulletin of Irkutsk State Academy of Economics, 1*, 206–212.

Knyazeva, G. A., N. Yu., & Kirusheva. (2016). Transition to a green economy on the example of innovative development of the regional forest complex. *Bulletin of St. Petersburg University, 2*, 119–145.

Kuttubaeva, T. A., Ischuk, L. V., Adarina, R. T., Glotko, A. V., & Fedosova, T. V. (2017). Applications of the concept of territorial marketing in forming and implementing competitive advantages for tourist areas. *Journal of Environmental Management and Tourism, 8*(4(20)), 713–726.

Kuzminov, I. F., Bakhtin, P. D., & Khabirova, E. E. (2018). *Global agriculture is going green: A reflection of the trend in scientific and industry periodicals*. Science Technology Innovations. Retrieved from https://issek.hse.ru/data/2018/02/14/1162674536/NTI_N_81_14022018.pdf.

Lipina, S. A., Agapova, E. V., & Lipina, A. V. (2016). *Green economy. Global development*. Prospect.

Lyaskovskaya, E. A., & Grigoryeva, K. A. (2018). Formation of a "green" economy and sustainability of development of the country and regions. *Bulletin of the South Ural State University. Ser. Economics and Management, 12*(1), 15–22.

Medvedeva, A. (2018, March 15). *Agriculture of the future—Bet on Organic Farming*. Retrieved from https://www.agroxxi.ru/stati/selskoe-hozjaistvo-buduschego-stavka-na-organicheskoe-zemledelie.html. Accessed July 29, 2018.

Pavlov, F. (2020, June 8). *The largest environmental disasters in Russia in the last five years*. FederalCity. Retrieved from https://federalcity.ru/index.php?newsid=8353. Accessed June 10, 2020.

Popov, E. V., & Semyachkov, K. A. (2014). Typology of green economy institutions. *Bulletin of Ural Federal University. Series: Economics and Management, 5*, 4–14.

UNEP. (2011). *Toward a green economy: Toward sustainable development and poverty eradication*. Retrieved from http://old.ecocongress.info/5_congr/docs/doklad.pdf.

World Bank. (2001). *World development indicators*. Retrieved from http://documents.worldbank.org/curated/en/988701468781815855/World-development-indicators-2001.

Yashalova, N. N. (2014). Development of green economy indicators at the regional level. *National Interests: Priorities and Security, 40*(277), 26–34.

Zakharova, T. V. (2015). Green economy and sustainable development of Russia: Contradictions and prospects. *Tomsk State University Journal of Economics, 2*(30), 116–126.

Environmental Factors for Measuring Service Performance

Larisa I. Nekhvyadovich, Raisa T. Adarina, and Daria S. Robets

Abstract Service managers need metrics for evaluating service performance. Performance metrics provide benchmarks for evaluating methods and improving the use of labor, automation, and logistics. Moreover, they allow for developing fair and motivational compensation systems for employees. Without such systems, the competitive improvements available in production would not be available to services. Therefore, these services would be substituted by competing goods and services. Productivity metrics for manufacturing are well studied and widely used in specific service areas, but they develop at slow rates. Such a slow development of service performance indicators is caused by intangibility, labor intensity, rapidity, simultaneity, complexity, entrepreneurial independence, and lack of attention from professional communities.

Keywords Environmental factors · Service performance · Customization · Aggregation · Disaggregation

JEL Codes Q5 · Q55 · Q57

1 Introduction

Considering productivity measures expressing the relationship between the results of service processes and the resources or inputs required to operate them, we note that appropriate definitions of results and inputs are critical to a meaningful analysis of production performance. Meaningful analysis cannot be successful without the correct specification of inputs and outputs derived from a thorough analysis of each

L. I. Nekhvyadovich (✉)
Altai State University, Barnaul, Russia

R. T. Adarina
Gorno-Altai State University, Gorno-Altaysk, Russia

D. S. Robets
Altai State Technical University Named After I. I. Polzunov, Barnaul, Russia
e-mail: darya_robec@bk.ru

E. G. Popkova and B. S. Sergi (eds.), *Geo-Economy of the Future*,
https://doi.org/10.1007/978-3-030-92303-7_4

facility's processes and alignment with the correct methods for measuring service performance. This is caused by the fact that there are four areas of performance measurement for services that need to be considered in the analysis. These areas are as follows:

- Inherent problems in performance measurement;
- Classification of the service processes;
- Available methods for measuring service performance;
- Suggestions for managers to solve the problems of measuring service performance.

Let us break down the intangible aspects of service. The analysis shows that the quality of customer contact and its personalization are highly variable, reflect personal values, and that most residents have to be dealt with in real-time. Therefore, measuring performance and service quality is a challenging task (Shenshinov, 2012). Cost and service output must be quantifiable to measure productivity. Although intangibility complicates measuring performance, managers should do their best to conduct this complicated task. We propose the following ways of separating direct results from indirect ones:

- Separate controlled and uncontrolled intangible assets;
- Measure controlled intangible assets;
- Tangibilize as many intangible assets as possible.

While intangible assets may be an inherent problem, they are not an excuse to evade performance analysis (Adarina et al., 2019).

The organization of a quality service must strive to maximize performance in line with system goals, which always include quality. This is nothing new. Service performance is always measured against a certain level of quality (Shenshinov, 2012). The problem is that it is easy to measure the tolerances or functionality of the finished product in manufacturing, but the quality of the final product in the service industry lies in the consumer's satisfaction.

2 Materials and Methods

Quality of service is most often defined as the difference between expectation and perceived performance. With this in mind, it is also necessary to add specific performance indicators for a particular service. Measuring the performance works effectively in services only if they are accompanied by a clear definition of service strategy and associated performance characteristics, as well as by an effective parallel quality measurement system based on those characteristics. The measurement of performance can begin once this quality has been directly or indirectly determined (Merdesheva et al., 2020).

In manufacturing, a company's productivity is related to planned production, not to consumer orders. In service, the demand must usually be satisfied at the moment of its occurrence. In fact, headcount and labor productivity are a function of the projected demand, which further complicates productivity analysis. One of the main dilemmas in measuring the performance of a service unit is to know if it operates at the time of measurement (Shenshinov & Al-Ali, 2020). If an enterprise in the service sector does not control the rate of arrival of customers directly or through an appropriate reservation system, performance becomes highly variable and more difficult to measure. In this case, the classification of the service process is carried out, which includes the complexity of measurement in the service industry, which is varied in different situations that must be evaluated. To make any meaningful generalizations about which measurement approaches are appropriate for which situation, it is necessary to develop a classification scheme for classifying situations and providing a basis for comparing them. These situations can be divided into three dimensions.

- Customization and consumer engagement. Customization affects the definition of the product and introduces several changes to the final result. User outputs are difficult to group discretely for analysis. The variability of results in the final version can be challenging to determine since it can significantly impact the choice of performance measures. Moreover, there can be a need to perform the analysis much more frequently to move to a higher aggregation level. Consumer involvement introduces variability. This in itself does not affect the product as much as the customization that often accompanies it. However, consumer involvement affects the design of a performance measurement program, requiring larger samples or longer measurement timescales to cope with process variability (Vukovich et al., 2018).
- The complexity of input and output data. A service process may have one, several, or plenty input factors and one, several, or plenty output factors. The measurement of service productivity requires a specific methodology, which can consider the totality of the rendered services, where the combination of services is considered a problem. In this regard, the service industry implements the following approaches:
- Inputs and outputs for each component are measured and compared separately;
- The use of a model simultaneously dealing with multiple inputs and outputs.
- Aggregation and disaggregation. Measuring the effectiveness of any activity requires careful specification of the appropriate unit of analysis, the choice of which depends on the ultimate use of the performance indicator. Aggregated (e.g., solid or single) indicators are best suited for the following processes:
- Evaluating general economic policy;
- Making decisions on producing or rendering products or services;
- Deciding on the relative costs of labor and capital;
- Defining pricing strategies and policies.

However, these indicators do not always help identify specific sources of inefficiency and cannot significantly improve the enterprise's efficiency and effectiveness (Nekhvyadovich, Kostin, et al., 2020).

Disaggregated (e.g., product or process) indicators are helpful under the following conditions:

- Making operational decisions with the obligatory choice of methods of work;
- The existence of an approved system of remuneration of employees.

However, these conditions are not always met since it is challenging to develop provisions for complex services. Therefore, the focus of the research on service performance should be made on disaggregated measures.

Various performance metrics are available for services. Nevertheless, it is essential to remember that the service organization has effectively used some methods of measuring productivity in production. The authors consider strengths and weaknesses for each type of service.

Output and input coefficients can be used at both aggregate and disaggregate levels. They are regularly performed by government agencies (e.g., Federal State Statistics Service) and used at the aggregate level to evaluate economic policies (e.g., deregulation). One of the problems with these ratios is that they tend to consider only one input and one output at a time. As a result, costs are often tracked at a disaggregated level, especially headcount ratios. However, they are often criticized for their narrowness.

Let us consider this using the example of a hospital, where the ratio of home visits and one home health aide assistant as a standard can be questioned as having the potential to compromise quality standards and negatively impacting morale since too much emphasis is placed on the raw numbers (Tsvetkov et al., 2019). One might also question the appropriateness of comparing disaggregated coefficients under co-production conditions, variability in the range of services, capacity constraints, consumer interaction, quality measurement problems, and individualization. Therefore, such ratios seem to work best at the disaggregated level with the simplest services.

Most of the methods for measuring services based on time reporting (e.g., stopwatch timing and job sampling) are highly disaggregated ratio methods. They require that the work be identified, recorded, and standardized before any significant measurements are made. Therefore, they seem most appropriate for services where outputs and inputs are simple and the easiest to measure (Dragunov & Shenshinov, 2020).

We call these systems *plus* quality systems because they try to make quality a measurable outcome of service. There are many attempts to integrate quality and outcome while showing performance in the hospitality industry. The Service Assessment Matrix (SAM) is an interesting approach developed to incorporate quality aspects into service product measurement. In this approach, individual clients identify a potential set of service quality and performance criteria, which are further assembled into a matrix to guide the standards developed.

The *Quality Plus methods* incorporate some types of expert quality assessment and provide objective validity for those using this indicator. This method represents the future for disaggregated assessment of the performance of complex services. However, there is still research to be done in this area (Ragulina et al., 2019).

3 Results

With a classification scheme in hand and knowledge of available measurement methods, one can always consider how a service manager solves the problem of productivity in the service industry, which implies the following:

- First, the service manager must state the reason for studying the performance. This will force a reconsideration of the strategic questions behind the analysis, which will help focus the analysis on the level of aggregation important to the desired solution.
- Second, the service manager must analyze the existing system for delivering service decomposed into process steps and key decision areas. This analysis will focus on critical and specific operational areas to measure performance.
- Next, it is necessary to specify the service characteristics of strategic importance at each stage of the service process and key decision areas. This means that the management team must be absolutely clear about the services it wants to provide (Alekseev et al., 2018).
- After that, it is necessary to select and investigate the methods of performance measurement that seem most appropriate for the analytical goals. Selection criteria should include concern for complexity, individualization, and degree of aggregation.
- The service manager must be prepared for all sorts of objections about quality. Nothing can replace quality measurement and analysis. As the saying goes, "Do not expect what you do not check."
- Finally, the service manager should involve performers all along the way. The acceptance of any proposed measures is critical to employees' ability to improve performance because performance systems affect the effort expended and the rewards offered since they are an essential part of the psychological contract controlling workers' behavior.

The most significant gaps in the ability to measure service performance arise in disaggregated and individualized services. Aggregated approaches are not as sensitive to the detailed issues of quality measurement, joint results, and cost allocation. The problems associated with these issues also make it difficult to provide useful information to management at a disaggregated level. Additionally, managers must be trained to use available performance metrics in the service industry.

The measurement phase is the second vital component of service quality management, which provides coherence to the supporting concepts. Once the measurement methods have been developed and the results obtained, the investigated process can be placed in this measurable context and decisions made. The other aspects of the management of service quality are more complex than in production (Stroiteleva et al., 2019).

Quality originated in a manufacturing environment, and its terminology and methods were developed mainly in that environment. Nevertheless, some researchers have found that similar methods can be applied to service organizations as well. The application of these methods in a service environment requires adapting ideas to different circumstances. However, there have arisen some problems in meeting the growing need to define and specify the nature of service quality so that they can be addressed more effectively. The ambiguity of the definition and specification of service quality and the intangible nature of service quality is further complicated by its ephemeral and personal nature, which is unlikely to coincide in different organizations (Shenshinov, 2012). This makes the measurement of service quality and productivity a significant challenge for applying service quality management in all industries.

Because of the difficulty in determining service quality, the primary approach is to compare the gap between delivery and expectations. This led to the development of customer-oriented methods, such as the SERVQUAL scale, which measures customer satisfaction (Nekhvyadovich, Borodin, et al., 2020).

It is important to note that performance measures effectively work in services only if they are first accompanied by a clear definition of service strategy and related performance characteristics. Nevertheless, the relationship between performance and quality of services in the housing sector can be broadly defined as the ratio of results to costs in the quality equation (Avkopashvili et al., 2019).

4 Conclusion

It is generally believed that the responsibility for the success in implementing service quality management rests squarely on the shoulders of management and that effective service quality management is particularly important. It is necessary to use individualized production approaches designed to improve service. The technology of service efficiency, perceived by customers as an essential component, can help in this issue. Alternatively, the latent energy of frontline personnel can be implemented to advance quality management. However, in every organization, senior management must take the lead in identifying gaps and provide the structure and authority for the effective delivery of services. Service organizations must take a proactive and strategic approach to service quality management.

Despite the importance of the quality of service, quality management in services is still in its infancy. It can be argued that the incomplete understanding on the part of academics and the reactive approach of some service managers has led to what is now called the *pragmatic gap* between theory and practice. The remedy for these problems lies in a partnership between academic and business communities in exploring suitable service quality management approaches.

References

Adarina, R., Gazukina, Y., & Yankovskaya, K. (2019). Indicators of the "green economy" as a tool for monitoring the regional economy. *IOP Conference Series: Earth and Environmental Science, 395*, 012107. https://doi.org/10.1088/1755-1315/395/1/012107

Alekseev, A. N., Savenkov, D. L., Shvakov, E. E., Lebedev, N. A., & Ragulin, A. D. (2018). Financial problems of regional development and the ways of their solution in contemporary Russia. *Quality—Access to Success, 19*(S2), 71–75.

Avkopashvili, P. T., Polukhin, A. A., Shkodinsky, S. V., & Poltarykhin, A. L. (2019). The fundamental provisions of the concept of knowledge economy. In E. Popkova, Y. Ragulina, & A. Bogoviz (Eds.), *Industry 4.0: Industrial revolution of the 21st century* (pp. 57–64). Springer. https://doi.org/10.1007/978-3-319-94310-7_5.

Dragunov, A., & Shenshinov, Y. (2020). Effective development of innovative technologies as a method for increasing labor productivity in industry. *E3S Web of Conferences, 208*, 03001. https://doi.org/10.1051/e3sconf/202020803001.

Merdesheva, E. V., Titova, O. V., & Avkopashvili, P. T. (2020). Methodological approach to the classification of digital economies by environmental efficiency and sustainability criterion. In E. Popkova, & B. Sergi (Eds.), *Scientific and technical revolution: Yesterday, today and tomorrow* (pp. 1360–1368). Springer. https://doi.org/10.1007/978-3-030-47945-9_148.

Nekhvyadovich, L. I., Borodin, V. A., & Avkopashvili, P. T. (2020). A systemic approach to the study of the sustainability of the global digital economy: Economies of scale, ratchet effect and hysteresis effect. In E. Popkova, & B. Sergi (Eds.), *Scientific and technical revolution: Yesterday, today and tomorrow* (pp. 1369–1377). Springer. https://doi.org/10.1007/978-3-030-47945-9_149.

Nekhvyadovich, L., Kostin, K. B., & Mironov, V. V. (2020). Government and public monitoring of entrepreneurship in a multicultural region: New perspectives in the conditions of the digital economy. In E Popkova, & K. Vodenko (Eds.), Public Administration and Regional Management in Russia. *Challenges and Prospects in a Multicultural Region* (pp. 147–154). Springer. https://doi.org/10.1007/978-3-030-38497-5_17.

Ragulina, Y. V., Semenova, E. I., Avkopashvili, P. T., Dmitrieva, E. A., & Cherepukhin, T. Y. (2019). Top-priority directions of implementing new internet technologies on the territories of rapid economic development. In E. Popkova, & V. Ostrovskaya (Eds.), *Perspectives on the use of new information and communication technology (ICT) in the modern economy* (pp. 182–188). Springer. https://doi.org/10.1007/978-3-319-90835-9_21.

Shenshinov, Y. V. (2012). Labour quality management. *The Journal "North-East Asia academic Forum"*.

Shenshinov, Y., & Al-Ali, A. (2020). The tools of increasing efficiency of human resource in the lean production environment: Conceptual study. *International Journal of Core Engineering & Management, 6*(7), 1–18.

Stroiteleva, T. G., Kalinicheva, E. Y., Vukovich, G. G., & Osipov, V. S. (2019). Peculiarities and problems of formation of industry 4.0 in modern Russia. In E. Popkova, Y. Ragulina, & A. Bogoviz, (Eds.), *Industry 4.0: Industrial revolution of the 21st century* (pp. 145–153). Springer. https://doi.org/10.1007/978-3-319-94310-7_14.

Tsvetkov, V. A., Loginov, E. L., & Raikov, A. N. (2019). Integrated network interaction of education with science and production. *Educational Technologies and Society, 19*(1), 389–398.

Vukovich, G. G., Makuschenko, L. V., Bateykin, D. V., Titova, O. V., & Dobrosotskiy, V. I. (2018). Support of the territories of advanced economic development on human capital: Theory and practice. *Quality—Access to Success, 19*(S2), 157–160.

Environmental Audit as a Tool for Quality Management System

Irina V. Chernyaeva, Zalina M. Ivanova, and Yulia G. Gazukina

Abstract The environmental management system is particularly useful in the environmental work of companies. An environmental management system is a framework for guiding an organization to achieve or maintain good environmental performance by supporting company goals or meeting regulatory standards. More conservational work is necessary to improve the quality of the environment, but limited resources will constrain companies' efforts to protect the environment. Therefore, the planning stage should set the priorities for environmental goals.

Keywords Environmental audit · Quality management · Audit planning · Result · Reporting

JEL Codes Q5 · Q52 · Q56

1 Introduction

A rational management system consists of the following main interrelated elements:

- Planning;
- Organization;
- Implementation;
- Control.

Planning. The primary function of environmental indicators is to ensure that certain goals or standards are met. Therefore, goals and policies are needed at the planning stage. To determine the best goal and policy, it is necessary to determine

---

I. V. Chernyaeva (✉)
Altai State University, Barnaul, Russia

Z. M. Ivanova
Kabardino-Balkarian State Agrarian University Named After V.M. Kokova, Nalchik, Russia
e-mail: magda.808@list.ru

Y. G. Gazukina
Gorno-Altai State University, Gorno-Altaysk, Russia

E. G. Popkova and B. S. Sergi (eds.), *Geo-Economy of the Future*,
https://doi.org/10.1007/978-3-030-92303-7_5

the company's activities and position concerning environmental issues. In addition to identifying risks, the company needs to identify available resources, including the capital, time, human resources, and equipment (Shenshinov, 2012).

Organization and implementation. Organization and implementation are the detailed work following the initial planning of environmental policies and goals. It is necessary to assign responsibilities. A special environmental department or organization is designated to coordinate all environmental programs and resource allocation. Allocation of responsibility facilitates the implementation of any planned environmental program. The implementation of this program includes risk management and program management on specific issues that complete the environmental results.

Management. All of the above elements are essential for an effective and collaborative environmental program. To achieve an effective and cost-effective environmental program, controlling plays an essential role in the environmental management system (EMS). Without an ongoing analysis and monitoring of the environmental program, the company or management team cannot understand the adequacy and effectiveness of the programs being implemented. Based on information from the monitoring process, policies or goals may be revised and changed, and the structure or organization responsible for environmental work may be reorganized (Tashkulova & Kletskova, 2020). In general, controlling is a critical element in optimizing company resources to achieve the best environmental performance. An environmental audit is an effective tool for the management of controlling.

Environmental auditing first began in the US in the 1970s. European and Asian countries began auditing in the 1980s and 1990s. Compared to the history of the environmental movement, environmental auditing is still in its infancy. There is no universal format or definition. Different people interpret environmental audits differently. However, an audit differs from the assessment of environmental impact; it is essential to have a clear distinction between these two phenomena. An environmental assessment evaluates the potential environmental impact of the proposed facility based primarily on professional judgment. In turn, an audit is the systematic monitoring of the environmental performance of the existing operations and practices of a company by collecting data and backing it up with evidence (Alekseev et al., 2018).

There is no fixed definition for this phenomenon. However, the definition of environmental auditing given by the International Chamber of Commerce (ICC) is widely accepted.

2 Materials and Methods

An environmental audit is a voluntary program established by a company to measure the environmental performance of a company. An audit scope can vary from one company to another or from facility to facility, depending on the nature of the business and the organization's expectations. An audit can only be useful to a company when it is appropriate and well-planned. Otherwise, an environmental audit will require considerable resources but will only produce an unmanageable mass of data capturing non-compliance and providing no help in making management decisions.

The scope of environmental auditing ranges from the whole enterprise and all of its activities to just a short and small audit of the performance of a piece of equipment. An audit may include an evaluation of the past, current, and planned activities of the organization, as well as its products and services (Stroiteleva et al., 2020). The scope of environmental auditing can be divided into four streams:

- Geographical;
- Functional;
- Divisional;
- Compliance.

It is difficult for any company to initiate an audit program in all of its facilities or divisions where too little is known about an audit approach. The initial environmental audit scope may be limited to a single site for a global company or a multifunctional company. An audit may first be performed in a developed country where resources and auditing support are more available. The government or society usually puts much pressure to improve environmental performance (Glotko et al., 2019). The benefits of a positive attitude toward the environment are greater when the audit is viewed as a test for the entire organization. A prototype structure can be developed based on the environmental audit currently being implemented at the site, which would then be suitable for implementation at other company sites.

A company may choose to conduct an audit on a sample basis so that the audit can focus on specific activities and operations. Particular public attention is given to the environmental impact of manufacturing sectors and industrial activities. Environmental audits can focus on the discharge of pollutants into the environment, namely, air, water, and land. Environmental audits can be limited to examining air pollutant emissions or focusing on monitoring either a specific facility or the entire company (Stroiteleva, Kletskova, et al., 2020). Thus, organizational activities and their impact on the environment must be identified in developing the audit scope as soon as possible. The selection of the functional group assigned to conduct the audit depends on age, the scope of the audit location (i.e., closeness to public areas), number of past non-compliance issues, the type of involved processes, possible expansion of operations, and possible sales.

The nature of the company will determine the scope of the environmental audit. The previous commitment of the managing group in the company can increase the scope of the audit area (e.g., from the office to the company). An international company cannot have operations and activities of the same type, in the same geographic area, or the same country. The audit could start with one unit and then apply the amended environmental audit to other units (Adarina et al., 2019). The company conducting the environmental audit does not strive to ensure that the environmental indicators are met immediately and by any means. The company conducts an audit to see if standards and regulations are being met and if the provision and compliance are going well enough to avoid legal obligations and penalties. Compliance with regulatory requirements can be the first step in a company's long-term environmental program. An audit can be applied to verify compliance with specific regulations instead of all identified requirements due to the limited resources of a company and the urgency and importance of a particular process. There may be more immediate benefits to meeting one standard than another.

Developing an environmental program within a company can go beyond compliance with regulatory norms and have its own standards and compliance goals. The audit may inspect the company's policies and objectives, which can be more stringent to ensure environmental management (Shenshinov & Abdulsattar, 2020).

Goals of the environmental audit.

All industrial, commercial, and domestic activities are connected with the consumption of natural resources. People always forget about the value of natural resources, especially those that are of great value but have little financial costs to business. Some natural resources are commonwealth (e.g., fresh air, clean water, and land). These resources have no agreed and defined values. The value of such abundant resources is ignored or neglected because there are subsidies from the state.

The government cannot reflect the actual cost of resources. The audit of the activities of the company's site and the collection of actual data provide information on the consumption of resources related to any processes or activities of the company and allow implementing a more effective control of consumption (Dragunov & Shenshinov, 2020).

The help in managing and improving environmental performance comes after forming an in-depth understanding of the company's operations. Assisting the audited facility or production manager in improving environmental performance is given by the head of the auditing organization based on the information collected and the analysis performed. Deficiencies identified during the environmental audit are corrected. The recommendations contained in the audit report must necessarily be implemented.

Reduction of costs and optimization of resources. Careful monitoring and auditing of litigation and non-compliance will facilitate immediate remediation and save the company from prosecution and fines. Additionally, careful control of production efficiency (i.e., the amount of inputs per unit of output) will help optimize resources.

Demonstration of management's commitment to environmental action. The audit includes the presentation of results showing the company's commitment to environmental issues. The audit assesses the following groups:

- Shareholders;
- Investors;
- General public.

Various groups can form an idea of the extent of the environmental work done and the results achieved.

Educating and motivating the workforce. Employees involved in internal audits, whether auditors or those undergoing audits, will increase the company's environmental awareness. The reporting on progress in news letters will also increase support for environmental programs.

Planning an environmental audit.

The approaches used in environmental auditing are not definitive. However, there are common elements among most auditing programs (Karataev et al., 2020). Gathering information, evaluating findings, analyzing facts, and reporting to stakeholders are the main stages of an environmental audit. These activities can also be classified as pre-audit activities, on-site audit activities, and post-audit activities. The details of these activities may vary depending on the different methods of auditing, consultation, and cooperation of the organization.

The environmental auditing process includes the following:

- Revising the company's internal environmental control system;
- Understanding of management;
- Evaluation of the control system using tests such as measurements;
- Discussion of the findings with management;
- Making judgments on the information collected;
- Reporting results and making recommendations.

The following *six steps* are used to plan an audit of the understanding of environmental management systems:

- Assessment of the system's strengths and weaknesses;
- Collection of factual data;
- Evaluation of findings;
- Accountability;
- Action planning;
- Subsequent check.

A successful environmental audit program requires careful planning of the audit program at an early stage. It is especially important to select and clearly define goals and objectives appropriate to the organization. The goals will depend on the company's nature, management philosophy, and size. With certain goals in mind, one can then identify the scope of the proposed audit. In addition to determining the goals and scope of the research, the selection of the audit team is also important to an effective program. Responsibility should be assigned to each team member. The audit protocol, which lists the step-by-step procedures to be followed during the audit, is prepared during the planning stage to provide guidelines for implementing the audit program (Avkopashvili et al., 2019; Vukovich et al., 2018).

Before the beginning of the evaluation of ESM, auditors must understand its operations and activities. Background information should be collected from site managers or people responsible for the audited areas. This will facilitate the on-site audit process at later stages and maximize the scope and reliability of information collection.

EMS is the basis for guiding the company in the field of sustainable environmental development. The audit is the management tool in the field of performance control. Therefore, understanding the environmental management system of the audited company is necessary to assess the performance and identify strengths and weaknesses.

3 Results

When auditing the company's operations, processes, internal controls, responsibilities for those businesses within the company, or control devices, and past and current problems that have not been resolved, it is necessary to take the following steps:

- Evaluate the strengths and weaknesses of the system;
- Evaluate the validity of the control and program;
- Identify any potential control failures associated with mismanagement.

Some components are essential for effective ESM, and auditors should pay special attention to them. Let us indicate some of the questions that auditors should answer:

- Is the staff of the facility experienced and well-trained?
- Are the responsibilities clearly defined?
- Is there a conflict between job responsibilities?
- Are there any standard procedures for accidents and proper accounting of environmental performance, such as compliance with established regulations?

All of these findings are important for evaluating ESM and recommending improvements.

In addition to any inherited weaknesses in environmental management structures or programs, it is important to learn if the phases of the environmental program are executed as intended. The presence of a well-developed environmental program and clearly defined responsibilities with the incomplete implementation or without it will not help in reducing environmental impact. Evidence of successful implementation can come from a record of continued compliance with rules and standards, interviews and questionnaires with facility personnel, or on-site observation of the facility. This stage tests strengths and weaknesses (Ragulina et al., 2019).

The meeting between members of the audit team is followed by the collection of information. This information will be summarized and evaluated for any conclusions or comments. A preliminary assessment of the information can be performed on-site to see if there is sufficient information to evaluate the results. The audit team can also discuss the results with the company manager.

4 Conclusion

Reporting is a significant component of an environmental audit because the report captures all the information: required compliance, facility activities, environmental structure, programs, audit processes, findings, and supporting evidence.

The management team, facility manager, and operations staff can be validated with current facility performance or operations from the reports. Recommendations included in the report can help the manager of the company to improve productivity (Tsvetkov et al., 2019).

Reporting and recommendations are not the end of the environmental audit. Recommendations should be implemented by assigning responsibility for corrective actions and setting a timetable for their implementation. After planning and implementing actions, it is necessary to conduct a follow-up audit of the action program. A follow-up program could be another environmental audit program.

Although environmental audit programs evolve differently according to the nature and needs of organizations, an effective audit program has some common aspects, such as (Costanza, 1996):

- Clear support and commitment of senior management at all levels;
- Obligations with respect to follow-up on audits and reports;
- Clear goals, scope, and frequency of the audit;
- Adequate qualified audit team;
- Independence of the audit team from the audited activity;
- A clear procedure for recording information and reporting to reasonably assess and achieve the audit objectives;
- Subsequent systems to ensure the implementation of corrective actions;
- Quality assurance of the accuracy of the environmental audit.

References

Adarina, R., Gazukina, Y., & Yankovskaya, K. (2019). Indicators of the "green economy" as a tool for monitoring the regional economy. *IOP Conference Series: Earth and Environmental Science, 395*, 012107. https://doi.org/10.1088/1755-1315/395/1/012107

Alekseev, A. N., Savenkov, D. L., Shvakov, E. E., Lebedev, N. A., & Ragulin, A. D. (2018). Financial problems of regional development and the ways of their solution in contemporary Russia. *Quality—Access to Success, 19*(S2), 71–75.

Avkopashvili, P. T., Polukhin, A. A., Shkodinsky, S. V., & Poltarykhin, A. L. (2019). The fundamental provisions of the concept of knowledge economy. In E. Popkova, Y. Ragulina, & A. Bogoviz (Eds.), *Industry 4.0: Industrial revolution of the 21st century* (pp. 57–64). Springer. https://doi.org/10.1007/978-3-319-94310-7_5

Costanza, R. (1996). Ecological economics: Reintegrating the study of humans and nature. *Ecological Application, 6*(4), 978–990.

Dragunov, A., & Shenshinov, Y. (2020). Effective development of innovative technologies as a method for increasing labor productivity in industry. In *E3S Web of Conferences*, 208, 03001. https://doi.org/10.1051/e3sconf/202020803001.

Glotko, A. V., Sycheva, I. N., Dunets, A. N., Poltarykhin, A. L., Zhuravlev, P. V., & Tubalets, A. A. (2019). Integration processes in the agro-industrial complex of the regions: Development and environmental problems. *Journal of Environmental Management and Tourism, 10*(3(35)), 613–621.

Karataev, E. M., Merkuryev, V. V., & Titova, O. V. (2020). Integrating the "green economy" into the model of digital future of the modern socio-economic systems. In E. Popkova, & B. Sergi (Eds.), *Scientific and technical revolution: Yesterday, today and tomorrow* (pp. 1402–1410). Springer. https://doi.org/10.1007/978-3-030-47945-9_153.

Ragulina, Y. V., Semenova, E. I., Avkopashvili, P. T., Dmitrieva, E. A., & Cherepukhin, T. Y. (2019). Top-priority directions of implementing new internet technologies on the territories of rapid economic development. In E. Popkova, & V. Ostrovskaya (Eds.), *Perspectives on the use of new information and communication technology (ICT) in the modern economy* (pp. 182–188). Springer. https://doi.org/10.1007/978-3-319-90835-9_21.

Shenshinov, Y. V. (2012). Labour quality management. *The Journal "North-East Asia academic Forum"*.

Shenshinov, Y., & Abdulsattar, A.-A. (2020). The tools of increasing efficiency of human resource in the lean production environment: Conceptual study. *International Journal of Core Engineering & Management, 6*(7), 1–18.

Stroiteleva, T. G., Kletskova, E. V., & Balaeva, S. I. (2020). Ontogenesis of green development model of digital economy: Essence, stages, and prospects. In E. Popkova, & B. Sergi (Eds.), *Scientific and technical revolution: Yesterday, today and tomorrow* (pp. 410–417). Springer. https://doi.org/10.1007/978-3-030-47945-9_46.

Stroiteleva, T. G., Petrova, E. A., & Yungblyudt, S. V. (2020). Responsible information society as a social environment for creation of the "green" digital economy: Formation and monitoring. In E. Popkova, & B. Sergi (Eds.), *Scientific and technical revolution: Yesterday, today and tomorrow* (pp. 402–409). Springer. https://doi.org/10.1007/978-3-030-47945-9_45.

Tashkulova, G. K., & Kletskova, E. V. (2020). The model of organization of "green" digital production and consumption based on the internet of things and AI. In E. Popkova, & B. Sergi (Eds.), *Scientific and technical revolution: Yesterday, today and tomorrow* (pp. 329–336). Springer. https://doi.org/10.1007/978-3-030-47945-9_37.

Tsvetkov, V. A., Loginov, E. L., & Raikov, A. N. (2019). Integrated network interaction of education with science and production. *Educational Technologies and Society, 19*(1), 389–398.

Vukovich, G. G., Makuschenko, L. V., Bateykin, D. V., Titova, O. V., & Dobrosotskiy, V. I. (2018). Support of the territories of advanced economic development on human capital: Theory and practice. *Quality—Access to Success, 19*(S2), 157–160.

Mechanisms for Implementing the Green Economy Model at the Regional Level

Svetlana V. Lepeshkina, Toskanay A. Kuttubaeva, and Svetlana A. Temmoeva

Abstract The paper explores the possibilities of substantiating economic and institutional mechanisms for introducing the model of the green economy in the constituent entity of the Russian Federation. At the theoretical level, the authors expanded their understanding of the green economy model at the regional level, its individual components, and the logic of the transition to the green economy model. More than that, the authors identified management opportunities for monitoring the implementation of the green economy in the region due to a set of indicators demonstrating the success of the transition to the green economy model in the region. The authors also developed approaches for analyzing the regional regulatory system contributing to the transition to the green economy model. Furthermore, the authors formulated recommendations for transforming the regional economy to the principles of green production through the program-targeted approach.

Keywords Green economy model · Sectoral approach · Region · Indicator system

JEL Codes Q5 · Q51 · Q56

1 Introduction

The concept of a green economy was first introduced in 1989 during the discussion of the UN Environment Program (UNEP), which introduced a *cleaner production* program. Since then, interest in the green economy has not faded away, and derivatives of the concept of *green market* and *green production* have appeared. Attention to the green economy is explained by attempts to solve environmental problems and the problem of poverty, which is characteristic of developed countries. For these

S. V. Lepeshkina (✉)
Altai State University, Barnaul, Russia

T. A. Kuttubaeva
Gorno-Altai State University, Gorno-Altaysk, Russia

S. A. Temmoeva
Kabardino-Balkarian State Agrarian University Named After V.M. Kokova, Nalchik, Russia

E. G. Popkova and B. S. Sergi (eds.), *Geo-Economy of the Future*,
https://doi.org/10.1007/978-3-030-92303-7_6

countries, the green economy model is a strategic alternative to gaining competitive advantages in manufacturing environmentally friendly products (Bobylev, 2015; Costanza, 1996; Daly & Townsend, 1993; Dixon et al., 2000; Lipina et al., 2016).

Despite the theoretical elaboration of the essence of the green economy model, this phenomenon remains a *good intention* and does not work in current conditions.

Numerous attempts to use the principles of ecological and economic development of territories are not crowned with success due to the lack of scientifically substantiated economic and institutional mechanisms for the transition to the green economy.

Scholars identify the following reasons impeding the implementation of the green economy model:

- Lack of resources;
- Inaction on the part of organizations;
- Ineffectiveness of government regulation and policy.

When considering these reasons and the need to eliminate them, the authors believe that the scientific problem consists of the lack of detailed mechanisms for introducing the green economy in the region.

In the mid-twentieth century, the understanding of the *society–nature* relationship lay in creating gentle conditions for the environment that required protection from human influence. Nowadays, the green economy is an economic order in which everything (from consumption to production) must be built on the principle that the economy depends on the environment. Simultaneously, the ideas of the green economy are considered as alternatives to the existing economic order. One of the reasons for this is the theorization of the green economy model in the absence of practical examples of the functioning of the green economy as a model, at least at the level of a separate territory.

In the second half of 2017, the Republic of Altay (Russia) adopted a draft "*Strategy for the Socio-Economic Development of the Republic of Altay until 2035.*" The draft was formed, taking into account the principles of the green economy.

The project became a logical stage in searching for prospects for the development of the region. The Republic of Altay had an experience of the functioning of the ecological-economic zone. The results of this experiment are interpreted ambiguously. However, it certainly brought some results. The principles of activity of the ecological-economic zone *Altai* were based on the regulatory and legal foundations of the concept of ecological and economic sustainability, which were aimed to preserve the natural environment of the Republic of Altay and ensure ecological balance and harmonious development of social infrastructure on its territory.

2 Materials and Methods

Currently, foreign and Russian scholars have sufficiently studied the problems of forming the green economy model. According to B. N. Porfiriev, this definition considers the green economy as an economic activity, and such an interpretation of the green economy practically does not distinguish it from the concept of sustainable development (Porfiriev, 2012). Based on the analysis and generalization of existing approaches to the definition, E. M. Zomonova offers her understanding of the green economy in her paper "Definition and Principles of the Green Economy." She defines a green economy as a model of economic development based on sustainable development, internationalization of externalities, an integrated approach to decision-making, and improved quality of life in conditions of resource conservation and resource efficiency by using new technologies and innovations (Zomonova, 2016).

N. A. Vukovich considers the green economy as an economy of sustainable growth with the domination of environmentally friendly industries using alternative energy and resource-saving technologies, where the environmental and economic policy of the government actively stimulates economic growth and the development of the ecological culture in the population (in order to preserve human civilization on Earth) (Vukovic, 2018).

The development of the green economy model in the region presupposes the definition of the model's structure, which should contain the following components:

- Reasons and prerequisites for the transition to the green economy model in the region;
- Targets of the green economy model at the regional level;
- Basic principles for implementing the green economy model in the region;
- Sectoral priorities of the green economy at the regional level;
- Resource potential, mechanisms, and tools for implementing and developing the green economy model at the regional level;
- Expected results and indicators of the effectiveness of the green economy model in the region.

The formation of an effective green economy model in the region must be performed by analyzing and assessing the socio-economic and natural resource components of the territory. This analysis of the region's characteristics is a harbinger for the green economy model. At the same time, it is necessary to assess and analyze the complex of indicators of the region in the following areas:

- Availability, level, and state of the natural capital of the region;
- Socio-economic situation in the region;
- Environmental situation in the region;
- Institutional environment of the region (Ischuk et al., 2019).

Justification of the need to transit to the green economy model in Russia necessitates consideration of the current level of its development. Simultaneously, differentiation in the development of Russian regions makes it necessary to consider territorial specifics in their forecasts when making such an assessment.

Every region has its initial conditions for the formation of a green economy. Therefore, researchers are interested in assessing the level of the green economy in the Russian regions. This assessment allows researchers to develop and use various greening indices, including:

- Ecological rating of regions according to the All-Russian public organization *Green Patrol* (All-Russian public organization "Green Patrol", 2018);
- Ecological and economic index of regions, developed with the support of WWF Russia, *RIA Novosti*, and the All-Russian public organization *Russian Geographical Society* (Bobylev et al., 2012).

The proposed indicators are developed according to the type of indicators aggregated based on the integration of indicators reflecting the economic, industrial and technological, environmental, institutional, and social components. Indicators differ in their composition and methods of their processing. Nevertheless, an attempt to comprehensively assess the development of the region (including the economic potential, the degree of degradation of natural resources, and the level of human capital) is what they have in common.

The ecological index is calculated regularly since it is measured four times a year.

The second rating was developed once within the grant project. Despite the differences in the methodology for calculating both indices, the Republic of Altay occupies the leading position in both ratings. The All-Russian public organization *Green Patrol* assigns the Republic of Altay from second to fourth place during 2008–2018 in its environmental ratings. The ecological and economic index, formed in 2012, shows the leadership of the Republic of Altay in the rating formed on its basis. Moreover, in 2013, the administrative center of the Republic of Altay, Gorno-Altaysk, took second place (giving primacy to the city of Moscow) in the rating of cities developed by the consulting company *Ernst & Young*.

3 Results

The Altai Mountains are a mountain range located in the center of Asia and spanning from about 48° to 53° N and 82° to 90° E. In latitude, the territory of the Altai Mountains is at the level of the steppes and semi-deserts of the European part of Russia (Ministry of Economic Development & Tourism of the Republic of Altai, 2019).

The Republic of Altay does not generate electricity on its territory, which would cover the needs of the economy of the republic. The developments of renewable hydropower resources in the Republic of Altay, and the widespread use of solar and wind power generation are priorities in developing the regional fuel and energy complex.

The authorities of the Republic of Altay consider the development of small and autonomous generation through the use of solar energy as one of the ways to solve socio-economic issues in the region.

Currently, a large investment project aimed at developing solar power in the region is implemented in the Republic of Altay. The Republic of Altay is one of the sunniest places in Russia. There are 300 cloudless days a year. This geographical feature allows creating conditions for the active construction of solar power plants.

Solar power plants (SPP) allow to provide the energy-deficient region with a sufficient volume of capacities, reduce the flow of electricity from neighboring regions, and increase the reliability of the power supply.

By 2022, the authorities of the Republic of Altay plan to become self-sufficient in electricity. This will become possible after the commissioning of all new SPP. Their total capacity will equal 65 MW.

Solar power plants provide 30% of the energy consumption in the Republic of Altay. The second stage of the Ininskaya SPP with a capacity of 15 MW and the Ust-Koksinskaya SPP with a capacity of 40 MW started to transmit electricity to the regional power grid on December 1, 2019. Another station is planned to be commissioned in January 2021.

Earlier, investors built and put into operation the following SPPs in the Republic of Altay:

- Two stations in Kosh-Agach with a capacity of 5 MW each;
- Stations in Ust-Kan and Ongudai with a capacity of 5 MW each;
- Maiminskaya SPP with a capacity of 25 MW;
- First stage of the Ininskaya SPP with a capacity of 10 MW.

In January 2021, the Chemal SPP with a capacity of 10 MW will begin operating. This will ensure the reliability of the power supply to the Chemal district, which is a popular tourist destination and has many tourist centers. Thus, the installed capacity of solar generation in the region equaled to 120 MW in 2020.

The capacities of the SPP built in 2014–2015 are enough to cover the daily needs for electricity in the Kosh-Agach district and partially in the neighboring Ongudai and Ulagan districts.

Nowadays, both stages of the Kosh-Agach SPP provide a stable power supply to more than 1000 households. The total generation of the station from the beginning of electricity supply to the grid until January 2020 amounted to more than 65 million kWh.

In total, eight SPPs are put into operation in the Republic of Altay: in Maima, Ongudai, Ust-Kan, Chemal, In, Amur, and the Kosh-Agach district (two SPPs).

Thus, the development of alternative energy in the Republic of Altay is an example of the effective use of the geographical advantages of the region without compromising the ecological state. This approach allows the region to maintain its relatively high environmental rating and implement the main provisions for forming the green economy.

Municipal solid waste negatively impacts the environment and the health of the local population. This waste is heterogeneous in composition, which significantly complicates its collection and processing.

According to statistics, one person emits an average of 50 kg of waste per month (Tapyshpan & Khovalyg, 2018). A considerable part of municipal solid waste falls on packaging containers, which practically do not decompose for many centuries.

The following methods of disposal of solid household waste exist nowadays:

- Storage at a landfill;
- Incineration in specialized ovens;
- Waste processing using a separate waste collection.

A perfect disposal method is not elaborated yet; each of the methods has its advantages and disadvantages. The incineration of municipal solid waste generates heat energy that can be used for the needs of the local population. The storage of poorly sorted waste can cause enormous harm to the environment. The authors note that developed countries try to significantly reduce the percentage of solid household waste that will be subject to storage. Incineration and storage do not consider the possibility of recycling municipal solid waste. The environmental hazards of municipal solid waste and economic losses are often interrelated. Disposal of unsorted waste leads to an irrecoverable loss of up to 90% of the products demanded in the market for secondary raw materials (Yashalova & Gridnev, 2013).

There is a concept of zero waste in the world. The principles of this concept largely coincide with the ideas of the green economy concept. The concept implies that all waste must be considered as a resource; that is, it must be recycled or reused. According to the UNEP classification, waste belongs to the main sectors of the green economy. Therefore, the goal is set to reduce waste disposal by at least 70% (Knyazeva & Kirusheva, 2016). Based on the studied Russian and international experience in managing solid household waste, the authors formulated the main conditions for their minimization with consideration of the green economy concept.

Every year, this problem becomes increasingly urgent for the Republic of Altay as a tourist and recreational region. The authors note the main barriers to the effective solution to the problem of solid waste disposal in the Republic of Altay. These barriers are as follows:

- Imperfection of legislation;
- Low level of ecological culture of the local population and tourists;
- Poor development of enterprises for the collection, processing, and disposal of solid household waste;
- Disinterest of business and public authorities.

By Order No. 55-VD of August 17, 2018, the Committee on Tariffs of the Republic of Altay established the average monthly standard for accumulating solid municipal waste, which is as follows: 0.1189 m^3 per one person in apartment buildings and 0.13673 m^3 per one resident in detached houses. The average standard for one person is 0.127815 m^3. As of January 1, 2018, the population of the Republic of Altay was 218,063 people. According to the reference data, the density of municipal solid waste is on average 200 kg/m^3. Consequently, the accumulation of solid household waste by the population of the republic is, on average, 66.9 thousand tons per year. This figure does not consider legal entities operating on the territory of the Republic of Altay and the tourist flow. Currently, on the territory of the Republic of Altay, there is only one solid waste processing plant with a capacity of 32.5 thousand tons per year. The capacity of the plant for processing municipal solid waste is insufficient. It is necessary to significantly increase the capacity of the existing plant or build new facilities to process solid waste in the Republic of Altay.

The use of technologies for the recycling of solid household waste can stimulate the economic activity of the region within the green economy model. In the Republic of Altay, a business for collecting and processing secondary raw materials (paper, glass, plastic, electronic waste, etc.) looks promising. According to experts, the waste recycling industry is profitable for business representatives and can create additional jobs in the region. The combination of economic, social, and environmental benefits is a fundamental aspect of the green economy concept.

The problem of municipal solid waste in the region cannot be solved without raising the ecological culture and awareness of the region's population. It is important to convey the benefits, which the region and every resident will receive from solving the problem of solid household waste, and outline their financial and economic benefits and benefits for the environment in an accessible form. Special educational programs that can correct consumers' and producers' daily behavior are effective tools for enhancing the ecological culture of the population.

Legislative tightening of requirements and control in the field of collection, processing, and disposal of solid household waste will give the Republic of Altay tangible results (Kolb, 2018). For example, in many European countries, it is not profitable for the population to throw away unsorted garbage since the fines for this violation are substantial. Effective administrative measures and propaganda can form the sustainable environmental behavior of the population and provide significant assistance in solving the problem of solid waste disposal.

According to the experience of foreign countries, the system of preferential tariffication of services for the removal of solid household waste has shown efficiency along with the system of fines. For instance, the Swiss population, who sort their garbage, has a reduced tariff for garbage disposal. Germany, one of the leading countries in waste management, has achieved positive results in this area mainly due to the voluntary sorting of waste by the population. The presence of specialized buckets with compartments for various types of garbage in houses and apartments is also an essential step towards solving the problem of solid household waste. Nevertheless, the proposed system will only work if containers for sorting waste are ubiquitous in the region. Nowadays, in the Republic of Altay, there is practically no separate waste collection system.

The minimization of municipal solid waste is another significant step towards the greening of the regional economy. It is easier to minimize the volume of municipal solid waste in the region than spend significant financial and administrative resources on solving the existing problem.

Attracting funds to the Republic of Altay for organizing waste processing facilities is an important task for the region. In particular, subsidies, concessional lending, and tax breaks can significantly help.

The problem of municipal solid waste, with consideration of the damage they cause to the region's environment, deserves special attention. Overcoming barriers to a rational solution to the problem will require the Republic of Altay to develop a holistic approach to solid waste management, taking into account the principles of the green economy.

4 Conclusion

The analysis of the strategic planning documents of the Republic of Altay has shown that despite adhering to the principles of the green economy, it is one of the three tasks in the implementation of the "Strategy for the Socio-Economic Development of the Republic of Altay Until 2035", which is not solved in practice. The managerial gap between strategic objectives and the methods of solving them is the main reason for this. State programs are the immediate tool, the use of which should give the desired result. However, their analysis has shown that among the goals and objectives of the 14 State Programs of the Republic of Altay, they do not include the target indicators necessary for forming the green economy model. These state programs should include green investments, green industries, green employment, organic agriculture, organic land use, and recycling and its organization in the branches of regional specialization, such as agriculture, tourism, processing industry, energy, and housing and communal services.

The absence of the above-mentioned priorities in government programs leads to the absence of measures and resources necessary to implement the green economy model.

Five of the fourteen programs play a crucial role in forming the green economy model in the Republic of Altay:

- Development of the agro-industrial complex and regulation of markets for agricultural products, raw materials, and food;
- Environmental safety and improvement of environmental performance;
- Development of economic potential and entrepreneurship;
- Further development of the housing and communal services and the transport complex;
- Improving the quality of water supply in the Republic of Altay.

These programs require reformation through their goals, objectives, and developments of target indicators and activities structured in terms of the objectives of a specific state program.

After solving these conceptual provisions of state regulation, it is necessary to take the following steps:

- Create conditions for obtaining positive effects from consumers of alternative energy by reducing electricity tariffs;
- Export electricity from solar power plants;
- Develop the regional system of organic animal husbandry and provide state support for agricultural producers;
- Preserve the resource potential of organic animal husbandry and organic farming;

- Introduce recycling standards in the agro-industrial complex.

The implementation of the developed recommendations will require the use of tools, such as green taxes, subsidies, state and municipal procurement, certification, and licensing, considering regional characteristics.

References

All-Russian public organization "Green Patrol". (2018). *Ecological rating of Russian regions*. Retrieved from https://greenpatrol.ru/ru/stranica-dlya-obshchego-reytinga/ekologicheskiy-reyting-subektov-rf?tid=351.

Bobylev, S. N. (2015). *Green economy. A new paradigm for the country's development*. SOPS.

Bobylev, S. N., Minakov, V. S., Solovyova, S. V., & Tretyakov, V. V. (2012). Ecological and economic index of the regions of the Russian Federation. Methodology and indicators for calculation. In A. Ya. Reznichenko, E. A. Schwartz, & A. I. Postnova (Eds.). WWF Russia, RIA Novosti. Retrieved from https://wwf.ru/upload/iblock/dc8/index.pdf.

Costanza, R. (1996). Ecological economics: Reintegrating the study of humans and nature. *Ecological Application, 6*(4), 978–990.

Daly, H., & Townsend, K. (1993). *Valuing the Earth: Economics, ecology, ethics*. The MIT Press.

Dixon, J., Backes, J., Hamilton, K., Kant, A., Lutz, E., Pejiola, S., & Hee J. (2000). Expanding the measure of wealth. Indicators of environmentally sustainable development. In V. N. Sidorenko, & T. A. Glushko (Trans. from English; S. N. Bobylev, & V. N. Sidorenko) (Eds.). Center for Preparation and Implementation of International Technical Assistance Projects; Project of Global Environmental Facility "Biodiversity Conservation".

Ischuk, L., Kuttubaeva, T. A., & Shvakova, O. (2019). Prerequisites for implementing the principles and tools of "green" economy in the Altai Republic. *IOP Conference Series: Earth and Environmental Science, 395*, 012099. https://doi.org/10.1088/1755-1315/395/1/012099

Knyazeva, G. A., & Kirusheva, N. Y. (2016). Transition to green economy on the example of innovative development of the regional forestry complex. *Bulletin of St. Petersburg University, 2*, 119–145.

Kolb, A. A. (2018). Administrative responsibility for environmental and other offenses in the field of waste management, including solid municipal waste. *Agrarian and Land Law, 11*(167), 20–25.

Lipina, S. A., Agapova, E. V., & Lipina, A. V. (2016). *Green economy. Global development*. Prospekt.

Ministry of Economic Development and Tourism of the Republic of Altai. (2019). *Tourist Passport of the Republic of Altay*. Retrieved from https://altai-republic.ru/tourism/tourist_passport.pdf

Porfiriev, B. (2012). *The "green" economy: New tendencies and trends in the development of the world economy*. Scientific works: Institute of Economic Forecasting of the Russian Academy of Sciences, 10, 9–33. Retrieved from https://cyberleninka.ru/article/n/zelyonaya-ekonomika-novye-tendentsii-i-napravleniya-razvitiya-mirovogo-hozyaystva. Accessed April 16, 2020.

Tapyshpan, A. M., & Khovalyg, A. A. (2018). The problem of household waste. *Bulletin of Tuva State University. Technical and Physical and Mathematical Sciences, 3*(38), 70–74.

Vukovic N. A. (2018). Green economy: Definition and main directions of development. *Bulletin of Ural Federal University. Series Economics and Management, 17*(1), 128–145. https://doi.org/10.15826/vestnik.2018.17.1.006.

Yashalova, N. N., & Gridnev, A. E. (2013). Ecological and economic problems of waste recycling. *National Interests: Priorities and Security, 9*(43(232)), 28–36.

Zomonova, E. M. (2016). Definition and principles of green economy. *ASR: Economics and Management, 1*(14), 13–17. Retrieved from https://cyberleninka.ru/article/n/ponyatie-i-printsipy-zelenoy-ekonomiki. Accessed April 16, 2020.

A Retrospective Overview of the Definition of Industrial Ecology in the Theory of Industrial Management

Larisa I. Nekhvyadovich, Yulia G. Bozieva, and Igor Y. Pushkarev

Abstract This paper is an overview of the field of industrial ecology, which notes a significant contribution to research and applied developments, leading to the position at which it is now. A short overview of the term is given, and scholars and practitioners who helped defining this field as an important contribution to the efforts on environment protection and management are discussed. Industrial ecology takes energy from various sources, and the concepts that enriched this field and made it attractive have to be analyzed. This discussion sheds light at certain aspects that constitute the field of industrial ecology's competence and how it looks today. The paper contains the main characteristics of industrial ecological studies, which should be observed and evaluated in order to implement the methods for a constructive analysis and criticism of these characteristics in the future. The historical approach to development of industrial ecology helps understanding the dynamic formation of the field, which still consists of participants and ideas that are rooted in various disciplines, including almost all fields of natural and technical sciences, and certain fields of social sciences. Understanding of its historical development means that the concepts that lie in the basis of industrial ecology and the components of its competencies could be evaluated from the point of view of their substantiation (Glotko et al. in Journal of Environmental Management and Tourism 10, 3(35): 613–621 2019).

Keywords Industrial management · Industrial ecology · Industrial symbiosis · Industrial metabolism · Biological eco-system of analogy

JEL Codes Q5 · Q55 · Q57

L. I. Nekhvyadovich (✉)
Altai State University, Barnaul, Russia

Y. G. Bozieva
Kabardino-Balkarian State Agrarian University named after V.M. Kokova, Nalchik, Russia

I. Y. Pushkarev
Branch of Kuzbass State Technical University, Novokuznetsk, Russia

E. G. Popkova and B. S. Sergi (eds.), *Geo-Economy of the Future*,
https://doi.org/10.1007/978-3-030-92303-7_7

1 Introduction

Industrial ecology—as an industrial term—attracts attention of the wide pubic. Different studies help to mobilize the scientific society and draw attention to scarce efforts on this topic. Such work as "Production strategies" are studied by managers of industrial production and are used in restructuring and reorganization, which eventually leads to significant reduction of the burden on the environment. An industrial system that is obtained as a result of the restructuring will have a lot of common characteristics with biological and natural ecosystems. According to Erkman, intuitive compassion regarding ecological development, especially with industrial production, within the imposition of ecological order, allows for the cooperation between industrialists and the public.

"Production strategies" are based on the idea that the traditional linear method of industrial production (expenditures-processing-issue) for improvement of the ecological indicators of industry have to be replaced with the cyclical method, at which one manufacturer's products are used to the full—so all expenditures that accompany production become byproducts, which could be sold to other manufacturers. Thus, two problems are solved at the same time—deficit natural resources for entering the production produce waste—as a result of processing—which then is delivered to special grounds and put in drains in the biosphere (pollution). Closure of the linear flows of resources is an elegant solution that requires technological and organizational changes for manufacturers (Karataev et al., 2020).

It is believed that industrial ecosystems are created for preserving valuable resources, especially metals (e.g., turnover of iron and precious metals from the platinum metals). These resources should be moved from waste deposits back to production processes (Nekhvyadovich et al., 2020).

Other forms of pollution—such as depositions of various materials in water and air require further research, preferably at the intersection of business and ecology.

2 Methodology

The attempts to create a new discipline at the intersection of business and ecology started in 1970, when the journal "Industrial ecology" was created—however, it was short-lived. During this time, Japan and certain countries of Europe made attempts to consider and evaluate industrial production from the point of view of eco-system (Glotko et al., 2019). German scholars emphasized that apart from the mention of a project in Austria and a conference in Germany, industrial ecology was present in the agenda since the late 1960s, when the Center for research and information of socio-political life of an independent research center in Brussels started its work in Belgium. Belgian economists at that time were focused on the flows of materials and energy.

These efforts were included in the following projects, so over the recent three decades there was a necessity to constantly develop and change the directions of the activities in the fight for the issue of ecologically safe product. Instead of staying in the independent position, industrial manufacturers have to get involved in the research programs of industrial technologies development (Avkopashvili et al., 2019).

Promoting industrial ecology as the main conceptual basis, which includes and strengthens ecological management is a task of each industrial production, which will provide positive result between industry and ecology; however, for this it is necessary to distance oneself and industrial ecology from the dominating environmentalism, which is traditionally treated by industry as passive, regressive, anti-growing, and anti-technological.

Let us consider seven fields, in which the application of the concepts and practices of industrial ecology could be useful (Vukovich et al., 2018):

- creation of industrial ecosystems (discussed below);
- balancing of industrial expenditures and issue in view of the natural ecosystems' potential;
- dematerialization of industrial production;
- improvement of metabolic ways of industrial processes;
- use of materials;
- creation of the systemic models of energy use;
- coordination of policy and the long-term perspective of industrial systems' evolution.

According to Tibbs, industrial ecology could ensure the perspective and measure for industrial development in the long-term, where industrial and natural systems co-exist in harmony—the former do not destroy the latter (Adarina et al., 2019).

The flagship application of industrial ecology is networks of waste exchange. Discussion of the evolutionary system of waste exchange becomes popular, which has led to a lot of studies.

In order to make industry pay attention to the concept of industrial ecology and start developing interest in environment management, it is necessary to focus on technical solutions; there are also moral obligations, which traditionally pose certain problems for business companies.

The research field of activities in this direction could be considered well-developed and full from the point of view of active participants in the fight for issue of ecologically safe products, which requires engineering and materials research, and economic justifications on business management. Conceptual documents on ecological development have expanded and developed the basic structure that forms a scientific foundation; however, they pay more attention to application of these tools primarily in industry. Certain tools that constitute the industrial ecological work in research and practice are considered after and in view of the discussion of the central plan of industrial ecology: its observance over economic systems through the lens of ecological systems (natural ecosystems) (Shenshinov, 2012).

It should be noted that this sphere is obliged to—consciously or not—perform the systematization of ecological studies, which could be implemented not only in industrial production.

The idea that lies in the basis of industry is expressed by the large number of participants of this field, which could be understood mainly in three various forms, which are drawn to each other and obtain power from association with each other.

Though these three formulates come from different points of view and focus on different characteristics, it is possible to see that they use the same approach, viewing industrial and economic activities from the ecological and biophysical points of view.

In these three cases, the central idea in industrial ecology is as follows:

- "industrial metabolism";
- "biological ecosystem of analogy";
- "industrial symbiosis".

These definitions are representative, since all of them describe the process activities on the use of materials and energy—though in different scales (Dragunov & Shenshinov, 2020).

In biology, metabolic activity belongs to substance and energy processing at the level of a separate organism. Due to this, metabolic activity in industrial systems is considered as the activities from cradle to grave, being parallel to biological organisms, which swallow nutritional elements and produce droppings. The systemic limit for observing metabolic activity lies here, at the individual level, which could make transition of the notion to large objects—e.g., economies or industrial systems—more complex (Bogoviz et al., 2020).

Symbiosis manages the activities on substance and energy processing at the level between organisms, while the "ecosystem analogy" describes processes at the highest level—e.g., ecosystems that include relations between separate factors and processing activities within a certain factor. Results of the eco-systems' analysis are transferred by industrial ecologists to the transformation of the whole industrial systems or even the whole socio-economic system. Another reason for discussing industrial metabolism and industrial symbiosis as the main notions in industrial ecology is the fact that both notions have existed before the industrial ecology was created as a separate field. The latter, however, covered the former pretty soon, and they began to be treated as being in the center of the field.

Let us consider the terms that are used for further understanding of not only industrial ecology as a scientific discipline but also the policy that creates conditions for restructuring of industry. Industrial ecology is aimed at performing the descriptive and directing roles.

Industrial metabolism. This term was created and disseminated by an engineer who became a professor of ecological management. "Industrial metabolism" became a predecessor in the sphere of industrial ecology, but, since the emergence of the latter, it has been closely related to and included in industrial ecology—as a direction of research that is coordinated with the common goal of this field.

Industrial metabolism defines the comprehensive totality of physical processes that transform resources and energy, as well as labor, in final products and waste. It should be noted that according to this definition, metabolic activities of industry and their analysis stop in the final product (Alekseev et al., 2018).

The concept of industrial metabolism takes value out from the idea of dissipation in systems: like all complex self-organizing systems—natural or industrial—they dissipate materials and energy, losing them in the process of use. The theory of self-organizing systems provides the means for the transfer of understanding of the kingdom of a natural system into one of systems that are created by human.

Studies called "industrial metabolism as a part of industrial ecology" are connected to mapping of materials and energy flows through the whole industrial system (mainly with the use of the analysis of materials—balance and expenditures—issue) (Shenshinov, 2020).

This idea draws attention to the main character of an economic system—i.e., processing of materials and energy through their transformation in production, which leads to circular money flows, primarily between manufacturers and households. Metabolic activities are usually associated with biological organisms, which are based on continuous consumption of nutritional elements, partially assimilating them in body mass and partially using the energy of nutritional elements, and producing droppings in the solid, liquid, and gas forms. Assimilation of metabolism of biological organisms to the activities of an industrial system is the use of a metaphor (Stroiteleva et al., 2020). Industrial metabolism requires further development, since the concept neglects the space–time aspects of material flows and the ties with social and human measures. Until now the concept of industrial metabolism has not been able to ensure the comprehensive analysis for the purpose of reduction of general influence on the environment, especially as to the influence connected to consumption and trade. It must be acknowledged that the engineering shift of industrial ecology has to be overcome.

3 Results

Thus, the notion "Industrial ecology" has the normative qualities that predetermine its final state, but does not describe its present state, as envisaged by the term "industrial metabolism" and its use. The idea that industrial ecology covers "analogy" and becomes the directing tool deserves and requires a more comprehensive analysis.

Comparison of systems that are created by human and natural systems constitutes the attractiveness of industrial ecology for various participants—engineers, ecologists, business management, physicists, and related academic researcher and practitioners (Tashkulova & Kletskova, 2020).

Time demands that the general concepts of ecology—like biodiversity—be used and moved to an industrial organization. An earlier or more inclusive attempt of movement aims at stimulating industrial ecological development, understanding it from the point of view of dynamics of natural ecosystems (Ragulina et al., 2019).

This field is based on the natural and biological "ecosystem analogy", since it is mentioned almost in every publication devoted to the term "Industrial ecology". Due to this reason, "analogy" deserves close attention and evaluation, since it informs and forms applications in industrial ecology.

Industrial symbiosis. This defining term for industrial ecology is a contribution not from the studies of biological systems but from application of an industrial organization, which became well-known as the most successful implementation of industrial ecology and which had appeared before the concept of industrial ecology emerged. This is a system of waste exchange (or rather by-products exchange).

4 Conclusion

Industrial symbiosis describes individual relations between the economic subjects of an industrial park and substance and energy exchange for their mutual benefit, in particular by using the flows of waste and energy resources. The whole industrial park of productions, which consist of separate symbiotic relations was called an "industrial ecosystem". Dissemination of industrial symbiosis leads to optimization of the effectiveness of material and energy flows within large-scale industrial processes.

References

Adarina, R., Gazukina, Y., & Yankovskaya, K. (2019). Indicators of the "green economy" as a tool for monitoring the regional economy. IOP Conference Series: Earth and Environmental Science. In *International Conference on Sustainable Development of Cross-Border Regions 2019*, SDCBR 2019, 012107.

Alekseev, A., Savenkov, D., Shvakov, E., Lebedev, N., & Ragulin, A. (2018). Financial problems of regional development and the ways of their solution in contemporary, Russia. *Quality—Access to Success, 19*(2), 71–75.

Avkopashvili, P., Polukhin, A., Shkodinsky, S., & Poltarykhin, A. (2019). The fundamental provisions of the concept of knowledge economy. *Studies in Systems, Decision and Control, 169*, 57–64.

Bogoviz, A., Shvakov, E., Tretyakova, O., Zakharov, M., & Abramov, A. (2020). *Globalization of education in the conditions of formation of the global knowledge economy: regularities and tendencies*. Growth Poles of the Global Economy: Emergence, Changes and Future Perspectives. Series "Lecture Notes in Networks and Systems" Plekhanov Russian University of Economics, Luxembourg (pp. 993–1000).

Dragunov, A., & Shenshinov, Y. (2020). *Effective development of innovative technologies as a method for increasing labor productivity in industry*. E3S Web of Conferences.

Glotko, A., Sycheva, I., Dunets, A., Poltarykhin, A., Zhuravlev, P., & Tubalets, A. (2019). Integration processes in the agro-industrial complex of the regions: Development and environmental problems. *Journal of Environmental Management and Tourism, 10, 3*(35), 613–621.

Glotko, A., Voronkova, O., Miheeva, N., Skoblikova, T., Alekhina, N., & Zhminko, A. (2019). The economic development of a transboundary region under the conditions of the Eurasian economic union. In *Proceedings of the 33rd International Business Information Management Association Conference, IBIMA 2019: Education Excellence and Innovation Management through Vision 2020*. 33, Education Excellence and Innovation Management through Vision 2020 (pp. 7957–7968).

Karataev, E., Merkuryev, V., & Titova, O. (2020). *Integrating the "green economy" into the model of digital future of the modern socio-economic systems*. Lecture Notes in Networks and Systems, 129 LNNS (pp. 1402–1410).

Nekhvyadovich, L., Kostin, K., & Mironov, V. (2020). *Government and public monitoring of entrepreneurship in a multicultural region: new perspectives in the conditions of the digital economy*. Public Administration and Regional Management in Russia. Challenges and Prospects in a Multicultural Region. Series "Contributions to Economics" Switzerland (pp. 147–154).

Ragulina, Y., Semenova, E., Avkopashvili, P., Dmitrieva, E., & Cherepukhin, T. (2019). Top-priority directions of implementing new internet technologies on the territories of rapid economic development. *Advances in Intelligent Systems and Computing, 726*, 182–188.

Shenshinov, Y. (2012). *Labour quality management*. North-east Asia academic forum.

Shenshinov, Y. (2020). Abdulsattar Al-Ali–PhD. The tools of increasing efficiency of human resource in the lean production environment: Conceptual study. *International Journal Of Core Engineering & Management, 6*(7).

Stroiteleva, T., Petrova, E., & Yungblyudt, S. (2020). Responsible information society as a social environment for creation of the "green" digital economy: Formation and monitoring. *Lecture Notes in Networks and Systems, 129*, 402–409.

Tashkulova, G., & Kletskova, E. (2020). Place and role of human in the system of circular reproduction in the digital regional economy. *Lecture Notes in Networks and Systems, 129*, 943–951.

Vukovich, G., Makuschenko, L., Bateykin, D., Titova, O., & Dobrosotskiy, V. (2018). *Support of the Territories of Advanced Economic Development on Human Capital: Theory and Practice Quality—Access to Success, 19*(2), 157–160.

Development of Tools for Managing Economic Risks When Designing Well Pattern to Implement the Sustainable Development Concept

Nelly N. Galeeva, Anastasia M. Rogacheva, Yuliya R. Rudneva, Lubov I. Vanchukhina, and Anton Y. Gavrilov

Abstract The paper aims to develop a method for assessing the economic risks associated with designing the optimal density of wells of oil facilities through adaptational geological and field models. The theoretical and methodological basis of the research includes scientific works devoted to the theory and practice of risk assessment in subsoil use, including in the field of oil and gas production. The methodological basis includes a systematic approach to the studied processes, methods of comparative analysis, systematization and synthesis of doctrinal provisions and practical materials, analytical procedures, and other methods. The paper identifies and analyzes the risks accompanying the design of the optimal density of the well pattern. These risks include geological (the difference of heterogeneity coefficient, sandiness, oil saturation, permeability, porosity, etc.), physical and chemical (the difference of oil viscosity, oil density, tar and asphaltene content in oil, etc.), and economic (changes in crude oil prices, electricity tariff, capital costs, and other operating costs). Of the listed risks, the authors identify the factors predetermining the emergence of economic risks and determine their importance. The authors propose a comprehensive methodology for assessing the risks associated with designing the optimal density of well pattern based on sensitivity analysis and the method of scenarios. The proposed methodology allows one to identify the most significant risks and determine the likelihood of different project development scenarios.

Keywords Reservoir engineering · Risk management · Sustainable development · Economic risks

JEL codes L710 · M190 · M140 · O130

N. N. Galeeva (✉) · A. M. Rogacheva · Y. R. Rudneva · L. I. Vanchukhina · A. Y. Gavrilov
Ufa State Petroleum Technological University, Ufa, Russia

A. M. Rogacheva
e-mail: izabo@bk.ru

A. Y. Gavrilov
e-mail: anton.gavrilov1997@mail.ru

E. G. Popkova and B. S. Sergi (eds.), *Geo-Economy of the Future*,
https://doi.org/10.1007/978-3-030-92303-7_8

1 Introduction

As a rule, the most important and riskiest oil and gas operations are concentrated in the processes of the upstream sector (Suda et al., 2015), which includes the exploration and production of hydrocarbons.

In the context of implementing the concept of sustainable development and in connection with the progressive depletion of the resource base, the rationalization of the development of oil and gas fields is of particular relevance. The most urgent areas of rational use of oil and gas resources are the introduction of new methods allowing for rapid business decisions to ensure a high level and rate of oil production and increase oil recovery. In this case, a rational system of developing the oil and gas field will be considered when the system will provide the highest possible level and rate of oil production at minimum costs (labor, material, and financial). This fact prompts to increase the responsibility of research organizations for design decisions and decision-making in business systems based on the feasibility evaluation of design studies (Leybert & Khalikova, 2020a, b).

One of the most important factors determining the rational system of field development is selecting the optimal density of the well pattern—a time-consuming process associated with various risks. Identifying risks and factors causing them will reduce unintended costs and increase the payback of designing the optimal density of wells of oil facilities through adaptational geological and field models.

In terms of the economic security of the country, industry, and individual companies using subsoil, the greatest importance is represented by the economic risks of appropriate levels. The development of tools for economic risk management, taking into account industry specifics and peculiarities of production stages, is in the sphere of interest of modern risk management. This fact determines the relevance of this research.

2 Methodology

Many scientists are engaged in studying various aspects of risk assessment and management in subsoil use. Their works lay down the theoretical and methodological foundations of this problem. The scientific community considers branch specificity of risks in the oil and gas sphere, both by types of subsoil use (Khatataeva, 2011; Murygin & Zhulanov, 2019; Murzin & Rogova, 2017; Rose, 2011; Tolstonogov & Kiforenko, 2014; Vasiliev et al., 2014) and by their individual stages (Kotlyarov & Petrov, 2015; Kraynova, 2009; Lazarenko, 2003; Suda et al., 2015).

The analysis of the development of the research topic allows us to conclude that, despite the considerable number of scientific works on the assessment of economic risks in the development of oil fields, insufficient attention is paid to the issues of assessing economic risks, especially when designing the optimal density of wells pattern.

There is practically no universal structured system of factors in the activities of oil and gas production companies (Zubarev et al., 2016). Analyzing publications on the subject of risks in the oil industry, we considered their main types: geological risks, infrastructure risks, risks associated with government policy, economic risks, business partnership, technical, and production risks. The considered types of risks may be caused by each other; some risks may be included in several types of risks.

One of the elements of risk management is risk assessment.

Since the development of oil field can be presented in the form of an investment project, the combined approach (including sensitivity analysis and the method of scenarios with the elements of the analysis of probability distributions) is most applicable for risk assessment when designing the optimal density of wells pattern through adaptational geological and field models. This will allow us to form a significant set of risks and obtain a visual picture of the development of events, taking into account the most significant factors.

The method of adaptational geological and production models is a statistical method of forecasting the oil recovery factor of oil objects (at different values of water cut of production fluid) based on the calculation of oil recovery factor using mathematical relationships. These mathematical relationships include the coefficients, selected experimentally by results of a large group of fields, and geological, physical, and technological parameters multiplied by these coefficients.

The density of the well pattern characterizes the ratio of the field area within the oil-bearing contour to the number of wells drilled. On different fields, both sparse and frequent patterns can be the optimal density. It depends on the geological and physical parameters of the reservoir and its geological structure.

The methodology of designing the optimal density of well pattern density through adaptational geological and field models consists of 12 steps:

1. Finding principal components by object parameters;
2. Identification of the object of development;
3. Prediction of a water–oil factor of the object at different water cut;
4. Forecasting specific oil production per one well that was in operation;
5. Forecasting the total development time for different values of the stock of wells that have been in operation;
6. Forecasting the accumulated volume of water injection over the entire development period;
7. Finding the value of volumetric reserves attributable to one well that has been in production at different points in time;
8. Forecasting the final oil recovery factor according to the adaptational model for different values of the well stock that has been in production;
9. Filtering values of volumetric reserves attributable to one well that has been in production, at different points in time for different values of the overstock, which are within the permissible limits;
10. Calculation of well pattern density at different values of the stock that has been in operation;

11. Calculation of net cash flow values at different values of the stock that has been in operation;
12. Plotting the dependence of profit from developing the object and the oil recovery factor on the density of the well pattern and the value of the volume of reserves per one well, and analyzing the results.

This methodology was used on the example of a development object related to terrigenous sediments exploited in the fields located in the Volga-Ural oil and gas province. It was found that the maximum net cash flow is achieved with a density of wells from 35 to 38 ha per well.

Since the methodology for designing optimal well pattern density is based on selecting the density of well pattern that will provide the greatest net cash flow during the development of the facility, the design process can be considered a payback process. With the rational use of resources and competent and detailed design of the development, the profit from the sale of oil exceeds the cost of well construction and operation of development facilities.

The methodology includes many economic indicators, changes in which cause economic risks. The paper concludes that economic risks should be assessed at stages 11 and 12 of the methodology since these stages involve the calculations of economic indicators.

3 Results

The research investigates the risk of oil production enterprises, including environmental, geological, environmental, foreign policy, transport, technological, and economic risks. This research allowed us to form the author's approach to classifying the factors of risk events in the design of the optimal density of wells pattern through adaptational geological and field models. It is proposed to divide these risks into the following groups:

- Geological (difference of heterogeneity coefficients, sandiness, oil saturation, permeability, porosity, etc.);
- Physical and chemical (difference of oil viscosity, oil density, tar and asphaltene content in oil, etc.);
- Economic (change of crude oil price, electric power tariff, and capital and other operation costs).

The paper proposes a comprehensive methodology for assessing the risks associated with designing the optimal density of well pattern, consisting of two stages.

The first stage of a comprehensive assessment of economic risks includes analyzing the sensitivity of the investment project of oil field development with the design of the optimal pattern of wells through adaptational geological and field models using the identified set of factors of risk events.

The next step in the complex methodology is conducting the scenarios of the development project with the design of the optimal density of the well pattern through adaptational geological and field models with elements of probability distributions, taking into account the most significant risks identified by the sensitivity analysis.

Sensitivity analysis will be performed by changing the indicators (decrease and increase by 5%, 10%, and 15% of their estimated value) and calculating the change in net cash flow for a particular change in the indicator by the following formula (1):

$$\Delta NCF_{i\%,j} = \frac{NCF_{i\%,j}}{NCF_0} \times 100\% - 100\%, \quad (1)$$

where:

$\Delta NCF_{i\%,j}$—percentage change in net cash flow at the value of risk factor *j* changing by *i* percent;

$NCF_{i\%,j}$—value of net cash flow in rubles at the value of risk factor *j* changing by *i* percent;

NCF_0—value of net cash flow in rubles at the correct determination of the optimal density of wells on an oil facility.

Let us consider the methodology on the example of risk assessment when designing the optimal density of oil and gas wells.

Sensitivity analysis of changes in geological parameters showed that the most significant negative risk factor is the error in determining the heterogeneity coefficient. The influence of error in the measurement of physical and chemical parameters on the NCF is, on average, greater than the influence of geological parameters. The most significant negative risk factor is the error in determining the oil's viscosity, and the only positive factor is the error in the change in relative viscosity.

Table 1 shows the results of the analysis of the sensitivity of net cash flow to changes in economic indicators.

The most significant risk factors are changes in the price of crude oil and changes in electricity tariffs.

The ranges of changes in the indicators were identified, and statistical indicators were calculated based on the market analysis results.

Table 1 Dependence of changes in net cash flow (NCF) of the project on changes in economic risks

Risk factor: parameter change	Changes in NCF at optimum engineering, procurement, and construction in % with an error of the index's measurement						
	−15%	−10%	−5%	0%	+5%	+10%	+15%
Electricity price	**1.70**	**1.13**	**0.57**	**0.00**	**−0.57**	**−1.13**	**−1.70**
Other operating costs	0.42	0.28	0.14	0.00	−0.14	−0.28	−0.42
Capital costs	0.52	0.34	0.17	0.00	−0.16	−0.33	−0.48
Price of crude oil	**−18.62**	**−12.43**	**−6.22**	**0.00**	**6.25**	**12.51**	**18.78**

Source Compiled by the authors

From 2017 to 2019, the electricity tariff ranged from 3.80 to 5.78 rubles/kWh. The mathematical expectation is 4.69 rubles/kWh. The root-mean-square deviation is 2.87 rubles/kWh. The coefficient of variation is 61.2%. Thus, the probability of a change in the electricity tariff is high.

From 2017 to 2019, the oil price varies from 20.74 to 39.81 million rubles per thousand tons. The mathematical expectation is 29.08 million rubles per thousand tons. The root-mean-square deviation is 30.10 million rubles per thousand tons. The coefficient of variation is 103.6%. Thus, the price of oil is unstable and changes frequently.

We determined the probabilities of reaching different values of the electricity rate and oil price by rounding the electricity rate for the past three years to one decimal unit and rounding the oil price for the past three years to the nearest integer whole units. Next, we counted how many times the electricity rate and oil price reached each value obtained and then divided the number of reaching each value by the total number of values. The results of the probability calculations are shown in Figs. 1 and 2.

The graphs show that the probability of reaching the average price of oil is quite high.

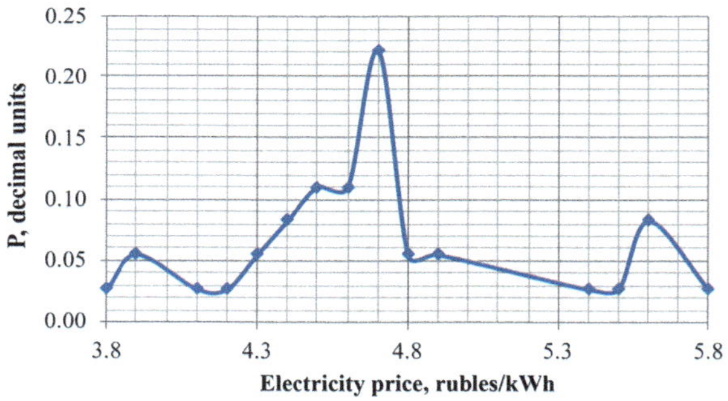

Fig. 1 Probabilities of reaching the values of the electricity price. *Source* Compiled by the authors

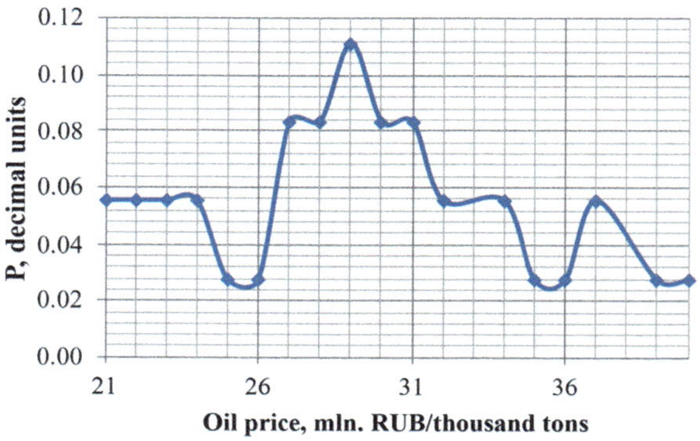

Fig. 2 Probabilities of reaching the values of the oil price. *Source* Compiled by the authors

The next stage of risk assessment is a scenario analysis of the development project with the design of the optimal density of wells pattern through adaptational geological and field models, considering the sensitivity analysis of the most significant risk factors. The algorithm of using the method of scenario analysis in assessing the risks associated with the design of optimal well pattern density includes five consecutive steps:

(1) Probabilities of achieving different values of risk factors;
(2) Assessment of the probability of scenarios of simultaneous achievement of specific values of risk factors;
(3) Calculation of the net cash flow under different scenarios;
(4) Calculation of optimal well pattern density under different scenarios;
(5) Determination of the probability of obtaining the values in the specified range.

In this paper, we calculated the probability of achieving different values of the two indicators simultaneously. The results are presented in Table 2.

The obtained values allow us to determine the probability of achieving certain ranges of net cash flow and values of well pattern density. For example, it is necessary to define the ranges of the optimal density of wells pattern and net cash flow at a possible change of oil price from 25 to 34 million rubles per thousand tons and at a possible change of electricity tariff from 4.4 to 4.8 rubles per kWh. Additionally, it is necessary to define the possibility of getting the obtained ranges of the well pattern density and net cash flow.

For this purpose, from the data presented in Table 2, we filter the values of the indicators included in the necessary ranges (Table 3). The lower limit of the net cash flow range and, consequently, the upper limit of the optimal well pattern density will be considered the scenario with the maximum electricity price (4.8 rubles/kWh) and the lowest oil price (25 million rubles per thousand tons). Substituting the values in the methodology calculation, we obtain NCF = 29,643 million rubles and $S' =$ 33.0 ha per well.

The upper limit of the net cash flow range and, consequently, the lower limit of the optimal well pattern density range will be considered the scenario with the lowest electricity price (4.4 rubles/kWh) and the highest oil price (34 million rubles/kWh). Substituting the values in the methodology calculation, we obtain NCF = 43,553 million rubles and $S' = 28.3$ ha per well.

The probability that the optimal density of well pattern and net cash flow will fall within the obtained ranges is equal to the sum of the probabilities of all scenarios within limits. In this case, P equals 0.356.

Thus, the range of optimal density of well pattern at a possible change of oil price from 25 to 34 million rubles per thousand tons and at a possible change of electricity price from 4.4 to 4.8 rubles per kWh looks like [28.3 … 30.0] ha per well. The NCF range is [29,643 … 43,553] million rubles. The probability that the optimal density of well pattern and net cash flow will fall within these ranges is 0.356.

Table 2 Probabilities of reaching different values of the electricity price and oil price simultaneously

Electricity price		Oil price, million rubles per thousand tons																	
		21	22	23	24	25	26	27	28	29	30	31	32	34	35	36	37	39	40
		P, decimal units																	
Value, rubles per kWh	P, decimal units	0.056	0.056	0.056	0.056	0.028	0.028	0.083	0.083	0.111	0.083	0.083	0.056	0.056	0.028	0.028	0.056	0.028	0.028
3.8	0.028	0.002	0.002	0.002	0.002	0.001	0.001	0.002	0.002	0.003	0.002	0.002	0.002	0.002	0.001	0.001	0.002	0.001	0.001
3.9	0.056	0.003	0.003	0.003	0.003	0.002	0.002	0.005	0.005	0.006	0.005	0.005	0.003	0.003	0.002	0.002	0.003	0.002	0.002
4.1	0.028	0.002	0.002	0.002	0.002	0.001	0.001	0.002	0.002	0.003	0.002	0.002	0.002	0.002	0.001	0.001	0.002	0.001	0.001
4.2	0.028	0.002	0.002	0.002	0.002	0.001	0.001	0.002	0.002	0.003	0.002	0.002	0.002	0.002	0.001	0.001	0.002	0.001	0.001
4.3	0.056	0.003	0.003	0.003	0.003	0.002	0.002	0.005	0.005	0.006	0.005	0.005	0.003	0.003	0.002	0.002	0.003	0.002	0.002
4.4	0.083	0.005	0.005	0.005	0.005	0.002	0.002	0.007	0.007	0.009	0.007	0.007	0.005	0.005	0.002	0.002	0.005	0.002	0.002
4.5	0.111	0.006	0.006	0.006	0.006	0.003	0.003	0.009	0.009	0.012	0.009	0.009	0.006	0.006	0.003	0.003	0.006	0.003	0.003
4.6	0.111	0.006	0.006	0.006	0.006	0.003	0.003	0.009	0.009	0.012	0.009	0.009	0.006	0.006	0.003	0.003	0.006	0.003	0.003
4.7	0.222	0.012	0.012	0.012	0.012	0.006	0.006	0.019	0.019	0.025	0.019	0.019	0.012	0.012	0.006	0.006	0.012	0.006	0.006
4.8	0.056	0.003	0.003	0.003	0.003	0.002	0.002	0.005	0.005	0.006	0.005	0.005	0.003	0.003	0.002	0.002	0.003	0.002	0.002
4.9	0.056	0.003	0.003	0.003	0.003	0.002	0.002	0.005	0.005	0.006	0.005	0.005	0.003	0.003	0.002	0.002	0.003	0.002	0.002
5.4	0.028	0.002	0.002	0.002	0.002	0.001	0.001	0.002	0.002	0.003	0.002	0.002	0.002	0.002	0.001	0.001	0.002	0.001	0.001
5.5	0.028	0.002	0.002	0.002	0.002	0.001	0.001	0.002	0.002	0.003	0.002	0.002	0.002	0.002	0.001	0.001	0.002	0.001	0.001
5.6	0.083	0.005	0.005	0.005	0.005	0.002	0.002	0.007	0.007	0.009	0.007	0.007	0.005	0.005	0.002	0.002	0.005	0.002	0.002
5.8	0.028	0.002	0.002	0.002	0.002	0.001	0.001	0.002	0.002	0.003	0.002	0.002	0.002	0.002	0.001	0.001	0.002	0.001	0.001

Source Compiled by the authors

Table 3 Values of the indicators within the required ranges

Electricity price		Oil price, million rubles per thousand tons								
		25	26	27	28	29	30	31	32	34
		P, decimal units								
Value, rubles per kWh	P, decimal units	0.028	0.028	0.083	0.083	0.111	0.083	0.083	0.056	0.056
4.4	0.083	0.002	0.002	0.007	0.007	0.009	0.007	0.007	0.005	0.005
4.5	0.111	0.003	0.003	0.009	0.009	0.012	0.009	0.009	0.006	0.006
4.6	0.111	0.003	0.003	0.009	0.009	0.012	0.009	0.009	0.006	0.006
4.7	0.222	0.006	0.006	0.019	0.019	0.025	0.019	0.019	0.012	0.012
4.8	0.056	0.002	0.002	0.005	0.005	0.006	0.005	0.005	0.003	0.003

Source Compiled by the authors

4 Conclusion

The most appropriate method of risk assessment when designing an optimal density of well pattern through adaptive geological and field models is a combined approach to risk assessment that includes sensitivity analysis and scenario method with the elements of the analysis of probability distributions, which will allow forming a meaningful set of risks and getting a visual picture of the development of events, taking into account the most significant risks.

Since the methodology for designing optimal density of well pattern is based on selecting the density that will provide the greatest net cash flow during the development of the facility, the design process can be considered a payback process. With the rational use of resources and competent and detailed design of the development, the profit from the sale of oil exceeds the cost of the construction of well and the operation of development facilities. The methodology includes many economic parameters, changes in which cause economic risks. Economic risks are present in stages 11 and 12 of the methodology because these stages involve the calculations of economic indicators. The paper revealed that the most significant economic risks are changes in oil and electricity prices.

The authors proposed a comprehensive methodology for assessing the risks associated with designing the optimal density of well pattern, based on the use of sensitivity analysis and the method of scenarios, which allows us to identify the most significant risk factors, and then, using these factors to determine the likelihood of different scenarios of a project on oil field development.

References

Khatataeva, RSh. (2011). The factors influencing level of economic risk. *Terra Economicus, 9*(4–3), 46–49.

Kotlyarov, I. D., & Petrov, S. V. (2015). Risk consideration in geological, economical and cost evaluation of deposits. *Mine Surveying and Subsurface Use, 1*(75), 8–13.
Kraynova, E. A. (2009). Economic mechanism of project risk management in the development of oil and gas resources. *Journal of Mining Institute, 184*, 144–149.
Lazarenko, M. A. (2003). Assessment of investment risk in the oil industry in designing the development of oil fields of poorly explored areas. *Journal of Mining Institute, 155*(1), 237–239.
Leybert, T. B., & Khalikova, E. A. (2020). Economic evaluation of the effectiveness of design solutions for the installation of a compressor station for the preparation and transportation of associated petroleum gas. *SOCAR Proceedings, 5*, 70–78. https://doi.org/10.5510/OGP20200100425
Leybert, T. B., & Khalikova, E. A. (2020). Current tendencies of transformation of the Russian practice of decision making in business systems. In E. G. Popkova & A. V. Chesnokova (Eds.), *The leading practice of decision making in modern business systems* (pp. 13–26). Emerald Publishing.
Murygin, A. V., & Zhulanov, E. E. (2019). Method of evaluating the impact of industrial risks on the efficiency of investment projects for developing missed hydrocarbon deposits. *PNRPU Sociology and Economics Bulletin, 4*, 275–289. https://doi.org/10.15593/2224-9354/2019.4.21
Murzin, A. D., & Rogova, T. M. (2017). Identification of ecological and economics risks of mining industry. *Mining Informational and Analytical Bulletin, 5*, 97–104.
Rose, P. R. (2011). *Risk analysis and management of petroleum exploration ventures* (V. I. Poroskun Transl. from English; N. A. Malyshev Ed.). Moscow-Izhevsk, Russia: Research Center "Regular and Chaotic Dynamics"; Izhevsk Institute for Computer Research. (Original work published 2001).
Suda, K. A., Rani, N. S. A., Abdul Rahman, H., & Wang, C. (2015). A review on risks and project risks management: Oil and gas industry. *International Journal of Scientific and Engineering Research, 6*(8), 938–943.
Tolstonogov, A. A., & Kiforenko, I. K. (2014). The principles of investment project formation of oil field development considering the influence of risks. *Fundamental Research, 6–3*, 577–580.
Vasiliev, Yu. N., Gimadeyeva, R. N., & Ilnitskaya, V. G. (2014). Uncertainties and risks of gas field development design and management. *Vesti Gazovoy Nauki, 4*(20), 16–22.
Zubarev, A. A., GluKhova, M. G., & Makovetskaya, E. G. (2016). Systematization of the risk factors of oil and gas companies. *Proceedings of the Voronezh State University of Engineering Technologies, 3*(69), 370–374. https://doi.org/10.20914/2310-1202-2016-3-370-374

Impact of the COVID-19 Pandemic on the World Steel Market

Irina N. Belova, Ravil M. Chusmakaev, Filipp D. Belov, and Elena A. Egorycheva

Abstract The article analyzes the activities of steel companies during the COVID-19 pandemic in 2020. The COVID-19 pandemic and the subsequent economic crisis have become one of the most serious challenges for world steel companies. The consequences, which are expressed in a decrease in demand for steel products from the key consumer sectors—automotive industry and construction, and other consumers, have caused a high level of uncertainty in the international steel market, and have identified new development vectors that should help overcome the crisis in the steel industry. The article assesses the impact of the pandemic and current global economic crisis on the global steel industry, including Russia; identifies trends and prospects for the development of the global steel market; identifies key areas for the recovery of the steel industry after the COVID-19 pandemic.

Keywords World market · Steel · Rolled metal · COVID-19 pandemic · Crisis · Economy · Efficiency · Industry · Demand

JEL Codes F01 · F14 · L10 · L11

I. N. Belova (✉) · R. M. Chusmakaev · E. A. Egorycheva
Peoples' Friendship University of Russia (RUDN University), Moscow, Russia
e-mail: belova_in@rudn.ru

R. M. Chusmakaev
e-mail: Chusmakaev@inbox.ru

E. A. Egorycheva
e-mail: egorycheva-ea@rudn.ru

F. D. Belov
Russian Research Institute of Economics, Politics and Law in Science and Technology, Moscow, Russia

E. G. Popkova and B. S. Sergi (eds.), *Geo-Economy of the Future*,
https://doi.org/10.1007/978-3-030-92303-7_9

1 Introduction

The metallurgical industry is one of the main branches in the structure of industrial production and belongs to the basic global branches of heavy industry. A well-developed and competitive heavy industry plays a crucial role in ensuring stable and efficient economic growth for any country.

Mechanical engineering, construction, metalworking industry, oil and gas industry and railway industry are among the main consumers of steel products. It is worth noting that the metallurgical industry is closely related to the chemical industry, as the latter makes it possible to create new chemical compositions that are parts steel products, which makes it possible to launch new steel grades on the market or improve the quality of existing grades (PROMZN, 2018).

Due to the continuous improvement of production processes, as well as increasing the efficiency of technological processes, Russia is becoming an increasingly competitive player in the global ferrous metals market, becoming the top 10th leading country in the production of steel products (Fig. 1).

It is worth noting that the world market of steel products is influenced by certain factors that may have both positive and negative impacts on the metallurgical industry.

To begin with, let's consider the factors that favourably affect the metallurgical industry and contribute to its development:

- Stable growth from key consumers;
- Favourable economic climate;
- Non-aggressive trade policy-easing anti-dumping regulation;
- Uninterrupted delivery of steel products by land and sea;

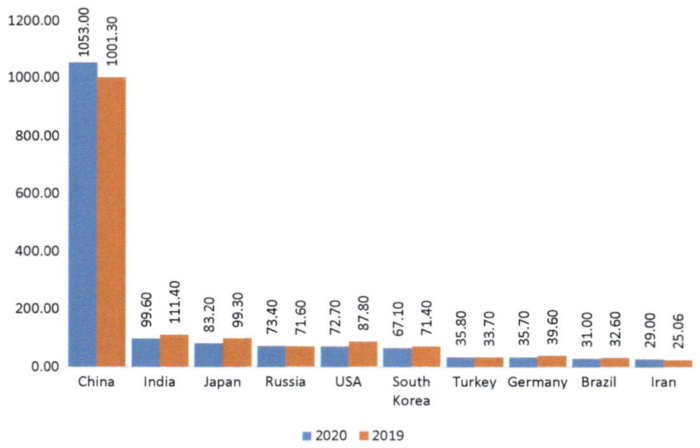

Fig. 1 Top 10 steel-producing countries. *Source* https://www.worldsteel.org/media-centre/press-releases/2021/February-2021-crude-steel-production.html

- Stability in the development of related industries: oil and gas industry, automotive industry and others;
- Relatively cheap iron ore and coal.

At the same time, the realities of the modern economy are such that any producer is in a state of uncertainty and instability, which is caused by the following factors:

- Strict trade regulation of product exports;
- Introduction of duties on certain types of products;
- Anti-dumping regulation;
- Economic crisis;
- Reduced activity in related sectors of the economy;
- Falling demand from key consumers as a result of economic instability.

Since the beginning of 2020, the COVID-19 pandemic, which has spread globally, has led to a decline in international trade, due to the introduction of a quarantine regime, transport restrictions, reduced demand, and business activity of manufacturers.

In this regard, it is quite relevant to analyze the impact of the COVID-19 pandemic on the world market of steel products and the activities of metallurgical enterprises in these conditions.

2 Methodology

To achieve this goal, the authors relied on the work of Russian economists, specialists in the field of metallurgy and trade: Zainullin (2021); information and analytical resources and websites such as World Steel Associations (2021), RIA Rating (2020), TASS (2020), and Deloitte in the CIS (2020).

The article uses quantitative, qualitative, and empirical research methods.

3 Results

According to the World Steel Association, the total volume of steel produced in 2020 was 1864 million tons of steel, which is only 0.9% less than in 2019 (Worldsteel Association, 2021).

As a result of the beginning of the crisis caused by the COVID-19 pandemic, in 2020, the metallurgical industry faced one of the main challenges—a drop in demand for steel, a reduction in production and the subsequent shutdown of production capacities, a break in production and technological chains.

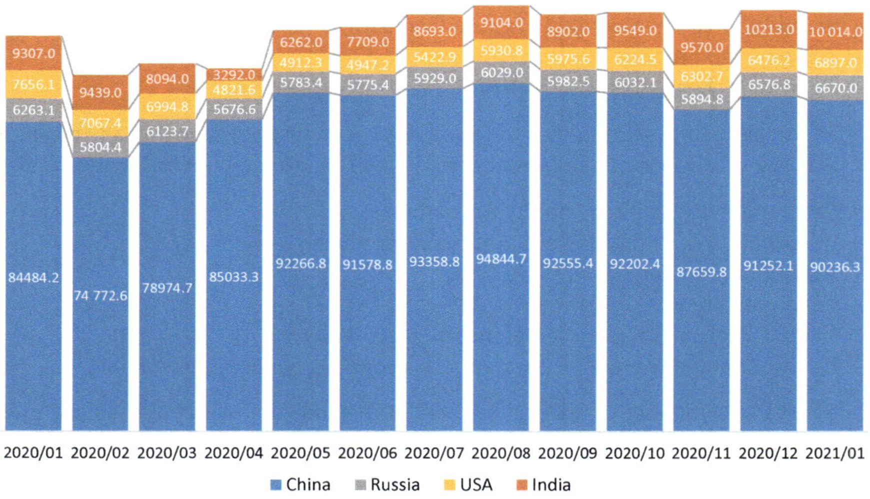

Fig. 2 Steel production volumes, thousand tons, monthly. *Source* https://extranet.worldsteel.org/statistics/steel-data-viewer/MCSP_crude_steel_monthly/CHN/IND

Since mid-May 2020 (Fig. 2), economic activity has started to recover, which can be described as a V-shaped recovery.[1] However, it is worth noting that this was not enough to compensate for the losses that occurred during the world lockdown. Most of the sectors of the economy that consume steel products remained below their level of activity than before the COVID-19 epidemic (Worldsteel Association, 2020). Due to the onset of the second wave of COVID-19, the recovery of global markets remained fragile and uncertain, as social restrictions were hardly eased, which in turn contributed to the continued growth of unemployment, weak confidence of industrial sectors in the public sector, as well as the recovery of demand for products from the most important sectors of the economy.

China, the world steel leader played a major role in the recovery and maintenance of steel demand, as it first of the other countries to lift restrictions and thus prevented the shutdown of steel production and a critical decline in activity in the steel industry. Since the end of February 2020, China has shown a very rapid recovery, which has been steady and projected China's GDP growth in 2020, despite GDP shrinking by 6.8% in the first quarter of 2020 (Worldsteel Association, 2020).

It is worth noting that in 2019, the Chinese steel industry accounted for more than 50% of global steel production, but the COVID-19 pandemic certainly had a negative impact on the Chinese steel industry, but in the 3rd quarter of 2020, China managed to increase its production by 3.4% (Deloitte in the CIS, 2020). China managed to demonstrate positive dynamics in steel production not only due to the support from the state but also since in 2016 China upgraded its old capacities (steelmaking units)

[1] The V-shaped recovery reflects an economic situation where a major downturn in the markets meets an equally strong upturn in the markets.

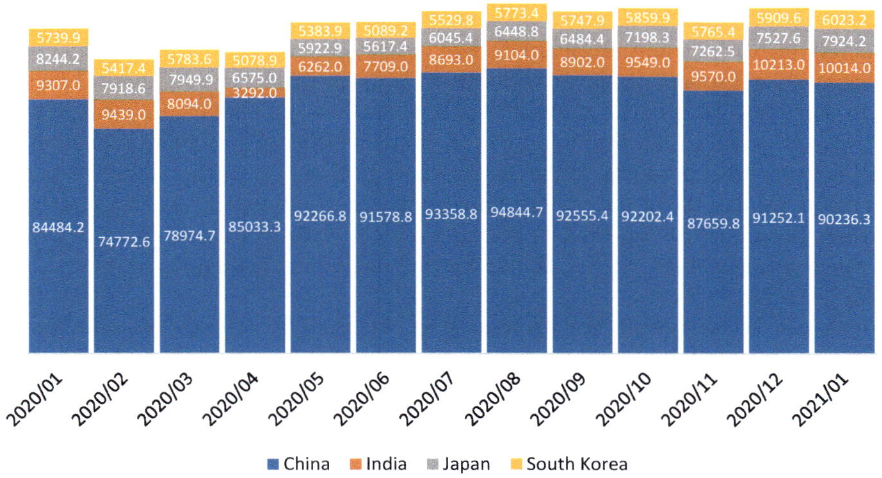

Fig. 3 Steel production in Asia, thousand tons, monthly. *Source* https://extranet.worldsteel.org/statistics/steel-data-viewer/MCSP_crude_steel_monthly/CHN/IND

to more modern and high-tech ones, which allowed maintaining the total production capacity of about 1.2 billion tons of steel per year since 2016. It is worth noting that thanks to the actions of the Chinese government, in 2020, China, as the state that first introduced restrictions in connection with the beginning of the COVID-19 pandemic, was the first to lift these restrictions, thereby supporting the demand for metal products in the world.

In the diagram above (Fig. 3) the dynamics of the volume of steel products produced in Asian countries is presented. It is worth noting that due to China, as well as its government support programs and the growth of domestic consumption, the Asian market grew by 1.5% in 2020. In China, growth was 5.2%, and the global production share of the Asian market increased from 53.3% to 56.5% in 2020.

In March 2021, the National Bureau of Statistics of the People's Republic of China (NBS) adjusted data on the volume of steel production in China for 2020. The adjusted figure for smelting volumes was 1065 million tons, compared to the initial figure of 1054 million tons. Compared to the volumes of smelting in 2019, the growth in 2020, taking into account the adjusted indicator, was 7% (AIS "Metal Supply and Sales", 2021).

At the same time, the remaining major producers of the Asian market, such as India (second largest in the world) and South Korea (sixth largest in the world) were able to restore their pre-crisis production volumes only by the end of the year. In Japan (the third place in the world), the drop was 16.2% and the end of the year failed to restore the pre-crisis level. Given that China is the leader not only in the world but also in the Asian steel market, the overall dynamics, taking into account the decline in production in some Asian countries, did not affect the positive overall annual dynamics of production in Asian countries as a whole.

European countries were hit pretty hard even before the COVID-19 pandemic started. In 2019, there was a sharp decline in steel production by 5.4% at the largest steel mills in Europe, namely, ArceloMittal, USS Slovakia, Arvedi, SSAB, Liberty, and Salzgitter. The gradual recovery that began at the beginning of 2020 was slowed down again due to the beginning of the crisis, which led to a sharp reduction in demand for steel products, to the shutdown of steelmaking units, which led to the postponement of the restart of blast furnaces and their subsequent shutdown, as well as the closure of rolling lines at the largest steel mills in Europe (ArcelorMittal, Thyssenkrupp, Libery, SSAB, British Steel, Tata and many others) (Deloitte in the CIS, 2020).

As can be seen from the diagram (Fig. 4), there is a V-shaped recovery in the economic activity of the metallurgical sector in Europe, characterized by the instability of production volumes.

The North American steel market was characterized by a decline and then a recovery process, but it was not possible to achieve pre-crisis indicators by the beginning of 2021 (Fig. 5).

Currently, it is difficult for the US government to stabilize the market due to the uncontrolled situation with the spread of COVID-19, and if this dynamics is stable, there may be a decline in the recovery momentum (Metals of Eurasia, 2020). According to analysts, it is expected that 2021 will not be less optimistic, as there is a high probability of a repeated lockdown of the world economy and the beginning of the third wave of COVID-19, and a new vector of development of the American economy as a result of the choice of the new US President John Biden, as well as rather pessimistic prospects for the development of the most important sectors of the US economy, which in turn determine the demand for American steel products and reflect the state of the American economy (Metals of Eurasia, 2020).

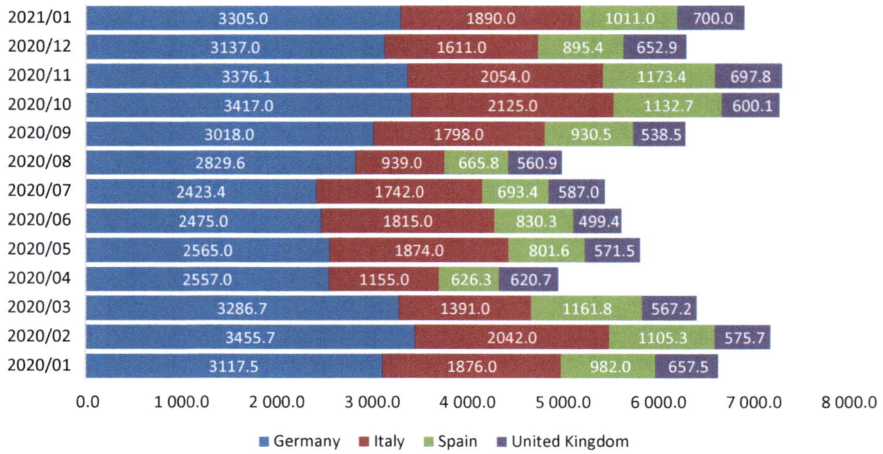

Fig. 4 Steel production in Europe, thousand tons, monthly. *Source*: https://extranet.worldsteel.org/statistics/steel-data-viewer/MCSP_crude_steel_monthly/CHN/IND

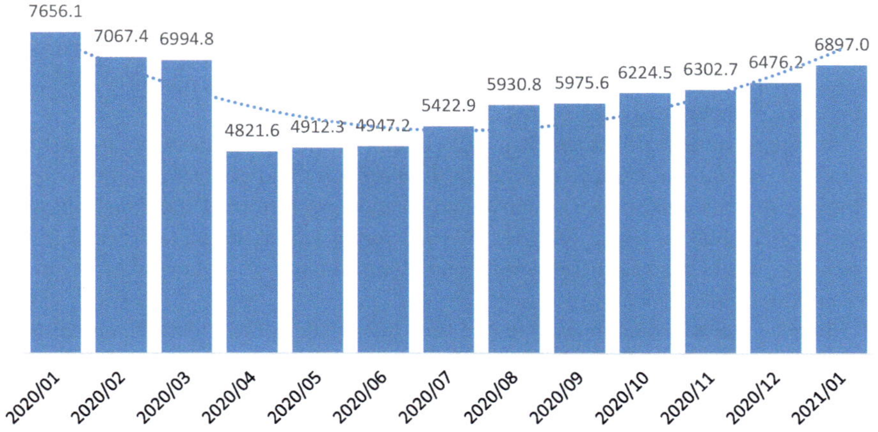

Fig. 5 Steel production in the United States, thousand tons, monthly. *Source* https://extranet.worldsteel.org/statistics/steel-dataviewer/MCSP_crude_steel_monthly/CHN/IND

In 2021, new capacities of steel companies will be put into operation in North America: Nucor, Steel Dynamic, and Big River Steel, some of which were postponed until 2023–2025 (Deloitte in the CIS, 2020). This step was made due to the forced and critical situation in the North American steel industry.

The global crisis and COVID-19 also affected the Russian steel products market (Fig. 6), causing a decline in domestic demand in Russia, exports, as well as the suspension of one of the largest consumers of steel pipes—the Nord Stream project.

It should be noted that Russian steel companies have demonstrated a high level of technological development, as well as a large accumulated experience in optimizing

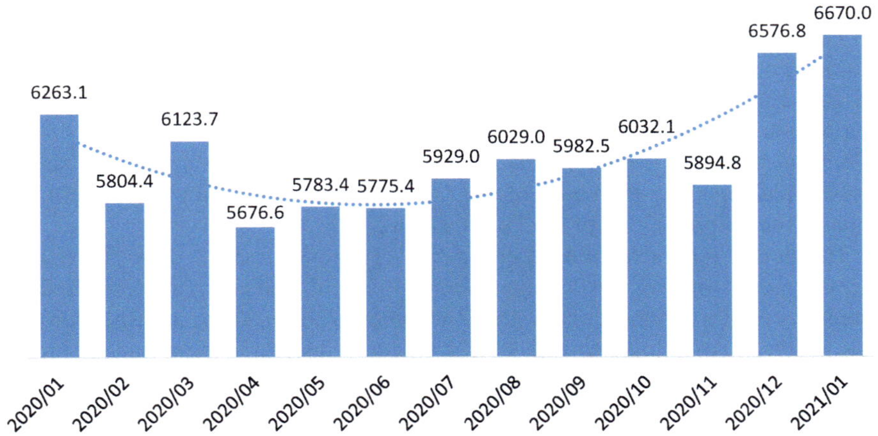

Fig. 6 Steel production in Russia, thousand tons, monthly. *Source* https://extranet.worldsteel.org/statistics/steel-data-viewer/MCSP_crude_steel_monthly/CHN/IND

the business process in a period of uncertainty. The combination of these qualities allowed us to quickly adapt to the changing situation during the crisis, as well as due to the expanded presence of several Russian companies abroad, diversify risks, and compensate for losses in some areas.

In general, in the global economy, the process of restoring demand and activity from key consumers, namely the automotive industry and construction, due to the removal of strict restrictions, contributed to the restoration of the functioning of metallurgical plants. Since October 2020, a rapid restart of metallurgical plants around the world has begun, and more than 30 million tons of hot metal has been put into production.

Most steel companies in Europe and the United States have already started their work, but there are still production delays, as it is necessary to increase the lost capacity during the pandemic. At the same time, this was not sufficient to meet the increased demand in key markets, as stocks, which are one of the prerequisites for the effective functioning of the metallurgical plant, fell to historical lows. Restocking across the entire steel value chain in Europe and the United States created additional demand that steel companies initially could not cope with, and this was the result of a sharp increase in prices in all regions by the end of 2020. Fitch Ratings expects prices to peak by the end of Q1 2021.

4 Conclusion

The impact of COVID-19 on the steel industry is one of the most important issues for shaping the future strategies of steel companies. The steel industry is one of the few among other heavy industries that have survived 2020 with almost minimal losses compared to 2019 and entered a new year 2021. Global steel production for 2020 decreased by 3.2%, while the Chinese steel market grew by 4.5% (Ria Rating, 2020).

The worst result for 2020 was shown by the titanium and pipe rolling industry. It is worth noting that a significant decline in pipe production was caused by the global crisis, a decline in economic activity in the oil industry, as well as a result of the completion of major pipeline projects.

It is worth noting that the timely response of the Chinese government to the new global challenge was the result of the fact that China became the most efficient country for the production of steel products in 2020.

The United States experienced strong recovery growth in 2020, driven by government support. However, the downturn was less painful than previously expected and forecast, even though most of the steel companies in North America suspended their production due to falling demand, which caused a reduction in the volume of orders. In the third quarter of 2020, steel production in North America decreased by 19.8% (Deloitte in the CIS, 2020).

Due to reduced demand for steel products and an actual shortage of orders that would require the use of all available units of metallurgical companies, about 30%

of metallurgical capacities worldwide, except for China, were suspended or were not operating at full capacity (Fitch Wire, 2021).

The recovery of demand and activity from key consumers, namely the automotive industry and construction, due to the lifting of strict restrictions, contributed to the restoration of the functioning of metallurgical plants. At the same time, this was not sufficient to meet the increased demand in key markets, as stocks, which are one of the prerequisites for the effective functioning of the metallurgical plant, fell to a historic low. Restocking across the entire steel value chain in Europe and the United States created additional demand that steel companies initially could not cope with, and this was the result of a sharp increase in prices in all regions by the end of 2020.

Since October 2020, a rapid restart of metallurgical plants around the world has begun, and more than 30 million tons of hot metal has been put into production. Most steel companies in Europe and the United States have already started their work, but there are still production delays, as it is necessary to increase the lost capacity during the pandemic.

According to most experts, metal prices depend on the pace of economic recovery in three key metallurgical regions: China, the United States, and Europe. To date, China has recovered and already in February 2021 showed positive growth in the volume of steel produced. The United States is still in a serious state due to high mortality rates and COVID-19 infections, and active vaccination of the population is currently underway to stabilize the situation with the pandemic. In the EU, the situation is unstable, most European countries are closing their borders again due to the beginning of the third wave of COVID-19, and there is also a weak financial incentive.

Although the COVID-19 pandemic has become a global challenge for the entire world, including the steel industry, the Russian ferrous metallurgy market has demonstrated high resilience and flexibility, as well as the ability to make non-standard decisions in a short period. In addition to sustainability, the Russian steel industries demonstrated good performance in 2020, which allows Russian steel companies to remain competitive in the domestic and international markets.

Acknowledgements This paper has been supported by the RUDN University Strategic Academic Leadership Program and Grant of RUDN University No. 060119-0-000.

References

AIS "Metal supply and sales". (2021). *China recalculated steel production for 2020*. https://www.metalinfo.ru/ru/news/123754. Accessed 04.03.2021.

Deloitte Research Center in the CIS. (2020). *Overview of the ferrous metallurgy market*. https://www2.deloitte.com/ru/ru/pages/research-center/articles/overview-of-steel-and-iron-market-2020.html. Accessed 03.03.2021.

Fitch Wire. (2021). Fitch ratings. *The global steel price rally will be short-lived*. https://www.fitchratings.com/research/corporate-finance/global-steel-price-rally-will-be-short-lived-02-02-2021. Accessed 08.02.2021.

Metals of Eurasia. (2020). *Ahead-recovery and recovery.* http://www.eurasmet.ru/magazine/. Accessed 10.02.2021.
PROMZN.RU (Industrial portal). (2018). *The role of ferrous metallurgy in industry and economy.* https://promzn.ru/metallurgiya/chernaya.html. Accessed: 05.04.2021.
Ria Rating. (2020). *Has prepared an overview of the metallurgy industry for 9 months of 2020.* https://riarating.ru/macroeconomics/20201222/630193046.html. Accessed 12.03.2021.
TASS. (2020). *Expert: Russian metallurgists are more confident in 2020 than the rest of the world.* https://tass.ru/interviews/10017735. Accessed 13.03.2021.
Worldsteel Association. (2021). *Global crude steel production will decline by 0.9% in 2020.* https://www.worldsteel.org/media-centre/press-releases/2021/Global-crude-steel-output-decreases-by-0.9--in-2020.html. Accessed 11.04.2021.
Worldsteel Association Press Release. (2020). *Worldsteel Short Range Outlook.* https://extranet.worldsteel.org/. Accessed 04.12.2020.
Worldsteel Association Press Release. (2021). *February 2021 crude steel production.* https://www.worldsteel.org/media-centre/press-releases/2021/February-2021-crude-steel-production.html. Accessed 05.04.2021.
Zainullin, E. (2021). *Steel is easily cast. The largest metallurgists of the Russian Federation reduced output by 2.4%.* https://www.kommersant.ru/doc/4671379. Accessed 12.03.2021.

Transformation of Consumer Behavior During the COVID-19 Pandemic

Evgeniya K. Karpunina, Natalia V. Ruzhanskaya, Oksana N. Podorova-Anikina, Natalia N. Zubareva, and Roman S. Luchaninov

Abstract The purpose of the research is to study the features of consumer behavior during the pandemic, to establish its current trends, as well as to determine the vector of transformation of certain segments of entrepreneurial activity under the new model of consumer behavior. The authors assessed the structure of household consumption in Russia before the pandemic and during the pandemic. It is determined that the decline in real monetary incomes of the population during the pandemic and the conditions of uncertainty caused negative changes in consumer behavior in Russia. The authors identified changes in the structure of consumption due to the pandemic (a reduction in discretionary spending, an increase in the share of household spending on household food, a change in the ratio between food and non-food products, a reduction in the volume of paid services to the population), and also justified the emergence of new categories of consumers with a certain type of consumer behavior. The article identifies new trends in consumer behavior and suggests mechanisms for their reflection in the activities of enterprises.

Keywords Consumer behavior · Pandemic · Consumer income · Preferences · Consumption · Food products · Non-food products · Consumer habits

JEL Codes D11 · D12 · D13

E. K. Karpunina (✉)
G.R. Derzhavin Tambov State University, Tambov, Russia

N. V. Ruzhanskaya
The Komi Republican Academy of State Service and Administration (Krassa), Syktyvkar, Russia

O. N. Podorova-Anikina
A. I. Herzen Russian State Pedagogical University, St. Petersburg, Russia

N. N. Zubareva · R. S. Luchaninov
Belgorod National Research University, Belgorod, Russia

E. G. Popkova and B. S. Sergi (eds.), *Geo-Economy of the Future*,
https://doi.org/10.1007/978-3-030-92303-7_10

1 Introduction

Consumer behavior in the market is the starting point for the organization of the economic activities of enterprises. The needs and preferences of potential consumers, the factors influencing the change in requests, the causes of unsatisfied needs determine the nature of internal and external cause-and-effect relationships between the consumer and the manufacturer (Karpunina et al., 2020a). Changes in the processes of real consumption, as well as purchases of goods of various types by population groups, signal the need to adjust the strategy of the behavior of enterprises and reorient their production activities under new conditions. Tracking the changes that are taking place and promptly responding to them determine the success of the enterprise in the long term. Currently, it is difficult to predict the behavior of consumers in the new post-market realities. Indeed, the COVID-19 pandemic has not only created non-standard business conditions for enterprises, contributed to the disruption of traditional economic ties, but also transformed the daily life of ordinary people. The introduction of a lockdown, the transfer to remote work, increased attention to hygiene and health issues are just some features of the COVID reality (Karpunina et al., 2020b). In such conditions, people's attitude to shopping is changing, a new behavioral economy is being created and new trends are being formed. The study of the peculiarities of consumer behavior under the influence of the pandemic is important for predicting the processes of economic recovery and the organization of economic activity in conditions of uncertainty.

2 Literature Review

Consumer behavior is a reflection of the nature of the activities of individuals, groups or organizations related to the purchase, use and disposal of goods and services, including their emotional, mental and behavioral reactions preceding or following these actions (Parsons, 2017, 2020). The behavior of consumers and the structure of their purchases are determined by a set of factors. The basic factor is the level of consumer income, which defines the possibilities of increasing consumer spending, strengthening their confidence in the future and trust in the state, expanding access to credit resources and forming savings (Belyaevsky, 2013). In turn, the growing consumer demand stimulates the business activity of producers and provides positive economic dynamics of the growth of production volumes, employment, and credit activity of the population (Harrod, 2004; Kapelyushnikov, 2013; Kudinova, 2014; Takmakova, 2010; The World Bank, 2020; Zubarevich & Safronov, 2019). The pandemic has made adjustments to the usual patterns of consumer behavior due to restrictions on the movement of people during lockdown and isolation, the transition to remote work, rethinking established values, and drawing attention to issues of maintaining their health and family well-being (Karpunina et al., 2021). In turn, this was the impetus for the transformation of the production activities of enterprises and the optimization of the processes of delivery of finished products (Korolyuk et al., 2021).

3 Methodology

The research is aimed at studying the features of consumer behavior during the pandemic, establishing its current trends, as well as determining the vector of transformation of certain segments of entrepreneurial activity under the new model of consumer behavior.

Research objectives: (1) to assess the structure of household consumption in Russia before the pandemic and during the pandemic; (2) to investigate the impact of the pandemic on consumer behavior in Russia, to identify changes; (3) to determine the new trends of consumer behavior that define the nature of the production activity of enterprises.

Research methods: theoretical analysis, comparative analysis, graphical method, economic analysis, systematization method, system approach.

4 Results

Until 2013, the World Bank classified Russia as a group of high-income countries. In 2019, countries with a gross national income per capita of US $ 12,536 or more were classified as high-income countries (Zombart, 2005). In Russia during this period, the value of the indicator was US $ 11,260, that is, the country is included in the group of countries with an average level of income per capita (from US $ 1036 to US $ 12,615). From 2014 to 2017, the real incomes of the population in Russia decreased, in 2018 they showed a near-zero growth (+0.1%), and by the end of 2019 they increased by 1% in annual terms (Ovcharova, 2020, 2017). In general, in 2020, the financial situation of citizens worsened due to the pandemic. In the first half of 2020, the difficult situation regarding the dynamics of living standards was aggravated by emergency measures to contain the spread of coronavirus infection. They provided for the suspension of activities in many sectors of the economy. Despite the efforts to support the monetary income of the population, their reduction was inevitable (Galieva et al., 2020). In 2020, the real monetary income of the population decreased by 3.0% compared to 2019, real disposable monetary income declined by 3.5% (Burdiak, 2021). The level and quality of life of the population can be judged by the structure of spending earned income. The patterns of consumer behavior of the population are also revealed here. In the period 2013–2019, there was a decrease in the share of purchases of goods and services in the total amount of used income and an increase in household savings due to negative consumer expectations and a decline in consumer confidence. During the periods of stabilization of the socio-economic development of the country (2018–2019), there was an increase in the

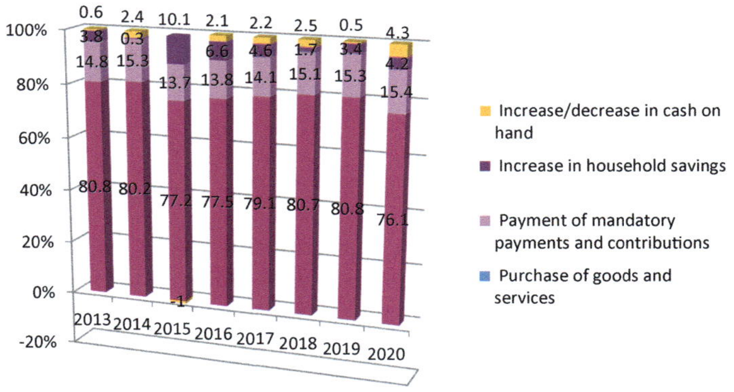

Fig. 1 Structure of the use of monetary income of the population in the Russian Federation, 2013–2020. *Source* Compiled by the authors based on (Federal State Statistics Service of the Russian Federation, 2020)

share of purchases of goods and services, as well as a decrease in the savings of the population (Galieva et al., 2020) (Fig. 1).

The figure shows that in 2020, during the COVID-19 pandemic, the share of purchases of goods and services in the total amount of used income decreased by 5.8%. The instability and unpredictability of the pandemic situation forced the population to save 1.2 times more.

If we consider the structure of consumer spending of the population, we can see that most of the population's spending is spent on the purchase of goods (in 2019—74%). In addition, Russia is characterized by a low share of services (on average, 22% for the analyzed period), much lower than developed countries. This indicates that the offer of services is not developed. The reasons for this situation may be the high cost of services; saving the population on services for the sake of buying goods; the availability of a wide range of free services provided by the state; a developed shadow sector in the service sector. In general, the share of services in total consumer spending has increased since 2014 (Fig. 2).

The dynamics of payments for goods and services made abroad for cash and using plastic cards indicates a problematic 2015 when the national currency of the Russian Federation collapsed. This year, the maximum value of this indicator was 4.1%. In 2016, the sharp weakening of the ruble caused a reduction in spending on the purchase of goods and services abroad to 2.9% of the total consumer spending of the population and led to a decrease in the effective demand of the population. The consumer behavior of the Russian population has pronounced features. The most significant is that a significant increase in GDP and household consumption per capita does not lead to positive changes in the structure of consumption (Snob, 2021). By contrast, a natural phenomenon for developed countries is an increase in the standard of living in the country and a reduction in the share of the resources of its population spent on food.

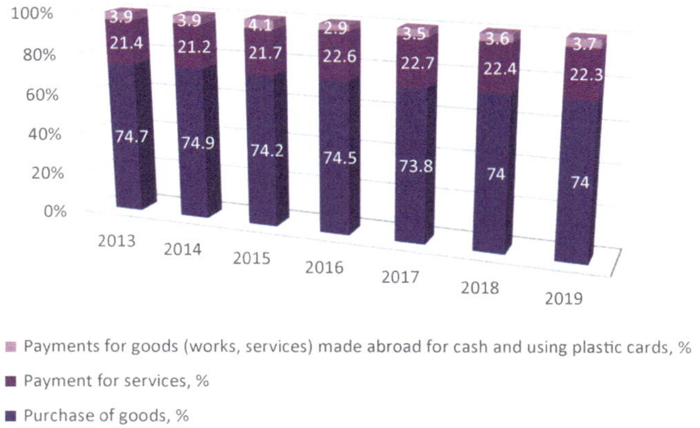

Fig. 2 Structure of consumer spending of the Russian population, in%, 2013–2019. *Source* Compiled by the authors based on (Federal State Statistics Service of the Russian Federation, 2019)

For example, the share of food products in the total consumption structure of the Russian population is disproportionately large in comparison with developed countries: for the period 2013–2018, the average value of the indicator was 37.8% (Fig. 3).

For comparison, in 2017, the share of spending on food in Germany was 8%, in the UK-13.5%, in the US—5.9%, in Russia-more than 31.2% (Shirov & Potapenko, 2020). In Turkey, Lithuania, and Estonia, food costs are traditionally at the level

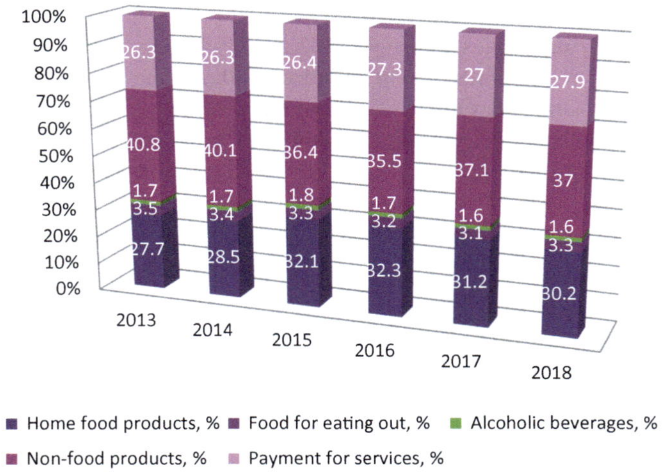

Fig. 3 Structure of population consumption, %, 2013–2018 *Source* Compiled by the authors based on Federal State Statistics Service of the Russian Federation (2020)

of 20%, in other Eastern European countries and Greece—in the range of 15–20%. And this is taking into account the fact that in 1998–2018, the physical volume of consumption in Russia increased by 2.7 times (the physical volume of GDP—by 2.1 times). This phenomenon, which is atypical for other countries, is called "the paradox of Russian consumption" (Snob, 2021).

According to the Deloitte report (2020), the key characteristics of consumer behavior of the Russian population in 2019 were (Deloitte, 2020):

- the way of shopping: 31% of consumers make purchases in stores, even if there is no need; 96% of Russians make purchases via the Internet; 70% of consumers make purchases via the Internet at least once a month;
- the most popular categories of purchases during the sales period: (1) in regular stores: 61%—food, 55%—household chemicals, 54%—alcoholic beverages; (2) in online stores: 37%—clothing and shoes, 37%—household appliances and electronics, 31%—cosmetics;
- categories of goods on which Russians save: 68%—ready-made food, 66%—cosmetics, 38%—medicines, 36%—children's goods and toys;
- when buying food, 55% of Russians pay attention to the label, 27% of consumers pay attention to the environmental friendliness of the product, 84% of buyers regularly check the shelf life of goods, 63% regularly check the composition of the product, 45% of consumers track the country of production;
- features of purchasing perfumes and cosmetics: 69% of Russians prefer natural cosmetics, even at a higher price, 64% of consumers are interested in environmental friendliness of products and biodegradable packaging, 44% of buyers need information about the absence of testing of cosmetics on animals;
- features of alcohol consumption: a high degree of consumer concern about counterfeit products, 46% of consumers are confident in their choice of alcoholic products;
- features of buying clothes and shoes: 40% of Russians prefer to buy in shopping centers, 34% of consumers favor branded clothing and shoes, 69% of buyers choose clothes and shoes made of natural materials;
- features of purchasing electronics and household appliances: 33% of consumers prefer shopping in shopping centers, 9% of Russians make purchases through online marketplaces, 45% of Russians buy equipment of the same brand, 61% of consumers make their choice based on collecting reviews on the Internet (Deloitte, 2020).

In times of crisis, the structure of consumption changes: discretionary spending is reduced, the share of household spending on household food increases, the ratio between food and non-food products, paid services of the population changes (Galieva et al., 2020). The pandemic crisis of 2020 led to a sharp decline in economic activity and shock compression of demand in large sectors of the economy (trade, services, tourism, transportation, etc.), their consequences were a significant increase in unemployment and a decrease in real incomes of the population (Rosstat, 2017). In the second half of 2020, the situation worsened due to the acceleration of inflation

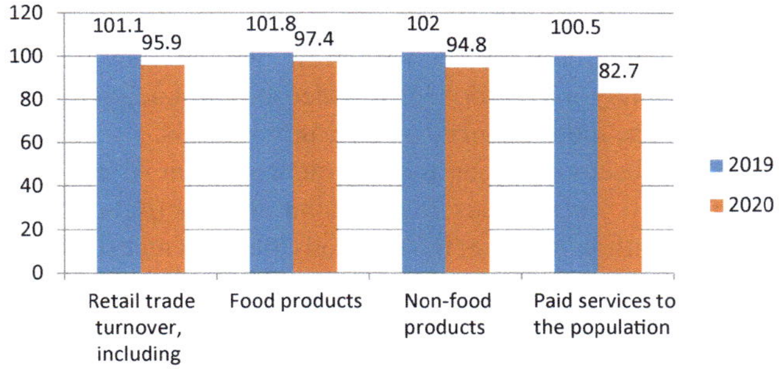

Fig. 4 Annual retail trade turnover in comparable prices and the index of the physical volume of paid services to the population, 2020% by 2019. *Source* Compiled by the authors based on (Federal State Statistics Service of the Russian Federation, 2020)

and the devaluation of the ruble, which led to an increase in the cost of imports and a decrease in the availability of goods for the Russian population.

In the first quarter of 2020, the share of spending on goods and services increased significantly against the background of a decrease in savings. This happened under the influence of rising inflation expectations, a further decrease in the attractiveness of foreign currency deposits, the desire to "fix profits" in the face of an increase in the exchange rate, as well as preparation for the upcoming period of self-isolation (Galieva et al., 2020). In the second quarter of 2020, a lockdown, a decrease in economic activity, as well as increased anxiety of the population led to a reduction in the share of consumer spending, as well as the activation of distance trading and the development of non-cash payments. According to the Federal State Statistics Service of the Russian Federation, the introduction of quarantine measures was reflected in a noticeable uneven drop in retail trade turnover—by 4.1% in general in annual terms (including by 2.6% for food products and 5.2% for non-food products) and a decrease by 17.3% in annual terms in the volume of paid services (Fig. 4).

During the pandemic, the automotive market became the most sensitive segment of the non-food market. In March 2020, sales of new cars in Russia increased by 4% compared to March 2019, and in April 2020 they sharply decreased by 72% (Aebrus.ru, 2020). According to the results of January–December 2020, the car market decreased by 9.1% compared to the same period of 2019. The demand is also negatively affected by the increase in the cost of cars: in the first quarter of 2021, they increased by an average of 2–8% (Forbes, 2021). The maximum decline in the retail trade and paid services sectors to the population was achieved in April 2020. This is due to the introduction of full lockdown and quarantine restrictions, the suspension of the work of enterprises. Starting from May 2020, a gradual recovery in demand began, but significant deviations in the structure of effective demand of the population were observed until the end of the year. In particular, during the pandemic, the growth of purchasing activity was ensured by purchases of food and children's goods, which the Russian consumer began to make more often on the Internet (Deloitte, 2020). During the same period, the segment of consumers who purchase household appliances through Internet applications increased by 20%. The segment of ready-made food purchased through online delivery services grew by 13%. 89% of Russians who used food delivery, ready-made food, and video communication services during self-isolation plan to use these services on an ongoing basis.

The pandemic has become a source of the development of new brands and the creation of new products and services. In particular, some activities, such as ordering products in apps and online training, have already become an established consumer habit. Studies based on the analysis of consumer behavior after previous crises show that people form an emotional connection with those brands that supported the audience during a difficult time for them (Netology, 2020, 2021). During the pandemic, there was a noticeable decrease in consumer activity in ordinary Russian stores, purchases of clothing and shoes decreased by 24%, electronic and household appliances—by 18%, cosmetics and perfumes—by 16%. At the same time, the cost of buying food in offline stores has increased (21% of Russians surveyed noted) (Deloitte, 2020). The most severe impact of the pandemic was on the restaurant business and cafes. Consumers' expenses for paying for food in public catering establishments have decreased by more than 20%. Purchasing activity for alcoholic beverages also decreased (by 11% compared to the same period of the previous year).

During the pandemic, the population changed not only their traditional consumer behavior but also their savings and credit behavior. In 2020, it began to save less, and redirect free cash to current consumption. In addition, there was a sharp decline in consumer lending due to the expected changes in cash income and related economic difficulties. Despite the pandemic, the majority of Russians have retained their consumer habits. In particular, the habit of buying goods in the same stores has remained, depending on the category of goods (from 38 to 62% of consumers). Also, the habit of Russians to purchase certain groups of goods from Russian and foreign manufacturers has remained unchanged: the Russian consumer prefers food, ready-made food, alcohol, medicines, and household chemicals of domestic production; children's goods, household appliances, perfumes and cosmetics, clothing and shoes of foreign production.

Due to the pandemic, Russians have become more aware of participating in sales. So, in 2019, 18% of Russians made purchases during online sales, while in 2020 the value of this indicator reached 25% (due to the sale of clothing and shoes, electronics and household appliances, cosmetics, children's goods, and perfumes). In ordinary stores during the sales period, food products, household chemicals, and alcoholic beverages are most often purchased.

In general, during the pandemic, new categories of consumers with a certain type of consumer behavior were formed (Khalilov, 2020). The first category includes consumers who have not changed their lifestyle, habits, who remain calm and carry out expenses in the same volumes (36% of Russians) (Struchenevsky, 2018). The second category includes consumers who have significantly reduced their expenses to save money during the pandemic (31% of Russians surveyed) and are pessimistic. The third category includes 18% of Russians surveyed who have stocked up on goods for the pandemic period and make purchases online. The fourth category is consumers who are most affected by the situation with the pandemic, who are forced to sharply reduce spending and spend money only on essential goods (15% of respondents). The current situation was the result of a change in the public consciousness of the Russian population. The main characteristic features of the perception of the world and society among Russians were the internal fear of losing the meaning of life; the lack of understanding of the vector of their actions in the conditions of a radical change in the traditional way of life; the division of people into those who are ready to act and those who have decided to wait; rethinking the value of real relations between people, the role of the state and individual social institutions in human life (PPCworld, 2020, 2021).

The identified trends in consumer behavior should be taken into account in the process of strategic planning of enterprises ' activities (Table 1).

5 Conclusion

Firstly, the authors assessed the structure of household consumption in Russia before the pandemic and during the pandemic. A feature of consumer behavior in Russia before the pandemic is the absence of positive changes in the structure of consumption due to a significant increase in GDP and per capita consumption. The authors concluded that the real monetary income of the population decreased during the pandemic, which caused negative changes in the structure of consumption in Russia. Secondly, the authors investigated the impact of the pandemic on consumer behavior in Russia. They determined that the structure of consumption has changed during the pandemic: discretionary spending has decreased, the share of household spending on household food has increased, the ratio between food and non-food products has changed, the volume of paid services to the population has decreased. The authors stressed that the decline in economic activity, as well as increased anxiety of the population, led to a reduction in the share of consumer spending. The authors identified the consumer habits that remained during the pandemic. The authors proved that the pandemic contributed to the formation of new categories of consumers with a certain type of consumer behavior. Thirdly, the authors identified new trends in consumer behavior and proposed appropriate mechanisms for reflection in the activities of enterprises.

Table 1 Current trends in consumer behavior of Russians during the pandemic and their reflection in the activities of enterprises

The trend of consumer behavior	Before the pandemic	During the pandemic	The mechanism of reflection on the activity of the enterprise
The growth of the share of food products in the total structure of population consumption	−	+	Regular monitoring of market changes, expansion of the range of food products, revision of the pricing policy for non-manufactured goods
Increase in the share of purchases made via the Internet	+	+	Using the potential of the Internet to promote goods and services, developing mobile applications Maintaining a high level of consumer confidence by ensuring information security standards
Reduction of the volume of paid services for the population	−	+	Identification and coverage of new market segments, adjustment of pricing policy taking into account the fall in consumer income
Increasing the propensity to participate in sales	−	+	Using the sales mechanism as a tool for promoting goods and services of the enterprise
Purchase of goods in the same stores, depending on the category of goods	+	+	Tracking consumer habits and relevant adjustment of sales channels of manufactured products
The emergence of new formats of product lines and brands	−	+	Conducting regular market research and analysis of consumer preferences to promptly adapt to changing conditions and maintain a position in a competitive market. Flexibility is an opportunity to see how the audience's insights change, what guides them when making decisions
Purchase of certain groups of goods from Russian and foreign manufacturers	+	+	Implementation of the principles of import substitution in the planning of production activities of enterprises, identification of new product niches
Redistribution of consumer activity in regular stores / in online stores	−	+	Development of company accounts in social networks and websites on the Internet, expansion of the online presence of enterprises, and the formation of feedback channels with consumers
Formation of new categories of consumers with a certain type of consumer behavior	−	+	Research of changes in consumer behavior, timely monitoring of changes, and implementation of the principles of flexibility in the implementation of the assortment policy of enterprises

Source Compiled by the authors

References

Aebrus.ru. (2020). *Statistics of car sales in the Russian Federation for 2020*. https://лада.онлайн/auto-news/autovaz/13249-aeb-statistika-prodazh-avtomobilej-v-rf-za-2020-god-itogi-91.html. Data accessed 15.06.2021.

Belyaevsky, I. (2013). Monetary incomes of the population and consumer spending: Level, trends and differentiation. *Economics, Statistics and Informatics, 2*, 108–118.

Burdiak, A. (2021). Consumption of goods and services in 2020: a pandemic model. Monitoring of the economic situation in Russia. *Trends and Challenges of Socio-Economic Development, 3*(135), 4–7.

Deloitte. (2020). *Consumer sector in Russia*. https://www2.deloitte.com/content/dam/Deloitte/ru/Documents/consumer-business/russian/CBT-2020-RU-NEW.pdf. Data accessed 15.06.2021.

Federal State Statistics Service of the Russian Federation. (2019). *Structure of monetary income and expenses of the population of the Russian Federation*. https://rosstat.gov.ru/folder/13397?print=1. Data accessed 15.06.2021.

Federal State Statistics Service of the Russian Federation. (2020). *Household income, expenditure and consumption*. http://old.gks.ru/wps/wcm/connect/rosstat_main/rosstat/ru/statistics/publications/catalog/doc_1140096812812. Data accessed 15.06.2021.

Forbes. (2021). *Car sales in Russia continued to fall*. https://www.forbes.ru/newsroom/biznes/425465-prodazhi-avtomobiley-v-rossii-prodolzhili-padenie. Data accessed 15.06.2021.

Galieva, G., Il'ina, L., Ponomarev, S., Khashir, B., & Karpunina, E. (2020). Consumption as a growth factor of the Russian economy. In *Proceedings of the 36rd IBIMA conference*, 4–5 November 2020, Granada, Spain (pp. 3763–3772).

Harrod, K. (2004). *Towards the theory of economic dynamics. World economic thought*. Through the prism of centuries: In 5 volumes. Mysl, Moscow.

Kapelyushnikov, R. (2013). *Behavioral economics and the new paternalism*. HSE Publishing House.

Karpunina, E., Butova, L., Sobolevskaya, T., Badokina, E., & Pliusnina, O. (2021). The impact of the Covid-19 Pandemic on the Development of Russian National Economy Sectors: Analysis of dynamics and search for stabilization measures. In *Proceeding of the 37th IBIMA conference*, 1–2 April 2021, Cordoba, Spain (pp. 1213–1226).

Karpunina, E., Yashin, S., Molchan, A., Lisova, E., & Derkacheva, E. (2020). The price of pleasure: The challenges and threats of the digital economy. In *Digital future: Economic growth, social adaptation, and technological perspectives*. Lecture Notes in Networks and Systems (pp. 363–373).

Karpunina, E., Zabelina, O., Galieva, G., Melyakova, E., & Melnikova, Y. (2020). Epidemic threats and their impact on the economic security of the state. In *Proceeding of the 35th IBIMA conference*, 1–2 April 2020, Seville, Spain (pp. 7671–7682).

Khalilov, D. (2020). *Future consumer index: How COVID-19 affects consumer behavior in Russia*. https://www.ey.com/ru_ru/consumer-products-retail/ey-future-consumer-index-russia. Data accessed 15.06.2021.

Korolyuk, E., Rustamova, I., Kuzmenko, N., Khashir, B., & Karpunina, E. (2021). Diagnostics of regional economic security problems during the 2020 crisis. In *Proceeding of the 37th IBIMA conference*, 1–2 April 2021, Cordoba, Spain (pp. 5248–5257).

Kudinova, A. (2014). The structure of consumption and economic development: the nature of the relationship. *Bulletin of the Novosibirsk State University. Series: Socio-Economic Sciences, 14*(1), 5–14.

Netology. (2020). *How the pandemic is changing brands and consumers: new topics and formats*. https://netology.ru/blog/06-2020-digital-posle-pandemii. Data accessed 15.06.2021.

Ovcharova, L. (Ed.). (2020). *Monitoring of the socio-economic situation and social well-being of the population*. October 2020, HSE, Moscow.

Parsons, E., Maclaran, P., & Chatzidakis, A. (2017). *Contemporary issues in marketing and consumer behaviour*. Routledge.

PPCworld. (2020). *People and the pandemic: What determines consumer behavior at the end of 2020*. https://ppc.world/articles/lyudi-i-pandemiya-chto-opredelyaet-pokupatelskoe-povedenie-v-konce-2020-goda/. Data accessed 15.06.2021.

Rbc. (2021). *Rosstat estimated the scale of the decline in the real disposable income of Russians*. https://www.rbc.ru/economics/28/01/2021/60129a749a7947cf1ca85d53. Data accessed 15.06.2021.

Rosstat. (2017). *The social situation and standard of living of the population of Russia*. 2017: Statistical collection, Rosstat, Moscow.

Shirov, A., & Potapenko, V. (2020). The paradox of Russian consumption. *ECO, 6*, 8–25.

Snob. (2021). *Trends in behavioral economics: How the crowning of Russians changed the consumer market*. https://snob.ru/entry/202138/. Data accessed 15.06.2021.

Struchenevsky, A. (2018). Why consumption will continue to grow faster than GDP. *Effective Anti-Crisis Management. Practice, 1*(5), 42–47.

Takmakova, E. (2010). Purchasing power of household income as a factor of economic growth. *Scientific Notes of the Orel State University. Series: Humanities and Social Sciences, 3–1*(37), 45–52.

The World Bank. (2020). *Gross National Income per Capita 2020*. https://gtmarket.ru/ratings/gross-national-income-ranking. Data accessed 15.06.2021.

Zombart, V. (2005). Why the United States is no socialism. World economic thought. Through the prism of centuries, Mysl, Moscow.

Zubarevich, N., & Safronov, S. (2019). The share of food in total consumption of the population of Russian regions as an indicator of the standard of living and consumption modernization. Bulletin of the Moscow University. Series 5. *Geography, 2*, 61–68.

Regional Survey of Health Workers' Opinions on the COVID-19 Pandemic

Vladimir V. Shkarin, Victoria V. Ivasheva, Svetlana Yu. Soboleva, Olga S. Emelyanova, and Irina V. Dneprovskaya

Abstract The emergence, spread, and development of COVID-19 has become a serious challenge and a test of the professional, organizational, and technological capabilities of the global medical community. The specific nature of the profession puts health workers at the forefront of responding to a COVID-19 outbreak and exposes them to a particularly high risk of infection. Hazards include contact with the pathogen, long work hours, psychological distress, fatigue, occupational burnout, and several others. Simultaneously, as members of society, health workers are also exposed to the general risks and influences of a difficult epidemiological situation (e.g., anxiety from uncertainty, lack of information, and general social and economic problems and constraints). As professionals, health workers must demonstrate the unity of opinion, attitude, and action and act as experts for patients on various issues, implementing the state's health care policy.

Keywords COVID-19 pandemic · Survey · Health workers · Healthcare · Anti-epidemic measures · Vaccination

JEL Codes I10 · I11 · I12 · I18

1 Introduction

Based on the survey of health workers, the authors aim to assess their professional perspective on the following aspects related to the pandemic:

- Validity and effectiveness of introduced anti-epidemic measures;
- Degree of confidence in the COVID-19 statistics;
- Degree of anxiety associated with the situation;
- Forecasts of post-pandemic realities;
- Attitude to vaccination;
- Difficulties encountered due to the epidemiological situation, etc.

V. V. Shkarin · V. V. Ivasheva · S. Yu. Soboleva · O. S. Emelyanova · I. V. Dneprovskaya (✉)
Volgograd State Medical University, Volgograd, Russia

E. G. Popkova and B. S. Sergi (eds.), *Geo-Economy of the Future*,
https://doi.org/10.1007/978-3-030-92303-7_11

2 Methodology

The authors developed an original questionnaire consisting of 39 questions (mostly closed). Subjective and direct questions revealing the socio-psychological attitude of the respondents and their attitude towards the subject of the survey prevailed among the questions. Several questions provided an opportunity to choose several answers and make a rating conclusion during the analysis.

The survey was conducted among health workers of medical institutions of Volgograd in April 2021 based on the principle of voluntariness.

The authors gave out 125 questionnaires, of which 122 were returned fully completed (97.6%). A considerable number of questions allowed multiple answers and contained the position "No answer." The responses within the position "No answer" were also considered when analyzing the percentages of meaningful responses chosen by respondents.

3 Results

The majority of the respondents are women (102 persons or 83.6%) aged 35–54 years (53 persons or 43.4%), working in polyclinics (77 persons or 63.1%), and having more than ten years of medical experience (75 persons or 61.5%). Most health workers (61.5%) did not work in covid hospitals; 28.7% are working in the "red zone," and 9.8% have worked in the "red zone" for some time.

The majority of surveyed medical workers had already had COVID-19 (70% of men and 53.9% of women). The proportion of COVID-19 survivors by age group was as follows:

- Aged 18–34 (youth group)—57.5%;
- Aged 35–54 (middle group)—60.4%;
- Over the age of 55 (senior group))—48.3%.

When assessing the interest of health workers in the pandemic situation in Volgograd, Russia, and abroad, the authors found that the majority of the respondents are interested in the regional situation (91%). Russian and international statistics are of interest to 88.5% and 77.9% of the respondents, respectively.

As the epidemic evolved, the opinion of the medical community on the degree of its danger changed (Shkarin et al. Shkarin et al., 2021). The dynamics of changes in the attitudes of health workers toward the pandemic were determined as follows:

- 60.7% believed and continue to believe that the situation is very serious;
- 4.1% believed and continue to believe that situation is not so serious;
- 11.5% changed their view on the situation from non-serious to the awareness of the danger;
- 8.2% decided that things were not as bad as they thought they thought at the beginning;

- 15% had difficulty answering.

More than half of the respondents (52.5%) assess the pandemic abroad as alarming; 26.2% say the situation improved, but there are still some concerns. Only women expressed fear and uncertainty about the situation.

Seventy-five percent of those who have had COVID-19 have certain anxiety and concern about the global pandemic situation. Among those who did not suffer the disease, there are 51% of people concerned. Only those who did not have COVID-19 (10% of those who were not ill) assess the situation as very difficult and feel fear.

In assessing the situation in Russia, there is a tendency to take a more positive view of the current situation. Thus, 54.1% of the respondents believe that the situation is improving although there are certain concerns (in relation to the foreign situation—26%); 32% of the respondents consider the situation difficult and feel anxiety (for foreign assessments there are 52.5%).

Among those who have had COVID-19 themselves or have relatives who have had it, the proportion of people considering the situation to be improving is 50.7% and 49.3%, respectively. However, 34.8% and 38.4% are anxious and worried. Among those who have not been ill themselves or whose relatives have not been ill, these proportions are 58.5% and 61.2% (optimists), and 28.3% and 22.4% (anxious).

In assessing the situation in Volgograd, there is a slightly more optimistic attitude compared to the situation in Russia. Thus, 14.8% find no reason to worry, 54.9% think the situation is improving, and 27.9% assess it as complex and alarming.

Answering the question about the world's future after the pandemic, about 50% of the respondents indicated that life would no longer be the same; a third believe that there will be some non-critical changes; about 10% hope that everything will be as before; about 11% found it difficult to answer.

Of all respondents, 57.4% are afraid of contracting COVID-19. However, there is some dependence on gender—45% among men and 60% among women. Fifty-five percent of men and 36.2% of women hope or are confident that they will not be infected. The degree of concern depends on whether the respondent has suffered COVID-19. Thus, about 43% of those who have recovered from the disease are afraid of getting the disease again, and more than 52% believe that they will not get the disease anymore; the ratio was 75.5 and 22.6% among those who were not sick.

The degree of fear also depends on the initial attitude toward the pandemic. Among those who think the pandemic is dangerous, more than 66% fear contracting the disease, and about 30% do not have such concerns. Among those who believe that the pandemic situation is not dangerous, the corresponding estimate is almost the opposite—40% and 60%.

Fifty-four percent of the respondents are afraid of infecting others. In this case, there is also some dependence on gender—35% of men and 58% of women are afraid of becoming the source of infection. Health workers in polyclinics assess their risks of becoming a source of infection for others higher compared with the risks of doctors in hospitals (about 60% and about 43%, respectively). The fear of infecting others also depended on whether the respondent had suffered COVID-19—about 46% of those who had the disease and more than 64% of those who had not. Among those who have already been vaccinated, 75% are afraid of infecting others. Among those who do not want to be vaccinated, only 25% are afraid of infecting others, while 67% are confident that they will not infect anyone.

Assessing the measures of protection against COVID-19, respondents cited (multiple answers allowed) the following most effective measures:

- Compliance with the rules of personal hygiene (about 92%);
- Keeping a social distance (about 88%);
- Wearing a mask in public places (86%).

The least effective measures are relocation outside the city (39%), staying at home (41%), and reduced working time (52.5%).

Women indicate the effectiveness of maintaining social distance more often than men (90.2% vs. 75%). However, they indicate the effectiveness of moving out of town less frequently than men (36.3% vs. 55%).

Restriction of going outdoors is considered an effective measure by 52.5% of the "youth" group (18–34 years old), 51.7% of the "senior" group (over 55 years old), and only 26.4% of the "middle" group (35–54 years old).

Those respondents who consider the situation dangerous also think that personal hygiene, face masks, and social distance are the most effective protective measures (94.6%, 89.2%, and 89.2%, respectively). Those who do not consider the pandemic situation dangerous most often indicate the social distance (80%). Personal hygiene and face masks accounted for 60% and 40%, respectively.

"Going out without necessity" (41.8% of the respondents), moving out of town (41% of the respondents), and reduced work hours (32.8% of the respondents) were indicated as the most ineffective measures. In the anti-rating, men more often indicated a reduction in work time (45%) and women—moving out of town (43.1%).

In general, the answer "staying at home" was at the top of the anti-rating in all combinations of the respondents (in relation to the pandemic, in terms of willingness to vaccinate, among those who were sick and those who were not, etc.).

The top three measures that respondents took to avoid getting sick were a more careful attitude to their health, taking vitamins, and eating fruits and vegetables (49.2%, 47.5%, and 45.1%, respectively). With slightly different ratios, these three positions prevailed in all groups formed according to different characteristics (gender, age, relation to the pandemic, whether the disease was suffered or not, etc.). Employees of hospitals additionally pointed out physical exercises. This was significantly noted as a measure taken by those who do not plan to get vaccinated and those who do not consider the pandemic dangerous.

The measures adopted in the region and rated by respondents as the most effective were self-isolation for people over 65 years old (83.6% of the respondents), obligation to wear masks (82%), and closing of public places (entertainment, restaurants, etc.) (77%). The group of people over 55 years of age found the transfer of students and schoolchildren to distance learning (a measure that has been applied in the whole country) to be the most effective measure (Shkarin et al., Shkarin, Ivasheva, et al., 2021).

Among the "anti-rating" measures, the most ineffective were "compulsory wearing of gloves" (45.1%), the abolition of social transport passes (43.4%), and the introduction of fines for violations of the rules (41%). The "senior" group (over 55 years old) included in the top-three of "anti-rating" the suspension of some businesses instead of imposing fines, which is consistent with the general concern of the population about the problems of government support for the economy during a pandemic (Lomovtseva et al., 2021).

Among the negative consequences of the pandemic for themselves personally (multiple answers were allowed), respondents ranked the abandonment of vacations (trips) (75.4%), abandonment of usual leisure activities (65.6%), and increased costs for food and medicine (55.7%). Additionally, 34.4% of the respondents noted a deterioration in their general perception of life (tension and fear). Other adverse effects were noted in less than 20% of cases.

When asked about the possible positive effects of the pandemic, 49.2% of the respondents said "nothing good," 24.6% said there was time for self-development, and 22.1% said there was more time for family.

The greatest share of positive sides of the pandemic was noted by respondents who do not currently view the situation as dangerous. The answer "nothing good" was given by only 30% of them. As positive sides, they noted communication with family (60%) and communication with friends (30%). Among those who assess the situation as very dangerous, about 80% see "nothing good" in the COVID-19 pandemic and note no positive consequences other than "an increased use of social media."

In assessing changes at work, about 75% of the respondents indicated no changes, about 10% noted a decrease in pay, and 4.9% of workers were sent on a leave of absence or at an inconvenient time.

Speaking on social support, 62.3% of the respondents reported receiving additional payments to health care workers, 19.3% of the respondents reported receiving cash payments for children under the age of 8, 11.8% reported receiving cash payments for children under the age of 16, and 21.1% received no support. Additional payments were made to 75.6% of those who worked in hospitals and 54.5% of those who worked in polyclinics.

When asked about helping others during the pandemic, over 60% of the respondents said they helped relatives, about 40% helped friends and acquaintances, about 24% helped co-workers, and 18% helped no one.

The questionnaire paid special attention to the degree of trust in official information about COVID-19 (abroad, in Russia, and Volgograd) and the most trusted sources of information. Around 46% of the respondents trust the information about the number of cases abroad, 40% expressed distrust, about 10% had difficulty answering, and 4% did not follow the information. In response to a similar question about the situation in Russia, the figures were 44.3% (yes), 46% (no), 8.2% (no answer), and 1.5% (do not follow the information). As for Volgograd, 42.7% of the respondents trust the information, 46.7% do not trust it, and 9% found it difficult to answer. Thus, there is a slight decrease in the proportion of people who trust information about the local situation.

In assessing official sources of information about the pandemic, 41% of the respondents indicated that they trusted federal television channels, 24.6% trusted official Internet sources, 11.5% trusted social networks and bloggers, and 18% noted other sources. Only 3.3% of the respondents trust the print media and newspapers, and 1.6% trust the local TV channels.

When asked about the degree of trust in other sources of information about the epidemiological situation, respondents noted the following (multiple answers were allowed): 82.8% of the respondents trust doctors and nurses they know, 85.2% trust epidemiologists, 77.8% trust general practitioners, 61.5% trust nurses and hospital staff, 56.6% trust hospital administrators, 55.8% trust representatives of Rospotrebnadzor, 48.4% trust the federal authorities, 46.8% trust regional authorities, 45% trust representatives of the Ministry of Health of the Russian Federation, 41.8% trust independent medical experts, 23.8% trust popular medical experts, and 12.3% trust bloggers.

In assessing the actions of the Russian authorities during the pandemic, 77% of the respondents evaluated them as adequate and timely, 11.5% found it difficult to answer, and another 11.5% gave a negative assessment.

Assessing how the Russian health care system is handling the epidemiological situation compared with other countries, 40.2% of the respondents indicate that the Russian system is doing better, 34.4% believe that it is doing as well as other countries, 7.4% believe that it is as bad as in other countries, 2.5% believe it is worse than in other countries, and 15.6% give no answer. Comparing the health care system of the Volgograd Region with other regions of Russia, 15.6% of the respondents favor the Volgograd Region, 46.7% feel that the system is doing as well as in other regions, 11.5% say it is equally bad, and 23% have difficulty answering.

During the COVID-19 pandemic, 57.4% of the respondents sought COVID-19-related care; 83% of those who sought care rated it positively, 17% rated it negatively, and 2.8% did not receive the care necessary. According to the questionnaire, 75.5% of those who were not ill and 17.4% of those who suffered COVID-19 did not seek help. The number of the respondents who sought help unrelated to COVID-19 was 50.8%; 79% of those who sought help evaluated it positively, and the rest evaluated it negatively.

The respondents were asked whether they are ready to get vaccinated; 16.4% of the respondents said they had already been vaccinated, 41% said they plan to get vaccinated, 24.6% said they were not ready, and 18% had difficulty answering. Thus, almost one in four of those surveyed did not express a willingness to be vaccinated.

Those respondents who do not plan (in the near future) to get vaccinated indicate various reasons for their opinion, such as the desire to wait until all tests are completed (53.8%), expecting the authorities to be vaccinated first (19.2%), and expecting the majority of the population to be vaccinated (15.4%).

The respondents used the following services during the COVID-19 pandemic: express tests for COVID-19 (56.6%), checkup by the local therapist (50%), doctor home visits (34.4%), computed tomography (CT) of the lung (34.4%), making an appointment through the call center (around 28%), calling an ambulance to the house (27%). Of all respondents, 15.6% did not use any services.

When assessing the quality of medical and social services (from 1 to 10 points), the maximum average score was given to "COVID-19 vaccination"—7.7 points. Next were "performing express tests for COVID-19" (7.59 points) and "free lung CT" (7.31 points). The last three places were given to "examination by narrowly focused specialists in polyclinics" (5.81 points), "examination by local therapists" (5.8 points), and "doctor home visits" (5.29 points).

Thus, the lowest scores were given to the positions associated mainly with the work of the primary level due to the maximum workload and the maximum expression of staffing shortages.

Noting the main difficulties faced by medical organizations during the pandemic, the respondents pointed to staffing shortages (82%), health risks for health workers (72.1%), and problems with drug supply (55.7%). The respondents also indicated the lack of lung ventilators (38.5%) and funding (37.7%).

Inefficient work of medical organizations and inefficient management were noted by 17.2% of respondents (this is the lowest percentage of all positions). Representatives of hospitals pointed to problems of drug supply (75.6% vs. 44.2%), shortage of lung ventilators (55.6% vs. 28.6%), and lack of funding (51.1% vs. 29.9%) more often than representatives of polyclinics. In turn, representatives of polyclinics were somewhat more likely than hospital workers to face staffing shortages (87% vs. 73.3%), a lack of transportation (33.8% vs. 26.7%), and health risks for medical professionals (76.6% vs. 64.4%).

4 Conclusion

The sociological survey of health workers of Volgograd regarding the COVID-19 situation showed their high professional involvement and interest. Health workers are most interested in the specific situation in the region, with an overwhelming number of workers following the pandemic situation in Russia and abroad.

Since the emergence of the infection and up to the present time, most medical professionals consider COVID-19 a dangerous disease. However, almost one in four is not ready to be vaccinated, and one in five finds it challenging to answer this question.

Several contradictory evaluations and vague responses indicate the absence of consensus on professional issues and general problems of the current situation among health workers. The most trusted source of information was the opinion of colleagues that health workers knew personally.

In assessing protection measures against COVID-19, respondents cited personal hygiene, social distance, and wearing a mask in public places as the most effective measures. The greatest negativism was caused by the necessity to wear gloves and the restriction of movement. The worst negative consequences of the pandemic were the inability to rest on vacation (not providing the vacation or providing it at an inconvenient time) and restriction of movement.

With a relatively high proportion of people who do not trust official (regional and federal) sources of statistical information and are skeptical about the real severity of the situation and the adequacy of measures taken, the vast majority of respondents assessed the actions of the authorities as adequate, timely, and consistent with global practices (Soboleva & Sobolev, 2021). This fact can be seen as a positive signal representing the potential and basis for a "growth zone"—increased confidence and consolidation of society (including professional communities) in cases of systemic threats.

Thus, the research demonstrated the need of the professional medical community for reliable, comprehensive, timely, and objective information, primarily at the level of official sources. This information is needed to reduce anxiety, uncertainty, and mistrust in stressful epidemiological conditions. One of the tools to build such trust could be regular thematic monitoring of the opinions, expectations, and needs of specific professional communities and society as a whole to improve the effectiveness of management decisions taken by all branches of government.

References

Lomovtseva, O. A., Soboleva, S. Yu., & Sobolev, A. V. (2021). Comparative characteristics of models of state support for national economies of the world countries in the conditions of the COVID-19 pandemic. *Bulletin of the University, 1*, 91–98. https://doi.org/10.26425/1816-4277-2021-1-91-98

Shkarin, V. V., Ivasheva, V. V., Yemelyanova, O. S., Chepurina, N. G., Simakov, S. V., & Danilenko, E. E. (2021). Organization of educational and medical activities of a medical university in the conditions of the COVID-19 pandemic—The experience of the region. *Current Problems of Healthcare and Medical Statistics, 1*, 193–207. https://doi.org/10.24411/2312-2935-2021-00012

Shkarin, V. V., Orlov, D. V., Dneprovskaya, I. V., Moskalev, Yu. A., Nevezhina, E. A., & Emelyanova, O. S. (2021). Improving coronavirus infection diagnostics with video surveillance during the COVID pandemic. *Health Care Standardization Problems, 1–2*, 27–33. https://doi.org/10.26347/1607-2502202101-02027-033

Soboleva, S. Yu., & Sobolev, A. V. (2021). Comparative analysis of the effectiveness of anti-epidemic measures in the countries of the world during the period of COVID-19. *Management in healthcare: Challenges and risks of the XXI century* (pp. 287–293). Volgograd State Medical University.

Crisis of the Statehood Culture and the Lessons of the Pandemic

Ivan D. Afanasenko

Abstract The research subject is the statehood culture. Transformations in the spiritual and material worlds are directly reflected in the system of state-building. Transformations in the spiritual and material worlds are directly reflected in the system of state-building. Never before has a crisis of social order been accompanied by such a devastating pandemic. Crisis phenomena have affected the main human-made attributes—the economy and the government. The government provides security, social protection, law and order, and justice as the common good. The COVID-19 crisis confirmed that even in the face of global shocks, the national economy becomes the guarantor of the well-being of its people and that there is a certain limit to economic security that cannot be left open. The destructive forces of transnational monopolies, who perceive national borders with their customs supervision and trade policies as a hindrance, have descended upon nation-states. The American way of democracy is realized through the destruction of nation-states and the forcible imposition of a single form of statehood on all people. This state of affairs necessitates the reconsideration of the assessment of the experience of countries using other forms of economic and social orientation and getting good results. It is necessary to turn to the historical experience of Russia, whose peoples have an unprecedentedly high national instinct, which allows overcoming the crisis of statehood.

Keywords State · Crisis of public administration · Statehood culture · Historical experience of Russia · Economic thought

JEL Code B30

1 Introduction

The universal organization doctrine includes two kinds of organization: the organization of the natural world and the organization of humanity. The result of the

I. D. Afanasenko (✉)
Saint Petersburg State University of Economics, St. Petersburg, Russia

E. G. Popkova and B. S. Sergi (eds.), *Geo-Economy of the Future*,
https://doi.org/10.1007/978-3-030-92303-7_12

new organization of the natural environment is the world economy. The result of the organization of human communities is a statehood culture.

Let us highlight the basic scientific provisions of the statehood culture:

- Each nation creates its own special forms of state structure;
- Main features of statehood are developed by an ethnos at the highest stage of its development and unity;
- Natural form of statehood is self-sufficient and remains stable, rational, and progress-oriented throughout the life of an ethnos (Afanasenko, 2016).

The signs of the state's independence are realized through the culture of forms of political organization of society and are manifested in the independence of the movement of people in time. The Russian researcher Ogarev, back in the nineteenth century, proved that the form "expresses the entire internal structure of national public life enclosed within known geographical limits. Its formation is influenced by the specific conditions determining this internal composition of the people. Moreover, its formation is also influenced by the psychology and 'spirit' of the nation-builder, which give an appropriate direction to its state-building instinct" (Ogarev, 1993, p. 99).

In examining the state's nature, we proceeded from the following methodological assumptions:

- State has a national origin;
- Natural form of the statehood culture is an integral part of the general culture of the people;
- Informal institutions, customs, traditions, and moral norms of society are conditioned by the national culture of statehood.

2 Materials and Methods

The Russian cultural and historical world has a single historical space filled with the events of civil and ethnic history. It is marked with the unity of the people's spirit with all confessional diversity. This allows for an interdisciplinary approach to research (Afanasenko, 2017).

From an interdisciplinary perspective, an attempt has been made to justify the fact that the choice of state form is not arbitrary and that the culture of state-building is part of the general culture created by a particular nation. It is subject to certain objective laws. Among all objective laws, we should mention the law of triune development established by K. N. Leontiev, the Russian philosopher and diplomat (Leontiev, 1996). As early as the middle of the nineteenth century, he established and substantiated that the evolution of the state system and the laws of state development are consonant with the evolution of all things on earth. Leontiev proved the movement of state forms in time and identified the stages of their development within the framework of the life cycle of 1000–1200 years (Leontiev, 1996).

The conclusion of the Russian scholar Solonevich, explaining the cause and origins of the creation of a statehood culture and state instinct by a particular nation, is of particular interest. "If the people do not have a state instinct, then these people will not create a state under no geographical, climatic, or other conditions. If the nation has a state instinct, the state will be created in spite of geography, climate, and, if you like, even in spite of history. Of all the Slavs, only Poles and Russians lived a long independent state life" (Solonevich, 1991, p. 147).

Scholars have proven that the empire is the highest type of state. Negative attitudes toward empire come from those countries that pursued colonial policies and collapsed, losing their colonies. It has been established that not all nations can create an empire where different peoples live on equal terms. Any attempt to build a state that focuses only on one nation in a multi-ethnic society is doomed to failure. One of the examples is the public policy of modern Ukraine.

3 Results

Thus, in the world people created for themselves, the most significant common goods are the economy and the state. The economy creates everything necessary for human life. The most demanded function of the state is the protection of the population from dangers of any kind (Afanasenko, 2016).

People were disturbed when they discovered that the liberal state could not protect them from epidemics and the threat of starvation. Russian scholars have substantiated the key features of the state system. It has been proven that the state is the highest form of organization of a people who have achieved self-consciousness. This form is not prompted by nature; it is an invention of human genius. It is established that the statehood culture is part of the overall culture created by a particular nation. Only natural forms of the state are self-sufficient, sustainable, and ensure evolutionary progress. The choice of the form of government is not arbitrary (contrary to current practice). The framework of the statehood culture is formed by national traditions, customs, and moral norms uniting the organizational (institutional) structure of society. Therefore, each nation consciously recognizes its national form of state (Ryazanov, 2011).

Statehood forms are subject to the objective law of the triune development process. In their movement through time, they pass through the stages of formation, prosperity, and destruction (Leontiev, 1996).

One more condition is that the nation-builder must have a state instinct (Solonevich, 1991).

Russian scholars have paid much attention to determining the social purpose and historical place of democracy in the movement of state form over time. The state form is refined in the process of ethnogenesis. Democracy has been shown to assert itself in "the final phase of the life cycle of the state, which is inevitably followed by decay and death" (Leontiev, 1996).

The Western super-ethnos is in the final phase of ethnogenesis. It is more than 1200 years old. The duration of this phase of ethnogenesis is 3–4 centuries. We can recall this historical fact that the Slavic super-ethnos also established a democratic form of government in the final phase of ethnogenesis. Veliky Novgorod, Pskov, and Vyatka were democratic republics.

In the Western world, democratization has been mistakenly perceived as progress. In fact, "it is not the progress of development, it is the process of secondary, blending simplification, decomposition" (Ogarev, 1993).

The imposition of the institutions of the American model of democracy is dangerous since it interrupts the natural process of developing the statehood culture in those nations that are in other phases of ethnogenesis. This was on full display during the COVID-19 pandemic, which highlighted a particular form of injustice—the failure of the liberal state to protect the population from the epidemic, unemployment, and starvation.

Not all modern countries, claiming to be the standard of "civility," have proven the ability to function effectively in an emergency. There was a public demand for a social-type state.

There emerged a new concept of a "functioning state." Russia is classified as one of these "functioning" countries. The Russian people are endowed with a state instinct of great power. Modern Russia has retained the inherited ability to act effectively in emergency situations. Fortunately, the politicians of the perestroika period did not have time to destroy the rich practice of the USSR in protecting the population from disasters. During the aggravation of the COVID-19 pandemic in Russia, the sanitary and epidemiological service worked successfully.

The combination of the socio-cultural crisis and the COVID-19 pandemic has revealed realities that lead to conclusions of a profound nature. This includes the attitude toward the individual. In the Western world, "personality" is not an abstract category of philosophy. It is the primary social value and is reduced to the most sacred thing for Western people—the possession of property (Kamyshova, 2014). In fact, it turns out that an American living on a salary has virtually nothing to protect him or her in an emergency. "Just a month of downtime, and a person is left with nothing. Without state assistance, he or she goes to the abyss. Americans were proud of their medicine, but it was only available to the rich people, and the likelihood of being cured depended on the cost of health insurance."

People suddenly realized that an individual's life resource was limited to a month's production downtime and a month's wages. Everyone understood that it is impossible to live like this. Nevertheless, no one knows the right way to live.

Ignorance of state nature was reflected in the government's competence. Intellectually and morally unprepared for such an occupation, people get access to the government (Afanasenko, 2019). The power they are endowed with exceeds their consciousness.

The freedom of the individual was revealed in a special way by the digital format of life. A person is dual—he or she is a material and a spiritual being. As a material being, a person lives according to the laws of cell division of living matter. The COVID-19 pandemic reminded us of the biological dependence of people.

The digital format of life bypasses human spirituality. A digit cannot be a measure of morality. Without a word, it is impossible to formulate a scientific concept, and it is impossible to preserve and transmit new knowledge to posterity.

Human is built on the mind's power. Thinking is a basic human strength. Digitalization multiplies human information power. Nevertheless, information is not yet knowledge. It becomes knowledge by going through thinking.

Thus, digital life is not only positive. Digital transformation is not limited to job losses and worsening employment problems (Kovalev, 2018; Plotnikov et al., 2018). The increased attention goes to the doppelganger (a digital avatar) and reduces interest in the real person.

There is an attempt to "refine" the human being so as to digitize him or her, gain control over his or her emotions, and correct the stereotypes of human social behavior. A great danger lurks here. In the human's essence, biological (natural) and social (spiritual) states complement each other. In an effort to improve humans, genetic engineering cannot change the spiritual genetic basis of humans.

4 Conclusion

Scientific discussions on the statehood culture raise questions on the problems of creativity and the spiritual world of humans. Creativity and spirituality are not adapted to a genetically advanced person. To achieve a creative state, a person needs not only the appropriate tension of the brain but also inspiration, that is, a state of mind.

Researchers of the human brain have found that learning increases brain capacity, but digitization does not. Most of the brain is not digital but analog. The digit is not capable of distinguishing the good from the bad. The word is the source of human social life.

Our research covered interesting scientific and practical issues, thereby reducing uncertainty about the future development of individuals, society, and the state in the context of digitalization. At the same time, we recognize the relevance and lack of solutions to such issues as spirituality (a socially valuable aspect of education), corporate legal relations in the digital age, features of the formation of legal consciousness in the information society, modernization of human rights and freedoms in the context of digitalization of society, etc. Thus, we call on scientists and experts to conduct research aimed at their solution.

Acknowledgements The article was prepared within the initiative research work "Formation of economic systems of the Eurasian type: Dynamics, contradictions, efficiency," of the Department of General Economic Theory and History of Economic Thought of the Saint Petersburg State University of Economics No. 121042600253-4.

References

Afanasenko, I. D. (2016). *Economy and state in the system of the Russian world*. St. Petersburg State University of Economics.

Afanasenko, I. D. (2019). Russian and Eurasianism, dialectics of living conditions and basic values. In DYu. Miropolsky & A. V. Kharlamov (Eds.), *Economic system of the Eurasian types: Problems of economic uncertainty* (pp. 171–183). St. Petersburg State University of Economics.

Afanasenko, I. D. (2017). Interdisciplinarity: Methods and practices in the use of total knowledge. *Bulletin of the St. Petersburg State University of Economics, 5*(107), 172–182.

Kamyshova, A. V. (2014). Debatable issues of government control techniques in economic systems global instability environment. *Journal of Legal and Economic Studies, 4*, 92–98.

Kovalev, S. G. (2018). Digitalization of socio-economic space: Challenge and chance for breakthrough development of Russia. *Philosophy of Economy, 5*(119), 79–92.

Leontiev, K. N. (1996). *The East, Russia, and Slavs*. Respublika.

Ogarev, N. P. (1993). State property. In K. Isupov & I. Savkin (Eds.), *Russian philosophy of property* (p. 99). Hansa.

Plotnikov, V., Vertakova, Y., & Moroz, N. (2018). Differentiation of the economic space in the context of ensuring economic security. In *Proceedings of IBIMA 2018: Innovation management and education excellence through vision 2020* (pp. 1588–1596). Milan, Italy.

Ryazanov, V. T. (2011). Political economics of the particular: Principles of Russian research tradition. *Russian Economic Journal, 5*, 22–48.

Solonevich, I. L. (1991). *People's monarchy*. Moscow, Russia: Feniks.

Developing a Sustainable System of Natural Resource Management

Elena V. Kletskova, Oksana V. Klimova, and Alexander A. Zhidkikh

Abstract Since the mid-twentieth century, scholars have recognized the importance of traditional knowledge for the sustainable use of natural resources. Several in-depth studies have demonstrated how traditional knowledge can contribute to the sustainable management of natural resources. Researchers discovered that communities that depend on fishing, horticulture, or subsistence agriculture had, through trial and error, accumulated knowledge on the sustainable use of natural resources. Studies clearly indicate that native subsistence methods and techniques promote biodiversity.

Keywords Sustainable system · Natural resources · Environmental issues · Effective management

JEL Codes Q51 · Q52 · Q56

1 Introduction

Environmental problems became especially prominent in almost all countries. Development concepts underwent significant changes, and the development at the expense of nature is now construed as unsustainable. Today, environmental protection and the maintenance of a balanced ecosystem are the most important tasks for all countries. Urban environmental problems are especially serious; they require cooperation between government, environmental managers, stakeholders, researchers, and policymakers. Environmental problems vary from city to city and region to region, depending on such variables as city size, income growth rate, local geography, climate, and institutional potential. Accordingly, the job of an economist or an environmental manager is more challenging in larger cities.

E. V. Kletskova (✉)
Altai State University, Barnaul, Russia

O. V. Klimova
Gorno-Altai State University, Gorno-Altaysk, Russia

A. A. Zhidkikh
Altai State Technical University Named After I.I. Polzunova, Barnaul, Russia

E. G. Popkova and B. S. Sergi (eds.), *Geo-Economy of the Future*,
https://doi.org/10.1007/978-3-030-92303-7_13

The Earth has limited waste processing capacity; if too much waste is released into the atmosphere rather than being recycled and reused, the environment cannot process and assimilate produced waste. This can result in pollution, resource degradation, and consequent economic damage (Bogoviz et al., 2020; Stroiteleva et al., 2020).

As urbanization continues, solid waste management is becoming a major concern both in cities and in villages. Urban environmental problems are substantial and require the involvement of government, environmental managers, stakeholders, and policymakers.

Solid waste production per capita continues to grow globally. Waste management issues pose a serious environmental threat in the developing world. In recent years, the Ministry of Natural Resources and Environment of the Russian Federation has passed more solid waste management bills than any other topic. Solid waste management is a complex and multifaceted process. Historically, waste management has always been a function of engineering due to the evolution of a technology that not only provided the benefits of mass production but also created waste disposal problems. Noxious fumes can be a nuisance, especially during the active life of a landfill. Such gases as hydrogen sulfide can be produced by sulfate-reducing bacteria if the landfill contains sulfate waste or decaying plant material (Tashkulova & Kletskova, 2020a, 2020b).

One of the main responsibilities of municipal authorities is managing the collection and disposal of solid waste so as to ensure sanitary security and maintain the aesthetic conditions of the environment. Improperly managed solid waste often undergoes bacterial decomposition due to its organic and nutrient components, thereby creating foul odors, propagating infectious diseases and pests, and contaminating the water. Improper disposal of solid waste negatively affects the ecosystem and the environment. Scientifically unsubstantiated waste disposal attracts birds, rodents, flies, and other detrivores, contributing to sanitary and aesthetic problems.

2 Methodology

Growing environmental awareness has increased the attention paid to solid waste management; in recent years, environmental issues have taken center stage. This sector concentrates on intensive human activity and prioritizes appropriate and safe solid waste management systems, providing healthy living conditions for the population.

The development concept underwent significant changes, and non-eco-friendly development is considered to be unsustainable. The main environmental problems faced by urban areas include motor vehicle pollution, noise pollution, solid waste disposal, industrial pollution, and municipal wastewater.

Municipal solid waste (MSW) generally consists of non-hazardous components but sometimes include hazardous waste—packaging, clothing waste, glass, dirty bottles, paint, batteries, industrial dust, ash, tires, metal cans and containers, animal carcasses, medical waste, abandoned cars, insulators, electrical conductors, sewage, and sludge.

The municipal authorities usually dispose of solid waste in low-lying areas called sanitary landfills. The municipal authorities usually dispose of solid waste in low-lying areas called sanitary landfills. Regulations specify many solid waste management instances, including proper separation of solid waste, biodegradable waste, recyclable waste, and non-recyclable types of waste. These are usually separated into color-coded containers (green for biodegradable waste, blue for non-degradable waste, and black for hazardous waste) at the waste source so that they can be properly recycled or disposed of (Adarina et al., 2019; Merdesheva et al., 2020).

The accumulation of waste is growing much faster than the birth rate in cities. The growth of waste production per capita is becoming a very serious problem, forcing the country to take immediate steps towards managing its waste production.

Since the dawn of time, humans and animals have been using the Earth's resources both to sustain life and to dispose of waste. In earlier times, waste management did not pose a significant problem because the population was very small compared to the size of the land available for waste disposal. But today, the issue of solid waste is regarded as a serious issue everywhere. Rapid population growth and uncontrolled industrial development are degrading the urban and semi-urban environment in many developing countries, putting tremendous pressure on natural resources and undermining sustainability.

Uncontrolled solid waste threatens public health and poses serious risks to the environment, society, and the economy. A complete and environmentally-sound policy requires the effective participation of all those involved in the problem. Each person is a part of the problem, and each person should also be a part of the solution, meaning that solving the problem of waste management depends on the collective actions and efforts of people.

Poor waste management is related to the growing health problems in all social strata, ranging from waste-borne diseases to groundwater contamination. Solid waste can affect all components of the environment—air, water, and soil. Massive volumes of waste with limited financial resources of local governments further exacerbate the problem. Waste takes a wide variety of forms and comes from various sources, including social, economic, and industrial activities (Karataev et al., 2020; Shenshinov & Abdulsattar Al-Ali, 2020).

According to the UN Development Programme research, the second most serious problem that urban residents face (after unemployment) is inefficient solid waste management. Typically, one-third to two-thirds of all generated solid waste goes uncollected, indiscriminately dumped outside of houses.

To address the growing solid waste production, environmental policies should employ a comprehensive, hierarchical approach to waste management with the four components: prevention, recycling, incineration, and disposal.

Waste management has a positive impact on both the environment and the economy. The goal of recycling is to use a combination of all three methods to ensure the safe and effective management of MSW. The Ministry of Natural Resources and

Environment of the Russian Federation recommends that the communities use the systems to address their individual needs, focusing on the following:

1. Preventing or reducing pollution;
2. Reusing and recycling, rather than incinerating or disposing.

Improper solid waste management is a problem in urban communities, especially in urban areas where more and more people move from rural areas due to unemployment and poorer quality of life. The lack of proper municipal amenities exacerbates the environmental threats. Sudden population surges force the municipal authorities to take special measures, making the problem even worse. This situation must be turned for the better, which requires an organized and thoughtful approach. In short, the main reasons for these problems are as follows:

- Population growth;
- Industrialization and urbanization;
- Lack of an integrated and comprehensive approach to solid waste management;
- Disregard for the important financial, political, and social aspects.

Solid waste is currently posing a major challenge to city managers. Optimal solid waste management strategies are being developed mainly in developing countries that employ the Clean Development Mechanism, which generates significant revenues from solid waste for local governments. These optimal strategies involve identifying and analyzing existing problems in order to effectively implement all the essential aspects of urban solid waste management (Costanza, 1996; Tsvetkov et al., 2019).

3 Results

MSW management aims to improve poor waste management practices that are prevalent in many low-income countries, where little to no attention is paid to this issue compared to other aspects of infrastructure, water supply, and transportation. Waste management is an interdisciplinary field that encompasses waste collection, transfer, and disposal. Due to its tremendous impact, it is important to have a broad outlook on the issue, not confining oneself to a specific technology. Currently, special attention is paid to building the existing capacity of waste management agencies, the main objectives of which are the following:

1. Assessment of waste production volumes, type of waste, and waste disposal regime in a given area;
2. Identification of measures taken by local authorities and assessment of their operational efficiency;
3. Survey of environmental quality and human health hazards caused by MSW to environmental protection workers;
4. Investigation into the feasibility of participating in the Clean Development Mechanism and creating a comprehensive framework for effective solid waste management policies.

All countries have adopted environmental management system as a tool for analyzing the intended goals—in other words, to look at the problem from start to finish for a meaningful solution to MSW management issues. These measures require involving experts from various disciplines, besides economists and environmental workers (Avkopashvili et al., 2019; Ragulina et al., 2019).

Current hypotheses are as follows:

1. Waste produced by households depends on the income and size of households.
2. Public attitude towards solid waste contributes to proper solid waste management.
3. Improper waste management at various stages promotes many diseases.
4. Solid waste landfills contaminate the surrounding groundwater.

The analysis encompasses both descriptive and inferential statistics, including linear regression analysis. While descriptive statistics are used to describe the background information collected from sample sanitation workers, inferential statistics allow analyzing the relationship between the behavior of respondents and the amount of waste generated.

The analysis employs standard statistical software for the social sciences. Analysis attributes include monthly household income, waste production per capita, family size, current utility fee for best available waste management service, and daily waste weight. The hypothesis of the study states that waste generation is directly related to income, family size, and the current utility fee for waste management (Vukovich et al., 2018; Glotko et al., 2019).

Economists and environmental scientists run a variety of statistical analyses, including bivariate and linear regression analysis, to determine the relationships between the determinant independent variables (such as income, family size, and waste disposal fees) and the dependent variable (such as waste production). The resulting correlation coefficient values help identify the attributes that significantly affect the weight of household waste. Starting with stepwise regression analysis, the scholars conducted linear regression analysis. All insignificant determinant variables were omitted, leaving behind only the following four parameters: income, waste production per capita, waste disposal fees, and family size.

Existing waste management systems are not always effective, resulting in pollution, health hazards, and economic losses and causing environmental and social problems. Approximately 90% of solid waste is disposed of in open dumps, which significantly increases health hazards and environmental degradation. Solid waste should be treated as a manufacturing resource, while non-recyclable plastic bags should be banned. Additionally, people usually suffer from diseases due to a lack of information about the health hazards associated with improper waste disposal and unsanitary conditions.

The scientific management of solid waste in rural areas remains largely neglected. Therefore, it requires prioritized attention in order to improve the rural environment and socio-economic development. However, most studies focus on urban waste. Current research focuses on the following areas: modern waste management practices, human behavior in terms of waste management, factors affecting waste quality and quantity, recovery of plant nutrients from household biodegradable waste via vermicomposting, recycling, and developing a sustainable model for rural areas (Nekhvyadovich et al., 2020). Furthermore, waste can be transformed into valuable resources by creating environmental management among the population by introducing science-based environmental education.

MSW generally consists of the following:

- Biodegradable waste—any organic material that microorganisms can break down into simpler stable compounds;
- Non-biodegradable waste—any waste that microorganisms cannot break down into simpler stable compounds.

Non-biodegradable waste is further divided into commercial recyclable waste and domestic hazardous waste, depending on its origin and use.

Solid waste generation in rural areas may increase rapidly due to population growth, increased consumption, changing food prices, the use of non-biodegradable products (single-use plastics and thermocell packaging), encouraging packaging practices, etc. The scientific management of solid waste in rural areas requires serious attention, given its importance for improving environmental sanitation, public health, and sustainable economic growth.

Disposal of household and other toxic waste in streets, ponds, drains, and other open spaces pose serious health and environmental hazard.

4 Conclusion

Solid waste management is currently too casual in both rural and urban areas. Most waste is disposed of without proper separation or treatment. Poor MSW management leads to diseases, foul odors, public nuisances, pollution (water, soil, and air), aesthetic nuisances, and economic losses to residents. The lack of adequate facilities for the safe disposal of solid waste in metropolitan areas negatively affects human health and ecosystems.

There is an urgent need for a mechanism to ensure 100% separation of solid waste into organic, inorganic, and household hazardous waste. Currently, there are no proper measures for the control over environmental pollution caused by MSW. Solid waste, including plastic waste, is incinerated or dumped into sewers, posing a serious threat to life.

In most cases, MSW is dumped in low-lying areas on the outskirts of cities with no compliance with established regulations for landfill sites. Most landfills no longer use compaction, leveling, soil cover, gas filtration, and monitoring systems.

Sustainable management of natural resources refers to an adequate infrastructure, which is now considered to be key. Solid waste is a potential resource; its extraction in the form of soil nutrients (compost), materials (recyclables), and energy (biological gas) is essential for the effective implementation of a natural resource management system. The political agenda for sustainable natural resource management should encourage behavioral changes in citizens, elected representatives, and policymakers so as to minimize waste production and maximize reuse and recycling. Sustainable natural resource management is a problem of managing people; overemphasis on technological solutions to these problems is less than ideal.

References

Adarina, R., Gazukina, Y., & Yankovskaya, K. (2019). Indicators of the "green economy" as a tool for monitoring the regional economy. *IOP Conference Series: Earth and Environmental Science, 395*, 012107. https://doi.org/10.1088/1755-1315/395/1/012107

Avkopashvili, P. T., Polukhin, A. A., Shkodinsky, S. V., & Poltarykhin, A. L. (2019). The fundamental provisions of the concept of knowledge economy. In E. Popkova, Y. Ragulina, & A. Bogoviz (Eds.), *Industry 4.0: Industrial revolution of the 21st century* (pp. 57–64). Cham, Switzerland: Springer. https://doi.org/10.1007/978-3-319-94310-7_5

Bogoviz, A. V., Shvakov, E. E., Tretyakova, O. G., Zakharov, M. Y., & Abramov, A. N. (2020). Globalization of education in the conditions of formation of the global knowledge economy: Regularities and tendencies. In E. Popkova (Ed.), Growth poles of the global economy: Emergence, changes and future perspectives (pp. 993–1000). Cham, Switzerland: Springer. https://doi.org/10.1007/978-3-030-15160-7_99

Costanza, R. (1996). Ecological economics: Reintegrating the study of humans and nature. *Ecological Application, 6*(4), 978–990.

Glotko, A. V., Voronkova, O. Y., Miheeva, N. M., Skoblikova, T. V., Alekhina, N. A., & Zhminko, A. E. (2019). The economic development of a trans boundary region under the conditions of the Eurasian economic union. In *Proceedings of IBIMA 2019: Education excellence and innovation management through vision 2020* (pp. 7957–7968). Granada, Spain.

Karataev, E. M., Merkuryev, V. V., & Titova, O. V. (2020). Integrating the "green economy" into the model of digital future of the modern socio-economic systems. In E. Popkova, & B. Sergi (Eds.), *Scientific and technical revolution: Yesterday, today and tomorrow* (pp. 1402–1410). Cham, Switzerland: Springer. https://doi.org/10.1007/978-3-030-47945-9_153

Merdesheva, E. V., Titova, O. V., & Avkopashvili, P. T. (2020). Methodological approach to the classification of digital economies by environmental efficiency and sustainability criterion. In E. Popkova, & B. Sergi (Eds.), *Scientific and technical revolution: Yesterday, today and tomorrow* (pp. 1360–1368). Cham, Switzerland: Springer. https://doi.org/10.1007/978-3-030-47945-9_148

Nekhvyadovich, L., Kostin, K. B., & Mironov, V. V. (2020). Government and public monitoring of entrepreneurship in a multicultural region: New perspectives in the conditions of the digital economy. In E. Popkova, & K. Vodenko (Eds.), *Public administration and regional management in Russia* (pp. 147–154). Cham, Switzerland: Springer. https://doi.org/10.1007/978-3-030-38497-5_17

Ragulina, Y. V., Semenova, E. I., Avkopashvili, P. T., Dmitrieva, E. A., & Cherepukhin, T. Y. (2019). Top-priority directions of implementing new internet technologies on the territories of rapid economic development. In E. Popkova, & V. Ostrovskaya (Eds.), *Perspectives on the use of new information and communication technology (ICT) in the modern economy* (pp. 182–188). Cham, Switzerland: Springer. https://doi.org/10.1007/978-3-319-90835-9_21

Shenshinov, Y., & Al-Ali, A. (2020). The tools of increasing efficiency of human resource in the lean production environment: Conceptual study. *International Journal of Core Engineering & Management, 6*(7), 1–18.

Stroiteleva, T. G., Kletskova, E. V., & Balaeva, S. I. (2020). Ontogenesis of green development model of digital economy: Essence, stages, and prospects. In E. Popkova, & B. Sergi (Eds.), *Scientific and technical revolution: Yesterday, today and tomorrow* (pp. 410–417). Cham, Switzerland: Springer. https://doi.org/10.1007/978-3-030-47945-9_46

Tashkulova, G. K., & Kletskova, E. V. (2020a). The model of organization of "green" digital production and consumption based on the internet of things and AI. In E. Popkova, & B. Sergi (Eds.), *Scientific and technical revolution: Yesterday, today and tomorrow* (pp. 329–336). Cham, Switzerland: Springer. https://doi.org/10.1007/978-3-030-47945-9_37

Tashkulova, G. K., & Kletskova, E. V. (2020b). Place and role of human in the system of circular reproduction in the digital regional economy. In E. Popkova, & B. Sergi (Eds.), *Scientific and technical revolution: Yesterday, today and tomorrow* (pp. 943–951). Cham, Switzerland: Springer. https://doi.org/10.1007/978-3-030-47945-9_101

Tsvetkov, V. A., Loginov, E. L., & Raikov, A. N. (2019). Integrated network interaction of education with science and production. *Educational Technologies and Society, 19*(1), 389–398.

Vukovich, G. G., Makuschenko, L. V., Bateykin, D. V., Titova, O. V., & Dobrosotskiy, V. I. (2018). Support of the territories of advanced economic development on human capital: Theory and practice. *Quality—Access to Success, 19*(S2), 157–160.

Strategic Foundations of Environmental Management

Alexander A. Zhidkikh, Yulia G. Gazukina, and Irina V. Chernyaeva

Abstract Since ancient times, humanity has understood that nature and the human race form an integral part of the life cycle system. This system consists of five elements: air, water, land, flora, and fauna that are interrelated and interdependent, have evolved and developed. Deterioration in one element inevitably affects the other. Mainly the needs of the poor, the avarice of the rich, and the reckless use of technology are the main causes of environmental degradation in the world. Humanity is at a crossroads right now. The population is growing rapidly by the day, requiring using the finite resources of nature. The dangers are clearly visible. If no concentrated and united effort is applied, tomorrow it may be too late. Not all is yet lost, despite the damage that has already been done to the Earth. With careful environmental management, much can be fixed, and much can be avoided in the future. The remarkable thing about the current decade is that people have begun to unravel and understand the complexity of ecosystems and that people's attitudes toward environmental protection have changed.

Keywords Ecological management · Development strategy · Environment · Interdisciplinary approach · Environmental management

JEL Codes Q5 · Q551 · Q56

1 Introduction

Environmental problems caused by industrial pollution and urban congestion are discouraging; however, the situation has not reached the point of no return. By restoring proper environmental management, integrating the inextricable economic

A. A. Zhidkikh (✉)
Altai State Technical University Named After I.I. Polzunova, Barnaul, Russia

Y. G. Gazukina
Gorno-Altai State University, Gorno-Altaysk, Russia

I. V. Chernyaeva
Altai State University, Barnaul, Russia

E. G. Popkova and B. S. Sergi (eds.), *Geo-Economy of the Future*,
https://doi.org/10.1007/978-3-030-92303-7_14

and ecological systems, and creating environmental awareness among people, humanity can choose environmentally compatible development paths.

Not every person can be an environmental manager. Moreover, effective environmental management must be based on a scientific-technological approach that takes full account of socio-economic parameters and needs.

Recent global experience clearly shows that environmental management, as a means of social change, must be fully supported by an appropriate political system. Fortunately, the Russian government is committed to conservation and environmental protection.

The main challenges that humanity faces are as follows:

- Population management and health care;
- Complex land use planning and water management;
- Restoration of marginal land vegetation to stop deforestation for firewood, fodder, and long-term ecological security (i.e., greening of uncultivated lands);
- Combating water pollution in river systems;
- Combating air pollution in industrial centers;
- Development of clean, renewable energy;
- Control over municipal solid waste via recycling;
- Preserving biodiversity;
- Development of new types of settlements that promote the well-being of people, especially through reduction of congestion;
- Environmental education at all levels of society.

Countries are realizing that environmental problems can best be solved through regional and local cooperative efforts, since these problems recognize geographic but not political boundaries. This fact is reflected in various government programs (Dragunov & Shenshinov, 2020; Glotko et al., 2019a).

Humanity's technical skills in dealing with environmental problems are not without reproach. There is hope that the prospects for the environment in the next decade are bright. If people all unite in environmental conservation, the goals of sustainable development may be attainable. Major producers, such as businesses, play a key role in maintaining the environment and life cycles. Since the future of the human race on land devoid of vegetation is bleak, there has been a resurgence in attention to plants and vegetation. Therefore, the least people can do is green the country as much as possible. This will create a healthy environment, leading to a healthy economy for both present and future generations.

As concern for environmental protection and conservation grows worldwide, the harmful effects of pollution on natural ecosystems are becoming increasingly important to resource management. There is no doubt about the progress and prosperity brought about by industrialization, but materialistic progress has damaged nature itself and threatens the environment. Deforestation has led to the destruction of wildlife, soil erosion, loss of fertility, flooding, and desertification. Urbanization and unplanned urban expansion have created problems of waste management, sanitation, and clean water.

2 Methodology

The ecological problems of developing countries are caused not only by excessive industrialization but also by incomplete knowledge of the possible consequences of scientific development. Progress has become synonymous with an attack on nature. Industrialized countries experienced much more severe pollution and began to realize its negative impact on the lives of people who have sought to profit from industrial production. The United Nations Conference on the Human Environment, held in Stockholm in June 1972, began an era of growing awareness of the demand for environmental protection. Fortunately, Russia has taken this issue seriously. The constitution was amended to raise awareness about environmental protection.

The pursuit for quality of life puts tremendous pressure on everything the environment can provide, so much so that part of the environment has begun to succumb. These phenomena can be seen in desertification, soil diseases, floods and droughts, urban congestion, threat to countless plant and animal species, and the widespread pollution of land, water, and air. Nevertheless, the need for food, shelter, clothing, energy, and other basic needs of society has not diminished, nor has the realization that environmental resources, even renewable ones, are finite.

Therefore, environmental management is the proper use and management of resources. Environmental management refers to an interdisciplinary approach to the use and processing of resources that act as a regulating force against the human desire for resource exploitation and waste. Environmental management involves reducing or minimizing the anthropogenic impact on the physical and ecological environment in an attempt to avoid overuse, misuse, and abuse of environmental resources (Tashkulova & Kletskova, 2020a, 2020b).

Many countries have now initiated programs to improve the environment and have set up organizational structures to address environmental problems. The global conservation strategy is based on nature-friendly development and the concept of protecting ecosystems to provide the basis for sustainable development. The Russian government was sensitive to environmental problems from the very beginning. It created ministries of environmental protection, which are entrusted with all responsibilities to ensure the protection and preservation of the environment.

The environment is a resource that includes the land surface, the underground, the water, the air, and everything they contain. This means that the government is concerned with the sensible management of resources, which is synonymous with environmental management. These environmental management functions include management of agricultural land and food production, which is controlled by energy resources. Russia depends on the anthropogenic impact on the security and sovereignty of the land, water, and air resources. Government is the ultimate environmental manager; its primary task is to preserve what must be preserved, protect what must be protected, and regulate the use of natural resources.

Environmental management requires access to the potential of the environment so as to support processes and provide resources for the development. This is vital for the continued progress of humanity. Population management is a necessity for environmental management. It is a very similar problem to urban management. As the city grows goes beyond a certain optimal size, the cost of providing basic urban services becomes exorbitantly high, leading to a deterioration of the quality of life. Land can be maintained at optimal levels so that quality of life does not fall lower than the established standards. Although cities drive economic development, failure to manage rapid urbanization threatens human health, environmental quality, and urban productivity. The most pressing environmental problem is the lack of safe drinking water. Water is a source of health, and its scarcity weighs heavily on the current generation, especially in urban areas, who suffer the most from poor health, low productivity, and reduced quality of life (Karataev et al., 2020; Shenshinov & Abdulsattar Al-Ali, 2020).

The goal of rapid urbanization is to sustain economic growth while addressing related environmental and social equality issues. Addressing environmental challenges offers a unique opportunity to improve health and living conditions, as well as the macroeconomic performance in rapidly growing cities.

People who set environmental goals should aspire to a higher quality of life. The ultimate goal of environmental management is to achieve environmental goals. The environment is represented by the soil, on which terrestrial life depends, water that hosts aquatic life, and air, without which nothing can survive. To prevent the degradation of these areas, environmental management techniques must be adopted now. Conservation is the goal of environmental management; it implies the rational use of environmental resources.

The following aspects are especially important to the global conservation strategy:

- Maintaining basic ecological processes and life cycle systems;
- Preserving genetic biodiversity;
- Sustainable use of species and ecosystems;
- Promoting the environment's ability to support a growing population;
- Improving the sustainability of natural systems, since its reduction leads to higher costs of goods and services;
- Lowering the resource use of major industries.

The world needs reliable sources of food and energy. To ensure such security, the following important steps must be taken:

- Supporting genetic diversity;
- Combating soil degradation;
- Combating deforestation;
- Stopping the spread of deserts;
- Limit the exploitation of land and water resources;
- Reducing pollution.

Environmental management is the optimal allocation of limited resources between different possible uses. Ecological criteria and economic concerns suggest that such an allocation must be efficient. At the same time, available resources must be protected from degradation, and finite scarce resources must be conserved (Alekseev et al., 2018; Glotko et al., 2019b).

Conservation of water bodies, air, and forests in such activities that use a natural resource like mining, manufacturing imply the following actions:

- Avoiding excessive exploitation of resources;
- Preserving the land.

The following activities are the cornerstones of environmental management:

- Environmental planning;
- Environmental status evaluation;
- Environmental impact assessment;
- Environmental legislation and administration.

The concept of environmental planning is based on integrating environmental considerations into the economic development planning process, as practiced so far.

Environmental planning process must establish policies, priorities, and methods in a way that can be easily reviewed and modified according to feedback from the system. Therefore, it should:

- Be flexible;
- Maintain sensitivity to the exploitation of natural resources, the impact on employment, prices, and quality of life;
- Be able to identify and promote alternative development projects;
- Promote public participation at various levels of decision-making;
- Stimulate the development of regional and local potential;
- Be creative and modal at the regional and local level;
- Understand the political basis for achieving a dynamic balance between environmental, social, political, and economic concerns.
- Environmental planning cannot be successful without the following steps:
- Adopting the improvement of the quality of life as a basic element of social policy;
- Taking proactive steps to integrate an ecological perspective into economic and social decision-making;
- Encouraging public participation in conflict resolution and negotiation between various development options;
- Promoting cooperation between the government, industrial structures, and academia for the development of a new way of thinking about the systematic interrelations of the economy, energy, and the environment;
- Protecting and preserving the environment in a manner compatible with the socio-economic objectives of a society already burdened by an energy crisis, rising unemployment, and spiraling prices.

Environmental planning seeks to promote economic development by avoiding or minimizing collateral environmental damage.

3 Results

Environmental management needs a comprehensive planning approach that seeks to manage human activities so as to maintain the balance between the quality of life and the natural environment. This obviously depends on societal values, which are by no means consistent and can vary greatly depending on circumstances and time.

The system of environmental status evaluation consists of various processes. Since the environment is a complex and dynamic system, the evaluation is extremely difficult and fraught with a number of uncertainties. There is no single metric or status index. Furthermore, any suitable computational index using several critical parameters would be inaccurate and too difficult for the current methods. In these circumstances, one possible approach is to estimate the status of the subsystems.

One can attempt to compare the status of the environment between two points in time and space using one or more parameters of an environmental sub-component. However, future needs require urgent, more extensive, and thorough research into the application of systems analysis methods to assess environmental status (Stroiteleva et al., 2020a; Stroiteleva, et al., 2020b).

Environmental impact assessment (EIA) is one of the main tools available today to the environmental manager, serving as a useful guide for decision-making. EIA is a procedure for identifying the potential anthropogenic impact on ecological systems. In some cases, such as pollution exposure assessment, the impact can be quantified to some degree, but for long-term environmental and social impacts, an acceptable degree of quantification is currently unattainable. The most significant result of the EIA is the comparison of development options and the selection of alternative sites for the placement of development projects. EIA identifies, measures, and evaluates both positive and negative impacts of development projects on the environment.

The methodologies emphasize the enumeration of possible impacts and the collection of data, rather than their relative importance. The latter, namely the assessment of the relative weight of values, is the main goal of the assessment. But current EIA procedures cannot be called the most up-to-date in terms of scientific accuracy.

EIA criticism includes the following:

- High time consumption and costs;
- Inefficiency in integrating environmental considerations into development assessment;
- Its low usability for medium-sized, routine development projects (since it mainly focuses on unique large-scale projects).

Environmental legislation, backed by well-developed regulations, meticulous enforcement, and administrative mechanisms, is an important component of environmental management. The legislative domain covers a wide range of issues, including land use and water rights, aiming to prevent and combat air pollution, conserve forests, and protect the wildlife and the environment. It also includes urban planning, licensing rules regarding toxic chemical production, sales, resource use, waste, food expiration and contamination, mining claims, patenting plants and organisms, etc.

Legislation alone is not enough to enforce and implement environmental measures. Persuasion and education are also equally important. Incentive and deterrent packages are considered persuasive methods to achieve environmental goals (Merdesheva et al., 2020; Ragulina et al., 2019; Shenshinov, 2012).

Environmental management activities require the following support systems:

- Information system of environmental management;
- Ecological monitoring and supervision system;
- Environmental research.

An information system provides information on the status of the environment for its further management. An adequate environmental information system is indispensable to the environmental management system.

4 Conclusion

Regardless of the data accumulation process, an environmental information system must have built-in subsystems for data analysis and interpretation, as well as for broad dissemination of relevant data to users, whomever they may be.

Environmental monitoring systems are equally important for the successful management of the environment. The subsystems of environmental change monitoring are presented further.

First, a network of volunteer environmental quality monitors is needed to alert interested agencies about gross violations of environmental quality.

Second, the subsystem of impact monitoring includes periodic assessment of environmental status, including human health, population dynamics, plant, and animal yields, soil productivity, etc. (Avkopashvili et al., 2019; Vukovich et al., 2018).

Research, both basic and applied, is certainly a basic prerequisite for initiating environmental management programs.

There is also a need for research on environmental assessment methodology, improving environmental monitoring methods, finding solutions to unavoidable environmental problems, controlling environmental degradation, reversing environmental trends, and restoring environmental health.

References

Alekseev, A. N., Savenkov, D. L., Shvakov, E. E., Lebedev, N. A., & Ragulin, A. D. (2018). Financial problems of regional development and the ways of their solution in contemporary Russia. *Quality—Access to Success, 19*(S2), 71–75.

Avkopashvili, P. T., Polukhin, A. A., Shkodinsky, S. V., & Poltarykhin, A. L. (2019). The fundamental provisions of the concept of knowledge economy. In E. Popkova, Y. Ragulina, & A. Bogoviz (Eds.), *Industry 4.0: Industrial revolution of the 21st century* (pp. 57–64). Cham, Switzerland: Springer. https://doi.org/10.1007/978-3-319-94310-7_5

Dragunov, A., & Shenshinov, Y. (2020). Effective development of innovative technologies as a method for increasing labor productivity in industry. *E3S Web of Conferences, 208*, 03001. https://doi.org/10.1051/e3sconf/202020803001

Glotko, A. V., Sycheva, I. N., Dunets, A. N., Poltarykhin, A. L., Zhuravlev, P. V., & Tubalets, A. A. (2019). Integration processes in the agro-industrial complex of the regions: Development and environmental problems. *Journal of Environmental Management and Tourism, 10*(3), 613–621.

Glotko, A. V., Voronkova, O. Y., Miheeva, N. M., Skoblikova, T. V., Alekhina, N. A., & Zhminko, A. E. (2019a). The economic development of a transboundary region under the conditions of the Eurasian economic union. In *Proceedings of IBIMA 2019: Education excellence and innovation management through Vision 2020* (pp. 7957–7968). Granada, Spain.

Karataev, E. M., Merkuryev, V. V., & Titova, O. V. (2020). Integrating the "green economy" into the model of digital future of the modern socio-economic systems. In E. Popkova, & B. Sergi (Eds.), *Scientific and technical revolution: Yesterday, today and tomorrow* (pp. 1402–1410). Cham, Switzerland: Springer. https://doi.org/10.1007/978-3-030-47945-9_153

Merdesheva, E. V., Titova, O. V., & Avkopashvili, P. T. (2020). Methodological approach to the classification of digital economies by environmental efficiency and sustainability criterion. In E. Popkova, & B. Sergi (Eds.), *Scientific and technical revolution: Yesterday, today and tomorrow* (pp. 1360–1368). Cham, Switzerland: Springer. https://doi.org/10.1007/978-3-030-47945-9_148

Ragulina, Y. V., Semenova, E. I., Avkopashvili, P. T., Dmitrieva, E. A., & Cherepukhin, T. Y. (2019). Top-priority directions of implementing new internet technologies on the territories of rapid economic development. In E. Popkova, & V. Ostrovskaya (Eds.), *Perspectives on the use of new information and communication technology (ICT) in the modern economy* (pp. 182–188). Cham, Switzerland: Springer. https://doi.org/10.1007/978-3-319-90835-9_21

Shenshinov, Y. V. (2012). *Labour quality management*. North-East Asia academic forum.

Shenshinov, Y., & Al-Ali, A. (2020). The tools of increasing efficiency of human resource in the lean production environment: Conceptual study. *International Journal of Core Engineering & Management, 6*(7), 1–18.

Stroiteleva, T. G., Kletskova, E. V., & Balaeva, S. I. (2020a). Ontogenesis of green development model of digital economy: Essence, stages, and prospects. In E. Popkova, & B. Sergi (Eds.), *Scientific and technical revolution: Yesterday, today and tomorrow* (pp. 410–417). Cham, Switzerland: Springer. https://doi.org/10.1007/978-3-030-47945-9_46

Stroiteleva, T. G., Petrova, E. A., & Yungblyudt, S. V. (2020b). Responsible information society as a social environment for creation of the "green" digital economy: Formation and monitoring. In E. Popkova, & B. Sergi (Eds.), *Scientific and technical revolution: Yesterday, today and tomorrow* (pp. 402–409). Cham, Switzerland: Springer. https://doi.org/10.1007/978-3-030-47945-9_45

Tashkulova, G. K., & Kletskova, E. V. (2020a). Place and role of human in the system of circular reproduction in the digital regional economy. In E. Popkova, & B. Sergi (Eds.), *Scientific and technical revolution: Yesterday, today and tomorrow* (pp. 943–951). Cham, Switzerland: Springer. https://doi.org/10.1007/978-3-030-47945-9_101

Tashkulova, G. K., & Kletskova, E. V. (2020b). The model of organization of "green" digital production and consumption based on the internet of things and AI. In E. Popkova, & B. Sergi (Eds.), Scientific and technical revolution: Yesterday, today and tomorrow (pp. 329–336). Cham, Switzerland: Springer. https://doi.org/10.1007/978-3-030-47945-9_37

Vukovich, G. G., Makuschenko, L. V., Bateykin, D. V., Titova, O. V., & Dobrosotskiy, V. I. (2018). Support of the territories of advanced economic development on human capital: Theory and practice. *Quality—Access to Success, 19*(S2), 157–160

Economic Foundations of Integrated Environmental Monitoring

Yulia V. Ragulina, Elena V. Merdesheva, and Olga V. Titova

Abstract Monitoring has traditionally been carried out along single disciplinary lines, even though it is a basic process to resource management and can never rely on a single discipline. The single disciplined monitoring led to the impossibility of predicting several environmental problems (e.g., forest damage from air pollution). A comprehensive understanding of the interaction of environmental elements requires an integrated approach to environmental monitoring, not just the consideration of physical, chemical, biological, or socio-economic aspects. This paper discusses the issue related to integrated environmental monitoring. It defines integrated monitoring, explores why integrated monitoring is usually not carried out. Some examples are provided. The authors discuss the role of long-term ecological research and monitoring. The focus is made on the role of the regions and the problems of regional monitoring. The authors outline the requirements necessary for comprehensive monitoring.

Keywords Environmental monitoring · Resource management · Environment · Integration · Marine environment · Environmental conditions

JEL Codes Q5 · Q51 · Q55

Y. V. Ragulina (✉)
Peoples' Friendship University of Russia (RUDN University), Moscow, Russia
e-mail: julra@list.ru

E. V. Merdesheva
Gorno-Altai State University, Gorno-Altaysk, Russia
e-mail: merdesheva@bk.ru

O. V. Titova
Altai State University, Barnaul, Russia
e-mail: otitova82@icloud.com

E. G. Popkova and B. S. Sergi (eds.), *Geo-Economy of the Future*,
https://doi.org/10.1007/978-3-030-92303-7_15

1 Introduction

It is necessary to develop a system of indicators reflecting the state and management of resources in accordance with the long-term goals of sustainability. A more holistic approach to environmental issues is required to focus on sustainability. Biophysical and socio-cultural considerations must be integrated as closely as possible.

One of the reasons for the lack of integrated monitoring is that the methods of traditional environmental management are not sufficiently adapted to accept such a holistic approach. Legislation existing in many countries tends to focus on one environmental document at a time, for example, the Clean Air Act of the USA (Washington State Department of Ecology & United States Environmental Protection Agency, 1990). Another example is the Mineral Workings (Offshore Installations) Act 1971, which indicates the integration of laws relating to resource management as one of its goals (Parliament of the UK, 1971).

The reason for the lack of a comprehensive approach is that, before the restructuring of local government, each organization (e.g., watershed or port) possessed its own monitoring program and had few formal mechanisms to ensure useful interaction and sharing of information with other local bodies. A lot depends on cooperation between individuals. Some joint boards conduct formal cross-organizational monitoring as mandated by the regional planning scheme and issue-specific monitoring reports. However, a narrow interpretation of the available authority makes the process random and not uniform throughout the territory (Bogoviz et al., 2020; Tashkulova & Kletskova, 2020b).

Despite efforts to restructure and better define functions with the formation of regional councils, monitoring that occurred before restructuring continues mainly on an ad-hoc basis. Information bases are often structured without regard to how this data can be used in the future. The attitudes of managers are influenced by the role of their organizational unit without regard to the data gathered by others.

It is necessary to develop and implement an integrated monitoring system. This system will not simply arise and become universally applied. Individuals tend to resist integrating their uniform procedures of discipline control. They find it challenging or even annoying if the benefits do not become clear to them.

Integration requires conscious and systematic consideration of the many diverse elements of a resource management problem in search of optimal solutions. This requires preventive, proactive, and reactive policies.

2 Methodology

When designing an integrated monitoring system or modifying an existing system, it is necessary to take into account the following parameters:

- Monitoring purpose;
- Required resolution in time and space;

- Accuracy requirement;
- Measurement errors (random or biased);
- Spatial variability of the aridity field and strength of spatial correlations (problems of heterogeneity of an ecosystem);
- Temporal variations in the data;
- Practical constraints;
- Design flexibility, allowing for changes in the development of understanding of processes;
- Possibility of a mismatch between the program's goals and the system's capabilities;
- Competing environmental models, which may involve different monitoring strategies, for example, the information needed for management ends on whether the stress response model or the input–output model (where flows from one location go to another);
- Problem of historical monitoring, in which data are fragmentary or impossible.

It is impossible to ignore socio-cultural and political aspects when considering examples of integrated monitoring since the methods used for this type of monitoring, by definition, involve communication across disciplinary boundaries, which may need to be facilitated in the decision-making process.

In the presented examples of integrated monitoring, there is a range of integration processes, from integrated monitoring of the biological, chemical, and physical environment, as implied in the definitions, to integrated monitoring of the biophysical and socio-economic aspects of the environment (Tashkulova & Kletskova, 2020a; Stroiteleva et al., 2020).

Let us discuss concentrations of pollutants in the marine environment. Contaminants can be measured in the water column, sediments, and biota. Their potential effect cannot be assessed without recognizing that these three carriers form an interconnected system. The integration of data from these carriers is necessary to provide a comprehensive picture of marine environmental quality. No biological indicator will provide all the information needed to determine the effects of stress on an ecosystem. Thus, several biological, physical, and chemical monitoring indicators may be required. Indicators in the water column may include nutrients, pathogens, oil, and organogenic wastes. Indicators in bottom sediments and invertebrates, fish, birds, and mammals may include trace metals and synthetic organic compounds.

Assessment of marine pollution requires chemical surveys and environmental studies. Determining concentrations of contaminants provides no information about their bioavailability or their potential for adverse effects because many metals are bound to sediments in the water, so they are mostly inert. In contrast, other compounds (e.g., chlorophenols) may be present in relatively low concentrations but can significantly impact flora and fauna. Chemical analysis of tissues can solve the problem. Nevertheless, some contaminants are metabolized in the body to amounts that standard methods would not normally detect. Also, some chemicals accumulate more in some tissues than in others.

Measurement of the structure of ecological community alone is not sufficient to determine pollutant exposure because the presence or absence of specific species may be due to factors other than pollution, such as variations in temperature, salinity, dissolved oxygen, sediment structure, water depth, or biotic factors (e.g., reproductive cycles, competition, or predation). Therefore, along with additional studies, the integration of chemical and environmental studies is required to verify the results (i.e., bio testing of samples from the field) (Adarina et al., 2019; Karataev et al., 2020).

Physical, chemical, and biological monitoring in terrestrial ecosystems is also necessary to adequately understand terrestrial ecosystems. For this purpose, a network of stations is being created for the comprehensive monitoring of pine forests. Stations must be in protected areas where no management changes are planned for the next decade, and the landscape is typical of that part of the country. A minimum monitoring program for these ecosystems should include the following:

- Recording the flow of major pollutants through vital parts of the ecosystem. For example, in the ecosystem of a pine forest, sulfur and heavy metals are examined in needles, lichens, litter, topsoil, and earthworm bodies;
- Conducting measurements of production and degradation of organic matter by comparable standard methods. For example, organic decline, wood growth, and litter decomposition rates are measured in a pine forest.
- Carrying out a sufficient description of the soil and vegetation cover, allowing us to link the studied ecosystem with the larger biological and geophysical classifications.

The obtained results will allow us to compare the functioning of forest ecosystems across the country and lay the foundation for their rational management.

Registration of environmental conditions in background regions (whose land, water, and air are relatively unaffected by pollution and other disturbances) is carried out to quantify the content of pollutants in the air, water, and land. Comprehensive monitoring is conducted in or near small watersheds across the country, usually selected in national parks or preserves. Watersheds represent different climatic, topographic, and vegetation conditions. The selection process considers watersheds where research has already been conducted and those near existing research stations. Coordination and integration in the development of several projects involve separate ecosystems combining plant and animal research with analysis of physical and chemical parameters (Costanza, 1996; Tsvetkov et al., 2019).

The desire to focus on the social aspects of managing natural wildlife and fisheries systems in addition to biological and physical resources leads to an integrated monitoring system. This system contains indicators of resource management of wildlife and fisheries that include a comprehensive range of biological, ecological, social, and institutional aspects. These indicators are represented by the following:

- Raw materials that resource managers have to work with (e.g., land, labor, time, and money);

- Features of social groups seeking to use the resource base (e.g., population density and current level of demand for wildlife viewing opportunities);
- Actions or activities usually performed by managers of wildlife resources and outputs of direct management activities (e.g., number of new participants after a young hunter training course);
- Impact on the results of production or consumption of products (e.g., reduced user conflicts caused by inexperienced hunters or heightened perceptions of management agency).

Indicators of performance of resource management were determined based on three criteria:

- Calculation of indicators provided by the annual information on the given elements of resources;
- Availability of these indicators in regularly collected agency data archives, available government documents, or secondary research literature;
- Type of source, which includes agency documents describing the use of identifiers of indicators, reports demonstrating the validity or bias of a particular indicator, and reports suggesting that a particular indicator may be useful to measure.

Of the identified indicators, institutional indicators were found to be the most accessible and acceptable, while social indicators were the least accessible. This lack of social indicators points to a lack of monitoring capabilities reflecting the social dimension, that is, people's demand for and use of wildlife resources, as compared to the biological and institutional dimensions. It is believed that the use of a system of indicators will lead to more comprehensive and clearer resource management decisions and help identify new promising areas of research, especially in the human dimension (Ragulina et al., 2019; Vukovich et al., 2018).

Long-term environmental research and monitoring inevitably go together and greatly contribute to ecology. The challenges currently facing long-term research are discussed, given the funding and limitations of research institutions and the brevity of professional careers. Purposeful leadership by one or more project managers is important, along with clear goals, simple research design, protection and management of research sites, selection and measurement of variables, and data collection and management. A good monitoring program for long-term studies is expected to consist of the following:

- Carefully chosen initial sample design and the variables to be measured;
- Interested scientists capable of interpreting the selected data, examining and processing this data as it is collected, so that the structure of the monitoring program can be modified as knowledge of the ecosystem increases;
- Flexible monitoring program allowing to change sampling frequencies, measurement areas, or measured parameters;
- Main monitoring program should not be so large that it takes up all the time and resources of the researcher. It is necessary to have the time and resources to conduct short-term research to answer the questions proposed in a long-term study that uses a filtering approach.

The filtering approach includes the following goals:

- Creating a comprehensive long-term observation program to identify changes;
- Conducting research on ecosystem processes to understand systems with the development of integrative conceptual and predictive models.

The main research areas include the following:

- Primary productivity models;
- Population dynamics of organisms of different trophic levels;
- Structure and control of organic accumulations in soils of arid sediments;
- Nutrient flow patterns;
- Disturbance patterns;
- Data managers.

Filtration sites are located in ecosystems representative in terms of climate, landscape, biota, and soil. The research areas cover a range of spatial scales from plots to global sites. Nevertheless, most studies focus on the plot and landscape level and are located in regions of significant geographic importance, such as large river valleys, erosion zones, and productively managed plantations (Glotko et al., 2019a, 2019b).

3 Results

Nowadays, there is an interest in expanding the disciplines, including social studies.

Scholars interested in long-term ecological research have suggested that the filter network would be as follows:

- Providing a focus for comprehensive research and monitoring;
- Easier comparison of results;
- Using common skills and technology;
- Ensuring long-term continuity of individual priority research projects;
- Availability of training and educational opportunities.

In this case, it is necessary to create a network of filtering sites, which will serve as focal points for integrating research and monitoring.

A broader type of integrated monitoring takes place in the global change program. Global change is defined as changes in global climate, biodiversity, land use, and pollutants. Currently, four areas have been identified in which the study of global change is important:

- Major experiments;
- Modeling and synthesis projects;
- Environmental monitoring;
- Development of technology.

The items of the action plan include an analysis of the interaction between land use and global change, which include the following goals:

- Predicting the impact of land-use changes on key biota and processes at different spatial scales;
- Planning a comprehensive research program, including modeling, collaborative research, and large-scale experiments;
- Predicting the impact of altered disturbance regimes (e.g., fires, hurricanes, and floods) on ecosystems and their biota;
- Creation of a comprehensive environmental monitoring program using existing environmental research sites.

This monitoring will include biotic and abiotic components, interagency collaboration, comparable methodology across sites, comparable data management systems, data synthesis, and modeling (Alekseev et al., 2018; Stroiteleva et al., 2020).

Monitoring and modeling are interdependent and interactive processes. An integrated renewable natural resource modeling system is currently in operation to provide a clearinghouse for renewable natural resource databases and models relevant to Europe. This system links integrated biotic and abiotic data along with socio-economic activities and policy questions regarding renewable resources. It continually seeks to answer the "what if" questions associated with policy options analysis.

At the local level, comprehensive monitoring may be required for land use planning and environmental impact assessment. Basic monitoring of individual disciplines is often done at this level. However, it is necessary to understand ecosystem processes to assess the impact of development or a new management regime. For example, information may be needed on predator–prey relationships or the habitat requirements of a particular species. Local and regional authorities will carry out monitoring at the local scale.

Integrated monitoring can be conducted at the regional level, for example, for watershed management or forestry. On a broader scale, a selection of key indicators may be needed along with baseline data on water, soils, and climate that need to be monitored for resource use. This level of monitoring will be carried out by regional agencies (Shenshinov & Abdulsattar Al-Ali, 2020).

Comprehensive monitoring on a national scale requires a nationally coordinated system with comparable monitoring at the regional level, either by a national agency that will collect and process data centrally or by regional agencies that process their own data. Data collected at this scale include developments for the reports on the global environmental state. The collected data may include data needed for the reports on the state of the environment. At this level, careful selection of indicators is required to ensure that key variables, species. Processes are monitored at the national level.

Being an adequate early warning of environmental change, environmental indicators should be used for a monitoring system, along with socio-economic indicators, so that policymakers can assess environmental conditions in conjunction with socio-economic quality. Sustainability indicators can be broad enough to reflect the environmental consequences of economic activity and environmental management (e.g., pollutants, inputs, and spatial requirements) and changes in environmental conditions (e.g., stocks of environmental assets, environmental quality, and biodiversity).

The selection of early warning indicators is based on well-articulated models of the impact of socio-economic stresses on the environment.

4 Conclusion

Direct information about the future state and development of relevant socio-economic and environmental variables forms the basis for proactive planning and management. Its predictive power is based on mathematical models of the human–environment system.

Retrospective indicators include traditional policy assessments and historical trend indicators: energy, natural resources, agriculture, and pollution levels, which provide information on the effectiveness of existing policies or autonomous developments. From these indicators, decision-makers can learn and improve the effectiveness of the implemented policy. They can provide indirect information about future sustainability. As a rule, information is quantified using a combination of measured data and reference values (e.g., historical situations, economic goals, and health care standards).

The formation of a policy is not locked in a "trial–error-evaluation-new" trial cycle. Nevertheless, scientifically reliable information is obtained that is not based on risky forecasting methods. By allocating resources to both predictive and retrospective strategies, modeling would benefit from increased data collection and evaluation of policy effectiveness.

References

Adarina, R., Gazukina, Y., & Yankovskaya, K. (2019). Indicators of the "green economy" as a tool for monitoring the regional economy. *IOP Conference Series: Earth and Environmental Science, 395*, 012107. https://doi.org/10.1088/1755-1315/395/1/012107

Alekseev, A. N., Savenkov, D. L., Shvakov, E. E., Lebedev, N. A., & Ragulin, A. D. (2018). Financial problems of regional development and the ways of their solution in contemporary Russia. *Quality—Access to Success, 19*(S2), 71–75.

Bogoviz, A. V., Shvakov, E. E., Tretyakova, O. G., Zakharov, M. Y., & Abramov, A. N. (2020). Globalization of education in the conditions of formation of the global knowledge economy: Regularities and tendencies. In E. Popkova (Ed.), *Growth poles of the global economy: Emergence, changes and future perspectives* (pp. 993–1000). Cham, Switzerland: Springer. https://doi.org/10.1007/978-3-030-15160-7_99

Costanza, R. (1996). Ecological economics: Reintegrating the study of humans and nature. *Ecol. Applicat, 6*(4), 978–990.

Glotko, A. V., Sycheva, I. N., Dunets, A. N., Poltarykhin, A. L., Zhuravlev, P. V., & Tubalets, A. A. (2019). Integration processes in the agro-industrial complex of the regions: Development and environmental problems. *Journal of Environmental Management and Tourism, 10*(3), 613–621.

Glotko, A. V., Voronkova, O. Y., Miheeva, N. M., Skoblikova, T. V., Alekhina, N. A., & Zhminko, A. E. (2019b). The economic development of a transboundary region under the conditions of the

Eurasian economic union. In *Proceedings of IBIMA 2019: Education excellence and innovation management through vision 2020* (pp. 7957–7968). Granada, Spain.

Karataev, E. M., Merkuryev, V. V., & Titova, O. V. (2020). Integrating the "green economy" into the model of digital future of the modern socio-economic systems. In E. Popkova, & B. Sergi (Eds.), *Scientific and technical revolution: Yesterday, today and tomorrow* (pp. 1402–1410). Cham, Switzerland: Springer. https://doi.org/10.1007/978-3-030-47945-9_153

Parliament of the UK. (1971). Mineral Workings (Offshore Installations) Act 1971 (July 27, 1971 chapter 61). London, UK.

Ragulina, Y. V., Semenova, E. I., Avkopashvili, P. T., Dmitrieva, E. A., & Cherepukhin, T. Y. (2019). Top-priority directions of implementing new internet technologies on the territories of rapid economic development. In E. Popkova, & V. Ostrovskaya (Eds.), Perspectives on the use of new information and communication technology (ICT) in the modern economy (pp. 182–188). Cham, Switzerland: Springer. https://doi.org/10.1007/978-3-319-90835-9_21

Shenshinov, Y., & Al-Ali, A. (2020). The tools of increasing efficiency of human resource in the lean production environment: Conceptual study. *International Journal of Core Engineering & Management, 6*(7), 1–18.

Stroiteleva, T. G., Kletskova, E. V., & Balaeva, S. I. (2020). Ontogenesis of green development model of digital economy: Essence, stages, and prospects. In E. Popkova, & B. Sergi (Eds.), Scientific and technical revolution: Yesterday, today and tomorrow (pp. 410–417). Cham, Switzerland: Springer. https://doi.org/10.1007/978-3-030-47945-9_46

Stroiteleva, T. G., Petrova, E. A., & Yungblyudt, S. V. (2020). Responsible information society as a social environment for creation of the "green" digital economy: Formation and monitoring. In E. Popkova, & B. Sergi (Eds.), *Scientific and technical revolution: Yesterday, today and tomorrow* (pp. 402–409). Cham, Switzerland: Springer. https://doi.org/10.1007/978-3-030-47945-9_45

Tashkulova, G. K., & Kletskova, E. V. (2020a). Place and role of human in the system of circular reproduction in the digital regional economy. In E. Popkova, & B. Sergi (Eds.), *Scientific and technical revolution: Yesterday, today and tomorrow* (pp. 943–951). Cham, Switzerland: Springer. https://doi.org/10.1007/978-3-030-47945-9_101

Tashkulova, G. K., & Kletskova, E. V. (2020b). The model of organization of "green" digital production and consumption based on the internet of things and AI. In E. Popkova, & B. Sergi (Eds.), Scientific and technical revolution: Yesterday, today and tomorrow (pp. 329–336). Cham, Switzerland: Springer. https://doi.org/10.1007/978-3-030-47945-9_37

Tsvetkov, V. A., Loginov, E. L., & Raikov, A. N. (2019). Integrated network interaction of education with science and production. *Educational Technologies and Society, 19*(1), 389–398.

Vukovich, G. G., Makuschenko, L. V., Bateykin, D. V., Titova, O. V., & Dobrosotskiy, V. I. (2018). Support of the territories of advanced economic development on human capital: Theory and practice. *Quality – Access to Success, 19*(S2), 157–160.

Washington State Department of Ecology and United States Environmental Protection Agency. (1990). *Clean Air Act of the USA (104 Stat. 2468).* Washington, DC.

Regional Focus of Environmental Monitoring

Maria G. Sukhova, Olga V. Zhuravleva, and Yulia G. Gazukina

Abstract To focus on sustainable development, an integrated approach to resource management offers the best potential for understanding and monitoring environmental change. The catchment basis for comprehensive monitoring is the appropriate spatial scale used in some long-term studies. At the regional or catchment level, the focus can be made on urban and rural sustainability. It may be appropriate to consider the functional planning of regions based on ecological criteria, such as climate and vegetation structures, soil classification, and watershed boundaries. Perhaps, sustainable regional development requires explicit recognition of the environmental limits of human activity and an acknowledgment that the problem lies in people, not in other elements of the environment. The success of such an approach will depend on a shift in public perception.

Keywords Environmental monitoring · Regional development · Ecosystem focus · Environmental effects

JEL Codes Q5 · Q51 · Q55

1 Introduction

A regional or ecosystem focus is adopted in the reforms of legislation on local self-government and resource management, which provide a more prominent role in environmental management to the regional government. The Resource Management Act of 1991 allows for closer integration of land, water, air, and coastal management processes into policy statements and plans. A regional board may prepare regional plans in the context of a regional policy statement in response to environmental management issues. These plans can be requested by the public and must include goals to be achieved. The reasons for their adoption and the expected environmental results are critical policy statements and plans containing environmental standards to

M. G. Sukhova (✉) · O. V. Zhuravleva · Y. G. Gazukina
Gorno-Altaisk State University, Gorno-Altaysk, Russia

E. G. Popkova and B. S. Sergi (eds.), *Geo-Economy of the Future*,
https://doi.org/10.1007/978-3-030-92303-7_16

be achieved. Monitoring these environmental or performance standards allows one to check whether the indicated goals are achieved.

Planning is necessary for every region. It includes a comprehensive list of environmental resources and a list of human activities that could put the quality and quantity of those resources at risk. Using the framework basis as a guide, an environmental monitoring program can be planned and implemented. Various frameworks (e.g., those used in environmental reporting) have been developed to guide integrated environmental monitoring (Adarina et al., 2019; Merdesheva et al., 2020).

Currently, regional monitoring is carried out by several agencies, including regional councils, territorial bodies, the Department of Nature Conservation, District Boards of Health, and the Department of Commerce. The central state agencies have their own goals and objectives, while the regional councils are still in the ad-hoc monitoring phase with few formalized monitoring goals and objectives. Much of the current monitoring is a continuation of some past response to problems; it also may be conducted since researchers think it is crucial. Information is collected without much thought about the reason for its collection or how it should be used. One variable at a time is often tracked with little or no integration with other collected data. There are also problems with access to information between organizations. The integration between people and between the methods of data collection is required.

However, some changes in monitoring procedures have reflected increased public awareness of environmental trends and issues. For example, monitoring development projects (e.g., environmental impact reports) increasingly considers a range of social, economic, natural, and physical variables, where interrelationships are noted (Karataev et al., 2020; Shenshinov & Abdulsattar Al-Ali, 2020).

An informal survey conducted by the Resource Management Center raised concerns about the monitoring methods and regional policies that the regions could use to collect data. Moreover, some of these methods have different priorities, such as regular measurement of indicators related to water quality. The program provides a database allowing for estimating significant changes. In this program, all sampling is done by field water survey teams using standard sampling methods.

Nowadays, there is no unified body that coordinates regional monitoring or sets monitoring standards. There is a need to link regional monitoring activities and the national monitoring system, requiring regional monitoring by central government agencies and local governments. A template is needed to facilitate this commitment at the national level. This template is now presented under the Resource Management Act. It provides a comprehensive, holistic view focusing on the sustainable use of resources at the regional level. This approach leads to a series of comparable regional state-of-the-environment reports that can be used at the national level.

In addition to the nationally required and regionally coordinated monitoring system, the different nature of the regions means that monitoring will need to address specific environmental issues (Tashkulova & Kletskova, 2020a; Vukovich et al., 2018).

Currently, there are two categories of monitoring requirements:

- A template for coordinating monitoring carried out at the regional level;

- A regional monitoring program focusing on current or potential environmental problems specific to the region.

These two monitoring programs can and should be integrated wherever possible.

The following steps must be taken to allow local authorities to carry out comprehensive monitoring at the regional level:

- *Change of focus.* In addition to focusing on resource issues, there is a need for a more holistic approach to the environment and sustainable use of resources.
- *Clear objectives.* There must be clear coordination of the monitoring program between the two different organizations so that the status of the entire resource is considered. For example, a change of authority at the administrative boundary should not affect the overall management objectives and resource monitoring.
- *Inventory of resources.* It is necessary to conduct an inventory of natural resources in each region along with a list of human activities that could threaten the quality or quantity of resources. It is advisable to create a structure to guide this process.
- *Indicators of resource management.* Monitoring the resource's state involves repeated measurement of specific environmental and social variables. On this basis, a careful selection of the indicators of resource management will allow policymakers and the public to focus on environmental policy and resource status without scrutinizing all measured environmental and social variables (Glotko et al., 2019a, 2019b).

Most existing monitoring systems are designed for early warning. Therefore, some of the current major environmental problems are foreseen early enough to take preventive measures. Growing technological development makes unexpected environmental consequences more probable.

The cumulative environmental and social effects of human activities at all spatial scales are of concern when considering the sustainable use of resources. Most of the best-known environmental problems (e.g., deforestation, damage from acid rains, ozone depletion, and global climate change) are the cumulative result of expanding economic activity.

2 Methodology

Cumulative impacts result from the additive or synergistic effects of multiple incremental actions. One of the examples of such impacts is delayed-action chemical bombs, that is, the gradual accumulation and sudden release of harmful human-made chemicals into the environment. This process occurs with changes in land use and climate and the continued accumulation of toxic substances in soils and sediments, resulting in lower emissions of these substances. For example, fish tend to contain increased amounts of mercury.

Society often pays little attention to gradual changes in environmental parameters until it is too late for effective mitigation measures. Since homeostasis and ecosystem resilience often absorb incremental responses over long periods with no apparent adverse effect, there is a false sense of security when, in fact, humanity is being led into an ecological trap.

The first comprehensive attempt to quantify the effects of acidic air pollution on forests in Europe shows early warnings on the future state of forests. It has been suggested that only forest pollution costs Europe at least $30 billion a year. Moreover, it is almost impossible to calculate the additional costs in terms of adverse health effects from heavy metals in groundwater and widespread damage to buildings. There is growing evidence that it would be cheaper to clean the air of pollution than to continue repairing the damage caused.

At the beginning of the twenty-first century, geologists warned about erosion problems in the Eastern Cape. They said that the destruction of forest cover on the Eastern Cape would lead to widespread erosion. Nevertheless, their warnings, like many others, are left unheeded.

Nowadays, there are problems connected with the characterization of electronic systems and predicting the response of ecosystems to disturbances. On the analogy of characterizing human health, they point out that the lack of knowledge makes it harder to define the health of ecosystems and their dynamics becomes less manageable. It is suggested to use sets of indicators to characterize an ecosystem. Each indicator should relate to specific parts of the ecosystem, although no single group of indicators adequately represents the entire ecosystem.

To predict changes in an ecosystem, it is necessary to understand the different time and space scales. External abiotic events significantly impact ecosystems but are mediated by strong biological interactions within ecosystems. The spatial scale is defined by the dispersion distance of the most mobile key biological variables. The period of up to several hundred years is defined by the longest-lived or slowest-acting key biological variables. The minimum spatial scale for analysis is extremely large, and the slowest variable is the lifetime of trees, which requires a timescale for analysis of about two centuries. The slowest variable that should have been tracked 30 years ago to predict acidification was the buffering capacity of the watershed soil. Other variables (e.g., lake water or sulfur emissions) could not have predicted changes in lakes.

If sustainable development of the biosphere is a long-term goal, monitoring must become its integral part. We need appropriate indicators showing that something is going wrong. Moreover, we need enough time to make policy decisions to remedy the situation. For the above purpose, we need to do the following:

- Look at historical analogies to identify indicators that could be used;
- Focus on early warnings of environmental shifts;
- Identify quantification and monitoring of the balance between positive and negative feedback mechanisms in biological, geophysical, ecological, and socio-economic systems.

Several approaches, which can interact with each other, can be used to implement these recommendations.

Monitoring can provide baseline data and trends in environmental variables over a long period to understand how ecosystems may change in the future given specific environmental conditions (e.g., climate or certain types of management). It is necessary to be careful when interpreting the data since it is necessary to distinguish between natural background variation and changes caused by catastrophic events or anthropogenic causes to identify true changes in the environment.

In the absence of long-term ecological monitoring, it is sometimes possible to gather a picture from previous ecosystems using historical evidence (e.g., tree rings), climate data, and flow records available long before any processes. Part of the problem of collecting data from past monitoring can be solved by using various methods of data collection and analysis.

By replacing space with time, long-term ecological processes can be understood in short-term studies. For example, to study plants in old fields, it is possible to study fields abandoned one, two, five, ten, and thirty years ago and assume that continuity among plant communities at different sites when combined is the same as plant community succession at the same site within the first three decades after abandonment. This process assumes a smooth and consistent environment with no sudden or abrupt changes caused by spasmodic climatic or other events (Alekseev et al., 2018; Stroiteleva et al., 2019).

The use of models to predict long-term ecosystem behavior will not provide a detailed picture of change because of limited knowledge of the real functioning of ecosystems. However, these models are useful in designing, performing, and interpreting long-term studies.

3 Results

Significant research is currently being performed on climate change prediction. The role of the ocean in determining climate is often underestimated. Nevertheless, understanding and predicting ocean behavior can lead to understanding and predicting climate. Thus, the role of monitoring and modeling becomes fundamental to this understanding. Additionally, it is necessary to establish the link between climate change and rising sea levels, which requires large-scale monitoring and modeling systems.

The models are useful for predicting an irreversible trend in an environmental variable. However, the predictions will be quite uncertain because the models are calibrated to current or historical data. Additionally, the early stages of the trend will be difficult to detect due to the great natural variability in environmental conditions. It is important to try to optimize early warning monitoring systems, for example, carefully choose indicators, monitoring locations, and averaging times.

In recent years, disturbances familiar to engineers and physicists (e.g., slaps and skips) have been applied to ecological and socio-economic systems, where the system quickly transitions to a new and completely different state or even collapses. Models that simulate random behavior can be used to identify key elements of an early warning monitoring system. To compare several management strategies, the model tracks certain ecosystem indicators to provide the earliest possible assessment of the effectiveness of a particular management strategy.

Another method of predicting environmental change is the method of material balance, which tracks the flow of hazardous chemicals through the industrial economy from production and processing to end-use and disposal. Estimates of the accumulation of trace substances can often be obtained more accurately with this method than with standard monitoring because the focus is made on specific hazardous chemicals and their breakdown products. However, this approach requires a list of relevant historical data supplemented by indirect indicators of past conditions.

It is necessary to keep detailed records of the sources and flows of particular toxic materials. Next, models can be built to generate a range of scenarios of possible material use that complement the numerous existing scenarios of the use of population and energy (Avkopashvili et al., 2019; Ragulina et al., 2019).

This approach can be used to track the flow of pollutants through ecosystems. For example, sulfur and heavy metals can be traced through needles, litter, topsoil components, and earthworms in a pine forest ecosystem. In aquatic ecosystems, heavy metals from industry, stormwater runoff, and wastewater can be traced from their source through the water column to sediment and biota.

The change in capacity demonstrates an ecosystem that can serve as an early warning signal of environmental change. The understanding of the carrying capacity provides a functional definition of sustainability. From an environmental perspective, any development or environmental activity that does not exceed a region's carrying capacity is sustainable (Bogoviz et al., 2020; Tashkulova & Kletskova, 2020b).

4 Conclusion

Regional carrying capacity is the maximum rate of resource consumption and waste discharge that can be maintained indefinitely over a certain period.

This applies only to renewable resources (e.g., water, agriculture, and forestry). The division of the total ecosystem capacity by the average per capita resource consumption and waste production will provide the maximum size of the human population that can be sustained. An early warning of environmental damage is provided by monitoring the carrying capacity.

The impacts of recreationists will continue to be an active and growing threat to conservation unless a comprehensive framework that includes all relevant biophysical, social, and management conditions is given responsibility for making judgments (Stroiteleva et al., 2020).

The ability to identify the redox potential of soil and sediments and the regulation of the retention and release of chemical contaminants are the potentials that, along with the accumulation of toxic chemicals, should serve the purposes of environmental assessment. However, it is preferable, where possible, to identify not only primary relationships between development activities and environmental impacts but also secondary relationships. That is, an activity may affect another indicator through an intermediate influence since a pollution source may affect the environment through an intermediate step, which may include depletion of some capacity due to the accumulation of another pollutant. This depletion can be detected by monitoring, which must be performed to maintain the normal condition of our environment.

Acknowledgements The research was financially supported by the Russian Foundation for Basic Research and the Republic of Altay under the scientific project No. 20-45-040016.

References

Adarina, R., Gazukina, Y., & Yankovskaya, K. (2019). Indicators of the "green economy" as a tool for monitoring the regional economy. *IOP Conference Series: Earth and Environmental Science, 395*, 012107. https://doi.org/10.1088/1755-1315/395/1/012107

Alekseev, A. N., Savenkov, D. L., Shvakov, E. E., Lebedev, N. A., & Ragulin, A. D. (2018). Financial problems of regional development and the ways of their solution in contemporary Russia. *Quality—Access to Success, 19*(S2), 71–75.

Avkopashvili, P. T., Polukhin, A. A., Shkodinsky, S. V., & Poltarykhin, A. L. (2019). The fundamental provisions of the concept of knowledge economy. In E. Popkova, Y. Ragulina, & A. Bogoviz (Eds.), *Industry 4.0: Industrial revolution of the 21st century* (pp. 57–64). Cham, Switzerland: Springer. https://doi.org/10.1007/978-3-319-94310-7_5

Bogoviz, A. V., Shvakov, E. E., Tretyakova, O. G., Zakharov, M. Y., & Abramov, A. N. (2020). Globalization of education in the conditions of formation of the global knowledge economy: Regularities and tendencies. In E. Popkova (Ed.), *Growth poles of the global economy: Emergence, changes and future perspectives* (pp. 993–1000). Cham, Switzerland: Springer. https://doi.org/10.1007/978-3-030-15160-7_99

Glotko, A. V., Sycheva, I. N., Dunets, A. N., Poltarykhin, A. L., Zhuravlev, P. V., & Tubalets, A. A. (2019). Integration processes in the agro-industrial complex of the regions: Development and environmental problems. *Journal of Environmental Management and Tourism, 10*(3), 613–621.

Glotko, A. V., Voronkova, O. Y., Miheeva, N. M., Skoblikova, T. V., Alekhina, N. A., & Zhminko, A. E. (2019b). The economic development of a transboundary region under the conditions of the Eurasian economic union. In *Proceedings of IBIMA 2019: Education Excellence and Innovation Management through Vision 2020* (pp. 7957–7968). Granada, Spain.

Karataev, E. M., Merkuryev, V. V., & Titova, O. V. (2020). Integrating the "green economy" into the model of digital future of the modern socio-economic systems. In E. Popkova, & B. Sergi (Eds.), *Scientific and technical revolution: Yesterday, today and tomorrow* (pp. 1402–1410). Cham, Switzerland: Springer. https://doi.org/10.1007/978-3-030-47945-9_153

Merdesheva, E. V., Titova, O. V., & Avkopashvili, P. T. (2020). Methodological approach to the classification of digital economies by environmental efficiency and sustainability criterion. In E. Popkova, & B. Sergi (Eds.), *Scientific and technical revolution: Yesterday, today and tomorrow* (pp. 1360–1368). Cham, Switzerland: Springer. https://doi.org/10.1007/978-3-030-47945-9_148

Ragulina, Y. V., Semenova, E. I., Avkopashvili, P. T., Dmitrieva, E. A., & Cherepukhin, T. Y. (2019). Top-priority directions of implementing new internet technologies on the territories of rapid economic development. In E. Popkova, & V. Ostrovskaya (Eds.), *Perspectives on the use of new information and communication technology (ICT) in the modern economy* (pp. 182–188). Cham, Switzerland: Springer. https://doi.org/10.1007/978-3-319-90835-9_21

Shenshinov, Y., & Al-Ali, A. (2020). The tools of increasing efficiency of human resource in the lean production environment: Conceptual study. *International Journal of Core Engineering & Management, 6*(7), 1–18.

Stroiteleva, T. G., Kalinicheva, E. Y., Vukovich, G. G., & Osipov, V. S. (2019). Peculiarities and problems of formation of industry 4.0 in modern Russia. In E. Popkova, Y. Ragulina, & A. Bogoviz (Eds.), *Industry 4.0: Industrial revolution of the 21st century* (pp. 145–153). Cham, Switzerland: Springer. https://doi.org/10.1007/978-3-319-94310-7_14

Stroiteleva, T. G., Petrova, E. A., & Yungblyudt, S. V. (2020). Responsible information society as a social environment for creation of the "green" digital economy: Formation and monitoring. In E. Popkova, & B. Sergi (Eds.), *Scientific and technical revolution: Yesterday, today and tomorrow* (pp. 402–409). Cham, Switzerland: Springer. https://doi.org/10.1007/978-3-030-47945-9_45

Tashkulova, G. K., & Kletskova, E. V. (2020a). Place and role of human in the system of circular reproduction in the digital regional economy. In E. Popkova, & B. Sergi (Eds.), *Scientific and technical revolution: Yesterday, today and tomorrow* (pp. 943–951). Cham, Switzerland: Springer. https://doi.org/10.1007/978-3-030-47945-9_101

Tashkulova, G. K., & Kletskova, E. V. (2020b). The model of organization of "green" digital production and consumption based on the internet of things and AI. In E. Popkova, & B. Sergi (Eds.), *Scientific and technical revolution: Yesterday, today and tomorrow* (pp. 329–336). Cham, Switzerland: Springer. https://doi.org/10.1007/978-3-030-47945-9_37

Vukovich, G. G., Makuschenko, L. V., Bateykin, D. V., Titova, O. V., & Dobrosotskiy, V. I. (2018). Support of the territories of advanced economic development on human capital: Theory and practice. *Quality—Access to Success, 19*(S2), 157–160.

Financial and Economic Policy, Bank Financing, Investment, and Project Financing for Sustainable Development of Future Geoeconomics

Aleksey A. Mironov, Olga V. Koneva, Sergey E. Kasatkin, Andrey P. Balakin, and Aleksey Yu. Popov

Abstract The paper aims to substantiate the significant contribution of public and corporate financing factors in achieving sustainable development of the future geoeconomics and identify prospects for improving this direction. Using the methods of correlation and regression analysis, the authors prove that financial and economic policy, bank financing, investment, and project financing play an essential role in the sustainable development of the geoeconomics of the future. Financial factors determine efficient and sustainable use of water, efficient and sustainable use of land, and efficient use of materials. The authors propose scenarios for the sustainable development of the geoeconomics of the future up to 2025, depending on the management of financial factors. Specifically, the authors obtained a balanced scenario that assumes an increase in the sustainability of water usage to 92.73 points (up 38.53%), land use efficiency and sustainability to 79.38 points (up 0.09%), and material efficiency to 83.21 points (up 0.29%). There is also a scenario for the sustainable use of land and water resources and a scenario for the sustainable management of materials. The authors provide recommendations for the practical implementation of the proposed scenarios. The obtained results laid the scientific and methodological basis for the transition to a new financial approach to ensure sustainable development of the geoeconomics of the future, relying on financial and economic policy, bank financing, investment, and project financing.

Keywords Financial and economic policy · Bank financing · Investment and project financing · Sustainable development · Geoeconomics

JEL codes C58 · D92 · E65 · G17 · G21 · G32 · Q01 · O13

A. A. Mironov (✉) · O. V. Koneva · S. E. Kasatkin · A. P. Balakin · A. Yu. Popov
Volga State University of Technology, Yoshkar-Ola, Russia
e-mail: kbua@inbox.ru

E. G. Popkova and B. S. Sergi (eds.), *Geo-Economy of the Future*,
https://doi.org/10.1007/978-3-030-92303-7_17

1 Introduction

Sustainable geoeconomics is a set of sustainable development goals (SDGs) and their outcomes related to environmental protection, including the following:

- Efficient and sustainable (clean) energy (SDG 7);
- Efficient and sustainable use of water (SDG 6);
- Sustainable use of land (SDG 2);
- Material use efficiency (SDG 11 and SDG 12);
- Environmental quality, reduction of greenhouse gas emissions (SDG 13);
- Biodiversity and protection of ecosystem (SDG 14 and SDG 15).

In the face of the COVID-19 pandemic and crisis, the world's attention is focused on accelerating economic growth (SDG 8), developing an innovative economy (SDG 9), and building global partnerships (SDG 17). The focus on these SDGs has resulted in insufficient attention to the sustainable development of the geo-economy. The problem lies in the fact that the current approach to sustainable geoeconomics relies on non-financial measures of public (e.g., environmental standardization and control) and corporate (environmental responsibility) management. The lack of funding can be (and, according to this research, is) the reason for the incomplete realization of the potential of sustainable development of the geoeconomics of the future.

This paper hypothesizes that financial factors play an essential role in the sustainable development of the geoeconomics of the future. The research proposes a new financial approach to ensuring the sustainable development of the geoeconomics of the future, which is based on financial and economic policy, bank financing, investment, and project financing. The paper aims to substantiate the significant contribution of the factors of public and corporate financing in achieving sustainable development of the future geoeconomics and identify prospects for improving this management.

2 Literature Review

The issues of financing the sustainable development of the geoeconomics of the future, including financial and economic policy, bank financing, investment, and project financing, are disclosed in the works of such scholars as Chen and Zhao (2021), Morozova et al. (2018), Popescu et al. (2021), Popkova et al. (2021), Pyka and Nocoń (2021), Quatrini (2021), and Rodrigo-González et al. (2021). The literature review shows that the existing publications give a secondary role to financial factors in the sustainable development of the geoeconomics of the future. In general, financial factors are poorly understood, and their contribution to the sustainable development of the geoeconomics of the future has not been quantified or clearly defined.

The research gap to be filled in this paper is the uncertainty of the importance of financial and economic policy, bank financing, investment, and project financing for the sustainable development of future geoeconomics.

3 Research Methodology

The authors determine how financial factors (e.g., financial and economic policy, bank financing, investment, and project financing) (Table 1) influence the manifestation of sustainable development of the future geoeconomics. Based on UNDP materials (2021), the authors formed a sample of countries with the highest level of sustainable development (Table 2).

4 Findings

To test the hypothesis and determine the influence of financial factors on the sustainable development of the geoeconomics of the future, using data from Table 1 and Table 2, the authors obtained the following correlations and regression relationships (Table 3).

Let us check the reliability of the statistical results obtained in Table 3 using the Fisher test. For 10 observations and 4 factor variables, the tabulated value of Fisher's F-criterion is 5.19. This value is exceeded in the models G_2 (10.93), G_3 (7.11), and G_4 (7.11). Consequently, only three regression models are reliable at the 0.05 significance level. Thus, the authors obtained the following three equations are obtained:

- $G_2 = -41.06–0.63f_1 + 0.91f_2 + 1.20f_3 - 0.02f_4$. The equation shows that only bank financing and investments positively impact the efficient and sustainable use of water, the change of which by 94.73% is explained by these financial factors;
- $G_3 = -132.34–1.68f_1 - 0.36f_2 + 3.53f_3 + 0.44f_4$. The equation shows that only financial and economic policy and bank financing positively impact the efficient and sustainable use of land, the change of which by 92.22% is due to these financial factors;
- $G_4 = 133.59 + 0.41f_1 + 0.16f_2 - 0.88f_3 - 0.14f_4$. The equation shows that only investments and project financing positively affect the efficient use of materials, a change of which by 90.62% is due to these financial factors.

According to the obtained regression dependencies, the authors compiled alternative scenarios for the sustainable development of the future geoeconomics up to 2025, depending on the management of financial factors (Table 4).

According to Table 4, the balanced scenario, which allows simultaneously increasing all three selected manifestations of sustainable development of the future geoeconomics, involves increasing the sustainability of water use to 92.73 points (by

Table 1 Financial factors for sustainable geoeconomics of the future in the sample countries in 2020

Country	Criterion for sampling	Financial and economic policy	Bank financing	Investments	Project financing
	Sustainable development index, points 1–100	Facilitation of the creation of "markets of tomorrow," especially in the areas requiring the public–private partnership, points 1–100	Domestic credit to the private sector, % of GDP	Increasing incentives to placing financial resources on long-term investments, strengthening stability and expanding inclusion, points 1–100	Venture capital deals, bn PPP$ GDP
		f_1	f_2	f_3	f_4
Finland	85.90 (1 place)	59.5	41.80	95.4	26.10
Sweden	85.61 (2 place)	52.2	59.70	89.0	37.00
Denmark	54.86 (3 place)	46.7	73.90	84.6	44.60
Germany	82.48 (4 place)	48.1	33.80	79.3	19.90
Belgium	82.19 (5 place)	49.3	29.90	81.2	26.00
Austria	82.08 (6 place)	47.3	36.80	88.3	18.00
France	81.67 (8 place)	50.1	46.40	83.0	36.20
Estonia	81.58 (10 place)	44.9	26.70	81.1	64.50
Netherlands	81.56 (11 place)	50.4	46.90	79.9	31.70
Russia	73.75 (46 place)	n/d	32.90	55.3	3.70

Source Compiled by the authors based on (UNDP, 2021; WIPO, 2021; World Economic Forum, 2021)

38.53%), the efficiency and sustainability of land use to 79.38 points (by 0.09%), and the efficiency of material use to 83.21 points (by 0.29%). For this purpose, the following is recommended:

- Facilitating the creation of "markets of tomorrow," especially in the areas requiring the public–private partnership (by 107.36%);

Table 2 Manifestations of sustainable geoeconomics of the future in the sample countries in 2020, points 1–100

Country	Efficient and sustainable energy	Efficient and sustainable use of water	Sustainable use of land	Material use efficiency	Environmental quality	Reduction of GHG emissions	Biodiversity and ecosystem protection
	G_1	G_2	G_3	G_4	G_5	G_6	G_7
Finland	74.16	70.90	93.04	76.48	88.43	62.58	63.63
Sweden	87.34	86.27	98.63	79.88	89.65	85.30	60.63
Denmark	79.52	100.00	81.02	85.39	81.69	70.38	67.35
Germany	55.89	66.08	76.69	86.55	84.76	76,21	73.78
Belgium	45.79	46.00	69.52	85.81	89.25	75.54	74.51
Austria	73.11	69.15	98.10	79.37	86.10	77.46	69.53
France	53.36	65.98	70.93	87.70	87.71	79.12	71.36
Estonia	57.71	52.50	99.13	75.36	91.46	61.52	74.32
Netherlands	469.55	58.66	55.03	83.66	87.14	70.58	59.20
Russia	26.55	53.85	51.02	89.50	87.72	42.23	49.05

Source Compiled by the authors based on (Global Green Growth Institute, 2021)

Table 3 Correlation and regression dependence of sustainable geoeconomics of the future on financial factors

Regression statistics	G_1	G_2	G_3	G_4	G_5	G_6	G_7
Multiple correlation (%)	55.28	94.73	92.22	90.62	70.30	80.94	78.01
F observed	0.55	10.93	7.11	5.74	0.41	0.18	1.94
Constant	933.07	−41.06	−132.34	133.59	83.00	69.71	71.36
Regression coefficient at f_1	12.65	−0.63	−1.68	0.41	−0.09	0.94	0.45
Regression coefficient at f_2	2.61	0.91	−0.36	0.16	−0.13	0.19	−0.22
Regression coefficient at f_3	−18.09	1.20	3.53	−0.88	0.15	−0.56	−0.24
Regression coefficient at f_4	−1.05	−0.02	0.44	−0.14	0.07	−0.15	0.14

Source Compiled by the authors

Table 4 Scenarios for the sustainable development of the geoeconomics of the future depending on the management of financial factors up to 2025

Variable	Baseline values in 2020, points 1–100	Balanced scenario		Scenario of sustainable use of land and water resources		Scenario of sustainable management of materials	
		value, points 1–100	increase, %	value, points 1–100	increase, %	value, points 1–100	increase, %
f1	44.85	93.00	107.36	44.85	0.00	80.66	79.84
f2	42.88	82.00	91.23	56.16	30.97	56.70	32.23
f3	81.71	100.00	22.38	99.21	21.42	81.71	0.00
f4	30.77	100.00	224.99	30.70	−0.23	30.77	0.00
G2	66.94	92.73	38.53	100.00	49.39	56.78	−15.18
G3	79.31	79.38	0.09	136.43	72.02	14.27	−82.01
G4	82.97	83.21	0.29	69.76	−15.92	100.00	20.53

Source Compiled by the authors

- Increasing domestic credit to the private sector to 82% of GDP (by 91.23%);
- Increasing incentives to place financial resources on long-term investments, enhancing stability, and expanding inclusiveness to 100 points (by 22.38%);
- Increasing venture capital deals to $100 billion (by 224.99%).

The scenario of sustainable land and water use improves the sustainability of water use to 100 points (by 49.39%) and the efficiency and sustainability of land use to 100 points (by 72.02%). For this purpose, the following is recommended:

- Increasing domestic credit to the private sector to 56.16% of GDP (by 30.97%);
- Increasing incentives to placing financial resources on long-term investments, enhancing stability, and expanding inclusiveness to 99.21 points (by 21.42%).

The scenario of sustainable use of materials provides an increase in material efficiency up to 100 points (by 20.53%). For this purpose, the following is recommended:

- Facilitating the creation of "markets of tomorrow," especially in the areas requiring the public–private partnership to 80.66 points (by 79.84%);
- Increasing domestic credit to the private sector to 56.70% of GDP (by 32.23%).

5 Conclusions

Thus, the authors proved that financial and economic policies, bank financing, investment, and project financing play an essential role in ensuring the sustainable development of the geoeconomics of the future. Financial factors determine efficient and sustainable use of water, efficient and sustainable use of land, and efficient use of materials.

The authors offered several scenarios for the sustainable development of geoeconomics up to 2025, depending on the management of financial factors. Moreover, the authors provide recommendations for the practical implementation of the proposed scenarios. It is advisable to choose among the proposed scenarios based on the possibility of management of non-financial factors in addition to financial management.

The limitation of this research lies in the fact that such manifestations of the future geoeconomics as environmental quality, reduction of greenhouse gas emissions, protection of biodiversity and ecosystem, and efficient and sustainable energy were found to be independent of financial factors. The authors propose to devote further research to studying the prospects for implementing these areas of sustainable development of the geoeconomics of the future, going beyond the management of financial factors.

References

Chen, Y., & Zhao, Z. J. (2021). The rise of green bonds for sustainable finance: Global standards and issues with the expanding Chinese market. *Current Opinion in Environmental Sustainability, 52*, 54–57. https://doi.org/10.1016/j.cosust.2021.06.013

Global Green Growth Institute. (2021). Green Growth Index 2020. Retrieved from https://greengrowthindex.gggi.org/ (Accessed September 8 2021)

Morozova, I. M., Litvinova, T. N., Przhedetskaya, N. V., & Sheveleva, V. V. (2018). The problems of financing of entrepreneurship infrastructure in developing countries and their solutions. In E. Popkova (Ed.), The impact of information on modern humans (pp. 277–283). Cham, Switzerland: Springer. https://doi.org/10.1007/978-3-319-75383-6_35

Popescu, I.-S., Hitaj, C., & Benetto, E. (2021). Measuring the sustainability of investment funds: A critical review of methods and frameworks in sustainable finance. *Journal of Cleaner Production, 314*, 128016. https://doi.org/10.1016/j.jclepro.2021.128016

Popkova, E. G., Oudah, A.-M. M. Y., Ermolina, L. V., & Sergi, B. S. (2021). Financing sustainable development amid the crisis of 2020. A research note. In E. G. Popkova, & B. S. Sergi (Eds.), Modern global economic system: Evolutional development vs. revolutionary leap (pp. 773–780). Cham, Switzerland: Springer. https://doi.org/10.1007/978-3-030-69415-9_88

Pyka, I., & Nocoń, A. (2021). Banks' capital requirements in terms of implementation of the concept of sustainable finance. *Sustainability, 13*(6), 3499. https://doi.org/10.3390/su13063499

Quatrini, S. (2021). Challenges and opportunities to scale up sustainable finance after the COVID-19 crisis: Lessons and promising innovations from science and practice. *Ecosystem Services, 48*, 101240. https://doi.org/10.1016/j.ecoser.2020.101240

Rodrigo-González, A., Grau-Grau, A., & Bel-Oms, I. (2021). Circular economy and value creation: Sustainable finance with a real options approach. *Sustainability, 13*(14), 7973. https://doi.org/10.3390/su13147973

UNDP. (2021). Sustainable Development Report. Retrieved from https://dashboards.sdgindex.org/rankings (Accessed September 7 2021)

WIPO. (2021). Global Innovation Index 2020. Retrieved from https://knoema.ru/GII2018Aug/global-innovation-index (Accessed September 8 2021)

World Economic Forum. (2021). Global competitiveness report special edition 2020: How countries are performing on the road to recovery. Retrieved from https://www.weforum.org/reports/the-global-competitiveness-report-2020 (Accessed September 8 2021)

Modelling the English-Language Course for Personnel of Healthcare Organizations Within the Framework of International Medical Tourism Enhancement

Vladimir V. Shkarin, Tatiana P. Reznik, Dmitry V. Orlov, and Marina V. Polyakova

Abstract This paper features basic aspects of designing and implementing a course of medical English to personnel of selected healthcare organizations in the Volgograd region within the framework of the regional governmental project "Export of Medical Services". Attention is focused on difficulties arising due to its specialist nature, effective solutions to the problems are also discussed. The authors created the model of the English-language course for personnel of healthcare organizations within the framework of international medical tourism enhancement. Of the 63 students enrolled on the course, 100% fulfilled the programme successfully and received a certificate of its completion. Among these 50% of the students proved to have the excellent level of the course material acquisition, 34%—medium, whereas 16%—the low-level results. It is concluded that the elaborate approach and carefully worked out model, embracing the organizational and teaching aspects of the course, as well as its technological character, should provide strong likelihood of positive educational results in future multiplication of this project.

Keywords Export of medical services · Teaching a foreign language for specific purposes · Foreign language communicative competence · Continuous professional development programmes · Foreign-language communicative situation

1 Introduction

One of the national state projects in Russia "Healthcare" (2019–2024) embraces a number of federal projects, one of them being "Export of Medical Services", which is being developed in the Volgograd region in the status of a key-priority

V. V. Shkarin
Volgograd State Medical University, Volgograd, Russia

T. P. Reznik · M. V. Polyakova
Volgograd State Socio-Pedagogical University, Volgograd, Russia

D. V. Orlov (✉)
SBHCI City Children's Clinic №. 2, Volzhsky, Russia

E. G. Popkova and B. S. Sergi (eds.), *Geo-Economy of the Future*,
https://doi.org/10.1007/978-3-030-92303-7_18

regional project. According to its roadmap the year of 2019 was taken up by initial preparatory measures, including implementing courses of medical English to personnel of selected healthcare organizations (HO). This paper features a case of such specialist-targeted courses implementation as well as perspectives of the project. The article is second in a series: in the first one (Orlov et al., 2019) the authors covered general approaches to developing the regional model of the "Export of Medical Services" project, and also described how language-level placement testing of medical personnel was realized. It is noteworthy that, based on the results of 71 medics tested, we found that 7 individuals had "zero" level of communicative competence in English, 52 had Basic-user Level (A1–25 people, A2–27 people), and 12 people—Independent-user Level (B1–9 and B2–3 people). It should be borne in mind that the B1 level is the minimum level at which medical professionals are able to handle successful communication with their "inophonic" medical tourist patients. Here we define "inophonic" as a native speaker of a foreign language and consequently a representative of the corresponding mentality (Azimov & Shukin, 2009).

2 Methodology

In 2019 within the framework of the abovementioned project a number of leading medical organizations of the Volgograd Region decided to take a professional development course of medical English for the staffers involved in treating and servicing inophonic patients. A significant share of such courses was provided by the educational centre "World Languages and Cultures" and the Institute of Modern Languages of the Volgograd State Socio-Pedagogical University (VSSPU). Prior to the active realization phase the specialists of the centre carried out a careful and comprehensive designing of the course, the latter having been compiled anew. Its foundation was laid on the baseline principles, which had been worked out at the very beginning of the project: teaching "on site", that is on the premises of the clinic; syllabus based directly on the medical profile of the HO; short-term course duration (from 16 to 48 academic hours); going by selected typical communicative scenarios of 'doctor-patient' interactions to build the necessary communicative skills and abilities; and, consequently, taking a *communicative situation* as the basic unit of the training process. Following the analysis of the specialist areas of the HOs, taking part in the project, three main variations of the training programme were produced: "English in medical tourism: Inpatient treatment", "English in medical tourism: Dentistry" and "English in medical tourism: medical consulting and diagnostic services", which in their turn were customized for the range of communicative situations to meet the demands of each particular customer HO (Akopova Przhedetskaya Przhedetsky Borzenko, 2020a, 2020b, 2020c, 2020d, 2020e; Inshakova et al., 2020).

One more specificity of arranging the training in question was a challenging and important task of putting the trainees into group. Besides the input language level, we had to take into consideration some other objective and, we should say, far from productive conditions and factors, such as shifts-based working schedule, students occupying managerial positions in their HO staying away from lessons due to their work commitments and so forth. Moreover, in a number of cases all the trainees of the HO assigned to take the course had to be put into one small group, which then turned out to unite students of three different language levels. This being the case, the teacher had to apply a range of individualized learning techniques: so the lesson plan was derived in at least two variations. The difference between them would be that A-level (Basic-level user according to the CEFR) students were offered more simple reproductive language-level tasks exercising basic speech cliches, employing a lot of drill and repetition to work out the necessary automatic speech habits in trainees. Meanwhile, B-level students were to do tasks based on transformation, as well as tasks of semi-communicative and communicative-proper types (Azimov & Shukin, 2009). The latter strategy was supposed to allow the students to apply their input scope of foreign language competence to the current new learning aims and tasks and for them to form new productive speech stereotypes as a result.

It is noteworthy that these fluctuations in the language level homogeneity of the groups caused us to tailor the initial module structure of the course. Originally, (Orlov et al., 2019) the course comprised two modules: *Basic*, i.e., an introductory part for revision to activate the necessary basic language knowledge and habits, and *Specialist*, aiming in particular at mastering the necessary communicative means to successfully manage interaction with inophonic patients. However, in a number of cases when teaching heterogeneous language level groups, we had to blend the contents of the two abovementioned modules, allowing more time and additional practice to recycle the material for revision. This, on the one hand, served to meet the demands of weak students to acquire stronger, steadier, more numerous basic level language skills and knowledge, and on the other hand, to provide more advanced language level students with an opportunity to study at their full capacity starting from the very first lesson.

Returning to the basic unit of the training process under description—communicative situation, below we present the complete inventory of such from our programme "English in medical tourism: Inpatient treatment":

- First visit
- Calling to make an appointment; At the reception desk
- Finding your way around the clinic
- Daily ward rounds by the attending doctor; weekly rounds by the head of the department
- Daily treatment procedures
- Accident and emergency service
- Tests and diagnosis: explaining the aim and the procedure; doing the manipulations; facilitating the patient in consultations with specialists, laboratory testing
- Prescriptions and recommendations
- Picking up a treatment programme
- Recovery and rehabilitation

- Disease prevention and healthcare recommendations
- Conducting and taking part in the medical council
- Dealing with psychological barriers, patient's fears and refusals
- Dismissal procedures
- Dealing with the patient's medical history
- Dealing with the patient's relatives

Studying each communicative situation was to take three academic hours. Besides, as practice showed, the contents of some situations (for example, "First visit" and "Prescriptions and recommendations") served a good base for many of the subsequent ones. Consequently, thorough work out of these basic situations ensured success in mastering the ones to follow of a congruent nature.

Undoubtedly, in the scope of one article it as impossible to describe all the components of the model, presenting the organizational and teaching aspects of the course in question, which was developed and piloted during the year of 2019. Other aspects:

- building up the topical essential vocabulary list,
- approaches to designing a vocabulary practice exercise unit, as well as such units targeted at acquiring grammar structures and speech cliches,
- the forms and content of the in-course progress and final assessment,
- the common and the specific of course design as varied by the profile and type of the HO,
- correlation of the approaches applied in teaching and the corporate culture of the HO are to be covered in the publications to follow.

3 Results

In the final part of the paper we will consider the results of the course implementation, as well as their analysis, and perspectives. Of the 63 students enrolled on the course, 100% fulfilled the programme successfully and received a certificate if its completion. Among these 50% of the students proved to have the excellent level of the course material acquisition, 34%—medium, whereas 16%—the low-level results. In 25% of the students, who were initially motivated by the external factors (motives of compulsion), we fixed evidence of their motivation transition to the internal type (being interested in the subject of study itself, targeting at self-development, personal realization in the professional sphere). To conclude, realization of the project should be considered as successful as the training course reached its goal of building in the trainees a certain type of communicative competence—being psychologically ready and linguistically equipped to interact efficiently with inophonic patients.

Nevertheless, we must remark and describe some specific difficulties, which we—both the teachers and the students—came across in and around the teaching process. The teaching–learning process was noticeably impeded by the doctors' overloaded working schedule, which made independent study mode—a crucial component of foreign language acquisition—practically excluded. Under the circumstances the teachers of the course had to slow down the pace and to include additional activities for the previous material revision. Besides that, they made intensive use of additional teaching–learning aids, hand-out material, visual aids to secure a sufficient amount of support to learn the vocabulary and other language material, both during the course and after its completion. Such techniques were seen as being helpful if thereafter the staffers had to refresh their communicative skills prior to any kind of international event or just before receiving an English-speaking patient. To say more about the teaching aids and learning materials, we should also note another problem which was practically impossible to overcome—finding proper audio material to go with each particular situation under study, which as mentioned above was considered the basic unit of the training process. As the choice of the situations was authentic, made by the developers of the course, it is nothing surprising that we couldn't find ready-made recordings to match the topics ideally and fully. A partial solution to this problem was using the recorded material of some well-known courses of medical English (Glendinning & Beverly, 2008), but these had to be shortened or taken by fragments, which, of course, decreased their teaching and communicative worth. Tackling this listening material shortage is still to be dealt with and outlines the perspectives of the project.

To get an informative enough feedback on the training conducted we asked the students to do a questionnaire which helped to assess how satisfied they were with the studies in general and some of its particular aspects, for example: "Did the contents of the course meet your professional demands?", "Have you used any of the skills and training experience at your work?", "Was the pace of the course convenient to you?", "Was the course length sufficient to reach its goals? Please, give an extended answer", "What is your opinion on the forms and teaching methods implemented during the course (types of tasks, approaches)". The students' answers testified to high appreciation of the course design and performance, done by the educators of the "World Languages and Cultures" Centre (VSSPU), proved the high demand of the foreign language among medical professionals, who are involved in the "Export of Medical Services" project.

Below is the graphical representation of the process discussed. Under Fig. 1 there are symbol designations and comments.

The Model of the English-Language Course for Personnel of Healthcare Organizations within the Framework of International Medical Tourism Enhancement is broken down by 4 stages, each being divided into several steps. The use of colour signifies the agent taking part in the process of the corresponding stage and step:

- those carried out by the health organizations are marked with the yellow colour; besides, they have the corresponding label on the left—the red cross;
- those carried out by the educational centre are marked with the green colour; they have the corresponding label on the right—the book;
- those which have both the colours blended and the labels at both sides are realized in collaboration of the health organization and the educational centre. The horizontally paired blocks symbolize parallel processes of the same stage of the model.

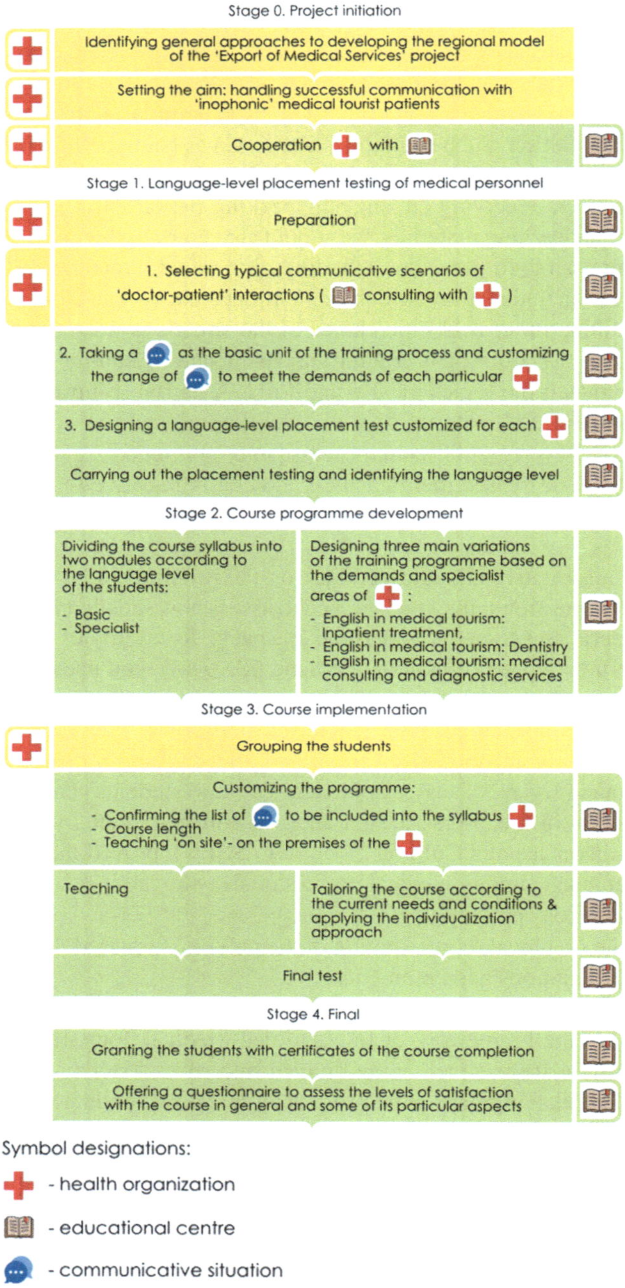

Fig. 1 Model of the English-language course for personnel of healthcare organizations within the framework of international medical tourism enhancement. *Source* Developed and compiled by the authors

4 Conclusion

To draw an intermediary conclusion and to outline the perspectives of further realization of the course of English in medical tourism, we should note that the elaborate approach and carefully worked out model, embracing the organizational and teaching aspects of the course, as well as its technological character, should provide strong likelihood of positive educational results in future multiplication of this project.

References

Akopova, E., Przhedetskaya, N., Przhedetsky, Yu., & Borzenko, K. (Eds.) (2020). The conceptual provisions of the theory of marketing of healthcare organizations. marketing of healthcare organizations: Technologies of Public-Private Partnership. A volume in the series Popkova, E. (ed.) Advances in Research on Russian Business and Management, Charlotte, NC, USA: Information Age Publishing. Available at: https://www.infoagepub.com/products/Marketing-of-Healthcare-Organizations

Akopova, E., Przhedetskaya, N., Przhedetsky, Yu., & Borzenko, K. (Eds.) (2020). Public-private partnership as a mechanism of increase of effectiveness of marketing activities. marketing of healthcare organizations: Technologies of public-private partnership. A volume in the series Popkova, E. (Ed.) Advances in Research on Russian Business and Management, Charlotte, NC, USA: Information Age Publishing. Available at: https://www.infoagepub.com/products/Marketing-of-Healthcare-Organizations

Akopova, E., Przhedetskaya, N., Przhedetsky, Yu., & Borzenko, K. (Eds.) (2020). Tendencies and problems of development of marketing of healthcare organizations in Modern Russia. marketing of healthcare organizations: technologies of public-private partnership. A volume in the series Popkova, E. (Ed.) Advances in Research on Russian Business and Management, Charlotte, NC, USA: Information Age Publishing. Available at: https://www.infoagepub.com/products/Marketing-of-Healthcare-Organizations

Akopova, E., Przhedetskaya, N., Przhedetsky, Yu., & Borzenko, K. (Eds.) (2020). Preconditions and Perspectives of development of marketing of healthcare organizations in modern Russia. Marketing of Healthcare Organizations: Technologies of Public-Private Partnership. A volume in the series Popkova, E. (Ed.) Advances in Research on Russian Business and Management, Charlotte, NC, USA: Information Age Publishing. Available at: https://www.infoagepub.com/products/Marketing-of-Healthcare-Organizations

Akopova, E., Przhedetskaya, N., Przhedetsky, Yu., & Borzenko, K. (Eds.) (2020). The algorithm of applying the mechanism of development of marketing of healthcare organizations with the help of public-private partnership in Modern Russia. Conclusions. About the Authors. Marketing of Healthcare Organizations: Technologies of Public-Private Partnership. A volume in the series Popkova, E. (Ed.) Advances in Research on Russian Business and Management, Charlotte, NC, USA: Information Age Publishing. Available at: https://www.infoagepub.com/products/Marketing-of-Healthcare-Organizations

Azimov, E., & Shukin, A. (2009). *New dictionary of teacher methods terms and notions (Theory and practice of language education)*. IKAR Publishing House.

Glendinning, E., & Beverly, A. (2008). *English in medicine* (3rd edn). Holmstrom. Moscow: CUP.

Inshakova, A., Goncharov, A., Kazachenok, S., & Sevost'yanov, M. (2020). Med-arb: concept, features of regulation, and prospects for law enforcement. In: A. Inshakova & A. Bogoviz (Eds.), *Alternative Methods of Judging Economic Conflicts in the National Positive and Soft Law*. A volume in the series Popkova, E. (ed.) Advances in Research on Russian

Business and Management, Charlotte, NC, USA: Information Age Publishing. Available at: https://www.infoagepub.com/products/Alternative-Methods-of-Judging-Economic-Conflicts-in-the-National-Positive-and-Soft-Law

Orlov, D., Shkarin, V., & Reznik, T. (2019). Maintaining communicative interaction with foreign medical tourists in Russian clinics. Modern tourism in medical treatment and healthcare. In *Proceedings of the 5th all-Russia scientific and practical conference with international participation* N. Sedovoy & G. Schekina (Eds.), Volgograd, pp. 107–120.

Specifics of Marketing Technologies in Healthcare Organizations

Natalia N. Zubareva

Abstract The purpose of the study is to identify the specifics of modern marketing technologies for promoting services in healthcare organizations, as well as to form recommendations for improving digital marketing methods in this area. The author highlighted the prerequisites for the development of marketing of healthcare organizations in Russia (the increase in the incidence of the country's population, the change in the ratio of public and private organizations in the field of healthcare, the tightening of competition conditions in the market of paid medical services, the growing confidence of the population in paid medical care), and also determined its industry specifics. The article highlights the existing limitations of the introduction of marketing technologies for the positioning of healthcare organizations and the promotion of medical services. The author systematized the marketing technologies used by healthcare organizations (marketing research, marketing concept formation, internal and external marketing, organization branding). It is proved that healthcare organizations should actively implement digital marketing technologies and adapt medical content to a dynamic virtual environment.

Keywords Marketing · Marketing technology · Healthcare · Medical service · Digital marketing · Promotion · Marketing communication

JEL Codes I15 · M31

1 Introduction

Currently, the conditions of the economic activity of organizations are changing. This is a consequence of the consumer preferences transformation (Payne et al., 2008; Prahalad & Ramaswamy, 2010; Vargo & Lasch, 2006), the emergence of new factors in the competitive environment development, as well as the appearance of advanced digital technologies that can ensure the success of the organization at a new level (Gusev et al., 2020; Karpunina, et al., 2020; Molchan et al., 2019).

N. N. Zubareva (✉)
Belgorod National Research University, Belgorod, Russia

E. G. Popkova and B. S. Sergi (eds.), *Geo-Economy of the Future*,
https://doi.org/10.1007/978-3-030-92303-7_19

The healthcare sector, which is certainly one of the most important for all segments of society, does not stand aside. Today, patients want to receive high-quality medical services in the shortest possible time, to have guarantees after receiving them and services for continuous interaction within the framework of the treatment, as well as to choose a medical institution and a price offer (Babenko, 2017; Borzenko, 2018; Halfin et al., 2017).

In this regard, new horizons of activity and growth opportunities are opening up for healthcare organizations. The success of the organization now depends on a competent development strategy and implementing a marketing mix. In Russia, there are prerequisites for the development of marketing healthcare organizations. Firstly, it is the growing demand for medical goods and services due to the increase in the incidence of the country's population by 150.21% over the past 25 years (from 1995 to 2020), as well as the deterioration of the environmental and epidemiological situation and the increasing attention of Russians to their health (Borzenko, 2018). Secondly, the number of outpatient clinics in Russia has risen (by 268.54% over the past 25 years) due to the increase in private organizations, as well as growing competition among healthcare organizations that provide paid medical services. Third, the growing trust of Russians in private healthcare organizations and their preference for paid medical services due to the low quality of free medical services provided (17% of patients surveyed in 2017 noted the poor quality of treatment, the professionalism of doctors, and their attitude to patients) (All-Russian Center for the Study of Public Opinion, 2018). Thus, the search for effective marketing technologies to promote the services of healthcare organizations and improve their brand is currently an urgent research task.

2 Literature Review

Tarasenko (2010) defines the marketing of a healthcare organization as a system of measures aimed at achieving the organization's goals related to meeting the needs of patients in various types of preventive, diagnostic, therapeutic, and rehabilitation services. This definition allows us to conclude that the main element of marketing in healthcare organizations is the promotion of medical services (Babenko, 2017). Another element of marketing in this area is the formation of the brand of a healthcare organization, that is, a set of brand attributes that distinguish it from its competitors (Sewell & Brown, 2017).

Researchers distinguish between internal marketing (aimed at creating a high degree of satisfaction with the quality of medical services provided to existing patients of the organization and the education of loyal consumers) and external marketing (external marketing communications to attract potential consumers-advertising, PR communications) (Borzenko, 2018; Halfin et al., 2017; Xandopulo, 2017). Customer orientation is an integral part of the successful work of the organization, and when promoting paid medical services, it becomes a key determinant (Kaliyeva et al., 2016; Voronina & Bereza, 2016).

Healthcare organizations use modern digital technologies, such as digital marketing, CRM systems, and SMM promotion, which determine the opportunities for them to achieve competitive advantages in the industry (Nikolaeva, 2017; Rakhmanova, 2017; Rostovsky, 2017).

3 Methodology

The purpose of the study is to reveal the specifics of marketing and modern marketing technologies for promoting services in healthcare organizations, as well as to form recommendations for improving the methods of industry digital marketing.

Research objectives: (1) identify the prerequisites for the development of marketing of healthcare organizations in Russia; (2) highlight the features of marketing in healthcare organizations; (3) determine the problems associated with the promotion of medical services in Russian healthcare organizations; (4) offer a set of effective marketing technologies to attract and retain patients in healthcare organizations.

Research methods: the method of theoretical analysis, the method of comparison, the method of economic and statistical analysis, the graphical method, the system approach.

4 Results

4.1 Prerequisites for the Development of Marketing of Healthcare Organizations in Russia

Health care ensures the preservation and maintenance of life and health of the population at the target level, is one of the most priority areas of the national economy, which has special features.

The analysis of the indicators of the health sector development in Russia allows us to conclude that the positive dynamics of the increase in the incidence of the country's population in the period from 2005 to 2018 by 15%. In addition, the total number of outpatient organizations in Russia has multidirectional dynamics: a decrease in the period 2005–2010 (by 27.8%), followed by a gradual increase of 28.6%, starting from 2010 to 2018 (Table 1).

On the one hand, the activities of healthcare organizations are directly related to the provision of medical services and the production of goods that are among the public goods (Borzenko, 2018). This means that the production and provision of medical services are unprofitable for private businesses due to the high costs to ensure their quality and the presence of government restrictions in the field of pricing for them to maintain their mass availability. On the other hand, the shortcomings of

Table 1 Dynamics of indicators of the healthcare sector development in Russia, 2005–2018

Indicators	2005	2010	2015	2016	2017	2018
Morbidity of the population, thousand people	100.306	106.328	115.326	115.326	114.382.2	114.840.8
Morbidity of the population, thousand people per 1 thousand people of the population	676	780.1	778.2	780.1	778.9	782.1
Number of outpatient organizations	21.783	15.732	18.564	19.126	20.217	20.228
Number of state organizations	17.172	12.173	13.985	14.117	14.465	14.424
Share of state organizations (%)	78.8	77.3	75.3	71	71.5	71.3
Number of private organizations	–	2.753	3.749	4.168	4.168	4.866
Share of private organizations (%)	–	17.5	20.2	21.8	20.6	24.1

Source Compiled by the author based on (Federal State Statistics Service of the Russian Federation, 2020)

the provision of medical goods and services by public organizations are compensated by the expansion of the range of paid medical services to the population provided by public organizations or private organizations. This pattern is confirmed by the change in the ratio of public and private organizations providing health services. If in 2005 the share of public organizations in the healthcare sector was 78.8%, in 2018 it decreased to 71.3%. The number of private organizations in the study period increased by 1.8 times.

During the same period, household spending on paid medical services per household member increased almost sixfold on average (Fig. 1).

The volume of paid medical services provided to the population increased by 567.930 million rubles in the period 2005–2018 (Fig. 2).

Fig. 1 Monetary expenditures of households on paid medical services, on average per household member, rubles *Source* Compiled by the author based on (Federal State Statistics Service of the Russian Federation, 2020)

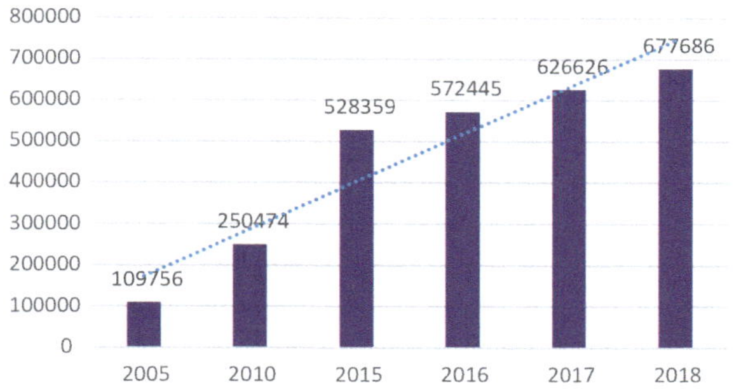

Fig. 2 Paid medical services to the population, million rubles, 2005–2018. *Source* Compiled by the author based on (Federal State Statistics Service of the Russian Federation, 2020)

This situation can be a consequence of both the deterioration of the quality of medical services provided by public organizations in the health sector and the expansion of the population's opportunities to receive the necessary range of paid services due to a more attentive attitude to their health and improving the quality of life.

Thus, in Russia, there is an increase in the number of private organizations and an expansion of the range of paid services provided by public health organizations. This is a prerequisite for applying a marketing approach to the promotion of medical services as commercial goods.

Marketing in the healthcare sector has specifics that distinguish it from marketing in other areas of the economy (Malakhova, 1998). Firstly, the possibilities of implementing paid medical services for the population are limited due to the parallel coexistence of the sphere of free services for many types of medical care (Krivenko, 2005). This creates a limited demand for paid services, in contrast to the almost limitless demand for free medical services provided at the expense of the budget or compulsory health insurance. Secondly, the uneven distribution of medical services between the paid and free markets leads to a price disparity. On the one hand, high prices for paid services are caused by the impossibility or difficulty of obtaining some services for free (expensive services, non-traditional types of services, etc.), on the other hand, budget-funded types of services are provided free of charge in most health care organizations in the amount of allocated funding, usually without the use of modern technologies and methods.

At the same time, in Russia there are certain restrictions for the implementation of the marketing package by healthcare organizations: (1) problems of the marketing budget deficit for conducting high-quality marketing research; (2) the complexity of processing marketing information due to the lack of patient feedback; (3) regulatory and legal restrictions related to the need for full geographical coverage of the territory with state medical services and their provision in full free of charge (under the terms of mandatory medical insurance) create problems with market segmentation and the implementation of their pricing policies of healthcare organizations; (4) the problem of expanding the product range due to the unavailability of new technologies and equipment and limited marketing communication channels; (5) low flexibility of healthcare organizations due to the need for strict compliance with state norms and standards (Przhedetsky et al., 2020).

4.2 Marketing Technologies Used by Healthcare Organizations

In our research, we understand the marketing of a healthcare organization as the process of forming the competitive advantages of a healthcare organization and its positioning in the medical services market through the use of a set of specialized technologies aimed at conducting marketing research, developing a marketing concept for positioning an organization and promoting medical services, and forming the organization's brand (Fig. 3).

The main final marketing goals of the healthcare organization are retaining patients and increasing the number of patients; expanding the range of services; achieving more favorable positions in the competitive market; increasing brand awareness; grow profits (Fedulova, 2019). Marketing research is a separate block of marketing aimed at studying the competitive market, finding free niches for future development, and choosing the most appropriate marketing technologies for promoting medical services at a specific stage of the organization's activity. The most applicable methods of marketing research are: benchmarking (analysis of positive innovations and shortcomings of competing organizations); conducting mass surveys and in-depth interviews and focus groups to assess the quality of medical services provided and the image of the organization by consumers; analysis of documentation and medical records of patients; a special program to assess the quality of medical care in a particular medical institution or region "secret patient" (Zakonguru, 2020); included observations.

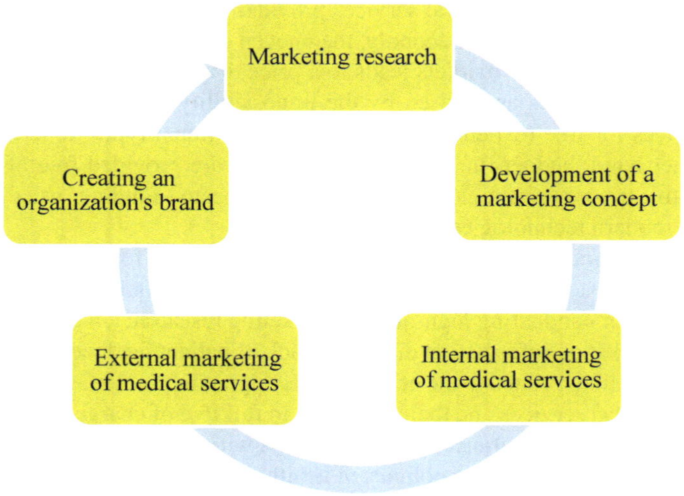

Fig. 3 Complex marketing technologies for healthcare organizations. *Source* Compiled by the author

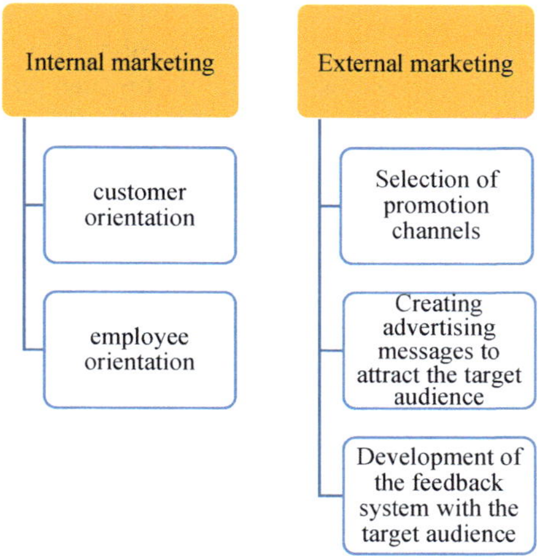

Fig. 4 Structural components of marketing of medical services of the healthcare organization. *Source* Compiled by the author

The formation of the marketing concept is the fundamental stage of the implementation of the marketing complex, which determines the goals in quantitative terms (expected results—patients, revenue, profit); the timing of the implementation of marketing activities; the marketing budget; a phased plan for the promotion of the organization and the stages of control and adjustment.

Marketing of medical services includes all types of activities of a healthcare organization aimed at communicating information about the content, purpose, and benefits of medical services to potential consumers and creating their desire to use this type of service. External and internal marketing technologies are used to promote medical services (Fig. 4).

Internal marketing is aimed at encouraging patients to make new requests and spreading positive information about the high quality of services, which helps to expand the target audience. In addition, the objects of internal marketing are the organization's personnel (the appearance of the staff; the style of communication of medical personnel with patients), the conditions of waiting and receiving patients, which determine their satisfaction with the quality of the service and the desire to return to this medical organization (patient-friendly registry; the presence of a Call Center; comfortable waiting conditions for admission; cleanliness and convenience; availability of modern information stands), availability of patient loyalty programs. The key to the successful promotion of medical services of a healthcare organization is also systematic work with negative reviews of patients (Tarasenko, 2010).

External marketing is based on the choice of media carriers, the development of an advertising "message" to attract the target audience, and the construction of a feedback system with a high level of convertibility with the target audience, taking into account the specifics of the organization and the medical services provided. Currently, there are quite a lot of options for promoting advertising messages about medical services, they include both traditional promotion channels and a corresponding set of marketing communications (outdoor advertising, advertising in print media, distribution of invitations-flyers, mailing lists, PR—active participation of the organization's staff in conferences, writing and publishing articles), as well as advertising on the Internet (Tarasenko, 2010). Healthcare organizations cannot ignore the current trends and technological development of medical services. The healthcare sector is actively introducing and using modern digital technologies and adapting medical content to a dynamic virtual environment (Lugovskaya & Zubareva, 2020). Healthcare organizations have access to a range of modern digital marketing technologies, including contextual advertising, SMM, and SEO optimization for effective promotion of services on the Internet and social networks, as well as maintaining constant organic traffic (Table 2).

Promotion of medical services should be carried out taking into account the requirements of the law (for example, it is forbidden to contact minors; give any guarantees on the outcome of treatment, including talking about the absence of side effects) (Blogingate, 2020). In addition, healthcare organizations in the practice of promoting medical services and brand formation use the following marketing technologies: advertising from bloggers, hashtags, live broadcast with doctors (similar to a webinar), teasers, promo posts, carousel, advertising records, promotion of community posts, YouTube publications, Google CMS, reputation marketing (working with negative reviews of patients) (Niejournal, 2020).

In modern conditions, healthcare organizations have new opportunities to form feedback with the target audience. For example, telemedicine and artificial intelligence technologies used in the development of the "connected patient" concept (monitoring and provision of medical services using embedded intelligent devices and telemedicine) are essentially advanced marketing communication technologies (Karpov et al., 2017). According to the report of the international KPMG "Healthcare insiders: Taking the temperature of artificial intelligence in healthcare", the vast majority of representatives of healthcare organizations (89%) believe that artificial intelligence already increases the efficiency of their work (Rbc, 2020). In 2019, 385 thousand patient consultations were conducted in Russia using telemedicine technologies (46.3% of them in real-time). Compared to 2018, the number of online consultations increased 1.7–2 times (Abdrakhmanova et al., 2020). As a rule, this technology is used for control consultations that do not require visualization of the patient after discharge from the hospital, for psychotherapeutic and neuropsychological consultations, as well as for diagnostic and preventive purposes (Lugovskaya & Zubareva, 2020). Telemedicine technologies have enabled healthcare organizations to expand their coverage during the COVID-19 pandemic (Karpunina, et al., 2020). If in 2017–2019 the growth of the Russian telemedicine market amounted to $20 million, then in the period April-December 2020 it increased by $35 million (Rg, 2021). Telemedicine advisory centers at the federal level remain unloaded due to the lack of a system of rules and mechanisms for implementing new opportunities. Meanwhile, private healthcare organizations are currently focusing on creating Internet services and wearable devices to use this channel to promote medical services (Rg, 2021). The marketing technologies described above have a direct impact on the formation of the organization's brand. The key components of the brand logo,

Table 2 Digital marketing technologies for the promotion of healthcare organizations services

Marketing technology	Application goals	Content	Result
Contextual advertising	Creating and maintaining advertising campaigns in Yandex. Direct, maintaining the display in the top block and attracting new customers	Collecting key requests for the service provided up to a frequency of 30 and dividing them into several groups Creating advertising campaigns in Yandex. Direct (for search, for priority requests, for retargeting) Enabling Yandex. Metrica and setting goals for counting conversions Optimization of advertising campaigns during the entire time of their operation	High clickability of ads, increasing the average cost per click. Reduction of budget expenditures. Increase in the flow of customers for this type of medical services
SEO optimization	Improve the position of the site pages in the target search queries and increase the constant organic flow of visitors	Conducting an audit of the site to identify and eliminate errors Completion of the site content for the necessary requests Performing optimization work for low-frequency queries	Access to the first positions of the output for low-frequency queries. Increase the regular organic flow of visitors
SMM-promotion	Facebook Instagram/Twitter profile management, increase in the number of subscribers, raise the number of applications from social networks	Identifying topics of interest to subscribers Development of the design of posts for the relevant topics Introduction of permanent categories: announcement of the organization's services with an analysis of indications and contraindications, question–answer with doctors, acquaintance with doctors, important news, congratulations on holidays, the announcement of promotions, promo posts of medical institutions Heating the audience's interest in promotional procedures	Improve the overall appearance of your accounts with a single color scheme of posts. Increase the number of subscribers, grow the number of clicks, and raise the minimum price per click. An increase in the number of applications from social networks. The growth of brand awareness on the Internet

Source Compiled by the author

corporate identity, repair, service, quality of services, reviews and recommendations, unique sales offer must meet the needs of the target audience and ensure their maximum satisfaction.

5 Conclusion

Firstly, the necessity of applying a marketing approach to the promotion of services of healthcare organizations in Russia is justified, and its specifics are revealed. Secondly, the limitations of the introduction of marketing technologies for the promotion of medical services in Russian healthcare organizations are identified. Thirdly, the content and features of marketing technologies used by healthcare organizations (marketing research, the formation of a marketing concept, internal and external marketing, branding of the organization) are disclosed and highlighted. Fourthly, the necessity of active use of digital marketing technologies (contextual advertising, SEO optimization, SMM) and digital marketing communications (telemedicine and artificial intelligence technologies used in the development of the "connected patient" concept) for the promotion of medical services by healthcare organizations is justified.

References

Abdrakhmanova, G., Kovaleva, & G., Utyatina, K. (2020). Digital maturity of healthcare. https://issek.hse.ru/mirror/pubs/share/385932983.pdf. Accessed: May 01 2021

All-Russian Center for the Study of Public Opinion. (2018). Press release No. 3277. The problem background of the country: The results of the year. https://wciom.ru/index.php?id=236&uid=1160141. Accessed: May 01 2021.

Babenko, A. (2017). Specificity of the organization of marketing activities in the field of paid medical services. *Glavvrach, 1*, 73–77.

Blogingate. (2020). Promotion of medical services: An overview of the main advertising tools. https://blog.ingate.ru/detail/prodvizhenie-meditsinskikh-uslug-obzor-osnovnykh-reklamnykh-instrumentov/. Accessed: May 01 2021.

Borzenko, K. (2018). Prerequisites and prospects for the development of marketing of healthcare organizations in modern Russia. *State and Municipal Administration. Scientific Notes, 2*, 208–214.

Federal State Statistics Service of the Russian Federation. (2020). Healthcare in Russia 2019. https://rosstat.gov.ru/storage/mediabank/Zdravoohran-2019.pdf. Accessed: May 01 2021.

Fedulova, N. (2019). External and internal marketing. Does he need a state medical organization? *Healthcare management: News, Views, Education. Bulletin of VSHOUZ, 5*(3), 15–23.

Gusev, V., Smyslova, O., Ioda, J., Karpunina, E., & Shevtsov, N. (2020). Industry 4.0 technologies: a tool for overcoming contradictions of socio-economic development or a source of new Threats? *Proceeding of the 35th IBIMA Conference 1–2 April 2020* (pp. 7654–7671) Seville, Spain.

Halfin, R., Syrtsova, L., & Lvova, D. (2017). Methodology for identifying and analyzing contact points for implementing patient-oriented and participatory approaches to marketing medical services. *Problems of Standardization in Healthcare, 7–8*, 67–74.

Kaliyeva, O., Taranukha, I., & Sheptukhin, M. (2016). Evolution of relations with consumers in a client-oriented organization. *Bulletin of Modern Science, 11–1*(23), 90–96.

Karpov, O., Subbotin, S., Shishkanov, D., & Zamyatin, M. (2017). Digital healthcare. Necessity and prerequisites. *Doctor and Information Technologies, 3*, 6–22.

Karpunina, E., Yashin, S., Molchan, A., Lisova, E., & Derkacheva, E. (2020). The price of pleasure: the challenges and threats of the digital economy. In *Digital Future: Economic Growth, Social Adaptation, and Technological Perspectives. Lecture Notes in Networks and Systems* (pp. 363–373). Springer.

Karpunina, E., Zabelina, O., Galieva, G., Melyakova, E., & Melnikova, Y. (2020). Epidemic threats and their impact on the economic security of the state. In *Proceeding of the 35th IBIMA Conference, 1–2 April 2020* (pp. 7671–7682) Seville, Spain.

Krivenko, D. (2005). Features of the concept of marketing in healthcare. https://elar.urfu.ru/bitstream/10995/86767/1/science_upi_2005_1_216.pdf. Accessed: May 01 2021.

Lugovskaya, M., & Zubareva, N. (2020). Telemedicine as a tool for remote marketing communications in the healthcare sector. *Practical Marketing, 12*(286), 29–34.

Malakhova, N. (1998). *Marketing of medical services*. ICFER.

Molchan, A., Karpunina, E., Kochyan, G., Petrov, I., & Velikanova, L. (2019). Effects of digitalization: New challenges for economic security systems. In *Proceedings of the 34th IBIMA conference 13–14 November 2019* (pp. 6631–6639). Madrid, Spain.

Niejournal. (2020). Promotion! Marketing! Medical services. https://niejournal-ru.turbopages.org/niejournal.ru/s/product/promotion-medical-services/. Accessed: May 01 2021.

Nikolaeva, E. (2017). Implementation of the client-oriented approach with the help of CRM system implementation. *Science and Education Today, 6*(17), 73–75.

Payne, A., Storbacka, K., & Frow, P. (2008). Managing the co-creation of value. *Journal of the Academy of Marketing Science, 36*, 83–96.

Prahalad, K., & Ramaswamy, V. (2010). *The Future of Competition. Creating a unique value together with consumers*. Publishing house "Williams", Moscow.

Przhedetsky, Yu., Przhedetskaya, N., & Sheveleva, V. (2020). Marketing of healthcare organizations in the conditions of implementation of national projects (on the example of the oncological service). *State and Municipal Management. Scientific Notes, 1*, 92–96.

Rakhmanova, A. (2017). The role of "digital marketing" in the development of medical institutions. In *Proceedings of the II All-Russian Scientific and Practical Conference. Postgraduates and undergraduates* (pp. 175–181). Yugorsk State University-Khanty-Mansiysk.

Rbc. (2020). According to the report of the international KPMG "Health-care insiders. https://plus.rbc.ru/news/5f6e921e7a8aa9bebb4f7e6b Accessed: May 01 2021.

Rg. (2021). In 2020, the number of telepatients in Russia increased to five million. https://rg.ru/2021/01/15/v-2020-godu-chislo-telepacientov-v-rossii-vyroslo-do-piati-millionov.html. Accessed: May 05 2021.

Rostovsky, Y. (2017). Medical Marketing on the Internet. *Glavvrach, 2*, 20–26.

Sewell, K., & Brown, P. (2017). Clients for life.Trans. from the English. M. Ivanov, M. Ferber-9th ed., Mann, Ivanov and Ferber, Moscow.

Tarasenko, E. (2010). Marketing and marketing communications in the practice of medical institutions: modern methods and technologies for effective promotion of medical services. https://www.hse.ru/data/2011/12/21/1261575462/Маркетинг.pdf. Accessed: May 01 2021

Vargo, S., & Lasch, R. (2006). Development of a new dominant logic of marketing. *Russian Journal of Management, 4*(2), 73–106.

Voronina, E., & Bereza, N. (2016). Client orientation as an integral component of the successful work of the company. Problems of economics and management in the XXI century: Current issues, trends, prospects (pp. 146–156). Penza

Xandopulo, G. (2017). Marketing in the field of healthcare, education, and science without borders: social and humanitarian sciences, 7, 196–200.

Zakonguru. (2020). A secret patient. https://zakonguru.com/medicina/tayniy-patsient.html. Accessed: May 01 2021.

Environmental Quality and Ecological Efficiency of Economic Practices

Tools for Overcoming the Crisis Phenomena of the Pandemic in the Socio-economic Development of the Countries

Julia V. Shurchkova, Irina S. Zinovyeva, Natalja V. Polujanova, Yulia N. Galitskaya, and Ruslan N. Berlizev

Abstract *The purpose* of the study is to identify the specifics of the socio-economic development of the countries during the COVID-19 pandemic, as well as to determine effective tools for overcoming the crisis phenomena in the economy. The study of the growth rate of gross domestic product calculated by the expenditure method and the number of cases of coronavirus diseases were the grounds for sampling countries for the following analysis. The analysis of socio-economic indicators and identification of the development's specifics during the pandemic is carried out on the example of the United States, France, Great Britain, Spain, Italy, and Germany. Based on the analysis, the authors concluded that the least effective implementation of anti-COVID measures is in Italy and Spain, as a result of the lack of effective programs to support lending to the real sector, state support for companies in terms of maintaining employment, and making investments to support business. The German public administration system effectively used a package of anti-crisis measures based on the balance of increasing budget and extra-budgetary infusions into the economy, easing monetary policy, so the country managed to maintain investment activity at the pre-crisis level and create a serious basis for the subsequent recovery from the global economic crisis.

Keywords COVID-19 pandemic · Crisis · Socio-economic development · Fiscal stimulus · Economic activity · Monetary policy · Mitigation of the consequences

JEL Codes E63 · E65 · O11

J. V. Shurchkova (✉)
Voronezh State University, Voronezh, Russia

I. S. Zinovyeva
Voronezh State Forest Engineering University Named After G. F. Morozov, Voronezh, Russia

N. V. Polujanova
Belgorod National Research University, Belgorod, Russia

Y. N. Galitskaya · R. N. Berlizev
Kuban State Technological University, Krasnodar, Russia

E. G. Popkova and B. S. Sergi (eds.), *Geo-Economy of the Future*,
https://doi.org/10.1007/978-3-030-92303-7_20

1 Introduction

The COVID-19 pandemic, like a snowball at the beginning of 2020, engulfing more and more countries, leads to a socio-economic crisis unprecedented in recent history, a force majeure sharp decline in economic activity, and a drop in the incomes of economic organizations and households. The response to the new challenges has been a massive shift in public administration to address the challenges of containing and mitigating the spread of the new virus, with an emphasis on initial disease reduction, reducing pressure on the global health system (testing, improving the treatment of people with severe symptoms and subsequent mass vaccination of the population) and, if the pandemic is further curbed, focusing on economic recovery. The massive introduction of fiscal stimulus packages by government authorities in developed and developing countries was aimed at mitigating the impact of the sudden decline in economic activity in the real sector and households, as well as at preserving the productive capacity of countries. These fiscal and economic packages differed not only in their size but also in their unprecedented application. However, the OECD estimates that most of the measures implemented by government and regulatory authorities to contain the effects of the pandemic were not timely, failed to offset the massive decline in economic activity, and had profound economic consequences for the global economy. Thus, the relevance of the chosen topic is confirmed by the need to analyze and evaluate the adopted economic and fiscal measures, as well as to develop the most effective solutions to maintain the cash flow of business and the monetary and financial systems of the world.

2 Literature Review

In 2020, the global collapse due to the COVID-19 pandemic forced the scientific community to turn to the study of its causes, features of its course, and assessment of the consequences. The authors try to make forecasts of the future development of the economy and its sectors, taking into account the ongoing social and economic changes (Ashraf, 2020; Boone, 2020; Drobot, 2020; Erokhina & Dzhergenia, 2020; Lev & Kolpakova, 2020; McKibbin & Fernando, 2020; Pechatkin, 2020; Phan & Narayan, 2020; Torkanovsky, 2020; Wagner, 2020; Zhang et al., 2020). For example, McKibbin and Fernando (2020) consider seven scenarios for the possible development of the global economy after a pandemic. Researchers are also interested in effective tools to overcome negative consequences for the economy and society; they analyze the situation in different countries to find optimal management solutions (Ali et al., 2020; Baldwin & Weder di Mauro, 2020; Berg, 2020; Cochrane, 2020; Drobot et al., 2020; Lisova et al., 2020; Mejokh et al., 2020; Nicola, 2020; Salisu, 2020).

3 Methodology

The purpose of the study is to identify the specifics of the socio-economic development of the countries during the COVID-19 pandemic, as well as to determine effective tools for overcoming the crisis phenomena in the economy.

The sample of countries for analysis was based on the study of the growth rate of gross domestic product, calculated by the expenditure method according to OECD (2020). The authors' selected countries for the analysis that, after a significant drop in the second quarter of 2020 compared to the first quarter of 2020, provided GDP growth in the third quarter of 2020. The second sampling criterion is the number of coronavirus cases (Gisanddata, 2021). The authors excluded developing countries (India, Russia, Brazil, Turkey) from the analysis. Thus, the following countries were selected for the analysis: the USA, France, Great Britain, Spain, Italy, and Germany.

4 Results

The analysis of GDP dynamics data for the selected countries, taking into account the number of cases of coronavirus diseases, is shown in Fig. 1 (horizontally: the IMF forecast of GDP per capita for 2020, at current prices, obtained by converting GDP in

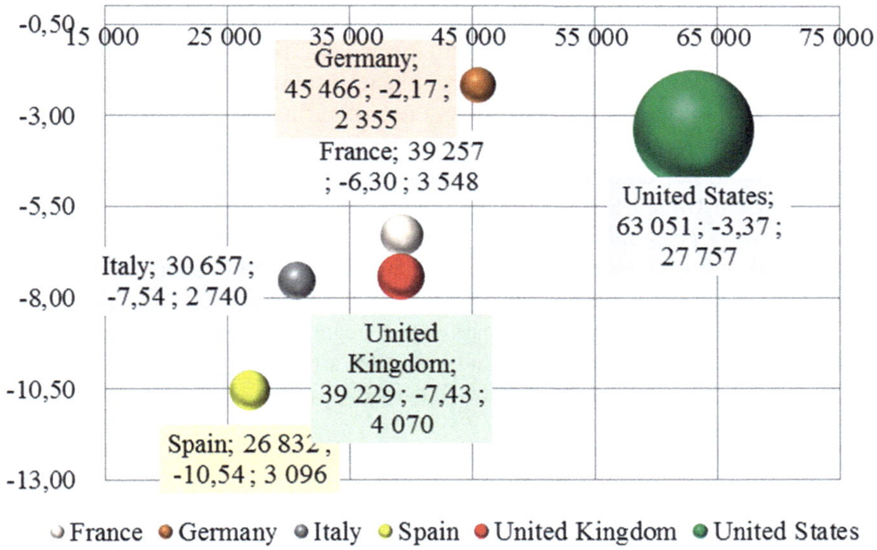

Fig. 1 GDP dynamics of the studied countries, taking into account the number of cases of coronavirus diseases. *Source* Compiled by the authors based on International Monetary Fund (IMF) (2020)

national currency into US dollars and dividing it by the total population; vertically: the IMF forecast GDP per capita growth rates for 2020 relative to 2019 data, at current prices, in US dollars; the size of the circle reflects the degree of damage to the selected countries by coronavirus) (International Monetary Fund (IMF), 2020).

The data suggest that Germany and the United States have a lower rate of decline in GDP per capita in 2020 compared to the data for 2019 −2.17% and −3.37%, respectively. These countries also have the highest GDP per capita for 2020: in the United States—63.05 thousand US dollars, in Germany—45.5 thousand US dollars. The rate of decline in GDP in the analyzed countries does not strongly depend on the total number of detected diseases but is due to the effectiveness of the adopted sets of measures to overcome the negative consequences of the pandemic. In Germany, 2354 million diseases were detected; the country has the lowest rate of decline in GDP per capita in 2020 compared to 2019. In Italy, 2.74 million people are registered, diseases and GDP per capita in 2020 decreased by 7.54%. The United States has the highest rate of COVID-19 (almost 27.7 million cases), the country managed to effectively resist the pandemic, which led to a drop in GDP per capita by only 3.37%. France and the United Kingdom reduced their GDP per capita by 6.3% and 7.43%, respectively. The most inefficient is Spain, which has about 3 million diseases and the highest rate of decline in GDP per capita −10.54%. Regulators that used macroprudential measures before the pandemic (measures aimed at minimizing the systemic risk of financial sector insolvency) have implemented their mitigation. Thus, the UK and Germany have completely released the countercyclical capital buffer, reducing it to zero. Some countries have taken temporary easing measures to reduce allowances and "Basel" standards, and at the same time, they extended orders to postpone the payment of dividends until the end of 2020 (USA, Spain, and Italy). The implementation of the policy of global financial support against the background of a reduction in key rates led to a drop in the cost of financial resources (Fig. 2).

A negative value of long-term interest rates in December 2020 is observed in Germany (−0.62%) and France (−0.34%), a positive value of the indicator was achieved in the United States (+0.93%), Italy (+0.58%), Great Britain (+0.32%), Spain (+0.04%). The implementation of financial and regulatory measures to support the financial and real sector also led to a reduction in the cost of short-term financial rates in the analyzed countries (Fig. 3).

France, Germany, Italy, and Spain have negative short-term interest rates (−0.54%), while positive short-term interest rates are typical for the UK (+0.03%) and the US (+0.17%). However, in the UK and the US, short-term rates fell by 0.71% and 1.48%, respectively.

The analyzed countries actively used budget and other economic incentives to restore their economies after the lockdown and mitigate the risks of the pandemic. The authors estimate the amount of total state budget expenditures to assess budget incentives (Fig. 4). According to the IMF, the maximum value of public spending in 2020 is observed in France (63.09% of the country's GDP), in 2020 they increased by only 7.53% compared to 2019, this is the minimum value among all the analyzed countries.

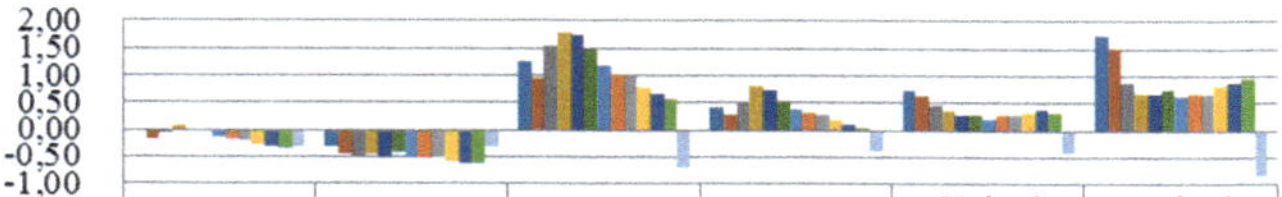

	France	Germany	Italy	Spain	United Kingdom	United States
Jan-2020	-0,01	-0,31	1,28	0,42	0,72	1,76
Feb-2020	-0,18	-0,47	0,96	0,27	0,63	1,50
Mar-2020	-0,06	-0,54	1,55	0,51	0,46	0,87
Apr-2020	0,06	-0,45	1,80	0,82	0,36	0,66
May-2020	-0,03	-0,52	1,76	0,75	0,27	0,67
Jun-2020	-0,04	-0,43	1,46	0,52	0,28	0,73
Jul-2020	-0,15	-0,52	1,20	0,38	0,21	0,62
Aug-2020	-0,17	-0,52	1,03	0,30	0,27	0,65
Sep-2020	-0,21	-0,52	0,98	0,28	0,27	0,68
Oct-2020	-0,30	-0,61	0,77	0,17	0,31	0,79
Nov-2020	-0,33	-0,62	0,66	0,09	0,38	0,87
Dec-2020	-0,34	-0,62	0,58	0,04	0,32	0,93
Change in long-term interest rates, December 2020 to January 2020, % per annum	-0,33	-0,31	-0,70	-0,38	-0,40	-0,83

Fig. 2 Change in long-term interest rates for the 12 months of 2020, in %. *Source* Compiled by the authors based on OECD (2020)

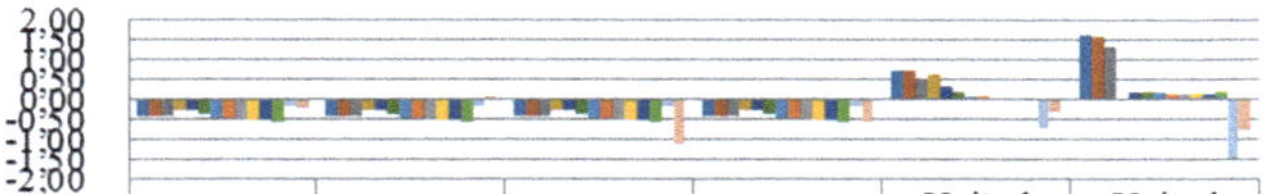

	France	Germany	Italy	Spain	United Kingdom	United States
Jan-2020	-0,39	-0,39	-0,39	-0,39	0,74	1,65
Feb-2020	-0,41	-0,41	-0,41	-0,41	0,75	1,59
Mar-2020	-0,42	-0,42	-0,42	-0,42	0,53	1,35
Apr-2020	-0,25	-0,25	-0,25	-0,25	0,65	
May-2020	-0,27	-0,27	-0,27	-0,27	0,33	0,17
Jun-2020	-0,38	-0,38	-0,38	-0,38	0,19	0,20
Jul-2020	-0,44	-0,44	-0,44	-0,44	0,10	0,18
Aug-2020	-0,48	-0,48	-0,48	-0,48	0,07	0,15
Sep-2020	-0,49	-0,49	-0,49	-0,49	0,06	0,13
Oct-2020	-0,51	-0,51	-0,51	-0,51	0,05	0,12
Nov-2020	-0,52	-0,52	-0,52	-0,52	0,04	0,16
Dec-2020	-0,54	-0,54	-0,54	-0,54	0,03	0,17
Change in short-term interest rates, December 2020 to January 2020, % per annum	-0,15	-0,15	-0,15	-0,15	-0,71	-1,48
The difference between short and long-term rates in December 2020, %	-0,20	0,08	-1,12	-0,58	-0,29	-0,76

Fig. 3 Change in the value of short-term financial rates in the analyzed countries for 12 months of 2020, in %. *Source* Compiled by the authors based on OECD (2020)

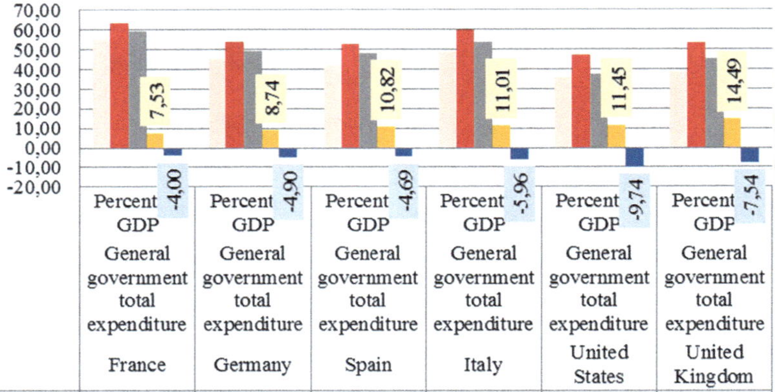

	France	Germany	Spain	Italy	United States	United Kingdom
2019	55,56	45,17	41,91	48,71	35,74	38,61
Forecast for 2020	63,09	53,91	52,73	59,72	47,19	53,09
Forecast for 2021	59,09	49,02	48,04	53,76	37,45	45,55
2020-2019	7,53	8,74	10,82	11,01	11,45	14,49
2021-2020	-4,00	-4,90	-4,69	-5,96	-9,74	-7,54

Fig. 4 Government spending, % of GDP for the period 2019 and 2020–2021. *Source* Compiled by the authors based on International Monetary Fund (IMF) (2020)

Ranking the growth of public spending for 2020 relative to 2019 shows that the maximum increase in public spending in the UK (by 14.5%), in the United States (by 11.45%), in Italy (by 11%), in Spain (by 10.8%), a slight increase is also projected in Germany (by 8.7%). The IMF forecasts a decline in government spending in 2021 in all the countries analyzed.

An effective tool for supporting the global economy is the implementation of a policy of budgetary and extra-budgetary incentives aimed at supporting income and ensuring credit growth, including government guarantees on loans (the most significant relative to GDP in Italy, Germany, France, and the United Kingdom); special government business support programs (grants to small and medium-sized enterprises, targeted support for healthcare, airlines); direct social payments to the population (subsidies for salary payments (UK), increase in unemployment benefits (USA), easing of requirements for unemployment benefits (Europe), direct lump-sum payment of money to the population (USA)); various tax breaks (tax holidays for individuals and businesses with deferred tax payments for 90 days (USA, Spain, France). An estimate of the level of additional expenditures or lost government revenue in the analyzed countries is shown in Fig. 5.

The maximum amount of additional spending will be in the United States (16.72% of GDP), the United Kingdom (16.26% of GDP), Germany (11% of GDP), France (7.71% of GDP), Italy below the average level (6.82% of GDP), and the minimum value of budget support is expected in Spain (4.13% of GDP). The maximum level of additional spending and lost revenue directed to the health sector is projected in the UK (5.34% of GDP), and the minimum—in Spain (0.4% of GDP). The maximum

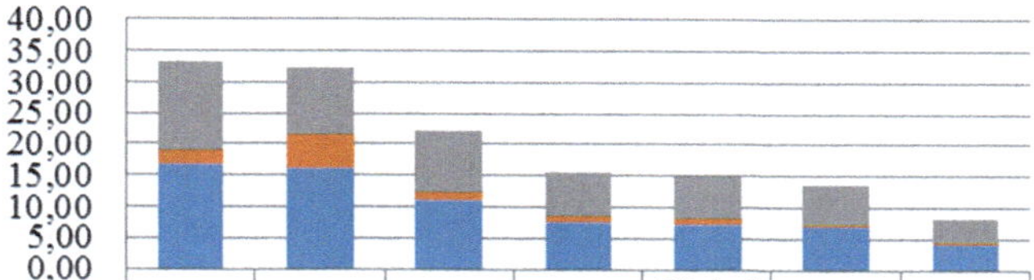

	USA	Great Britain	Germany	France	Globally in 191 countries	Italy	Spain
including in the non-medical sector	14,41	10,92	9,83	6,92	6,36	6,25	3,74
including in the field of healthcare	2,31	5,34	1,20	0,79	0,96	0,58	0,39
Additional expenses or lost revenue	16,72	16,26	11,03	7,71	7,44	6,82	4,13

Fig. 5 Estimate of the level of additional expenditures or lost government revenue, % of GDP. *Source* Compiled by the authors based on International Monetary Fund (IMF) (2021)

support for the rest of the economy, except for health care, is projected in the United States (14.41% of GDP), and the minimum in Spain (3.74% of GDP). An estimate of the level of extra-budgetary infusions is presented in Fig. 6. The maximum value of extra-budgetary infusions in 2020, according to the IMF forecast, will be in Italy (35.5% of GDP), Germany (27.85% of GDP), Great Britain (16.12% of GDP), France (15.75% of GDP), Spain (14.44% of GDP), and the minimum value will be in the

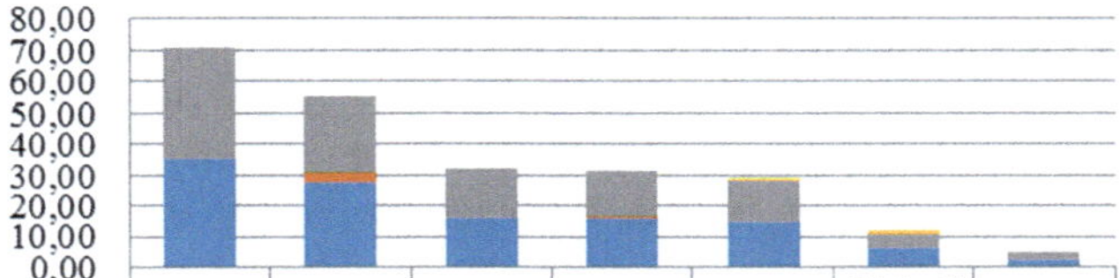

	Italy	Germany	Great Britain	France	Spain	Globally in 191 countries	USA
Quasi-fiscal operations	0,00	0,00	0,00	0,00	0,90	1,62	0,00
Guarantees	35,28	24,84	16,07	14,82	13,49	4,04	2,17
capital injections, loans, asset purchases, or debt assumptions.	0,20	3,01	0,05	0,93	0,06	0,42	0,27
Extra-budgetary injections	35,48	27,85	16,12	15,75	14,44	6,09	2,43

Fig. 6 The level of extra-budgetary infusions in the analyzed countries at December 2020, % of GDP. *Source* Compiled by the authors based on International Monetary Fund (IMF) (2021)

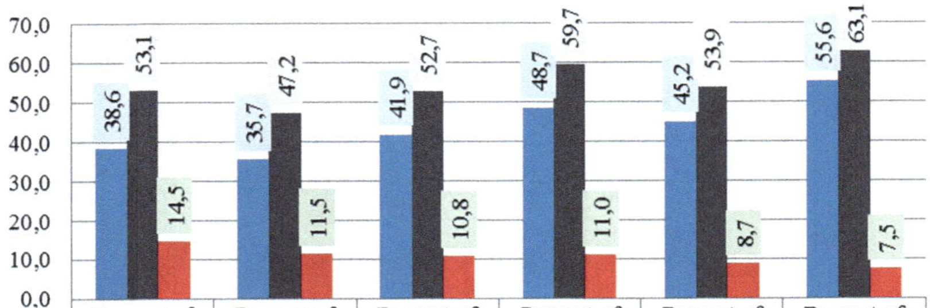

	Percent of GDP United Kingdom	Percent of GDP United States	Percent of GDP Spain	Percent of GDP Italy	Percent of GDP Germany	Percent of GDP France
2019	38,6	35,7	41,9	48,7	45,2	55,6
2020	53,1	47,2	52,7	59,7	53,9	63,1
2020-2019	14,5	11,5	10,8	11,0	8,7	7,5

Fig. 7 Gross public debt in the analyzed countries, as a % of GDP. *Source* Compiled by the authors based on International Monetary Fund (IMF) (2020)

United States (only 2.43% of GDP, which is lower than the average for 191 countries (6.1% of GDP)).

The xpansion of financial assistance by the analyzed countries did not lead to an increase in gross public debt (Fig. 7).

The United Kingdom (+14.5%), the United States (+11.5%), Italy (+11%), and Spain (+10.8%) increased the volume of public debt most significantly in 2020. Germany (+8.7%) and France (+7.5%) controlled it more effectively in 2020. The maximum amount of public debt at the end of 2020 was formed in France (63.1% of GDP), in Italy (59.7% of GDP), in Germany (53.9% of GDP), in the UK (53.1% of GDP), in Spain (52.7% of GDP), and its minimum value was reached in the United States (47.2% of GDP). The next element of support for the economy is the growth of total government net lending/borrowing, which measures the extent to which governments provide financial resources to other sectors of the economy and nonresidents (net lending), or use financial resources generated by other sectors and non-residents (net borrowing) (Fig. 8).

In 2020, the total amount of net lending and borrowing decreased for all the analyzed countries due to a decline in economic activity. The largest declines occurred in the United Kingdom (−14.26% of GDP), the United States (−12.37% of GDP), Italy (−11.34% of GDP), Spain (−11.27% of GDP), and the smallest declines were in Germany (−9.7% of GDP) and France (−7.77% of GDP). The decline in business activity during the pandemic could not but affect the investment activity of the state and economic entities (Fig. 9).

In all the analyzed countries, there was a decrease in investment activity. Germany was an exception, where, in 2020 the volume of investment remained at the level of 2019. The maximum decrease in investment occurred in Italy (−1.7% of GDP), the United Kingdom (−1.6% of GDP), France (−1.47% of GDP), the United States

	Percent of GDP General government net lending/borrowing United Kingdom	Percent of GDP General government net lending/borrowing United States	Percent of GDP General government net lending/borrowing Italy	Percent of GDP General government net lending/borrowing Spain	Percent of GDP General government net lending/borrowing Germany	Percent of GDP General government net lending/borrowing France
2019	-2,204	-6,349	-1,639	-2,826	1,521	-2,998
2020	-16,464	-18,718	-12,981	-14,092	-8,175	-10,769
2020-2019	-14,26	-12,37	-11,34	-11,27	-9,70	-7,77

Fig. 8 Total government net lending/borrowing, % of GDP. *Source* Compiled by the authors based on (International Monetary Fund (IMF), 2020)

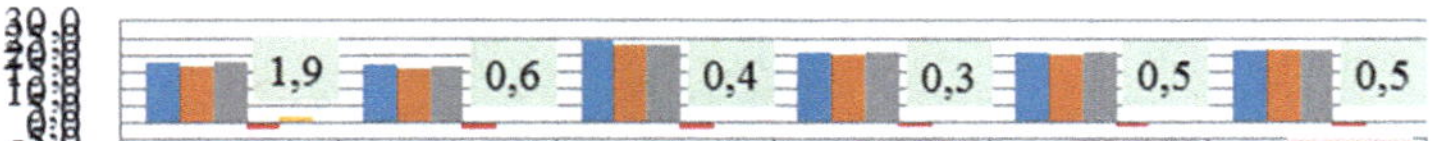

	Per GDP Total investment volume Italy	Per GDP Total investment volume United Kingdom	Per GDP Total investment volume France	Per GDP Total investment volume United States	Per GDP Total investment volume Spain	Per GDP Total investment volume Germany
2019	18,0	17,3	24,2	21,0	20,8	21,4
2020	16,3	15,7	22,7	20,3	20,3	21,4
Forecast for 2021	18,2	16,3	23,2	20,6	20,8	21,9
2020-2019	-1,67	-1,60	-1,43	-0,69	-0,58	-0,01
Forecast 2021-2020	1,9	0,6	0,4	0,3	0,5	0,5

Fig. 9 The total value of gross fixed capital formation, % of GDP. *Source* Compiled by the authors based on International Monetary Fund (IMF) (2020)

(−1.47% of GDP), Spain (−1.47% of GDP). Summary changes in the economic indicators analyzed by the authors are presented in Table 1.

5 Conclusion

The analysis allowed authors to conclude that the least effective use of a wide range of anti-COVID measures is Italy and Spain, where there was a maximum decline in

Table 1 Summary table of the ranking of countries on the effectiveness of economic measures aimed at reducing the negative consequences of the COVID-19 pandemic

Indicators	France	Germany	Italy	Spain	United Kingdom	United States
Percentage change in GDP per capita for 2020 to 2019,%	6.3	−2.17	−7.54	−10.54	−7.43	−3.37
Change in long-term interest rates, December 2020 to January 2020, % per year	−0.33	−0.31	−0.7	−0.38	−0.4	−0.83
Change in short-term interest rates, December 2020 to January 2020, % per year	−0.15	−0.15	−0.15	−0.15	−0.71	−1.48
Dynamics of changes in government spending in 2020 relative to 2019, % of the country's GDP	7.53	8.74	11.01	10.82	14.49	11.45
The total amount of additional government spending or the number of lost profits in December 2020, as a % of GDP	7.71	11.03	6.82	4.13	16.26	16.72
The total amount of extra-budgetary injections in December 2020, as a % of GDP	15.75	27.85	35.48	14.44	16.12	2.43
Change in total government gross debt for 2020 relative to 2019, % of GDP	7.53	8.74	11.01	10.82	14.49	11.45
Total government net lending/borrowing, at the end of December 2020, in % of GDP	−7.77	-9.7	−11.34	−11.27	−14.26	−12.37
Dynamics of the total investment volume, 2020 to 2019 in % of GDP	−1.43	−0.01	−1.67	−0.58	−1.60	−0.69

(continued)

Table 1 (continued)

The difference in the dynamics of the volume of exports of goods and services in 2020 and 2019, % per year	−18.68	−12.97	−18.99	−28.06	−19.63	−12.51
The difference in the dynamics of the volume of imports of goods and services in 2020 and 2019, % per year	−13.91	−11.01	−18.65	−23.57	−24.61	−13.39
The difference in the magnitude of the unemployment rate in 2020 and 2019, %	0.41	1.13	1.1	2.7	1.55	5.22
The difference in the value of inflation, in average consumer prices in 2020 and 2019, % per year	−0.84	−0.85	−0.5	−0.93	−1.02	−0.29
Dynamics of the gross national savings, in % of GDP	−2.68	−1.33	−1.40	−2.01	0.36	−0.94
The number of positive changes in economic indicators assessed by the authors in response to the COVID-19 pandemic	7	9	3	3	5	5
The number of changes in economic indicators negatively assessed by the authors in response to the COVID-19 pandemic	7	5	11	11	9	9
The difference between positive and negative indicators	0	4	−8	−6	−4	−6

Source Compiled by the authors

GDP per capita. Despite a significant increase in public spending, the implementation of a policy of reducing long-term and short-term rates, a decrease in the level of inflation against the background of a significant increase in the amount of extra-budgetary injections into the economy, and a significant increase in public debt, the state and regulatory authorities of Italy and Spain failed to effectively counter the COVID-19 pandemic. This was due to insufficient growth in additional expenditures or lost revenue, a significant decline in net lending and borrowing, a serious decline in investment activity, ineffective export, and import support policies, and poor performance in employment support measures in Italy and Spain. This was due to insufficient growth in additional expenditures or lost revenue, a significant decline in net lending and borrowing, a serious decline in investment activity, ineffective export, and import support policies, and poor performance in employment support measures in Italy and Spain. According to the authors, the main reason for the ineffectiveness of economic measures aimed at smoothing the negative consequences of the coronavirus pandemic in Italy and Spain lies in the emphasis on extra-budgetary sources of funding for the necessary economic measures.

This is also a consequence of the lack of effective programs to support lending to the real sector, state support for companies in terms of maintaining employment and making investments to support business, as well as serious underfunding of smoothing measures at the expense of state sources. Germany allowed a minimal decline in GDP per capita during the difficult period, this was achieved thanks to the effective use of a package of anti-crisis measures based on a policy of cheap interest rates, effective programs to support the real sector due to a significant increase in extra-budgetary infusions, an increase in additional expenditures, a slight decrease in the amount of net lending and borrowing, maintaining investment activity at the pre-crisis level, an increase in import operations, a serious decrease in inflation in the country, and a decrease in gross savings. According to the authors, the main reason for the effectiveness of economic measures in Germany lies in the balance of increasing budget and extra-budgetary infusions into the economy, easing monetary policy.

References

Ali, M. O. Moon, D., & Jiha, F. (2020). Response to the COVID-19 pandemic: Economics and health of economically developed and developing countries. *Economic Related, 4*. https://doi.org/10.18334/eo.10.4.111083

Ashraf, B. (2020). Economic impact of government interventions during the COVID-19 pandemic: International evidence from financial markets. *Journal of Behavioral and Experimental Finance, 27*, 100371. https://doi.org/10.1016/i.ibef.2020.100371

Baldwin, R., & Weder di Mauro, B. (2020). *Economics in the time of COVID-19*. CEPR Press.

Berg, J. (2020). The labour market policy response to COVID-19 must leverage the power of age. Available at: https://voxeu.org/article/labour-market-policy-response-COVID-19-must-leverage-powerage. Accessed 01 April 2021

Boone L. (2020). *Tackling the fallout from COVID-19. Economics in the Time of COVID-19* London: CEPR Press.

Cochrane, J. (2020). Flatten the coronavirus curve at a lower costs. *The Wall Street Journal.*
Drobot, E. (2020). Global economy in the context of the COVID-19 pandemic: Results of 2020 and prospects for recovery. *Economic Relations, 10*(4), 937–960.
Drobot, E., Makarov, I., Nazarenko, V., & Manasyan, S. (2020). The impact of the COVID-19 pandemic on the real sector of the economy. *Econ, Entrepren Law, 8*, 2135–2150.
Erokhina, E., & Dzhergenia, Yu. (2020). Threats to the implementation of the national interests of the Russian Federation in the post-pandemic period. *Economic Security, 3*, 259–272.
Gisanddata. (2021). COVID-19 dashboard by the center for systems science and engineering (CSSE) at Johns Hopkins University (JHU). Available at: https://gisanddata.maps.arcgis.com/apps/opsdashboard/index.html#/bda7594740fd40299423467b48e9ecf6. Accessed 01 April 2021
International Monetary Fund (IMF). (2020). World economic outlook database. Available at: https://www.imf.org/en/Publications/WEO/weo-database/2020/October/download-entire-database Accessed 01 April 2021
International Monetary Fund (IMF). (2021). Fiscal monitor: Database of country fiscal measures in response to the COVID-19 pandemic. Available at: https://www.imf.org/en/Publications/FM/Issues/2021/01/20/fiscal-monitor-update-january-2021. Accessed 01 April 2021
Lev, M., & Kolpakova, I. (2020). Budgetary factors of implementation of state programs and national projects in the system of economic and social security of Russia. *Economics of Entrepreneurship and Law, 5*, 1521–1534.
Lisova, E., Petrov, I., Koryakina, T., Kalombo Mulamba, V., & Karpunina, E. (2020). The well-being of Russian regions under threat: economic and social impact assessment of COVID-19. In *Proceeding of the 36th IBIMA Conference in Granada* (pp. 7603–7616) Spain, 4–5 November 2020.
McKibbin, W., & Fernando, R. (2020). *The Global Macroeconomic Impacts of COVID-19: Seven Scenarios*. CAMA working paper, 19. https://doi.org/10.2139/ssrn.3547729
Mejokh, Z., Korolyuk, E., Sozaeva, D., Pilipchuk, N., & Karpunina E. (2020). Economic Security of Russian Regions: Risk Factors and Consequences of the Covid-19 Pandemic. In *Proceeding of the 36th IBIMA conference in Granada* (pp. 8197–8205) Spain, 4–5 November 2020
Nicola, M. (2020). The socio-economic implications of the coronavirus and COVID-19 pandemic: A review. *International Journal of Surgery, 78*, 185–193.
OECD (2020). OECD Statistic. Available at: https://stats.oecd.org/index.aspx?lang=en Accessed 01 April 2021
Pechatkin, V. (2020). Formation and development of the digital economy in Russia as a strategic priority for the development of territories in the context of pandemics. *Innovation Economy Issues, 2*, 2020.
Phan, D., & Narayan, P. (2020). Country responses and the reaction of the stock market to COVID-19: A preliminary exposition. *Emerging Markets Finance and Trade, 56*(10), 2138–2150.
Salisu, A. (2020). Revisiting oil-stock nexus during COVID-19 pandemic: Some preliminary results. *International Review of Economics and Finance, 69*, 280–294.
Torkanovsky, E. (2020). Autarky 2.0: The global environmental agenda, the COVID-19 pandemic and the New normal. *Economic Relations, 3*, 663–682.
Tsvetkov, V., & Dudin, M. (2020). Media activity of domestic politicians and statesmen during the COVID-19 pandemic as a tool for strengthening the trust of the electorate. *Leadership and Manager, 2*, 89–122.
Wagner, A. (2020). What the stock market tells us about the post-COVID-19 world. *Nature Human Behaviour, 4*(5), 440–450.
Zhang, D., Ji, Q., & Hu, M. (2020). Financial markets under the global pandemic of COVID-19. *Finance Research Letters, 36*, 1–6.

Assessment of the Industrial Impact on the Environment

Elena S. Polovnikova, Madina A. Kyarova, and Yulia G. Gazukina

Abstract Industry is a human-made process; however, as time goes on, it moves further away from the ecological symbiosis to establish the current ecological imbalance, where all forms of natural resources are either depleted or exposed to ongoing adverse impacts. The industrial sector consumes colossal water resources, although this fact is not an environmental impact issue. The environmental impact of using such resources is both direct and indirect. Some aspects of the industry leading to controversial effects are mainstream and, therefore, immediately recognizable. Other aspects are less apparent but equally harmful. Economic success depends on the environmental processes; it will soon be primarily determined by many environmental factors (cultural, political, social, physical, aesthetic, financial, and legal).

Keywords Industrial impact · Environment · Industrial sector · Resource use · Measurable direction

JEL Codes Q5 · Q56 · Q57

1 Introduction

The main types of environmental impacts of the industrial sector can be classified into the following sections:

E. S. Polovnikova (✉)
Altai State University, Barnaul, Russia

M. A. Kyarova
Kabardino-Balkarian State Agrarian University Named After V.M. Kokova, Nalchik, Russia

Y. G. Gazukina
Gorno-Altaisk State University, Gorno-Altaisk, Russia

E. G. Popkova and B. S. Sergi (eds.), *Geo-Economy of the Future*,
https://doi.org/10.1007/978-3-030-92303-7_21

1.1 Land Use

Land use has the most significant environmental impact in the industrial sector. Industrial projects consistently occupy land, consume space above ground, exploit underground areas, and spread various effects that directly or indirectly affect the environment.

An adverse effect of this industry is the prevalence of developers who prefer to use new green areas. Although this situation is understandable from the point of view of ease of development and cost-effectiveness, the environmental impact is maintaining existing buildings, structures, and facilities in an abandoned or destroyed condition. Moreover, the situation is not ameliorated by the current trend towards stimulating external and internal (urban) shopping centers, industrial complexes and parks, and business parks. Although all these facilities are becoming accessible to a broader range and number of people, this happens due to the renovation of the inner city.

Habitat destruction undoubtedly lies at the heart of many processes affecting natural goods, landscape, and wildlife. The wasteful disregard for nature conservation often contributes to such hostility in the public perception (Costanza, 1996).

1.2 Use of Natural Resources

In addition to land use, the industrial sector is notorious for its use of natural resources. The environmental impact of using such resources is both direct and indirect. For instance, deforestation is a direct environmental effect of using natural timber. Mining is another direct effect of using natural stone and aggregates in production with an environmental effect resulting from extraction processes. Indirect environmental impact occurs due to manufacturing materials, components, and products negatively affecting the environment.

1.3 Energy Consumption

It is generally recognized that about half of the total energy consumption in the country is accounted for by the construction, operation, and demolition of buildings. Most energy is invariably wasted. Furthermore, natural resources, such as fossil fuels, are even more depleted for minor reasons.

1.4 Air Emissions

Perhaps there is no need to think of industry as synonymous with air emissions. However, the use of installations and equipment often leads to the appearance of atmospheric pollutants, such as diesel fuel vapor. Moreover, some manufacturing operations emit smoke and other toxic wastes into the air. However, if we speak about chlorofluorocarbons [CFCs] and similar gaseous emissions, the environmental impact is staggering. As an indicator of the severity of air emissions, about half of CFC emissions come from buildings in various forms, such as insulation materials, refrigerants, and fire-fighting equipment. One cannot deny the global environmental impact on the ozone layer as a result of such emissions. To a large extent, industrial enterprises are to blame for this situation due to their direct and indirect activities (Bogoviz et al., 2020).

1.5 Use of Water Resources

The industrial sector, directly and indirectly, consumes enormous water resources (although this is not a problem of environmental impact). However, buildings with final products do have environmental impacts due to the need for water and wastewater treatment facilities, as well as water supply and storage facilities with water maintenance for new and developing infrastructures. This situation inevitably results in specific side effects that require new solutions and, therefore, ultimately perpetuate their cycle of demand for ample water resources (Tashkulova & Kletskova, 2020b).

1.6 Water Discharges

The impact of the industrial sector on the environment is both direct and indirect. At the production level, the consequences are related to improper production practices. This situation leads to the direct discharge of pollutants (e.g., oil waste) into the water or the indirect entry of pollutants into the land that feeds natural water sources. In 2020, there were two cases of oil spills in Russia, one of which was unintentional, and the other was recognized as negligent. These disasters indicate that such aspects of production are quite frequent (Glotko et al., 2019).

1.7 Waste

Waste generated in the industrial sector is an urgent problem with harmful consequences for the environment. Construction and demolition waste includes waste resulting from any excavation or construction, civil engineering, site clearance, demolition, road works, and building renovation. These phenomena include various types of construction waste, rubble, earth, concrete, wood, and mixed materials for site clearance. Household waste and construction and demolition waste are the main components that account for about 84% of all solid waste in the world.

The environmental consequences of production lead to some violations of comfort for people who live and work near any industrial facility. Such consequences comprise the following:

- Noise from construction work and equipment;
- Dust from production processes and traffic;
- Inconveniences in the form of temporary housing and construction traffic jams;
- Hazardous pollution in the form of toxic waste;
- Visual impairments associated with signs, billboards, and other objects.

The apparent environmental effect of production is that its activities pose a danger to both employees and residents of the given territory (Stroiteleva et al., 2019).

2 Methodology

Currently, the attitude of residents of megacities to a clean environment is becoming increasingly strict. Many environmental protection events are held every year worldwide. Besides, there is a growing trend of complaints about environmental pollution in industry, when most people believe that rapid urbanization and industrialization make the city one of the noisiest and most crowded and polluted places in the world.

Additionally, many government departments increasingly regulate new construction by requiring the inclusion of a wide range of environmental protection measures, such as the Air Pollution and Noise Abatement Ordinance or the Waste Disposal Ordinance. Although these measures are taken mainly under public pressure, they have a positive response and raise the awareness of the government on the environment. Meanwhile, the employees of the construction industry see the environment in different ways. Many builders or suppliers consider an environmental problem only from the perspective of business benefits. They often consider protecting building components or resources from environmental impacts but rarely think deeply about the impact of construction on people and nature (Adarina et al., 2019). This situation is related to the fact that traditional industrial management operates in three dimensions:

- Cost;
- Time;
- Quality.

All business activities must consider these three dimensions, but not the environmental impact. In this regard, the industry management is ready to introduce environmental protection measures and meet the requirements, but with limited space, for people to expand the scope of work to protect the environment.

Indeed, economic success is dependent on environmental processes; it will soon be largely determined by many environmental factors: cultural, political, social, physical, aesthetic, financial, legal, and many others. An environmentally friendly production and the operation process will be recognized as successful if the environmental processes are measurable (Tashkulova & Kletskova, 2020a).

Measurement is a measurable direction or content with variable environmental dimensions. Environmental issues are widely discussed in all sectors of society. Therefore, this tradition is becoming less likely to lead the project to success.

In the twenty-first century, industry management must integrate the environmental dimension with cost, time, and quality management.

The environmental process should be assessed in four dimensions: the environment should be perceived as an essential problem in the management of any organization, along with issues of cost, time, and quality. Under the new management system, cost, time, and quality remain relevant. Nonetheless, a significant contribution will be made to the study of environmental factors, their significance for production and operation, and the adoption of measures to combat any causes and consequences of pollution.

The environmental pollution concerns associated with manufacturing are widely recognized, and many measures addressing these issues are technically studied and legally formalized. However, practicing manufacturers still do not pay much attention to these issues, the number of which is constantly growing. Although it is difficult to give the exact causes of this situation, the increase in production activity abroad and within the country is undoubtedly one of these reasons.

3 Results

In general, the problems of implementing an environmental management system in the industrial sector are defined as follows:

- The environment is interpreted differently in terms of cost, time, and quality. From a business perspective, measurement goals are set on behalf of industry leaders, but the goal of protecting the environment is imposed by external bodies, such as government agencies. In terms of the impact on economic development, they are micro-factors affecting the economy at the enterprise level in the short term, but the environment is a macro-factor affecting the economy and its development on a global scale in the long term (Nekhvyadovich et al., 2020; Ragulina et al., 2019). Such differences lead to the fact that, despite the acute problems of environmental management on the part of the industry leaders, environmental

protection measures during construction are carried out only to a limited extent (Vukovich et al., 2018).
- In most cases, result-oriented policies neglect environmental regulations (Merdesheva et al., 2020). The measures are taken only in the case of environmental pollution. In this regard, if elementary environmental rules are taken into account, much less effort will be made to identify the causes of pollution and study ways of preventing them. Some countries establish environmental audit departments. In environmental regulation, the environmental protection agencies must perform an audit based on the developers' requirements. The auditors are not independent but controlled by the client; therefore, they speak for the client regarding environmental responsibility (Shenshinov, 2012).

Besides, there are too many environmental regulations in the form of a decree, a law, or a white paper. Different issues cover different aspects, such as air and water pollution, waste and waste recycling, hazardous substances, health, and security. Not everyone can easily remember so many rules, especially if they do not directly benefit them. Thus, too many rules can annoy people (Avkopashvili et al., 2019). There is an urgent need to summarize all environmental regulations by industry in a consistent form that can be easily understood and implemented.

These cost, time, and quality management systems are particularly widely used in the industrial sector. Managers traditionally consider the need to take care of the environment as an unnecessary burden in time and cost. Today, such inappropriate attitudes and behavior are observed in a small number of managers. However, those who refuse to work with environmental indicators know exactly what indicators need to be fulfilled so that their business is competitive, and these indicators are as follows:

- Competitive economic environment;
- Environmental awareness;
- Environmental management costs;
- Sense of personal responsibility;
- Academic degrees obtained.
- Attitudes to labor and production methods driven by strong business competition.

Environmental management costs are a new element in the implementation of environmental regulations. These costs are mainly related to environmental protection, the search for new environmental protection methods, and the improvement of the damaged environment. For example, the costs associated with the use of (1) *dust screen*; (2) *water spray*; or advanced tools or equipment involving environmentally friendly materials (Stroiteleva et al., 2020).

4 Conclusion

A large amount of environmental management costs affects the expenses of the organization, such as staff welfare and savings for future development.

Any person can make the most of the environment without paying for it. Noise, dust, and waste can be eliminated by investing money. In this case, the person will appreciate it and the efficiency of this method, which will be reflected in the additional profit (Tsvetkov et al., 2019). Humanity should never forget that the environment is a unique resource that does not belong to any person or business. Anyone who uses it and damages it must pay for it. The value of the environment must be measured in terms of the benefit of the entire society.

References

Adarina, R., Gazukina, Y., & Yankovskaya, K. (2019). Indicators of the "green economy" as a tool for monitoring the regional economy. *IOP Conference Series: Earth and Environmental Science, 395*, 012107. https://doi.org/10.1088/1755-1315/395/1/012107

Avkopashvili, P. T., Polukhin, A. A., Shkodinsky, S. V., & Poltarykhin, A. L. (2019). The fundamental provisions of the concept of knowledge economy. In E. Popkova, Y. Ragulina, & A. Bogoviz (Eds.), *Industry 4.0: Industrial revolution of the 21st century* (pp. 57–64). Springer. https://doi.org/10.1007/978-3-319-94310-7_5

Bogoviz, A. V., Shvakov, E. E., Tretyakova, O. G., Zakharov, M. Y., & Abramov, A. N. (2020). Globalization of education in the conditions of formation of the global knowledge economy: Regularities and tendencies. In E. Popkova (Ed.), *Growth poles of the global economy: Emergence, changes and future perspectives* (pp. 993–1000). Springer. https://doi.org/10.1007/978-3-030-15160-7_99

Costanza, R. (1996). Ecological economics: Reintegrating the study of humans and nature. *Ecological Application, 6*(4), 978–990.

Glotko, A. V., Voronkova, O. Y., Miheeva, N. M., Skoblikova, T. V., Alekhina, N. A., & Zhminko, A. E. (2019). The economic development of a transboundary region under the conditions of the Eurasian economic union. Proceedings of the IBIMA 2019: Education excellence and innovation management through vision 2020 (pp. 7957–7968). Granada, Spain.

Merdesheva, E. V., Titova, O. V., & Avkopashvili, P. T. (2020). Methodological approach to the classification of digital economies by environmental efficiency and sustainability criterion. In E. Popkova, & B. Sergi (Eds.), *Scientific and technical revolution: Yesterday, today and tomorrow* (pp. 1360–1368). Springer. https://doi.org/10.1007/978-3-030-47945-9_148

Nekhvyadovich, L. I., Borodin, V. A., & Avkopashvili, P. T. (2020). A systemic approach to the study of the sustainability of the global digital economy: Economies of scale, ratchet effect and hysteresis effect. In E. Popkova, & B. Sergi (Eds.), *Scientific and technical revolution: Yesterday, today and tomorrow* (pp. 1369–1377). Springer. https://doi.org/10.1007/978-3-030-47945-9_149

Ragulina, Y. V., Semenova, E. I., Avkopashvili, P. T., Dmitrieva, E. A., & Cherepukhin, T. Y. (2019). Top-priority directions of implementing new internet technologies on the territories of rapid economic development. In E. Popkova, & V. Ostrovskaya (Eds.), *Perspectives on the use of new information and communication technology (ICT) in the modern economy* (pp. 182–188). Springer. https://doi.org/10.1007/978-3-319-90835-9_21

Shenshinov, Y. V. (2012). Labour quality management. The Journal "North-East Asia Academic Forum."

Stroiteleva, T. G., Kalinicheva, E. Y., Vukovich, G. G., & Osipov, V. S. (2019). Peculiarities and problems of formation of industry 4.0 in modern Russia. In E. Popkova, Y. Ragulina, & A. Bogoviz (Eds.), *Industry 4.0: industrial revolution of the 21st century* (pp. 145–153). Springer. https://doi.org/10.1007/978-3-319-94310-7_14

Stroiteleva, T. G., Petrova, E. A., & Yungblyudt, S. V. (2020). Responsible information society as a social environment for creation of the "green" digital economy: Formation and monitoring. In E.

Popkova, & B. Sergi (Eds.), *Scientific and technical revolution: Yesterday, today and tomorrow* (pp. 402–409). Springer. https://doi.org/10.1007/978-3-030-47945-9_45

Tashkulova, G. K., & Kletskova, E. V. (2020). Place and role of human in the system of circular reproduction in the digital regional economy. In E. Popkova, & B. Sergi (Eds.), *Scientific and technical revolution: Yesterday, today and tomorrow* (pp. 943–951). Springer. https://doi.org/10.1007/978-3-030-47945-9_101

Tashkulova, G. K., & Kletskova, E. V. (2020). The model of organization of "green" digital production and consumption based on the internet of things and AI. In E. Popkova, & B. Sergi (Eds.), *Scientific and technical revolution: Yesterday, today and tomorrow* (pp. 329–336). Springer. https://doi.org/10.1007/978-3-030-47945-9_37

Tsvetkov, V. A., Loginov, E. L., & Raikov, A. N. (2019). Integrated network interaction of education with science and production. *Educational Technologies and Society, 19*(1), 389–398.

Vukovich, G. G., Makuschenko, L. V., Bateykin, D. V., Titova, O. V., & Dobrosotskiy, V. I. (2018). Support of the territories of advanced economic development on human capital: Theory and practice. *Quality—Access to Success, 19*(S2), 157–160.

Benefits of Environmental Audits as a Means of Achieving the Corporate Goals of High-Tech Companies

Tamara G. Stroiteleva, Oksana V. Klimova, and Olga V. Leushina

Abstract Environmental audits are successfully implemented in a constructive way; there are many benefits from conducting them in a high-tech company. They are beneficial both for the companies themselves and for the environment. In this paper, the authors list some of the advantages of environmental audits for high-tech companies. First, an environmental audit ensures no litigation since it checks compliance with regulatory standards and helps avoid lawsuits and consequences such as fines or imprisonment. Second, environmental audit provides increased security for the management team since it can increase the credibility of the management team. Third, identifying and documenting the state of compliance with regulatory standards within the company or individual operations is communicated to senior management. Finally, the results and performance evaluation in the organization show a positive result and give confidence to the management teams so that they can focus on the activities of the company.

Keywords Environmental audit · High-tech companies · Resource optimization · Improving the image · Equipment installation

JEL Codes Q5 · Q51 · Q55

1 Introduction

Even if the results are negative (i.e., the standards are not met sometimes or constantly), the management team can appoint responsible individuals to correct the situation and ensure compliance in the future. The management team can get

T. G. Stroiteleva (✉)
Altai State University, Barnaul, Russia

O. V. Klimova
Gorno-Altai State University, Gorno-Altaysk, Russia

O. V. Leushina
Novosibirsk State University of Economics and Management, Novosibirsk, Russia
e-mail: lov@bk.ru

E. G. Popkova and B. S. Sergi (eds.), *Geo-Economy of the Future*,
https://doi.org/10.1007/978-3-030-92303-7_22

back to its job rather than being intimidated by sudden incidents and lawsuits, but it requires work on the following directions:

- Increasing operational efficiency, which is shaped by the fact that an environmental audit program can provide information to help managers improve their operations, understand current and new regulatory requirements, point out problems in the company or the activities being audited, and provide information on control methods. With all necessary information, managers can be more responsible in performing their job;
- Optimization of resources conducted using the results of an environmental audit designed to identify the company's environmental problems and prior problems. A list of priorities can help allocate company resources to the most urgent and pressing problems to achieve the most significant benefit (Shenshinov & Al-Ali, 2020). Identifying potential reductions in waste production, possible recycling, and energy concentration through audits will ultimately result in cost savings (Tashkulova & Kletskova, 2020a);
- Improving the competitiveness and sustainable development of the company by improving the public image. The benefit is also derived from the fact that the company has already started an environmental audit or has not begun to pollute the environment. An effective environmental audit significantly reduces unnecessary environmental impact. Additionally, by addressing and predetermining potential risks and taking immediate action to prevent them, a company can better protect itself from accidents. These actions significantly increase the confidence of banks and insurance companies to lend money and cover insurance amounts. Nowadays, more companies are taking a close look at environmental work; green companies are more competitive in the high environmental awareness market. The products of the food industry have great potential in the market. The greener the products, the more opportunities to develop and conquer new markets (Merdesheva et al., 2020);
- Improving the image by ensuring a systematic and documented environmental audit, as well as recording and reporting all results. The available and open information about the company's environmental performance is the only way for the public to understand the company better. The presentation of environmental indicators can gain public confidence and prevent the public from being suspicious of anything confidential, where the worst cases are always assumed and where development and business may be hampered;
- Training of personnel to perform environmental audits involving cooperation at all levels, regardless of whether the audit is conducted internally or with external consultants. The audit involves gathering information, site visits, interviews, or questionnaires. All of these processes require interaction with personnel in the audited area. Therefore, workers can better understand work and other regulatory requirements. In some cases, the auditors may be employees of the company. Therefore, the audit may serve as a training program that provides a better understanding of the organization, an opportunity to process deficiencies with respect

to environmental performance, and a chance to make improvements (Karataev et al., 2020).

Environmental audits are attractive to industries and businesses because of the benefits achieved within companies. Moreover, the environment also benefits through audits that improve and ensure environmental performance within companies. Compliance with environmental standards and legislation ensures the quality of the environment. The environment is further protected by increasing the audit scope from problem solving or compliance management to risk management. By identifying potential risks during audits and a high degree of precaution during operation, the potential risk of accidents and environmental impacts are significantly reduced. Therefore, the environment is protected from unnecessary pollution and accidents due to various mistakes. The optimization of the production resources of companies will save the land from overexploitation and allow achieving sustainable development.

2 Materials and Methods

Environmental audit is a new concept, and there is no accepted practice or format for it. Audits can be used extensively for a variety of purposes. The use of environmental audits can be grouped into the following categories:

- Environmental protection;
- Development of monitoring programs;
- Performance of the equipment;
- Physical risks and hazards;
- Business tools;
- Mergers and acquisitions;
- Products and markets;
- Law and policy;
- Compliance with laws and standards.

The purpose and results of the audit are versatile. An environmental audit aimed at ensuring an environmentally friendly product to maintain business and continuous development will also protect our environment from degradation (Shenshinov, 2012). It is believed that human-made chemicals destroy the ozone layer causing the so-called greenhouse effect and climate change. To make it easier to understand and demonstrate the role of auditing, the following conditions are defined:

- It is always cheaper to prevent contamination than to fix it. A well-designed monitoring program can indicate pollution trends at an earlier stage so that immediate action can be taken and irreversible consequences avoided;
- The relevant parameters with the desired accuracy in the monitoring program are essential. The primary purpose of an audit of a monitoring program is to determine

the required parameters and the appropriate degree of accuracy required for a particular research or project;
- Environmental audits are important in developing cost-effective monitoring programs and analyzing the effectiveness of existing monitoring programs;
- Installing pollution control or monitoring equipment is a standard measure to reduce the release of excess pollutants into the environment and ensure compliance with the established norms. There is a wide range of available pollution control equipment to meet a variety of requirements. Different methods of environmental auditing differ in performance and technical characteristics. To have the most appropriate model and ensure that the money is well spent on a particular company or project, the effectiveness of the used control measure must be monitored during its operation on site. This can be done by auditing the operation of the equipment.

In complying with these conditions, a company must conduct an environmental audit in several stages:

1. Determine the technical characteristics of the relevant equipment;
2. Measure the actual performance of the equipment (the input and output concentrations of pollutants under operating conditions covering the entire design range, not just the optimum performance);
3. The performance of the relevant equipment must be carefully analyzed under various operating conditions. Technical advice from suppliers or specialists and engineers is desirable when taking any action, whether applying other mitigation measures or reconfiguring equipment to improve performance (Dragunov & Shenshinov, 2020).

Doing business nowadays is more complicated than just the demands of improving business performance, that is, increasing profits. The stringency and number of regulations increase as public awareness of environmental issues and the need to protect the environment increases. Fines related to violations of environmental legislation are also increasing. The productivity of production in the field of environmental protection must also be high. Otherwise, the company or manager could face considerable financial problems or regulatory obligations. Therefore, auditing potential environmental impacts and financial liabilities arising from poor environmental management is an important application of environmental auditing.

When one company plans to take over another or intends to buy one of its sites or subsidiaries, it must make sure that it does not acquire environmental liabilities together with assets. Otherwise, a company intending to sell its stock must ensure its environmental performance to make a successful transaction (Alekseev et al., 2018). Sellers or purchasers may conduct an environmental audit determining the legal and financial obligations associated with the past and present environmental impacts of takeover targets.

The manufacturing sector plays an important role in protecting the environment and promoting sustainable development. A considerable number of resources are used to produce consumer goods. In addition to resource exploitation, waste is generated during production, final disposal, and packaging. The manufacturing process also contributes to environmental pollution, which includes: water, air, noise, and environmental impacts. The sources of raw materials used for production can also significantly affect the environmental performance of manufacturers. Using a source of raw materials from unfriendly processes can indirectly harm the environment.

An environmental audit can be designed to assess the degree to which products are environmentally friendly by identifying the manufacturing details of products (from raw materials, through production technology, to packaging and retailing). An environmental audit covers the following areas:

- Raw material sources;
- Production processes;
- Package.

During the audit, the first question to ask is what raw materials are used to produce the product and in what quantities. Listing the details of all raw materials for each product will help identify any products and materials that are hazardous and adverse to the environment.

The audit includes suggestions for substitutes for raw materials that are found to be detrimental to the environment. A well-presented report on all this information, prepared by the audit team (type of materials and their environmental impact), can make it easier for company managers to develop new production methods and environmental programs to improve environmental performance.

Most of the pollution and environmental impacts come from production processes, with audits revealing production processes and the environmental consequences of those processes. Questionnaires, checklists, and interviews can be used to get a complete picture of the production process. Measurement and monitoring programs are sometimes included in audit programs to quantify the environmental impact. Continuous monitoring can keep a manufacturer on the right track in managing manufacturing cite to prevent pollution (Stroiteleva et al., 2019).

Packaging is becoming a part of our lives. Plastic and paper are two types of materials often used in packaging. It turned out that plastics are not biodegradable and cause significant problems in disposal; they produce toxic gases when incinerated and occupy landfills. The production of paper from trees also damages the environment since fewer and fewer trees are left for a balanced ecosystem. Manufacturers need to do something to help reduce packaging and eliminate waste generation. How can manufacturers best contribute to improving the environmental situation? How do they know they are producing too much packaging? In this problem, the best assistant is an audit of the amount of resources (plastic and paper) used for packaging, as well as the provision of recommendations for packaging design (Stroiteleva et al., 2020).

3 Results and Discussion

In order to protect the environment, legislation is enacted, and standards and guidelines are set for compliance with established standards. For example, the allowable noise levels are established in a technical memorandum for construction noise and commercial activities. Water quality regulations are established for water quality control areas to protect water quality in the harbor. The purpose of setting these standards is to control the degree of pollution. Therefore, careful monitoring of pollutant compliance is essential to protect the environment (Glotko et al., 2019).

The verification of compliance of a particular operation, process, facility, company, or project with relevant laws and standards can be conducted either by regulators (e.g., EPD) or project operators, or by external audit groups such as community groups, non-governmental organizations, or consultants. The verification includes the following steps (Adarina et al., 2019):

- Determination of applicable rules and standards for obtaining monitoring data;
- Comparison of monitoring data with standards;
- Determination of the real state of compliance with all standards and regulations, as well as the frequency and severity of their violations;
- Follow-up on recommendations (Tashkulova & Kletskova, 2020b).

Environmental auditing is recognized as a very effective management tool and benefits the environment and companies. However, this development is hindered by the lack of a standard format and case reports. The government is the largest employer and the most significant organization—it has departments dealing with a wide range of activities, from administration to science and engineering. Initiating environmental audits in government agencies may be appropriate to develop an audit standard and guidance for the private sector. Additionally, the experience of audit in different areas provides valuable information for the private sector since government activities cover most areas.

Nevertheless, benefits are the only driving force for companies to respond positively to environmental work (including audits). Legislation is sometimes not an effective tool for getting people to behave properly. However, social pressure and market forces are always effective. The direct impact on sales caused by a boycott of environmentally unfriendly products by consumers (Russian and international) will force management to respond quickly to how this will affect their business performance. The new ecolabel, developed in Europe, will become a new international trend in the manufacturing sector. Stressing the importance of systematically accounting for environmental performance through environmental audits can encourage all companies to become more involved in environmental audits.

Companies are also prohibited from conducting environmental audits because any weaknesses in companies will be exposed (e.g., non-compliance with regulatory standards), which will lead to litigation against the company and damage its image. Moreover, the public will be extremely nervous about the consequences of the audit (Nekhvyadovich et al., 2020). However, some of the irregularities and consequences can be corrected. Revealing the consequences to the public would cause unnecessary frustration. Additionally, the context of an audit can be commercially sensitive (e.g., a development strategy). It is desirable to take a systematic approach to the disclosure of relevant information in audit reports.

The requirements to produce an environmentally friendly product force companies to conduct environmental audits, starting with franchisees and listings, since they are easier to evaluate. An environmental audit can be a pre-qualification for applying for a franchise or participating in a company listing.

4 Conclusion

Public awareness of environmental issues in megacities is still not very high. The pressure to improve environmental performance usually comes from only a few green groups. An immature public monitoring system is one of the reasons why companies do not feel the need to improve their efficiency. Therefore, education is essential. The education of people must be supported and organized through all channels (schools, daily activities, posters, TV commercials, or environmental functions and parties) (Bogoviz et al., 2020). As public environmental awareness grows, more and more companies will voluntarily perform environmental audits to improve their company image and secure their business. The flow of education and information should be carried out through state and environmental organizations and technical associations.

To increase the role of environmental audit and its importance in companies, it is necessary to provide and improve the quality of the audit program and auditors (Vukovich et al., 2018). The chartering of engineers to ensure the quality of engineering projects and engineers has long been established. The same scheme is necessary when conducting an environmental audit.

References

Adarina, R., Gazukina, Y., & Yankovskaya, K. (2019). Indicators of the "green economy" as a tool for monitoring the regional economy. In *IOP conference series earth and environmental science* (Vol. 395, p. 012107). https://doi.org/10.1088/1755-1315/395/1/012107

Alekseev, A. N., Savenkov, D. L., Shvakov, E. E., Lebedev, N. A., & Ragulin, A. D. (2018). Financial problems of regional development and the ways of their solution in contemporary Russia. *Quality—Access to Success, 19*(2), 71–75.

Bogoviz, A. V., Shvakov, E. E., Tretyakova, O. G., Zakharov, M. Y., & Abramov, A. N. (2020). Globalization of education in the conditions of formation of the global knowledge economy: Regularities and tendencies. In E. G. Popkova (Ed.), *Growth poles of the global economy: Emergence, changes and future Perspectives* (pp. 993–1000). Springer. https://doi.org/10.1007/978-3-030-15160-7_99

Dragunov, A., & Shenshinov, Y. (2020). Effective development of innovative technologies as a method for increasing labor productivity in industry. In *E3S Web of Conferences* (Vol. 208, p. 03001). https://doi.org/10.1051/e3sconf/202020803001

Glotko, A. V., Sycheva, I. N., Dunets, A. N., Poltarykhin, A. L., Zhuravlev, P. V., & Tubalets, A. A. (2019). Integration processes in the agro-industrial complex of the regions: Development and environmental problems. *Journal of Environmental Management and Tourism, 10*(3(35)), 613–621.

Karataev, E. M., Merkuryev, V. V., & Titova, O. V. (2020). Integrating the "green economy" into the model of digital future of the modern socio-economic systems. In E. Popkova, & B. Sergi (Eds.), *Scientific and technical revolution: Yesterday, today and tomorrow* (pp. 1402–1410). Springer. https://doi.org/10.1007/978-3-030-47945-9_153

Merdesheva, E. V., Titova, O. V., & Avkopashvili, P. T. (2020). Methodological approach to the classification of digital economies by environmental efficiency and sustainability criterion. In E. Popkova, & B. Sergi (Eds.), *Scientific and technical revolution: Yesterday, today and tomorrow* (pp. 1360–1368). Springer. https://doi.org/10.1007/978-3-030-47945-9_148

Nekhvyadovich, L., Kostin, K. B., & Mironov, V. V. (2020). Government and public monitoring of entrepreneurship in a multicultural region: New perspectives in the conditions of the digital economy. In E. G. Popkova, K. V. Vodenko (Eds.), *Public administration and regional management in Russia. Challenges and prospects in a multicultural region* (pp. 147–154). Springer. https://doi.org/10.1007/978-3-030-38497-5_17

Shenshinov, Y., & Abdulsattar Al-Ali. (2020). The tools of increasing efficiency of human resource in the lean production environment: Conceptual study. *International Journal of Core Engineering & Management, 6*(7), 1–18. Retrieved from http://ijcem.in/wp-content/uploads/2020/06/THE-TOOLS-OF-INCREASING-EFFICIENCY-OF-HUMAN-RESOURCE-IN-THE-LEAN-PRODUCTION-ENVIRONMENT-CONCEPTUAL-STUDY2.pdf

Shenshinov, Y.V. (2012). Labour quality management. *The Journal North-East Asia academic Forum.*

Stroiteleva, T. G., Kalinicheva, E. Y., Vukovich, G. G., & Osipov, V. S. (2019). Peculiarities and problems of formation of industry 4.0 in modern Russia. In E. Popkova, Y. Ragulina, & A. Bogoviz, (Eds.), *Industry 4.0: Industrial revolution of the 21st century* (pp. 145–153). Springer. https://doi.org/10.1007/978-3-319-94310-7_14

Stroiteleva, T. G., Kletskova, E. V., & Balaeva, S. I. (2020). Ontogenesis of green development model of digital economy: Essence, stages, and prospects. In E. Popkova, & B. Sergi (Eds.), *Scientific and technical revolution: Yesterday, today and tomorrow* (pp. 410–417). Springer. https://doi.org/10.1007/978-3-030-47945-9_46

Tashkulova, G. K., & Kletskova, E. V. (2020). Place and role of human in the system of circular reproduction in the digital regional economy. In E. Popkova, & B. Sergi (Eds.), *Scientific and technical revolution: Yesterday, today and tomorrow* (pp. 943–951). Springer. https://doi.org/10.1007/978-3-030-47945-9_101

Tashkulova, G. K., & Kletskova, E. V. (2020). The model of organization of "green" digital production and consumption based on the internet of things and AI. In E. Popkova, & B. Sergi (Eds.), *Scientific and technical revolution: Yesterday, today and tomorrow* (pp. 329–336). Springer. https://doi.org/10.1007/978-3-030-47945-9_37

Vukovich, G. G., Makuschenko, L. V., Bateykin, D. V., Titova, O. V., & Dobrosotskiy, V. I. (2018). Support of the territories of advanced economic development on human capital: Theory and practice. *Quality—Access to Success, 19*(2), 157–160. Retrieved from http://case.asu.ru/files/form_312-32290.pdf

Determining the Quality of Services and Measuring It via Environmental Indicators

Andrey A. Bezhovets, Elena V. Gubkina, and Tologon T. Omoshev

Abstract Determining the quality of service consists of assessing the importance of said quality and establishing the measurement system for its evaluation. The main task of service is satisfying the needs of customers. For example, waiting is a service provided at a restaurant. However, this logic does not provide any insight into what makes service a service. Service is not the same as a good. Therefore, any theory of goods cannot be applied to services. To understand service, one must first distinguish between service products (e.g., food or clothes) and service delivery. The main parameters that separate a delivery from a product are: intangibility, simultaneous production and consumption, and heterogeneity.

Keywords Quality of services · Environmental indicators · Service characteristics · Intangibility · Heterogeneity

JEL Codes Q5 · Q55 · Q56

1 Introduction

This is called a *continuum* in the definition of service quality. Services must decide which of their components deserve attention. The components include the result of service (e.g., a dress that a person bought) and the delivery process (e.g., trying on the dress, paying, or the store's atmosphere). Currently, the delivery belongs to two processes of service:

- Technical service;

A. A. Bezhovets (✉)
Altai State University, Barnaul, Russia
e-mail: bezhovets@mc.asu.ru

E. V. Gubkina
Gorno-Altai State University, Gorno-Altaysk, Russia
e-mail: helenvl@bk.ru

T. T. Omoshev
International University. K.Sh. Toktomamatova, Jalal-Abad, Kyrgyzstan

E. G. Popkova and B. S. Sergi (eds.), *Geo-Economy of the Future*,
https://doi.org/10.1007/978-3-030-92303-7_23

- Service quality, connected to the process and the functional aspect of service.

Technical and functional aspects are important factors influencing the overall perception of service quality. Most secondary literature in marketing and service management focuses on service quality, especially on service delivery. However, this does not mean that the delivery (functional aspect) is more important than the result (technical aspect). For example, the technical competence of a doctor can be much more important than their conduct. Nevertheless, most of the cost in a restaurant bill goes to service delivery (the atmosphere and waitstaff service) rather than the service result (the food). The importance of technical and functional maintenance depends on the type of service and the consumer's interest (Bogoviz et al., 2020).

2 Materials and Methods

Three parameters of service are mainly applicable only to the functional component. These are:

- Intangibility;
- Continuity;
- Heterogeneity.

These parameters are not the only way to define service; different definitions serve different purposes. The definitions below are useful for distinguishing between different services and their purposes.

Relative intangibility is work, processes, and actions. The defining characteristic of pure service is intangibility. This means that pure service cannot be seen, touched, held, or retained and brings no physical sensation. Pure service is essentially a process and an experience. Services provide more psychological experiences than physical possessions.

One example of pure service is watching a concert. The received service is the performance, the environment, and personal experience. All of these have no tangible attributes (Glotko et al., 2019). The intangibility of experience is best explained by comparing going to a concert with buying a music player. A music player can be brought home and compared to other goods, while the experience of a concert cannot be brought home.

In real case scenarios, not all services are pure. For example, a restaurant dinner includes both food (tangible aspect) and the waitstaff's work (intangible aspect). Or, buying a computer consists not only of the computer itself (tangible) but also of delivery and after-sales services (intangible).

In practice, most services have both tangible and intangible aspects. In the property management industry, services provided to residents include both pure service (i. e., greeting and attending to the guests) and mixed service (i. e., providing after-sales service and ordering food in the clubhouse). However, according to our calculations and experience, the intangible service is more important than the mixed service in the property management industry.

Relative continuity is when the production and consumption of a service occur simultaneously. For example, when someone asks a theater attendant to call them a cab, the service is both rendered and received at the same time. In the case of pure service, as mentioned earlier, production and consumption processes are inseparable (Alekseev et al., 2018).

Another peculiar feature of continuity is the fact that services can be rendered to several persons at the same time. For example, theater performance is a service aimed at all people in the audience. Since production and consumption of a service are inseparable, it makes sense to study how to maximize the use of a service by involving the largest possible number of people who receive the service (Shenshinov, 2012).

Relative heterogeneity is predetermined by the human factor, meaning that each rendered service can be potentially personalized and unique (Shenshinov & Al-Ali, 2020). This stems from the fact that different customers may have different demands, or different attendants may meet the same customer demands in different ways at different points in time. For example, one resident calls the customer service office to inquire about a management fee, while another wants to place an order for repair work; one is in a hurry, and another is just passing the time; one has several inquiries, but another has only one inquiry and hopes to get the answer promptly. Both residents get customer service from the same office, but they have different requirements, expectations, and desires, and staff must constantly adapt to these differences. Then again, even the same request can be handled differently depending on the people involved. A customer service worker may be able to respond to a simple inquiry promptly and very politely but struggle with orders. This relative heterogeneity can complicate the measurement of services and quality checks to ensure that they meet a common standard.

3 Results

Pure service depends on intangibility, continuity, and heterogeneity. The higher the service on the continuum ladder, the more pronounced these parameters are. In the property management industry, most services are pure, meaning that the influence of the human factor is great. As mentioned earlier, pure services are extremely challenging to evaluate and measure. To measure and evaluate service quality, we should introduce the following indicators:

- Definition of quality.
- Quality is an ambiguous term.

There are different approaches to defining quality:

- Philosophical;
- Technical;

- User-oriented.

Philosophical approach. In this approach, quality denotes inherent superiority that cannot be defined or analyzed further. This approach defines quality as follows: "Quality is achieving the highest standard, not trying to satisfy the customer with deliveries or by scamming. Quality does not allow for second-rate compromises." While this definition of quality may have its supporters, there is no point in treating quality as something transcendent and unmeasurable (Tsvetkov et al., 2019).

Technical approach. This approach mainly concerns itself manufacturing quality and quality standards. It measures quality objectively in terms of deviations from standards or defects. This approach is best suited to easily measurable things, such as mass-produced goods.

User-oriented approach. In this approach, the customer subjectively judges the quality of a product (Vukovich et al., 2018). Since different clients have different expectations and feelings, the same product or service may be rated differently by different clients. This approach is best suited for defining the quality of pure service. However, service can be measured using objective criteria so as to obtain accurate data.

Since service quality is subjective, the user-oriented approach provides the best definition of quality for services. To establish the standard of service and determine what aspect affected service quality the most, service providers should ask for customer feedback. However, getting high accuracy data would also entail applying a technical approach—including some definite criteria in a customer questionnaire (Nekhvyadovich et al., 2020).

Studying service quality is important due to several reasons:

Emergence of service quality sector.

Since the early 1920s, statistical quality control was developed in Europe and even implemented during World War II to control the production in countries suffering from war. Quality control grew both in Western and Asian countries due to the definite need for controlling product quality and reducing defects (Costanza, 1996).

In the early 1970s, manufacturing competition became global; many companies had to impose stricter quality control to remain competitive. The goal of high-quality engineering is to develop reliable technologies and methods that increase the competitiveness of new products by reducing their cost and improving their quality (Dragunov & Shenshinov, 2020).

Due to this idea, quality control received wide acceptance—most enterprises understand that quality control is essential to profitability.

Service industry growth.

The reason for the growing interest in service quality is the service sector boom, which is especially prominent in Western countries. The development of new service-oriented quality models is starting to dominate in economics (Avkopashvili et al., 2019).

The "Services" category includes hotel, recreational, health, legal, educational, social, and property management services. The review of a table illustrates the dominant role that services have come to play in the economy. Since development is tied to returns on investment, service quality received its own methodology (Merdesheva et al., 2020).

Competition and profit-making.

Service is a big part of the economy. The study of service quality can provide a competitive advantage over other service providers. To provide high-quality services, an employee must first obtain the necessary tools and resources (Adarina et al., 2019). For example, they must have sufficient training and the appropriate amount of work that turns them into confident, loyal workers, ready to provide an outstanding level of service. If management wants the workers to provide high-quality customer service, they must provide high-quality training to employees.

Even though workplace design and staff training are essential to the service industry, these two things are not enough (Ragulina et al., 2019). Corporate culture is equally important because an organizational climate is defined as the employees' perception of the support and rewards for providing high-quality service (Tashkulova & Kletskova, 2020). When an organization has a strong climate, its work tactics, purpose, practices, and procedures will reflect the importance of service, and employees will see service as a valuable asset to the company.

4 Conclusion

To be profitable, a company must have:

1. A detailed job design and ensure that employees are properly trained before starting out on the position;
2. A developed corporate culture and growth strategy that would influence future development.

The success of a company depends on these two factors since they directly determine customer loyalty and satisfaction. Therefore, researching the factors that influence service quality is essential for companies that look into the future.

References

Adarina, R., Gazukina, Y., & Yankovskaya, K. (2019). Indicators of the "green economy" as a tool for monitoring the regional economy. In *IOP Conference Series: Earth and Environmental Science* (Vol. 395, p. 012107). https://doi.org/10.1088/1755-1315/395/1/012107

Alekseev, A. N., Savenkov, D. L., Shvakov, E. E., Lebedev, N. A., & Ragulin, A. D. (2018). Financial problems of regional development and the ways of their solution in contemporary Russia. *Quality—Access to Success, 19*(S2), 71–75.

Avkopashvili, P. T., Polukhin, A. A., Shkodinsky, S. V., & Poltarykhin, A. L. (2019). The fundamental provisions of the concept of knowledge economy. In E. Popkova, Y. Ragulina, & A. Bogoviz (Eds.), *Industry 4.0: Industrial revolution of the 21st century* (pp. 57–64). Springer. https://doi.org/10.1007/978-3-319-94310-7_5

Bogoviz, A. V., Shvakov, E. E., Tretyakova, O. G., Zakharov, M. Y., & Abramov, A. N. (2020). Globalization of education in the conditions of formation of the global knowledge economy: Regularities and tendencies. In E. Popkova (Ed.), *Growth poles of the global economy: Emergence, changes and future perspectives* (pp. 993–1000). Springer. https://doi.org/10.1007/978-3-030-15160-7_99

Costanza, R. (1996). Ecological economics: Reintegrating the study of humans and nature. *Ecological Application, 6*(4), 978–990.

Dragunov, A., & Shenshinov, Y. (2020). Effective development of innovative technologies as a method for increasing labor productivity in industry. In *E3S Web of Conferences* (Vol. 208, p. 03001)

Glotko, A. V., Voronkova, O. Y., Miheeva, N. M., Skoblikova, T.V., Alekhina, N.A., Zhminko A.E. (2019). The economic development of a transboundary region under the conditions of the Eurasian economic union. In *Proceedings of the IBIMA 2019: Education Excellence and Innovation Management through Vision 2020* (pp. 7957–7968). Granada, Spain.

Merdesheva, E. V., Titova, O. V., & Avkopashvili, P. T. (2020). Methodological approach to the classification of digital economies by environmental efficiency and sustainability criterion. In E. Popkova, & B. Sergi (Eds.), *Scientific and technical revolution: Yesterday, today and tomorrow* (pp. 1360–1368). Springer. https://doi.org/10.1007/978-3-030-47945-9_148

Nekhvyadovich, L., Kostin, K. B., & Mironov, V. V. (2020). Government and public monitoring of entrepreneurship in a multicultural region: New perspectives in the conditions of the digital economy. In E. Popkova, & K. Vodenko (Eds), *Public administration and regional management in Russia* (pp. 147–154). Springer. https://doi.org/10.1007/978-3-030-38497-5_17

Ragulina, Y. V., Semenova, E. I., Avkopashvili, P. T., Dmitrieva, E. A., & Cherepukhin, T. Y. (2019). Top-priority directions of implementing new internet technologies on the territories of rapid economic development. In E. Popkova, & V. Ostrovskaya (Eds.), *Perspectives on the use of new information and communication technology (ICT) in the modern economy* (pp. 182–188). Springer. https://doi.org/10.1007/978-3-319-90835-9_21

Shenshinov, Y. V. (2012). Labour quality management. *The Journal North-East Asia academic Forum.*

Shenshinov, Y., & Al-Ali, A. (2020). The tools of increasing efficiency of human resource in the lean production environment: Conceptual study. *International Journal of Core Engineering & Management, 6*(7), 1–18.

Tashkulova, G. K., & Kletskova, E. V. (2020). Place and role of human in the system of circular reproduction in the digital regional economy. In E. Popkova, & B. Sergi (Eds.), *Scientific and technical revolution: Yesterday, today and tomorrow* (pp. 943–951). Springer. https://doi.org/10.1007/978-3-030-47945-9_101

Tsvetkov, V. A., Loginov, E. L., & Raikov, A. N. (2019). Integrated network interaction of education with science and production. *Educational Technologies and Society, 19*(1), 389–398.

Vukovich, G. G., Makuschenko, L. V., Bateykin, D. V., Titova, O. V., & Dobrosotskiy, V. I. (2018). Support of the territories of advanced economic development on human capital: Theory and practice. *Quality—Access to Success, 19*(S2), 157–160.

Increasing the Effectiveness of Environmental Management Based on the Use of Standards

Irina S. Dzakhmisheva, Natalya A. Shchetinina, and Natalya O. Derkach

Abstract The environmental management system is a generally accepted concept of management in any organization that considers the impact of the company's products, services, and processes on the environment. It provides a systematic methodology for allocating resources, assigning responsibilities, and continuously assessing environmental processes following the appropriate procedures. The standards began in England in 1992. The testing of environmental management standards took place at UK enterprises. Environmental management standards were formed based on the indicators of audit, strategic management process, and control. The Commission of the European Communities and its fifth Environmental Action Program created Eco-Management and Audit Scheme for businesses in the area of voluntary environmental management system certification and work on continuous improvement of environmental performance. Later, the International Organization for Standardization (ISO) developed a series of environmental management system standards known as the ISO 14000 series. The series was agreed upon in 1991 and was developed by the new technical committee ISO/TC 207—Environmental management in 1993. ISO 14001—Environmental Management Systems: Specifications with Guidance for Use was published September 1, 1996. ISO 14001 is the only system by which a company can become a benchmark and get a certification. With the introduction of ISO 14001, the previously developed standard BS7750 was withdrawn.

Keywords Environmental management · Using standards · Environmental management system (EMS) · Economic activity · ISO 14001 Standard

JEL Codes Q5 · Q52 · Q58

I. S. Dzakhmisheva (✉)
Kabardino-Balkarian State Agrarian University Named After V.M. Kokova, Nalchik, Russia
e-mail: irina_dz@list.ru

N. A. Shchetinina
Novosibirsk State University of Economics and Management, Novosibirsk, Russia

N. O. Derkach
Altai State University, Barnaul, Russia

E. G. Popkova and B. S. Sergi (eds.), *Geo-Economy of the Future*,
https://doi.org/10.1007/978-3-030-92303-7_24

1 Introduction

There is a common agreement that the Environmental Management System (EMS) is an essential tool for sustainable industrial development. ISO 14001 and Eco-Management and Audit Scheme (EMAS) have different types of EMS standards. There is a difference in the process and content of the two systems. EMAS contains several elements that are missing in ISO 14001. The difference lies precisely in the areas critical to sustainable development, namely, public access to the media and the accompanying legal procedures for environmental protection, as well as environmental indicators. Both systems have different concerns and interests (Avkopashvili et al., 2019; Shenshinov, 2012; Shenshinov & Al-Ali, 2020; Tsvetkov et al., 2019).

One of the key differences between ISO 14001 and EMAS is their environmental policy requirements. EMAS is based on performance, which requires continuous improvement in environmental performance to the level set by the economically viable best available technology. EMAS is based on performance, which requires continuous improvement in environmental performance to the level set by the economically viable best available technology. EMAS requires a company to provide a public report that informs stakeholders. In turn, the publicity ensures that companies are proactive in their approach.

The definition of ISO 14001 focuses on EMS itself—the organization must commit to continuous improvement. ISO deals with the organizational system and concludes that improving the environmental management system will improve environmental performance. ISO 14001 and EMS follow the "Plan-Do-Check-Act" cycle, designed to continually improve EMS to help improve environmental performance. The ISO 14001 standard is based on the following components:

- Industrial ecology;
- Planning;
- Implementation and adherence to processes;
- Analysis of the fulfillment of environmental indicators and the corresponding adjustment;
- Managerial decisions.

The rest of the indicators are just recommendations. ISO 14001 standards focus on the process and not on manufacturing performance. There is a growing trend of awareness of environmental problems and the likely problems associated with environmental degradation. However, much confusion, disagreement, and uncertainty prevail. Scientists in many countries present a comprehensive view of the impact of industrial activities on the environment. These representations of land use include impacts that destroy habitat, as well as various forms of pollution, waste, energy consumption and the use of natural resources to support health and safety. Many natural areas are irreversibly damaged by industrial activities that alter their ecological integrity. The main impact of industry on the environment is related to the consumption of materials, many of which are non-renewable. Nowadays, buildings account for about 40% of the materials entering the world economy each year and

25% of the world's wood use. The exploitation of forest resources and the extraction of minerals influence the environment. The construction of facilities produces many atmospheric pollutants.

2 Materials and Methods

Most of the services and products used in the production consume resources, many of which are non-renewable. If these resources continue to be produced, some of them may contribute to pollution. For example, the construction of buildings consumes about two-thirds of all energy used. Additionally, the construction of new buildings accounts for about 5% of the total energy consumption in its production process. The buildings and structures in use consume an enormous amount of natural resources and are most responsible for many undesirable effects on everything surrounding that building. Manufacturing and environmental impact are synonymous.

If society could better control every production process, people could enjoy a more comfortable living condition without much change to their natural environment; in other words, sustainable production is valued (Glotko et al., 2019; Nekhvyadovich et al., 2020; Ragulina et al., 2019).

Recently, environmental management in the industry saw a growing trend in terms of positive indicators. There is an increased number of studies on the successes of environmental management. Nowadays, the means by which the industry can achieve the objectives of ISO 14000 are allocated. All industrial enterprises are advised to consider environmental policies, goals, and objectives at each stage of the development project. Almost all enterprises implement the interaction between the created environmental policy, natural environment, and possible remedies in terms of social responsibility, environmental responsibility, environmental and energy auditing, sustainable design, energy standards, and labeling in the constructed buildings. After considering the environmental consequences of industrial operations, practitioners suggested that the client must accept *the environment* as the goal of the project to develop a culture of environmental protection in the industry. The corresponding business opportunities define the potential for the manufacturing industry to contribute to environmental protection positively.

Current environmental management concepts and their importance in the industry are the basis for a structured environmental management program. The environmental management program is based on BS 7750, and EMS is essential in industrial organizations to guide production, operation, and decommissioning. The example of industrial production served as the basis for building the EMS structure for other industrial companies. It is assumed that environmental management should be specified as a requirement in contract specifications and quantitative statements.

The Industrial Research and Information Association has produced a series of industrial management guidelines on implementing environmentally responsible practices. These guides aim to:

- Provide a checklist of environmental considerations and guidance on them at various stages of the project;
- Provide a basis for identifying existing information and current best practices;
- Provide a framework to help companies compile environmental impact registers and adopt appropriate environmental management procedures.

The difficulties that individual entrepreneurs face in managing the environment in production are summarized. The traditional approach to production management, which focuses only on cost, time, and quality, needs improvement since humanity enters a greener and more energy-dependent age. However, it is a time when environmental protection becomes an obligation not only for the producer but also for the consumer. Production management must integrate cost, time, quality, and environment.

These four dimensions should assess any development of the project. Moreover, practitioners have concluded that parts of the manufacturing process (including customers, designers, and contractors) must have great potential to make significant contributions and play an active role in environmental campaigns. Specialists in manufacturing and environmental management focus on the following areas:

- Efficient use of energy and natural resources;
- Proper selection and specification of environmentally friendly materials and control of toxic chemicals and hazardous waste;
- Pollution control, clean technologies, and recycling and waste management in the industry;
- Environmental education during education and training.

Additionally, there were some suggestions from other researchers. For example, the researchers proposed the following measures to solve environmental problems:

- Halting deforestation and desertification (by saving on the use of wood, finding alternatives to it, and exploring the role of industry in reforestation) and addressing the depletion of other resources (by saving on the use and recycling of materials and using renewable varieties);
- Addressing the problem of concentrated heavy rains and sea-level rise (by designing protection systems and developing materials and methods suitable for construction and allowing satisfactory performance under such conditions);
- Preventing pollution (by handling and controlling the waste, developing and using safe and non-polluting materials and appropriate construction, maintenance, and demolition methods);
- Taking care of the effects of pollution (by developing and using materials that can withstand harsh atmospheric conditions, such as the effects of acid rain and other toxic substances);
- Combating inadequate drainage, soil erosion, and excessive salinity of seawater (by implementing the appropriate design of constructed facilities and design and construction of protective systems);
- Reducing pressure on existing land and encroachment on green areas and places of natural beauty (by developing technologies for rehabilitating existing buildings

and commissioning abandoned, polluted, or naturally vulnerable land, as well as assessing the environmental impact of construction projects);
- Finding reliable and stable energy sources for mining, material production, production activities, and the use and maintenance of constructed facilities (Bogoviz et al., 2020; Costanza, 1996; Tashkulova & Kletskova, 2020).

3 Results

EMS is the formal structure of an organization that implements environmental management. BS 7750 requires an organization to develop, implement, and maintain EMS to ensure that its activities are consistent with the established environmental policies, strategies, goals, and objectives. Additionally, the system must comply with all applicable environmental laws governing its economic activities. To create an EMS, an organization must address the following issues:

- Develop an EMS system manual—a documented set of procedures and instructions that meet the environmental standards;
- Implement procedures and work instructions during its activities;
- Maintain procedures and work instructions and keep them up to date.

These three requirements must be met since they are fundamental to the internal development and implementation of EMS in the enterprise. Moreover, the heads of production facilities must understand their role and responsibility for implementing environmental management.

Typically, the level of EMS complexity is proportional to the complexity of the production process with its technological processes, as well as the financial situation of the enterprise, taking into account the human potential and the social component of the organization. The development of EMS requires a comprehensive structure, procedures, and an understanding of the range of available skills, abilities, and commitments within the enterprise itself.

Some companies can develop EMS from scratch, but they have to put more effort into the production process. However, other businesses may have their management system. They can only use the ISO 9000 quality management system (QMS).

BS 7750 establishes and describes the basic requirements for EMS in key sections. Within these sections, each requirement is formulated in terms of organizational responsibility around which the QMS of the organization can develop the system according to its business activities (Adarina et al., 2019; Stroiteleva et al., 2020a, Stroiteleva et al., 2020b). Thus, the personality and core business of an enterprise may need to develop some aspects of the system more than the minimum requirements specified in the standard. When creating an EMS, businesses may be required to refer not only to the standard but also to other arbitration guidance documents. The documents must contain the requirements of industry experts and specifications established by the regulatory authorities.

4 Conclusion

To maximize the use of EMS, an enterprise should consider the concepts of environmental management and its application from a broader perspective. At the initial level, EMS can contribute to improvements within the enterprise itself (e.g., energy savings in the work environment). At the secondary level, EMS can highlight procedures and practices that can be improved (e.g., the reduction of waste or encouragement of recycling of materials and resources) (Adarina et al., 2019; Dragunov & Shenshinov, 2020). At the tertiary level, EMS can act as a marketing and public relations tool to promote the enterprise's image in the commercial market. Therefore, to maximize the potential of EMS, the company must consider both internal and external aspects while complying with the international standard ISO 14000, which sets the requirements for EMS and allows the company to formulate policies following the legal requirements and necessary information on the positive impact on the environment, as well as to control the implementation of environmental aspects that companies can regulate and influence.

The application of these aspects depends on factors such as environmental policies, taking into account the nature of the company's activities and the conditions in which these activities are formed.

References

Adarina, R., Gazukina, Y., & Yankovskaya, K. (2019). Indicators of the "green economy" as a tool for monitoring the regional economy. In *IOP Conference Series: Earth and Environmental Science* (Vol. 395, p. 012107). https://doi.org/10.1088/1755-1315/395/1/012107

Avkopashvili, P. T., Polukhin, A. A., Shkodinsky, S. V., & Poltarykhin, A. L. (2019). The fundamental provisions of the concept of knowledge economy. In E. Popkova, Y. Ragulina, & A. Bogoviz (Eds.), *Industry 4.0: Industrial revolution of the 21st century* (pp. 57–64). Springer. https://doi.org/10.1007/978-3-319-94310-7_5

Bogoviz, A. V., Shvakov, E. E., Tretyakova, O. G., Zakharov, M. Y., & Abramov, A. N. (2020). Globalization of education in the conditions of formation of the global knowledge economy: Regularities and tendencies. In E. Popkova (Ed.), *Growth poles of the global economy: Emergence, changes and future perspectives* (pp. 993–1000). Springer. https://doi.org/10.1007/978-3-030-15160-7_99

Costanza, R. (1996). Ecological economics: Reintegrating the study of humans and nature. *Ecological Application, 6*(4), 978–990.

Dragunov, A., & Shenshinov, Y. (2020). Effective development of innovative technologies as a method for increasing labor productivity in industry. In *E3S Web of Conferences* (Vol. 208, p. 03001).

Glotko, A. V., Sycheva, I. N., Dunets, A. N., Poltarykhin, A. L., Zhuravlev, P. V., & Tubalets, A. A. (2019). Integration processes in the agro-industrial complex of the regions: Development and environmental problems. *Journal of Environmental Management and Tourism, 10*(3(35)), 613–621.

Nekhvyadovich, L. I., Borodin, V. A., & Avkopashvili, P. T. (2020). A systemic approach to the study of the sustainability of the global digital economy: Economies of scale, ratchet effect and hysteresis effect. In E. Popkova, & B. Sergi (Eds.), *Scientific and technical revolution: Yesterday, today and tomorrow* (pp. 1369–1377). Springer. https://doi.org/10.1007/978-3-030-47945-9_149

Ragulina, Y. V., Semenova, E. I., Avkopashvili, P. T., Dmitrieva, E. A., & Cherepukhin, T. Y. (2019). Top-priority directions of implementing new internet technologies on the territories of rapid economic development. In E. Popkova, & V. Ostrovskaya (Eds.), *Perspectives on the use of new information and communication technology (ICT) in the modern economy* (pp. 182–188). Springer. https://doi.org/10.1007/978-3-319-90835-9_21

Shenshinov, Y. V. (2012). Labour quality management. *The Journal North-East Asia academic Forum.*

Shenshinov, Y., & Abdulsattar Al-Ali. (2020). The tools of increasing efficiency of human resource in the lean production environment: Conceptual study. *International Journal of Core Engineering & Management, 6*(7), 1–18. Retrieved from http://ijcem.in/wp-content/uploads/2020/06/THE-TOOLS-OF-INCREASING-EFFICIENCY-OF-HUMAN-RESOURCE-IN-THE-LEAN-PRODUCTION-ENVIRONMENT-CONCEPTUAL-STUDY2.pdf

Stroiteleva, T. G., Kletskova, E. V., & Balaeva, S. I. (2020). Ontogenesis of green development model of digital economy: Essence, stages, and prospects. In E. Popkova, & B. Sergi (Eds.), *Scientific and technical revolution: Yesterday, today and tomorrow* (pp. 410–417). Springer. https://doi.org/10.1007/978-3-030-47945-9_46

Stroiteleva, T. G., Petrova, E. A., & Yungblyudt, S. V. (2020). Responsible information society as a social environment for creation of the "green" digital economy: Formation and monitoring. In E. Popkova, & B. Sergi (Eds.), *Scientific and technical revolution: Yesterday, today and tomorrow* (pp. 402–409). Springer. https://doi.org/10.1007/978-3-030-47945-9_45

Tashkulova, G. K., & Kletskova, E. V. (2020). Place and role of human in the system of circular reproduction in the digital regional economy. In E. Popkova, & B. Sergi (Eds.), *Scientific and technical revolution: Yesterday, today and tomorrow* (pp. 943–951). Springer. https://doi.org/10.1007/978-3-030-47945-9_101

Tsvetkov, V. A., Loginov, E. L., & Raikov, A. N. (2019). Integrated network interaction of education with science and production. *Educational Technologies and Society, 19*(1), 389–398.

Sustainable Development of Industrial Innovation in the Context of Environmental Safety

Ksenia G. Yankovskaya, Olga I. Ergardt, and Zalimkhan R. Kudaev

Abstract This study focuses on the sustainable development of society and economy, namely the future relations between industrial production and the environment. Current ways of industrial development have negative ecological and social consequences. Industrial ecology is a theory of developing new structures and functions for an industry that are environmentally compatible, economically beneficial, and community-oriented. We aim to explore the practical application of industrial ecology theory in the development of eco-industrial parks. We intend to prove that local governments have a significant role in determining the sustainability of their regions, even though the main forces of change lie outside of their immediate sphere of influence. Planning and development decisions made at local levels can still impact the global environment. Local policies and regulatory tools that foster innovation in production-consumption cycles are fundamental to making progress toward sustainable development.

Keywords Sustainable development · Environmental security · Effective policies · Consumption levels · Eco-industrial parks

JEL Codes Q5 · Q52 · Q56

1 Introduction

Humanity faces severe and lasting issues stemming from pollution and cultural upheavals caused by materialism, rising unemployment, and generational poverty. German scholars have summarized current studies on the issues of sustainable development (Vukovich et al., 2018).

K. G. Yankovskaya (✉)
Gorno-Altai State University, Gorno-Altaysk, Russia

O. I. Ergardt
Altai State University, Barnaul, Russia

Z. R. Kudaev
Kabardino-Balkarian State Agrarian University Named After V.M. Kokova, Nalchik, Russia

E. G. Popkova and B. S. Sergi (eds.), *Geo-Economy of the Future*,
https://doi.org/10.1007/978-3-030-92303-7_25

The planet supports around 8 billion people now, but this only possible through rapid deterioration of natural resources, namely arable land, karst groundwater, and biodiversity. Moreover, humanity uses around 40% of energy among all terrestrial animals. The current scale of the human economy is too large to sustain the same standard of living for future generations. The human population, which has nearly doubled in the last century, now stands at 7.5 billion, and is expected to reach 10 billion by 2050 (Tashkulova & Kletskova, 2020).

Since World War II, resource consumption has increased in parallel with population growth. This is now reinforced by declining natural resource reserves that cannot be replaced in any meaningful timeframe. Trade and technology have allowed humans to take resources from virtually every ecosystem on the planet. There is plenty of evidence of environmental degradation resulting from human activities, including loss of biodiversity, deforestation, soil degradation, fish stock collapse, groundwater contamination, and general pollution. Humanity transforms the Earth at a level that rivals natural phenomena: machines move twice as much material as geological forces.

The current situation suggests that there is a maximum level of stress that the environment can sustain without taking irreversible damage. Studies of material and energy flows show that the stress already exceeds the long-term global capacity. The *Sustainable Europe* campaign suggests that achieving sustainability will require roughly halving the use of material resources in the global economy in the coming decades (Glotko et al., 2019). To allow the developing countries to claim their fair share of global capacity, the developed countries must reduce their use of non-renewable resources by 88–94%. The Geneva-based World Business Council for Sustainable Development confirms this fact. It urges industrialized countries to reduce material production, energy consumption, and emissions by more than 90% by 2040 to meet the needs of the growing global population while staying within the Earth's ecological limits.

Due to the rising need to bring material consumption and waste generation in line with the Earth's regenerative capacity, the social transformation toward sustainability has become a global imperative (Nekhvyadovich et al., 2020).

Many alternative definitions and interpretations of sustainable development have now been proposed. Sustainability literally refers to the need to live justly within the natural limits. This requires certain approaches to sustainable development. It is very important to understand that *development* does not include growth in annual economic output and gross national products. Unlimited growth of these indicators is impossible in the situation of limited resources. Expansion requires a constant increase in material and energy flows in the human economy. Instead, *development* means maximizing human potential by increasing efficiency and quality.

2 Materials and Methods

Sustainable development requires having balanced economic and social goals within ecological limits. This proves that society, as a socio-economic system, is extremely dependent on the throughput of material resources (Adarina et al., 2019). Promoting nature-compatible production, goods, consumption patterns, material policies, and lifestyles in a healthy environment will require fundamental changes in existing social values and developmental frameworks. This will take nothing less than social transformation, encompassing technological innovation, social ingenuity, and the collective will to refocus existing production patterns and consumption. The industrial analysis of development points to several inherent "industrial gaps. At certain branching points, technological change can go in several directions. The choice of direction depends on political and institutional factors. In the theory of complex systems, this is called path dependence.

Due to growing environmental pressures, fundamental changes occur in society. This creates a need for viable alternatives to the manufacturing status quo. Industrial ecology is an alternative that aims for a fundamental paradigm shift in the theory of relations between industry and ecology. This thesis presents industrial ecology as a local planning tool for aligning human activities with the Earth's regenerative capacity (Karataev et al., 2020).

This study focuses on the role of municipal government in the development of eco-industrial parks, a key strategy for implementing industrial ecology. We aim to discover whether municipal policies and regulations are conducive to eco-industrial development. The general political climate and the current regulatory framework applicable to design can only be used when the characteristics of eco-industrial parks are being analyzed. New concepts of industrial development and operation must be supported in order to provide insight into the capacity of municipal policies. This results in a definite understanding of how local governments can promote eco-industrial development (Bogoviz et al., 2020). This study can provide information on the circumstances in other municipal governments that have not yet adopted an industrial ecology (Avkopashvili et al., 2019).

Industrial ecology theory requires that the industrial system be modified to mimic natural ecosystems in their general functioning. Ecology is the study of the structure, functions, and behavior of the natural systems that make up the biosphere. An ecosystem is a community of organisms interacting with each other and with the physical environment in a particular area in a somewhat self-sustaining manner. The ecosystem structure consists of both the living and nonliving components of the community. Communities of different species in an ecosystem are usually structured as chains and webs of producers, consumers, etc.

Diversity, inertia, sustainability, and feedback loops are important attributes of ecosystems. These are some of the features of mature ecosystems that allow them to work efficiently, recycle materials, and produce as little waste as possible. This is what industrial production must strive for. In this case, the industrial ecosystem should:

- Minimize fossil fuel use for energy; reduce waste generation and consumption of primary resources;
- Recycle and use industrial waste and emissions from manufacturing in other industrial processes "...similar to the nutrient cycles in an ecological food chain";
- Create a diverse and sustainable system to absorb and recover from unexpected shocks.

There is one fundamental difference between natural and eco-industrial systems. Biological systems evolve toward a local equilibrium (not necessarily an optimal state) via evolutionary mechanisms of variation, selection, and reproduction at the level of organisms. In biological processes, population adaptation is the result of natural selection, not of conscious design. Industrial systems adapt to a changing environment (the market), adjusting different processes to each other and to their natural environment. However, this can also be achieved by deliberate action through planning and design. The word "industrial" in "industrial ecology" should be understood more broadly than "pertaining to production enterprises" (Dragunov & Shenshinov, 2020).

This set of production processes that integrate technology, materials, and energy should be seen as an interacting network of resources, processes, and wastes. To cause a paradigm shift in understanding the relationship between industry and ecology, one must understand that industrial ecology believes that industry is the most significant environmental factor and, as such, should be the focus of analysis. Industrial ecology has a longer-term perspective and seeks comprehensive solutions (Shenshinov, & Abdulsattar Al-Ali., 2020).

The broad vision of industrial ecology can be conceptualized by several attributes widely recognized by scholars in this field.

Industrial ecology recognizes two main areas that distinguish the flow and transformation of materials and energy.

The first is the technical area, which focuses on infrastructure and technology for resource flow within separate production processes and between enterprises.

The second area deals with the less strict behavior of the network and includes other types of relationships between enterprises, such as the shared use of services, transportation, and equipment, rather than just material and energy flows. This second area has expanded to include community-business interactions aimed at developing symbiotic networks between businesses, the community, and the public sector.

In this area, social relationships and creativity take precedence over technological solutions for more sustainable management of water, materials, and energy (Shenshinov, & Abdulsattar Al-Ali., 2020). These flows are controlled by traditional management tools, such as hierarchy, the market, and law.

3 Results

The actual reality of industrial ecology is studied at two different levels. One concerns the micro- and meso-scale application of ecosystem principles in industrial systems. Virtual development of an eco-industrial park involves work within a single industrial park or with all industrial enterprises within a settlement or a region (Stroiteleva et al., 2020). These two systems are not mutually exclusive, and the advantages and disadvantages of each are complementary. For example, eco-industrial parks are not closed systems and will prove more effective if integrated into a communal or regional exchange network. Macro-scale deserves a different type of attention, aimed at reducing the resource intensity of production and consumption at a societal level (dematerialization).

Industrial ecology is an emerging field. It is multidisciplinary and has been described as "a wide range of concepts rather than a single theoretical construct." Industrial ecology includes a range of approaches, such as:

- Environmental design;
- Life cycle analysis;
- Pollution prevention;
- Industrial metabolism;
- Resource recovery or by-product exchange;
- Green construction;
- Extending the service life of products;
- Multi-stakeholder processes and services economy concept.

Our study focuses on community-business interactions at the micro-level and examines industrial ecosystems within geographic boundaries or eco-industrial parks.

4 Conclusion

Industrial business development requires clear planning of the output volumes, minimizing its production costs (raw materials, electricity, and heat), and creating sustainable economic, environmental, and social relations.

Today, eco-industrial parks highlight the focus of industrial ecology on developing networks between businesses and communities so as to optimize resource use and minimize economic and environmental costs. Implementing this is a complex task of coordinating multiple economic actors, local governments, and community members with the ultimate goal of integrating economic development with the local ecosystem (Merdesheva et al., 2020; Tsvetkov et al., 2019). Like other environmental management initiatives, these measures are all united by traditionally diverging political perspectives (Costanza, 1996). Effective policies require public participation in planning and decision-making and solving complex inter-relationships between social, technological, and environmental issues. Taken as a whole, these measures create environmental security.

References

Adarina, R., Gazukina, Y., & Yankovskaya, K. (2019). *Indicators of the "green economy" as a tool for monitoring the regional economy*. In *IOP Conference Series: Earth and Environmental Science* (Vol. 395, p. 012107). https://doi.org/10.1088/1755-1315/395/1/012107

Avkopashvili, P. T., Polukhin, A. A., Shkodinsky, S. V., & Poltarykhin, A. L. (2019). The fundamental provisions of the concept of knowledge economy. In E. Popkova, Y. Ragulina, & A. Bogoviz (Eds.), *Industry 4.0: Industrial revolution of the 21st century* (pp. 57–64). Springer. https://doi.org/10.1007/978-3-319-94310-7_5

Bogoviz, A. V., Shvakov, E. E., Tretyakova, O. G., Zakharov, M. Y., & Abramov, A. N. (2020). Globalization of education in the conditions of formation of the global knowledge economy: Regularities and tendencies. In E. Popkova (Ed.), *Growth poles of the global economy: Emergence, changes and future perspectives* (pp. 993–1000). Springer. https://doi.org/10.1007/978-3-030-15160-7_99

Costanza, R. (1996). Ecological economics: Reintegrating the study of humans and nature. *Ecological Application, 6*(4), 978–990.

Dragunov, A., & Shenshinov, Y. (2020). Effective development of innovative technologies as a method for increasing labor productivity in industry. In *E3S Web of Conferences* (Vol. 208, p. 03001). https://doi.org/10.1051/e3sconf/202020803001

Glotko, A. V., Voronkova, O. Y., Miheeva, N. M., Skoblikova, T. V., Alekhina, N. A., & Zhminko, A. E. (2019). The economic development of a transboundary region under the conditions of the Eurasian economic union. In *Proceedings of the IBIMA 2019: Education Excellence and Innovation Management through Vision 2020* (pp. 7957–7968). Granada, Spain.

Karataev, E. M., Merkuryev, V. V., & Titova, O. V. (2020). Integrating the "green economy" into the model of digital future of the modern socio-economic systems. In E. Popkova, & B. Sergi (Eds.), *Scientific and technical revolution: Yesterday, today and tomorrow* (pp. 1402–1410). Springer. https://doi.org/10.1007/978-3-030-47945-9_153

Merdesheva, E. V., Titova, O. V., & Avkopashvili, P. T. (2020). Methodological approach to the classification of digital economies by environmental efficiency and sustainability criterion. In E. Popkova, & B. Sergi (Eds.), *Scientific and technical revolution: Yesterday, today and tomorrow* (pp. 1360–1368). Springer. https://doi.org/10.1007/978-3-030-47945-9_148

Nekhvyadovich, L., Kostin, K. B., & Mironov, V. V. (2020). Government and public monitoring of entrepreneurship in a multicultural region: New perspectives in the conditions of the digital economy. In E. Popkova, & K. Vodenko (Eds.), *Public administration and regional management in Russia* (pp. 147–154). Springer. https://doi.org/10.1007/978-3-030-38497-5_17

Shenshinov, & Abdulsattar Al-Ali. (2020). The tools of increasing efficiency of human resource in the lean production environment: Conceptual study. *International Journal of Core Engineering & Management, 6*(7), 1–18.

Stroiteleva, T. G., Kletskova, E. V., & Balaeva, S. I. (2020). Ontogenesis of green development model of digital economy: Essence, stages, and prospects. In E. Popkova, & B. Sergi (Eds.), *Scientific and technical revolution: Yesterday, today and tomorrow* (pp. 410–417). Springer. https://doi.org/10.1007/978-3-030-47945-9_46

Tashkulova, G. K., & Kletskova, E. V. (2020). Place and role of human in the system of circular reproduction in the digital regional economy. In E. Popkova, & B. Sergi (Eds.), *Scientific and technical revolution: Yesterday, today and tomorrow* (pp. 943–951). Springer. https://doi.org/10.1007/978-3-030-47945-9_101

Tsvetkov, V. A., Loginov, E. L., & Raikov, A. N. (2019). Integrated network interaction of education with science and production. *Educational Technologies and Society, 19*(1), 389–398.
Vukovich, G. G., Makuschenko, L. V., Bateykin, D. V., Titova, O. V., & Dobrosotskiy, V. I. (2018). Support of the territories of advanced economic development on human capital: Theory and practice. *Quality—Access to Success, 19*(S2), 157–160.

The Scientific Nature of Industrial Ecology: A Theoretical Analysis

Elena V. Merdesheva, Kseniya A. Melekhova, and Madina V. Blieva

Abstract Researchers and practitioners of industrial ecology believe that the current form of development of humankind will lead—in the long-term—to a catastrophe for homo sapiens and, probably for a lot of forms of flora and fauna. Recognition of the fact that the current form of economic production and consumption, together with the growing population, contradicts the natural system of the biosphere, makes industrial ecologists who want to study nature and development of life in the recent periods to development of companies, with further elaboration of the central idea of industrial ecology, which concerns the adaptation of socio-economic processes of human and especially industrial production, as well as to natural limitations at the global level (Bogoviz et al. in Growth Poles of the Global Economy: Emergence, Changes and Future Perspectives, Plekhanov Russian University of Economics. Luxembourg, pp. 993–1000, 2020 (Bogoviz et al. 2020)). Industrial ecology is a part of the important movement from human society's release from natural limitations (which could be treated as the basis of the whole technological development, starting from the Stone Age). "Learning from nature", industrial ecology tries to reverse this growing isolation of humankind from its resource base (including the potential of waste assimilation).

Keywords Industrial ecology · Theoretical analysis · Global system · Sustainable society · Creation of services · Learning from nature

JEL Codes Q5 · Q551 · Q56

E. V. Merdesheva (✉)
Gorno-Altai State University, Gorno-Altaysk, Russia
e-mail: merdesheva@bk.ru

K. A. Melekhova
Altai State University, Barnaul, Russia

M. V. Blieva
Kabardino-Balkarian State Agrarian University Named After V.M. Kokova, Nalchik, Russia

E. G. Popkova and B. S. Sergi (eds.), *Geo-Economy of the Future*,
https://doi.org/10.1007/978-3-030-92303-7_26

1 Introduction

"Studying nature" is the main philosophy of industrial ecology. This is still a new discipline, which tries to find its place in the canon of sciences. Industrial ecologists distinguish several situations in which nature is mainly treated as natural ecosystems, which were used as the basis for organizing and using various resources. Industrial ecologists worked at companies, rebuilding production processes. Ecological disciplines demonstrated a shift in favor of technological solutions, which are created by engineers of industrial ecology. "Learning from nature, however, is not a trivial company on search and copying of the best practices' situation, especially regarding "collection and mixture". When the motto "learning from nature" is formulated seriously, the question envisages more than just technological importance: what is the image of "nature"? Whence does human understanding of nature come? What is "learning"? What could be learned from nature and where exactly does behavior in natural systems leads to a delusion as an example of socio-economic behavior? Thorough work in the sphere of industrial ecology should take into account not only engineering but also scientific perspectives. In this sense, the discipline reflects the general discourse on economic issues—technological, economic, biological / economic, or social and philosophical—where the understanding is necessary and which are used more frequently where the complex perspectives of these spheres is necessary (Tashkulova & Kletskova, 2020).

"Learning from nature" is further understood as an attempt of a transfer between natural and social systems. In case of a sphere with the growing volume of research, but without a proper understanding of the moving forces, this thesis should be supplemented by applied research. The main results of this practice consist in the attempt to confirm industrial ecology as a comprehensive scientific sphere, which applications are the reflection of its main ideas and values, and establishment of "learning from nature" as a scientific study. This is necessary because industrial ecology is still in the process of formation and is far from being sustainable. Scholars from dozens of various academic circles ensure at least a theoretical basis for an active discussion (Glotko et al., 2019).

2 Methodology

Reorganization of socio-economic systems could be performed by science in any serious direction. Due to this, the communication between scholars from different spheres reminds rather of "children's mess" than informative work.

The current attempts to understand and solve the conflict between short-term human and long-term economic well-being require breaking the limits between various sciences, especially natural and social (to which here humanitarian sciences are assigned) (Merdesheva et al., 2020). A strict division and definition of scientific disciplines is a recent phenomenon, which has passed the following stages of development:

- from establishment of modern sciences after the humanistic philosophy in early sixteenth century to late nineteenth century (e.g., the influence of the Darwin's "On the Origin of Species" in the British Victorian society);
- scientific discourses in the sphere of education and understanding;
- dissemination of knowledge since those times, when, together with quick technological development and specialization of research, this wider perspectives was made very difficult for achievement and understanding;
- calling itself a "theory of sustainability", industrial ecology shares the blame for this tunnel vision, together with any other monoculture of mind;
- combination of ideas from a range of the sphere of sciences, which is written in a reflexive manner (which largely owes to essays of humanitarian writers and which supposedly supports and stimulates this early modern form of an integrative scientific dialogue).

Such approach to the research object allows us treating the object and the work completely differently than would have been possible in case of the empirical approach:

- firstly, industrial ecology has a lot of thematic studies and applied works—it is said that this "science" "operationalizes" sustainability;
- secondly, drawing various spheres of sciences, which stimulate eco-economic substantiation and its application, applied research in the form of thematic studies could be easily performed by other projects, which are more limited in their coverage. Each of them cannot get far from the object that they study (Stroiteleva et al., 2020).

Besides, in the course of discussion of such notions as globalization, sustainable development, economic growth, and well-being—ideas that also have the political nature, not only economic or technological one. All these terms are difficult to define in the way that has been widely accepted by most of the participants of the scientific and public discourse, since any attempt of definition faces a discussion of the political motives of the person who offers the definition. This political position could be found in the selective approach to collection and analysis of data: the nature of the research object and the surrounding "world system", which have a lot of data, require a research approach, which, for most purpose, has to replace the volume and topicality of the focus and strictness. For example, this became topical for life cycle analysis. As for the topic of globalization, it emerged in the course of this process (Alekseev et al., 2018).

The complex character of the studied issue allowed adapting the selected proofs to the preconceived theory, not to an—allegedly—independent research. The fact that industrial ecology gains popularity among researchers, companies, and government officials, means that the discipline should be critically considered in view of the fact what has to be achieved (sustainable economic system) for it not to become the last tool for "conquer of the environment". This requires not only additional thematic studies (which were considered in industrial ecology as low hanging fruit" of short-term benefits) but also the systemic study of all aspects.

Therefore, it is necessary to study not only the influence of product manufacture and services creation on the environment but also the social consequences of their distribution (e.g., scale effect and competition) and consumption. The main attention to industrial ecology is paid by the whole systems of production and consumption, which are embodies in infrastructures, not by a separate product (Adarina et al., 2019).

Such position requires the movement away from the offer, which is preferred by commercial companies—the nature of demand is treated as an exogenous factor of economic analysis, and the change of demand that envisages a critical view at growth and consumption. However, in view of the impossibility to fully dematerialize the economic activities, this is the place to establish sustainability and, therefore, industrial ecology. Industrial ecology—due to the metaphor that lies in its basis—should be a common concept for the whole sustainable economic system, not just a technological solution that is used in production. Inclusion of the analysis of distribution and consumption systems in the industrial ecological work reunites the discipline with its conceptual roots and usefulness of considering nature as an example for studying industrial systems (Dragunov & Shenshinov, 2020).

The main concept of the paper, which requires adoption of the interdisciplinary approach, is the effect of transfer of notion from one sphere to another—i.e., activities of "learning" in the motto of industrial ecology. It is necessary to understand not only the nature of the source and object of the transfer but also the process of transfer. In the history of science, this process was regularly used, and false ideas of one subject were adopted as a dogma in another subject. This transfer becomes the basis of industrial ecology as an "ecological metaphor", which is used for understanding a social system in the terms of natural sciences (Shenshinov, 2012).

This producing is the main goal of creation of preanalytical vision", which includes the following:

- fundamental recognition which considers the ecological view at human society and socio-economic system, which is built into the totality of all ecosystems, i.e., biosphere of the Earth;
- preanalytical vision includes recognition of the biophysical limits of resources and, at last, resource saving.

It is important that the final ability of the biosphere is assimilated with pollution and belongs to the category of biophysical limits. This is a physical perspective, which limits the economic system that grows in the absolute physical expression.

Industrial ecology was created at the merging of several academic and professional disciplines, of which the most vivid is engineering. Since this sphere is based on the ideas and concepts of various spheres, it has to be evaluated with the help of analytical tools from these spheres and has to be logically substantiated (Costanza, 1996). This thesis is devoted to studying the substantiation of the contribution of industrial ecology to sustainable development. This claim to reality becomes more important in the course of strengthening of a belief that the theoretical foundations of industrial ecology have already been set, and further work will have to concentrate on real restructuring and informing the directive bodies of the goals. Besides, this sphere claims to be the most important basis for creating structures that stick to sustainable development goal. Industrial ecology is described as a "science of sustainability" and as an "operational approach to sustainability".

That's why here it is offered to evaluate industrial ecology as compared to adjacent disciplines and academic research and practical application (Stroiteleva et al., 2019). In the course of the history of science, the growing number of research efforts led to the fundamental basis, which is common for all sciences, becoming fragmentary, which led to a rudimentary understanding of the scientific concepts outside of any selected sphere.

3 Results

For the subject of industrial ecology, which is largely based on the views of other scientific disciplines, a conceptual transfer should be performed with conscious application—if it needs wide respect in the scientific community and formation of government and corporate decision, not a green cloak for a predetermined type of adoption of the policy, which is not able to perform fundamental changes.

Due to this, the paper consists of three parts:

- 1st—introduction to the sphere and the circle of problems;
- 2nd—evaluation of industrial ecology from the point of view of natural and social sciences;
- 3rd—reconsideration and reformation of its application.

While theoretical and applied research is being conducted in the sphere of industrial ecology, the first part is the material for distinguishing the problem spheres and the most important limits of the considered system (Stroiteleva et al., 2020).

Industrial ecology is considered as a potentially useful attempt to create a sustainable human society; it cannot be treated as a panacea: success of this concept is research and its application largely depend on acknowledgement of the fact how actively it is used.

Considering industrial ecology, which is based on the eco-economic thinking, as a part of a sustainable economic system, researchers observe the system of industrial ecology in the corresponding—i.e., complex and systemic—context, where the ecological elements are a central aspect of the research, for they are aimed at creating the basis for proper scientific efforts in the sphere of industrial ecology (Tsvetkov et al., 2019).

It is proved that any system that is created by a human has to be considered as a self-organizing dissipative system—like non-human animate or inanimate systems. The analysis focuses on a dissipative system and, in particular, a system created by human, which lies in the basis of energy and material flows. These conclusions from the works in the sphere of thermodynamics, in particular the nature of all complex systems that dissipates energy, will provide the discipline of industrial ecology with substantiation and guide for the transfer between natural and social systems, to which it strives.

Having determined the reason for comparing the characteristics of natural ecosystems and industrial systems, let us give the rules of this transfer, which, up to now, was performed in the chaotic and intuitive way at isolated examples (Karataev et al., 2020).

Discussing the metaphor that lies in the basis of industrial ecological thinking and evaluating it from the point of view of quality of scientific thinking, we see that a lot of scientific works in all spheres of science are actually managed with the help of metaphors, but with different levels of success. Validity of the main notions of one discipline in this case is based on the suppositions that were obtained as a result of the transfer of ideas from other spheres.

Let us show to which consequences the reorientation of industrial ecological analysis and restructuring, when the whole economic system is treated as an observed system, will lead. Industrial ecology is pierced with social measuring, which is yet absent: inclusion of the models of distribution of market and consumption in the industrial ecological analysis, which shows that this sphere cannot be based only on the technological or engineering foundation (Vukovich et al., 2018). These issues have to be studied in order to find out whether they stimulate the process of distribution of the resources and energy market.

4 Conclusion

This paper is devoted mainly to criticism of the modern state of scientific studies in the sphere of industrial ecology and creation of a more proper scientific basis for studies in the sphere of industrial ecology; the paper considers new concepts that have been taken from the already set foundations. As a political offer, implementation of the reform of ecological tax is performed—in the understanding of gradual increase of taxes on natural fuel in the course of a timeframe, which is considered to be the basis of industrial ecology, with serious attitude towards sustainable development of territories.

References

Adarina, R., Gazukina, Y., & Yankovskaya, K. (2019). Indicators of the "green economy" as a tool for monitoring the regional economy. In *IOP Conference Series: Earth and Environmental Science. International Conference on Sustainable Development of Cross-Border Regions 2019, SDCBR 2019* (p. 012107).

Alekseev, A., Savenkov, D., Shvakov, E., Lebedev, N., & Ragulin, A. (2018). Financial problems of regional development and the ways of their solution in contemporary Russia. *Quality—Access to Success, 19*(S2), 71–75.

Bogoviz, A., Shvakov, E., Tretyakova, O., Zakharov, M., & Abramov, A. (2020). Globalization of education in the conditions of formation of the global knowledge economy: regularities and tendencies. In: *Growth poles of the global economy: emergence, changes and future perspectives.* Series "Lecture Notes in Networks and Systems", Plekhanov Russian University of Economics. Luxembourg (pp. 993–1000).

Costanza, R. (1996). Ecological economics: Reintegrating the study of humans and nature. *Ecological Application, 6*(4), 978–990.

Dragunov, A., & Shenshinov, Y. (2020). Effective development of innovative technologies as a method for increasing labor productivity in industry. In *E3S Web of Conferences.*

Glotko, A., Sycheva, I., Dunets, A., Poltarykhin, A., Zhuravlev, P., & Tubalets, A. (2019). Integration processes in the agro-industrial complex of the regions: development and environmental problems. *Journal of Environmental Management and Tourism, 10, 3*(35), 613–621.

Karataev, E., Merkuryev, V., & Titova, O. (2020). Integrating the "green economy" into the model of digital future of the modern socio-economic systems. *Lecture Notes in Networks and Systems, 129*, 1402–1410.

Merdesheva, E., Titova, O., & Avkopashvili, P. (2020). Methodological approach to the classification of digital economies by environmental efficiency and sustainability criterion. *Lecture Notes in Networks and Systems, 129*, 1360–1368.

Shenshinov, Y. (2012). *Labour quality management*. North-east Asia academic forum.

Stroiteleva, T., Kalinicheva, E., Vukovich, G., & Osipov, V. (2019). Peculiarities and problems of formation of industry 4.0 in modern Russia. *Studies in Systems, Decision and Control, 169*, 145–153.

Stroiteleva, T., Kletskova, E., & Balaeva, S. (2020a). Ontogenesis of green development model of digital economy: Essence, stages, and prospects. *Lecture Notes in Networks and Systems, 129*, 410–417.

Stroiteleva, T., Petrova, E., & Yungblyudt, S. (2020b). Responsible information society as a social environment for creation of the "green" digital economy: Formation and monitoring. *Lecture Notes in Networks and Systems, 129*, 402–409.

Tashkulova, G., & Kletskova, E. (2020). Place and role of human in the system of circular reproduction in the digital regional economy. *Lecture Notes in Networks and Systems, 129*, 943–951.

Tsvetkov, V., Loginov, E., & Raikov, A. (2019). Integrated network interaction of education with science and production. *Educational Technologies and Society, 19*(1), 389–398.

Vukovich, G., Makuschenko, L., Bateykin, D., Titova, O., & Dobrosotskiy, V. (2018). Support of the territories of advanced economic development on human capital: Theory and practice. *Quality—Access to Success, 19*(S2), 157–160.

Production Quality Management Within the Increase of Product Safety and Environmental Friendliness

Larisa B. Nyurenberger, Irina V. Chernyaeva, and Milana A. Dzuganova

Abstract Due to the growing complexity of products and stricter standards of safety, product recall is inevitable; however, it damages a company's work and requires effective management. An important decision related to recall of products is selection of the time of recall, which influences future sales of a company and the risks of product safety. Companies make a lot of temporary decisions during the whole process of recall. Soon after a potential defect is spotted, a company has to study whether the defect requires a wide recall and, if this is justified, to initiate the recall. Prompt recall (right after companies spot products' defects) could lead to large losses in sales, while postponed recall could lead to large expenditures for correction of defects, large safety risks for consumers, and large fines from government bodies (Shenshinov in Labour quality management. North-East Asia Academic Forum, 2012 (Shenshinov, 2012)). Product safety is of the utmost importance for consumers and society.

Keywords Environmental friendliness · Quality management · Increase of safety · Increase of expenditures · Defect

JEL Codes Q5 · Q55 · Q57

1 Introduction

In the current conditions, a company can manipulate its temporary decisions in two ways:

- during the process of recall—to postpone the negative influence of the recall, especially for products with a long life cycle;

L. B. Nyurenberger (✉)
Novosibirsk State University of Economics and Management, Novosibirsk, Russia
e-mail: l.b.nyurenberger@nsuem.ru

I. V. Chernyaeva
Altai State University, Barnaul, Russia

M. A. Dzuganova
Kabardino-Balkarian State Agrarian University Named After V.M. Kokova, Nalchik, Russia

E. G. Popkova and B. S. Sergi (eds.), *Geo-Economy of the Future*,
https://doi.org/10.1007/978-3-030-92303-7_27

- company can deliberately ignore the results of the research and postpone the recall until the end of the research.

When a company models the process of recall for products with a long life cycle, it is possible to observe how the corresponding decisions on the timeframe of the recall are made, and how the tools are given to governments to prevent long delays of the recall. In such cases, the following questions arise.

Firstly, considering the fact that companies spotted a defect (which requires a recall), which types of companies will probably initiate the postponed recall? In this case, we consider trade operations, which stimulate companies to postpone the recall.

Secondly, if a company considers the postponed recall, how exactly does it delay (deliberately ignoring the research results and manipulating the duration of the research) the delivery and how long does it take? At last, after understanding the driving forces of delays with the recall, how governments could use various tools for preventing (long) delays with the recall?

To model the process of the recall of products with (long) cycles of issue, the Bass model is used—it maximizes profit, when product sales are influenced by external and internal aspects (e.g., marketing buzz). It is supposed that potential defect of product is revealed in the exogenous way at a certain time and could be spotted only by the company and its products' users, thus causing negative influence on the internal influences and product sales through marketing buzz before the recall. Then the company starts studying the potential effect, and after that decides when to recall the defective goods (Vukovich et al., 2018). A quick recall leads to negative coverage in mass media (i.e., external influence), while postponed recall leads to negative marketing buzz among the existing users (i.e., internal influence). Both cases lead to future losses of sales for companies. Besides, delay of the recall inevitable leads to increase of expenditures for elimination of the defect (since more items have been sold) and to a larger fine for the delay of the recall (including reputation risks for the company, legal fees for safety failures, and government fines). That's why the company makes a decision on the time of the recall, concluding an agreement between losses from sales after the recall (i.e., negative influence of external aspects) and general expenditures for the recall (e.g., expenditures for the recall of all sold products, late fees, and negative influence of internal aspects) (Avkopashvili et al., 2019).

2 Methodology

Companies cannot manipulate the duration of investigations due to the following reasons:

- a potential defect revealed at a late stage of the product processing;
- negative incremental influence of the recall on the internal channels, which dominate over the external ones.

Knowing this, the company will show that when the share margin of the product is much higher than the cost factor of the recall, it will be always delaying the recall. If the cost of the recall of a product unit grows, the company will still be delaying the recall—if the defect is revealed at an early stage of the production cycle. In case of further increase of the cost of the recall of a product unit, the company might not consider the postponed recall at all stages (Tsvetkov et al., 2019).

For companies that could freely manipulate the duration of their investigation (e.g., complex product that requires professional knowledge during the research), there is a model which results have been described above (without manipulation of the research)—the only exception is that companies that make decisions on a postponed recall will extend the duration of the investigation in order to avoid responsibility for quick recall, and the product with a higher margin / recall cost ratio will have a larger duration of manipulation. Postponed recall of a product (with too large duration of the investigation) does not influence the companies' sales much, so companies are not too eager to postpone the recall (Costanza, 1996).

This model envisages political consequences for governments, if the recall, which is postponed for too long, is discouraged due to the following reasons.

Firstly, government can develop fees for making companies react quickly to safety defects. We consider and compare three schemes of fees for restraining stimuli of companies that have large delay or large duration of manipulated investigation. However, the changes of policy or law might be time-consuming and complex. The offered model also envisages the inspection of potential defects of a product as a short-term correction (Bogoviz et al., 2020). This model shows that review of complaints with large margin as compared to the recall's cost, as well as products that are at an early stage of the product cycle, could oppose serious delays of the recall more effectively. Besides, it takes in the account the fact that internal effect of marketing buzz could effectively deal with the recall's delays due to investigation of defects, which, in its turn, could reduce the duration of the delay (Tashkulova & Kletskova, 2020).

The negative influence of products' recall on the company's cost, demand for products, and the marketing activities does not allow reducing the risks of a recall, and companies cannot always use various strategies of operations—e.g., operations of geographical distribution of products and contracts for joint use of expenditures for a recall. As soon as there emerges a crisis, related to damage to a product that requires a recall of the products, companies will have to take into account various factors, when they decide when to announce the recall. At present, scholars demonstrate the companies strategically postpone recalls according to their need for attraction of external financing and study the factors the correlated with delays of recall, ignoring the time marks in the process of recall. Besides, studies of the government policy on regulation of product recalls focus on responsibility of products, not on the government monitoring or fees for the delay of product recall (Glotko et al., 2019).

3 Results

To reduce the delays with product recall and the related social damage, the following measures have to be taken:

- the theoretical model has to describe—in detail—the product's cycle and the emergence of product recall, which would allow understanding how the short-term and long-term influence of recall impacts the company's sales;
- explain how companies make decisions on product recall. In particular, when product's defect is revealed at an early stage, a company could strategically delay the recall in order to receive profit, despite the multiple fines and penalties caused by the recall (Nekhvyadovich et al., 2020).

Thus, though product recall is a necessary business solution, it is also an important social solution. To analyze a compromise that lies in the basis of a company's decision on the times of recall, where it offers potential solutions that allow avoiding serious delays of recall and the corresponding social costs.

The practice of outsource raises the effectiveness of production and decreases production costs, but also increases the risk of emergence of defected products. Since more and more components are transferred to outsource, management of crises that are related to product defects has become an important function of product quality management (Merdesheva et al., 2020). Product recall is the most frequently used, but double-edged tool, for managing such crises, since the recalls send a bad signal regarding products' quality and lead to the heterogeneous influence on demand for products for recalls with different structure of incomes and expenditures and recalls at different stages of product cycles. In particular, we have shown here that if the recall negatively influences consumers via the internal channels (e.g., marketing buzz), as compared to the external channels (e.g., mass media), companies might ignore the possibility of delay of their corresponding recalls due to potential large reputation risks (Alekseev et al., 2018). However, if the recall is influenced more negatively by external rather than internal factors, the recall that is connected to defects of products with a higher margin (as compared to recall's cost) or defects that are revealed at an early stage for products with a moderate margin is usually postponed. It is no wonder that a higher margin also leads to long delays, so companies can deliberately postpone the time of the recall after they have revealed the defects, or can manipulate the duration of the investigation but pretend that the recall is on time (Shenshinov, 2020).

To reduce the consequences of such unlawful behavior, there are schemes of penalties and various mechanisms of information dissemination. Let us consider three schemes of fines:

- fine that is based on the delay's duration;
- fine that is based on the delay's duration and the duration of the manipulated investigation;
- fine that is based on the manipulation's behavior.

Depending on the needs, capabilities, and resources of politicians, they could select the one that restrains the most undesirable behavior. This also agrees with scientific works that study government sanctions or a monitoring form for preventing harmful actions (Dragunov & Shenshinov, 2020). Besides, disclosure of information that is to raise the public awareness in the part of improvement of the actions to reveal defects emphasizes the reputation risks for companies that could also prevent the long delays of recall.

4 Conclusion

Since we study company's decisions on the time of product recall (i.e., whether it should be delayed, how and for how long) during the long cycle of the product, we understand how companies make such decisions and that this is only the first step. We expand this understanding, providing counteractions that might be taken by government from the point of view of social provision (Karataev et al., 2020; Stroiteleva et al., 2019). As for companies' decisions on the time of product recall during the whole production cycle and at the last state of product quality management, we think that this direction could be further developed by works that consider the dynamics between consumers, companies, and regulating bodies.

References

Alekseev, A., Savenkov, D., Shvakov, E., Lebedev, N., & Ragulin, A. (2018). Financial problems of regional development and the ways of their solution in contemporary Russia. *Quality—Access to Success, 19*(S2), 71–75.

Avkopashvili, P., Polukhin, A., Shkodinsky, S., & Poltarykhin, A. (2019). The fundamental provisions of the concept of knowledge economy. *Studies in Systems, Decision and Control, 169*, 57–64.

Bogoviz, A., Shvakov, E., Tretyakova, O., Zakharov, M., & Abramov, A. (2020). Globalization of education in the conditions of formation of the global knowledge economy: regularities and tendencies. In *Growth poles of the global economy: Emergence, changes, and future perspectives. Series "Lecture Notes in Networks and Systems"* (pp. 993–1000). Plekhanov Russian University of Economics.

Costanza, R. (1996). Ecological economics: Reintegrating the study of humans and nature. *Ecological Application, 6*(4), 978–990.

Dragunov, A., & Shenshinov, Y. (2020). Effective development of innovative technologies as a method for increasing labor productivity in industry. In *E3S Web of Conferences.*

Glotko, A., Sycheva, I., Dunets, A., Poltarykhin, A., Zhuravlev, P., & Tubalets, A. (2019). Integration processes in the agro-industrial complex of the regions: development and environmental problems. *Journal of Environmental Management and Tourism, 10, 3*(35), 613–621.

Karataev, E., Merkuryev, V., & Titova, O. (2020). Integrating the "green economy" into the model of digital future of the modern socio-economic systems. *Lecture Notes in Networks and Systems, 129*, 1402–1410.

Merdesheva, E., Titova, O., & Avkopashvili, P. (2020). Methodological approach to the classification of digital economies by environmental efficiency and sustainability criterion. *Lecture Notes in Networks and Systems, 129*, 1360–1368.

Nekhvyadovich, L., Kostin, K., & Mironov, V. (2020). Government and public monitoring of entrepreneurship in a multicultural region: new perspectives in the conditions of the digital economy. In *Public administration and regional management in Russia. Challenges and prospects in a multicultural region. Series "Contributions to Economics"* (pp. 147–154).

Shenshinov, Y. (2020). Abdulsattar Al-Ali–PhD. The tools of increasing efficiency of human resource in the lean production environment: conceptual study. *International Journal of Core Engineering & Management, 6*(7).

Shenshinov, Y. (2012). *Labour quality management*. North-East Asia Academic Forum.

Stroiteleva, T., Kalinicheva, E., Vukovich, G., & Osipov, V. (2019). Peculiarities and problems of formation of industry 4.0 in modern Russia. *Studies in Systems, Decision and Control, 169*, 145–153.

Tashkulova, G., & Kletskova, E. (2020). Place and role of human in the system of circular reproduction in the digital regional economy. *Lecture Notes in Networks and Systems, 129*, 943–951.

Tsvetkov, V., Loginov, E., & Raikov, A. (2019). Integrated network interaction of education with science and production. *Educational Technologies and Society, 19*(1), 389–398.

Vukovich, G., Makuschenko, L., Bateykin, D., Titova, O., & Dobrosotskiy, V. (2018). Support of the territories of advanced economic development on human capital: Theory and practice. *Quality—Access to Success, 19*(2), 157–160.

Development of Key Concepts of the Theory of Industrial Ecology

Svetlana I. Balaeva, Olga V. Titova, and Nina A. Kocheeva

Abstract The research process consists of a correspondence between the following phases: the personal context of the researcher, the interpretive paradigm, the research strategy, and the methods of data collection and analysis. This research is based on theoretical and applied analysis. In contrast to quantitative research, the focus is made on the processes operating in the socially constructed reality rather than on measuring and analyzing the relationship between variables. The research does not claim to be within a value-free framework, since any research can never be morally neutral, and all research results have political implications. Skilled researchers also address the relationship between the researcher and what is being studied, taking into account the situational constraints that shape the study in question. The current model of international economic development is based on a neoliberal worldview. The research strategy is a path of discovery connecting the research paradigm to the empirical world. The ecological worldview seeks political, social, economic, and institutional transformation to change behavior, values, and attitudes toward economic growth. Such research focuses on conducting case studies, which will preserve the holistic and meaningful features of real events and allow for a design logic unique to particular circumstances.

Keywords Qualitative analysis · Situational limitations · Ethical position · Ethical position · Ecological worldview

JEL Codes Q5 · Q51 · Q59

S. I. Balaeva (✉)
Kabardino-Balkarian State Agrarian University Named After V.M. Kokova, Nalchik, Russia
e-mail: balaeva.s@list.ru

O. V. Titova
Altai State University, Barnaul, Russia
e-mail: otitova82@icloud.com

N. A. Kocheeva
Gorno-Altai State University, Gorno-Altaysk, Russia

E. G. Popkova and B. S. Sergi (eds.), *Geo-Economy of the Future*,
https://doi.org/10.1007/978-3-030-92303-7_28

1 Introduction

Qualitative research is multi-method in orientation. It assumes an interpretive and naturalistic approach to its subject with a complex historical development of its current genres. The biographically situated subject lies behind each focus of the research activity. In our understanding, research is an interactive process shaped by the personal history, biography, gender, social class, race, and ethnicity of the researcher and people in a particular setting (Bogoviz et al., 2020; Glotko et al., 2019). In general, the research occurs within an interpretive community that incorporates its own historical research traditions into a separate point of view. These factors have influenced the subject of this research.

Our interest in developing scientific research on the theory of industrial ecology is a result of our intrinsic concern for nature. Our current view is that no single method can fully capture the subtle variations in the current human experience, which underscores the value of using a wide range of interrelated interpretive methods (Tashkulova & Kletskova, 2020a).

The ethical position that can be applied to this study is the contextual-consequentialist model. It is based on a statement of trust, an attitude to the ecological situation of the country, the health of people living in the area, and their ethical attitude and authority.

2 Materials and Methods

A paradigm can be understood as a basic set of values, beliefs, and assumptions about the world surrounding us. This research is interpretive since it is guided by individual paradigms that determine the way the world should be understood and studied. The study is based on our view of the nature of reality and our relation towards the received knowledge about the world in the social sciences and ecology in the country, which, in essence, is a critical theoretical approach. Beyond factors that are real in space and time, reality exists and has been shaped by social, political, cultural, economic, ethnic, and gender factors that are now accepted as *real*. Over time, the past structures that perceived these factors as something natural and immutable became flawed. The relationship between the subject and object of cognition is transactional and subjective (Stroiteleva et al., 2020).

The dialogue between subject and researcher must be dialectical to transform historically mediated structures accepted as immutable into a more informed consciousness.

Since this study focuses on environmental politics, the theoretical rationale should be expanded into the realm of factors that are real in space and time and fixed, like the physical sciences. It is essential to recognize the role of values, beliefs, and assumptions in the way cultures interact with nature. The paradigm we ascribed to this is known as the ecological worldview (Tashkulova & Kletskova, 2020b).

It harks back to the physics and biology of the twentieth century, nonequilibrium thermodynamics, complex systems theory, deterministic chaos, and systems ecology. The central scientific assumptions lie in the following:

- Behavior of natural systems is unknowable at a system-wide level;
- Uncertainty is large and irreducible across broad scales;
- Holistic approaches provide a better understanding of global change.

Nevertheless, regardless of our research position, people are an integral part of the ecosphere (Glotko et al., 2019).

The nature perspective is the one that recognizes the obligatory dependence of people on the ecosphere and the fact that resources control people. It is important to note that nature has an intrinsic value exceeding the value of production in markets.

The ecological worldview stands in stark contrast to the dominant neoliberal paradigm in Western culture, which is becoming a global worldview. The prevailing neoliberal worldview is largely mechanistic and rooted in the Enlightenment and the subsequent scientific revolution, particularly the determinism of Newtonian analytic mechanics. A perspective is anthropocentric and separates the observer from the observed object. One of the results is that nature is objective, and human is perceived as something separate and dominant over the natural world. In essence, nature is adapted to human needs and requirements. It is valued primarily as a source of resources and waste flow. The combination of mechanical philosophy and anthropocentrism explains the technological approach to nature. This neoliberal (expansionist) paradigm views the economy as a self-sustaining circular flow between production and consumption, ignoring the links between commodities and ecological processes (Costanza, 1996). The current model of international economic development is based on a neoliberal worldview.

The ecological worldview attempts to reconcile the environmental destruction caused by economic activity by incorporating physical and biological principles. Similar to other organisms, people survive by extracting energy and materials from the ecosystem in which they live; they are limited by the existence of biophysical constraints. This fact fundamentally connects all human activity (including the economy) to the entire ecosphere. The economy must be understood as an open and growing subsystem fully dependent on a finite, non-expanding, and materially closed ecosphere (Tsvetkov et al., 2019).

All transformations of energy and materials (including material economic production processes) are subject to thermodynamics laws, which impose biophysical constraints. Ecosystems and, therefore, the economy are regulated and governed by the laws of thermodynamics.

The first law defines the conservation of energy during transformation—the total amount of energy remains unchanged.

The second law concerns the quality of energy and indicates that the entropy (disorder or chaos) of a system is constantly increasing. Ecosystems are open systems that have evolved a set of *dissipative structures* to direct high-quality solar energy sources to self-organize and create new, complex structures and *pump out* the entropy production associated with the second law. The latter function is essential to the survival of organisms and ecosystems; it imposes a significant burden on systems in terms of the need for sustainable energy (Nekhvyadovich et al., 2020). The struggle for existence is essentially a struggle for high-quality available energy. The key process in this struggle to utilize incoming solar radiation is the process of photosynthesis by autotrophs. It is a significant source of high-quality energy for ecosystems and the natural capital used by the human economy. Constraints on energy availability impose limits on the growth and activity of the ecosystems (Stroiteleva et al., 2020; Vukovich et al., 2018).

Thermodynamic considerations are essential to the economic and environmental, social, and technological processes of which it is composed. While natural systems depend on the sun for energy, human systems used significant reserves of chemical energy (primarily fossil fuels). Mocking nature, people are not very open to other forms of energy. This access has made humans free from the laws governing the development of other species, and it seems that we have gained control over the conditions regulating life. However, the economy is an inherently dissipative structure since no transformation process can operate at 100% efficiency (the first law of thermodynamics). Low-entropic energy and materials must be converted into high-entropic energy and materials to build and maintain complex structures. The economy managed to expand to a global scale using reserves of chemical thermodynamic potential. Thermodynamically, this leads to the following problems (Adarina et al., 2019; Merdesheva et al., 2020; Shenshinov & Abdulsattar, 2020):

- Reliance on energy conversion from chemical storage facilities has no mechanism for exporting entropy from the global ecosystem;
- There is currently virtually no environment outside the economy to export entropy.

The ecological worldview recognizes the inherent unsustainability of the current socio-economic system and advocates a biophysical view of the economy that recognizes its relationship to the ecosphere. The material transformations from which we currently derive economic wealth are the result of unstable linear flows (Dragunov & Shenshinov, 2020). Historically, biophysical constraints seemed inconsequential because the economic subsystem was small in relation to the global ecosystem. However, the current scale of material and energy use makes these restrictions increasingly binding.

3 Results

Based on the limitations imposed by the laws of nature, the principles that must be observed by people include the following:

- Consumption of the products and services of nature (natural capital) cannot exceed the rate of their production in the ecosphere;
- Waste production cannot exceed the assimilative capacity of the ecosphere;

- Socio-economic activities must not endanger the basic life-support systems of the ecosphere.

The development and orientation of production processes toward integration with natural systems and adherence to thermodynamic principles is a step in the social transformation toward a sustainable biophysical economy.

The research strategy is a path of discovery, connecting the research paradigm to the empirical world. The ecological worldview seeks political, social, economic, and institutional transformation to change behavior, values, and attitudes toward economic growth. On this basis, our research challenges the consideration of local development models for production purposes. However, instead of identifying the full range of alternatives for local industrial development with improved environmental performance, let us turn to the viability of one theory representing a new field. Such research focuses on the ability to conduct case studies, which will preserve the holistic and meaningful features of real events and allow for a design logic that is unique to particular circumstances. The research method has one focus. The unit of analysis is the regulatory framework and policies of local government regarding the design features of eco-industrial parks (Tashkulova & Kletskova, 2020b). The case study is expected to demonstrate the way traditional regulatory frameworks and local policies affect the development potential of industrial ecosystems, which includes the following issues:

- What are the main goals of the environmental effectiveness of each element?
- What means were used to achieve these goals in existing eco-industrial parks (EIP)?
- How favorable are the subject's policy and regulatory framework for achieving these goals?
- What can we learn from any barriers to EIP implementation?

This research used primarily libraries, computerized indexes and databases, and the Internet, which facilitated the identification of key works. Moreover, the research included excerpts from journals and used websites, as well as official studies (Glotko et al., 2019).

4 Conclusion

Design features specific to industrial ecology operating to achieve environmental performance goals were drawn from case studies. The authors compiled a list of regulatory barriers to implementing key elements of the competencies of industrial ecology. Then, the authors considered the cases where it is applicable to implement this list to determine the following:

- Which rules govern each practice;
- The level of responsibility for developing a conceptual theory of industrial ecology;

- Whether there is a program that facilitates or hinders the implementation of design elements.

References

Adarina, R., Gazukina, Y., & Yankovskaya, K. (2019). Indicators of the "green economy" as a tool for monitoring the regional economy. In *IOP Conference Series: Earth and Environmental Science* (Vol. 395, p. 012107). https://doi.org/10.1088/1755-1315/395/1/012107

Bogoviz, A. V., Shvakov, E. E., Tretyakova, O. G., Zakharov, M. Y., & Abramov, A. N. (2020). Globalization of education in the conditions of formation of the global knowledge economy: Regularities and tendencies. In E. Popkova (Ed.), *Growth poles of the global economy: Emergence, changes and future perspectives* (pp. 993–1000). Springer. https://doi.org/10.1007/978-3-030-15160-7_99

Costanza, R. (1996). Ecological economics: Reintegrating the study of humans and nature. *Ecological Application, 6*(4), 978–990.

Dragunov, A., & Shenshinov, Y. (2020). Effective development of innovative technologies as a method for increasing labor productivity in industry. In *E3S Web of Conferences* (Vol. 208, p. 03001). https://doi.org/10.1051/e3sconf/202020803001

Glotko, A. V., Sycheva, I. N., Dunets, A. N., Poltarykhin, A. L., Zhuravlev, P. V., & Tubalets, A. A. (2019). Integration processes in the agro-industrial complex of the regions: Development and environmental problems. *Journal of Environmental Management and Tourism, 10*(3(35)), 613–621.

Glotko, A. V., Voronkova, O. Y., Miheeva, N. M., Skoblikova, T. V., Alekhina, N. A., & Zhminko, A. E. (2019). The economic development of a transboundary region under the conditions of the Eurasian economic union. In *Proceedings of the IBIMA 2019: Education Excellence and Innovation Management through Vision 2020* (pp. 7957–7968). Granada, Spain.

Merdesheva, E. V., Titova, O. V., & Avkopashvili, P. T. (2020). Methodological approach to the classification of digital economies by environmental efficiency and sustainability criterion. In E. Popkova, & B. Sergi (Eds.), *Scientific and technical revolution: Yesterday, today and tomorrow* (pp. 1360–1368). Springer. https://doi.org/10.1007/978-3-030-47945-9_148

Nekhvyadovich, L. I., Borodin, V. A., & Avkopashvili, P. T. (2020). A systemic approach to the study of the sustainability of the global digital economy: Economies of scale, ratchet effect and hysteresis effect. In E. Popkova, & B. Sergi (Eds.), *Scientific and technical revolution: Yesterday, today and tomorrow* (pp. 1369–1377). Springer. https://doi.org/10.1007/978-3-030-47945-9_149

Shenshinov, Y., & Abdulsattar, A.-A. (2020). The tools of increasing efficiency of human resource in the lean production environment: Conceptual study. *International Journal of Core Engineering & Management, 6*(7), 1–18.

Stroiteleva, T. G., Kletskova, E. V., & Balaeva, S. I. (2020). Ontogenesis of green development model of digital economy: essence, stages, and prospects. In E. Popkova, & B. Sergi (Eds.), *Scientific and technical revolution: Yesterday, today and tomorrow* (pp. 410–417). Springer. https://doi.org/10.1007/978-3-030-47945-9_46

Stroiteleva, T. G., Petrova, E. A., & Yungblyudt, S. V. (2020). Responsible information society as a social environment for creation of the "green" digital economy: Formation and monitoring. In E. Popkova, & B. Sergi (Eds.), *Scientific and technical revolution: Yesterday, today and tomorrow* (pp. 402–409). Springer. https://doi.org/10.1007/978-3-030-47945-9_45

Tashkulova, G. K., & Kletskova, E. V. (2020). Place and role of human in the system of circular reproduction in the digital regional economy. In E. Popkova, & B. Sergi (Eds.), *Scientific and technical revolution: Yesterday, today and tomorrow* (pp. 943–951). Springer. https://doi.org/10.1007/978-3-030-47945-9_101

Tashkulova, G. K., & Kletskova, E. V. (2020). The model of organization of "green" digital production and consumption based on the Internet of things and AI. In E. Popkova, & B. Sergi (Eds.), *Scientific and technical revolution: Yesterday, today and tomorrow* (pp. 329–336). Springer. https://doi.org/10.1007/978-3-030-47945-9_37

Tsvetkov, V. A., Loginov, E. L., & Raikov, A. N. (2019). Integrated network interaction of education with science and production. *Educational Technologies and Society, 19*(1), 389–398.

Vukovich, G. G., Makuschenko, L. V., Bateykin, D. V., Titova, O. V., & Dobrosotskiy, V. I. (2018). Support of the territories of advanced economic development on human capital: Theory and practice. *Quality—Access to Success, 19*(S2), 157–160.

Current Problems of the Effective Use of Resource Potential in the Development of Kyrgyzstan's Regions: Social and Investment Aspect

Chinara R. Kulueva, Ayganysh A. Sayakbaeva, Kubanych K. Toktorov, Cholpon A. Nuralieva, and Kalmira Dz. Ismailova

Abstract The paper focuses on the regional economy, modern problems of the effective use of the resource potential of Kyrgyzstan's territories in terms of social and investment aspects. The authors define the role of the development and improvement of the theory of regional development in the context of adopted relevant programs, concepts, and territorial objectives focused on stabilizing and creating conditions for achieving positive dynamics in improving the level and quality of the population's life through sustainable and balanced reproduction of the regional resource potential. The authors also clarify terminology, which is the basis for expanding further philosophical and economic research objects of the regional economy and regional development, which occupy a worthy place in modern economic science. Methods of observation and theoretical research allowed us to reliably determine the role and importance of the regional economy in the sustainable development of the country and create the model of a developed country. In recent years, the development of regions is studied and disclosed by the Kyrgyz scientific schools as a policy direction. Considerable attention is paid to applied research with the identification of constraints, their diagnosis, analysis, and problem-solving. A special place in this research is given to the problems of labor resources and issues of effective use of mineral resources (minerals) of the regions. The authors propose to create special conditions for the development of labor and mineral resources by working out optimal long-term scenarios of socio-economic development of territories taking into account the risks and resource opportunities in the regions. The authors consider the role and usefulness of the development and production of minerals and gold deposits to develop regions and rural areas. The role of accounting and financial control over compliance with the coefficient of social effectiveness and usefulness of long-term investments is defined; their calculation methodology is provided. The authors propose specific recommendations on government regulation to strengthen

C. R. Kulueva (✉) · K. K. Toktorov · K. Dz. Ismailova
Osh State University, Osh, Kyrgyzstan

A. A. Sayakbaeva
Kyrgyz National University Named After Zhusup Balasagyn, Bishkek, Kyrgyzstan

C. A. Nuralieva
Kyrgyz-Russian Slavic University Named After B. Yeltsin, Bishkek, Kyrgyzstan

E. G. Popkova and B. S. Sergi (eds.), *Geo-Economy of the Future*,
https://doi.org/10.1007/978-3-030-92303-7_29

the social significance of long-term investment projects in the development of the regions and the Kyrgyz Republic.

Keywords Human resources · Regions · Gold deposits · Investment · Social effect · Government regulation

JEL Codes J18 · R11 · O15 · L78 · P45

1 Introduction

In the theory of regional development and the regional economy, an important place is taken by the resource potential, which is formed in a single territory, actively used in public production. Resource potential reveals its essence as a set of all types of resources (Faculty of International Economic Relations of Belarus State Economic University, n.d.). Few people know how to develop regions so that it would be comfortable to live, work, rest, provide family, friends, and relatives, and not to move to another place.

These and other questions became the object of our research. Many research works are devoted to the issues of regional economic development (Zhang, 1999). Our task is to find the solution to the problem based on the conducted theoretical analysis, where the interests of the region, people, and government are placed at the center.

It is a well-known fact that the development of regional economies largely depends on the size and efficiency of the use of the regional resource potential, especially labor and rich deposits of minerals, which predetermines the level of income of the population and is reflected in the standard of living. Prominent economic scientists (Stighlitz et al., 2016) compared the socio-economic indicators of the countries by the level of GDP and critically assessed their impact on the systemic level of development of the regional economy and the population's life. We agree with the opinion of L. L. Yakovleva that regional program documents contain the unevidenced conclusions about the expected positive dynamics in the standard of living caused by the implementation of various measures (Yakovleva, 2018).

In the late 1980s and early 1990s, the key topics of economic theory were the problems of regional development, where the dynamics of economic evolution (in our case, the dynamics of regional development) were investigated. Thus, the scholars were busy modeling regional dynamics. Some scholars studied sudden and unpredictable disturbances in the continuous development of regions (Andersson & Batten, 1987; Wilson, 1981). We agree with Andersson's (Andersson, 1986) beliefs about the increasing role of logistics and logistics systems and their manifestations as variable values in enriching the theory of regional development, where one can argue the existence of the "logistics—region" dilemma, which has recently been the subject of studies and detailed research. Nowadays, it can be argued that the supply chain and global changes in the world economy are closely interrelated and interact with each other. In our opinion, the activation of the logistics system mirrors the level of territorial development. In our case, these are the investments included in the supply chain. These investments create and affect changes in the nature of social and labor relations, as well as in production, labor, production placement, use of natural and labor potential, culture, and social institutions.

This process should be carried out at a steady and slow pace. The occurring changes should not have a chaotic and discontinuous nature. Therefore, it is required to adopt effective (ambitious) program documents of regional development pursuing the qualitative aspects of territorial growth.

Summarizing the above, it should be noted that the level of income and quality of life in the areas rich in minerals (in our case, gold deposits (1/7 of Kyrgyzstan has deposits of gold and other precious non-ferrous metals)) is much lower than in regions without similar natural wealth. This "unfair" situation is the subject of lively discussion in the media and increased discontent on the part of the residents of these territories. It is worth noting that a considerable share of rural governments in the regions of the Kyrgyz Republic (80%) are subsidized regions, characterized by difficulty in predicting their revenues and the dominance of political factors. In this case, it seems appropriate to cite the opinion that the choice of the right strategy for budgeting depends on the comfort of living in the region (Yakovleva, 2018).

2 Materials and Methods

2.1 Theoretical Foundation of the Problems of Regional Labor Potential

Before turning to the problems of labor potential, let us recall the economic nature of labor potential. Based on the study of international and Kyrgyz experience, we conclude that labor potential expresses the qualitative side and potential capabilities of labor (human) resources. In turn, the essence of the labor resources determines quantitative features of labor potential and the structure of labor. Considering both sides of labor potential, it is necessary to add intellectual, social, and physical components. Therefore, theoretically, labor potential is constantly in flux. It is a changeable category requiring constant study and searching for ways and specific mechanisms to mitigate or reduce labor risks.

The use of labor potential depends directly on the state of the national economy of the region and the pace of development of the national economy. The situation is exacerbated when labor potential, namely the ability to work, cannot ensure the economic development of territories and meet the ever-growing and changing needs of the population.

In this case, it is appropriate to recall the American experience of managing innovation processes in economic sectors (the 1980s), especially in the agricultural sector. At that time, the management of territories and regions successfully used not only the process of developing innovative programs but also a competitive selection of the executors. This policy reflected the goal-setting of an effective innovation system, which included nine interrelated factors focused on the social position of the region (Toktorov et al., 2020).

The social nature of the rational use of labor potential explains the degree of involvement of the able-bodied population, especially young people, in objective production relations. The level of real income received by workers, the degree of satisfaction of the working population with the material and socio-cultural needs, purposeful motivation, etc., are extremely important (Kulueva, 2015).

Consequently, market relations determine the presence of the following problems in the sphere of social and labor relations and in the process of using the labor potential of the able-bodied population (Kulueva, 2009):

1. Achievement of the highest possible level of employment of the able-bodied population in the sectors and spheres of the national economy;
2. Rational distribution of the employed able-bodied population, taking into account the synchronicity and interaction between production rates and the degree of satisfaction of socio-cultural and material needs;
3. Achievement of a certain balance in the development of regional production and distribution of labor and material resources, which allows for the efficient use of labor potential for providing the most favorable conditions for labor and life of people;
4. Marketing research and analysis according to the law of supply and demand conducted in each subject of the economy;
5. Increase the attractiveness of the Kyrgyz economy for the most able-bodied population, especially young people, for starting a business to reduce the intensification of migration outflow and "brain drain" to the near and far abroad countries;
6. Growth of young people with higher professional education (about 40%–45%);
7. The presence of young people (13–15 years old) in regional labor markets is the main reason for competition in the service and trade sectors, where no special vocational training is required, and labor results are paid comparatively lower.

The seventh issue is a serious problem that requires an appropriate assessment from the part of the government. In our opinion, this issue must be handled with some caution since it has two contradictory sides. On the one hand, recognition of child (13–15 years old) labor at the legislative level can protect them from discrimination, which is especially common in rural and border areas (Kulueva, 2016). On the other hand, disadvantages, and lack of a family budget push many minors into the market, where they often do unsustainable work, do not attend school, and become sick, exhausted, and unfit for their age. The most disappointing fact is that there is a growing trend in the number of underage children in the unskilled labor market. The social pressures and insecurities of low-income families can cause illiteracy throughout their lives. So far, these are individual cases amenable to local solutions; otherwise, it could become a threat to the country and all humankind (Kulueva, 2015).

In this case, it is necessary to implement cardinal measures on the part of the government since the problems mentioned above are impossible to solve only by providing one-time financial assistance (Kulueva et al., 2019). These issues require a well-thought-out socio-economic policy for the future. This policy is reflected in state programs, concepts, etc. (Government of the Kyrgyz Republic, 2017), where special attention is paid to the comprehensive development of the human and, in particular, labor potential of the country. Undoubtedly, the future of the country is in the hands of our children, the younger generation.

2.2 Theoretical Justification of the Problems of Using the Natural and Mineral Potential of the Regions

The theory of regional development places particular importance on the issues of studying such important economic directions as effective and rational use of available resources. The main priorities are given to the problems of balanced placement and use of natural wealth and human resources, which requires effective territorial organization of the economy and labor aimed at sustainable development of regions and improving the welfare of their inhabitants.

The Kyrgyz Republic has significant reserves of deposits and minerals, in particular, gold deposits, which are being gradually developed as the attracted investment resources come in. The development of the reserves of deposits and minerals should significantly contribute to the socio-economic development of the national economy and positively impact the welfare of the people, thereby improving the quality of life and increasing the level of social welfare of territories, especially rural areas.

The problem of the efficient use of minerals and resources has long existed. It is the center of political, economic, and international discussions. There exist many opinions and disputes around this problem and a sufficient number of research and monographic studies.

In current conditions, particularly acute is the question of the usefulness and profitability of these deposits to the local population (i.e., the owners of these lands, whose ancestors lived and inhabited these territories for centuries).

The development of fields requires huge capital investments, which Kyrgyzstan cannot afford. The legal framework of Kyrgyzstan is convenient for the development of mineral deposits, including gold. Therefore, almost 90% of gold deposits are developed mainly by foreign capital. The Kyrgyz Republic agrees with this policy of development of these territories, but this does not mean that foreign capital should fully dominate the territory of Kyrgyzstan and its regions.

We would like to focus on the examples when the deposits do not acquire the promised original appearance after their development, which indicates a weak control or, even more, the improper governmental control over the activities of foreign firms and companies engaged in gold mining. Currently, there is no fact of compliance with the promised environmental standards, let alone the impact of the social effectiveness of foreign investment in the developed areas (Slovo Kyrgyzstana, 2018). This situation applies to Chinese companies working day and night along the Ala-Buka-Chatkal road and inside the mountain ranges.

As for the social part of investment projects, the successful foreign gold mining company Kumtor Gold Company is a case in point. The Kumtor deposit was discovered 40 years ago by the Kyrghyzgeologiya geophysical expedition. Nevertheless, the development of this deposit during the Soviet Union was postponed several times for two reasons: high project cost (around $1.46 bln. at that time) and the reluctance to attract foreign investors for production needs since it could undermine the reputation of the USSR throughout the world economy as the leader of developed socialism (Slovo Kyrgyzstana, 2018).

On August 31, 1991, Kyrgyzstan, like all other republics, declared its independence. The process of transformation of socio-labor and production relations into market relations began. The course of the socially-oriented market economy was taken, and new norms and mechanisms of development of the national economy began to adapt. There appeared the first problems and difficulties in financing. The Kyrgyz Republic set itself ambitious tasks, the solution of which required enormous resources. Thus, there emerged the question of developing the gold deposit by attracting long-term foreign investment. After a thorough study of several proposals, the government gave its preference to Cameco, Canadian corporation renown worldwide for the development of uranium deposits. A strategic document on developing the goldfield between Kyrgyzstan and Canada on implementing a joint project, "Kumtor Gold Company," was signed on December 4, 1992 (Kumtor Gold Company, n.d). By 1997, the construction of a gold extraction plant was completed in the deposit, financial documents were developed and approved, and work on the production of gold raw materials began. In 1998 alone, one million ounces of gold were smelted. In 2002, gold production at Kumtor exceeded 100 tons (3.2 million ounces) (Kumtor Gold Company, n.d).

Analyzing the company's contribution to the comprehensive development of the region, we can note that the efforts of the government and local communities have established constructive cooperation between the Kyrgyz Republic and the gold mining company. Although, as shown by the more than twenty years, this period has seen all sorts of things, in particular, rallies and discontent of the local population (the Issyk-Kul Region), which were expressed to the authorities and the Kumtor Operating Company on the issues of social and economic development of these territories, compliance with environmental standards, environmental protection, reasonable distribution of mined gold, etc. Consequently, these processes positively impacted the implementation of the Sustainable Development Strategy of the Issyk-Kul Region. This strategy includes the following benchmarks and priorities for the future:

- Support for agricultural producers and the development of agro-industrial production;
- Promotion of the development of business structures and SMEs;
- Finding ways to make rational use of the territory's human and natural potential;
- Pursuing a policy of compliance with environmental norms and standards.

All business projects and investment initiatives supported by Kumtor must be financially independent and sustainable, bringing financial benefit to local communities and the public by 2026 (the end of the company's operations). Looking at the company's contribution to the economy of the Kyrgyz Republic from 1994 to 2017, the payments exceeded $3 billion 472 million (Table 1).

The analysis of Kumtor's contribution to the economy of Kyrgyzstan showed a solid share in industrial production, which amounted to 21.1% of the total, and 9.7% of the country's GDP. This fact once again confirms the importance of this project in the development of not only the individual Issyk-Kul Region, where gold extraction works are carried out but also the entire Kyrgyz Republic since the official statistical

Table 1 The contribution of the Kumtor Gold Company to the economy of Kyrgyzstan from 1994 to 2017 ($ mln)

No	Name of indicators	For 2017	1994–2017
1	Tax, customs, and other mandatory payments	99.314	1083.889
2	Payments to the Social Fund of the Kyrgyz Republic	20.049	187.845
3	Payments to the Issyk-Kul Development Fund	6.429	62.634
4	Payments for licenses and permits	0.423	3.809
5	Environmental payments and contributions to the Environmental Protection Agency	0.310	5.299
6	Payments to Kyrgyzaltyn OJSC	0.521	13.340
7	Refining fee	4.364	55.265
8	Proceeds from the sale of Centerra shares	–	86.000
9	Dividends	–	93.315
10	Procurement in the Kyrgyz Republic:		
	– supplies and services	44.468	771.727
	– food	4.317	69.540
11	Payments related to the infrastructure of the Kyrgyz Republic:		
	– electricity	12.869	179.470
	– improving and maintaining roads outside the mine	1.742	44.487
	– Tamga-Kumtor power line	–	41.612
12	Net salary of the employees of the Kumtor Gold Company	66.428	695.627
13	Treatment in health centers	–	0.744
14	Educational support, scholarships, and training	0.581	6.398
15	Sponsorship and sustainability projects	0.916	29.039
16	Assistance to the Government in implementing social projects	–	31.000
17	Cancer Service Support Fund	7.000	7.000
18	Agreement on Community Payments	–	4.400
19	Total	269.731	3472.44

Source (Kumtor Gold Company, n.d)

data is currently presented in two versions: with the inclusion of Kumtor's indicators and without them.

At the same time, the social nature of the project required the participation of local personnel in the extraction of gold reserves. This issue was also solved with difficulty since foreign investors invited "gold experts," but the demands of residents did their job. Currently, 97% of the company's personnel are locals, and about 3% are foreign specialists. Nowadays, the number of employees reaches 3778, including 2676 permanent staff, of which 2618 people are citizens of Kyrgyzstan, and 58 people are foreign specialists. The number of employees of contractors working at the mine is 1102 (Kumtor Gold Company, n.d.).

3 Results and Discussion

Continuing our discussion on the social nature of investment projects, it is necessary to indicate that local and senior authorities should consider the social usefulness of long-term programs for the development of deposits of non-ferrous metals, in particular, gold. There are still about a hundred foreign mining companies operating on the territory of Kyrgyzstan. These companies do not consider or do not want to understand the need to strengthen policies to increase the social importance of investment. The officials may also be involved in this.

Thus, the investment policy of Kyrgyzstan does not exclude the attraction of foreign investment in conditions of shortage of financial and material resources (Sayakbaeva et al., 2019) since foreign capital is a more available financial source than domestic investment due to the lack of governmental guarantee on their part. In our opinion, the main attractive priorities for investment are the labor and natural potential of the territories and the transit routes and communications in the Kyrgyz Republic, which allow for the implementation of a comprehensive economic mechanism covering all aspects of regional development.

According to the formed practice, foreign investors show a particular interest in entrepreneurship in those sectors, the expected results of which are not long to wait. These sectors are mainly oil and gas, energy, and mining industries.

Analyzing the dynamics of long-term foreign investments in the Kyrgyz Republic for 2014–2018 (Fig. 1), we notice a comparative reduction of investments by 21.6%. The most attractive period for investment was in 2015 ($1573.2 million) when Kyrgyzstan became a full-fledged EAEU partner (Fig. 1). After this period, there was some reduction in the volume of foreign investment caused mainly by the persistence of the unfavorable shadow political situation in the country and the high level of corruption of the upper level of government, which affected the investment attractiveness and living standards of the people living in Kyrgyzstan.

Nevertheless, we should consider the role of foreign investment in radical structural changes during the formation and adaptation of market relations with the ensuing circumstances associated with a flexible business policy, equipment with

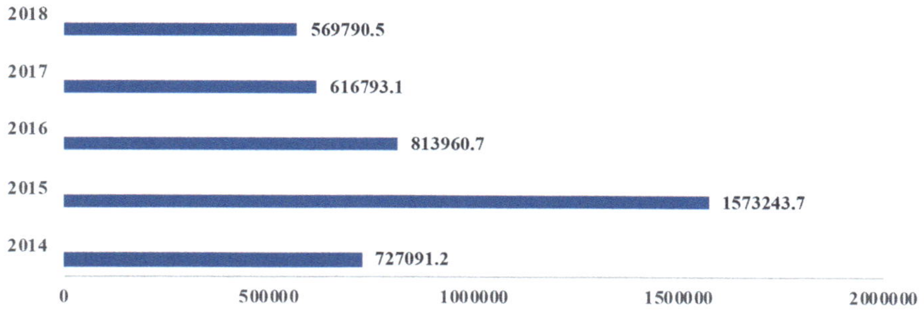

Fig. 1 Analysis of foreign investment capital inflow to the Kyrgyz Republic for 2014–2018, $ thousand. *Source* (Sayakbaeva et al., 2019)

the advanced technology of the latest generation, competent professionals, risk managers, and relevant challenges of the modern economy.

Methodology for assessing the social effectiveness of projects. There is currently a fairly approximate methodology for determining the social effectiveness, which can calculate the coefficient of social effectiveness of investment projects. In our opinion, no project, whether domestic or foreign, should be approved without such calculations reviewed at the state and local levels. Before an investment project is implemented, it must pass one of the important stages—a comprehensive performance evaluation. This stage involves a deep and detailed diagnosis of technical, economic, and financial information for the final decision (Stoyanova, 2000). Therefore, started from 2018, the Kyrgyz Republic began developing regions and border areas, where it is necessary to pay attention to the economic efficiency of attracted investments and their social side.

In practice, the diagnosis of the social nature of investment projects should be measured in monetary terms and used to calculate economic efficiency. In our case, it is necessary to strengthen control over compliance with social norms and standards, which should become mandatory conditions in implementing investments in mining regions, particularly regions with gold reserves.

An important indicator in assessing the coefficient of social effectiveness in implementing an investment project is the number of jobs created, which are adjusted for the situation in the regional or national labor markets in the areas of the implementation of investment projects. Let us give a methodology for determining the coefficients of social efficiency. The economic literature presents various methods, which provide a real opportunity to get a clear picture of the level of social or public utility of the implemented investment projects. We believe that the proposed formulas should be present for each investment project to be implemented. Unfortunately, these formulas are not taken into account in practice.

In the case of Kyrgyzstan and its regions, we consider it necessary to use methods calculating social efficiency with regional utility, taking into account the whole range of interrelated and interacting endogenous-exogenous effects (social, political, ecological, etc.), relevant costs, and characteristic results (Stoyanova, 2000), which require a competent accounting policy.

In turn, the policy of the organization of cost accounting is the basis for the formation of production accounting. This policy is influenced by serious factors determining the role of the responsibility center and cost carriers as a prerequisite for controlling. The practice of production accounting allows economic entities, in our case, the company, to calculate the structure of the production costs (Taigashinova & Nuralieva, 2018), as well as to diagnose the expected social effect of the production of valuable metal and maintain appropriate accounting control over their use in the regions.

Consequently, the territories are interested in the level of the positive impact of the investment project on their development and well-being. The proposed coefficient of regional utility allows us to determine the interaction of the following indicators:

1. Provision of the studied region with services, goods, and resources;
2. Provision of investment in fixed capital per one inhabitant of a particular territory;
3. Level of the ratio of the average annual values of prices for the services rendered, which are expressed in the Formula (1):

$$\boldsymbol{C}_{reg.ul.} = \left[\boldsymbol{\alpha}_{serv.av.} + \boldsymbol{\beta}_{invest.av.+serviceprice.}\right]/3, \tag{1}$$

where

$\boldsymbol{C}_{reg.\ ut.}$—coefficient of regional utility;

$\boldsymbol{\alpha}_{serv.\ av.}$—ratio of per capita availability of services provided in the course of project implementation (the Kyrgyz Republic to a specific region);

$\boldsymbol{\beta}_{invest.\ av.}$—ratio of the availability of investment in fixed capital per capita (the Kyrgyz Republic to a specific region);

$\boldsymbol{\delta}_{service\ price}$—ratio of average annual price levels for services provided during the project implementation, per one service (a specific region to the Kyrgyz Republic).

In practice, the social effect of a particular investment project is calculated using its integral indicator. This integral indicator is calculated according to the following formula:

$$\boldsymbol{J}_{soc.ef.} = \boldsymbol{I}_{soc.ef.} \times \boldsymbol{C}_{reg.ut.}, \tag{2}$$

where

$\boldsymbol{J}_{soc.\ ef}$—integral social effect of the investment project;

$\boldsymbol{I}_{soc.\ ef.}$—social performance indicator;

$\boldsymbol{C}_{reg.\ ut.}$—coefficient of regional utility.

It should be noted that the proposed calculation serves as the basis for the assessment of the expected social effect of investment, the social nature of which is mirrored in the quality of life of people and their well-being.

Thus, the use of the above methods cannot be definitive and universal due to their diversity. In our opinion, these methods can be used as mechanisms for selecting the most profitable variants of long-term investment projects from among the proposed. Moreover, these methods can be used for diagnostics or assessment of the degree of social investment of working projects.

The assessment of the social effectiveness of an investment project, namely, the calculation of the coefficient of regional utility ($\boldsymbol{C}_{reg.\ ut.}$) and integral social effect, gives us an ample opportunity for accounting and financial control over mining, particularly in the development of gold deposits, where the parties concerned can get transparent and accessible financial information. Nevertheless, such accounting and control are not carried out now, the reason for which may be the potential profit to foreign investors.

Diagnostics of economic and social utility and profitability of investment business projects is necessary to choose the most optimal and profitable option. It serves as a mechanism for preliminary minimization of the expected risk (Stoyanova, 2000). The assessment of the level of social nature of an investment project through a detailed study of unemployment and employment in a particular territory also requires particular attention (Chernyak et al., 2003).

The question of the social efficiency of the investment project is touched upon since the world economy considers investment projects in terms of their focus on improving the quality of life and ensuring sustainable growth of territories.

In turn, the social results of investment projects should be mirrored in the following areas (Kulueva, 2015):

- Increasing the number of jobs;
- Improvement of housing and communal conditions;
- Improvements in the sphere of cultural and social services;
- Changes in the direction of simplifying working conditions for employees and compliance with occupational safety (health and safety);
- Improving the level of logistics associated with the interrupted supply of residents of hard-to-reach areas with necessary products and resources (fuel, energy, food, etc.);
- Improving the health of local residents and workers by creating innovative equipment, advanced technologies, and medical health centers;
- Improving transportation and other means of transportation;
- Improving information services (Internet);
- Improving the location of the trade network, the creation of civilized trade centers;
- Digitalization of the region, increasing the availability and development of Internet communication, e-mail, and other types of communication (Kulueva et al., 2020).

In our opinion, the formation of new models of development of the information society with the help of digitalization will enable the development of a new type of economy—knowledge economy and regional economy, which will require the development and adoption of radical and adequate economic solutions to gain real access to the world economy (Ismailova & Mombekova, 2016).

In our case, when implementing investment projects (especially in 2019, which was declared by the President of the Kyrgyz Republic as the year of development of regions, digitalization of society and economy, and the beginning of the radical change in the regional development of the country), it is indicated that the long-term goals of strategic development of the Kyrgyz Republic are inextricably linked primarily with the objectives of improving the quality of life of citizens throughout the country (Kulueva, 2015). Based on the above, it is necessary to consider the existence of the following important indicators of social efficiency in investment:

1. Social efficiency determining the level of satisfaction of the population with the quality of life;
2. Social effect characterizing the degree of improvement in the quality of working life of people;
3. Indicator of socio-economic effectiveness, which gives a certain idea of the economic efficiency of long-term investment resources in the social sphere of rural and border areas, taking into account the positive result obtained and the social effect achieved (Kulueva, 2015).

Consequently, exploring this topic, we must understand that gold deposits are the wealth of the people and the country. We believe that there are great opportunities for solving the following problems at the state level:

- Proper and rational use of financial resources will restore the former glory of domestic industrial production;
- Covering the external debt (over $4.8 billion);
- Introduction of a separate section on the calculation of the coefficient of social efficiency in investment projects will ensure a decrease in the level of migration activity from rural and border areas, which will make it possible to maintain labor activity in the Kyrgyz Republic.

Nevertheless, such efforts should be supported by appropriate foreign and domestic public policies since Kyrgyzstan is at the initial stage of digitalization of society (Kulueva et al., 2019) and the economy on the way to globalization, where an essential role in addressing the problems of improving the quality of life of working people and the human resources of the country must be addressed by the hierarchy and respected in accordance with influencing factors at levels of the world economy, country, and regions (Table 2).

4 Conclusion

In our opinion, the government has no right to allow a unilateral decision in favor of investors. It should pursue a policy of protectionism in ensuring the socio-economic development of territories, improving the welfare of residents, and increasing the

Table 2 Hierarchical factors influencing the solution of problems of effective use of the Kyrgyzstan resource potential

At the world economy level
– Impact of globalization on the quality of labor and life in the national economy of Kyrgyzstan determined by the level of communications, external relations, and foreign trade – Openness of the Kyrgyz economy in the use of best foreign experience in the interaction and interconnection of resource potential structures of the Kyrgyz regions – Need for an integrated policy of scientific and educational cooperation and interaction between economies of the world economy – Degree of the response of the Kyrgyz economy and regions to exogenous interactions – Changing state of the values of labor law and investment cooperation – Intercountry division of labor and the degree of labor cooperation – Cyclical nature of the world economy, global economic crisis
At the Kyrgyz Republic level
– Development and perfection of labor legislation – State (antitrust) regulation of competition – Investment and tax policies supporting entrepreneurial and business structures – Tendency to concentrate and reorganize production – State support for entrepreneurship and business – Public social policy – Standard of living of the population, income, and the degree of real income – Development of labor relations and social partnership
At the level of regions and provinces
– Decent pay and regular income growth – Increasing the investment attractiveness of the real sector industries – Organization of competition for the selection of effective and profitable business projects for the development of regions (districts) – Development of sources of domestic investment – Regional development based on the project approach – Availability of social protection and employment guarantees, regulation of the rights and obligations of working people – Availability of a high-tech and productive workplace – Participation of employees in the management of business entities – Real and potential opportunities for qualification and professional growth – State of the moral and psychological climate in work collectives – Satisfaction of the economically active population with work – Degree of self-realization and self-expression in work

Source Compiled by the authors

quality of their life, which should positively impact the reduction of the migration activity of the able-bodied population.

Over the past two decades, Kyrgyzstan has directed all efforts toward implementing the socially-oriented market model chosen at the beginning of market transformations. The current model exacerbates the unanswered social problems. It is time to move from a "resource" country to the industrial model of development of the market economy with subsequent digitalization, proposed as a model of transforming the next development stage of the national economy (Ministry of Digital Development of the Kyrgyz Republic, n.d.). The taken cardinal measures under the globalization conditions will probably positively affect the formation and development of equal partnership relations during the production and sale of goods and services. Moreover, these measures may increase investment attractiveness, reduce migration outflow of able-bodied population, raise the position of labor migrants in the countries of near and far abroad to the civilized level, increase the attractiveness of domestic economy for its citizens, and improve the quality of villagers' life.

References

Andersson, A. E. (1986). The Four logistical revolutions. *Papers of Regional Science Association, 59*(1), 1–12.

Andersson, A. E., & Batten, D. F. (1987). Creative nodes, logistical networks, and the future of the metropolis. *Transportation, 14*, 281–293.

Chernyak, V. Z., Chernyak, A. V., & Dovdienko, I. V. (2003). *Business planning: Practical training manual*. Publishing House RDL.

Faculty of International Economic Relations of Belarus State Economic University. (n.d.). BSEU Blog. Retrieved from https://economy-web.org/

Government of the Kyrgyz Republic. (2017). Regional Policy Concept of the Kyrgyz Republic for the period 2018–2022 (Approved by decree of March, 31 2017 No. 194). Bishkek, Kyrgyz Republic. Retrieved from http://cbd.minjust.gov.kg/act/view/ru-ru/99907

Ismailova, K. J., & Mombekova, G. B. (2016). *Problems of development of a new educational direction Business informatics* (Vol. 2, pp. 40–46). Bulletin of Osh State University. Retrieved from https://www.oshsu.kg/univer/temp/url/ilim/2016-2.pdf

Kulueva, C. R. (2009). Problems of labor resources in the Kyrgyz Republic. In *International scientific-practical conference dedicated to the 15th anniversary of the Kazakh Academy of Labor and Social Relations: Theoretical and practical aspects of socio-economic and political development of the Republic of Kazakhstan, Central Asia and the CIS countries at the present stage* (pp. 34–42). Almaty, Republic of Kazakhstan.

Kulueva, C. R. (2015). Some issues of interaction of investment policy and labor in the regions of Kyrgyzstan. *Economy and Society, 6–3*(19), 499–509.

Kulueva, C. R. (2016). Does Kyrgyzstan need labor protectionism policy: Regional aspect. *Current Problems of the Humanities and Natural Sciences, 4–2*, 68–74.

Kulueva, C. R., Ubaidullayev, M. B., Ismanaliev, K. I., Kuznetsov, V. P., & Romanovskaya, E. V. (2020). Digitalization of Kyrgyz society: Challenges and prospects. In E. Popkova, & B. Sergi (Eds.), *The 21st century from the positions of modern science: Intellectual, digital and innovative aspects*. Springer. https://doi.org/10.1007/978-3-030-32015-7_26

Kumtor Gold Company. (n.d.). *History*. Retrieved from https://www.kumtor.kg/en/deposit/history/

Ministry of Digital Development of the Kyrgyz Republic. (n.d.). *The concept of the digital transformation of the Kyrgyz Republic "Digital Kyrgyzstan," 2019–2023*. Retrieved from http://ict.gov.kg/index.php?r=site%2Fsanarip&cid=27

Sayakbaeva, A. A., Akylbekova, N. I., & Nurmukambetova, J. K. (2019). Impact of foreign investment on the development of enterprises in the Kyrgyz Republic. *Eurasian Scientific Association, 10–4*(56), 334–338.

Slovo Kyrgyzstana. (2018, January 12). *2018 has been declared the year of regional development*. Retrieved from http://slovo.kg/?p=95213

Stighlitz, J., Sen, A., & Fitussi, J.-P. (2016). Evaluating our life incorrectly. In *Why doesn't GDP have sense?* Publishing House of the Gaidar Institute.

Stoyanova, E. S. (Ed.). (2000). *Financial management: Theory and practice. Textbook* (4th ed.). Publishing House Perspektiva; Urait.

Taigashinova, K. T., & Nuralieva, C. A. (2018). Aspects of the regulation of accounting in the Republic of Kazakhstan and Kyrgyzstan. In T. M. Mezentseva, & V. L. Nazarova (Eds.), *Current problems of accounting and reporting in Russia and the Commonwealth of Independent States* (pp. 264–269). Financial University under the Government of the Russian Federation; Almaty, Republic of Kazakhstan: Almaty Academy of Economics and Statistics.

Toktorov, K. K., Batyr uulu A., & Mametova, G. A. (2020). Experience of foreign countries on innovation activity of agricultural subjects. *Actual Issues of the Modern Economy, 3*, 116–125. https://doi.org/10.34755/IROK.2020.37.77.004

Wilson, A. G. (1981). *Catastrophe theory and bifurcation: Application to urban and regional systems*. University of California Press.

Yakovleva, L. L. (2018). Indicative planning model of regional social sector. *Economics of Contemporary Russia, 1*(80), 84–95.

Zhang, W.-B. (1999). *Synergetic economics: Time and change in non-linear economics* (N. V. Ostrovskaya Transl. from English; V. V. Lebedev, & V. N. Razhevaykin Eds.). Mir.

World Nuclear Power in the Context of Sustainable Development

Guzel G. Karachurina, Yulia P. Vassilyeva, Albina V. Gaisina, Irek A. Khisamutdinov, and Tatiana F. Mantserova

Abstract The paper provides a retrospective analysis of the nuclear power industry in the world over the past three decades in the context of sustainable development on the example of major electricity producers. Additionally, the paper outlines the role of nuclear power in other major types of carbon-free energy sources. The authors defined the dynamics of nuclear energy production, outlined the main factors of global changes in demand in the electric power industry, and identified the main flows of the world's electricity exports. The authors show that Europe, the Middle East, and East and South Asia are most active in exploiting nuclear capacity. Since nuclear power originated on the territory of the former Soviet Union, the authors pay particular attention to studying the dynamics and development of nuclear power in Russia. It is shown that nuclear power has always been in the spotlight in Russia since its progress contributes to implementing the innovative strategy in its economy. The paper aims to examine the state of the nuclear power industry in the world in the context of sustainable development. To achieve the set goal, the authors define the dynamics and trends of nuclear power in the international context and indicate the feasibility of its development in the socio-environmental and economic system. The research subject is the world's nuclear power industry. The research object is the countries of the world, namely the largest producers of electricity. Based on the conducted research, the authors offer recommendations to countries seeking to preserve the possibility of utilizing nuclear energy.

G. G. Karachurina (✉) · Y. P. Vassilyeva · A. V. Gaisina · I. A. Khisamutdinov
Ufa State Petroleum Technological University, Ufa, Russia
e-mail: karachurina@bk.ru

Y. P. Vassilyeva
e-mail: vasilevajulia@bk.ru

A. V. Gaisina
e-mail: a_gaisina@bk.ru

I. A. Khisamutdinov
e-mail: hisamutdinov07@bk.ru

T. F. Mantserova
Belarusian National Technical University, Minsk, Republic of Belarus

E. G. Popkova and B. S. Sergi (eds.), *Geo-Economy of the Future*,
https://doi.org/10.1007/978-3-030-92303-7_30

Keywords Nuclear power · Sustainable development · Export · Nuclear reactors · Electricity

JEL Codes F180 · Q420 · Q510

1 Introduction

The global reason that influenced the worldwide development of nuclear power is associated with the concept of the transition to Sustainable Development Goals, which emerged in the 1990s. These principles imply the achievement of economic performance and the improvement of the quality of life without compromising the environment, that is, "conflict-free" economic development. The concept of sustainable development was adopted at the UN Conference on Environment and Development held in Rio de Janeiro in 1992.

In the early days of nuclear power, there was a significant reduction in the negative impact on the environment in terms of "mitigating" processes that deplete the natural resource potential of the environment, since nuclear power does not use hydrocarbon raw materials (which are currently extracted in increasing amounts), and in terms of significantly reducing the pressure on the natural-assimilative capacity of the environment, since it minimizes the amount of waste.

Nuclear power contributes to many of the Sustainable Development Goals set by the global community, including achieving lasting long-term energy security.

2 Methodology

The works of Russian and international scholars dealing with the problems of the world economy, economic geography, geopolitics, energy security, energy policy, distribution of productive forces, and industrial relations significantly contributed to the study of the problem of the world nuclear industry (Ustinova et al., 2016; Manolov et al., 2020; Karachurina, 2011; Kudryavtseva et al., 2018; Vanchukhina et al., 2019; Peskova et al., 2019; Gaysina, 2020; Gagarinskii et al., 2005; Adamantiadesa & Kessides, 2009). In particular, Kudryavtseva and Velekhov consider the principles of sustainable development in the field of nuclear energy. Ponomarev-Stepnoi studies the role of nuclear power in the structure of world energy production. Adamantiades and Kessides highlight the current state of nuclear power, opportunities for its expansion, and recent advances in nuclear reactor technology.

Additionally, the collection of statistical information and analysis of the world's nuclear power industry is carried out by leading international (e.g., International Energy Agency (IEA), World Nuclear Association, International Atomic Energy Agency, Nuclear Energy Agency, etc.) and national organizations (e.g., U.S. Energy Information Administration, Agora, Federal State Statistics Service of the Russian

Federation, etc.) (IEA, n.d.; World Nuclear Association, n.d.; International Atomic Energy Agency, n.d.; U.S. Energy Information Administration, 2020; PJSC "INTER RAO," n.d.; Redl et al., 2021; Analytical Center under the Government of the Russian Federation, 2020; Khisamutdinov, 2013).

The methodological basis of the research is formed by such areas of economic science as economic theory, economic statistics, etc. The empirical, analytical, and comparative methods were used in solving the research tasks. The importance of nuclear power in the world is analyzed based on value judgments.

The authors reveal that the steady increase in society's needs for fuel and energy, with limited renewable resources, prompts the development of the world's nuclear power industry.

3 Results

3.1 Development of the World Nuclear Power Industry

The world energy space is in a constantly changing environment, which requires a constant review of the strategy for energy market development. The COVID-19 pandemic has reinforced the need to support Sustainable Development Goals and has turned the vector of the energy market in favor of low- or no-carbon energy. According to forecasts of leading international organizations, particularly the IEA, the global demand for primary energy by different types of fuel may change by 2030 (Fig. 1).

By 2030, the demand for coal, oil, and traditional biomass use is negative (respectively, -1531 Mtoe, -562 Mtoe, and -588 Mtoe). In turn, demand for renewable and nuclear energy is positive (respectively, 1514 Mtoe and 168 Mtoe). This prediction

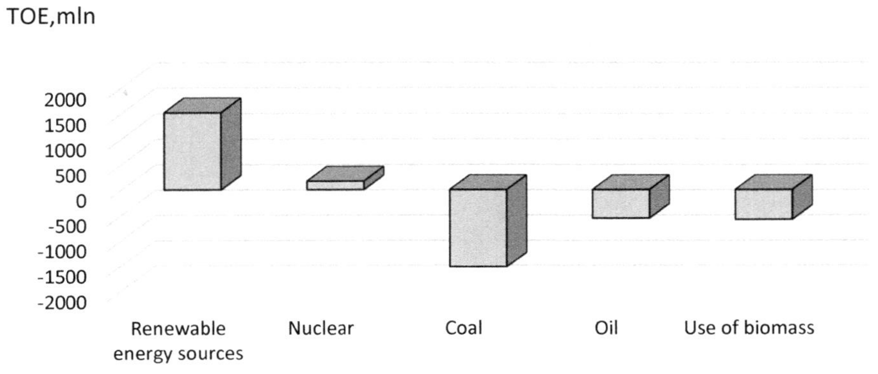

Fig. 1 Change in global demand for primary energy by fuel type in the sustainable development scenario, 2030 versus 2019. *Source* Compiled by the authors based on (IEA, n.d.)

confirms the importance and necessity of researching nuclear power and setting it apart from all other fuels.

It is advisable to consider nuclear power as part of the other major carbon-free sectors (e.g., hydropower, wind power, and solar power). Thus, on the example of OECD countries, we can see that nuclear energy has a large share, with hydropower in second place, wind power in third, and solar power in fourth place (Fig. 2).

These types of energy have their pros and cons. Nevertheless, they all serve the same purpose—to achieve the concept of sustainable development, a key indicator of which is the reduction of carbon dioxide emissions. Nowadays, 17% of the world's electricity production comes from nuclear power plants, which produce electricity using nuclear reactors (Fig. 3).

According to the World Nuclear Association (World Nuclear Association, n.d.), in 2019, nuclear reactors worldwide produced 2657 TWh of electricity. Worldwide nuclear power production increased by 4% due to the resumption of production in Japan and South Korea. The most significant increase in nuclear power production compared to 2018 was observed in the Asian region (+17%). Africa, South America, Eastern Europe, and Russia also showed significant growth (Fig. 4).

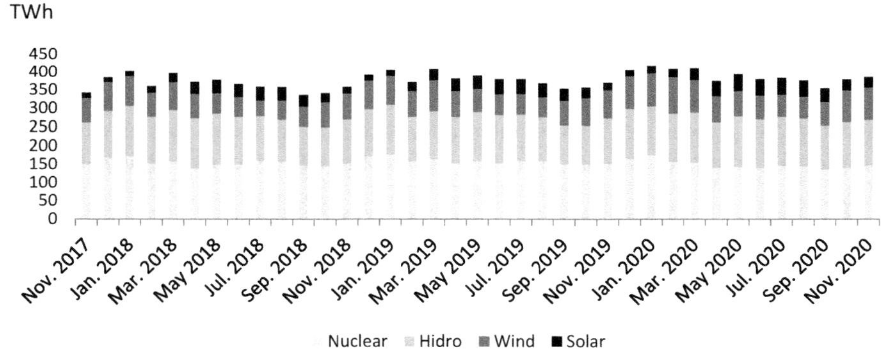

Fig. 2 Production of electricity in OECD countries by type of hydrocarbon-free fuel, TWh, (November 2017–November 2020). *Source* Compiled by the authors based on (IEA, n.d.)

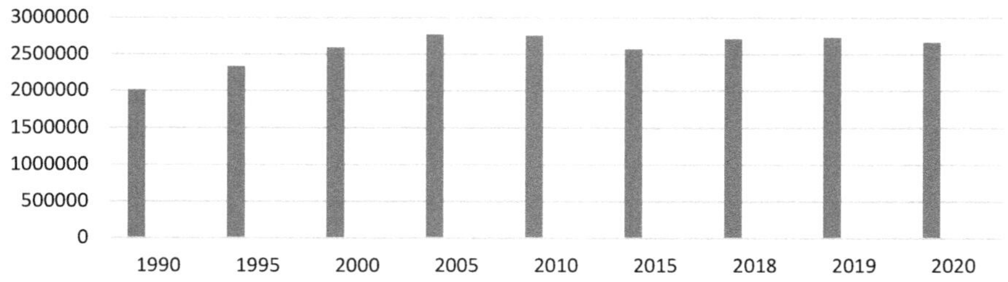

Fig. 3 Electricity production at nuclear power plants worldwide, GWh, 1990–2020. *Source* Compiled by the authors based on (World Nuclear Association, n.d.)

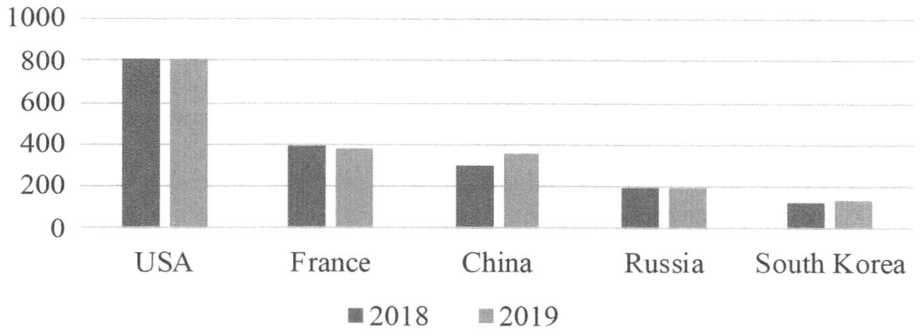

Fig. 4 Nuclear power generation, billion kWh. *Source* Compiled by the authors based on (World Nuclear Association, n.d.)

According to the World Nuclear Organization, 442 nuclear reactors worldwide will produce 393,064 MWh of electricity in 2021 (World Nuclear Association, n.d.). The largest producers of nuclear electricity are as follows:

1. With a 19.7% share of nuclear power, the USA generated 2657 billion kWh of electricity in 2019 (an increase of 0.17% compared with 2018). The USA is the largest producer of nuclear energy (a global share of 30%). This position is ensured by 95 reactors with a total capacity of 97.2 GW. According to the data of the Energy Information Administration of the United States (EIA) on the work of power plants, 9 out of 10 power plants in the USA that produced the highest amount of electricity in 2019 were nuclear plants. These ten power plants produced a total of 230 million MWh of electricity in 2019, representing 5.6% of all electricity production in the United States (U.S. Energy Information Administration, 2020). In 2020, the country exported 9.965 million kWh of electricity.
2. With a 70.6% share of nuclear power, France generated 382.4 billion kWh of electricity in 2019, reducing production by 3.41%. In 2020, the German think tank Agora calculated that nuclear power generation in France fell by 11% due to the shutdown of the main generating company EDF (Fig. 5). Across the EU,

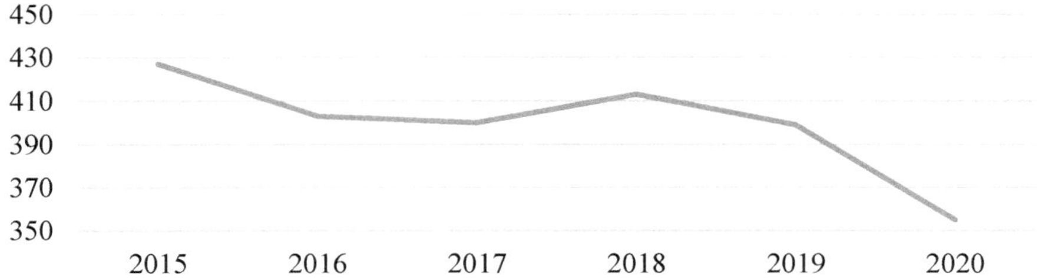

Fig. 5 Generation of nuclear electricity in France, TWh. *Source* Compiled by the authors based on (PJSC "INTER RAO," n.d.)

the reduction was 10%. The history of nuclear power in France dates back to October 1987. There are currently 56 operating nuclear reactors. The French government plans to reduce the share of nuclear power in the country's energy mix to 50% by 2035, which will require the closure of 14 reactors (International Atomic Energy Agency, n.d.).

France is currently the largest global exporter of electricity, with export revenues of more than 3 billion euros per year. In 2020, the country exported 61410 million kWh of electricity.

3. With a share of nuclear power of 18.3%, China generated 361.2 billion kWh of electricity, showing a growth of 18.89%. The first nuclear power plant in China was laid in 1970. Nowadays, China has 52 operating reactors, of which 48 are located in mainland China and four in Taiwan. The country plans to increase the share of nuclear power in the country's total energy mix to 28% by 2050. In 2020, the country exported 18,910 million kWh of electricity.
4. With a 19.7% share of nuclear power, Russia generated 208.8 billion kWh of electricity in 2019. Over the past decade, the generation of nuclear power has grown by 22.5%. The first nuclear reactor in Russia (USSR) was launched in 1964. Nowadays, there are 38 operating reactors at 11 nuclear power plants in Russia. According to Inter RAO Group, Russia exported 11.707 billion kWh of electricity in 2020. The largest export destinations include Lithuania (3.143 billion kWh), China (3.060 billion kWh), and Finland (2.637 billion kWh) (PJSC "INTER RAO," n.d.).
5. With 26.2% of nuclear power, South Korea generated 138.8 billion kWh of electricity in 2019 (+9.2%), which makes it the fifth-largest nuclear power producer in the world (Fig. 6). The first nuclear power plant was launched in 1977. Currently, South Korea has 24 reactors with a total capacity of 23.2 GWh.

The year 2020 impacted all areas of the global economy, including the electric power industry, where there has been a significant global drop in demand. Wind and solar PV production continued to grow at more than 10% and 20%, respectively, providing more than 9% of the world's electricity supply (about 8% in 2019). In

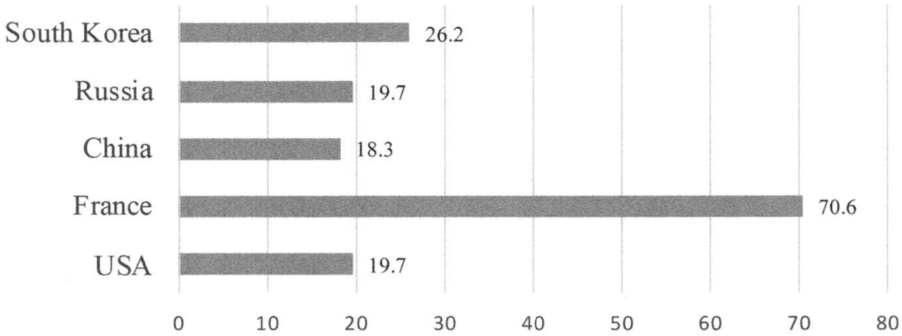

Fig. 6 Share of the nuclear power industry, 2019, %. *Source* Compiled by the authors based on (IEA, n.d.)

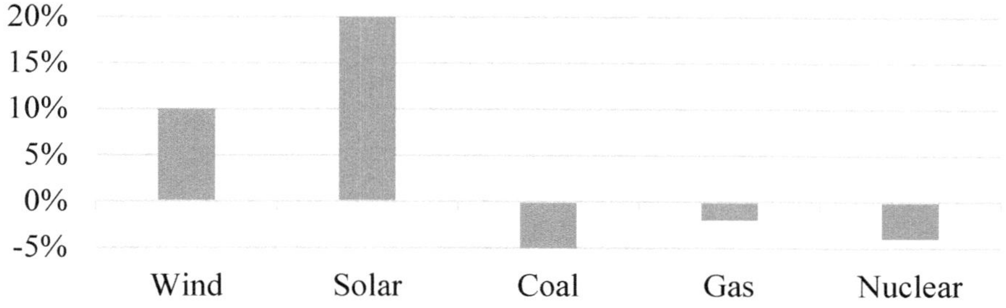

Fig. 7 Global power change in 2020. *Source* Compiled by the authors based on (IEA, n.d.)

total, renewable energy grew by about 7% in 2020 and provided 28% of the world's electricity (about 2% more than in 2019). The growth of renewable energy and the drop in demand have led to a decline in other generation technologies, namely coal, gas, and nuclear power (Fig. 7). Coal generation suffered the most. According to preliminary data from the International Energy Agency, in 2020, global energy production from coal was down more than 5%, compared to 3% in 2019. Energy production from gas decreased by about 2%. The decline in nuclear power was about 4%. The most considerable decline in the demand for nuclear power occurred in the first half of the year. The nuclear power supply also declined due to plant closures in France, Sweden, Germany, Switzerland, and the USA in 2019–2020. The exception to global trends was China, where nuclear power generation increased by about 6% due to the commissioning of new facilities. Oil, which accounts for only 3% of global electricity production, was also affected by the shrinking thermal generation market. Thus, many markets saw declines partially offset by some growth in the Middle East.

All over the world, there are programs aimed at the long-term operation and management of the aging of nuclear reactors. According to the International Atomic Energy Agency, by early 2020, more than 66% of the world's nuclear power capacity has been in operation for three decades or more, and 16% of the world's capacity has been in operation for more than four decades (Fig. 8) (International Atomic Energy Agency, n.d.).

In 2020, the nuclear power industry saw the commissioning of new capacity. These are new power units in China, India, Russia, Belarus, Korea, Slovakia (potentially postponed until early 2021), Turkey (construction of the third unit), and the United Arab Emirates. These power plants add more than 8 GW of new capacity (most of which is planned to launch by the end of 2020), offsetting the closure of units in France, Sweden, and the USA.

However, the need to commission new facilities and their safety is the subject of worldwide debate. One of the most striking examples of the above-mentioned countries is the Republic of Belarus, where the Belarusian Nuclear Power Plant (BelNPP) is in the final stage of construction. Currently, the readiness of Unit 1 of the BelNPP is 99%; it is going through the stage of pilot operation. Unit 1 has already produced the first billion kWh of electricity. The readiness of Unit 2 is 78%; it is going through the stage of pre-start-up adjustments. With the commissioning of

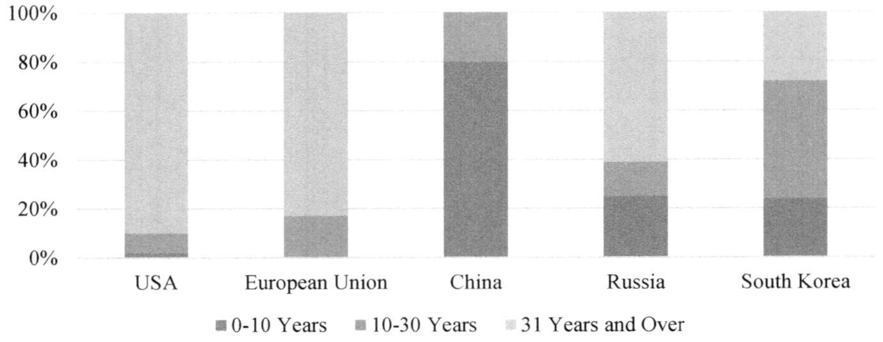

Fig. 8 Age of nuclear power capacity. *Source* Compiled by the authors based on (Redl et al., 2021)

two units, the electricity production will equal 18 billion kWh per year, which will replace 4.5 billion cubic meters of natural gas annually. As of January 1, 2021, the installed capacity of Belenergo's generating sources was 8947.31 MW. The share of nuclear power, taking into account the commissioning of the BelNPP into commercial operation, could reach about 25% in installed capacity and about 51% in electricity generation (Fig. 9).

However, in February 2021, the European Parliament passed a resolution demanding that the launch of the BelNPP be suspended because of safety concerns. Due to these circumstances, the Republic of Belarus is at risk of experiencing a surplus of capacity and production of excess electricity. Currently, in Belarus, several government decisions are being made to encourage domestic energy consumption, particularly electric boilers and electric cars.

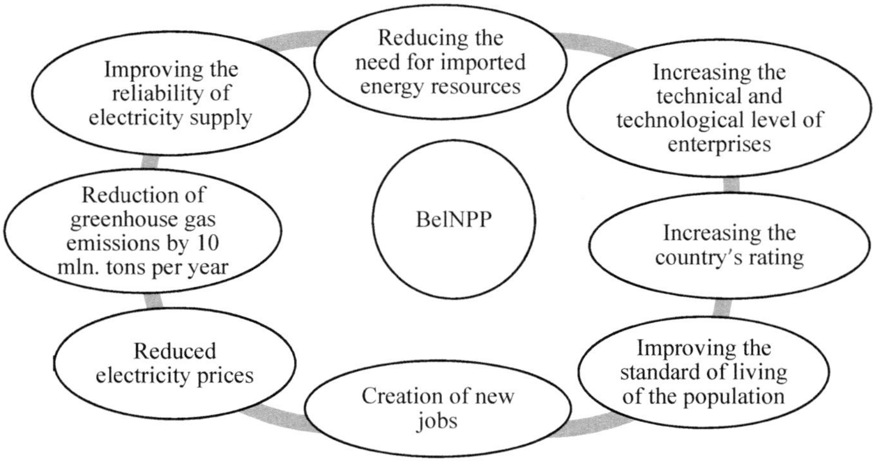

Fig. 9 Benefits of operating the BelNPP. *Source* Compiled by the authors

3.2 Development of Nuclear Power in Russia

The beginning of the emergence of nuclear power in Russia and its further development is associated with the decision of the Soviet government to start work on the "atomic project," adopted on August 20, 1945.

Together with the work on atomic weapons in the USSR, a considerable amount of work was performed on implementing atomic energy for peaceful purposes, including developing nuclear power and constructing nuclear power plants. In 1954, the world's first nuclear power plant was launched in Obninsk, the Kaluga Region.

In the 1970s, the USSR developed a large-scale program for the construction of nuclear power plants. By the end of the 1980s, the entire European part of the USSR was to be covered by a dense network of nuclear power plants. By 1990, the USSR had 46 power units at 15 nuclear power plants. However, a serious accident at a nuclear power plant in the USA in 1979 and an accident in Chernobyl in 1986 in the USSR significantly slowed down the development of nuclear energy in the USSR (Russia) and the world.

After the aforementioned accidents, the degree of safety of nuclear power plants left much doubt about their "conflict-free" nature to the environment and quality of life of the population. Thus, according to various expert estimates, the direct and indirect economic damage of the Chernobyl accident in current prices amounted to more than 950 billion rubles, not considering human losses (Khisamutdinov, 2013).

Over the last decade, nuclear power generation has grown by more than 20%. There are currently 36 power units at 10 nuclear power plants in Russia. Their electricity production has almost reached the level of the Soviet period of 205–215 billion kWh. According to the energy strategy of the Russian Federation, it is expected to increase energy production further and bring the share of nuclear power generation to 25% by 2035 (Government of the Russian Federation, 2020).

The optimism of Russia is based on the fact that it has always been strong in the field of energy, including nuclear energy. This optimism is also confirmed by real developments in modern domestic basic science. These developments allow the safe operation of nuclear reactors and meet the requirement of environmentally safe sustainable development.

4 Conclusion

Over the past five decades, nuclear power has reduced carbon dioxide emissions by more than 60 gigatons (almost two years of global energy emissions). However, in advanced economies, nuclear power has begun to decline, plants are closing, and almost no new investments are being made when the world needs more low-carbon electricity. To achieve a trajectory consistent with the Sustainable Development Goals, the expansion of clean electricity production must be three times faster than now. This requires about 85% of the world's electricity to come from clean sources by 2040.

In this context, countries that intend to retain nuclear power should consider the following measures:

- Allow extension of service life of existing nuclear power plants to the maximum safe life possible;
- Maintain the safe use of nuclear energy and accelerate innovation in the industry, manufacturers need to be subsidized from the state budget;
- The country needs to create a level playing field for nuclear power with other low-carbon energy sources;
- It is necessary to create mechanisms for risk management and financing of capital mobilization for the renovation of existing facilities and construction of new ones.

Acknowledgements The paper was prepared with the financial support of the Federal State Budgetary Educational Institution of Higher Professional Education "Ufa State Petroleum Technological University".

The authors would also like to express their gratitude to the International Energy Agency, the World Nuclear Association, and the International Atomic Energy Agency for providing access to the data information system, reports, bulletins, and other sources of statistical information.

References

Adamantiadesa, A., & Kessides, I. (2009). Nuclear power for sustainable development: Current status and future prospects. *Energy Policy, 37*(12), 5149–5166. https://doi.org/10.1016/j.enpol.2009.07.052

Analytical Center under the Government of the Russian Federation. (2020). Energy Bulletin, 2020. Retrieved from https://ac.gov.ru/publications/topics/topic/2290?page=3. Accessed February 22, 2021.

Gagarinskii, A. Yu., Ignatev, V. V., Ponomarev-Stepnoi, N. N., Subbotin, S. A., Tsibulskii, V. F. (2005). Role of nuclear power in world energy production in the 21st century. *Atomic Energy, 99*(5), 759–769.

Gaysina, A. V. (2020). The relationship between living standards and human capital development. *Eurasian Law Journal, 12*(151), 470–471.

Government of the Russian Federation. (2020). Energy Strategy of the Russian Federation to 2035 (June 9, 2020 No. 1523-r). Moscow, Russia. Retrieved from http://static.government.ru/media/files/w4sigFOiDjGVDYT4IgsApssm6mZRb7wx.pdf. Accessed February 22, 2021.

International Atomic Energy Agency. (n.d.). Official website. Retrieved from https://cnpp.iaea.org/. Accessed February 20, 2021.

International Energy Agency (IEA). (n.d.). Official website. Retrieved from https://www.iea.org/. Accessed February 20, 2021.

Karachurina, G. G. (2011). Factors and conditions for sustainable development of the region in the modern economy. *Innovation and Investment, 1*, 141–144.

Khisamutdinov, I. A. (2013). Socio-environmental and economic assessment of the region in the context of the transition to sustainable development. *Regional Economics: Theory and Practice, 4*(283), 9–15.

Kudryavtseva, O. V., Derkach, A. S., Manushko, S. V., Nesvetov, F. D., Pekarev, S. V., & Chetvertakov, V. S. (2018). Nuclear power in frames of sustainable development. *Scientific Research of Faculty of Economics. Electronic Journal, 10*(4), 33–49. https://doi.org/10.38050/2078-3809-2018-10-4-33-49

Manolov, G., Orlova, D., Khodkovskaya, J., Barkova, E., & Nazarov, M. (2020). A Smart specialization strategy for sustainable development of regions. In *E3S Web of Conferences* (Vol. 208, p. 08009). https://doi.org/10.1051/e3sconf/202020808009

Peskova, D. R., Vasileva, J. P., Nazarova, U. A., & Sharifyanova, Z. F. (2019). World and Russian fuel and energy market development perspectives and drivers. In *E3S Web of Conferences* (Vol. 124, p. 04017). https://doi.org/10.1051/e3sconf/201912404017

PJSC "INTER RAO." (n.d.). Official website. Retrieved from https://www.interrao.ru/. Accessed February 21, 2021.

Redl, C., Hein, F., Buck, M., Graichen, P., & Jones, D. (2021). *The European power sector in 2020*. Agora Energiewende. Retrieved from https://ember-climate.org/wp-content/uploads/2021/01/Report-European-Power-Sector-in-2020.pdf. Accessed February 15, 2021.

U.S. Energy Information Administration. (2020, September 25). In 2019, 9 of the 10 highest-generating U.S. power plants were nuclear plants. Retrieved from https://www.eia.gov/todayinenergy/detail.php?id=45276. Accessed November 30, 2020.

Ustinova, O. V., Rudov, S. V., Kostyleva, E. G., Grogulenko, N. V., & Kulishova, N. D. (2016). The processes of globalization in the Russians' views. *Man in India, 96*(7), 2165–2177.

Vanchukhina, L. I., Leybert, T. B., Khalikova, E. A., Rudneva, Y. R., & Rogacheva, A. M. (2019). Industry 4.0 and closed-loop economy in the context of solving the global problems of modern times. In E. Popkova, Y. Ragulina, & A. Bogoviz (Eds.), *Industry 4.0: Industrial revolution of the 21st century* (pp. 31–53). Springer. https://doi.org/10.1007/978-3-319-94310-7_4

World Nuclear Association. (n.d.). Official website. Retrieved from https://www.world-nuclear.org/. Accessed February 20, 2021.

Environmental Efficiency as an Evaluation Criterion for the Methods of Hydrocarbon Production Intensification

Elena A. Zakharova and Natalia A. Likhacheva

Abstract The authors select indicators to assess the environmental effectiveness of the methods of intensifying hydrocarbon production, taking into account greenhouse gas emissions into the atmosphere, discharges into the hydrosphere, and solid waste disposal. The authors analyze criteria for evaluating environmental effectiveness and mechanisms for selecting environmental indicators. It is proposed to use a new parameter—the intensity of release into the atmosphere of greenhouse gases generated during hydrocarbon production, both in applying methods of production intensification and during the burning of associated petroleum gases. The study uses not the absolute values of volumes but their ratio to the volume of products produced. The environmental effectiveness of the methods of intensifying hydrocarbon production are assessed using the following tools: (1) selection of criteria describing the process of impact on the environment according to the scheme "Impact – State – Response"; (2) determination of integral environmental features, taking into account all selected indicators for the sections "Impact," "State," and "Response"; (3) calculation of the integral indicator of the environmental efficiency of the methods of intensifying hydrocarbon production. For the first time, this system is used to evaluate a separate stage of hydrocarbon production—hydraulic fracturing.

Keywords Intensification of hydrocarbon production · Hydraulic fracturing · Environmental indicators · Environmental efficiency · Greenhouse emissions · Environmental pollution

JEL Codes Q35 · Q53 · Q56 · Q57

E. A. Zakharova (✉) · N. A. Likhacheva
Ufa State Petroleum Technological University, Branch of the University in the City of Salavat, Salavat, Russian Federation

E. G. Popkova and B. S. Sergi (eds.), *Geo-Economy of the Future*,
https://doi.org/10.1007/978-3-030-92303-7_31

1 Introduction

Increased attention is paid to the methods of intensification of hydrocarbon production in the extractive industry. This stimulates research to find economically and environmentally effective methods of increasing the volume of hydrocarbon production.

All stages of the technological chain of oil production (exploration, rig construction, drilling, intensification of production, extraction of fluid to the surface, and preparation of hydrocarbons for transportation and storage) are dangerous for the biosphere as a whole and its specific ecosystems.

The development of industrial production entails an increase in the anthropogenic impact on the biosphere and its ecosystems, which catastrophically deteriorates its conditions (Zakharova, 2020b). The conflict between the economic and ecological demands of production and society is a task that must be resolved immediately.

The shortage of hydrocarbons currently creates conditions for improving the production system. One of the primary methods of intensifying hydrocarbon production is hydrodynamic fracturing. Its use is often associated with high environmental risks manifested in environmental changes, high levels of water consumption, and changes in ecosystems and the biosphere. The most significant environmental risk associated with hydraulic fracturing is the contamination of aquifers with chemicals, the use of considerable volumes of water, the release of methane into the atmosphere, and an increase in seismic activity of the fractured rock. Flaring of associated petroleum gas is the primary source of greenhouse gases in the oil and gas industry (Kokarev & Semenovich, 2017; Zakharova, 2020a).

The assessment of the environmental impact from the extractive and processing industries is currently carried out according to different methodologies and criteria based on comparative analysis. The methodology is most often chosen depending on the research objectives and the area of production, the environmental efficiency of which is to be evaluated.

The classic indicators of the production impact on the environment are rational use of natural resources; gas emissions; wastewater discharge; generated solid waste; places of storage and disposal of solid waste; use of wasteless technologies; recycling of resources (Ridlington & Rumper, 2013). However, the control of the volume and reduction of greenhouse gas emissions are no less important.

2 Materials and Methods

Nowadays, it is necessary to assess the possible impact on living organisms and their habitat. One of the evaluation techniques is environmental performance evaluation. Environmental efficiency of production is determined following GOST R ISO 14031–2016 "Environmental Management. Assessment of environmental performance. General requirements" (Environmental Management, 2017).

The impact of the intensification of hydrocarbon production can be assessed by creating groups of environmental indicators formed, taking into account the main types of impact on the biosphere of a particular production or enterprise and activities aimed at leveling the harmful impact.

This research aims to improve the methodological approach and tools for assessing the environmental effectiveness of the methods of intensifying hydrocarbon production. The authors examined a system of environmental performance indicators widely recognized throughout the world. The basis of this system is the interaction scheme "Impact – State – Response."

The analysis of modern methods of the intensification of hydrocarbon production has shown that hydrodynamic fracturing (HF) is currently the most widespread and effective hydrocarbon production method. Hydraulic fracturing is used in conventional oil and gas production and the production of shale hydrocarbons. In this case, the impact on the environment significantly increases and requires a thorough assessment.

In this regard, we believe that it is particularly relevant to assess the environmental effectiveness of hydraulic fracturing using a set of environmental indicators and calculating characteristics according to the interaction scheme "Impact – State – Response." The secondary literature has various materials on the state of the environment after using single and multiple hydraulic fracturing in the field.

The following tools are proposed to carry out a comprehensive assessment of the environmental effectiveness of the methods of hydrocarbon production intensification:

- Selection and analysis of environmental indicators for groups "Impact," "State," and "Response," defined by the "Methodology and Criteria for Evaluating Environmental Performance" (Federal Service for Supervision of the Use of Natural Resources (n.d.));
- Calculation and analysis of summary environmental characteristics according to the scheme "Impact – State – Response" based on the recommended indicators normalized in the state standards;
- Evaluation of the integral indicator (I) of the environmental effectiveness of the methods of intensifying oil and natural gas production and its interpretation into points.

From the basic list of categories (emissions into the atmosphere, water consumption, and waste management (Federal Service for Supervision of the Use of Natural Resources (n.d.)), the authors selected environmental indicators to characterize the methods of intensification of hydrocarbon production (Table 1).

Based on the considered environmental criteria, the authors determined the environmental performance indicators and analyzed the basic and proposed additional indicators for the groups "Impact," "State," and "Response." This research will allow creating a unified database of environmental performance indicators of the methods of intensifying hydrocarbon production.

Table 1 Environmental indicators adapted to the evaluation of hydraulic fracturing

Group	Indicator
Impact	$y_{1.1}$—water supply of the production process (general) $y_{2.1}$—water consumption (from surface water bodies) $y_{3.1}$—emissions of pollutants (into the atmosphere) $y_{4.1}$—emissions of specific pollutants (into the atmosphere) $y_{5.1}$—formation of solid waste $y_{6.1}$—emissions of greenhouse gases, in particular, CO_2 (direct)
State	$y_{7.2}$—concentration of pollutants (in the atmosphere) $y_{8.2}$—concentration of pollutants (in water bodies) $y_{9.2}$—area of contaminated (disturbed) lands
Response	$y_{10.3}$—pollutant discharges (reduction) $y_{11.3}$—utilization of associated petroleum gas (APG) $y_{12.3}$—emissions of greenhouse gases, in particular CO_2 (emissions reduction) $y_{13.3}$—waste recycling $y_{14.3}$—share of restored land area in disturbed land

Source Compiled by the authors based on (Mayorova, 2017)

It is proposed to use an adapted methodology for assessing the effectiveness of the methods of hydrocarbon production intensification, which consists of the following stages:

1. Systematization of initial data of the adapted set of environmental indicators of hydrocarbon production intensification;
2. Rationing identified environmental indicators based on the established minimum and maximum threshold values of probable technogenic impacts;
3. Calculation of the total integral indicator of the studied process based on the previously calculated summary environmental indices for the proposed groups;
4. Final discussion of the features of environmental effectiveness of hydraulic fracturing identified during the calculation and analysis.

3 Results

Based on publicly available data on the amount and types of hydrocarbon fuels, electricity consumed, and the volume of production and flaring of associated petroleum gases, we calculate the proposed environmental indicator “Direct emissions of greenhouse gas” ($y_{6.1}$) (Analytical Center under the Government of the Russian Federation, 2013, 2019, 2020).

The volumes of greenhouse gases (CH_4, CO_2) generated during hydraulic fracturing were determined based on the data for 2014–2017 and emission factors. The volume of greenhouse gases from the combustion of associated petroleum gas was estimated.

It is proposed to use the integral environmental indicator (I) and the selected indicators to calculate the integral summary characteristics for the whole model "Impact – State – Response."

The calculation of indices I_1 for the group "Impact," I_2 for the group "State," and I_3 for the group "Response" were carried out following the "Methodology and Criteria for Evaluating Environmental Performance" (Federal Service for Supervision of the Use of Natural Resources (n.d.)).

In the "Impact" group, the ratio of the I_1 index and the indicators $y_{1.1}$—$y_{6.1}$ is determined by a monotonically decreasing relationship. High values of indicators $y_{1.1}$—$y_{6.1}$ show low values of I_1 and, consequently, low environmental efficiency.

In the "Impact" group, rationing is performed using indicators $y_{1.1}$—$y_{6.1}$. The threshold maximum is the indicator for the base year. The threshold minimum is the indicator formed based on standard or target values defined for a particular production or industry as a whole.

In the "State" group, the rationing is performed by indicators $y_{7.2}$—$y_{9.2}$. In this group, the relationship between the I_2 index and the indicators $y_{7.2}$—$y_{9.2}$ is also characterized by a monotonically decreasing dependence. High values of indicators $y_{7.2}$—$y_{9.2}$ give low values of I_2 and, consequently, low environmental efficiency.

Similarly, the threshold maximum is the indicator for the base year. The threshold minimum is the indicator formed based on standard or target values defined for a production, economic sector, or region.

The "response" group is marked with a monotonically increasing dependence of the I_3 index and the indicators $y_{10.3}$—$y_{14.3}$. The high values of indicators $y_{1.1}$—$y_{6.1}$ show a high value of I_3 and, consequently, high environmental efficiency.

The threshold minimum is the indicator for the base year. The threshold maximum is the indicator formed based on standard or target values defined for production, industry, or region (Analytical Center under the Government of the Russian Federation, 2019; Federal Service for Supervision of the Use of Natural Resources (n.d.); Mayorova, 2017).

The highest destructive impact on the biosphere is defined as a zero value (0) of the standardized indicator for the groups "Impact" and "State." The minimum impact is the highest value, in our case—one (1) (Federal Service for Supervision of the Use of Natural Resources (n.d.); Mayorova, 2017).

During the research, we calculated the composite indices of environmental effectiveness of hydraulic fracturing according to the scheme "Impact – State – Response," the private index of greenhouse gas emissions into the atmosphere, and the integral index of environmental effectiveness of hydraulic fracturing as one of the most used in the intensification of hydrocarbon production (Table 2). The obtained indicators were evaluated using a scale of performance levels (Table 3).

The analysis of the results of the study on environmental effectiveness (Table 4) allows us to assess the level of "Impact" of hydraulic fracturing on the environment as high (reduced environmental effectiveness—3 points). There is a significant contribution of such indicators as water supply and pollutant emissions. That is, hydraulic fracturing has the maximum impact on water resources and soil.

Table 2 Qualitative treatment of relative integral characteristics

Value range	Score	Qualitative assessment of environmental effectiveness
0–0.1	0 points	No production activity
0.1–0.3	1 point	Very low
0.3–0.5	2 points	Low
0.5–0.7	3 points	Reduced
0.7–0.9	4 points	Sufficient
0.9–1	5 points	High

Source Compiled by the authors based on (Mayorova, 2017)

Table 3 Assessment of environmental effectiveness of hydraulic fracturing

	Value	Level	Points
Composite index by group			
I_1 (Impact)	0.59	Reduced	3
I_2 (State)	0.55	Reduced	3
I_3 (Response)	0.71	Sufficient	4
I_{GH} (A partial index characterizing the dynamics of the emissions of greenhouse gas into the atmosphere)	1	High	5
Integral index	0.62	Reduced	3

Source Compiled by the authors

According to the "State" group, the composite index is also assessed as "reduced efficiency," which indicates a significant contribution of this method of intensification of hydrocarbon production to the state of the atmosphere and water resources.

For the "Response" group, the composite index (4 points) allows us to evaluate hydraulic fracturing as sufficiently effective in preventing and reducing the destructive impact on the biosphere.

The analysis of the structure of the composite indices confirms the earlier conclusions that hydraulic fracturing has the greatest impact on the atmosphere and a significant impact on water resources and soil (Fig. 1).

The index of greenhouse gas emission intensity shows high environmental efficiency, which may be related to the growing attention to the problem of utilization of associated petroleum gases and the increasing reduction in APG volume sent for combustion.

The integral index (I) allows us to summarize and evaluate the whole model "Impact – State – Response." Based on the value of the integral index (3 points), we can identify the reduced environmental effectiveness of hydraulic fracturing as a method to intensify hydrocarbon production.

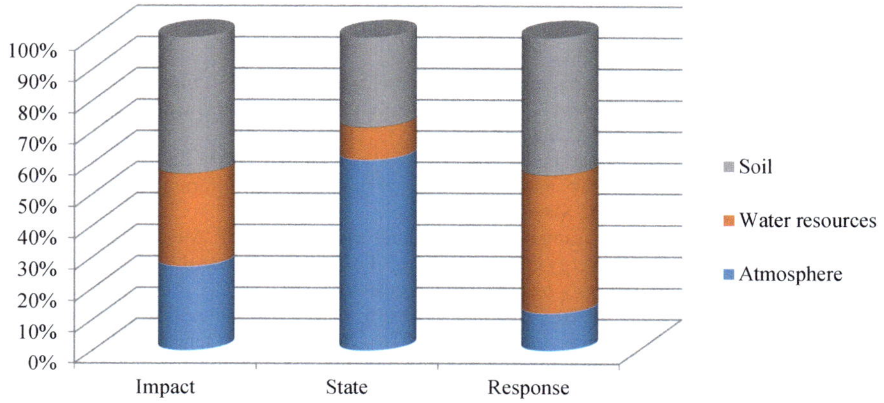

Fig. 1 Structure of consolidated indices of environmental efficiency. *Source* Compiled by the authors

4 Conclusion

Our calculations and research allowed us to formulate the following provisions:

1. The environmental state can be monitored by creating groups of environmental indicators formed, taking into account the main types of impact on the biosphere of a particular production or enterprise and activities to reduce the effects of the alleged impact.
2. Comprehensive assessment of environmental effectiveness can be carried out by indicators characterizing the interaction of indicators "Impact – State – Response."
3. Based on the calculations, it is concluded that hydraulic fracturing as a method to intensify hydrocarbon production has reduced environmental effectiveness.
4. The index of greenhouse gas emission intensity shows high environmental efficiency, which may be related to the growing attention to the problem of utilization of associated petroleum gases and the increasing reduction in APG volume sent for combustion.
5. Assessment of ecological efficiency of methods of intensification of hydrocarbon production using the scheme "Impact – State – Response" can help improve the system for assessing the impact of oil and gas complexes on the environment.

References

Analytical Center under the Government of the Russian Federation. (2013). Environmental threats to hydrocarbon production. *Energy Bulletin, 7*. Retrieved from https://ac.gov.ru/archive/files/publication/a/1105.pdf. Accessed 22 May 2020.

Analytical Center under the Government of the Russian Federation. (2019). Statistical collection: Fuel and energy complex of Russia—2018. Retrieved from https://ac.gov.ru/archive/files/publication/a/22922.pdf. Accessed 22 May 2020

Analytical Center under the Government of the Russian Federation. (2020). Measuring sustainable development in the fuel and energy complex. *Energy Bulletin, 82*. Retrieved from https://ac.gov.ru/uploads/2-Publications/energo_march_20.pdf. Accessed 22 May 2020.

Environmental management. (2020). Environmental performance evaluation. Guidelines on environmental performance evaluation (GOST R ISO 14031–2016, submitted June 1, 2017). Retrieved from https://docs.cntd.ru/document/1200142908. Accessed 12 Apr 2020.

Federal service for supervision of the use of natural resources. (n.d.). Methodology and criteria for assessing the environmental performance of enterprises, as well as environmental rating systems that meet the objectives of an objective reflection of the environmental situation on the subjects of the Russian Federation. Retrieved from http://rpn.gov.ru/sites/all/files/users/rpnglavred/filebrowser/docs/algoritm_dlya_predpriyatiy.doc. Accessed 12 Apr 2020.

Kokarev, M. O., & Semenovich, S. P. (2017). Layer hydraulic fracturing as issue source toxicants in underground waters. *Bulatov Readings, 4*, 160–164.

Mayorova, T. V. (2017). Methodological tools for assessing the effectiveness of environmental management in the implementation of the concept of low-carbon development (Synopsis of the Dissertation of Candidate of Economics). Yekaterinburg, Russia.

Ridlington, E., & Rumper, J. (2013). *Fracking by the numbers: Key impacts of dirty drilling at the state and national level.* Denver, CO: Environment America Research & Policy Center. Retrieved from https://environmentamerica.org/sites/environment/files/reports/EA_FrackingNumbers_scrn.pdf

Zakharova, E. A. (2020a). Assessment of the impact of hydraulic fracturing on the environment. In *Proceedings of the International scientific and methodical conference "Integration of Science and Education in Oil and Gas Universities"* (pp. 201–203). Ufa, Russia: Ufa State Petrolium Technological University.

Zakharova, E. A. (2020b). Determination of environmental efficiency as a mechanism for assessing the impact of production on the environment. In *Proceedings of the International scientific and methodical conference "Integration of Science and Education in Oil and Gas Universities"* (pp. 138–140). Ufa, Russia: Ufa State Petrolium Technological University.

The Impact of FDI on Algeria's Economic Growth

Vladimir Z. Chaplyuk, Fakhraddin N. Akhmedov, Mhd Shaker Zeitoun, Milana M.-S. Abueva, and Ahmad S. Al Humssi

Abstract This article highlights the impact of inward Foreign Direct Investment on Algeria's economic growth during the period 1990–2020. It uses many statistical regression models such as Linear model, Logarithmic model, Inverse model, Quadratic model, Cubic model, Compound model, Power model, S curve model, Growth model, Exponential model, and Logistic model to analyze the relationship between FDI and Algeria's economic growth. However, all models found a weak impact of FDI on economic growth in Algeria based on the correlation criteria R, Fisher Criterion F, and fitted trend lines. Therefore, the article concluded that there is no significant influence of FDI on Algeria's economic growth. The absence of a significant impact of foreign investment on Algeria's economic growth reveals a weakness in the Algerian government's performance in terms of attracting FDI. This problem has been addressed and diagnosed in this article finding that the Algerian government was unable to take strategic steps to attract investments amid the lack of a qualified investment environment. The paper suggests some policy recommendations on how to improve the investment environment in Algeria.

Keywords Algeria · Foreign direct investment · Doing business · Economic growth

JIT Codes F21 · F23 · O43 · O55

1 Introduction

The Foreign Direct Investment, as an external source of finance, plays an important role in achieving economic development for developing or developed countries. FDI is also considered an alternative source of external borrowing (Chung, 2010) and the

V. Z. Chaplyuk (✉) · F. N. Akhmedov · M. S. Zeitoun · M. M.-S. Abueva · A. S. Al Humssi
Peoples' Friendship University of Russia (RUDN University), Moscow, Russia

F. N. Akhmedov
e-mail: akhmedov_fn@pfur.ru

E. G. Popkova and B. S. Sergi (eds.), *Geo-Economy of the Future*,
https://doi.org/10.1007/978-3-030-92303-7_32

importance of this source crystalizes in periods of financial crises on the global and internal levels.

In contemporary conditions, many factors led to the increase in the volume of foreign direct investments as the global FDI reached more than 2.68 trillion US dollars in 2016 compared to 1990 when the volume of FDI reached 0.24 trillion dollars. Foreign investment reached 1.57 trillion dollars in 2000 (World Bank, 2021).

One of the most important factors that led to the increase in the volume of global investments is the technological and banking sophistication (Omisakin et al., 2012), the emergence of international organizations as the main regulator of the movement of capital between countries by setting up control to oblige countries to ensure the free movement of investments, and the emergence of international arbitration that guarantees the rights of investors.

Many factors negatively affect FDI in many countries. The most important one was the lack of a suitable environment for investment, such as the lack of a suitable infrastructure (Nguea, 2020) (Boga, 2019). Without the availability of such infrastructure, investment and economic activities cannot be carried out in any country. This infrastructure means the availability of water, electricity, roads, etc. Moreover, the Constitutional and Legislative Guarantees (Alomran, 2019) are very important factors to encourage foreign investors to invest capital in a specific country. Moreover, several other *factors* can also affect FDI such as weakness in the financial and banking structure in the country. (Sghaier & Abida, 2013), (Kamasa et al., 2020), war and armed conflicts (Li et al., 2017), disasters & global epidemics like Covid-19 (Duan et al., 2020), (Uygun & Gujrati, 2020), (Chaplyuk et al., 2021).

In modern conditions, wars and the spread of disasters are the main reason for the emigration of capitals, and this problem is distinguished from other investment obstacles.

There are positive and negative factors that affect the movement of FDI between countries. These countries can provide certain conditions to attract foreign direct investment. To improve the conditions of attracting FDI, many factors should be taken into account such as economic reforms, institutional quality, fighting corruption, government activeness, regulatory quality, enforcing rule of law, and accountability. These factors can positively influence growth, while political stability and the absence of violence/terrorism is not statistically significant. Moreover, SDGs such as poverty, income distribution, education, innovation, transport infrastructure, and information technology are noteworthy drivers of growth (Gherghina et al., 2019).

To determine the role of FDI in the economies of countries, many studies have identified the mechanism of direct and indirect influence of foreign direct investment, for example, M. Apostolov found that there are positive ties of foreign direct investments with gross domestic product in all of the domestic economies (Apostolov, 2016). The results of the research of (Okwu et al., 2020) showed some mixed growth effects of the variables in general. More specifically, FDI had a positive and significant effect on the economic growth of countries during the time. (Sarker & Khan, 2020) found short-run and long-run relationships and the causality running from GDP (economic growth) to FDI advocates for placing greater emphasis on policies that are appropriate to maintain a steady growth rate of GDP.

East African countries, such as Algeria, need external sources of funding, including attracting FDI to achieve economic development. Therefore, this research focuses on the role of FDI in the economic growth of the state of Algeria.

2 Methodology

Algeria is considered one of the most important countries in the African continent in terms of economy and geography. It is one of the developing countries located in the north of the African continent. Due to its closeness to the European Union and its large potential to attract foreign investment, such as the low cost of human capital and its large stocks of raw materials and water, Algeria is considered a vital space to attract investments.

In our research, we used economic data of the Republic of Algeria as reported by World Bank and the Arab Investment Guarantee Corporation during the period 1990–2020.

To serve the purpose of this research, the hypotheses for the test can be determined as follows:

- The null hypothesis, H0: there is no significant impact of FDI on Algeria's economic growth
- the research hypothesis, H1: there is a significant impact of FDI on Algeria's economic growth

To address the impact of FDI on economic growth, we use one of the most important models (IBM SPSS) to address the relationships between FDI & GDP growth are the following:

$$\text{Compound model} : Y1 = \beta 0 * (\beta 1^{x}) + \varepsilon i \tag{1}$$

$$\text{Quadratic model } Y2 = \beta_0 + (\beta_1 * x) + \left(\beta_2 * x^2\right) + \varepsilon i \tag{2}$$

$$\text{Cubic model} : Y3 = \beta_0 + (\beta_1 * x) + \left(\beta_2 * x^2\right) + \left(\beta_3 * x^3\right) + \varepsilon i \tag{3}$$

$$\text{Growth model} : Y4 = e^{(\beta 0 + (\beta 1 + x)} + \varepsilon i \tag{4}$$

$$\text{Exponential model } Y5 = \beta 0 + e^{(\beta 1 * x)} \tag{5}$$

$$\text{Power model} : Y6 = \beta 0 * (x^{\beta 1}) + \varepsilon i \tag{6}$$

$$\text{S curve model} : Y7\ e^{(\beta 0 + \frac{\beta 1}{x})} + \varepsilon i \tag{7}$$

$$\text{Logistic model}: Y8 = \frac{1}{\frac{1}{u} + (\beta 0 + (\beta 1^t))} + \varepsilon i \tag{8}$$

$$\text{Logarithmic model}: Y9 = \beta_0 + (\beta_1 * ln(x)) + \varepsilon i \tag{9}$$

$$\text{Inverse model } Y10 = \beta_0 + (\beta_1 / x) + \varepsilon i \tag{10}$$

$$\text{Linear model}: Y11 = \beta_0 + \beta_1 + \varepsilon i \tag{11}$$

In models (1, …, 11) where

$Y = (Y1, Y2, \ldots)$	is the value of the response (independent) variable.
$\beta = (\beta 0, \beta 1, \ldots)$	is the regression parameter.
x	a predictor (explanatory, independent) variable.
εi	an error term.
u	the upper boundary value.
e	Euler's Number

Depending on the objective of the research and using various mathematical models, first, there is a need to determine whether there is an impact of FDI on Algeria's economic growth (positive or negative). And then, we are going to address the reasons behind any possible impact whether positive or negative.

3 Results

In 2019, Algeria ranked 55th at the global level, in terms of the gross domestic product reaching USD 171.2 billion. (United Nations, 2019), but the COVID-19 led to a decrease in the Algerian gross domestic product to USD 147.3 billion in 2020. As a result, Algeria's classification by GDP declined to 57th on the global level (International Monetary Fund, 2020 estimates).

However, due to many factors in 2020, including COVID 19, Algeria achieved negative economic growth and a negative FDI as shown in the following figures (Figs. 1 and 2).

Having analyzed the data of Algeria's GDP growth, we found a positive growth from year to year but only from 1991 to 1993 there was a negative growth due to the Algerian civil war. After 1993, growth returned to be positive due to the rapid recovery of the economy after the end of the civil war. However, in 2020, Algeria's economic growth was negative due to the epidemic COVID-19.

The analysis of FDI data to Algeria during the period 1990–2020 shows that the volume of FDI coming into Algeria is at its best in two years, in 2001 and 2009, as the volume of FDI has reached USD 2.2 billion.

Fig. 1 Algeria's GDP growth % from 1990 to 2020 (2020 estimates). *Source* Compiled by the authors based on (World Bank, 2021), (The Arab Investment & Export Credit Guarantee Corporation, 2021)

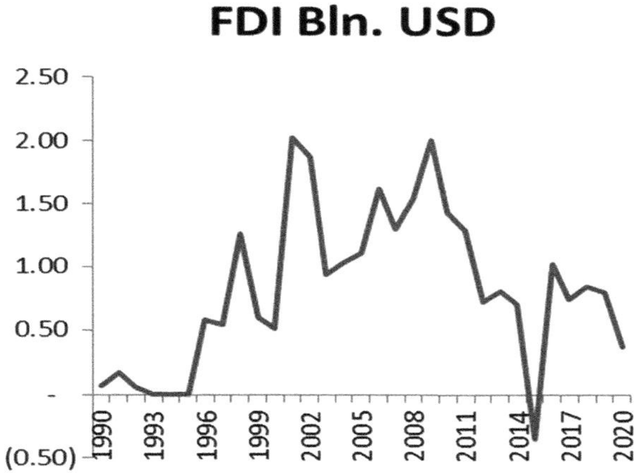

Fig. 2 Algeria's inward FDI from 1990 to 2020 (in billion current U.S. dollars) (2020 estimates). *Source* Compiled by the authors based on (World Bank, 2021), (The Arab Investment & Export Credit Guarantee Corporation, 2021)

By using the statistical program SPSS for analyzing statistical data, we have the following results (Table 1).

Data in Table 1 shows that there is no model, whether linear or non-linear, that represents the relationship between FDI and the economic growth in Algeria.

By analyzing the data, we found that F statistics in some models is near to F critical value. However, in other models, F statistics were less than the F critical value (**α = 0.05**), with 95% confidence accepting the null hypothesis, H0, as a reasonable possibility. For more details, by using linear and nonlinear mathematical models, we found that all of them confirmed that there is no impact of FDI on economic growth, as in the following figure (Fig. 3):

By analyzing the data in Fig. 3, we found that the correlation coefficient is less than 0.25, which can be explained by the fact that the impact of FDI on the economic growth of the Republic of Algeria is low. In other words, it has a weak impact, while the impact of other factors on the Algerian economic growth is more than 75%

Table 1 Coefficients of the relationship between FDI and economic growth in the Republic of Algeria

№.	Model	Chose the model Yes Y, No N	*R*	*P* value	*F*	*R* Square
1	Cubic	N	0.413	0.072	2.887	0.171
2	Quadratic	N	0.41	0.076	2.832	0.168
3	Inverse	N	0.393	0.029	5.284	0.154
4	Logarithmic	N	0.389	0.031	5.174	0.151
5	Linear	N	0.384	0.033	5.013	0.147
6	S curve	N	0.357	0.049	4.235	0.127
7	Power	N	0.356	0.05	4.197	0.126
8	Compound	N	0.353	0.052	4.116	0.124
9	Growth	N	0.353	0.052	4.116	0.124
10	Exponential	N	0.353	0.052	4.116	0.124
11	Logistic	N	0.353	0.052	4.116	0.124

Source Compiled by the authors based on IBM SPSS

To attract FDI, countries should provide appropriate conditions to attract investment (ease of doing business). The "ease of doing business" score data issued by the World Bank on an annual basis, helps to assess the absolute level of regulatory performance over time (World Bank) and reflects the extent to which countries provide the basic conditions for attracting foreign direct investments, as shown in charts 4 and 5 (Figs. 4 and 5).

By analyzing the indicator EoDB score (Ease of Doing Business score) between 2006 and 2020, we found that the indicator has gradually declined during that period.

To determine the extent of congruence and compatibility between the statements issued by the World Bank related to the introduction of business in Algeria and its reflection on the reality of foreign direct investment, we have studied the correlation between the two indicators of ease in doing business and FDI coming into Algeria as follows (Fig. 6).

Among different models, and as shown in Fig. 1 which describes the relationships between FDI and economic growth, we found that the linear curve is the best curve to represent the relationship between the two indicators (Table 2).

Thus, the regression equation becomes as follows:

$$\mathbf{Y_i} = 0.08798586 * \mathbf{X} - 2.55767608 + \varepsilon \mathbf{i} \quad (12)$$

Thus, one of the most common obstacles to attracting foreign investment to Algeria is the lack of appropriate conditions to attract FDI such as the difficulty and weakness in the business registration process, the process of obtaining building permits, the process of connecting to the power supply system, the process of property registration, lending, investor protection, taxation, international trade, the enforcement of contracts and the process of liquidation of enterprises.

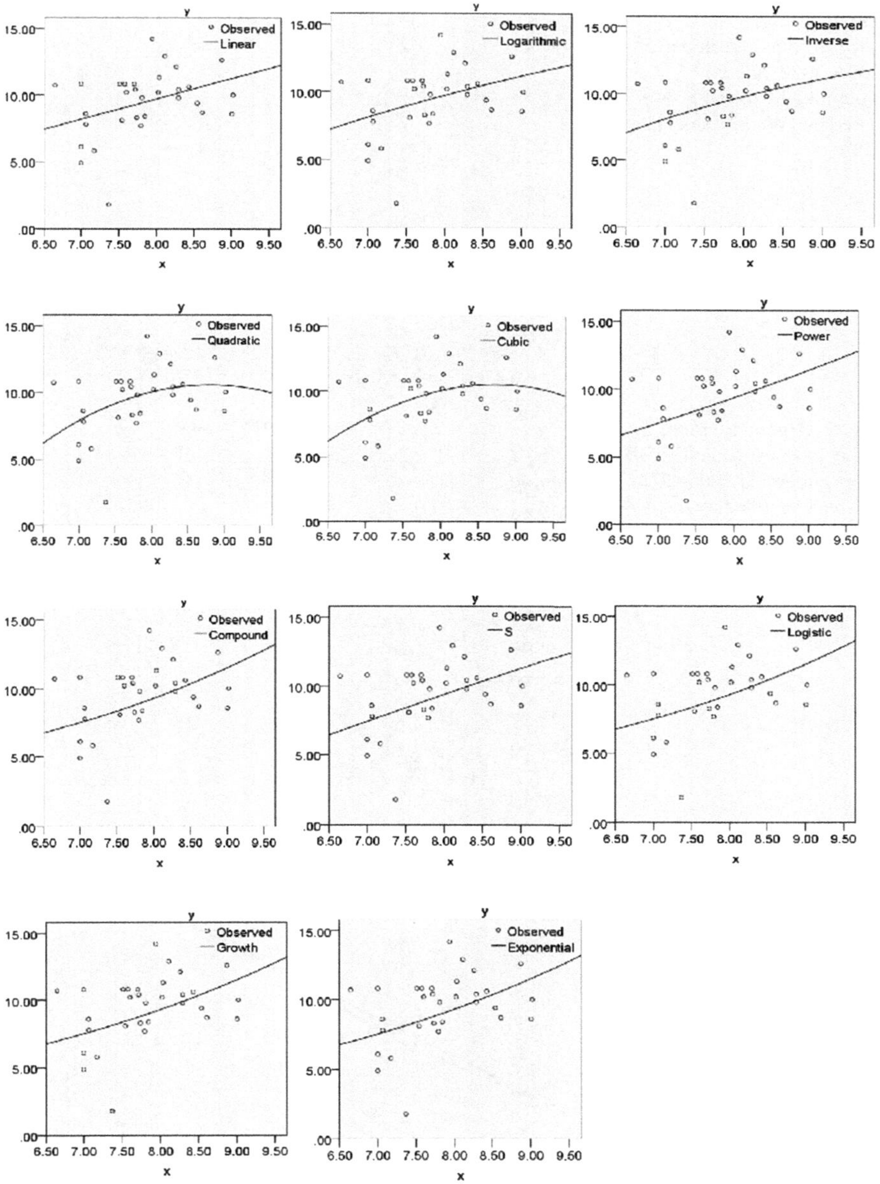

Fig. 3 The linear and non-linear relationship between the FDI and the economic growth in the Republic of Algeria during the period 1990–2020. *Source* Compiled by the authors based on IBM SPSS

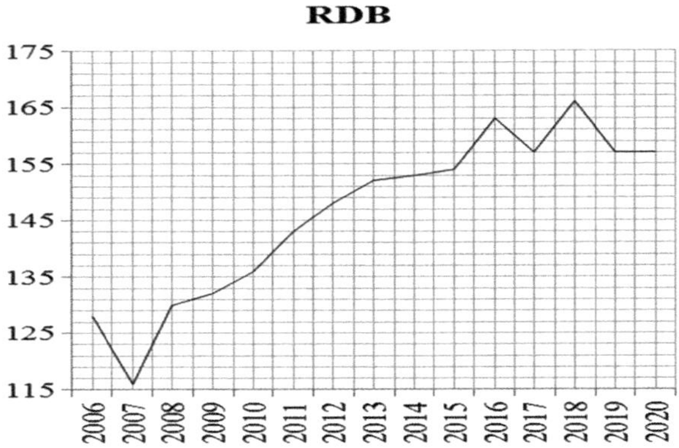

Fig. 4 Algeria's doing business rankings in period 2006–2020. *Source* Compiled by the authors based on (World Bank Data, 2021)

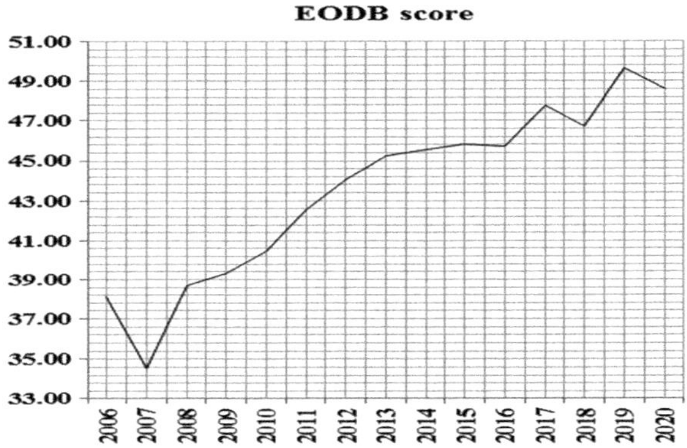

Fig. 5 Algeria's ease of doing business score (EoDB score) in period 2006–2020. *Source* Compiled by the authors based on (World Bank Data, 2021)

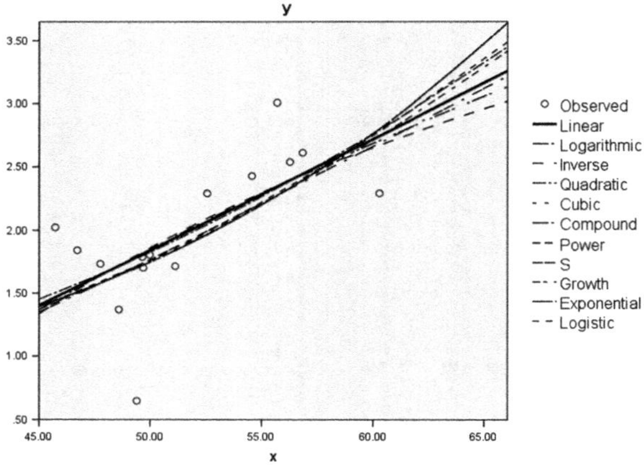

Fig. 6 Mathematical models that reflect the relationship between the ease of doing business and the volume of foreign investments input into Algeria during the period 2006–2020. *Source* Compiled by the authors based on IBM SPSS

Table 2 Model summary

R	R Square		Adjusted R Square	Std. Error of the Estimate	
0.647	0.419		0.374	0.453	
	Sum of Squares	**df**	**Mean Square**	**F**	**Sig.**
Regression	1.924	1	1.924	9.358	0.009
Residual	2.674	13	.206		
Total	4.598	14			
	Unstandardized Coefficients		**Standardized Coefficients**	**T**	**Sig.**
	B	Std. Error	Beta		
X	0.088	0.029	0.647	3.059	0.009
(Constant)	-2.558	1.490		-1.716	0.110

Source Compiled by the authors based on IBM SPSS

As for recommendations, there is a need for an effective government that can be able to efficiently attract and direct FDI to serve the state's economic objectives and thus achieve economic growth. The government in Algeria must reformulate the mechanisms associated with attracting foreign investors, mainly the laws governing the process of attracting investors, for example, facilitating the conducting of investments, rational tax exemptions, investment guarantee, fight corruption within the state's administrative apparatus, monitor investment activities and determine the reasons that hinder investors by investigating the investors themselves about the obstacles and working on a joint strategy between the government and investors to overcome these obstacles.

To improve the investment climate in Algeria, the economic team in the government must secure a good infrastructure, and this, in turn, must be done primarily by increasing the state's government spending on infrastructure. Developing the infrastructure can be financed from long-term internal borrowing from domestic investors by raising the interest rate, establishing investment partnerships between the government and the private sector, or external borrowing. Therefore, it is necessary to define strict and tightly controlled supervision over economic management to prevent corruption to achieve the desired economic goals.

4 Conclusion

The article highlights the impact of FDI on the economic growth of Algeria during the period 1990–2020. Many linear and non-linear mathematical models were used to study the impact of foreign direct investments on Algeria's economic growth.

The study concluded that there is no significant impact of FDI on the Algerian economic growth, and thus it reveals that Algeria does not capitalize on this important resource in achieving economic growth in the country, as Algeria occupies a late place on the global level to serve businesses.

The study found that the main reason behind the inability of the Algerian government to attract FDI is primarily the weak performance of providing adequate conditions to attract FDI. The article has addressed the main obstacles to attracting investments, the main of which was the poor and inadequate infrastructure, the spread of corruption, and weak governance.

To attract FDI and to capitalize on this source to achieve economic development, Algeria's economic governing must be reformulated and work to encourage investment and build appropriate infrastructure and appropriate controls to ensure that investment activities.

References

Alomran, N. M. S. (2019). Constitutional and legislative guarantees of foreign investment in light of Jordanian legislation. *Journal of Legal, Ethical and Regulatory Issues, 22*(4). https://www.abacademies.org/articles/constitutional-and-legislative-guarantees-of-foreign-investment-in-light-of-jordanian-legislation-8580.html. Data accessed 02 May 2021.

Apostolov, M. (2016). Effects of foreign direct investments. Evidence from Southeast Europe. *Cuadernos de Economía, 39*, 99–111. http://dx.doi.org/10.1016/j.cesjef.2015.10.003

Boga, S. (2019). Determinants of foreign direct investment: A panel data analysis for Sub-Saharan African Countries. *EMAJ: Emerging Markets Journal, 9*, 80–87. https://doi.org/10.5195/emaj.2019.175

Chaplyuk, V. Z., Alam, R. M. K., Abueva, M. M. S., Hossain, M. N., Al Humssi, S. A. (2021). COVID-19 and its impacts on global economic spheres. In: E. G. Popkova, B. S. Sergi (Eds.), *Modern Global Economic System: Evolutional Development vs. Revolutionary Leap*. ISC 2019. Lecture Notes in Networks and Systems, vol. 198. Springer. https://doi.org/10.1007/978-3-030-69415-9_94

Chung, K. (2010). Foreign Debt, FDI and Volatility. *International Economic Journal, 24*(2), 171–196. https://doi.org/10.1080/10168731003649628

Duan, W., Zhu, S., Lai, M., (2020). The impact of COVID-19 on China's trade and outward FDI and related countermeasures. *Journal of Chinese Economic and Business Studies*, 1–10. https://doi.org/10.1080/14765284.2020.1855395

Gherghina, S. C., Simionescu, L. N., & Hudea, O. S. (2019). Exploring foreign direct investment–economic growth nexus—empirical evidence from central and Eastern European Countries. *Sustainability, 11*, 5421. https://doi.org/10.3390/su11195421

International Monetary Fund. (2020). *World Economic Outlook Database*. https://www.imf.org/external/pubs/ft/weo/disclaim.htm. Data accessed 12 Apr 2021.

Kamasa, K., Mochiah, I., Doku, A. K., & Forson, P. (2020). The impact of financial sector reforms on FDI in an emerging economy: empirical evidence from Ghana. *Journal of Humanities and Applied Social Sciences, 2*, 271–284. https://doi.org/10.1108/JHASS-11-2019-0077

Li, C., Murshed, S. M., & Tanna, S. (2017). The impact of civil war on FDI flows to developing countries, *The Journal of International Trade & Economic Development, 26*(4), 488–507. https://doi.org/10.1080/09638199.2016.1270347

Nguea, S. M. (2020). The Impact of Infrastructure development on FDI in Cameroon. 2020. halshs-02446497. https://halshs.archives-ouvertes.fr/halshs-02446497/document. Data accessed 20 Apr 2021.

Okwu, A. T., Oseni, I. O., & Obiakor, R. T. (2020). Does FDI enhance economic growth? Evidence from 30 leading global economies. *Global Journal of Emerging Market Economies, 12*(2), 217–230. https://doi.org/10.1177/0974910120919042.

Omisakin, O. A., Oyinlola, A. M., & Adeniyi, O. A. (2012). Foreign direct investment, economic growth and financial sector development in small open developing economies. *Economic Analysis and Policy, 42*(1), 105–127. https://doi.org/10.1016/S0313-5926(12)50008-1

Sarker, B., & Khan, F., (2020). Nexus between foreign direct investment and economic growth in Bangladesh: an augmented autoregressive distributed lag bounds testing approach. *Financial Innovation, 6*.https://doi.org/10.1186/s40854-019-0164-y

Sghaier, I. M., & Abida, Z. (2013). Foreign direct investment, financial development and economic growth: empirical evidence from North African Countries. *Journal of International and Global Economic Studies, 6*(1), 1–13. https://www2.southeastern.edu/orgs/econjournal/index_files/Page410.htm. Data accessed 21 Mar 2021.

The Arab Investment & Export Credit Guarantee Corporation. (2021). *Annual Reports*. https://dhaman.net/corporate-publications/#annual-report. Data accessed 05 Apr 2021.

United Nations Statistics Division—National Accounts. https://unstats.un.org/unsd/snaama/Basic. Data accessed 12 Feb 2021.

Uygun, H., & Gujrati, D. R. (2020). COVID-19: Impact on foreign direct investment. In *Conference paper: COVID 19 Impact on FDI and Various Sectors*. Tradepreneur Global Academic Platform Southampton UK. https://www.researchgate.net/publication/343529491_COVID-19_Impact_on_Foreign_Direct_Investment. Data accessed 11 Apr 2021.

World Bank. (2021). *Foreign direct investment, net inflows (BoP, current US$)*. https://data.worldbank.org/indicator/BX.KLT.DINV.CD.WD. Data accessed 19 Apr 04 2021.

World Bank Data. (2021). *Doing Business in Algeria*. https://www.doingbusiness.org. Data accessed 22 Apr 2021.

Environmental and Ecological Determinants

Mikhail E. Tarasov, Galina A. Bulatova, and Marina N. Levkina

Abstract In examining issues of ecology and natural resource management, we should begin, for the sake of conceptual clarity, with a brief explanation of the term "ecology." The subject of ecology is the interactions between living systems and their environment. Ecology refers to the study of organisms in their natural habitat; its purpose is to discover and understand the relationships between living things and their environment. Webster's dictionary gives the following definition of ecology—"the totality or pattern of relations between organisms and their environment." The term may also be interpreted as the interaction of organisms with their chemical and physical environments. An environmental scientist deals with the effects of the environment on the organism and its interaction with other organisms under specific conditions. Autoecology and synecology are two different fields of ecology that study particular species or communities, respectively. There are also specialized fields of ecology: the ecology of forests, deserts, grasslands, freshwater, seawater, etc. But conservation, ecology, resource ecology, and pollution ecology have recently gained more importance for applied studies.

Keywords Ecological determinants · Environment · Exploitation of the ecosphere · Habitat · Consequences of degradation · Environmental debates

JEL Codes Q5 · Q55 · Q57

1 Introduction

Like other sciences, ecology also has its own principles and basic concepts:

M. E. Tarasov (✉)
Arctic State Agrotechnological University, Yakutsk, Russia

G. A. Bulatova
Altai State University, Barnaul, Russia

M. N. Levkina
Gorno-Altai State University, Gorno-Altaysk, Russia
e-mail: kumandina@mail.ru

E. G. Popkova and B. S. Sergi (eds.), *Geo-Economy of the Future*,
https://doi.org/10.1007/978-3-030-92303-7_33

1. All living organisms and their environment interact by affecting each other in different ways. Animal populations, flora, and vegetation are interdependent via their environment.
2. The environment is a dynamic complex of several interrelated factors. It works as a filter, selecting successful organisms as some of the environmental factors become critical for the life cycle.
3. A species strives to maintain its homogeneity in structure, function, reproduction, growth, and development by preserving its gene pool.
4. Not only does the environment affect the life of organisms—the organisms themselves change their environment as a result of their growth, propagation, reproduction, death, decay, etc. The dynamic environment and the organisms create pathways for different kinds of species to develop. This process continues until a somewhat stable community forms. The final stage of such a community is called the climax.

The biome includes plants, animals, and the environment. They may house more than one interconnected communities, some of which may reach climax stage. This complex of several communities, represented by a set of different species of plants and animals in a common climate, is called a biome (Tashkulova & Kletskova, 2020b; Bogoviz et al., 2020).

2 Methodology

At present, the basic concepts of ecology are explained mainly on a structural basis. However, with the introduction of the ecosystem concept in ecology, it becomes necessary to emphasize the functional aspects. These aspects define the role of the environment, where various factors interact with each other in the overarching ecosystem. The environment includes all living and non-living factors that work in complex ways. In modern ecology, there are several new key concepts:

1. Ecosystems are the basic structural and functional units of nature. Discrete biological units consisting of populations and communities occupy a certain niche, a unique functional position in relation to other organisms with which they interact.
2. There are various degrees of positive, negative, or even neutral interactions between organisms at the interspecific and intraspecific levels, which define, along with abiotic parameters, the success of a particular population in a habitat.
3. Energy is the driving force of ecosystems. Sunlight is captured by autotrophic organisms and transferred in the form of organic molecules to heterotrophic organisms. This energy flow is unidirectional and non-cyclic.
4. The chemical components in an ecosystem follow specific cycles.
5. Successful growth is determined by limiting factors.
6. In natural conditions, different populations undergo succession. Ecosystems orderly change over time, increasing in complexity. This process changes the

physical environment of the community. The terminal or stable state is known as the climax.

7. Natural events or human activities reduce the species diversity of an ecosystem. Human exploitation of ecosystems seeks to meet human needs by diverting productivity. Applied ecology (human ecology) is concerned with using ecological concepts to describe human activities and determine how people can best meet their needs through ecosystems. Ecosystems that are significantly altered by human activities are called managed ecosystems, whereas the others are called natural ones.

The environment is traditionally defined as the sum of forces and influences that affect the organism. The organism, in turn, is able to react to the environment and influence it. However, several environment types affect humans since they are socio-ecological organisms living in and interacting with a bio-ecological environment (Glotko et al., 2019b; Tashkulova & Kletskova, 2020a).

The environment consists of interacting and inseparable elements of physical, biological, and cultural nature that are connected both individually and collectively in myriad ways. The physical elements define the variable nature of the human habitat, its possibilities, and its limitations. The biological elements (plants, animals, microorganisms, and humans) make up the biosphere. The cultural elements (economic, social, political) are the human-made features that create the cultural environment.

The term "environment" refers to all surroundings, including people, flora, and fauna. Each of these influences the others, and all three influence the environment together. The nature and the human race form an inseparable part of the global habitat. This system consists of the following five elements: air, water, land, plants, and animals.

These elements are interconnected and interdependent; they evolved and adapted together. Deterioration of one element inevitably affects the rest.

Management involves conscious choice among multiple alternatives; moreover, it involves a deliberate commitment to the acknowledgment and achievement of goals. If general management involves the conscious adoption of strategies aimed at achieving a realistic short-term goal, ecological management is concerned with the interaction between humans and the environment, where biophysical and socio-cultural systems interact in a complex way.

Since 1990s, Russia has made significant progress in almost all spheres of life and economy. Food production and industrial manufacturing have increased. Old settlements have grown, and new ones have sprung up. Huge irrigation and power generation projects were built. Roads have connected remote areas of the country. People have been able to transform nature with the available technology. Humans have forgotten that natural resources are over-exploited in this mad race for progress, and some projects led to environmental degradation.

The harmful effects of environmental degradation manifest themselves mainly in water and air pollution, soil erosion, and deforestation. Industrialization is advancing rapidly, causing widespread water and air pollution.

People use water from rivers, streams, and other sources for their needs—eating, drinking, transportation, energy waste disposal, and so on. Water demand increases over time. Industry, agriculture, irrigation, mining, power generation, and the concentration of millions of people in cities contribute to pollution (Glotko, 2019a; Stroiteleva et al., 2020).

As serious as water pollution is, air pollution is even more of a threat to humans in many parts of the country. Air pollution is increasing due to motor vehicle traffic and industrial growth. Pollution has been proven to cause numerous respiratory illnesses. Hydrocarbons from asphalt, petrochemicals, and car exhausts have increased lung cancer rates, and children are the most affected by it. The effects of air pollution on trees, plants, crops, and people are evident. Property damage is another detrimental effect of pollution.

The growing need for energy has turned humanity's attention to nuclear power. However, radioactive waste produced by nuclear power plants today will have to be safely managed for periods far exceeding all recorded history.

Another problem is desertification and soil erosion. Experiments have shown that the process of desertification is anthropogenic, caused by the degradation of the land itself. Grazing of goats and other animals, removal of vegetation causes further expansion of deserts.

Human activities also harmed plants. In the last decades, the forests of the country depleted pretty rapidly.

Environment versus Development is an old discussion. It frequently asks whether a country that prioritizes development can have a good environment. Environmental conservation is not a luxury. Ecological debates took the Development's point, rather than the Environment's, stating that good environmental management is inherently a good thing.

Environmental management seeks to integrate nature with society without damaging the former or harming the interests of the latter. Rational use of the environment presupposes rational use of natural and scientific resources, thus requiring an interdisciplinary approach to the problem of resource use, reuse, and recycling. The focal point of environmental management is the minimization of anthropogenic impact on the environment.

Improving the quality of life requires promoting rapid and intensive development in many areas. However, such a development should not harm nature too much so as to preserve the ability of the planet to host life. No competent person would say no development should take place and that the environment should remain exactly as it is.

Many conservationists fear that too much time was lost on identifying the need for rational environmental management and that the best humanity can hope for is preventing further degradation of ecosystems, resources, and the environment.

Humans are an integral part of global ecosystems—they both depend on each other, even despite the fact that environmental management is inherently biased towards humanity's viewpoint. The decisions that people make in environmental management will decide whether humanity will survive as a part of the global ecosystem.

If humans were to be content with halting excessive development and strive for economic and ecological sustainability, the planet's population would naturally stabilize due to birth and death rates. Industrial growth should stop for the sole reason of ousting old and outdated industrial facilities (Alekseev et al., 2018; Stroiteleva et al. 2020). People's attitudes should also change towards preferring such non-material goods as leisure and education. What is the point of economic growth if the only thing it creates is a soul-crushing environment?

The tasks of environmental management are various and multifaceted. The goals of comprehensive ecological planning are usually divided into three groups:

- Protecting physical and mental well-being;
- Strengthening economic values;
- Preserving collective satisfaction.

These goals should aim to preserve life and its quality. However, the goals are often generic, dwelling mostly on the quality of the environment. The definition of "quality of environment" varies from person to person—it can mean zero population growth, being content with simple living, or eradicating poverty so as to preserve human existence and dignity.

There are two approaches to environmental management: conservational and conservative.

The first approach demands non-interference with the biophysical world and a complete adaptation of humans to it. Adaptation is interpreted as a somewhat limited tolerance, where the stress is acknowledged but accepted without changing the existing way of life, at least in the short term.

The second approach reserves room for human adaptation to the biophysical environment. Adaptation involves some positive and deliberate reaction, typically aimed at reducing exposure to harmful elements, thus serving a homeostatic purpose.

The biota (flora and fauna) of one habitat differs from that of another. The factors that limit the viability of a species in a particular habitat are called environmental factors. They are responsible for the growth, distribution, behavior, and survival of organisms. Physical and chemical factors combine to form the non-living (abiotic) environment, whereas the living (biotic) environment comprises the interrelationship of living communities aimed to obtain food, shelter, energy, etc. Moreover, abiotic and biotic factors often overlap, creating perfect conditions for a particular community.

Communities are collectively influenced by these factors. However, some factors have more influence than others do. Research into relative influences allowed the creation of certain rules and principles concerning the limitation of some factors (Adarina et al., 2019; Nekhvyadovich et al., 2020).

Any factor that tends to slow down the metabolic rate or potential growth in an ecosystem is called a limiting factor. However, any factor that controls survival is called regulating a factor. Many factors determine the success of a population or community. The limiting factor can be any condition that approaches or exceeds the tolerances of an organism. Although the quantity and quality of incoming energy, as well as the laws of thermodynamics, impose some limits, different ecosystems have various combinations of limiting factors that may constrain the biological structure and function further.

Climatic elements that affect plant growth are temperature, rainfall, and sunlight exposure. Edaphic factors are related to soil characteristics.

All of these factors play a role in plant growth and development. Soil acts as a source of moisture that plants need during growth.

The activity of certain environmental factors, such as temperature exceeding the maximum, will eventually terminate all life.

3 Results

Every living organism is affected by the environment in different ways. Those environmental or ecological factors govern the survival and growth cycles of life. Ecological laws regulate the reactions of organisms to the environment.

Basic materials available in amounts close to the critical minimum will tend to serve as limiting factors. This shows that plant growth depends on the volume of nutrients that are available in the minimum amounts. Therefore, plant growth is limited by the essential nutrients that are present in sufficient quantities relative to the needs of the plant. In other words, if any essential nutrient is present only in minimal quantities relative to the plants' needs, plant growth will be limited (Dragunov & Shenshinov, 2020; Merdesheva et al., 2020).

Light, temperature, nutrients, and basic elements are considered limiting factors. This can be exemplified by the absence of some plants in shaded areas or the absence of vegetation above certain heights in the Alps. This is the reasoning behind the distribution of plants in terms of insufficient light, temperature, or nutrients. Crop yields are often limited not by nutrients but by other factors that are needed in negligible amounts and are in short supply (e.g., zinc). Although this general concept is well-received by scholars, it is also criticized because it does not fully explain the concept of length. These factors operating, other than the minimum, can change the availability and utilization of resources. In some plants, zinc may act as a limiting factor only when grown in sunlight since these plants require less zinc when grown in the shade than in full sunlight.

In environmental science, three terms can be used to illustrate the responses of plants to an increase in the intensity of one or another environmental factor. These terms are referred to as absolute, harmonic, and ecological optima. They can be taken in a broader sense than the previously discussed cardinal points and their meanings, which can be expressed as follows:

- Absolute optimum corresponds to the highest activity of any plant function, such as transpiration or respiration. As the temperature rises to a certain point, plants can transfer more and more water at given intervals. Then, due to interference or destruction of certain complex parts of the organism, the activity rate decreases

sharply. Therefore, the absolute optimum point of activity can be defined as the point at which limiting or restraining factors begin to act.
- Harmonic optimum corresponds to the most favorable intensity of one function relative to the other functions of the plant. Whereas transpiration is a necessary function of the plant, excessive activity of any function (including it) would soon result in the death of the plant. The plant achieves the highest activity when the activity rates of various functions are properly coordinated and in harmony with each other.
- Theoretical ecological optimum is comprised of the sum of different harmonic optimums. Determining the average of the ecological optimum requires accounting for the relative importance of the various plant functions in their relation to the growth and behavior of the whole organism. However, it is difficult to sum different harmonic optimums or the exact location of the ecological optimum exactly (Avkopashvili et al., 2019; Shenshinov & Al-Ali, 2020).

The concept of limiting factors includes the limiting effect of maximums and minimums, meaning that such factors can be limiting in both small and excessive quantities. Any environmental factor that lies below the critical minimum or significantly above the critical maximum will limit the growth of the organism in that area. This is known as Shelford's law of tolerance.

Organisms have ecological maximum and minimum requirements for each environmental factor—the limits of an organism's tolerance to a given factor. Different environmental factors do not affect organisms independently. One factor is modified by another one, and thus organisms respond to the whole environment (Vukovich et al., 2018). According to the law of tolerance, each environmental factor has two different zones: a zone of tolerance and a zone of intolerance.

4 Conclusion

Not all possible factors are equally important in a given situation or for a given organism. Moreover, the physical environment can also have an important influence, both by limiting or promoting the activity of organisms.

Protecting and preserving the environment is one of the most pressing problems. Humanity has reached critical points in several areas of anthropogenic impact on nature. Progressive development and the very existence of modern civilization depend on a constructive solution to the problem of the relationship between humans and their environment. Modern environmental problems are industrialization and urbanization, the depletion of traditional energy and raw materials resources, the constant growth of demographic pressure on nature, the violation of the natural ecological balance, the economic elimination of certain plant and animal species, the negative consequences of pollution of nature by waste, etc.

The degradation of Earth's biosphere under the influence of increasing human activities is an alarming and dangerous sign of environmental crises. It forces the global scholarly community to seek solutions to this issue and to reconsider the very notion of rationality in our impact on nature. The key lies in understanding the nature of ecological equilibrium in terms of macro- and micro-ecosystems and the systematic management of natural resources in a truly rational and scientific manner.

References

Adarina, R., Gazukina, Y., & Yankovskaya, K. (2019). Indicators of the "green economy" as a tool for monitoring the regional economy. In *IOP Conference Series: Earth and Environmental Science* (Vol. 395, p. 012107).https://doi.org/10.1088/1755-1315/395/1/012107

Alekseev, A. N., Savenkov, D. L., Shvakov, E. E., Lebedev, N. A., & Ragulin, A. D. (2018). Financial problems of regional development and the ways of their solution in contemporary Russia. *Quality—Access to Success, 19*(S2), 71–75.

Avkopashvili, P. T., Polukhin, A. A., Shkodinsky, S. V., & Poltarykhin, A. L. (2019). The fundamental provisions of the concept of knowledge economy. In E. Popkova, Y. Ragulina, & A. Bogoviz (Eds.), *Industry 4.0: Industrial revolution of the 21st century* (pp. 57–64). Cham, Switzerland: Springer. https://doi.org/10.1007/978-3-319-94310-7_5

Bogoviz, A. V., Shvakov, E. E., Tretyakova, O. G., Zakharov, M. Y., & Abramov, A. N. (2020). Globalization of education in the conditions of formation of the global knowledge economy: Regularities and tendencies. In E. Popkova (Ed.), *Growth poles of the global economy: Emergence, changes and future perspectives* (pp. 993–1000). Cham, Switzerland: Springer. https://doi.org/10.1007/978-3-030-15160-7_99

Dragunov, A., & Shenshinov, Y. (2020). Effective development of innovative technologies as a method for increasing labor productivity in industry. In *E3S Web of Conferences* (Vol. 208, p. 03001). https://doi.org/10.1051/e3sconf/202020803001

Glotko, A. V., Sycheva, I. N., Dunets, A. N., Poltarykhin, A. L., Zhuravlev, P. V., & Tubalets, A. A. (2019). Integration processes in the agro-industrial complex of the regions: Development and environmental problems. *Journal of Environmental Management and Tourism, 10*(3), 613–621.

Glotko, A. V., Voronkova, O. Y., Miheeva, N. M., Skoblikova, T. V., Alekhina, N. A., & Zhminko, A. E. (2019b). The economic development of a transboundary region under the conditions of the Eurasian economic union. In *Proceedings of IBIMA 2019: Education Excellence and Innovation Management through Vision 2020* (pp. 7957–7968). Granada, Spain.

Merdesheva, E. V., Titova, O. V., & Avkopashvili, P. T. (2020). Methodological approach to the classification of digital economies by environmental efficiency and sustainability criterion. In E. Popkova, & B. Sergi (Eds.), *Scientific and technical revolution: Yesterday, today and tomorrow* (pp. 1360–1368). Cham, Switzerland: Springer. https://doi.org/10.1007/978-3-030-47945-9_148

Nekhvyadovich, L., Kostin, K. B., & Mironov, V. V. (2020). Government and public monitoring of entrepreneurship in a multicultural region: New perspectives in the conditions of the digital economy. In E. Popkova, & K. Vodenko (Eds.), *Public administration and regional management in Russia* (pp. 147–154). Cham, Switzerland: Springer. https://doi.org/10.1007/978-3-030-38497-5_17

Shenshinov, Y., & Al-Ali, A. (2020). The tools of increasing efficiency of human resource in the lean production environment: Conceptual study. *International Journal of Core Engineering & Management, 6*(7), 1–18.

Stroiteleva, T. G., Kletskova, E. V., & Balaeva, S. I. (2020). Ontogenesis of green development model of digital economy: Essence, stages, and prospects. In E. Popkova, & B. Sergi (Eds.), *Scientific and technical revolution: Yesterday, today and tomorrow* (pp. 410–417). Cham, Switzerland: Springer. https://doi.org/10.1007/978-3-030-47945-9_46

Stroiteleva, T. G., Petrova, E. A., & Yungblyudt, S. V. (2020). Responsible information society as a social environment for creation of the "green" digital economy: Formation and monitoring. In E.

Popkova, & B. Sergi (Eds.), *Scientific and technical revolution: Yesterday, today and tomorrow* (pp. 402–409). Cham, Switzerland: Springer. https://doi.org/10.1007/978-3-030-47945-9_45
Tashkulova, G. K., & Kletskova, E. V. (2020a). Place and role of human in the system of circular reproduction in the digital regional economy. In E. Popkova, & B. Sergi (Eds.), *Scientific and technical revolution: Yesterday, today and tomorrow* (pp. 943–951). Cham, Switzerland: Springer. https://doi.org/10.1007/978-3-030-47945-9_101
Tashkulova, G. K., & Kletskova, E. V. (2020b). The model of organization of "green" digital production and consumption based on the internet of things and AI. In E. Popkova, & B. Sergi (Eds.), *Scientific and technical revolution: Yesterday, today and tomorrow* (pp. 329–336). Cham, Switzerland: Springer. https://doi.org/10.1007/978-3-030-47945-9_37
Vukovich, G. G., Makuschenko, L. V., Bateykin, D. V., Titova, O. V., & Dobrosotskiy, V. I. (2018). Support of the territories of advanced economic development on human capital: Theory and practice. *Quality—Access to Success, 19*(S2), 157–160.

Ecotourism as a Basis for Sustainable Regional Development

Larisa I. Nekhvyadovich, Toskanay A. Kuttubaeva, and Nikita E. Petrenko

Abstract Tourism includes many elements, but the environmental aspects act as a key factor of tourism. These aspects include Inputs, Processes, Outputs, and Feedback. The Inputs include elements from demand—the tourism potential, motivation, travel opportunities, and destination resources—attractions, services, information, and lodging. The Processes include economic, social, and environmental interactions that can have positive and negative effects. The Feedback allows planning appropriate supervision measures, capacity, policies, and strategies for tourism growth while minimizing negative effects. The tourism system has been defined and modeled via several different perspectives. A functioning tourism system consists of several interrelated components. Demand consists of the tourism market and includes the interest of people to travel and their ability to travel.

Keywords Ecotourism · Sustainable development · Mass tourism · Environment · Impacts of tourism

JEL Codes Q5 · Q51 · Q56

1 Introduction

Supply components include transportation, attractions, services, and information. Attractions are quality resources designed to meet the needs of visitors. Services include the variety and quality of food, lodging, and other products. Information/promotion is essential to encourage tourists to use the available products. The systemic approach to tourism involves moving tourists from their origin area to their

L. I. Nekhvyadovich (✉)
Altai State University, Barnaul, Russia

T. A. Kuttubaeva
Gorno-Altai State University, Gorno-Altaysk, Russia

N. E. Petrenko
Novosibirsk State University of Economics and Management "NINH", Novosibirsk, Russia
e-mail: nepetrenko@ro.ru

E. G. Popkova and B. S. Sergi (eds.), *Geo-Economy of the Future*,
https://doi.org/10.1007/978-3-030-92303-7_34

destination through a transit region. This approach emphasizes the interdependence of the source environment and the destination environment. However, many tourism flows are hierarchical and may include many nested and overlapping destinations. The tourism system is complex in nature and includes several critical actors responsible for the functioning of tourism. This system becomes even more complex in environmentally fragile destinations, where conservation is of paramount importance. For this reason, studying the delicate relationship between the environment and tourism is especially important (Glotko et al., 2019; Tashkulova & Kletskova, 2020a).

The growing interest in conservation and environmental well-being has grown dramatically over the past two decades. At the same time, there has been a boom in global tourism, leading to a phenomenon called "mass tourism." With this unprecedented growth, the interactions between environmental conservation and tourism become inevitable and essential to study. Both favorable and unfavorable effects on the natural and socio-cultural environment can occur in areas where tourism already exists or may develop soon. There are two different views on the system of tourism:

1 Tourism harms the natural environment, meaning that tourism and conservation conflict.
2 Tourism and conservation can potentially form symbiotic relationships.

2 Methodology

The tourism system has been the topic of debate for the past three decades. The International Union for Conservation of Nature first raised the issue of relations between tourism and conservation when its Director-General published the study "Tourism and Environmental Conservation: Conflict, Coexistence, or Symbiosis?" Thirteen years ago, this question seemed to remain unanswered, but another question—"Tourism and the environment – agreement or disagreement?"—was posed. In this way, the tourism system is examined from one of two points of view—conflict or symbiosis. Either point can be accepted and defended. However, regardless of which one is supported, the way to reduce conflict or increase compatibility is to understand, plan, and manage the system based on knowledge and understanding of environmental conflicts. This approach will promote sustainable development.

The tourism system is based on the sustainable use of natural resources, supported by the following:

- World Conservation Strategy;
- Sustainable Development Strategy;
- World Commission on Environment and Development.

Tourism often serves as the link between the environment and development. This partnership is based on resource sustainability, and tourism must be fully integrated into the resource management process. That would require adopting resource conservation values along with more traditional development goals. Protection and maintenance of environmental quality are central to the goals of environmental conservation and resource sustainability. This primary goal relies on the awareness of the need to protect and maintain the environment while helping to realize the tourism potential. Tourism management usually follows a reductionist, deterministic, linear view on nature and conservation that includes several participants.

Ecotourism involves a huge range of stakeholders with different interests and goals. Some play a more prominent role than others, but almost all are represented in the development and management of ecotourism facilities. The key to the success of ecotourism lies in forming strong partnerships with stakeholders so that multiple conservation and equitable development goals can be achieved. However, partnerships can be difficult because of the number of stakeholders involved and their different needs. Ecotourism stakeholders are broadly classified into major decision-makers and supporting players. The key decision-makers are leaders, local communities, the tourism industry, government officials, and non-governmental organizations. While the funding agencies are the key supporting players, all stakeholders play an important role in the functioning of ecotourism, and it is the consensus and coordination among these critical stakeholders that determine the success of an ecotourism project (Alekseev et al., 2018; Stroiteleva et al., 2020).

Ecotourism is an interdisciplinary topic, often overlapping with several other disciplines. There is a shortage of secondary literature on the basics of ecotourism and impact assessments. Most publications are concerned with international ecotourism, while national studies are limited.

There are about 1500 scholarly studies published worldwide on various aspects of tourism and the environment, most of which have been published in the last 25 years. There are several volumes on ecotourism, varying greatly in approach and technical detail. There are also texts on definitions, principles, and types of ecotourism. Ecotourism follows two important principles of sustainability:

- Support the preservation of natural ecosystems and help maintain the local economy, which develops with varying intensity;
- Ecotourism can be soft or hard, depending on the strictness of the visitors' conduct. Some proponents of ecotourism believe that humans are part of nature and their impact is part of the natural process, while critics of ecotourism argue that people simply should not visit natural areas, as they invariably harm them;
- Mass tourism, where maximum income and performance are the most important factors;
- Alternative tourism, where sustainability (i.e., limiting the number of tourists) is the most important criterion for success.

Ecotourism can be further divided into passive (sightseeing), active (rafting, rock climbing, river crossing, etc.), or exploitative (staying at the lodge and using resources). Economic and environmental incentives result in different types of ecotourism.

In addition, the studies focus mainly on covering aspects of forest ecology, natural regeneration, forest fires, and wildlife. However, most of these studies include anthropogenic pressures, such as the dependence of rural residents on wild products, while the growth of tourism is considered a potential threat.

Methods of multicriteria decision-making, particularly the process of analytical hierarchy via geographic information systems and remote sensing, are among the most commonly used approaches to ecotourism decision-making criteria (Nekhvyadovich et al., 2020; Stroiteleva et al., 2020).

Recently, the use of social media for quantitative assessment has also gained importance in ecotourism research. Spatial, temporal, and financial limitations on the scope of surveys at the entrances to major attractions have been replaced by a new source of information available from social networking sites that can be used to approximate the number of visits to various destinations. Here, photo location data is used to estimate the number of visits to various recreational sites around the world, and information derived from photographer profiles is used to determine the origins of travelers. These estimates are then compared to empirical data at each site. Crowdsourced information can serve as a reliable proxy for empirical visitation rates. This new approach also allows understanding what elements of nature attract people to places worldwide and whether changes in ecosystems will change visitation rates. The integrated evaluation of ecosystem services and recreation trade-off model help determine the spatial distribution of visitors at a given destination, which can be further correlated with the underlying natural, built-up, and accessible features. Geographic information systems play a crucial role in the sustainable development of ecotourism destinations. They are used to understand the suitability of sites, visitor patterns, and their impact on social and ecological aspects.

Ecotourism initiatives have developed as a result of the effective participation of all relevant stakeholders. Despite geographic and other differences, the success of these initiatives has many common features. These features mainly include the full acceptance and support of local community members, consensus among community individuals who participate in ecotourism activities, benefit-sharing, and the resolution of potential conflicts (Adarina et al., 2019; Merdesheva et al., 2020).

Ecotourism has become a widely used tool for generating profit, developing communities, and preserving the environment. It can sometimes achieve all of these things, but it impacts the natural and cultural environment of the host destination in the process. In particular, many ecotourism products rely on technologies that harm the environment. As the pressures of ecotourism on protected areas continue to grow, stakeholders are finding it increasingly important to understand the effects of ecotourism and how to manage them. The impacts of ecotourism in the context of environmental and social aspects are presented below.

The first methods developed to deal with tourism impacts arose from the concept of carrying capacity, which originated in the field of rangeland management. However, the first analyses of tourist carrying capacity and impacts were done in the USA in the 1930s; most of them are debatable in every context.

Most definitions contain two aspects:

- The issue of "capacity"—"How many tourists can be accommodated before some adverse effects occur?"
- The issue of "capacity perception"—"How many tourists can be served before the visitor satisfaction declines?"

These cases simultaneously focus on the impact of the host country, population attitudes, and tourist satisfaction.

More recently, the concept of sustainable development acknowledged the need for a multidimensional approach that combines social, economic, and environmental aspects simultaneously (Costanza, 1996). The approach is used to determine the desirable upper limits of development—the optimal use of tourism resources. It presupposes that "the transit city cannot really be separated from the decision to limit use, because one is related to the other." Nevertheless, there are several carrying capacities, and "no single capacity can be attributed to the entire area." Therefore, the concept can be seen as the correct balance between the often conflicting goals of its various components.

According to ecotourism data, all fixed and flexible components of the natural environment (ecological potential, natural heritage potential, shoreline, climate, etc.) and the infrastructure systems (water supply, sewage, electricity, gas supply, transportation, utilities, etc.) should be measured first. Socio-demographic measurement has its own indicators, which are continuously monitored. Conclusions on socio-demographic improvement are based on the monitoring results.

Other aspects of monitoring visitor characteristics and experiences include four components, namely: biophysical, psychological, socio-cultural, and managerial. But some scholars add different components to determine the overall impact of an ecotourism project, such as the experiential and economic components (Avkopashvili et al., 2019; Ragulina et al., 2019; Vukovich et al., 2018).

3 Results

Ecotourism management is now seen mainly as a planning process and a systematic strategic policy tool for sustainable tourism development, rather than "a scientific measure." In fact, the need to limit tourism development is widely accepted so as to limit its negative impact on natural resources, social structures, cultural patterns, economic activities, and land use in local communities. In addition, "capacity may not be used as an absolute limit, but as a means of identifying critical thresholds that require attention, and thereby removing obstacles or applying control measures."

Since the phrase "carrying capacity" is still widely used to refer to the concept of limiting tourism to reduce negative impacts, it has raised the perceived importance of impact monitoring. In turn, this fact has led to the development of a more sensitive and specific method. These methods recognize that tourism will cause changes and that the key goal of managing visitors is to limit exposure to predetermined values.

Some of these methods include the following:

- Range of recreational opportunities;
- Limits of acceptable variation;
- Visitor impact management;
- Visitor experience and resource protection;
- Tourism Optimization Management Model (Bogoviz et al., 2020; Tashkulova & Kletskova, 2020b)

The ecotourism plan must:

- Identify environmentally sensitive areas surrounding protected areas to ensure the ecological integrity of the buffer zones and prevent their shrinking or destruction;
- Evaluate carrying capacity of visitors and vehicles at three levels: physical, real, and effective;
- Set the maximum number of visitors allowed to enter at any one time, based on the carrying capacity of the habitat;
- Specify the zone open to tourism, designated as an "ecotourism zone" (Stroiteleva et al., 2020);
- Develop, in cooperation with local authorities, a community-based tourism development strategy to ensure long-term regional benefit sharing and the promotion of regional activities;
- Develop codes and standards for private tourist sites located near critical wildlife habitats, ecologically sensitive areas, or buffer zones to provide benefits and revenues to these areas;
- Create monitoring mechanisms to assess the impact of tourism;
- Develop general guidelines for environmentally and culturally appropriate practices, as well as for all new designs.

4 Conclusion

In the context of the studied area, the number of visits has steadily increased without any assessment of the impact on regional ecology and socio-demographic environment. With the increase in the number of visitors, the number of vehicles coming from outside has also increased significantly, contributing to carbon dioxide emissions. In addition, the random expansion of guesthouses and hotels increases the pressure on limited water and land resources. Agricultural land is sold for construction purposes and results in a significant shift in the socio-economic structure of society. There is an urgent need to develop a set of measures for all interested stakeholders in ecotourism. It is impossible to maintain the ecological, cultural, and spiritual integrity of destinations without understanding their boundaries. However, stated intentions tended not to translate into concrete action, so there is a great need to develop checks and balances to ensure their implementation. Guidelines need to be based on an assessment of carrying capacity to be scientifically sound and implementable. This study aimed to develop a framework and methodology for evaluation, which can help to develop a practically implementable contextual policy, serving as a guide for the development of tourism destinations.

References

Adarina, R., Gazukina, Y., & Yankovskaya, K. (2019). Indicators of the "green economy" as a tool for monitoring the regional economy. In *IOP Conference Series: Earth and Environmental Science* (Vol. 395, p. 012107).https://doi.org/10.1088/1755-1315/395/1/012107

Alekseev, A. N., Savenkov, D. L., Shvakov, E. E., Lebedev, N. A., & Ragulin, A. D. (2018). Financial problems of regional development and the ways of their solution in contemporary Russia. *Quality—Access to Success, 19*(S2), 71–75.

Avkopashvili, P. T., Polukhin, A. A., Shkodinsky, S. V., & Poltarykhin, A. L. (2019). The fundamental provisions of the concept of knowledge economy. In E. Popkova, Y. Ragulina, & A. Bogoviz (Eds.), *Industry 4.0: Industrial revolution of the 21st century* (pp. 57–64). Cham, Switzerland: Springer. https://doi.org/10.1007/978-3-319-94310-7_5

Bogoviz, A. V., Shvakov, E. E., Tretyakova, O. G., Zakharov, M. Y., & Abramov, A. N. (2020). Globalization of education in the conditions of formation of the global knowledge economy: Regularities and tendencies. In E. Popkova (Ed.), *Growth poles of the global economy: Emergence, changes and future perspectives* (pp. 993–1000). Cham, Switzerland: Springer. https://doi.org/10.1007/978-3-030-15160-7_99

Costanza, R. (1996). Ecological economics: Reintegrating the study of humans and nature. *Ecol. Applicat, 6*(4), 978–990.

Glotko, A. V., Sycheva, I. N., Dunets, A. N., Poltarykhin, A. L., Zhuravlev, P. V., & Tubalets, A. A. (2019). Integration processes in the agro-industrial complex of the regions: Development and environmental problems. *Journal of Environmental Management and Tourism, 10*(3), 613–621.

Merdesheva, E. V., Titova, O. V., & Avkopashvili, P. T. (2020). Methodological approach to the classification of digital economies by environmental efficiency and sustainability criterion. In E. Popkova, & B. Sergi (Eds.), *Scientific and technical revolution: Yesterday, today and tomorrow* (pp. 1360–1368). Cham, Switzerland: Springer. https://doi.org/10.1007/978-3-030-47945-9_148

Nekhvyadovich, L., Kostin, K. B., & Mironov, V. V. (2020). Government and public monitoring of entrepreneurship in a multicultural region: New perspectives in the conditions of the digital economy. In E. Popkova, & K. Vodenko (Eds.), *Public administration and regional management in Russia* (pp. 147–154). Cham, Switzerland: Springer. https://doi.org/10.1007/978-3-030-38497-5_17

Ragulina, Y. V., Semenova, E. I., Avkopashvili, P. T., Dmitrieva, E. A., & Cherepukhin, T. Y. (2019). Top-priority directions of implementing new internet technologies on the territories of rapid economic development. In E. Popkova, & V. Ostrovskaya (Eds.), *Perspectives on the use of new information and communication technology (ICT) in the modern economy* (pp. 182–188). Cham, Switzerland: Springer. https://doi.org/10.1007/978-3-319-90835-9_21

Stroiteleva, T. G., Kletskova, E. V., & Balaeva, S. I. (2020). Ontogenesis of green development model of digital economy: Essence, stages, and prospects. In E. Popkova, & B. Sergi (Eds.), *Scientific and technical revolution: Yesterday, today and tomorrow* (pp. 410–417). Cham, Switzerland: Springer. https://doi.org/10.1007/978-3-030-47945-9_46

Stroiteleva, T. G., Petrova, E. A., & Yungblyudt, S. V. (2020). Responsible information society as a social environment for creation of the "green" digital economy: Formation and monitoring. In E. Popkova, & B. Sergi (Eds.), *Scientific and technical revolution: Yesterday, today and tomorrow* (pp. 402–409). Cham, Switzerland: Springer. DOI: https://doi.org/10.1007/978-3-030-47945-9_45

Tashkulova, G. K., & Kletskova, E. V. (2020a). Place and role of human in the system of circular reproduction in the digital regional economy. In E. Popkova, & B. Sergi (Eds.), *Scientific and technical revolution: Yesterday, today and tomorrow* (pp. 943–951). Cham, Switzerland: Springer. https://doi.org/10.1007/978-3-030-47945-9_101

Tashkulova, G. K., & Kletskova, E. V. (2020b). The model of organization of "green" digital production and consumption based on the internet of things and AI. In E. Popkova, & B. Sergi (Eds.), *Scientific and technical revolution: Yesterday, today and tomorrow* (pp. 329–336). Cham, Switzerland: Springer. https://doi.org/10.1007/978-3-030-47945-9_37

Vukovich, G. G., Makuschenko, L. V., Bateykin, D. V., Titova, O. V., & Dobrosotskiy, V. I. (2018). Support of the territories of advanced economic development on human capital: Theory and practice. *Quality—Access to Success, 19*(S2), 157–160.

Modern Scenario of Solid Waste Management

Nikolay S. Chasovskikh, Kseniya A. Melekhova, and Marianna M. Teryutina

Abstract Characterization of waste is critical to ensure the proper implementation of waste management projects. The composition of municipal solid waste varies from city to city. It depends on various factors, such as economic status, cultural norms, climate, location, and energy sources. Municipal waste typically includes compostable organic matter (e.g., food, fruits, vegetable peels, and garden waste), recyclable materials (e.g., plastic, paper, metals, glass, etc.), toxic materials (e.g., discarded medical ice, pesticides, etc.), and contaminated waste (e.g., sanitary napkins, cotton with bloodstains, etc.) The municipal solid waste was observed to contain 40–60% of organic matter, 30–40% of fine soil and ash, 3–6% of paper, and less than 1% of metals, plastics, and glass. Organic matter can be recycled into valuable compost.

Keywords Municipal solid waste · National resources · Efficient technologies · Decentralized recycling

JEL Codes Q5 · Q56 · Q57

1 Introduction

In low-income countries, the share of organics in municipal solid waste (MSW) is 64%, and the share of paper is only 5%. In high-income countries, the share of organics is 28%, and the share of paper is 31%. A 2009 World Bank report showed that global MSW with the majority of the organic fraction (46%) are followed by paper (17%), plastics (10%), glass (5%), metals (4%), and other waste (18%).

The characteristics of the waste depend on the following factors:

N. S. Chasovskikh (✉)
Gorno-Altaisk State University, Gorno-Altaysk, Russia

K. A. Melekhova
Altai State University, Barnaul, Russia

M. M. Teryutina
Arctic State Agrotechnological University, Yakutsk, Russia

E. G. Popkova and B. S. Sergi (eds.), *Geo-Economy of the Future*,
https://doi.org/10.1007/978-3-030-92303-7_35

- Way of life;
- Nutrition habits;
- Income status;
- Cultural practices;
- Season of the year;
- Climate;
- Type of commercial activity.

The amount of waste generally depends on social and economic status, the standard of living, climatic conditions, etc. The proportion of the biodegradable fraction tends to decrease with increasing economic status. However, the share of organic waste is higher in the waste of rural households than in urban households. The physical and chemical composition of waste is closely related to population density and the economic growth of cities. Rapid economic growth and urban overcrowding change living standards and other habits, ultimately increasing the generation of MSW per capita (Glotko et al., 2019b; Tashkulova & Kletskova, 2020).

Most MSW is dumped directly into open unsanitary landfills. It accounts for over 92% of the total waste generated nationwide. In most villages, MSW is also disposed of in a very unscientific way: in water bodies (particularly in village ponds), drains, open areas, and along roads running from the outer contour of the village. Additionally, waste, including plastics, is burned in the open air, creating potential pollution and health hazards. This practice of MSW disposal poses a serious threat to health, environmental degradation, and economic losses. There is an urgent need to pay special attention to the scientific study of MSW and find the reasons leading to the creation of environmentally friendly and economically viable decentralized models of MSW management.

Janitors collect solid waste from the streets and usually dump it in open areas near the outer contour of homes or burn it in open areas, streets, and landfills. In urban areas, janitors clean up roads and streets and put garbage in specially designated areas (e.g., street garbage cans in banks, markets, and offices or nearby unsanitary landfills where MSW is burnt in open areas).

There is a considerable shortage of the necessary waste collection system, including insufficient vehicles and budget for their maintenance. For several reasons, most local authorities cannot organize the necessary amount of waste disposal containers in all areas of the cities. The average MSW collection efficiency is more than 80% in cities and up to 30% in villages. Local authorities play a key role in improving the MSW collection efficiency. Vehicles designed to collect and transport MSW are not of the desired design and can create pollution on the routes. However, advanced hydraulic vehicles are being gradually introduced.

2 Methodology

In 2010, the National Green Tribunal (NGT) was created with authority to receive and seek final review of applications or appeals within six months of filing. Given the environmental hazards associated with MSW, NGT has released several guidelines since its inception. The amount of household and livestock waste is also increasing in rural areas. Most villages still lack the necessary functional mechanism for MSW disposal. The lack of involvement in the preservation of national resources is the reason why everyone produces waste, but not everyone is in favor of managing at least their waste (Nekhvyadovich et al., 2020; Shenshinov, 2012).

The unscientific management of MSW worldwide has made MSW one of the biggest problems in terms of environmental degradation and health and economic losses. Left unattended, MSW chokes the drains and sewers, which become perfect breeding grounds for mosquitoes and a major cause of urban flooding. In these cases, scientists play a key role in achieving purity. To improve livelihoods and public health, eliminate pollution, and protect the environment, especially in low- and middle-income countries, it is necessary to develop the most effective strategies, which include the principle of minimizing the amount of generated MSW and maximizing waste recycling and resource recovery. To use these strategies scientifically, it is necessary to create a functional mechanism so that it has a tangible impact on the local community. A comprehensive solid waste management mechanism should consist of a variety of environmentally friendly and cost-effective technologies and easily implemented effective mechanisms involving all interested parties. To solve the problem of unscientific disposal of solid waste, where possible, it is necessary to adhere to the principle of waste reduction in accordance with the established regulations (Glotko et al., 2019a; Stroiteleva et al., 2020).

MSW management in rural areas is the most critical component of environmental sanitation compared to urban waste. The problem is becoming even more acute due to the rapid growth of the population and the lack of a suitable mechanism for the scientific disposal of MSW in rural territories. The highest priority should be given to rural territories to conserve the environment, precious land resources, and management costs. The preference is given to decentralized recycling, which allows minimizing transportation costs and environmental impact (e.g., bio mechanization, microbial composting, vermicomposting, anaerobic digestion, or any other appropriate treatment to stabilize biodegradable waste). MSW has a higher percentage of organic waste, making it suitable for composting and producing biogas and biofuel.

The amount of MSW is constantly increasing and disposed of in an unscientific and uncontrolled manner, resulting in serious environmental and health losses in rural areas. Compared to urban areas, the standard of living and eating habits of rural residents are also changing due to the improved road and media connectivity across the country. MSW management must be put in place to improve environmental sanitation and life in rural areas. There is a common problem of open dumping in urban and rural areas, creating considerable health and environmental problems. Scientific and decentralized management of MSW, the closest to its source, is essential to achieving improved environmental sanitation.

3 Results

Rural areas strongly focus on safe and hygienic solid waste management, including source segregation, door-to-door collection, organics composting, materials recovery, and safe disposal of inert waste. Several preferences should be considered to minimize costs and ensure maximum resource extraction from MSW in the form of compost, biogas, and the recycling of fuel waste (Merdesheva et al., 2020; Shenshinov & Al-Ali, 2020).

The first preference is source reduction, segregation, reuse, and recycling. It can be accomplished by prohibiting the production or reducing the use of non-biodegradable items (plastic bags, plastic packaging, and single-use plastic thermocouples) and promoting the reuse of waste. In achieving clean and litter-free cities that include a ban on all types of plastic bags, mandatory source segregation is a key strategy for reducing the cost of any operating model or any technology. There is an urgent need to ensure 100% segregation of waste sources, maximum material recovery, minimal input to sanitary landfills, and scientific recycling within the economy by deploying qualified environmental scientists to achieve improved systems.

The second preference is recycling. Recycling commercial secondary waste can effectively utilize resources and reduce management costs and landfill burdens.

The third preference is composting waste. Converting organic waste into compost is a viable and scientific method of restoring plant nutrients. The application of compost derived from organic waste can improve soil health and crop yields. The use of compost will also reduce the consumption of chemical fertilizers, unnecessary transportation and management costs, and the burden on landfills.

The fourth preference is that waste energy can only be obtained from commercial non-recyclable waste. It should meet the required calorific properties for use as fuel in cement, biomass, and thermal power plants located within a radius of 100 km of the waste processing site.

The fifth preference is waste management. This is residual waste, consisting of inert waste such as street sweepings, fine earth, ashes, etc. They can be disposed of in landfills according to regionally accepted regulations.

Technological options for MSW management include several options. Community involvement, awareness, and low-cost technology are vital components to achieve a sustainable integrated MSW management system. The most preferred technologies should be chosen for their environmental relevance, socially acceptable, low capital operating costs, low space requirements, and ease of maintenance. Local authorities must be actively involved from the planning stage to the design, implementation, operation, and maintenance of the MSW management mechanism.

Scientific integrated and decentralized systems controlled by the community have proven to be more successful than others.

The health problem is the problem of unsafe solid waste disposal. It is critical to have complete and accurate information about the quantity and quality of MSW to plan an effective recycling system in rural areas. Many technological options are available in the market, including old and well-accepted low-cost options for the scientific treatment and safe disposal of solid waste across the country. Turning organic waste into compost and using it in agriculture is an effective, low-cost option for burial or burning in open landfills. Compost derived from organic waste acts as a soil conditioner and organic fertilizer since it contains a high percentage of organic content and moisture-holding capacity (90–95%). Nevertheless, this compost has a generally low percentage of macro-and micronutrients compared to commercial fertilizers (Costanza, 1996; Tsvetkov et al., 2019).

The biodegradable MSW consists of commercially recyclable, non-commercially recyclable, and non-recyclable household hazardous and inert waste.

- Commercially recyclable waste includes plastics, paper, cardboard, metals, rags, etc. Financial investment is necessary for scientific development to ensure maximum material recovery and marketing. Various studies have shown that the composition of generated MSW varies from city to city.
- Non-commercial recyclable materials (e.g., tetrahedral package, plastic bags, small pieces of packaging materials, rags, etc.) can be converted into high-calorie waste fuel that can be used as fuel in cement, biomass, and thermal power plants.
- Construction and demolition waste and fine-grained earth can be recycled and reused for road and building construction.
- Household hazardous waste and sanitary materials must be scientifically treated and disposed of following established environmental legislation standards.

Approximately 15% of municipal solid waste can be conveniently separated into 25 categories of potentially recyclable sources. Failure to segregate recycling at the source impairs its quality and marketing value. According to the established rules, it is prohibited to incinerate mixed waste since it will cause serious pollution of the environment by releasing toxic gases and heat into the atmosphere and damaging precious natural plant nutrients rich in organic matter and valuable recyclable resources.

The composition of biologically degradable waste is highly suitable for turning it into a nutrient-rich biological fertilizer since it has a high percentage of organic content and moisture. Rural waste has a higher proportion of organic waste than urban waste. Moreover, it is almost free of toxic contamination (Ragulina et al., 2019; Vukovich et al., 2018). Most of the organic fraction consists of food waste, fruit and vegetable peels, and organic products, which have a high concentration of organic carbon and other plant nutrients (e.g., nitrogen, phosphorus, zinc, iron, etc.). Using a variety of available technologies, these plant nutrients can be stabilized and collected in the form of compost. Composting and vermicomposting of organic waste is a perfect, sustainable, and cost-effective method for rural areas. The presence of heavy metals in urban compost beyond the permitted limits constrains its use for agricultural purposes.

Anaerobic microbial digestion of organic waste produces flammable biogas and compost rich in methane. Rural areas have a higher potential for biogas production since there is a huge availability of cattle manure and biodegradable kitchen waste. Biogas plants at the household and community levels can be put into operation depending on feedstock availability, area, and demand for biogas. The biogas suspension can be used as a nutrient-rich compost that is also free of pathogens and weed seeds.

The composting technology is old, environmentally friendly, the most appropriate, inexpensive, and sustainable for processing the organic fraction of solid waste in both rural and urban areas. Composting is the natural decomposition and stabilization of organic waste. Mechanical and manual composting units can be installed depending on the amount of raw material. During composting, the amount of MSW is reduced by up to 50–85%. Other studies have observed that about 30% of the volume of organic municipal waste is reduced during composting.

In recent years, various advanced scientific methods and approaches have been developed to produce good quality, odor- and pathogen-free compost. Composting of bio-soluble MSW can be done aerobically and anaerobically at both the household and community levels. During the natural composting of waste, temperatures are usually raised to kill pathogens. Currently, compost prepared from solid waste is being promoted to solve the problem of waste disposal and provide a supply of fertilizer for agricultural purposes.

There are many individual and commercial composting facilities and about 69% of MSW is converted to compost nationwide. Composting organic matter is a natural and local method used for centuries. The bio soluble waste fraction is suitable for compost production by the pit, windrow, heap, and vermicompost methods. Earthworms act as a bioreactor and turn organic waste into valuable, nutrient-rich organic fertilizer. This method is environmentally friendly and cost-effective.

Anaerobic microorganisms convert complex organic compounds into a stabilized form by a reduction process. It is well known that the anaerobic digestion of organic waste produces mainly methane and carbon dioxide, and the temperature inside the heap of organic waste does not increase much. Anaerobic composting takes longer and requires no flocking compared to aerobic composting. Anaerobic composting of organic waste can be carried out in compost pits, barrels, closed containers, waste heaps, etc.

Aerobic composting is an environmentally friendly, inexpensive, easy to operate and maintain, and socially acceptable technology for composting compostable waste (e.g., kitchen, household, park, garden, and horticultural waste, etc.). In aerobic composting, microorganisms oxidize the complex organic compounds in the waste and produce carbon dioxide gas. Thus, the compost is enriched with nitrites and nitrates. Aerobic composting is exothermic, the temperature rises to 650 °C, and the condition can become anaerobic. Waste must be flocked mechanically or manually at regular intervals to maintain an aerobic state and to control the temperature inside the waste heaps. Aerobic conversion of soluble waste into compost is faster than anaerobic one. Biological treatment of the organic fraction of MSW is the most environmentally acceptable method (Nekhvyadovich et al., 2020). Vermicompost acts as a biocontrol agent and suppresses the growth of fungal pathogens through volatile organic compounds released by symbiotic microbes. Compared to conventional compost, the activity of earthworms can effectively reduce the bacterial pathogen load in the organic fraction of MSW. However, both compost and vermicompost are salmonella-free, making them suitable for agricultural use and improvement of soil fertility. Vermicomposting is an effective technology to reduce the toxicity of industrial waste. It is considered suitable due to the availability of nutrients, low cost, and simplicity. Moreover, conventional composting and vermicomposting of organic waste is a safe and sustainable method of disposal (Adarina et al., 2019; Karataev et al., 2020). It is noted that the quality of vermicompost was better compared with garden compost.

Vermicomposting is a better option than classical composting with various means. Particles of vermicompost, which are typically less than 5 mm in size and more uniform, show better agricultural potential than coarser classical compost. The smaller particles have a larger surface area. Therefore, they are more accessible to microbes. Moreover, the finer compost releases more nutrients than the coarser compost.

4 Conclusion

Social, ecological, and technological advances have changed the daily activities and habits of the rural and urban communities, which becomes the key issue of scientific management of the huge amount of generated MSW with a significant proportion of non-biodegradable waste. Compared to urban areas, MSW management in rural areas remains the most neglected, despite the fact that it is a key component of various sanitation programs. Given the current scenario of MSW management in rural areas, there is tremendous scope for maximizing the extraction of valuable plant nutrients in the form of compost, including vermicompost, recyclable materials, and improved environmental sanitation by engaging the local community and implementing local sustainable technologies. The results of our research will help local governments develop and implement an environmentally friendly, socially acceptable, low-cost, easy-to-operate, self-sufficient, and employment-generating mechanism for MSW mechanism.

References

Adarina, R., Gazukina, Y., & Yankovskaya, K. (2019). Indicators of the "green economy" as a tool for monitoring the regional economy. *IOP Conference Series: Earth and Environmental Science, 395*, 012107.https://doi.org/10.1088/1755-1315/395/1/012107

Costanza, R. (1996). Ecological economics: Reintegrating the study of humans and nature. *Ecol. Applicat, 6*(4), 978–990.
Glotko, A. V., Sycheva, I. N., Dunets, A. N., Poltarykhin, A. L., Zhuravlev, P. V., & Tubalets, A. A. (2019). Integration processes in the agro-industrial complex of the regions: Development and environmental problems. *Journal of Environmental Management and Tourism, 10*(3), 613–621.
Glotko, A. V., Voronkova, O. Y., Miheeva, N. M., Skoblikova, T. V., Alekhina, N. A., & Zhminko, A. E. (2019b). The economic development of a transboundary region under the conditions of the Eurasian economic union. In *Proceedings of IBIMA 2019: Education Excellence and Innovation Management through Vision 2020* (pp. 7957–7968). Granada, Spain.
Karataev, E. M., Merkuryev, V. V., & Titova, O. V. (2020). Integrating the "green economy" into the model of digital future of the modern socio-economic systems. In E. Popkova & B. Sergi (Eds.), *Scientific and technical revolution: Yesterday, today and tomorrow* (pp. 1402–1410). Cham, Switzerland: Springer. https://doi.org/10.1007/978-3-030-47945-9_153
Merdesheva, E. V., Titova, O. V., & Avkopashvili, P. T. (2020). Methodological approach to the classification of digital economies by environmental efficiency and sustainability criterion. In E. Popkova & B. Sergi (Eds.), *Scientific and technical revolution: Yesterday, today and tomorrow* (pp. 1360–1368). Cham, Switzerland: Springer. https://doi.org/10.1007/978-3-030-47945-9_148
Nekhvyadovich, L. I., Borodin, V. A., & Avkopashvili, P. T. (2020) A systemic approach to the study of the sustainability of the global digital economy: Economies of scale, ratchet effect and hysteresis effect. In E. Popkova & B. Sergi (Eds.), *Scientific and technical revolution: Yesterday, today and tomorrow* (pp. 1369–1377). Cham, Switzerland: Springer. https://doi.org/10.1007/978-3-030-47945-9_149
Nekhvyadovich, L., Kostin, K. B., & Mironov, V. V. (2020). Government and public monitoring of entrepreneurship in a multicultural region: New perspectives in the conditions of the digital economy. In E. Popkova & K. Vodenko (Eds.), *Public administration and regional management in Russia* (pp. 147–154). Cham, Switzerland: Springer. https://doi.org/10.1007/978-3-030-38497-5_17
Ragulina, Y. V., Semenova, E. I., Avkopashvili, P. T., Dmitrieva, E. A., & Cherepukhin, T. Y. (2019). Top-priority directions of implementing new internet technologies on the territories of rapid economic development. In E. Popkova & V. Ostrovskaya (Eds.), *Perspectives on the use of new information and communication technology (ICT) in the modern economy* (pp. 182–188). Cham, Switzerland: Springer. https://doi.org/10.1007/978-3-319-90835-9_21
Shenshinov, Y. V. (2012). Labor quality management. North-east Asia academic forum.
Shenshinov, Y., & Al-Ali, A. (2020). The tools of increasing efficiency of human resource in the lean production environment: Conceptual study. *International Journal of Core Engineering & Management, 6*(7), 1–18.
Stroiteleva, T. G., Petrova, E. A., & Yungblyudt, S. V. (2020). Responsible information society as a social environment for creation of the "green" digital economy: Formation and monitoring. In E. Popkova & B. Sergi (Eds.), *Scientific and technical revolution: Yesterday, today, and tomorrow* (pp. 402–409). Cham, Switzerland: Springer. https://doi.org/10.1007/978-3-030-47945-9_45.
Tashkulova, G. K., & Kletskova, E. V. (2020). The model of organization of "green" digital production and consumption based on the internet of things and AI. In E. Popkova & B. Sergi (Eds.), *Scientific and technical revolution: Yesterday, today and tomorrow* (pp. 329–336). Cham, Switzerland: Springer. https://doi.org/10.1007/978-3-030-47945-9_37
Tsvetkov, V. A., Loginov, E. L., & Raikov, A. N. (2019). Integrated network interaction of education with science and production. *Educational Technologies and Society, 19*(1), 389–398.
Vukovich, G. G., Makuschenko, L. V., Bateykin, D. V., Titova, O. V., & Dobrosotskiy, V. I. (2018). Support of the territories of advanced economic development on human capital: Theory and practice. *Quality—Access to Success, 19*(S2), 157–160.

A Complex of Measures Aimed at the Preservation of the Unique Ecosystem of the Volga-Akhtuba Floodplain on the Territory of the Volgograd Region

Alexander I. Belyaev, Alexander P. Istomin, Alexander G. Zhikharev, Anna M. Pugacheva, and Alexey A. Sukhov

Abstract The Volga-Akhtuba floodplain in the arid semi-desert zone of southern Russia has a particular value. A necessary condition for the functioning of the Volga-Akhtuba floodplain is the annual optimal flood moisture and filling of the hydrographic network. However, due to the artificial regulation of the Volga river flow by a cascade of reservoirs, without taking into account the values of the Volga-Akhtuba floodplain, its natural diversity, over a long period, led to serious changes in the hydrological regime, landscape structure, species composition and productivity of key natural complexes and ecosystems. Within the framework of the federal project "Rehabilitation of the Volga" of the national project "Ecology", measures are being taken to build and reconstruct culverts, clean up and ecological rehabilitation of water bodies, and additional watering of the Volga-Akhtuba floodplain.

Keywords Volga-Akhtuba floodplain · Water body · Additional watering · Clearing · Ecological rehabilitation · Culverts

JEL Code 556

A. I. Belyaev (✉) · A. P. Istomin · A. G. Zhikharev · A. M. Pugacheva · A. A. Sukhov
Federal Scientific Centre of Agroecology, Volgograd, Russia
e-mail: director@vfanc.ru

A. P. Istomin
e-mail: istomin-ap@vfanc.ru

A. G. Zhikharev
e-mail: zhiharev-ag@vfanc.ru

A. M. Pugacheva
e-mail: pugachevaa@vfanc.ru

A. A. Sukhov
e-mail: suhov-a@vfanc.ru

E. G. Popkova and B. S. Sergi (eds.), *Geo-Economy of the Future*,
https://doi.org/10.1007/978-3-030-92303-7_36

1 Introduction

The territory of the Volga-Akhtuba floodplain has an international nature conservation significance, it is a large array of wetlands in the arid South-East of Russia (namely, in the Volgograd and Astrakhan Regions) and in 2000 was included in the List of objects recommended for inclusion in the list of water—wetlands protected by the Ramsar Convention, adopted to draw the attention of the international community to the topic of the accelerating loss of wetland habitats. Countries, including the Russian Federation, by acceding to the Convention, have committed themselves to promote activities aimed at reversing the loss and degradation of declining wetlands.

In addition, since 1999, the Volga-Akhtuba floodplain has the status of a key ornithological territory of international importance "Akhtubinskoye Poozerie", which confirms the need to preserve the habitat of rare and endangered bird species listed in the International Red Book, Red Data Books of the Russian Federation and the Volgograd Region.

In 2011, with the support of the Volgograd Region Administration, the Ministry of Natural Resources of Russia, the UNESCO Office in Russia, at the International Coordinating Council of the UNESCO "Man and the Biosphere Program" in Dresden, this nature park was included in the World Network of Biosphere Reserves.

Thus, the natural park became the 40th biosphere reserve of the Russian Federation and the first territory of regional importance, which was included in the list of international biosphere reserves, it imposes additional obligations for the conservation of biodiversity under international agreements and conventions in which the Russian Federation is a party.

2 Methodology

To fulfill the assumed international obligations to preserve the unique natural complex, some conditions must be met.

The proximity of the Volga-Akhtuba floodplain to Volgograd and Volzhsky, as well as the historically established features of using this territory, namely the development of housing, recreation, agriculture, leave their mark on the preservation of the unique and typical natural complexes of the floodplain.

One of the conditions for the functioning of the natural complex is the annual optimal flood moisture and filling of the hydrographic network. Filling the network of tracts, eriks, and lakes, on the territory of the floodplain in high water depends both on the volume of the special allowance and on the duration of the discharge of the maximum discharge through the Volgograd hydroelectric complex. Favorable watering of the floodplain area is achieved at maximum flow rates—26,000–28,000 m^3/s, lasting 6–12 days (Sheppel, 1986).

A change in the natural hydrological regime leads to the depletion and shallowing of water bodies of the Volga-Akhtuba floodplain, and as a consequence to a violation

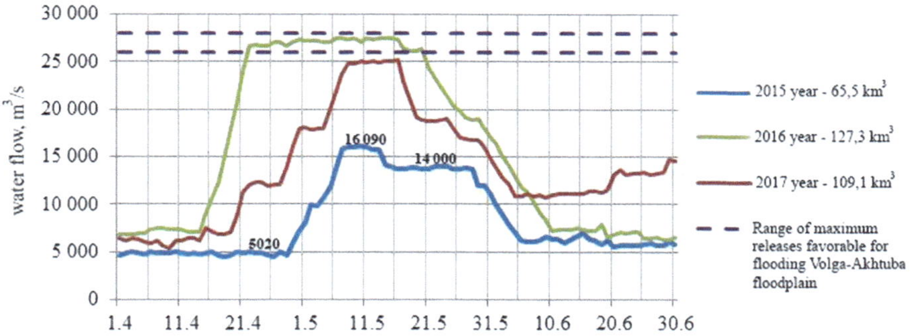

Fig. 1 Schedule of a special spring release through the Volgograd hydroelectric complex in the second quarter of 2015–2017. *Source* Developed and compiled by the authors

of the conditions for the existence of aquatic and terrestrial ecosystems. A striking example is the catastrophic low-water spring floods of 2006 and 2015, in which water practically did not enter the Volga-Akhtuba floodplain, thereby causing a significant blow to the unique natural territory and agricultural producers (Fig. 1).

This served as the basis for the inclusion of measures for the restoration and preservation of the floodplain in the fundamental documents for the development of the water sector of the Russian Federation, including the Water Strategy, the federal target program, and as a result, reflected in the federal and regional projects "Rehabilitation of the Volga" (Federal Project, 2018) of the national project "Ecology" (Federal Project, 2018).

To minimize the degradation of the floodplain, as well as its restoration, instead of the traditionally carried out hydro-mechanized works on clearing tracts, eriks, and lakes of the floodplain, a fundamentally new decision was made aimed at an integrated approach to the rehabilitation of the floodplain area, namely:

- to clean up the water body from reed vegetation and remove alien species of arboreal and shrubby vegetation that oppress the local floodplain flora;
- on the formation of a complex bottom topography with the required depths and shallow waters, planting trees and shrubs of typical floodplain flora;
- on the formation of the recreational qualities of the coastal strip.

In 2020, measures were completed for the restoration and environmental rehabilitation of water bodies on an area of 594 hectares, which is 40% of the planned target (State program of the Volgograd Region, 2013).

Taking into account the importance and relevance of the preservation of the unique ecosystem of the Volga-Akhtuba floodplain (Sazonov et al., 2015), the optimal regime of which depends on the duration and volume of incoming water, there is a need to maintain the operability of existing culverts and to build new culverts that fill the network of tracts, eriks, lakes and the floodplain territory (Istomin, 2017).

More than 150 culverts are located in the northern part of the Volga-Akhtubinskaya floodplain. The technical condition of the overwhelming majority of culverts is assessed as unsatisfactory and does not provide the maximum possible flow of water under the conditions of a special spring release through the Volgograd hydroelectric complex. Over 60% of culverts need major repairs and reconstruction.

In the first quarter of 2021, 17 culverts were completed and commissioned, which is 24% of the planned target.

It should be noted that measures for clearing and ecological rehabilitation of water bodies, as well as for the construction and reconstruction of culverts are preparatory measures necessary for the implementation of a large-scale project for additional watering of the Volga-Akhtuba floodplain.

The value of the annual runoff of the river. Over the period from 1881 to 2020, the Volga changed insignificantly, the most significant changes occurred in the intra-annual distribution of runoff (Gorelits et al., 2018), namely, the runoff of the spring flood decreased by almost 30% compared to natural conditions, the low-water runoff, especially the winter low-water period, increased more than 2 times.

Considering the above, favorable watering of the territory of the Volga-Akhtuba floodplain during the spring flood in modern conditions is not ensured.

To preserve the unique ecosystem of the Volga-Akhtuba floodplain, it is necessary to carry out comprehensive measures aimed at additional watering of the territory in dry years.

At present, design documentation has been developed for the facility "Complex of hydraulic structures providing additional watering of the Volga-Akhtuba floodplain" (Fig. 2). The main task of this complex of hydraulic structures is to transfer part of the flow of the Volgograd reservoir directly into the Akhtuba river, bypassing the Volgograd hydroelectric complex, creating an artificial hydrological regime on the

Fig. 2 Schematic diagram of additional water cut. *Source* Developed and compiled by the authors

Akhtuba, providing watering of the main eriks and lakes, as well as the floodplain area.

In addition to achieving the main goal of the project, improving the environmental situation in the northern part of the Volga-Akhtuba floodplain, solutions to auxiliary problems have been proposed. In particular, the generation of electricity using the potential energy of the water flow supplied from the Volgograd reservoir (the upper pool of the Volzhskaya HPP) to the Akhtuba River and the supply of water for irrigation of agricultural lands located along the banks of the projected supply canal.

The design solutions provide for the construction of a complex as part of the following main structures: a water intake supplying a culvert 32 km long, a hydroelectric power station building, hydraulic structures (dams-regulators) on the Akhtuba River, as well as a pumping station designed to supply water to the Krasnoslobodsky water tract of the Volga-Akhtuba floodplain, bank protection structures on the Akhtuba River.

The inlet culvert is capable of passing flow rates from 200 m^3/s (during the low-water season) to 1000 m^3/s during spring floods. The beginning of the canal—the village of Verkhneopogromnoye, the end is the village of Zayar (Sredneakhtubinsky district of the Volgograd Region).

During the low-water period, it is planned to pass 200 $m^3$of water through the supply channel, of which 35 m^3/s is directed to irrigate the steppe part of the Sredneakhtubinsky municipal district. At present, irrigation of this area is carried out through the existing irrigation canal with a throughput of 5 m^3/s.

Implementation of measures for additional watering of the Volga-Akhtuba floodplain will create up to 50 thousand hectares of irrigated land, 1750–1900 new jobs in the agro-industrial complex (based on the current number of people employed in the Volgograd region per 100 ha—3.5–3.8 units).

It is planned to pass 165 m^3/s through the hydraulic units, which are then directed to additional watering of the Volga-Akhtuba floodplain. In order to compensate for the operating costs of the complex of hydraulic structures, it is planned to construct a building of a hydroelectric power station with an installed capacity of 31.2 MW in the Zayar village.

The functioning of the complex of hydraulic structures will raise the water level in the Akhtuba section by 3 m from the low-water level, and during the flood period by an additional 4 m, which will ensure the gravity flow of water to the floodplain, accumulate water and redistribute it, if necessary, to optimize the hydrological regime. Volga-Akhtuba floodplain throughout the year.

To solve the main problem, in addition to the construction of hydraulic structures described above, it is planned to build pressure structures at the entrance to the main floodplain eriks (Pahotny, Bugrovaty, Staraya Akhtuba, and Bugai).

The throughput of each structure is 12 m^3/s. Thus, during the low-water period, it is possible to provide a stable water supply to the main waterways with a flow rate of 48 m^3/s and the direction of water downstream of the dam—the regulator at the village. Float flow rate 115–120 m^3/s, which corresponds to the low-water discharge of the Akhtuba river (about 100 m^3/s).

On the territory of the northern part of the Volga-Akhtuba floodplain, there are two main waterways (Kashirinsky and Krasnoslobodsky), with a total length of 150 km. In order to increase the water supply of the Kashirinsky water tract, the construction of an overflow dam on the Kashirin river near the Nevidimka village was completed in 2020. As a result, the water level in the waterway in the 37.8 km section from the overflow dam to the mouth of the erik (near the Akhtuba river) increased by 1.5 m to minus 6.50 m, which made it possible to increase the watering of lakes, eriks of the first and of the second order in the conditions of the discharge of maximum flows through the Volgograd hydroelectric complex in 2021, the accumulation capacity of the tract is 2.8 million m^3.

The maximum discharge flow through the Volgograd hydroelectric complex during the special spring release in 2021 was 25,000 m^3/s with a duration of 7 days (agricultural shelf), which corresponds to the lower limit of the basic rules (Basic rules for the use of water resources of the Volgograd reservoir on the river, 1983).

To cover the water deficit of the Kashirinsky and Krasnoslobodsky waterways during the low-water period, pumping of the existing head pumping station (HPS) with a capacity of 9 m^3/s is provided. It is planned to preserve this gas pumping station as part of measures for additional watering of the Volga-Akhtuba floodplain in order to supply water through the Krasnoslobodsky water tract system.

In order to create conditions for maximum watering of the floodplain area, it will be necessary to organize a unified service for the operation of all hydraulic structures located on the territory of the Volga-Akhtuba floodplain.

The construction of a complex of hydraulic structures will ensure the replenishment of the groundwater level, the main water bodies of the floodplain during the low-water period, as well as minimize the negative factors of low-water years, in which a standard spring flood with a maximum flow rate of 25,000–27,000 m^3/s lasting 5–7 days is difficult or not possible (Basic rules for the use of water resources of the Volgograd reservoir on the river, 1983).

3 Results

The result of the implementation of these measures will be a gradual restoration of the natural state of the floodplain territory, the creation of a balanced ecosystem with an effective self-cleaning mechanism, in addition, after the spring flood, it will make it possible to ensure the use of previously restored storage tanks of water bodies to increase the area of irrigated land for the cultivation of, for example, scarce vegetable crops.

References

Basic rules for the use of water resources of the Volgograd reservoir on the river. (1983). *The Volga* (pp. 261–283). Moscow.

Federal Project "Improvement of the Volga", approved by the minutes of the meeting of the project committee for the national project "Ecology" dated December 21, 2018, No. 3.

Gorelits, O. V., Ermakova, G. S., & Terskiy, P. N. (2018). The hydrological regime of the Lower Volga in modern conditions. *Meteorology and Hydrology, 10*, 27–39.

Istomin, A. P. (2017). Protection and conservation of water resources of the unique system of the Volga-Akhtuba floodplain. In *Ecological and reclamation aspects of rational nature management. Materials of the International Scientific and Practical Conference* (pp. 59–70). https://elibrary.ru/item.asp?id=29814203

Sazonov, V. E., Istomin, A. P., Kalyuzhnaya, N. S., & Kalyuzhnaya, I. Yu. (2015). Methodological and legal aspects of restoration and ecological rehabilitation of water bodies (on the example of the Volga-Akhtuba floodplain). *Facets of Knowledge, 4*(38), 9–19. https://elibrary.ru/item.asp?id=23437889

Sheppel, P.A. (1986). *Flood and floodplain Volgograd: Lower Volga book publishing house*, 240p.

State program of the Volgograd Region "Use and protection of water bodies, prevention of the negative impact of water on the territory of the Volgograd region", approved by the Government of the Volgograd Region of 30 Aug 2013 No. 453-p.

Sustainable Agriculture and Development of Rural Territories

Digitalization of Agriculture—The Path to the Future of Geoeconomics

Ilkhom M. Umarov, Zauresh B. Akhmetova, Nadira M. Turlybekova, Nazym T. Tulebayeva, and Shakizada U. Niyazbekova

Abstract In the era of globalization of the world economy, many sectors of the economy are being transformed, and agriculture is the engine of economic development in developing countries. Countries with developing economies are adopting the experience of developed countries and are beginning to apply innovative, digital technologies in agriculture, which is a key tool in increasing the productivity and efficiency of the industry. Of course, the digitalization of agriculture occupies one of the leading positions in this issue, because the economic stability of the state largely depends on the degree of development of the agricultural sector in the country. Agriculture in the world is turning from a traditional to a high-tech industry, which can create new markets for innovative solutions and developments. The results of the study: (1) the analysis of digital technologies in agriculture is presented; (2) methods and techniques of farming through digital technologies are disclosed; (3) the concepts reflecting various forms of digitalization in the agricultural sector are considered; (4) the economic efficiency of grain production in the regions of Kazakhstan is determined; (5) various applications of digital technologies and innovations in agriculture and farming are shown.

Keywords Digital technologies · Innovations · Agricultural sector · Digitalization · Sustainable agriculture · Strategic development · Economics · Innovative developments

JEL Codes O10 · O20 · O30 · P33 · P47

I. M. Umarov · Z. B. Akhmetova · N. M. Turlybekova · N. T. Tulebayeva
Al-Farabi Kazakh National University, Almaty, Kazakhstan

S. U. Niyazbekova (✉)
Moscow Witte University, Financial University Under the Government of the Russian Federation, Moscow, Russia

E. G. Popkova and B. S. Sergi (eds.), *Geo-Economy of the Future*,
https://doi.org/10.1007/978-3-030-92303-7_37

1 Introduction

The dynamics of the growth of agricultural production, and in particular grain farming, directly depend on the use and implementation of scientific research, modern technologies, and intensive innovations in the industry (Umarov, 2015).

Digital technologies are used to create virtual reality, using special robotics, 3D printing, touch devices, artificial intelligence, computer learning, and blockchain (Alm et al., 2016). Digitalization is expected to radically change everyday life (Yoo, 2010), and agricultural production processes and the associated food, fiber, and bioenergy supply chains and systems (Poppe et al., 2013) are already evident (the first signs of transformation).

In the agricultural sector, several concepts have emerged that reflect different forms of digitalization in the structure of production itself, in value chains, and, more broadly, in food systems. These include Smart Farming (Blok, 2018; Wolfert et al., 2017), Precision Farming (Eastwood et al., 2019; Leonard et al., 2017; Wolf & Buttel, 1996), Digital Farming (Keogh & Henry, 2016; Shepherd et al., 2018), and Agriculture 4.0 (Rose & Chilvers, 2018).

Regardless of the exact term used, digitalization implies that management tasks on and off the farm, in the broader value chain and food system, focus on different types of data (location, weather, landscape, sanitation, energy consumption, prices, and economic information, etc.) using sensors, machines, drones, and satellites to monitor animals, soil, water, plants, and people. The data obtained are used to interpret the past and predict the future, to make more timely or accurate decisions through constant monitoring or special requests (Eastwood et al., 2019; Wolfert et al., 2017).

Precision farming is an agricultural concept that includes new production and management methods that actively use digital data about the state of the field and the crop. It involves targeted point management of agricultural land (fields) with the help of special intelligent electronics.

To determine the economic efficiency of grain production, a system of natural and cost-based natural efficiency indicators is used, where the main indicator is the yield of wheat. The natural indicator is the basis for calculating cost indicators, such as profit, profitability, cost of production, the output gross output of production.

According to the Committee on Statistics, an analysis of wheat yield in Kazakhstan was carried out, where the data of the actual yield was compared with the equilibrium one. The sown area and the gross wheat harvest from 2004 to 2017 were also taken into account. According to these data, the fluctuation of wheat yield was approximated.

Taking into account the overall wheat yield in Kazakhstan as a whole, we would like to note that in the conditions of a sharply continental climate and the biological diversity of soil territories in the country, the Akmola, Kostanay, and North Kazakhstan regions of Kazakhstan remain the most favorable for wheat production. In these areas, it is possible to form 94% of the grain industry for winter crops and receive up to 96% of the crop, thus forming 100% of the export grain. In Northern Kazakhstan, you can grow soft and hard varieties of winter wheat, and in Southern Kazakhstan and Western, as well as in the southeast of the country, you can grow

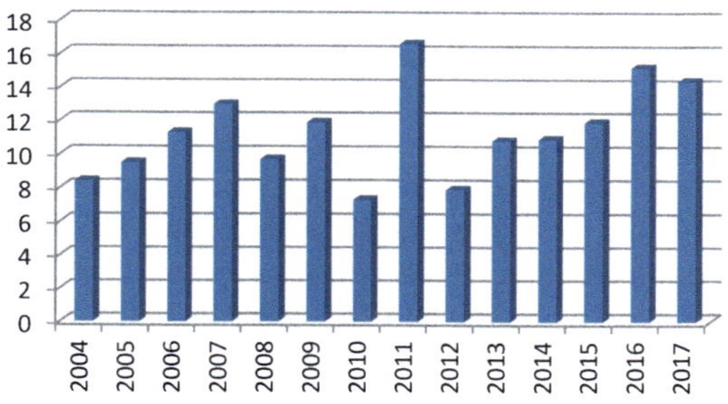

Fig. 1 Wheat yield in Kazakhstan for 2004–2017, c/ha. *Source* Compiled by the author based on (Committee on Statistics)

spring wheat. If we take into account this distribution of commercial wheat production for winter and spring crops, then such a concentration of the farm will bring more than half of the volume of grain produced in Kazakhstan. Of course, there is a need for state support and the development of the grain industry in these regions and regions of the country, which will produce export grain to world markets. As can be seen from Fig. 1, in 2011, the wheat yield reached its record high of 16.6 c/ha, the highest in 14 years. In 2017, the yield was 14.4 c/ha.

In the future, it is planned to increase the yield and production of wheat through the introduction of innovative technologies and soil cultivation by 1.5 times by 2021.

Aerial photographs taken using drones and quadrocopters (unmanned aerial vehicles) provide complete information about the state of the field, including indicators such as soil moisture, weeds, pests, and plant diseases. The data is received relatively quickly, and this allows you to take appropriate measures at the same time. Field robots are great for the careful tillage of soil and plants because they are very light. They are human-controlled and can be used to determine a specific sowing and fertilizing pattern.

Today, the use of information technologies (Farming 4.0) helps many agricultural enterprises to improve the technological process of the entire complex system of agro-industrial production. In this context, there is an integration of digital technologies with innovations that relate to the technical and technological process, namely, modern technologies of tillage for sowing crops and innovative technologies for storing and processing crops. Such integrated work from the moment of receiving the right information and the right decision on the part of farmers on the point impact on the problem areas of the fields helps to save the costs of agricultural work, equipment, inventory, fertilizers, and pesticides.

High-tech agriculture is focused on the work of the entire existing system of agriculture. This comes from the digital transmission of data via satellite systems to the computers, tablets, or phones of farmers who use special programs related to control electronics. Special programs developed for farmers are managed online through special applications installed on the devices. Data such as soil moisture, temperature, atmospheric pressure, and the amount of ultraviolet radiation are transmitted to the device every minute and recorded. Farmers analyze this data and compare it with the technical standards for growing this crop, and after receiving the results in the direction of the greatest deviation from the norm, make their adjustments and correct the situation. After taking preventive measures, the data is compared again and recorded in the report. Consequently, the number of activities carried out from the beginning of the fieldwork to the harvest takes into account the costs and quantity of the crop, as well as the efficiency of using all equipment from computers to machines related to digital data transmission.

Thus, it is expected that in the future, the digitalization of agriculture will become an integral part of the optimization of the economic mechanism of the agricultural production system and will bring this industry to a new level of development. Thus, this process will help to solve social problems related to agriculture, improve the quality of food and improve animal husbandry (Yeates, 2017), as well as influence the choice of agricultural methods that ultimately affect the environment (Balafoutis et al., 2017). Digitalization is expected to accelerate the sharing of agricultural knowledge and data (Baumüller, 2017) and improve the proctoring of problems and crises encountered in agricultural sectors and systems (Stevens et al., 2016). In the twenty-first century, the use of digital technologies is becoming ubiquitous in all countries, but they are most actively used in the cultivation of agriculture and viticulture. Digitalization is most often used in agriculture with the use of point-impact technology (Bramley, 2009), and to a lesser extent in poultry and animal husbandry, which in turn places high hopes on the transformation and development of this industry in the future (Rose et al., 2018).

2 Literature Review

Umarov and Walter (2015, 2016) considered in their works the technical aspects of the application of technologies for improving agricultural practices and increasing labor productivity, as well as changes in agriculture through digitalization. The application of digital technologies based on the introduction of sensors for monitoring animal health on farms was considered by Rutten et al. (2013). Most of the works of scientists are focused on the study of digital technologies (Balafoutis et al., 2017), innovations (Umarov, 2015), the practical application of digitalization (Alm et al., 2016; Butler & Holloway, 2015; Leonard et al., 2017), information and communication technologies (Poppe et al., 2013; Yoo, 2010), and the methodological approach of digitalization in agriculture (Ulezko et al., 2019).

The theoretical, methodological, and practical approach of many scientists in discussing digital technologies, innovations in agriculture and farming in this article will help young scientists and specialists in the field of natural and technical sciences, biology, and other interdisciplinary sciences in studying digitalization and the application of new digital technologies in agriculture.

In their work, Mitchell et al. (2018) consider the problem of agriculture, tillage with old equipment and technology that leads to the erosion of the upper soil layer and propose the introduction of new technologies of precision farming in agriculture, directly in crop production. Ulezko et al. (2019) in their work "Theoretical and methodological aspects of digitalization of agriculture" offer their theoretical and methodological aspects of the application of digitalization in agriculture. They propose to use digital robotics in agriculture by analyzing and modeling the data obtained to improve the efficiency of this industry.

3 Methodology

This study is based on the study and analysis of scientific approaches to the development and application of digital technologies and innovations in agriculture.

The introduction of new digital technologies in agriculture at first glance seems very expensive, but summing up the resulting profit with total expenses, you can deduce the percentage of net profit and only then understand how much this process is costly.

In addition, digitalization affects the cultural aspects of rural areas, as it sometimes contradicts traditional farming methods and affects the identity of the farmer, radically changing his direct purpose (Burton et al., 2012; Carolan, 2019). By changing the cultural characteristics of farming, based on practical application and experience (Butler & Holloway, 2015), digitalization changes the image of farmers and builds the entire system of work as a whole in the right rational direction. Ultimately, the integration of digitalization with such methods and approaches to agriculture raises discussions on the part of scientific and practical applications in agriculture.

Speaking about the digitalization of agriculture, it is impossible to miss such a moment as the use of digital technologies in animal husbandry, namely in dairy farming by the example of their use in the milking process (Burton et al., 2012) and the replacement of heavy manual labor with robotics and technologies of the full cycle of the cow milking process (Butler & Holloway, 2015). The use of robotic milking accelerated the process of collecting and processing milk, which ultimately led to the avoidance of problems with its souring and spoilage. This approach also has a social factor that has a positive impact on the health of animals and a more careful attitude towards them.

Digital technologies in agriculture can especially help small farms, where the staff is not so large and most of the proctoring is done by drones and robotics. Farmers can quickly use digital data obtained from satellite tracking or through drones, thus building cartography and ranking the data according to the time of year and weather conditions. Such data can be successfully used in planting and harvesting, as this is one of the most important measures in the production of crops.

4 Results and Reviews

It should be noted that the process of digitalization of agriculture would require solving some tasks of various types, such as:

- drawing up a plan for solving management tasks that are solved only with the help of digital technologies and the information obtained through them that is necessary for this solution;
- creation of specialized research centers that will collect information in a single database, build economic and mathematical models and functions that will allow you to create various options and scenarios for the development of high-level agricultural activities;
- integrate information and management activities between agricultural enterprises (farms) and government agencies at all levels;
- to ensure the relationship between agricultural producers by combining them into a single information space with the right to use data in digital ecosystems;
- creation of a single digital platform for the collection and processing of data obtained through satellite tracking, drones, and robotics;
- abandoning the old form of farming with the transition to high-tech management methods and the use of digital technologies and innovations;
- ensuring an uninterrupted flow of information, with the ability to quickly collect, analyze, store and systematize data and timely meet the information needs of agricultural producers and farmers;
- high degree of information security of agricultural enterprises and full protection of the database from unauthorized users and hackers.

In general, we can offer several scenarios for the development and implementation of the digitalization process in agriculture.

First, it is possible to develop pilot projects that will be aimed at testing the viability of new technologies and innovations in agriculture.

Second, it will be possible to create research centers with an equipped laboratory and storage for a database of data obtained using satellite navigation, drones, and robotics, where data with an estimated forecast for farmers will be analyzed and processed.

Third, it will be necessary to create a triple model of the spiral of innovation science–business–state.

Fourth, it will be necessary to create digital programs with the possibility of their use by farmers from phones, tablets, and laptops in such a way as to monitor the process of crop production in a continuous mode.

Fifth, the proposed programs and the best solutions from research centers will help to solve similar problems in agricultural enterprises at lightning speed and avoid possible risks in the future.

According to the study, another solution to the issue of the introduction and use of digital technologies in agriculture can be the creation of such services of integrated cloud services. In principle, the algorithm of such services is comparable to scientific centers, which in fact can also receive and process data received from digital devices and translate it into a report that is understandable for farmers to use. Although, some scientists believe that a single service will help to effectively use digital data for point impact on areas of fields, mapping the terrain, identifying problem areas in the fields, and using the weather forecast for planning agricultural work, etc. (Brönnimann et al., 2018).

Access to a common database will help agricultural producers to monitor not only data on the state of the soil, crop, weather conditions, etc., but also to build transport routes from the point of harvest to the storage and the movement of goods from the producer to the consumer directly using the logistics infrastructure. In addition, it will be possible to monitor the prices for the current period for agricultural products, reduce the cost of intermediary services and choose the most convenient and cheap supplier of transport services. The creation of such centers and services will certainly accelerate the process of digitalization of agriculture and increase the profit from each hectare of land.

It is important to note that such technologies are being implemented not only in the European Union, the United States, Canada, and other developed countries, but also in India, Uganda, the Philippines, Morocco, Kenya, South Africa, Zambia, Namibia, Mali, and Nigeria (Deichmann et al., 2016). In developing countries, modern technologies are focused on small–scale agricultural enterprises.

Another advantage of the introduction of digital technologies in agriculture is a small impact on the environment through pre-identified problems and immediate preventive measures to address them. This is one of the breakthroughs in the digitalization of agriculture, as it fully protects the state of crops and cultivated land without unnecessary environmental impact and with minimal use of chemicals and pesticides.

A striking example of the development of a new e– and digital agriculture together with FAO are such countries as Albania from 2019, Armenia from the second half of 2018, Moldova from 2014, the Russian Federation from 2018, Turkey from November 2019, Kyrgyzstan from 2019, Tajikistan and Uzbekistan from 2019 (ITU Publications, 2020). These countries collaborate with FAO on many agricultural projects, including the development of e– and digital agriculture strategies, digital roadmaps, and the introduction of information and communication technologies in agriculture.

Another example of the introduction of digital technologies is the use of special ultraviolet lamps with a special radiation spectrum close to sunlight for the normal photosynthesis of plants in cloudy weather (Vinther & Müller, 2018).

Recently, rooftop farms have become increasingly common in European countries, making it much easier to use the untouched area of the structure for growing edible agricultural plants. This simplifies, first of all, the availability of agricultural producers to consumers and reduces the cost of long-term transportation.

The joint use of digital technologies in agriculture will help solve the problem of food supply, due to the increase in the world's population. In this regard, FAO helps countries that need to finance agricultural projects in the field of digital technologies.

Farmers using digital technologies are under the close attention of consumers, who have the opportunity to monitor the transparency of the actions of agricultural companies and track information about the readiness of agricultural products. The ability to purchase products also includes information about the wholesale and retail prices of the product. This approach in the work of agricultural enterprises opens up new markets and makes agricultural products competitive and in-demand (Umarov et al., 2020).

By continuing to use new digital technologies, farmers are reaching the point where they are starting to use environmental resources, such as solar energy to charge the battery mechanism of agricultural machinery, without exhaust fumes and unnecessary noise. Thus, they save resources and take care of the surrounding flora and fauna adjacent to the cultivated fields.

Agricultural machinery equipped with digital sensors with the ability to process data obtained through digital field tracking tools can partially make autonomous decisions. The farmer mainly performs the functions of managing autonomous agricultural machinery. The autonomous decisions made by the harvesters include adjusting the tire pressure and balancing the system to save resources.

Next, we would like to consider the SWOT analysis of the strengths and weaknesses of the application of digitalization in agriculture (Table 1).

Let's also consider PEST—analysis (Table 2).

5 Conclusion

Many farmers today understand the benefits of using digital technologies in agriculture and strive to find ways to enter this process painlessly. Since the process itself is still knowledge-intensive and expensive, many farmers continue to use the old techniques and technology of agricultural cultivation. This is an erroneous opinion since the process of digitalization itself greatly facilitates the activities of agricultural enterprises and significantly increases profits. Digitalization of agriculture has found its direct application in the developed countries of Europe, the United States, and Canada. However, many emerging economies have already begun to adopt new digital and electronic technologies, together with the support of FAO, and have proven themselves to be the best in this regard. To date, many digital technologies, programs, and models for agriculture have already been developed. What is missing, is the integration of knowledge transfer from the producer to the consumer, as well as the spiral of relations: science—business—state. The trend of transition of many agricultural enterprises to high-tech systems is growing rapidly.

This leads to the conclusion that the main condition for the transition to high-tech systems of many agricultural enterprises is the expansion of the broadband network, the use of high-speed Internet, innovative and digital technologies in crop production, and animal husbandry. The focus should be primarily on the strategic approach of States to the use of digital technologies in agriculture. And for this purpose, subsidies should be allocated to small and medium-sized farms for the earliest possible development and application of digital technologies in agriculture. In turn, these farms will further become a strong foundation of the country's economy and ensure food security.

Table 1 SWOT–analysis

Strength: • Adaptability to extreme climatic conditions; • A well-established and balanced system of point impact on the field; • Increase of soil fertility; • Use of clean environmental resources; • Monitoring the state of the soil and crop at any time convenient for farmers • FAO's logistical support to countries in the application of digital technologies in agriculture	**Weakness**: • The absence of a triple spiral of innovative development: science—business—state; • Lack of an insufficient number of research centers and servers for collecting and processing digital data; • Not all sectors of agriculture are moving equally to the use of digital technologies; • Weak market communication and marketing information system
Opportunities: • Creation of research centers and integrated cloud services • Efficient use of digital data by all agricultural enterprises connected to digital data transmission programs • Development of small and medium-sized businesses • Creating your production chain for agricultural products	**Threats**: • Adverse changes in natural and climatic conditions (global warming and the associated increase in desert and semi–desert lands, increasing water scarcity, unstable weather conditions, etc.) • Critical deterioration of the infrastructure for transporting products to target markets and the associated increase in the cost of delivery • The spread of animal and plant diseases and pollution of the natural environment, parasitic species of plants, animals, fish, insects, will reduce the availability of land, water, and other resources and reduce the productivity of the industry as a whole, may reduce the country's export potential • Depletion of the potential of land, water, biological resources, and the genetic potential of animals, plants, and fish as a result of short-term profit orientation and lack of funding

Source (Balafoutis et al., 2017; Baumüller, 2017; Bramley, 2009; Brönnimann et al., 2018; Burton et al., 2012; Butler & Holloway, 2015; Carolan, 2019; Keogh & Henry, 2016; Mitchell et al., 2018; Rajeev & Nagendran, 2018; Rose & Chilvers, 2018; Rose et al., 2018; Rutten et al., 2013; Shepherd et al., 2018; Stevens et al., 2016; Sulimin et al., 2019; Ulezko et al., 2019; Walter, 2016; Yeates, 2017).

Table 2 PEST– analysis

Political factors: • Changing the country's development strategy • Changes in taxation and tax rates and benefits for agricultural enterprises • Wage legislation—Minimum wage and overtime • Political stability and the importance of the agricultural sector • Pricing rules	**Economic factors**: • Stability/instability of the economic system • State intervention in the economic activities of agricultural enterprises • Exchange rates and currency stability/instability in agricultural producing countries • The skill level of the workforce • Unemployment rate • Inflation rate • Economic growth rates
Social factors: • Demographics the population level • Social security in society • Gender equality • Level of education	**Technological factors**: • The latest technological developments in the field of agriculture • The contribution of technology to the development of the market • Development of the Internet, satellite data transmission system, mobile devices, and devices • Activities and development of scientific and technical centers and research in the field of digitalization of agriculture • The speed of technological progress • Innovative technologies to simplify actions in agriculture

Source (Balafoutis et al., 2017; Baumüller, 2017; Bramley, 2009; Brönnimann et al., 2018; Burton et al., 2012; Butler & Holloway, 2015; Carolan, 2019; Keogh & Henry, 2016; Mitchell et al., 2018; Rajeev & Nagendran, 2018; Rose & Chilvers, 2018; Rose et al., 2018; Rutten et al., 2013; Shepherd et al., 2018; Stevens et al., 2016; Sulimin et al., 2019; Ulezko et al., 2019; Walter, 2016; Yeates, 2017)

References

Alm, E., Colliander, N., Lind, F., Stohne, V., Sundström, O., Wilms, M., & Smits, M. (2016). *Digitizing the Netherlands: How the Netherlands Can Drive and Benefit From an Accelerated Digitized Economy in Europe*. Boston Consulting Group.

Balafoutis, A., Beck, B., Fountas, S., Vangeyte, J., Van Der Wal, T., Soto, I., Gómez-Barbero, M., Barnes, A., & Eory, V. (2017). Precision agriculture technologies positively contributing to GHG emissions mitigation, farm productivity, and economics. *Sustainability, 9*(8), 1339.

Baumüller, H. (2017). The little we know: An exploratory literature review on the utility of mobile phone-enabled services for smallholder farmers. *Journal of International Development, 30*, 134–154.

Blok, V. (2018). Technocratic management versus ethical leadership redefining responsible professionalism in the agri-food sector in the anthropocene. *Journal of Agricultural and Environmental Ethics, 31*(2–3), 583–591.

Bramley, R. (2009). Lessons from nearly 20 years of precision agriculture research, development, and adoption as a guide to its appropriate application crop & pasture science. *60*, 197–217.

Brönnimann, S., Brugnara, Y., Allan, R., & Brunet, M. (2018). A roadmap to climate data rescue services. *Geoscience Data Journal, 5*, 28–39.

Burton, R., Peoples, S., & Cooper, M. H. (2012). Building 'cowshed cultures': A cultural perspective on the promotion of stockmanship and animal welfare on dairy farms. *Journal of Rural Studies, 28*(2), 174–187.

Butler, D., & Holloway, L. (2015). Technology and restructuring the social field of dairy farming: Hybrid capitals, 'Stockmanship' and automatic milking systems. *Journal of the European Society for Rural Sociology, 56*, 513–530.

Carolan, M. (2019). Automated agrifood futures: robotics, labor and the distributive politics of digital agriculture. *The Journal of Peasant Studies,* 184–207.

Deichmann, U., Goyal, A., & Mishra, D. (2016). *Will Digital technologies transform agriculture in developing countries?* World development report team & development research group environment and energy team. Policy Research Working Paper, p. 7669.

Eastwood, C., Ayre, M., Nettle, R., & Dela Rue, B. (2019). Making sense in the cloud: farm advisory services in a smart farming future. *NJAS—Wageningen Journal of Life Science.*

Keogh, M., & Henry, M. (2016). The implications of digital agriculture and big data for Australian agriculture australian farm institute, Sydney, Australia.

Leonard, E., Rainbow, R., Trindall, J., Baker, I., Barry, S. Darragh, S., Darnell, R., George, A., Heath, R., Jakku, E., Laurie, A., Lamb, D., Llewellyn, R., Perrett, E., Sanderson, J., Skinner, A., Stollery, T., Wiseman, W., Wood, G., Zhang, A. (2017). *Summary report. Accelerating precision agriculture to decision agriculture: enabling digital agriculture in Australia.* Cotton Research and Development Corporation.

Mitchell, S., Weersink, A., & Erickson, B. (2018). Adoption of precision agriculture technologies in Ontario crop production. *Canadian Journal of Plant Science, 98*(4), 1384–1388.

Official website of the Ministry of National Economy of the Republic of Kazakhstan. *Committee on Statistics.* Available at: www.stat.gov.kz

Poppe, K., Wolfert, S., Verdouw, C., & Verwaart, T. (2013). Information and communication technology as a driver for change in agri-food chains. *EuroChoices, 12*(1), 60–65.

Rajeev, M., & Nagendran, P. (2018). Are gold loans glittering for agriculture? *Economic and Political Weekly, 53*(26), 32–38.

Rose, D. C., & Chilvers, J. (2018). Agriculture 4.0: Broadening responsible innovation in an era of smart farming. *Frontiers in Sustainable Food Systems, 87.*

Rose, D. C., Parker, C., Fodery, J., Park, C., Sutherland, W. J., & Dicks, L. V. (2018). Involving stakeholders in agricultural decision support systems: Improving the user-centered design. *International Journal of Agricultural Management, 6*, 80–89.

Rutten, C. J., Velthuis, A. G., Steeneveld, W., & Hogeveen, H. (2013). Invited review: Sensors to support health management on dairy farms. *Journal of Dairy Science, 96*(4), 1928–1952.

Shepherd, M., Turner, J. A., Small, B., & Wheeler, D. (2018). Priorities for science to overcome hurdles thwarting the full promise of the 'digital agriculture' revolution. *Journal of the Science of Food and Agriculture.*

Status of digital agriculture in 18 countries of Europe and Central Asia. *Food and Agriculture Organization of the United Nations.* ITU Publications.

Stevens, T. M., Aarts, N., Termeer, A. M., & Dewulf, A. (2016). Social media as a new playing field for the governance of agro-food sustainability. *Journal Current Opinion in Environmental Sustainability, 18*, 99–106.

Sulimin, V. V., Shvedov, V. V., & Lvova, M. I. (2019). Digitization of agriculture: innovative technologies and development models. In *IOP Conference Series: Earth and Environmental Science,* vol. 341.

Ulezko, A., Reimer, V., & Ulezko, O. (2019). Theoretical and methodological aspects of digitalization in agriculture. In *IOP Conference Series: Earth and Environmental Science*, (Vol. 274, pp. 012062). IOP Publishing.

Umarov, I., Shurenov, N., Kozhamkulova, Zh., Abisheva, & Kyz-Zh. (2020). Marketing and innovative aspects of the research of the competitiveness of countries in the grain market (for example, wheat). In *E3S Web of Conferences* (Vol. 159, p. 04003). https://doi.org/10.1051/e3sconf/202015904003. Access date 7 Apr 2021.

Umarov, I. M. (2015). Substantiation on innovative development prospects for grain production in Kazakhstan. *Journal of Advanced Research in Law and Economics (VI, Winter), 3*(13), 691–698.
Vinther, K. S., & Müller, S. D. (2018). The imbrication of technologies and work practices: The case of Google Glass in Danish agriculture. *Scandinavian Journal of Information Systems, 30*(1), 32–46.
Walter, A. (2016). How will digitalization change agriculture? *International Trade Forum, 2*, 28–29.
Wolf, S. A., & Buttel, F. H. (1996). The political economy of precision farming. *American Journal of Agricultural Economics, 78*, 1269–1274.
Wolfert, S., Ge, L., Verdouw, C., & Bogaardt, M. J. (2017). Big data in smart farming—A review. *Agricultural System, 153*, 69–80.
Yeates, J. W. (2017). How good? Ethical criteria for a 'Good Life' for farm animals. *Journal of Agricultural and Environmental Ethics, 30*(1), 23–35.
Yoo, Y. (2010). Computing in everyday life: a call for research on experiential computing. *MIS Quarterly*, 213–231.

Socio-economic Situation of the Russian Rural Population: Status and Problems

Arif G. Ibragimov, Nikolay G. Platonovsky, and Rafail R. Mukhametzyanov

Abstract The paper focuses on the problem of the socio-economic situation of the rural population in Russia. The paper examines the dynamics of the rural population, the number of rural settlements and their populations, and the level and quality of life in the Russian rural areas. The authors established that the total rural population in Russia has been decreasing from year to year for the last three decades. Small villages are being devastated. There is a large gap between the incomes of rural and urban populations. The paper also examines the levels of poverty, employment, and unemployment of people in rural and urban areas. It was found that the volume of unemployment in rural areas is two times higher than in cities. Moreover, the number of poor people in rural areas is 3–4 times higher than in the city. The authors note that the rural population of Russia is in a worse situation in terms of social facilities: the average radius of access to the most necessary social facilities in rural areas ranges from 14 to 90 km. To improve the situation, the Ministry of Agriculture and the Government of the Russian Federation have developed and implemented several Federal Target Programs over the past decade. However, the planned targets were not achieved. Therefore, the authors recommend fundamental changes in the government attitude toward agricultural labor and rural infrastructure development. In this case, the benchmark should be a comparison of rural and urban living standards.

Keywords Rural population · Rural settlement · Disposable income · Subsistence level · Average wage · Unemployment · Poverty · Social infrastructure

JEL Code Q180

A. G. Ibragimov (✉) · N. G. Platonovsky
Department of Management, Russian State Agrarian University–Moscow Timiryazev Agricultural Academy, Moscow, Russia

R. R. Mukhametzyanov
Department of World Economy and Marketing, Russian State Agrarian University–Moscow Timiryazev Agricultural Academy, Moscow, Russia

E. G. Popkova and B. S. Sergi (eds.), *Geo-Economy of the Future*,
https://doi.org/10.1007/978-3-030-92303-7_38

1 Introduction

Our research on the current situation shows that the rural population is significantly inferior to the urban population in terms of standard of living and quality of life (both economically and socially) practically throughout the Russian Federation. These factors lead to depopulation in rural areas, especially among the younger generation.

Credit must be given to the efforts of government agencies in their attempts to remedy the situation. Over the past decades, the Russian government and the Ministry of Agriculture, which oversees the agricultural sector of Russia, have developed and implemented several projects to develop rural areas to correct the adverse effects in this industry. In our opinion, the "Strategy for the sustainable development of rural areas of the Russian Federation for the period up to 2030" (Strategy) is one of the most elaborated programs (Ministry of Agriculture of the Russian Federation, 2015).

The Strategy involves two steps. The first step is preparation. It began in 2015. According to this stet, mechanisms and incentives must be developed and created to positively change the demographic situation in rural areas within five years. It is stated that special attention should be paid to such an essential problem as improving the legal regulation in this area. The legal regulation must address the complex issues hindering the sustainable development of rural areas in Russia. The first stage should also solve the problem of creating and implementing the mechanisms relative to interagency cooperation, which would be effective and would be able to provide a solution to the complex problem of developing Russian rural areas (Ministry of Agriculture of the Russian Federation, 2015).

The main stage of the Strategy is designed for ten years and will start in 2021. During this period, it is planned to implement the main activities. One of the essential tasks is to achieve the goal of forming and using a balanced set of incentives and mechanisms that can improve the demographic situation in rural areas (including an increase in the population). Additionally, this task should be closely linked to introducing new and cost-effective mechanisms in economic activity and the system of social service delivery (Ministry of Agriculture of the Russian Federation, 2015).

In general, the implementation of the proposed measures by 2030 in Russia should allow the country to achieve several critical indicators and guarantee the solution to the most acute social and demographic problems of rural areas. The budget of the Strategy totals 299.2 billion rubles, of which 90.4 billion rubles are allocated from the federal budget. Subjects of the Russian Federation are supposed to allocate 150.6 billion rubles. Another 58.2 billion rubles should be allocated from non-budgetary sources. Under the Strategy, it is planned to create 31.8 thousand jobs, build 5.4 million square meters of housing for rural residents, and build cultural and leisure facilities and schools with a total capacity of 9.9 thousand and 22.3 thousand places, respectively (Ministry of Agriculture of the Russian Federation, 2015).

The primary research purpose is to analyze the status, indicate problems, and identify reserves for improving the socio-economic situation of the rural population and developing rural areas in Russia.

2 Materials and Methods

To ensure the objectivity of the research results and avoid incorrect conclusions, the authors reviewed the data on the subject of the stated problem obtained from the relevant official statistical reports of the Federal State Statistics Service (Rosstat) and the Ministry of Agriculture. Additionally, the authors analyzed the scientific works of scholars leading in this area, and, based on a critical understanding of these works and their own observations, the authors made conclusions and provided recommendations.

3 Results and Discussion

As noted in some reports and secondary literature, several goals have been achieved as part of the practical implementation of the activities outlined in the federal program "Sustainable development of rural areas for 2014–2017 and until 2020" (Ministry of Agriculture of the Russian Federation, 2012). For example, 55 integrated compact development projects were completed. Nevertheless, there are also negative assessments of their implementation. Some people noted the poor quality of the materials used and many flaws that people had to fix at their own expense.

According to the reports, the necessary engineering facilities for constructing individual houses, quality living conditions in rural areas, and social infrastructure were created. However, these facilities are marked with the same disadvantages as outlined above. Still, these facilities have provided an advantage for workers in the agricultural and social sphere of the village, including young professionals.

Officials reported that they improved housing conditions for 265,700 rural families. In terms of one average, it gives a figure of one million inhabitants, which is not much compared to the real number of rural residents in need of these changes. The reports also note that the above Fig. (265,700 rural families) includes 91.7 thousand young professionals and young families, accounting for 34.5% of the total. Nevertheless, it does not seem quite correct to combine these two categories.

Let us note that such an indicator as the gasification of the rural housing stock increased from 33.1% to 56.5% (an increase of 24.4%). The supply of the rural population with drinking water also increased from 40.7% to 59.6%, which is undoubtedly an important achievement (Federal State Statistics Service of the Russian Federation (Rosstat), n.d.). However, even the achieved figures are perplexing. There is an impression that Russia is still being in the last century compared with the current level of these indicators in other developed countries.

The network of general education schools in rural areas added 105.8 thousand places for students. Nevertheless, not everything is unequivocally positive here either.

New rural schools are being built only in large communities. In some smaller communities, schools have simply been eliminated. Some students have to travel a considerable distance to school or move to relatives in large villages or urban settlements to receive a general secondary education.

Let us also note several other declared results within the fulfillment of the Strategy's targets. The network of regional and district hospitals increased by 6.1 thousand places and 751 medical and obstetric stations (Rosstat, n.d.). According to villagers, medical and obstetric stations are one of the most critical infrastructure facilities. According to reports, visits to outpatient clinics located in the rural areas of Russia increased by 7.5 thousand visits per shift. The area of cultural and leisure institutions grew by 24.6 thousand seats. The number of sports fields and facilities increased by 304 units (Rosstat, n.d).

Despite some positive aspects, the Strategy failed to stop the decline in the number of rural settlements and the emptying of small rural settlements. According to Rosstat, as of January 1, 2020, there were only 132.3 thousand rural settlements in Russia, which is 30 thousand units less than in 1989 (Rosstat, n.d). Let us note that Russia has been an agrarian country for most of its history. During different periods, most of the Russian population lived in the countryside despite changing state structures and controlled territories, which gave its undeniable advantages. However, since the October Revolution, throughout the Soviet period, and after the transition to a market economy in the early 1990s, the rural population has declined year by year (Fig. 1).

This trend is still observed nowadays. Nevertheless, this process has accelerated in the last three decades (Fig. 2).

Additionally, this process is accompanied by a decrease in the number of rural settlements, a decrease in rural population, and the aging of the average rural resident (Leshcheva et al., 2021). Thus, over the past 30 years, the number of rural settlements in Russia has decreased by 30 thousand units (18.4%). There were 162,200 rural settlements in 1989 and only 132,300 rural settlements in 2020. The number of settlements without population and with population from 1 to 10 people increased by 7.3 thousand (an increase of 6.9%) and by 1.6 thousand (an increase of 5.4%),

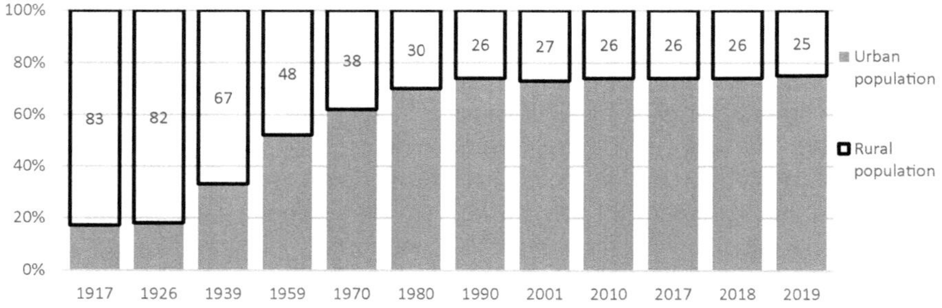

Fig. 1 Ratio of the urban and rural population in different years during 1917–2019, %. *Source* Compiled by the authors based on (Rosstat, n.d)

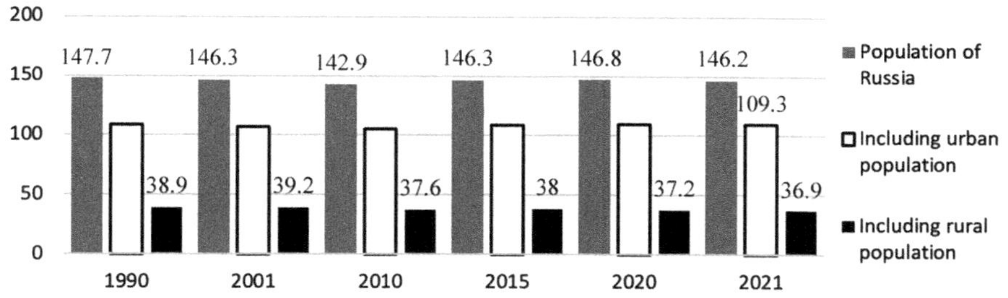

Fig. 2 Structure of Russia's population in different years during 1990–2021, million people (as of January 1 of a particular year). *Source* Compiled by the authors based on (Rosstat, n.d)

respectively (Skalnaya, 2018). Thus, for the total studied period, the total rural population in Russia decreased by 1.4 million people (3.6%) (Table 1). At the same time, the largest number (about 80%) of rural settlements with less than ten inhabitants and without population were concentrated in the regions of the Central and Northwestern Federal Districts (Rosstat, n.d).

It should be noted that this situation in the Russian rural areas is due to the fact that the standard of living in rural areas significantly lags behind the urban one. This can be judged by the average per capita disposable monetary income of a family, including all monetary and in-kind income (in monetary terms). As noted in the statistical publications of Rosstat, over the past 20 years, the size of total income achieved in rural households lags far behind the urban parameters of this indicator (Ibragimov, 2019; Ibragimov et al., 2020). In 2000, the average per capita disposable income of rural households was 1151.5 thousand rubles per month, while in urban households,

Table 1 Changes in Russia's rural settlements in the selected years during 1989–2020

Indicator	1989	2002	2010	2016	2020
Total rural population in Russia, million people	39.0	38.7	37.6	37.9	37.6
Share of rural population (%)	26.6	26.7	26.3	25.9	25
Number of rural settlements, thousand settlements	162.2	155.3	153.1	143.5	132.3
Number of empty rural settlements, thousand settlements	9.3	13.1	19.4	18.7	16.8
Share of empty rural settlements (%)	5.7	8.4	12.7	13.0	12.6
Number of disappearing rural settlements with a population of 1–10 people, thousand settlements	30.2	34.0	36.2	34.4	31.8
Proportion of disappearing rural settlements with a population of 1–10 people (%)	18.6	21.9	23.7	24.0	24.0
Number of rural settlements with a population of 11–50 people, thousand settlements	44.7	38.1	32.7	30.1	28.8
Share of rural settlements with a population of 11–50 people (%)	27.6	24.6	21.39	21	21.7

Source Compiled by the authors based on (Rosstat, n.d)

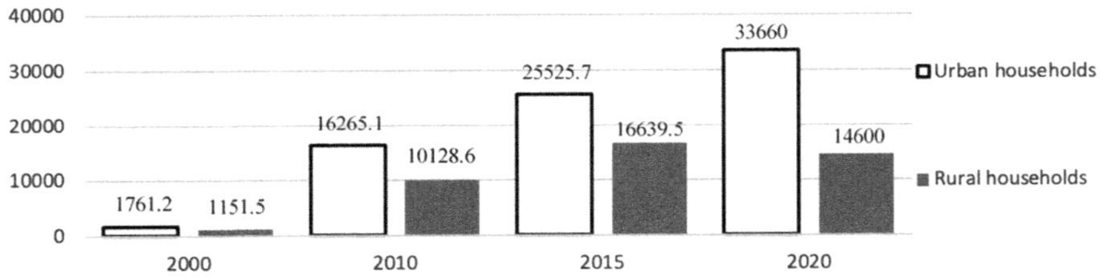

Fig. 3 Average per capita disposable income of households in some years during 1990–2020, rubles per month. *Source* Compiled by the authors based on (Rosstat, n.d)

it equaled 1761.2 thousand rubles per month. The indicator was 34.4% lower in the first category than in the second one (Ibragimov et al., 2020). On average, rural households had 37.7% less per capita disposable income relative to urban households in 2010, 34.8% in 2015, and 56.6% in 2020 (Fig. 3). That is, the situation has only worsened, and the gap has widened over 20 years.

The main source of disposable income of rural households is monthly wages. In this regard, there is a similar relationship as in the previous graph (Fig. 4). As can be seen, the year 1990 saw the best situation. At that time, the gap was minimal—2.7%. In 2000, the average monthly wage in the agricultural sector relative to the urban level was 40.1%, in 2010—48.7%, in 2015—56.5%, and in 2020—59.6%. That is, the gap compared to the 1990s is significant.

As shown in Table 2, in the Russian rural areas in 2018, the share of the population with incomes below the poverty line was 22%, which is 3.1 times more than in the city (7.1%). In addition, this indicator exceeds the 10% level more than twice, which, by international standards, is considered the maximum allowable. That is, every fifth rural resident of Russia is below the poverty line (Table 2).

In our opinion, one of the main reasons that led to such a gap in the financial situation of rural families is the lower wages in agriculture.

At the end of the 1980s, the average wage of workers in agricultural enterprises (state farms), were 5.4% higher than the wages of the urban population. At the same

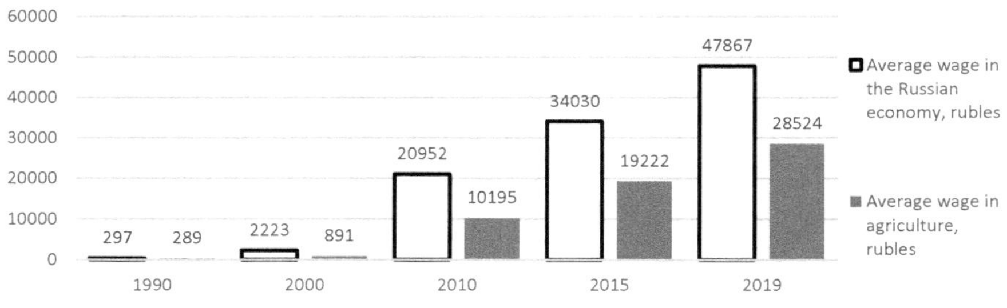

Fig. 4 Average wages in the Russian economy and agriculture in some years in 1990–2019. *Source* Compiled by the authors based on (Rosstat, n.d)

Table 2 The spread of poverty in Russia in selected years during 2013–2018

Indicator	2013	2014	2015	2016	2017	2018
Number of the poor in rural areas, million people	8.1	8.7	10.3	10.2	9.4	8.2
Share of the poor in the total population (%)						
Urban areas	5.2	5.5	8.4	8.3	8.1	7.1
Rural areas	22.0	23.4	27.8	27.1	24.9	22.0
Ratio of rural areas to urban areas, times	4.2	4.3	3.3	3.3	3.1	3.1

Source Compiled by the authors based on (Rosstat, n.d)

time, the average wage of collective farmers was at the level of 89% of the wages of factory workers and civil servants (Ibragimov et al., 2020; Ibragimov et al., 2020).

Other essential indicators characterizing the socio-economic situation of the population (including rural population) are the levels of employment, unemployment, and the accessibility of social facilities. During the studied period, in rural areas, the observed levels of these indicators are not satisfactory. According to Rosstat, the employment rate in rural areas is about 60%, and the unemployment rate is 8%–10%. For comparison, in the urban areas, these figures are 67% and 4%–6%, respectively (Rosstat, n.d) (Fig. 5).

The provision of social facilities for the rural population is worse than the urban population. Compared with 1990, this indicator only increased by 2019. Currently, the average radius of access to appropriate social infrastructure in rural areas is as follows: kindergartens—23 km, schools—16 km, clubs—14 km, hospitals—90 km, outpatient clinics—39 km, paramedic midwifery units (PMU)—15 km (Fig. 6).

In Russia, the standard of living in rural areas is steadily lower than in urban areas due to the differences in wages and the provision of residents with social infrastructure (Russkiy & Platonovskiy, 2019). According to some scholars, 65% of settlements in rural areas have no health care facilities, 70% have no consumer services, and 40% have no stores or retail outlets (Romanyuk et al., 2019). The

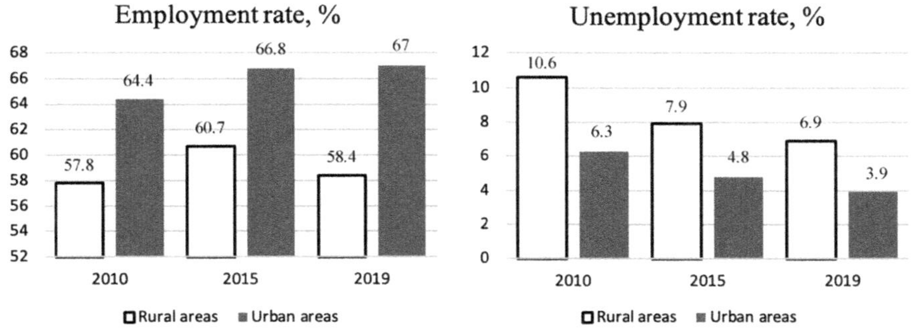

Fig. 5 Employment and unemployment rates among the able-bodied population in some years during 2010–2019 (annual average, %). *Source* Compiled by the authors based on (Rosstat, n.d)

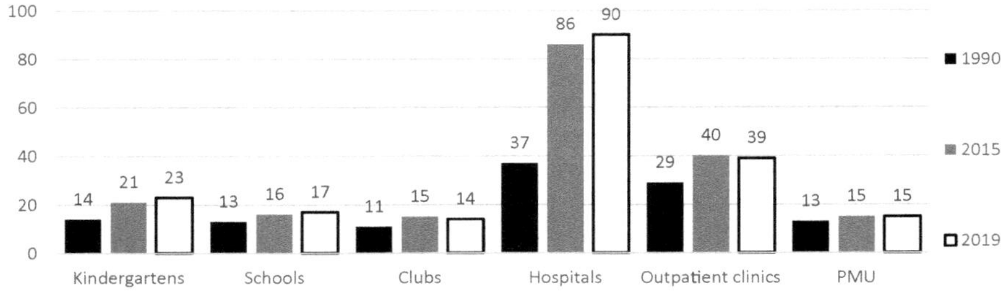

Fig. 6 Average radius of access to social facilities in rural areas, km. *Source* Compiled by the authors based on (Rosstat, n.d)

situation was similar in many European countries before the early 1990s. Rural areas had the repulsive image of an economically backward, under-equipped, aging, and isolated environment (Koteev et al., 2002; Vorontsova et al., 2021).

The current attractiveness of rural areas to European citizens is linked to government policies on improving living conditions in these territories. The EU has gradually shifted the approach to rural development from sectoral to territorial (Merlin, 2009).

4 Conclusion

Our research shows that rural areas are degrading in terms of population and the number of settlements, despite attempts of federal, regional, and local authorities to correct the situation. The challenging situation with the quality of life in most rural settlements and low accessibility of medical, educational, and social services is a negative effect of the reforms ongoing within the framework of the so-called optimization. Low wages, high hidden unemployment, and a shortage of well-paid jobs in rural areas are the current reality of the vast majority of rural communities. These objective conditions of modern Russia force the population, especially young people, to migrate to the cities.

In our opinion, possible solutions to the current problems are as follows:

1. Comprehensive governmental measures to increase the income of the rural population;
2. Encouraging the creation of new well-paid jobs in rural areas;
3. Improvement of transport, communal, medical, and social infrastructure in rural areas;
4. Strengthening state support for agriculture, with the special attention needed all (large, medium, and small) producers. Support for these categories of farms should be provided, including per one hectare of cultivated area.

References

Federal State Statistics Service of the Russian Federation (Rosstat). (n.d.). Official website. Retrieved from http://www.gks.ru/. Accessed 11 June 2021.

Ibragimov, A. G. (2019). To the question of effective land use in Russia. *Journal of Economy and Entrepreneurship, 9*(110), 1316–1319.

Ibragimov, A. G., Borulko, V. G., & Leshcheva, M. G. (2020). The effectiveness of dairy production in Russia. *Journal of Economy and Entrepreneurship, 8*(121), 255–258. https://doi.org/10.34925/EIP.2020.121.8.052

Ibragimov, A. G., Borulko, V. G., & Sergienkova, N. A. (2020). The effectiveness of beef cattle breeding in Russia. *Journal of Economy and Entrepreneurship, 8*(121), 233–236. https://doi.org/10.34925/EIP.2020.121.8.047

Ibragimov, A. G., Dzhancharova, G. K., & Leshcheva, M. G. (2020). Prospects and problems of digitalization of agriculture in Russia. *Journal of Economy and Entrepreneurship, 9*(122), 587–589. https://doi.org/10.34925/EIP.2020.122.9.114

Koteev, S. V., Mukhametzyanov, R. R., Nazranov, M. R., Agirbov, Y. I., & Lesnov, A. P. (2002). *Market relations in the agro-industrial complex and fruit and vegetable subcomplex*. Publishing house of the Russian Engineering Academy of Management and Agribusiness.

Leshcheva, M. G., Steklova, T. N., Uryadova, T. N., Ibragimov, A. G., & Borulko, V. G. (2021). Demographic problems with human resources for agricultural production in the region. In A. V. Bogoviz (Ed.), *The Challenge of Sustainability in Agricultural Systems* (pp. 518–527). Cham, Switzerland: Springer. https://doi.org/10.1007/978-3-030-72110-7_60

Merlin, P. (2009). *L'exode urbain: de la ville à la campagne*. Paris, France: La Documentation Française.

Ministry of Agriculture of the Russian Federation. (2012). *State program for the development of agriculture and regulation of markets of agricultural products, raw materials, and food from 2013 to 2020*. Moscow, Russia: Rosinformagrotech.

Ministry of Agriculture of the Russian Federation. (2015). *Strategy for sustainable development of rural areas of the Russian Federation for the period up to 2030*. Moscow, Russia: Rosinformagrotech.

Romanyuk, M. A., Platonovskiy, N. G., & Pavlova, I. M. (2019). About strategic planning in the Russian Federation. *Economics of Agriculture in Russia, 6*, 100–105.

Russkiy, V. G., & Platonovskiy, N. G. (2019). On the issue of rural poverty in Russia. *Investments in Russia, 11*(298), 9–15.

Skalnaya, M. M. (2018). Income of rural people as a factor of social stability of rural territories. AIC: Economics. *Management, 1*, 62–71.

Vorontsova, N. V., Merzlov, A. V., Mukhametzyanov, R. R., & Platonovskiy, N. G. (2021). Assessment of the attractiveness of rural areas in terms of internal migration in Russia and the EU countries. *Economy of Agricultural and Processing Enterprises, 6*, 40–47. https://doi.org/10.31442/0235-2494-2021-0-6-40-47

Economic Mechanisms for the Development of the Material and Technical Base of Agricultural Enterprises in the Region

Tatyana G. Gurnovich, Lyubov V. Agarkova, Elena M. Puchkova, Irina V. Sinitsyna, and Olga N. Nikulina

Abstract Purpose: improvement of economic mechanisms for renewing the material and technical base of agricultural organizations and ensuring its maximum adaptation to the conditions of resource constraints. Design/methodology/approach: The design of this article is an algorithm for developing a plan for the implementation of a designated goal by setting and step-by-step solution to a set of research problems:

- clarification of the economic content of the concepts of "resource potential" and "material and technical base (MTB) of agricultural enterprises";
- systematization of factors influencing the reproduction of the material and technical base of agricultural production and the processes of renewal of its elements;
- substantiation of the improvement of the economic mechanism for the development of the material and technical base of the agricultural industry in modern conditions, taking into account state support.

The research methodology is based on:

- the fundamental provisions and concepts of evolutionary and institutional economic theory, the theory of reproduction, modern scientific approaches to the study of subsystems and markets of the agro-industrial complex, mechanisms of state regulation;
- scientific works of domestic and foreign agricultural economists in the field of functioning and development of the agricultural economy;
- recommendations of research institutes and institutions of higher agricultural education on the reproduction of the material and technical base of agricultural organizations;

T. G. Gurnovich (✉)
Kuban State Agrarian University named after I.T. Trubilin, Krasnodar, Russia
e-mail: gurnovich@inbox.ru

L. V. Agarkova
Stavropol State Agrarian University, Stavropol, Russia

E. M. Puchkova · I. V. Sinitsyna · O. N. Nikulina
Nevinnomyssk State Humanitarian and Technical Institute, Nevinnomyssk, Russia

E. G. Popkova and B. S. Sergi (eds.), *Geo-Economy of the Future*,
https://doi.org/10.1007/978-3-030-92303-7_39

• directive and regulatory documents in the field of development of subsystems of the agro-industrial complex on an innovative platform.

Findings: The economic content of the categories "resource potential" and "MTB of agricultural production" is expanded; the factors influencing the processes of reproduction of MTB in agriculture are systematized, their general classification is given and the priorities of the current stage are highlighted; justified recommendations for improving the economic mechanism for the development of MTB based on the implementation of the calculation, regulatory and strategic approaches that complement each other. Originality/value: The theoretical value of the work lies in the development of the economic content of the categories "resource potential" and "material and technical base" based on the implementation of the dialectical approach to the interaction of the material and technical component and innovative technologies. The applied significance of scientific research lies in the development of methodological approaches to assessing the resource potential and material and technical base of agricultural production, as well as practical recommendations for improving the economic regulation of MTB renewal processes.

Keywords Resource potential · Material and technical base · Agricultural production · Economic mechanism · State support · Innovative development · Reproduction

JEL Codes O13 · Q10 · Q18

1 Introduction

Modern economic science has developed many views, concepts, and approaches to the interpretation of interrelated categories: "resource potential" and "material and technical base (MTB)". The issues of economic assessment of the elements of MTB of agricultural enterprises and the development of mechanisms for their modernization and development are also widely discussed in scientific circles. They are reflected in monographic studies, scientific articles by foreign and domestic scientists.

The works of foreign scientists-economists are devoted to the problems of formation, organization of use and development of the resource base of agricultural enterprises in the context of ensuring the efficiency of agricultural production (Banker & Natarajan, 2008; Battese & Coelli, 1992; Coelli, 1970) and others.

Among the scientific studies of domestic scientists on the composition and structure of the resource potential of agricultural enterprises, we note the works of Belokopytov and Zharova (2009); Davydkina, 2013; Demesheva & Tetyurkina, 2018; Leshchilovsky et al., 2007; Makarkin et al., 2019; Minakov, 2020; Petrikov, 2010; Piskunov, 2013; Potapov, 2014; Smagin, 2021; Smirnova, 2006; Udalykh, 2020) and others.

Various aspects of the development of economic mechanisms for the renewal of MTB were considered in the works of Bershitskaya et al. (2020); Bershitsky et al., 2015; Doronin et al., 2019; Krasnoshchekov, 2008; Lysenko & Lysenko, 2013; Orsik & Kormakov, 2008; Polukhin, 2013; Reznichenko et al., 2018) and other scientists.

However, despite the diversity of points of view on the economic content of the studied categories, it should be noted that there is a lack of integrity in the interpretation of the resource potential of agriculture, which is expressed in a weak emphasis on the use of technological innovations in economic practice, which provide a synergistic effect when interacting with elements of MTB and thereby influencing on the final results of production.

The working hypothesis of this study is based on the following provisions:

- increasing the efficiency of agricultural production and ensuring sustainable growth in the agricultural sector cannot be achieved without a material and technical base that meets modern requirements and innovative technologies;
- the ongoing changes in the process of reproduction of MTB form the need to modernize the economic regulation mechanism and the development of its components;
- the imperative of agrarian policy at present is a qualitative transformation of the forms and methods of state support for the renewal of MTB of agricultural organizations and a quantitative increase in its volumes.

2 Materials and Method

The theoretical and methodological basis for studying the composition and structure of material and technical resources of enterprises in the agrarian sector was the provisions of economic theory, which formed the basic concepts of the reproduction of resources in the process of their transformation.

The study was carried out based on the application of the systemic and structural-logical approaches to the development of the author's idea of the resource potential of agricultural organizations, the improvement of methods for assessing material and technical resources, the substantiation of the mechanisms for the development of MTB of agricultural enterprises in the region. A complex of methods was used that combines:

- theoretical methods-operations: analysis, synthesis, comparison, generalization, analogy;
- theoretical methods of action: dialectics, scientific theories, hypothesis construction;
- empirical methods-operations: monographic;
- empirical methods of action: monitoring, study and generalization of experience.

The information base was formed based on the results of monographic studies by domestic and foreign scientists on the economic content of the material and technical base of agricultural production, methodological approaches to assessing its components and adapting the reproduction mechanism to economic conditions.

The main conclusions and results of the study are formulated based on empirical factual data and relevant information obtained in the process of their collection and processing.

3 Results

The material and technical base is part of the resource potential of agricultural organizations, which also includes biological, technological, innovation, financial and other functional subsystems.

In some works studied (Piskunov, 2013; Potapov, 2014), the economic content of the resource potential of agricultural enterprises is revealed without fully taking into account the dialectics of the constituent elements. As one of the stages of the resource provision process, the calculation of the need for each type of resource is highlighted, while the authors put financial and organizational aspects at the forefront, ignoring the technological component of the resource base. Without diminishing the importance of the organizational and financial component, we consider it necessary to determine the use of innovative technologies in agricultural production as the main driving force for the reproduction of resource potential. At the same time, we emphasize that the subsystems of innovative potential and agricultural technologies interact, first of all, with the elements of MTB. An analytical approach to the study of the category of "resource potential", based on the research of predecessor scientists (Minakov, 2020; Potapov, 2014), made it possible to differentiate the main structural elements by the processes of formation and use (Fig. 1).

Technical units and mechanisms in the production process, interacting with biological objects, land resources, technologies used, are combined into a bio industrial system. Therefore, when assessing the resource potential, it is necessary to take into account the climatic and agro economic factors, the peculiarities and contradictions of the pricing policy in the agro-food and resource markets, as well as the impact on the functioning of the resource potential subsystems of the processes of adaptation of agricultural producers to the conditions of the innovative economy. In addition, we believe that to develop directions for improving the economic mechanisms for renewing MTB of agricultural organizations, it is advisable to clarify methodological approaches to assessing the components of MTB resources.

Improving the methodological approach to assessing the resource potential of agricultural production, we assumed that the diversity of the functional purpose of its elements requires the division of estimated indicators both by the components of the resource base (land resources, finance, personnel, fixed assets, material circulating assets) and by the nature of their use (structural, providing, effective) (Fig. 2). Based on the presented indicators, the level of resource availability and the efficiency of agricultural production is calculated. To assess the resource potential of agricultural enterprises in the region, one can use the index method, according to which the index is calculated for each indicator as to the ratio of the values of the territories of the analyzed region (edge, region, country) to the regional average value.

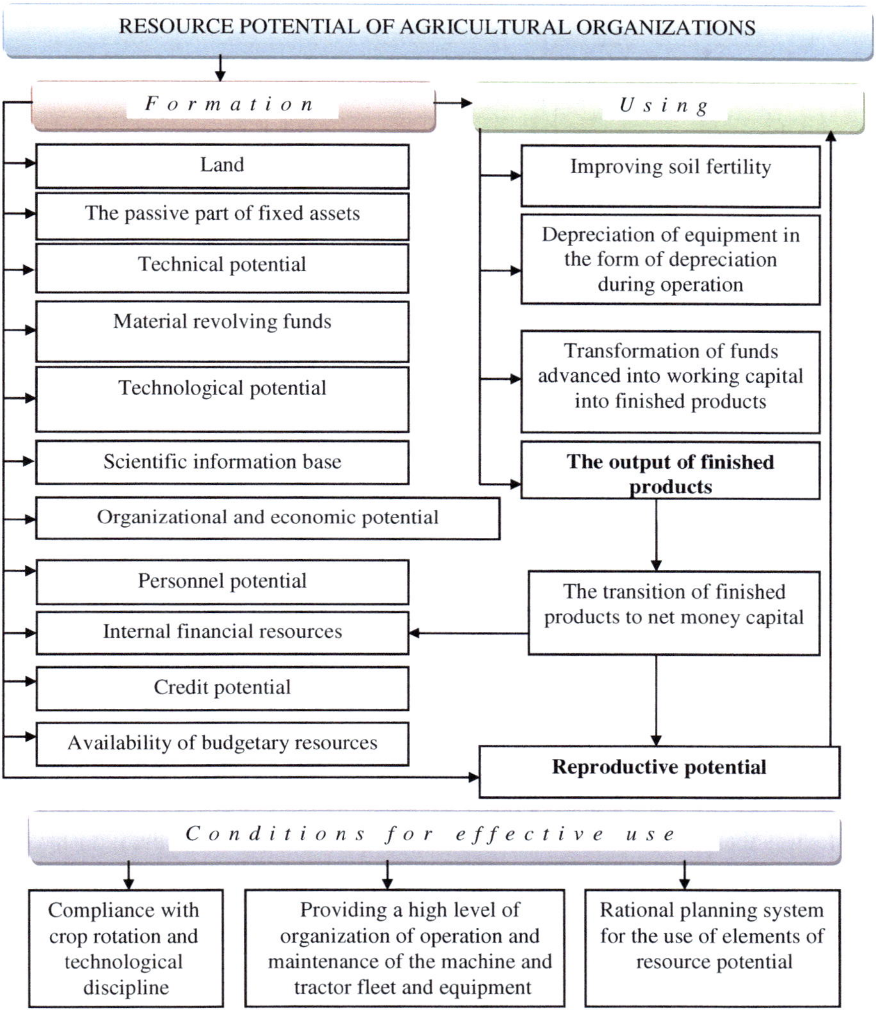

Fig. 1 Formation and use of the resource potential of agricultural organizations. *Source*: Developed and compiled by the authors based on sources (Gurnovich et al., 2018; Minakov, 2020; Potapov, 2014)

The process of material and technical renewal of agricultural production includes the implementation of a set of measures of a technical, technological and organizational and economic nature, aimed at saving resources and, ultimately, increasing the efficiency of agricultural enterprises (Bershitskaya et al., 2020).

The renewal of the technical base can be carried out extensively and intensively. In the first method, the equipment that has exhausted its resource is replaced with a similar one. The second method involves the commissioning of equipment of higher

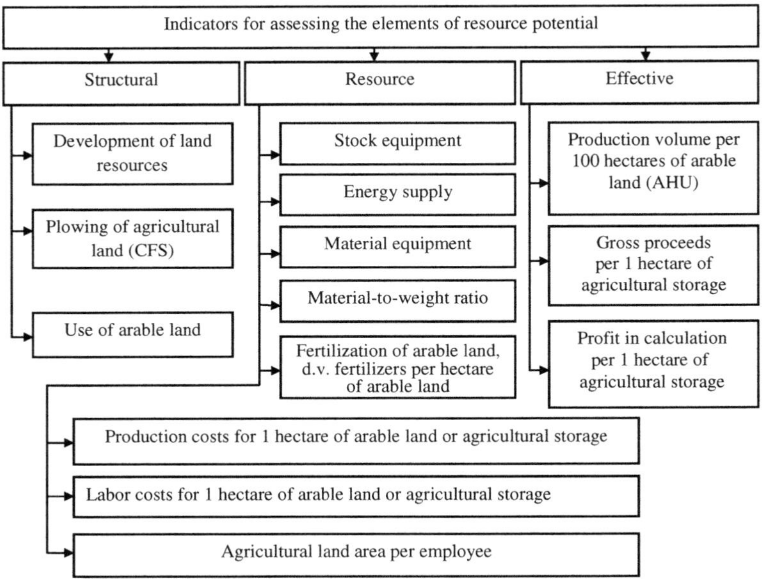

Fig. 2 Differentiation of indicators for assessing the resource potential of agricultural enterprises. *Source*: Developed and compiled by the authors

productivity while using progressive technologies for the production of cultivation of crops or raising animals.

Within the framework of improving the system of state support for agricultural producers and the selection of appropriate measures, it is important to differentiate the factors that determine the vector of innovative development of the industry into external and internal ones.

The system of measures for technical and technological renewal of agricultural production, aimed at increasing the economic efficiency of agricultural producers, is presented in Table 1.

The problem of modernizing the material and technical base of domestic agricultural production is associated with the development of two interrelated aspects: substantive (technical) and supporting (financial). The technical aspect is directly related to the establishment of general criteria for the nomenclature and technical characteristics of the machine and tractor fleet for plant growing and equipment for animal husbandry in terms of their compliance with the requirements of innovative development, the financial aspect is focused on the effective use of own and attracted investment resources in conditions of their limitedness. This indicates the need to improve regulatory mechanisms, including based on state support, the development of MTB in agricultural production.

The use of a computational and normative approach to determining the need for agricultural machinery and equipment is a well-known tool for solving the complex problem of increasing the efficiency of agricultural production.

Table 1 System of measures for technical and technological renewal of agricultural production

Character of activities	Name of the event
Exogenous	Improvement of scientific, technical and tax policy of the state
	Financial support for updating the material and technical base of agricultural producers
	EQUIPMENT insurance
	Price formation
	Determination of the procedure for calculating depreciation and write-off of equipment
	Updating the regulatory framework
Endogenous	Determination of counterparties that supply equipment at the lowest cost with a guarantee of its quality, other service companies that perform mechanized work (MTS, mechanized teams, etc.)
	Purchase of new (Russian or foreign) or used equipment
	Restoration of the equipment, which is subject to write-off
	Introduction of types of on-farm use of equipment
	Implementation of forecasting methods for the mechanization of agricultural production
	Material incentives for the introduction of new technology in agricultural production
	Justification of funding sources and options for the purchase of equipment

Source developed and compiled by the authors based on source (Bershitskaya et al., 2020)

Compliance with standards is the key to a rational organization of production processes on farms, compliance with agro technical and temporary requirements. At the same time, it should be emphasized the need for timely adjustments to the standards, taking into account the requirements of scientific and technical progress.

The strategic approach is associated with the establishment of indicators of state support for the renewal of MTB in strategies for the socio-economic development of municipalities, which is extremely important at the present stage. State support for the modernization of MTB, aimed at creating favorable conditions for agricultural producers for its sustainable development on an innovative basis, is most fully implemented in modern programs for the development of the agro-industrial complex.

We believe that the implementation of projects for the technical and technological renewal of agricultural production, taking into account the agricultural intensity of the region, should also be included in the development strategies of municipalities as priority areas.

We propose, within the framework of measures to modernize the economic mechanism and build forecasts for the development of MTB, to combine the calculation, normative and strategic approaches.

4 Conclusion

The study of various points of view on the economic content of the interrelated categories of "resource potential" and "material and technical base of agricultural enterprises" made it possible to conclude that it is necessary to define as the main driving force for the reproduction of the resource potential and, first of all, its material and technical component, the use of innovative technologies in agricultural production.

The author's proposal is of an applied nature that when assessing the resource potential of a particular agricultural enterprise, it is advisable to divide the estimated indicators both by the components of the resource base and by the nature of their use. The use of the index method is aimed at conducting an intraregional comparative assessment of the provision of agricultural producers.

The work carried out a systematization of factors affecting the reproduction of MTB in agricultural production and the processes of renewal of its elements, in accordance with which it is possible to choose the priority areas of external and internal regulation. Assessment of the dynamics of the provision of agricultural organizations with tractors and combines in the Russian Federation for 2013–2019 showed no positive trend. There is a decrease in the number of basic types of equipment and an increase in the load on it.

This requires the improvement of the economic mechanism for the development of MTB in the agricultural sector in modern conditions, taking into account state support. Within the framework of solving this problem, a set of measures of an internal and external nature is proposed, as well as the use of calculation, normative and strategic approaches to the regulation of the technical and technological renewal of MTB in agricultural production is justified.

Acknowledgements The article was prepared with the financial support of the Russian Fund for Fundamental Research. Grant of the RFFR No. 20-010-00079.

References

Banker, R. D., & Natarajan, R. (2008). Evaluating contextual variables affecting productivity. *Operations Research, 56*, 48–58. https://doi.org/10.1287/opre.1070.0460Accessed:10.05.2021

Battese, G. E., & Coelli, T. J. (1992). Frontier production functions, technical efficiency and panel data: With application to paddy farmers in India. *Journal of Productivity Analysis, 3*, 153–169. https://doi.org/10.1007/BF00158774Accessed:10.05.2021

Belokopytov, A. V., & Zharova, O. L. (2009). Economic problems of using the resource potential of agricultural enterprises in the Smolensk region. *Smolensk*, 105.

Bershitskaya, G. F., Gurnovich, T. G., Kuvarin, D. Yu., Matveev, A. S., & Pozoyan, D. P. (2020). Innovative development of the material and technical base of agricultural production in the context of import substitution. In *Proceedings of the Kuban State Agrarian University* (Vol. 86, pp. 15–20).

Bershitsky, Y. I., Kastidi, Y. K., & Tyupakov, K. E. (2015). Features of the economic assessment of grain harvesting equipment. *Polythematic Network Electronic Scientific Journal of the Kuban State Agrarian University, 111*, 287–298.

Coelli, T. J. (1970). Measurement of total factor productivity growth and biases in technological change in western Australian agriculture. *Journal of Applied Econometrics, 11*(1), 77–91.

Davydkina, O. A. (2013). Resource potential as a factor in increasing the competitiveness of agricultural production in the region. *Actual Problems of the Humanities and Natural Sciences, 11–1*, 201–204.

Demesheva, I. A., & Tetyurkina, E. V. (2018). *Resource potential of agricultural enterprises in the region and the effectiveness of its use* (101p). Belgorod State Agrarian University named after V.Ya. Gorin.

Doronin, B. A., Detistova, O. I., Gritsay, D. I., Alekseenko, V. A., & Serikov, S. S. (2019). The formation of the enterprise intangible assets is the basis to develop competitive regional agricultural machinery. *Research Journal of Pharmaceutical, Biological, and Chemical Sciences, 10*(1), 1406–1411.

Gurnovich, T. G., Agarkova, L. V., & Indyukov, A. I. (2018). *Perspective directions of innovative development of the material and technical base of agricultural production: Monograph* (175p). Kuban State Agrarian University.

Krasnoshchekov, N. V. (2008). *Innovations in machine use of the agro-industrial complex of Russia* (vol. 1, no. 2, 404p). FGNU "Rosinformagrotech".

Leshchilovsky, P. V., & Gusakov V. G. et al. (2007). *Economy of enterprises and branches of the agro-industrial complex* (574p). Belarus State Economic University.

Lysenko, M. V., & Lysenko, Y. V. (2013). Methodology for assessing the state of technical potential and the effectiveness of its use. *Basic Research, 4*, 202–206.

Makarkin, N. P., Potapova, L. N., Alferina, O. N., & Korneeva, N. V. (2019). Methodological approaches to the study of the economic essence of the resource potential of agricultural enterprises. *Bulletin of the Altai Academy of Economics and Law, 10–2*, 87–92.

Minakov, I. A. (2020). *Economics and management of enterprises, branches, and complexes of the agro-industrial complex: A tutorial* (404p). Lan.

Orsik, L. S., & Kormakov, L. F. (2008). Directions for solving the problem of technical equipment in agriculture. *Rural Machinery and Equipment, 4*, 7–10.

Petrikov, A. V. (2010). Agriculture of Russia: Potential, mechanisms of development, international cooperation. *Agricultural Economics of Russia, 11*, 15–20.

Piskunov, A. I. (2013). *Efficiency of resource provision of enterprises of the agro-industrial complex* (Ph.D. in Economy thesis). Ekaterinburg, 182p.

Polukhin, A. A. (2013). Economic assessment of the prospects for the technical modernization of agriculture in the Russian Federation in the context of the WTO. *Agrarian Russia, 11*, 33–36.

Potapov, A. P. (2014). Resource potential of agricultural production in the food security system. *Nikon Readings, 19*, 36–38.

Reznichenko, S. M., Bershitsky, Yu. I., Shichiyakh, R. A., Kurnosov, S. A., & Kuzmenko, O. V. (2018). Methodological characteristics for the efficient evaluation of investments in updating the technical basis of agricultural organizations dedicated to planting. *Espacios, 39*(31), 10.

Smagin, B. I. (2021). Algorithm for calculating the production potential of the agricultural sector of the economy. *Bulletin of Michurinsk State Agrarian University, 1*(64), 153–162.

Smirnova, A. V. (2006). Structure and methods of economic assessment of the resource and production potential of agricultural enterprises. *Agrarian Bulletin of the Urals, 6*(36), 13–20.

Udalykh, O. A. (2020). *Directions for improving the management of the production potential of agricultural enterprises.*(Vol. 20, pp. 225–234). Collection of scientific works of the series "Public administration".

Successful Global Practices in Responsible Agriculture for Sustainable Development

Aleksei V. Bogoviz, Lyubov V. Postnikova, Daria D. Postnikova, Svetlana V. Lobova, and Alexander N. Alekseev

Abstract The paper studies successful global practices of responsible agriculture for sustainable development. The authors develop an alternative methodological approach to studying responsible agricultural practices for sustainable development. The new approach involves using regression analysis to find the dependence of the index of corporate environmental responsibility and climate change on responsible agriculture. First, this approach allows clarifying the results on the Sustainable Development Goals related to environmental protection. Second, it allows for separating voluntary environmental initiatives of agricultural businesses from environmental requirements and standards developed by governments. The authors analyze a Monte Carlo scenario, which identifies the prospects for improving the practices of responsible agriculture for sustainable development in developed and developing countries up to 2030. The results showed that the impact of responsible agriculture on the index of corporate environmental responsibility and climate change is much more positive only in developing countries, though it is moderate. The optimistic scenario improves the index of corporate environmental responsibility and combating climate change in developing countries by 3.34%. The conducted research scientifically identified and accurately quantified the direct contribution of responsible agriculture to progress toward the environmental SDGs.

Keywords Successful global practices · Responsible agriculture · Sustainable development · Developed countries · Developing countries

JEL Codes Q13 · M14 · M15

A. V. Bogoviz (✉)
Moscow, Russia

L. V. Postnikova · D. D. Postnikova
Russian State Agrarian University—Moscow Timiryazev Agricultural Academy (RSAU—MAA named after K.A. Timiryazev), Moscow, Russia

S. V. Lobova
Altai State University, Barnaul, Russia

A. N. Alekseev
Plekhanov Russian University of Economics, Moscow, Russia

E. G. Popkova and B. S. Sergi (eds.), *Geo-Economy of the Future*,
https://doi.org/10.1007/978-3-030-92303-7_40

1 Introduction

Responsible agriculture involves the stability of the volume, prices, and quality of food products achieved by successful adaptation of agricultural production technologies to the natural and climatic conditions of the territory (which, in particular, involves import substitution in the agro-industrial complex) or reducing sensitivity to these conditions by creating autonomous (independent of external environmental) agrarian enterprises. Moreover, responsible agriculture involves rational and careful use of natural resources to preserve them for future generations.

The quantitative embodiment of agricultural responsibility is the "Natural Resources and Resilience" indicator (The Economist Intelligence Unit Limited, 2021). It is generally accepted that this indicator contributes to sustainable development. However, this contribution remains poorly understood and proven scientifically for two reasons. First, the indicator of sustainable development (calculated by UNDP) is aggregated and reflects the results for all Sustainable Development Goals (SDGs), not specifying the results for SDGs related to environmental protection (e.g., SDGs 13–15). Second, responsible agriculture implies a key role of private agrarian business and its voluntary environmental initiatives, while the "Natural Resources and Resilience" indicator (The Economist Intelligence Unit Limited, 2021) considers these initiatives inseparable from environmental requirements and standards developed by the government.

Thus, the direct contribution of responsible agriculture to progress toward the achievement of environmental SDGs has not been scientifically defined or measured. A gap in scientific knowledge represents a serious flaw in the current approach to studying these issues. It seems relevant to search for an alternative methodological approach beyond the standard generalized indicators to fill the identified gap. The development of such an approach is the subject of this research, which aims to explore successful global practices in responsible agriculture for sustainable development.

2 Literature Review

Selected national practices of responsible agriculture have been reviewed by Adriant et al. (2021), Chamuah and Singh (2020), Cvejić et al. (2019), Mayumi (2020), Pandey et al. (2020), Regan (2021), Rose and Chilvers (2018), Rose et al. (2021). The contribution of agriculture to protecting the environment and combating climate change as a goal of sustainable development is discussed by Akopova et al. (2020), Fokina (2020), Karanina (2020), Sozinova (2020).

However, the international experience of responsible agricultural practices is not systematized. The differences between developed and developing countries are not defined. The impact of these practices on environmental protection and combating climate change as one of the SDGs has not been defined or measured. These gaps define the field for further research and are filled in this paper.

3 Research Methodology

The author's (alternative) methodological approach to studying the practices of responsible agriculture for sustainable development has been developed. The new approach involves using regression analysis to find the dependence of the index of corporate environmental responsibility and climate change on responsible agriculture.

This approach allows us to clarify the results on the SDGs related to environmental protection. Moreover, it allows us to separate voluntary environmental initiatives of agricultural businesses from governmental environmental requirements and standards. Successful global practices of responsible agriculture for sustainable development are examined using the example of the top ten developed and top ten developing countries in terms of the value of the corporate environmental responsibility and climate change index in 2020 (Institute of Scientific Communications, 2021). The data for the research are summarized in Table 1.

Table 1 Statistics on responsible agriculture and corporate sustainability support in developed and developing countries in 2020, points 1–100

Developed countries			Developing countries		
Country	Corporate environmental responsibility and climate change index	Responsible agriculture (natural resources and resilience)	Country	Corporate environmental responsibility and climate change index	Responsible agriculture (natural resources and resilience)
	$Cerclch_{dd}$	$Ragr_{dd}$		$Cerclch_{ding}$	$Ragr_{ding}$
Belgium	48.7585	48.2	Sri Lanka	14.8923	46.3
Japan	33.8577	58.6	Panama	11.1038	53.0
Republic of Korea	32.3738	56.1	Oman	10.4438	43.8
France	27.3540	59.0	Argentina	8.5923	44.9
Switzerland	19.2486	64.2	Uruguay	8.2571	68.6
Austria	19.2413	61.8	Ecuador	7.7134	43.9
Italy	16.8835	50.7	Bangladesh	7.6867	35.8
Spain	16.7770	58.4	Kuwait	7.2099	37.2
New Zealand	14.1848	69.9	Pakistan	6.6171	42.1
Ireland	13.5118	73.2	Russia	6.2467	55.0
Arithmetic mean	24.22	60.01	–	8.8	47.06
Standard deviation	–	7.74	–	–	9.64

Source Calculated and compiled by the authors based on the materials of Institute of Scientific Communications (2021) and The Economist Intelligence Unit Limited (2021)

The authors also analyze a Monte Carlo scenario, which identifies prospects for improving the practices of responsible agriculture for sustainable development in developed and developing countries up to 2030.

4 Findings

The results of studying successful global practices of responsible agriculture for sustainable development are examined by regression analysis based on Table 1 (Fig. 1).

According to Fig. 1, a one-point increase in agricultural responsibility leads to a 0.0175-point increase in the corporate environmental responsibility and climate change index. In developing countries, there is the opposite effect. A one-point increase in agricultural responsibility leads to a 0.4517-point decrease in the corporate environmental responsibility and climate change index.

Based on the arithmetic mean and standard deviations calculated in Table 1, we generated 100 random numbers using the Monte Carlo method. These numbers reflect projections of agricultural responsibility for the period up to 2030 in developed and developing countries. Histograms of these projections are plotted in Fig. 2.

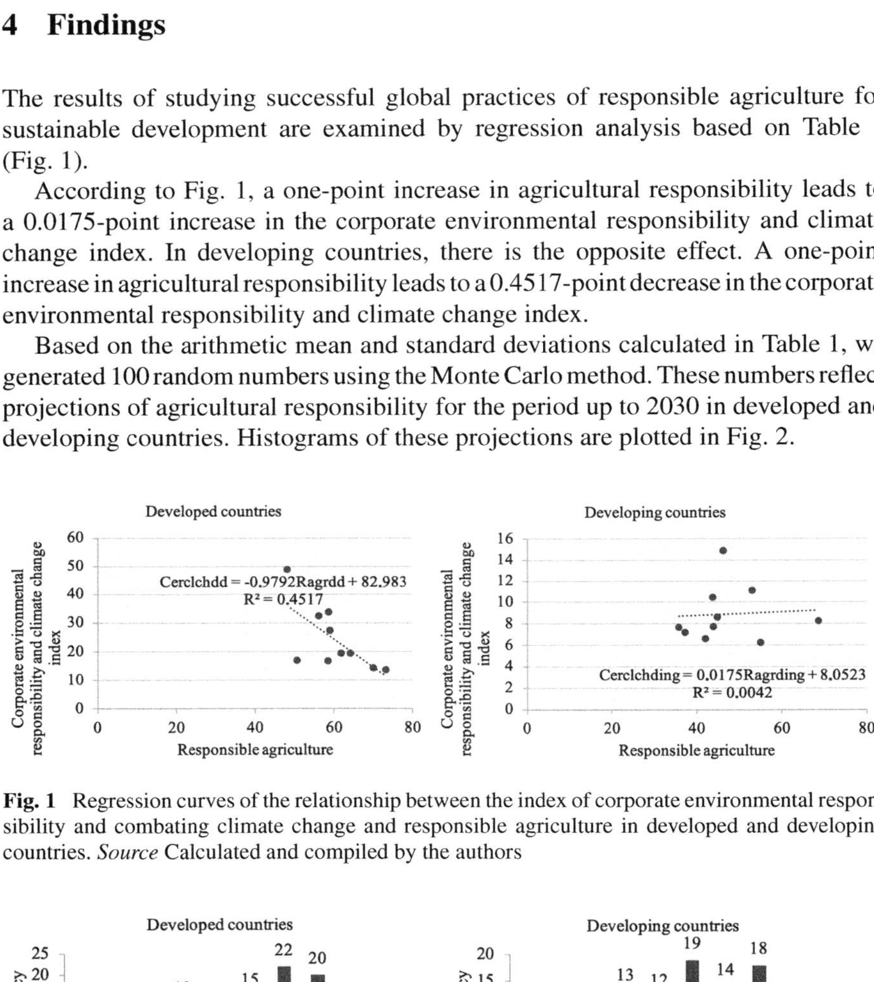

Fig. 1 Regression curves of the relationship between the index of corporate environmental responsibility and combating climate change and responsible agriculture in developed and developing countries. *Source* Calculated and compiled by the authors

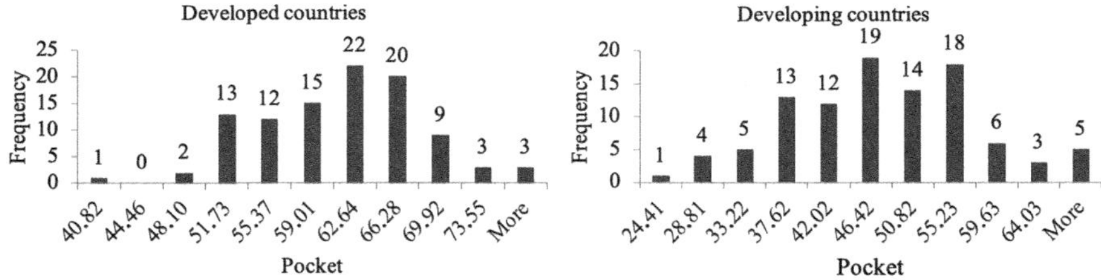

Fig. 2 Projections of agricultural responsibility for developed and developing countries up to 2030. *Source* Calculated and compiled by the authors

Table 2 Scenarios for responsible agriculture for sustainable development in developed and developing countries up to 2030

Characteristics of scenarios		Developed countries		Developing countries	
		Corporate environmental responsibility and climate change index	Responsible agriculture	Corporate environmental responsibility and climate change index	Responsible agriculture
Basic value		24.22	60.01	8.88	47.06
Optimistic scenario	Value, points 1–100	32.33	51.73	9.17	64.03
	Growth (%)	33.48	−13.8	3.34	36.06
Realistic scenario	Value, points 1–100	21.65	62.64	8.86	46.42
	Growth (%)	−10.62	4.38	−0.13	−1.36
Pessimistic scenario	Value, points 1–100	10.96	73.55	8.56	28.81
	Growth (%)	−54.73	22.56	−3.6	−38.78

Source Compiled by the authors

According to Fig. 2, we identified the optimistic (ensuring the maximum growth of the corporate environmental responsibility and climate change index), realistic (the most probable), and pessimistic (leading to the greatest decline of the corporate environmental responsibility and climate change index) scenarios. The consequences of these scenarios are presented in Table 2.

According to Fig. 2, under a realistic scenario, the index of corporate environmental responsibility and combating climate change in developed countries would decrease by 10.62%. In turn, it would remain virtually unchanged (decreasing by 0.13%) in developing countries. In the pessimistic scenario, the index of corporate environmental responsibility and climate change will decline by 54.73% in developed countries and by 3.6% in developing countries. In the optimistic scenario, the index of corporate environmental responsibility and climate change will increase by 33.48% in developed countries and by 3.34% in developing countries.

5 Conclusions

Thus, the conducted analysis of successful global practices of responsible agriculture for sustainable development showed that the influence of responsible agriculture on the index of corporate environmental responsibility and climate change is much more pronounced (though negative) in developed countries. In developing countries, the impact is moderate and positive.

The optimistic scenario allows increasing the index of corporate environmental responsibility and combating climate change. The index will increase by 33.48% in developed countries and by 3.34% in developing countries. This requires a 13.8% decrease in agricultural responsibility in developed countries and a 36.06% increase in developing countries.

Our research scientifically identified and accurately quantified the direct contribution of responsible agriculture to progress toward achieving environmental SDGs. The identified differences in this contribution in developed and developing countries require the design and application of different approaches to managing responsible agriculture for sustainable development in these groups of countries, taking into account their specifics.

References

Adriant, I., M.simatupang, T., & Handayati, Y. (2021). The barriers of responsible agriculture supply chain: The relationship between organization capabilities, external actor involvement, and supply chain integration. *Uncertain Supply Chain Management, 9*(2), 403–412.https://doi.org/10.5267/j.uscm.2021.2.003

Akopova, E. S., Przhedetskaya, N. V., Przhedetsky, Yu. V., & Borzenko, K. V. (Eds.). (2020). Marketing of nonprofit organizations in business-oriented economy: New challenges and priorities. In *Marketing of healthcare organizations: Technologies of public-private partnership* (pp. 15–24). Charlotte, NC: Information Age Publishing. Retrieved from https://www.infoagepub.com/products/Marketing-of-Healthcare-Organizations. Accessed 7 June 2021.

Chamuah, A., & Singh, R. (2020). Securing sustainability in Indian agriculture through civilian UAV: A responsible innovation perspective. *SN Applied Sciences, 2*(1), 106.

Cvejić, R., Istenič, M. Č, Glavan, M., Honzak, L., Klančnik, K., Kompare, K., & Pintar, M. (2019). Increasing climate change resilience in agriculture: Who is responsible? *Water Science and Technology: Water Supply, 19*(5), 1405–1412. https://doi.org/10.2166/ws.2019.009

Fokina, O. V. (2020). *Marketing management of projects for the introduction of "smart" learning technologies as a method of de monopolization of digital economy markets.* Retrieved from https://iscconf.ru/маркетинговое-управление-проектами/. Accessed 7 June 2021.

Institute of Scientific Communications. (2021*). Dataset "Corporate social responsibility, sustainability, and climate change: Simulation modeling and neural network analysis in regions of the world 2020.*" Retrieved from https://iscvolga.ru/dataset-climate-change. Accessed 7 June 2021.

Karanina, E. V. (2020). *Digital approach to the fight against shadow economy in the region: Advantages for the economic security and investment attractiveness of the territory.* Retrieved from https://iscconf.ru/цифровой-подход-к-борьбе-с-теневой-эко/. Accessed 7 June 2021.

Mayumi, K. T. (2020). Reconsidering agriculture, forestry, and fishery in Japan: Searching for a responsible development pathway. *Lecture Notes in Energy, 76*, 143–167. https://doi.org/10.1007/978-3-030-43225-6_7

Pandey, P., Valkenburg, G., Mamidipudi, A., & Bijker, W. (2020). Responsible research and innovation in the global south: Agriculture, renewable energy, and the pursuit of symmetry. *Science, Technology and Society, 25*(2), 215–222. https://doi.org/10.1177/0971721820902961

Regan, Á. (2021). Exploring the readiness of publicly funded researchers to practice responsible research and innovation in digital agriculture. *Journal of Responsible Innovation, 8*(1), 1–5. https://doi.org/10.1080/23299460.2021.1904755

Rose, D. C., & Chilvers, J. (2018). Agriculture 4.0: Broadening responsible innovation in an era of smart farming. *Frontiers in Sustainable Food Systems, 2*, 87. https://doi.org/10.3389/fsufs.2018.00087

Rose, D. C., Lyon, J., de Boon, A., Hanheide, M., & Pearson, S. (2021). Responsible development of autonomous robotics in agriculture. *Nature Food, 2*(5), 306–309. https://doi.org/10.1038/s43016-021-00287-9

Sozinova, A. A. (2020). *Comprehensive diagnostics of the shadow economy in the regions of Russia according to the criterion of marketing activity in entrepreneurship in the interests of sustainable development*. Retrieved from https://iscconf.ru/комплексная-диагностика-теневой-эко/. Accessed 7 June 2021.

The Economist Intelligence Unit Limited. (2021). *Global food security index*. Retrieved from https://foodsecurityindex.eiu.com/Resources. Accessed 7 June 2021.

The Assessment of Consumer Demand Potential for Organic Agricultural Products

Natalia V. Bannikova, Natalia V. Vorobyeva, Yuliya V. Orel, Svyatoslav S. Serikov, and Angelika R. Baicherova

Abstract The market of organic products in Russia is currently at the development stage; therefore, it is important to determine the factors, which influence consumers' willingness to buy eco-clean products. The purpose of the article is to define the consumers' attitude to these products, as well as to identify the motivation for purchasing and the driving forces that determine the behaviour of potential consumers of organic products. Particular attention was paid to organic fruit and vegetables. The research data is based upon the results of processing 192 questionnaires, received during the survey of residents of Stavropol Territory. The survey has revealed the rather high willingness of those surveyed to purchase organic products, which is limited by such factors as high cost, insufficient trust in the certification system, the deficiency of a well-established sales market and information support. High priority measures in the field of development of the market under consideration are providing the consumers with the corresponding information and marketing communications, improving the system of regulation and support for the manufacturers of organic products and the selection of the most efficient distribution channels. The research has revealed quite many concurrences with the results of similar researches by other authors in Russia and abroad.

Keywords Organic food · Local food · Demand · Consumer behaviour · Willingness to pay · Questionnaire

JEL Code M31

1 Introduction

The sector of organic food products occupies a relatively small share of agricultural production. However, the demand for these products is growing all over the world. According to the data, provided by the Organic Farming Unit (The Union of Organic Farming, 2021), the market of organic products has grown by more than 7 times

N. V. Bannikova (✉) · N. V. Vorobyeva · Y. V. Orel · S. S. Serikov · A. R. Baicherova
Stavropol State Agrarian University, Stavropol, Russia

E. G. Popkova and B. S. Sergi (eds.), *Geo-Economy of the Future*,
https://doi.org/10.1007/978-3-030-92303-7_41

within 2000–2019 (from 18 to 129 billion USD) and is expected to continue growing at the rate of 10–12% per year in the next few years.

The development of this market and the acquisition of traditional consumers into the sector of organic food products require performing researches, which describe consumers' attitude to these products. It is impossible to plan the production volume and adequate price policy without studying the potential consumer demand. Publications' review shows that researches in this field are relevant for different countries.

Primarily, the authors study the impact of demographic and psychological factors upon the consumers' willingness to pay. Thus, Winterstein and Habisch (Winterstein & Habisch, 2021) consider the impact of age upon consumer behaviour. Sandhu et al. (2019) pay particular attention to factors, which relate to environmental belief, whereas (Jose et al., 2020; Torres-Ruiz et al., 2018) focus on factors that associate with confidence in organic certification.

Shahabi Ahangarkkolaee and Gorton (2021), as well as some other researchers, point out that consumers frequently merge the concepts of "organic products" and "locally (home) produced food". Giving preference to locally produced food is a limiting factor for export development models of organic production.

According to Rodríguez-Bermúdez et al. (2020), the research of consumers' behaviour should be performed individually for different countries and regions, because the attitude to organic food products may reveal significant differences. However, the comparative analysis of our research and similar researches of our foreign colleagues shows that there exist quite many concurrences in the behaviour and perception of potential consumers, which will be further registered in the analysis of obtained results.

2 Methodology

In order to conduct the survey, there was developed a questionnaire with 21 questions, which related to respondents' attitude to organic products (primarily, fruit and vegetables), their willingness to purchase organic products and purchasing conditions, as well as to personal attributes of those surveyed.

The achievement of exact representativeness is not of paramount importance in course of performing the sampling for the estimation of potential demand in marketing. Exploratory researches allow using random sampling. Therefore, the present research applies simple random sampling with a sampling size of 150–200 respondents and a sampling error of 8%, as recommended by Efimenko (2019).

Thus, the general population of research involved the residents of Stavropol Territory and comprised 200 questionnaires. The research study was conducted in November and December 2019. Primary processing, which includes checking questionnaires for accuracy, sufficiency and absence of direct contradictions was followed by sorting. Consequently, 192 questionnaires were presented for processing using SPSS Statistics.

3 Results

192 respondents included 68.2% of women and 31.8% of men. Theoretically, this distribution corresponds to traditional roles in running a household, where women are more frequently involved in purchasing food products, than men.

The distribution of respondents by age was as follows: 18–24 years—35.6%, 25–39 years—30.9%, 40–59 years—23.0%, 60 years and older—10.5%. The certain bias of sampling towards younger age is compliant with the purpose of the research because organic production increase will take some time and the younger generation is more responsive to current trends in consumption.

The indirect method for the assessment of financial status is the response to the question on which share of respondents' financial income is spent on food products. The responses were distributed as follows: 9.9% of respondents spend up to 20% (referred to as "high-income" group in the research); 22.0% of respondents spend 20–30% ("above-average" income group); 40.8% of respondents spend 30–40% ("average-income" group); 27.2% of respondents spend 40% ("low-income" group).

70.2% of respondents reside in big cities, 16.2%—in towns (with less than 100 000 population), 13.6%—in rural areas.

The recent adoption of National standards for the production of organic agricultural products in Russia is an important step for the creation of a regulated market of organic products. The conducted survey has shown that only 48.4% of respondents are aware of the adoption of new standards; moreover, the older is the age, the lower is the awareness.

Notably, the certification mark of organic products is not the most decisive factor in deciding on purchasing food products (Fig. 1).

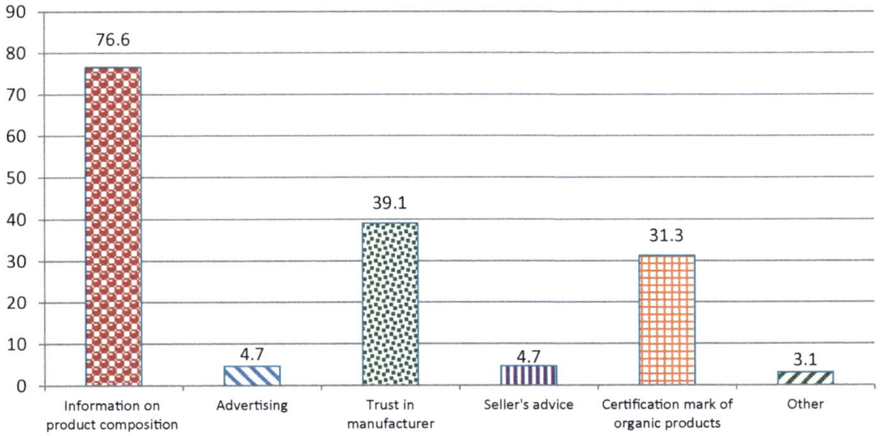

Fig. 1 The distribution of responses to the question "What other than price is significant in deciding on purchasing food products?" %. *Source* According to author's calculation

Interestingly, the significance of advertisement and seller's recommendations for purchasing food products is very low. This is consistent with the conclusions of other researchers (Nasir & Nasir, 2017).

Since the statutory definition of the "organic product" concept appeared quite recently, Russian citizens have different ideas on what belongs to these products. The respondents are most confident in the products that they have grown themselves (30%). The obtained data corresponds to the results of other surveys (Krasovskaya et al., 2018), which show that due to distrust in eco-labelling of organic products up to 40% of the urban population prefer to grow organic products on their pieces of ground, if possible.

Interestingly, the significance of advertisement and seller's recommendations for purchasing food products is very low. This is consistent with the conclusions of other researchers (Nasir & Nasir, 2017).

Since the statutory definition of the "organic product" concept appeared quite recently, Russian citizens have different ideas on what belongs to these products. The respondents are most confident in the products that they have grown themselves (30%). The obtained data corresponds to the results of other surveys (Krasovskaya et al., 2018), which show that due to distrust in eco-labelling of organic products up to 40% of the urban population prefer to grow organic products on their pieces of ground, if possible.

Also, consumers give preference to products labelled as "BIO, ORGANIC, ECO, etc." and products with "good composition" (free of food additives, genetically engineered components, etc.).

Every eighth of respondents deems products purchased from small producers ("farmers") to be organic. Despite the similarity of production technologies employed by farming enterprises and large-scale agricultural businesses, consumers already have the understanding that "farmers'" products are healthier and more natural. Thus, it can be said that there exists a spontaneously-formed brand of "farmers' products", although it may have no relation to organic products.

45.8% of respondents said that they had met and 21.4% had not met organic products in the market, based on their idea of what organic products are.

A significant indirect indicator of the unwillingness to purchase organic products is the index that shows that one-third of survey respondents has chosen the response "I pay no attention to that". Though, indifference to organic products decreases with age.

The survey data shows a very low level of trust in the existing certification system, which should ensure products' compliance with adopted standards. This may become a serious problem for the development of the organic food market in Russia. The problem of trust is also very important in foreign countries, therefore many authors indicate that consumers should be provided with corresponding information and marketing communications (Rodríguez-Bermúdez et al., 2020; Sandhu et al., 2019; Torres-Ruiz et al., 2018).

According to questionnaire results, only 4.7% of respondents fully trust, 59.9% rather trust than do not trust and 35.4%, (i.e. one-third of respondents) do not trust in the certification system.

The level of distrust is higher among men (44.3%) and rural residents (66.7%) and tends to increase with age. The better the financial status, the higher the trust level.

Besides, respondents are rather prone to trust in local manufacturers; which is proved by the responses to the question on preferences among several products similar in price and quality. 20.3% of respondents prefer imported products, 29.7% prefer domestic products, and 50.0% prefer local products.

Imported food products are mostly preferred by younger consumers; also, imported products are more often chosen with income increase.

The questionnaire paid particular attention to fruit and vegetables. The place of their purchase is an important factor from the viewpoint of regulating this market. The survey data has confirmed the hypothesis that consumers still prefer to purchase these products in the market (as specified by 57.3% of respondents).

Supermarkets have scored the second most important place—38.5% of respondents purchase fruit and vegetables here.

29.7% of respondents have chosen the response "I tend to consume fruit and vegetables, grown on my piece of ground".

Such sales channels as "Through advertisement from small producers" and "Delivery via Internet" were almost not used by the respondents in course of the survey.

The responses to the question "In your opinion, how much should organic products cost?" were distributed as follows:

- 10–20% more expensive than regular products—59.9%;
- 20–50% more expensive than regular products—31.8%;
- 50–80% more expensive than regular products—5.2%;
- At least twice as expensive—3.1%.

In general, the obtained results correspond to the data (Krasovskaya et al., 2018), which shows that 37% of urban residents are not willing to pay more than an extra 20% for the "ecological properties" of the products. This is also congruent with the data, provided in the research work of Rodríguez-Bermúdez et al. (2020).

Our research has shown that men more frequently than women noted higher price, recognizing the fact that the production of organic products cannot be low-cost. Many researchers (Önel et al., 2019; Shahabi Ahangarkolaee & Gorton, 2021; Winterstein & Habisch, 2021) mention the development of the regulatory system and the system of support for organic products' manufacturers as the major source for cost reduction and the improvement of the situation on the market of organic products.

Expectedly, respondents with higher income more often tend to acknowledge that organic products can be rather expensive (Fig. 2).

Rural residents also have a similar opinion, because their understanding of the chemical-free agricultural production process is better.

At the same time, according to the researches of other Russian authors (Mitrofanova et al., 2018), eco-clean production is more expensive than regular production by 30–100%, by different estimates.

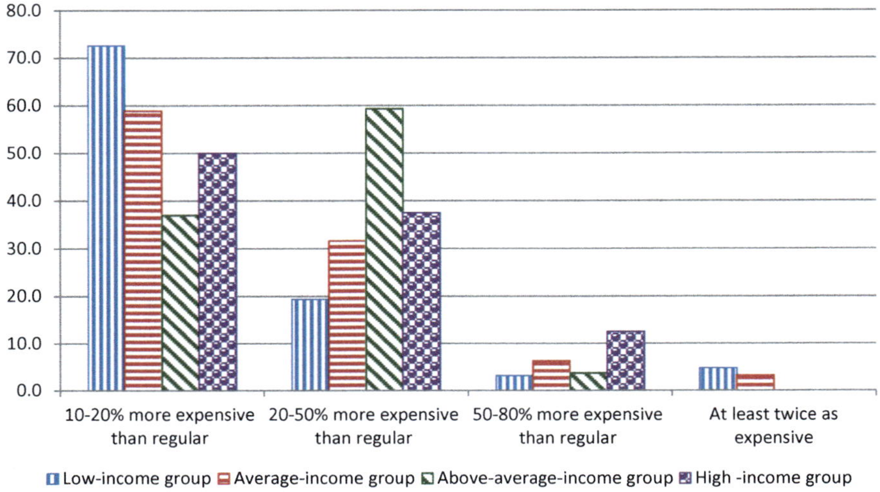

Fig. 2 The impact of financial status on guesses regarding the price of organic products, %. *Source* According to the author's calculation

One of the most important questions in the questionnaire was "Are you willing to purchase eco-clean fruit and vegetables, even though they can be significantly more expensive than regular products?" 58.9% of respondents answered in affirmative, which corresponds to the data from other researches (Vanyushina, 2019).

Women more often than men are willing to purchase organic products. The highest willingness to buy organic products was also noted in the age group of 25–39 years (72.9%).

Our research was carried out at the end of 2019; however, its data remains generally up-to-date. Despite the decrease of population income in 2020, experts notice significant growth in demand for organic products, in both Russia and other countries (The Union of Organic Farming, 2021). According to their opinion, the COVID-19 pandemic stimulated the demand for eco-clean products, as a "natural immunity" source.

Those unwilling to purchase eco-clean fruit and vegetables explained the reason for their unwillingness. Insufficient income (63.3% of respondents) was mentioned in the first place. In general, this data corresponds to the data of buyer behaviour research, performed by Non-profit organization Ecological Unit and Ecobureau GREENS in 2018 (Vanyushina, 2019).

The second major reason for the unwillingness to purchase organic fruit and vegetables was distrust: "I do not believe that these products are more eco-clean and eco-safe"; notably, the degree of distrust is higher among men and people of older age.

Finally, only 13.6% of respondents have chosen the response "Consuming such products will not significantly improve health condition". Younger respondents are more likely to see no relation between health condition and organic products'

Table 1 The significance of different factors for the selection of organic products

Selection factors	Rank
Good reputation of organic products' manufacturer	1
Special marking on the packaging	2
Reasonable price	3
Ease of purchase	4
Availability of buyer's advice	5
Good advertising and informing the buyer about the advantages of these products	6

Source According to author's calculation

consumption, whereas the older is the age, the more significant is the factor of maintaining health with the help of quality food.

Those willing to buy organic products provided answers to more specific questions, related to their purchase. The answers to the question "What is more significant for you when purchasing eco-clean fruit and vegetables?" (1—most significant, 6—least significant), were distributed as follows (Table 1).

When responding to the question "Where would it be more convenient for you to purchase eco-clean fruit and vegetables?" three response options have got almost equal preference (28–31%): "In organic food stores (separate areas in the market)"; "In agricultural manufacturers' brand stores"; "In hypermarkets".

One of the most significant questions for the development of the organic products market is as follows: "At what income per family member would you be willing to buy eco-clean fruit and vegetables?" The correlation-regression analysis allowed determining the relationship between the income per family member and the willingness to purchase organic fruit and vegetables. The resulting model of dual linear regression looks as follows: $Y1 = -13.52 + 4.44\ X1$. This means, that the increase of income per family member by 1 thousand rubles increases the share of consumers willing to purchase eco-clean fruit and vegetables by 4.44%.

Younger people (up to 40 years) and people with children are rather more willing to purchase eco-clean fruit and vegetables.

4 Conclusion

The analysis of questionnaire results has allowed making the following general conclusions and substantiating recommendations.

1. On average, 59% of respondents have shown their willingness to purchase organic products, although they can be more expensive than regular products.

The typical profile for a potential consumer of organic agricultural products is an urban resident 25–39 aged woman with higher or secondary vocational education, who spends 30–40% of her income on food products.

It should be noted that this "portrait" is, to a great extent, congruent with the typical profile of a consumer of agricultural organic products based on the results of research by Rodríguez-Bermúdez et al. (2020).

2. Consumers mentioned the price of organic fruit and vegetables as a major obstacle to their purchase. Besides, respondents suppose that organic products should not be expensive (60% are convinced that they should be only 10–20% more expensive than regular products).

The analysis has shown that per capita income increase by 1000 rubles (about 15 USD) increases the share of consumers willing to purchase organic products by 4.44%.

3. The development of the organic products' market requires carrying out explanatory work among potential consumers. One-third of respondents mentioned that they do not pay attention to whether a product is organic or not when purchasing it. Every fifth man and every fifth big city resident suppose that consuming such products will not significantly improve their health condition.

In addition to the above, experts associate the development potential of the organic products market with the progress in the field of motivation. According to the data of the National Organic Union of Russia (Mitrofanova et al., 2018), during 2013–2016 these products significantly moved from the luxury segment (comprising more than 90% of buyers in 2007–2008) to the segment for health-minded people. Experts note that COVID 2019 pandemic contributed to the increase in demand for organic products.

4. The degree of potential consumers' awareness about the regulation of organic products' market and accepted standards is not high. The significance of explanatory work is also associated with a high level of people's distrust in the system of products certification, which should ensure their compliance with standards. One-third of those unwilling to consume organic fruit and vegetables mention distrust as a reason for their unwillingness. Since the information on the composition of fruit and vegetables is not available on them (whereas, almost 80% of respondents consider the information on product composition as a reason for purchasing decision and 30%—as a reason to treat a product as being organic), it is necessary to facilitate the increase of trust in the certification system.
5. When choosing the way for informing the potential consumers it should be noted that standard advertising methods would not be efficient, because people generally do not trust advertising. Half of the respondents consider local production as a substantial argument for purchasing. One of the ways to enhance the potential consumers' trust in local manufacturers of organic products (whereas 40% of respondents consider trust in the manufacturer as a reason for purchasing decision) could be arranging access to enterprises for those willing.
6. Ease of purchase is the fourth significant factor for buying organic products. In addition, the potential consumers would prefer to purchase organic products in

organic food stores (separate areas in the market), in agricultural manufacturers' brand stores and supermarkets. Thus, arranging a specialized trade network with sales points in different locations is important for the development of the organic products market. Experts (Krasovskaya et al., 2018) also note that so far retail trade networks do not position improved-quality products separately, which makes it difficult for average consumers to find them.

Selling products via the Internet is currently a quite vacant market niche. Its potential has started to be more actively used during COVID 2019 pandemic. The growths of organic products' popularity, as well as sales growth, can provide new opportunities for implementation. In particular, if there are multiple purchase orders from one district, a manufacturer could arrange door-to-door delivery, using economy vehicles. This could additionally emphasize the value of organic products for consumer's perception, decrease demand elasticity for these products and distance them from regular products with adequate profits for the seller.

Acknowledgements The chapter was prepared with the financial support of the Ministry of Agriculture of the Stavropol Territory. Governmental contract No. 157/19 dated August 12, 2019.

References

Efimenko, Y. (2019). Sampling volume—Sampling method in sociological research. http://fb.ru/article/263333/obyem-vyiborki---vyiborochnyiy-metod-sotsiologicheskogo-issledovaniya. Accessed: 01.12.2019.

Jose, H., Kuriakose, V., & Koshy, M. P. (2020). What motivates Indian consumers' to buy organic food in an emerging market? *Asia-Pacific Journal of Business Administration, 12*(2), 97–113.

Krasovskaya, N. V., Deev, A. S., & Chernomorchenko, S. I. (2018). Russian market of organic food: Problems and prospects. *The Eurasian Scientific Journal*, *6*(10), 20–31.

Mitrofanova, I. V., Ivanova, T. B., Vishnevskiy, V. S., & Chestnova, M. A. (2018). Production of organic food products as a growth factor of stability of the agrarian and industrial complex in the south of Russia. *Regional economics. The South of Russia*, *4*, 182–195. https://doi.org/10.15688/re.volsu.2018.4.18

Nasir, V. A., & Nasir, S. (2017). Determining the role of communication and distribution channels for organic foods. In *Driving Agribusiness with Technology Innovations*, pp. 149–164.

Önel, G., Seidu, A., & Seale, J. L. (2019). Consumer demand for organic food groups and implications for farmers' revenues under the organic land subsidy scheme: The case of Denmark. *Journal of Food Distribution Research, 50*(2), 47–68.

Rodríguez-Bermúdez, R., Miranda, M., Orjales, I., Ginzo-Villamayor, M. J., Al-Soufi, W., & López-Alonso, M. (2020). Consumers' perception of and attitudes towards organic food in Galicia (Northern Spain). *International Journal of Consumer Studies, 44*(3), 206–219.

Sandhu, Y. A., Perumal, S. A. L., & Fauzi, W. I. M. (2019). The predictors and consequences of personal norms in the context of organic food among Pakistani consumers. *International Journal of Financial Research, 10*(3), 314–336.

Shahabi Ahangarkolaee, S., & Gorton, M. (2021). The effects of perceived regulatory efficacy, ethnocentrism and food safety concern on the demand for organic food. *International Journal of Consumer Studies, 45*(2), 273–286.

The Union of Organic Farming. (2021). Organic market in Russia and worldwide. https://soz.bio/opublikovana-svezhaya-mirovaya-statist/. Accessed: 05.04.2021.

Torres-Ruiz, F. J., Vega-Zamora, M., & Parras-Rosa, M. (2018). Sustainable consumption: Proposal of a multistage model to analyse consumer behaviour for organic foods. *Business Strategy and the Environment, 27*(4), 588–602.
Vanyushina, O. I. (2019). Organic agriculture in Russia: Peculiarities and development potential. In *The Collection of Articles of the 9th International Scientific and Practical Conference "Actual problems of contemporary society development"*, pp. 72–75.
Winterstein, J., & Habisch, A. (2021). Organic and local food consumption: A matter of age? Empirical evidence from the German market. *ABAC Journal, 41*(1), 26–42.

Clustering of the Economy of the Agro-Industrial Complex of the Region

Andrey V. Polyanin, Lyudmila I. Pronyaeva, Anna V. Pavlova, Olga A. Fedotenkova, and Irina A. Eremina

Abstract The main purpose of this research is to improve the methodology for identifying regional clusters when they are formed, adapted for the agro-industrial complex of the region. The authors substantiated the application of the cluster approach that improves the structure of the regional agro-industrial complex and ensures the strengthening of intraregional economic ties and the transition to an innovative type of development. The proposed method of identifying regional clusters in the agro-industrial complex during their formation is a combined approach based on the application of the method of calculating localization coefficients in the context of municipal areas, the use of the point-rating method, which allows to identify the territorial location of the cluster core, the cartographic method for determining the location and boundaries of clusters and the formalization method to substantiate the composition of participants and reveal their cluster potential. The use of the author's methodology will contribute to the effective implementation of the cluster approach in the regional agro-industrial complex and to leveling the shortcomings of existing methodologies for the agro-industrial sphere, characterized by the difficulty of conducting, the lack of opportunities to determine the specific boundaries of a potential cluster within the region and a clear justification for its specialization. The application of the proposed methodology for identifying regional agro-industrial clusters is characterized by the simplicity of calculations, which will systematically establish organizational work to make clusters, form a full-fledged composition of their members and develop strategic documents with a clear targeted focus on the existing cluster potential of the territory.

Keywords Cluster · Agro-industrial complex · Region · Economy · Territory potential

JEL Codes R11 · R12 · R58 · Q13 · Q18

A. V. Polyanin (✉) · L. I. Pronyaeva · A. V. Pavlova · O. A. Fedotenkova · I. A. Eremina
Central Russian Institute of Management—RANEPA Branch Orel, Orel, Russia

E. G. Popkova and B. S. Sergi (eds.), *Geo-Economy of the Future*,
https://doi.org/10.1007/978-3-030-92303-7_42

1 Introduction

In the structure of the economy of most regions of the European part of Russia, an important place is occupied by the agro-industrial complex, which is a multisectoral system, including agriculture, food and other industries that use agricultural products as raw materials. Its level of development ensures food security and economic independence for the region and the country as a whole. However, for many food groups in Russia, to date, it has not been possible to achieve food security thresholds. In addition, the agricultural industry is low-cost compared to other areas of activity, which hinders its technological and innovative development. The reason for this situation in the country is that during the years of reforms in the agro-industrial complex, many inter-economic ties were disrupted, intersectoral imbalances arose. The new territorial structure of the regional agro-industrial complex, based on a cluster approach that promotes more progressive methods of business interaction, production and exchange of innovations, can help to eliminate these problems.

In view of the special importance of AIC in ensuring food security and creating a multiplier effect of the development of related industries, research is needed in terms of developing a cluster identification methodology adapted for the agro-industrial sector of the region.

2 Materials and Methods

The scientific justification of the cluster approach as a way to increase the competitiveness of economic systems, methods of forming clusters and evaluating their activities, development of cluster policy has been gained in the works of a number of foreign scientists (Beaudry & Swan, 2009; Delgado et al., 2016; Lindqvist et al., 2013; Porter, 2010; Power & Asheim, 2019; Wennberg & Lindqvist, 2010). In studies of domestic scientists, models of agro-industrial clusters were proposed (Bogdanova, 2014; Ivanova & Menshchikova, 2018; Khukhrin et al., 2016). An important problem in the formation of clusters is the use of the most appropriate technique for identifying them. Among many such methods, it is necessary to note in our opinion the most successful: a method based on the calculation of localization and concentration coefficients (Feser, 2014); method based on establishing the effects of the territorial concentration of potential cluster members (Chatterji et al., 2013); a method of determining a regional cluster index (RCI) based on a system of economic performance criteria (Kovaleva, 2018). Methods of their identification can be noted for agro-industrial clusters: based on the integrated application of a sufficiently large number of methods of qualitative and quantitative assessment (Khukhrin & Nastin, 2012); based on the typology of the territories of the region by the level of stability (Tichyi & Koreva, 2019). Note the complexity of these techniques, the impossibility of justifying specialization and determining the boundaries of a potential cluster.

The methodological basis of this research was the general methods of analysis and comparative analysis aimed at identifying trends in the clustering processes of the agro-industrial complex based on the materials of the European and Russian Cluster Observatory. For the practical implementation of the proposed approach to identify regional agro-industrial clusters, a study of the resource potential of the municipal districts of the Orel region was conducted using the author's method, which consists in determining the possibilities for clustering based on the calculation of indicators that allow determining: the core of the cluster (based on the rating method, taking into account the localization of labor resources and organizations, the concentration of production and land resources, specialization of agricultural production); areas to be included in the cluster (based on setting the optimal range of cluster boundaries). A mapping method was used to demonstrate the locations of potential clusters, as well as a formalization method to substantiate the composition of cluster participants and the possibilities of their interaction.

3 Results

The use of the cluster approach is implemented in the agricultural sector through the development of stable production links between agricultural organizations and agricultural processors, the promotion of cluster products to new markets, including foreign ones, strengthening cooperation with small businesses, and the introduction of innovative technologies to create competitive advantages. The possibility of applying a cluster approach in the agro-industrial complex is confirmed by the experience of a number of foreign countries (Spain, France, Poland, Germany, Italy, Austria, etc.). Clusters have become a natural stage in the evolution of the agro-industrial complex and they also become a tool to strengthen the agricultural sector and increase its competitiveness in developed countries (Table 1).

In total, 89 clusters operate in Europe in the sphere of agriculture, which 15% are concentrated in Spain, 14% in France and 21% in Serbia.

A related industry closely linked to agriculture is the food industry, in which 109 clusters function, most of which function in Spain, France, Italy, Poland, the Netherlands and Romania.

Clusters of over 500 participants are very large in agriculture, including the following: Gebiedscoöperatie Westerkwartier (Netherlands); INBIOM—The Danish Innovation Network for Biomass (Denmark); Vitagora (France).

Assessing the experience of foreign agro clusters, it should be noted there is a strong base for innovation, for example, in the structure of the Vitagora cluster, the share of research organizations is about 20%. In foreign countries, the total number of employees in the economic sectors covered by clustering processes is more than 57 million people, which more than 2 million people in the agro-industrial sector (Ketels & Protsiv, 2020).

Experience in forming clusters in the agro-industrial sector is also available in Russia. Clusters in the agro-industrial complex are formed in two key specializations:

Table 1 Rating of European countries by the number and types of agro-industrial clusters

Country	Number of clusters in the country	Number of clusters in the country by industry specialization «Agricultural resources and services»	Share of clusters in the country by industry specialization «Agricultural resources and services»	Number of clusters in the country by industry specialization «Food Industry and Manufacturing»	Share of clusters in the country by industry specialization «Food Industry and Manufacturing»
Spain	160	13	8.1	16	10.0
France	106	12	11.3	14	13.2
Italy	80	8	10.0	10	12.5
Romania	52	6	11.5	4	7.7
Serbia	29	6	20.7	2	6.9
Netherlands	29	5	17.2	6	20.7
Belgium	41	4	9.8	3	7.3
Portugal	20	4	20.0	3	15.0
Ukraine	23	4	17.4	2	8.7
Slovenia	16	3	18.8	1	6.3
Germany	101	2	2.0	5	5.0
Austria	27	2	7.4	2	7.4
Poland	69	1	1.4	6	8.7
Hungary	25	1	4.0	3	12.0
Denmark	31	1	3.2	2	6.5
Great Britain	28	1	3.6	4	14.3

Source Compiled by the authors based on data from the European Cluster Observatory (European Observatory for Clusters and Industrial Change. https://www.clustercollaboration.eu/eu-initiatives/europeanclusterobservatory. Data accessed: 20.01.2021)

«agriculture and fish farming» and «food and beverage production». According to the Russian Cluster Observatory, clusters in the first specialization function in the Vologda, Rostov, Astrakhan and other regions, in the second—4 clusters (Tatarstan, Rostov, Leningrad regions and Krasnodar Territory) (Russian Cluster Observatory. https://cluster.hse.ru. Data accessed: 20.01.2021).

Most clusters in Russia have an initial level of organizational development. The number of participants in cluster structures ranges from 10 to 130. The largest in terms of the number of participants is the Agro-industrial cluster of the Kemerovo region, established in 2015.

An important difference among Russian agro-industrial clusters and foreign ones is the low share of research organizations in the structure of cluster participants.

To develop the methodology for identifying regional agro-industrial clusters, a combined approach should be proposed, based on the application of the methodology for calculating localization coefficients and the point-rating method, which allows identifying the territorial location of cluster nuclei. In order to determine the transboundary territories with which it is advisable to organize interaction with the core of the cluster to form its optimal boundaries, the proposed methodology recommends taking into account optimal distances and assessing the available resource potential.

The author's methodology for identifying regional agro-industrial clusters during their formation is represented by the following stages: analysis of the production and economic potential of the agro-industrial complex of the region; assessment of the contribution of a specific municipal region of the region to the production of agricultural products; analysis of production opportunities and agrarian specialization of municipal districts of the region; determination of opportunities for clustering in the agro-industrial complex of the region based on calculation of localization factors and concentration of production in the context of municipal districts; rating of the cluster potential of municipal districts; identification of the core area of the potential cluster, transboundary areas and the specialization of the potential cluster; substantiation of the composition of the participants in the potential cluster and the possibilities for their interaction; development of the Development Program and Strategy of the potential cluster.

We will conduct a more detailed disclosure of the essence and testing of the proposed methodology for identifying regional agro-industrial clusters when forming them on the materials of the agricultural sector of the Orel region, one of the typical agricultural regions of Russia.

Describing the Orel region, it should be noted that its economy is represented by various types of activities (industry, agriculture, trade, construction, transport). The region has sufficient potential for agricultural and industrial activities. In the structure of gross regional product, the largest share is occupied by 3 industries—agriculture (20%), manufacturing (17%) and trade (16%).

According to 2019 data, agricultural activity in the region is carried out on a total land area of 1.24 million hectares, of which 1.22 million hectares are for agricultural land. The sectoral specialization of the region is crop production, with a share in the total volume of agricultural products of about 70%. The crop industry of the region specializes in the production of cereals and sugar beets. The livestock industry specializes in the production of milk, cattle and pigs.

Administratively, the Orel region is divided into 24 municipal districts, different in the level of socio-economic development and specialization of agricultural activities (Table 2).

To determine the possibilities for clustering in the agro-industrial sphere of the region, we will calculate the factors of localization and concentration of production in the context of municipal districts (Table 3).

At the next stage of determining opportunities for clustering in the agro-industrial sphere of the region, using the score-rating method, points are determined by groups of coefficients in the context of municipalities. Points are assigned according to the occupied share of the region in the structure of the region for each indicator (Table 4).

Table 2 The structure of agricultural industry specialization of the municipal districts of the Oryol region according to the data of 2019

Districts	Production volumes at current prices, million rubles			Structure, %					Production	
	Total	Crop production	Livestock production	Total in the region		In the context of the district			per 1 employee	Per 1 agricultural organization
				Crop production	Livestock production	Total	Crop production	Livestock production		
Bolkhovsky	911.9	658.2	253.7	1.5	1.4	1.5	72.2	27.8	5.1	152.0
Verkhovsky	2077.8	1680.0	397.7	3.8	2.2	3.4	80.9	19.1	5.8	346.3
Glazunovsky	1306.4	1201.9	104.6	2.8	0.6	2.1	92.0	8.0	10.5	326.6
Dmitrovsky	1400.5	1350.7	49.8	3.1	0.3	2.3	96.4	3.6	16.9	200.1
Dolzhansky	1736.6	1688.7	48.0	3.9	0.3	2.8	97.2	2.8	4.5	289.4
Znamensky	701.1	696.1	5.0	1.6	0.0	1.1	99.3	0.7	46.7	233.7
Zalegoshchensky	2750.8	2750.8	0.0	6.3	0.0	4.4	100.0	0.0	9.8	687.7
Kolpnyansky	2701.9	2656.4	45.5	6.1	0.2	4.4	98.3	1.7	12.2	450.3
Korsakovsky	1065.2	1051.3	13.9	2.4	0.1	1.7	98.7	1.3	36.7	1065.2
Krasnozorensky	930.0	801.4	128.7	1.8	0.7	1.5	86.2	13.8	3.4	232.5
Kromskoy	1835.3	1834.1	1.1	4.2	0.0	3.0	99.9	0.1	12.7	183.5
Livensky	7915.9	5814.7	2101.2	13.3	11.5	12.8	73.5	26.5	3.3	359.8
Mtsensky	4891.6	3063.4	1828.2	7.0	10.0	7.9	62.6	37.4	5.0	815.3
Maloarkhangelsky	2360.2	2170.9	189.3	5.0	1.0	3.8	92.0	8.0	7.1	786.7
Novoderevenkovsky	1914.9	1883.5	31.3	4.3	0.2	3.1	98.4	1.6	33.0	383.0
Novosilsky	985.2	985.2	0.0	2.3	0.0	1.6	100.0	0.0	32.8	985.2
Orlovsky	3910.9	2248.4	1662.6	5.1	9.1	6.3	57.5	42.5	2.7	170.0
Pokrovsky	5278.0	3255.8	2022.2	7.5	11.0	8.5	61.7	38.3	9.1	659.7

(continued)

Table 2 (continued)

Districts	Production volumes at current prices, million rubles			Structure, %					Production	
				Total in the region		In the context of the district				
	Total	Crop production	Livestock production	Crop production	Livestock production	Total	Crop production	Livestock production	per 1 employee	Per 1 agricultural organization
Sverdlovsky	2455.1	2325.8	129.3	5.3	0.7	4.0	94.7	5.3	1.0	129.2
Soskovsky	2228.9	544.3	1684.6	1.2	9.2	3.6	24.4	75.6	557.2	2228.9
Trosnyansky	3227.8	1260.8	1967.0	2.9	10.7	5.2	39.1	60.9	124.1	645.6
Uritsky	2565.5	1386.9	1178.7	3.2	6.4	4.1	54.1	45.9	3.6	1282.8
Khotynetsky	1236.0	1160.1	75.9	2.7	0.4	2.0	93.9	6.1	14.5	412.0
Shablykinsky	5596.8	1189.1	4407.7	2.7	24.1	9.0	21.2	78.8	233.2	2798.4
Total	61,984.6	43,658.4	18,326.1	100.0	100.0	100.0	70.4	29.6	3.6	366.8

Source Calculated by the authors on the basis of Federal State Statistics Service data (Federal State Statistics Service, 2019)

Table 3 Factors of localization and concentration of agricultural production

Coefficient	Calculation Procedure
Agricultural industry concentration factor by volume of products shipped (CCP)	The share of agricultural products of the region in the total volume of shipped products in the region/The share of agricultural products of the region in the total volume of shipped products in the region
Agricultural industry localization factor by number of employees (LFE)	Share of agricultural workers in total employment in the region/Share of agricultural workers in the region in total employment in the region
Agricultural industry localization factor by number of organizations (AFO)	Share of the number of agricultural organizations in the total number of organizations in the region/Share of the number of agricultural organizations in the total number of organizations in the region
Agricultural localization factor by size of agricultural land (AFL)	Shares of agricultural land in the total land area of the region/Share of agricultural land in the region in its total land area
Agricultural Industry Specialization Ratio (AS)	Share of agricultural production in the region in total agro-industrial products of the region/Share of agricultural production in the region to gross regional product

Source Developed and compiled by the authors

The results of the calculations show that in the northeastern zone of the region Livensky district occupies a leading position, it concentrates sufficient potential for the formation of a cluster core that will specialize in the production and processing of crop products. In the southwestern zone, the Soskovsky district, which specializes in the production of livestock products, demonstrates strong positions. It borders on a number of areas also focused on this specialization. Estimating the combined potential of areas in points, it can be noted that potential agro clusters have fairly good starting positions (Table 5).

In identifying areas with which interaction is appropriate, the distances and resource potential between them are taken into account. The cluster boundary range is designed so that the distance from the cluster core does not exceed 70 km and logistical costs do not greatly affect the efficiency of cluster members (Fig. 1).

To justify the membership of potential clusters and the possibilities for their interaction, the following should be noted:

- The «Grain-sugar cluster", with its core in the Livensky region, brings together the potential of a number of transboundary areas that specialize in the production of grains and sugar beets using favorable climatic conditions. For the processing of grain products on the territory of a potential cluster there are two elevators, a feed mill and a bakery in Kolpnyansky and Verkhovsky districts. To produce sugar from sugar beets, 2 sugar plants in the Livensky and Kolpnyansky districts operate

Table 4 Rating of districts, potential participants of agro-industrial clusters in the Oryol region

	CCP	Point	LFE	Point	AFO	Point	AFL	Point	AS	Point	Total points
Bolkhovsky	1.18	0.6	0.63	0.2	3.31	0.6	0.96	0.8	1.15	0.6	2.8
Verkhovsky	1.01	0.4	1.38	0.2	2.33	0.4	0.94	0.6	0.99	0.4	2
Glazunovsky	0.89	0.4	2.25	0.4	2.51	0.4	0.83	0.4	0.87	0.4	2
Dmitrovsky	1.74	0.8	1.75	0.4	2.57	0.4	0.82	0.4	1.70	0.8	2.8
Dolzhansky	1.63	0.8	3.50	0.6	3.49	0.6	0.99	0.8	1.59	0.8	3.6
Znamensky	1.88	1	0.50	0.2	2.80	0.4	0.78	0.2	1.83	1	2.8
Zalegoshchensky	1.12	0.6	2.25	0.4	2.03	0.2	0.92	0.6	1.09	0.6	2.4
Kolpnyansky	0.78	0.4	2.25	0.4	2.78	0.4	0.90	0.6	0.76	0.4	2.2
Korsakovsky	1.97	1	1.00	0.2	2.58	0.4	0.84	0.4	1.92	1	3
Krasnozorensky	0.89	0.4	2.63	0.4	1.81	0.2	1.07	1	0.87	0.4	2.4
Kromskoy	0.39	0.2	1.75	0.4	2.62	0.4	0.98	0.8	0.38	0.2	2
Livensky	1.12	0.6	6.38	1	5.53	1	1.08	1	1.10	0.6	4.2
Mtsensky	0.89	0.4	4.13	0.8	2.66	0.4	0.90	0.6	0.87	0.4	2.6
Maloarkhangelsky	1.06	0.4	3.13	0.6	2.04	0.2	0.84	0.4	1.03	0.4	2
Novoderevenkovsky	1.29	0.6	2.88	0.6	2.83	0.4	1.03	1	1.26	0.6	3.2
Novosilsky	1.41	0.6	1.50	0.2	1.54	0.2	0.72	0.2	1.37	0.6	1.8
Orlovsky	0.56	0.2	1.75	0.4	5.16	1	0.99	0.8	0.55	0.2	2.6
Pokrovsky	1.42	0.6	4.38	0.8	2.91	0.4	1.10	1	1.38	0.6	3.4
Sverdlovsky	0.83	0.4	2.63	0.4	3.20	0.6	0.98	0.8	0.80	0.4	2.6
Soskovsky	2.03	1	0.50	0.2	1.94	0.2	1.03	1	1.97	1	3.4
Trosnyansky	1.72	0.8	3.88	0.6	2.59	0.4	0.95	0.8	1.68	0.8	3.4
Uritsky	0.94	0.4	0.75	0.2	2.88	0.4	1.10	1	0.91	0.4	2.4
Khotynetsky	1.02	0.4	3.00	0.6	2.71	0.4	0.94	0.6	0.99	0.4	2.4
Shablykinsky	2.09	1	3.13	0.6	3.13	0.4	0.91	0.6	2.03	1	3.6

Source Calculated by the authors on the basis of Federal State Statistics Service data (Federal State Statistics Service, 2019)

on the territory of the cluster. The existing industrial potential of the Livensky, Kolpnyansky and Verkhovsky districts for the production of high-grade products will contribute to the growth of competitiveness of the cluster;

- The «Meat cluster» with the core of the cluster in the Soskovsky district allows combining the potential of a number of transboundary areas, whose agrarian organizations specialize in raising cattle and pigs. In the Soskovsky, Kromskoy and Trosnyansky districts there are tribal mother reproducers and feedlot workshops for breeding, raising and slaughtering pigs of the "Exima" agricultural holding. In the Trosnyansky and Shablykinsky districts there are enterprises for the cultivation of meat breeds of livestock of the agricultural holdings "Miratorg" and "Bryansk

Table 5 Point score of the potential of agro-industrial clusters

Cluster name	Districts of the region	Number of points	Combined potential of cluster	Potential cluster specialization
Grain-sugar cluster	Livensky	4.2	17,80	Crop production
	Dolzhansky	3.6		
	Pokrovsky	3.4		
	Krasnozorensky	2.4		
	Kolpnyansky	2.2		
	Verkhovsky	2.0		
Meat cluster	Shablykinsky	3.6	17,60	Livestock production
	Soskovsky	3.4		
	Trosnyansky	3.4		
	Dmitrovsky	2.8		
	Uritsky	2.4		
	Kromskoy	2.0		

Source Developed and compiled by the authors

Fig. 1 Cartogram of the location of potential regional agro-industrial clusters in the Orel region. *Source* Developed and compiled by the authors

Meat Company". In addition, many livestock farms specialize in the cluster. In Uritsky and Kromskoy districts there are enterprises for the production of feed. The existing industrial potential of Dmitrovsky and Kromskoy districts for the production of high-grade products (meat processing plants) will strengthen the starting capabilities of the cluster.

The results of the research of the existing production and resource potential of identified agro-industrial clusters in the region will allow us to move to the last stage of the formation of a potential cluster—the development of the Cluster Strategy and other policy and organizational documents.

4 Conclusion

For the further progressive development of the agro-industrial complex of the regions of Russia, its structural restructuring is necessary on the basis of a cluster approach that contributes to increasing the competitiveness of agrarian organizations, their orientation towards innovative development and strengthening intraregional economic ties. The basis of this conclusion was the assessment of foreign and domestic clustering experience in the agro-industrial sector.

The effective implementation of the cluster approach in the agro-industrial complex of the region will be facilitated by the application of an improved methodology for identifying agro-industrial clusters when they are formed. The author's methodology proposed in the research makes it possible to offset the shortcomings of existing methods characterized by the difficulty of conducting, the lack of opportunities to determine the specific boundaries of a potential cluster and a clear justification for its specialization. The developed author's methodology is a combined approach based on the use of the method of calculating localization coefficients, the use of the rating, mapping and formalization methods, which allow you to identify the territorial location of the cluster core, its boundaries, justify the composition of cluster participants and evaluate their resource potential.

The application of the proposed methodology for identifying regional agro-industrial clusters will allow stakeholders in their formation to systemically build organizational work and develop strategic documents with a clear targeted focus on the existing cluster potential of the territories.

References

Beaudry, C., & Swan, G. (2009). Firm growth in industrial clusters of the United Kingdom. *Small Business Economics, 32*(4), 409–424. https://doi.org/10.1007/s11187-007-9083-9

Bogdanova, O. V. (2014). Methodological aspects of cluster formation in the agricultural sector of the regional economy. *Economic Research, 5*, 1.

Chatterji, A., Glaese, E., & Kerr, W. (2013). Clusters of entrepreneurship and innovation. In *NBER Working Paper*. https://scholar.harvard.edu/files/glaeser/files/clusters_of_entrepreneurship_nberwp. Data accessed: 20.01.2021.

Delgado, M., Porter, M., & Stern, S. (2016). Defining clusters of related industries. *Journal of Economic Geography, 16*(1), 1–38.

European Observatory for Clusters and Industrial Change. https://www.clustercollaboration.eu/eu-initiatives/europeanclusterobservatory. Data accessed: 20.01.2021.

Federal State Statistics Service. (2019). Regions of Russia: socio-economic indicators—2019. http://www.gks.ru/wps/wcm/connect/rosstat_main/rosstat/ru/statistics/publications/catalog/doc_1138623506156. Data accessed: 20.01.2021.

Feser, E. (2014). Clusters and your economy: an illustrated introduction. *Economic Development Quarterly, 28*(4), 379–380. https://doi.org/10.1177/0891242414555062

Ivanova, E. V., & Menshchikova, V. I. (2018). Prospects for the development of cluster structures in the agro-industrial complex of the Tambov region. *Bulletin of Michurinsk State Agrarian University, 3*, 151–156.

Ketels, C., & Protsiv, S. (2020). Cluster presence and economic performance: A new look based on European data. *Regional Studies*. https://doi.org/10.1080/00343404.2020.1792435

Khukhrin, A. S., & Nastin, A. A. (2012). Regional approaches to the formation of agricultural clusters. *Rationing and Wages in Agriculture, 9*, 24–31.

Khukhrin, A. S., Bundina, O. I., & Agnaeva, I. Yu. (2016). Agro-industrial clusters in Russia: Outlines of the future. *Economics of Agricultural and Processing Enterprises, 12*, 11–23.

Kovaleva, T. Yu. (2018). Theoretical and methodological foundations and results of assessing the effectiveness of cluster spatial development of regions of the Russian Federation. *Bulletin of the Astrakhan State Technical University, 2*, 101–111. https://doi.org/10.24143/2073-5537-2018-2-101-111

Lindqvist, G., Ketels, C., & Sölvell, Ö. (2013). *The cluster initiative Greenbook 2.0*. Ivory Tower Publishers.

Power, D. (2019). True clusters: A severe case of conceptual headache. In B. Asheim (Ed.), *Regional economic advantage* (pp. 50–67). Edward Elgar Publishing.

Porter, M. (2010). *Competition* (Updated & Expanded ed). William.

Russian Cluster Observatory. https://cluster.hse.ru. Data accessed: 20.01.2021.

Tichyi, V. I., & Koreva, O. V. (2019). Increasing the sustainability of the development of rural areas in the region through the formation of a territorial agro-industrial cluster. *InterCarto. Intergis, 25*(1), 206–218.

Wennberg, K., & Lindqvist, G. (2010). The effect of clusters on the survival and performance of new firms. *Small Business Economics, 34*(3), 221–241.

Agricultural Technologies as a Factor in the Development of Organic Farming: Regulation of Foreign Trade Turnover in Russia and the EAEU

Agnessa O. Inshakova, Aleksey P. Anisimov, and Denis E. Matytsin

Abstract The chapter substantiates the conclusion that an organic product is a food product produced as a result of the use of the latest environmentally friendly technologies. Without the use of pesticides and agrochemicals (including plant growth stimulants, hormonal and other drugs that pose a danger to the health of citizens), artificial food additives and GMOs, without refining, mineralization, the addition of artificial flavors, dyes, and other additives that change the natural properties of such a product, which is enshrined in national legislation. Having studied the agricultural legislation of Russia and other EAEU countries, the authors note that these legal acts contain a large number of common approaches and methods of state regulation of the production and turnover of organic products, although it differs in terms of individual production conditions, prohibitions, and restrictions. In addition, the EAEU countries differ in the duration of the transition period their laws do not mention the quality of other (other than land) natural objects as a condition that can negatively affect the quality of organic products grown. Although the adoption of laws on organic products in Russia and the EAEU countries was a great victory for agricultural producers and consumers, there is still a lot of work to be done. It is necessary to bring the EAEU standards in line with international requirements, to develop specific measures to support producers of organic agricultural products, to fine-tune control mechanisms, and to establish sanctions for violations in this area. In particular, it is advisable to establish legal liability for the use of organic labeling without proper grounds. As economic measures to support producers of organic products, it is necessary to establish a regulatory obligation of the state to purchase only organic products for military personnel, hospitals, orphanages, as well as other similar social needs. Producers engaged in environmentally friendly production should granted land tax benefits, as well as other taxes. These directions of the state agrarian policy are since the activity of growing organic agricultural products has certain features, is carried

A. O. Inshakova (✉) · A. P. Anisimov · D. E. Matytsin
Volgograd State University, Volgograd, Russia
e-mail: ainshakova@list.ru; gimchp@volsu.ru

D. E. Matytsin
e-mail: Dmatytsin@ya.ru; gimchp@volsu.ru

E. G. Popkova and B. S. Sergi (eds.), *Geo-Economy of the Future*,
https://doi.org/10.1007/978-3-030-92303-7_43

out with minimal human intervention, using the natural laws of the development of agricultural plants and animals.

Keywords Agriculture · Organic products · Crop production · Animal husbandry

JEL Code K32

1 Introduction

The current state of the market for environmentally friendly products characterized by rapid development, since where these technologies were introduced, there is an increase in yields and an improvement in soil fertility (Inshakova et al., 2020). According to experts, in the developed countries of the world, about 80% of consumers prefer environmentally-friendly (organic) products, and the incentive for the development of its market is the fact that programs to support organic agriculture belongs to the measures of the "green basket" of the WTO, and the costs are not limited to them (Voronin, 2013). But organic farming is not just a "green" project, but also a real tool for the development of modern competitive agriculture in the global food market (Zankovsky et al., 2020). Currently, 178 countries practice organic production, 87 of which have their regulatory framework. In 2018, Russia was included in the list of such states. There are 2.7 million producers of organic products in the world; the market volume is 89.7 billion dollars. Since 2000, it has increased 5 times, and despite all the world crises, it is growing.

At the moment, the development of Russian organic agricultural production is carried out (in contrast to a fairly common global trend), mainly by increasing the volume of products produced at large agricultural enterprises. According to available statistics, the average area of agricultural land on which such environmentally friendly products are produced is 3400 ha. At the same time, the market for organic agricultural products in Russia is not developed. Meanwhile, such products could be in demand in the EU countries. According to available data, the current demand for organic products of Russian production from EU consumers exceeds 100 thousand tons of cereals, legumes, and industrial crops (Organization of Organic Agricultural Production in Russia, 2018).

Consumer motives for purchasing organic products can be divided into two categories: the desire to participate in the regulation of the production process of agricultural products (organic food is grown using environmentally friendly technologies) or the desire to regulate the quality of the product (organic food does not contain pesticides). Most people who buy organic food cite health and safety as their strongest motivating factor. Indeed, organic production means avoiding additives, pesticides, toxins, genetically modified ingredients, hormones, or antibiotics. The environmental aspect of consumers' choice of organic technologies is also due to the fact that chemicals in the form of fertilizers and pesticides can pollute groundwater and streams through runoff cause algae blooms and oxygen depletion in water bodies, contribute to soil acidification, kill beneficial insects and potentially poison wildlife and its reproductive system (Czarnezki, 2011a). Along with this, consumers may have several additional considerations. Thus, the emerging local food movement often gives priority to the proximity of the food source, as this leads to a reduction in transport costs and emissions of harmful substances (Ong, 2008).

It follows from this that the production of environmentally friendly products using technologies that do not have a negative impact on the state of the environment solves not one, but several important tasks at once. In addition to meeting the demand for quality food and nature protection, the development of organic farming contributes to the economic development of rural areas, as well as the employment of rural residents, including due to the growth of manual labor (Margera et al., 2015).

Meanwhile, even in the face of the obvious advantages of the development of the organic sector of agriculture, many countries of the world do not take adequate measures to support it. This conclusion was also true for a long time concerning Russia, whose government did not pay due attention to the production of environmentally friendly products. The situation changed only after the adoption of Federal Law No. 280-FZ of August 3, 2018 "On Organic Products and Amendments to Certain Legislative Acts of the Russian Federation". The Law is important for the sustainable development of agriculture and ensuring food security. The unique territorial and climatic conditions of our country create an optimal springboard for the development of ecological agriculture and the accompanying market for organic (environmentally friendly) products.

2 Methodology

In the process of research, general scientific methods are used, such as formal-logical, dialectical, system-structural, critical cognition. Methods of synthesis, classification, and generalization were used to interpret the results of the study. The paper also uses private scientific methods: formal-legal, the principle of assessing legal processes, the method of comparative analysis, etc.

3 Results

Until recently, the Russian law did not have the term "environmentally friendly product" or any other concept that denotes it. Federal Law No. 280-FZ of August 3, 2018 "On Organic Products and on Amendments to Certain Legislative Acts of the Russian Federation" established that organic products are environmentally friendly agricultural products, raw materials, and food, the production of which meets the requirements established by this law. In addition, Article 1 of Federal Law No. 29-FZ of January 2, 2000 (as amended on 13.07.2020) "On the quality and safety of food products" contains the concept of "food quality". This is a set of characteristics of safe food products that meet the requirements established according to the legislation of the Russian Federation, the terms of the contract, the sample, standardization documents, technical documentation that determine their consumer properties, nutritional value, authenticity, grade (caliber, category, etc.), and meet the physiological needs of a person (Matytsin & Rusakova, 2021). According to Article 15 of Law No.29-FZ, food products intended for sale must meet the physiological needs of a person in the necessary substances and energy. They must meet the mandatory requirements established by the legislation of the Russian Federation for the permissible content of chemical (including radioactive), biological substances and their compounds, microorganisms, and other biological organisms that pose a danger to the health of current and future generations. Thus, Russian legislation suggests that there are two approaches to the assessment of agricultural products: it can be simply safe, not have negative consequences for health, and meet the physiological needs of a person, or it can be useful and grown in compliance with special environmental technologies (Matytsin, 2021).

This means that environmentally friendly production and environmentally safe production are two systems of agricultural activity that do not completely coincide, requiring a different approach in legal regulation. It seems that products that do not harm nature and human health and meet the safety requirements (sanitary standards, environmental regulations, or technical regulations) can be considered environmentally safe. Eco-friendly products are a more neutral term than organic products.

At the moment, Article 4 of Federal Law No. 280-FZ of August 3, 2018 "On Organic Products and Amendments to Certain Legislative Acts of the Russian Federation" provides for many requirements for the production of organic products.

Among them: the separation of the production of organic products from the production of products that are not related to organic products; a ban on the use of agrochemicals, pesticides, antibiotics, growth stimulants and fattening animals, and hormonal drugs. Except for those that are allowed for use in Russia by national, interstate, and international standards in the field of organic production; prohibition of the use of the hydroponic method of growing plants; prohibition of the use of ionizing radiation, etc. Along with this law, the quality of organic products must be guaranteed by technical regulations, but at the moment in Russia for organic products, technical regulations are not adopted, so its "confirmation of compliance" is made concerning GOST standards adopted by both the Russian authorities and interstate authorities.

The requirements of the Russian legislation regarding the concept and conditions of organic production are determined by the content of scientific discussions that have been going on in Russian and world science for a long time.

Thus, the first group of authors believes that organic agriculture is a technology of crop production and animal husbandry, based on generally recognized environmental principles that allow not only to guarantee the health of citizens and the preservation of natural objects but also to increase the quantity and quality of products produced (Morgera et al., 2012). Accordingly, the production of environmentally friendly products is one of the alternative methods of farming, which is the latest, adapted to the natural environment of the management system. This system is based on a holistic approach and consists in the management of agricultural production that ensures its sustainable development to meet the needs of present and future generations in organic products of crop production, animal husbandry, aquaculture, and agroforestry (Kurman, 2019).

Representatives of the second group of authors emphasize that environmentally-friendly (organic) products are characterized not so much by regulatory criteria and requirements (or their properties), but mainly by the characteristics inherent in the product itself. These authors indicate that such products should "be considered agricultural products characterized by high nutritional value, useful for human health, which does not contain toxic or other substances that can potentially negatively affect the health of citizens" (Yarandaikin, 1999).

Thus, in the first case, the authors, when describing environmentally friendly (organic) products, focus on the methods of production of such products. While the second scientific school focuses, on the final result of agricultural production, related to the quality of these products, and its representatives focus on the characteristics of the products themselves, putting legal criteria in the second place. It is difficult to agree with this because it is the law that fixes the quality properties of a particular product, based on medical or other criteria. Consequently, the Russian legislator supported the position of the first group of scientists.

Summarizing it seems that an organic product is a food product produced on basis of natural agriculture and animal husbandry, without the use of pesticides and other plant protection products, chemical fertilizers. Without growth stimulants and fattening of animals, antibiotics, hormonal and veterinary drugs, artificial food additives and GMOs, without refining, mineralization, the addition of artificial flavors, dyes, and other additives that change the natural properties of such a product, which is enshrined in national legislation.

The Russian legislation provides for several tools of state management in the field of organic production, including some technological prohibitions (for example, on the use of food additives, ionizing radiation, etc.), as well as means of state regulation: certification, labeling, and a single register of manufacturers.

This means that an ecological entrepreneur has more responsibilities than a traditional agricultural producer. However, at the same time, it receives certain advantages in the consumer market, allowing it to significantly increase profits.

The main question at the same time is how to guarantee the consumer high-quality organic products, with the help of which information means he should receive information about such a product, and how the state should control this process. For example, the U.S. Government controls the production and circulation of organic products through three main methods: certification, authentication, and labeling. Product certification is the first step to get any organic product on the market. This is a quality assurance procedure by which an organization independent of the manufacturer (seller, contractor) and consumer (buyer) certifies in writing that the products meet the specified requirements. Authentication means that after the farmer receives the organic certification document, it is subject to periodic verification. Marking means the application of conventional signs (letters, numbers, graphic signs, and inscriptions) on the packaging of the product for its further identification, an indication of its properties and characteristics (Harrison, 2008). There are two main functions of organic food labeling: first, informing the consumer, allowing him to make an informed choice based on full awareness, and, secondly, it is a tool of legal regulation, that reduces the risk of low-quality products appearing on the market. This means that labeling is the main lever of influence on the content of consumer preferences. In the EU, environmental certification (state-owned and produced voluntarily by non-profit organizations) is the most common way of environmental information, which allows you to maintain consumer confidence in organic products. However not environmental certification is given to the competence of private organizations. In the United States, Sweden, and some other countries, the state is responsible for controlling environmentally friendly production. The American National Organic Standard was developed by the US Department of Agriculture.

The certification itself is carried out by specialists of the US Department of Agriculture. Accordingly, environmental labeling in the United States is a state quality mark. The Swedish label "KRAV" is controlled by the Swedish Agriculture Authority and the Swedish National Food Department. The rest of the certification bodies and environmental logos, both national and international, are based on EU Regulations and do not raise doubts among consumers. In Russia, after the normative consolidation of the concept of "organic product", there is no mandatory certification of organic production or products, that is, it is made voluntarily. According to Article 21 of Federal Law No. 184-FZ of 27.12.2002 "On Technical Regulation", voluntary confirmation of compliance is carried out at the initiative of the applicant on the terms of the contract between the applicant and the certification organ.

Voluntary conformity assessment can be conducted to establish compliance with standardization documents, voluntary certification systems, and contract terms. The certification organ confirms the conformity of objects to these standards; issues certificates of conformity for objects; grants applicants the right to apply the mark of conformity; suspends or terminates the validity of the certificates of conformity issued to them. The system of voluntary certification can be created by a legal entity or an individual entrepreneur (their associations). The conformity mark informs the buyer of the product that the product is certified and meets the established quality standards. Currently, there are four national certification bodies in Russia: "Organic-Expert", "Roskachestvo", "Rosselkhoznadzor", and "Organic-Certification". At the same time, the analysis of the Russian market of organic products and certification procedures shows some problems including consumer distrust and low level of education of the population; irresponsibility of producers; the lack of state support for producers of organic products; the lack of a detailed and detailed list of sanctions; and the lack of a mechanism for mutual recognition of the equivalence of organic standards; political influence on the development of the organic products market; the lack of control over the movement of foreign products and control over enterprises for voluntary certification (Bydanova, 2020). Other countries face similar challenges, where the rapid growth in the number of environmental labels has led to problems of lack of transparency, clarity, and faith in labeling, which increases the confusion in the consumer market. Consumers want more detailed information about the product, as well as an improvement in the quality of its information. It is difficult to ensure that environmental labels are of high quality and that the products sold in the markets comply with them, which requires increased administrative regulation (Czarnezki, 2015). Given the importance of this area, independent control of the stages of the production cycle is required from the entrepreneur, for real compliance with all environmental standards and prevention of abuse. This is also important because, according to some experts, there is a threat of a collapse of the market for organic products, as the results of studies on the presence of residual amounts of GMOs or certain chemical elements in organic products increasingly made public in government reports and the media. In this connection, the consumer may stop paying extra money for products of questionable quality, requiring increased state control (Watnick, 2014).

That is why Russia created a special system for labeling organic products, which includes legal norms that establish mandatory requirements within the framework of state regulation, as well as a system of organs, that monitor the implementation of these norms. Under Article 7 of the previously mentioned Federal Law No. 280-FZ, producers of environmentally friendly (organic) products (after confirming their compliance with the standards) are entitled to put special markings on them—special inscriptions and/or graphic images. They are applied to containers, packaging, or in any other way not prohibited by law.

No less important than certification and labeling, an element of public administration in the Russian Federation was the introduction of the Unified State Register of Organic Producers. This register allows you to distinguish environmental entrepreneurs engaged in the production of organic products from the total mass of traditional industries. Accordingly, information about organic production receives an official status, and the producer receives an additional information platform provided by the state. The practice of creating and maintaining state registers is not new for the Russian Federation, where there is a Register of Hazardous Production Facilities, a State Register of Waste Disposal Facilities, and some others. The Ministry of Agriculture of the Russian Federation is responsible for maintaining this register.

In other EAEU countries, we see the following situation:

(a) The Republic of Kazakhstan has the Law of the Republic of Kazakhstan "On the production of organic products" dated November 27, 2015, No. 423-V (as amended on 28.10.2019). The Law contains rules and requirements comparable to the Russian legislation in the field of production and turnover of organic products. In particular, it provides (Article 11) conditions for the production of organic products (for example, including a ban on the use of synthetic substances, pesticides, hormones, antibiotics, and food additives). As in the Russian Federation, in the Republic of Kazakhstan, the confirmation of the conformity of organic production is voluntary (Article 12). The result of the confirmation of the conformity of organic production is a certificate of conformity. In addition, there are several requirements for labeling organic products, as well as maintaining a register of its producers. A distinctive feature of this law is the extension of its activities concerning the products of aquaculture (fish farming) and the products of wild plants.

(b) The Law of the Republic of Belarus of November 9, 2018, No. 144-Z "On the production and circulation of organic products", in contrast to the laws of other EAEU countries, regulates in sufficient detail the powers of various government organs (agriculture, health, etc.). A special feature of the law is also the regulation of seed production and beekeeping. In other respects, the law contains comparable requirements for the production of organic products, as well as other EAEU countries, as well as methods of public administration—labeling, registration, and voluntary certification of products.

(c) The Law of the Republic of Armenia of April 8, 2008, No. 23-N "On Organic Agriculture" aimed at the development of organic production, promotion of innovations, the introduction of modern technologies, export of organic products within the framework of interstate economic cooperation. An independent certification system for organic products provided. In Armenia, the issues of organization of production, processing, packaging, storage, transportation, sale, as well as marking of organic products of plant and animal origin, including beekeeping products, are regulated by law. Registration of economic entities engaged in organic agriculture carried out.

Certification of organic products at all stages of its production is carried out by a private certification body, Ecoglobe LLC, accredited under the American (USDA NOP 7 CFR 205) and German certification systems (DAkkS). There are no state standards in the field of organic agriculture in Armenia.

(d) Law of the Kyrgyz Republic No. 65 of May 18, 2019 "On Organic Agricultural Production in the Kyrgyz Republic". As in the rest of the EAEU countries, this law contains a standard set of management methods, including maintaining a register, voluntary certification, and labeling of organic products. However, unlike the Russian Federation, in the Republic of Kyrgyzstan, the law directly prescribes measures of state support for producers of organic products.

Thus, the legislation of Russia and other EAEU countries contains a large number of common approaches and methods of state regulation of the production and turnover of organic products, although it differs in terms of individual production conditions, prohibitions, and restrictions. In addition, the duration of the transition period differs in the EAEU countries, and their laws do not mention the quality of natural objects other than land as a condition that can negatively affect the quality of organic products grown by them (Navasardova & Zakharin, 2019). Further development of the legislation of the EAEU countries on organic products will be influenced by international standards, including in force in the EU countries. For example, the new EU Regulation No. 2018/848, adopted on 30 May 2018, on the production and labeling of organic products (entered into force on 01.01.2021), provides for tighter control of the supply chain (including to non-EU countries) of environmentally friendly (organic) products. As well as strengthening measures to counteract chemical contamination of agricultural soils, expanding the list of types of organic products, establishing bans on certain technologies (hydroponic method, nanomaterials) in the production of these products, etc.

Within the framework of the regulatory regulation of the production of organic (environmentally friendly) agricultural products in various countries of the EAEU, many issues remained unresolved that require the attention of representatives of agricultural and legal science. Among them, we note the following:

(1) Since there are three basic models of organizing agricultural production (traditional model; using GMOs; organic farming), the national authority must approve a special Provision regulating the coexistence of these models. The fact is that if a field where organic products are grown is located next to a field where GMO crops grow, bee pollination of both fields may occur, which will violate the purity of organic products. In Russia, the cultivation of agricultural products using GMOs is currently prohibited, although in other countries the issue of the distance between organic fields and fields with GM crops, the difference in crop rotation, information issues, etc. is being discussed (Meniv, 2016).The size of such "buffer zones" requires a separate discussion, as well as the issues of compensation for damage caused by the placement of GMO crops next to organic crops, since this can lead to losses for organic producers.
(2) The question follows from the previous one: can there be GMO impurities in the composition of organic products, and if so, in what amount. For the EAEU countries, the experience of the United States is of particular interest, where the labeling of organic products has four levels, setting the degree of "organic" products in the range from less than 70–100%. The general principle underlying such labeling rules is that the degree of labeling increases as the organic content of the product increases. Thus, the higher the organic content of the product, the more clearly this organic content is shown to the consumer (Czarnezki, 2011b). In the EU, there are three categories of "organic" goods: if the product contains more than 95% organic ingredients, it called organic; if there are 70–94% organic components, the word "organic" used in the list of ingredients; less than 70% organic ingredients-the word "organic" used on the package (Belyakova, 2018). A separate problem with organic product labeling raised by US scientists is that organic labeling is not suitable for animals and fish living in the wild. So, salmon is a migratory and carnivorous fish, and for the one who catches it, there is no way to determine what this caught fish ate during its life. As a result, it is not appropriate to label such products as "organic", since they are not always 100% organic, and this violates consumer expectations and undermines the integrity of organic labeling (Hass, 2010).
(3) The agricultural clusters can be created for the cultivation of organic products. The category "cluster" is widely used in Russian legislation (for example, concerning the creation of industrial clusters).

An agricultural cluster is "a set of agricultural infrastructure objects within the borders of the territory of one or more subjects of the Russian Federation (at the choice of the Government of the Russian Federation), as well as some agricultural producers in this territory who interact with each other and public authorities on issues of agricultural production" (Selivanova, 2016). Agricultural clusters in the Russian Federation and other EAEU countries should focus on strengthening their innovation orientation shortly. The introduction of organic production technologies within the framework of agricultural clusters will accelerate the formation of a competitive regional economy and ensure the expansion of its investment attractiveness. In turn, the attraction of foreign investment will positively affect the growth of the volume of production of organic products of the agricultural sector, which has a high demand (Isvaliev, 2017). This will require the addition of the Federal Law of 29.12.2006 (as amended on 15.10.2020) "On the development of Agriculture" with the special term "agricultural cluster", indicating the legal regime, incentives, and restrictions for agricultural producers engaged in the cultivation of organic products within its borders.

(4) The management of agricultural production, including the cultivation of organic products, is closely linked to climate problems. On the one hand, climate change strongly affects the efficiency of agricultural production (for example, due to droughts). On the other hand, the question arises about the impact of such production on climate change. In the US, a consumer group claims that the carbon released from transporting organic food can be so much that its impact on climate change outweighs the environmental benefits of these foods. Moreover, some experts argue that organic production of certain foods is more energy-intensive and requires more land than if it is grown traditionally (Ong, 2008). This statement of the question is justified. Any environmentally friendly production still affects the state of the environment. Even wind turbines and solar panels have environmental side effects (for example, danger to birds or interference to radio waves from the operation of wind turbines) that have to put up with it. Therefore, while welcoming the discussion of the environmental consequences of growing organic products, it noted that this is part of a system of more important discussions about the consequences of using a new generation of environmentally friendly technologies in general. While recognizing the existence of such consequences, we should note that their positive value for nature protection is incomparably greater.

(5) Today in many countries of the world (including Russia), the market of organic products has been captured by large producers. It already makes sense to think about a system of legal incentives and restrictions that will allow small farms that produce organic products to remain on the market. This requires a significant adjustment of the agricultural and tax legislation, which provides such farmers with additional tax benefits and state support measures. However, much can be done now. For example, in many countries of the world, there is a tradition of buying agricultural products produced in this area by the population. In this regard, information support of farmers from the local authorities is required both in terms of marketing and development of online trade in organic products, which would reduce their costs. The change in legislation could allow local authorities to purchase organic products for schools and other social facilities.

The promotion of organic products could also serve as an adjustment of school educational programs.

(6) On August 28, 2011, the state of Vermont (USA) experienced one of the worst natural disasters—tropical storm Irene (Irene). Many areas of Vermont were devastated by the flooding, and millions of dollars worth of damage were caused to citizens as a result. Floodwaters covered several agricultural fields, including organic fields, causing soil flooding with pollutants, foreign substances, and pesticides from upstream areas. Although crops that have come into direct contact with floodwaters are prohibited from entering the market due to their contamination, consumers still need to be sure that food grown in the same soil will meet the necessary organic production standards in the future. It is therefore important to ensure that public authorities properly regulate this aspect of organic production cultivation. Meanwhile, regulations in many countries do not adequately protect future organic crops from growing on previously flooded soils, and this raises important consumer protection issues for buyers of organic products. A separate question is whether the national government should provide more assistance to organic farmers than to non-organic farmers because of the difference in costs between these production methods (Robert, 2012).

4 Conclusions

An organic product is an environmentally friendly agricultural product grown in compliance with environmental principles and norms, without the use of pesticides and agrochemicals, various types of feed and food additives, refining or mineralization, which should be enshrined in national legislation. The legislation of Russia and other EAEU countries contains a large number of common approaches and methods of state regulation of the production and turnover of organic products, although it differs in terms of individual production conditions, prohibitions, and restrictions. In addition, the EAEU countries differ in the duration of the transition period, and their laws do not mention the quality of other (other than land) natural objects as a condition that can negatively affect the quality of organic products grown by them.

Although the adoption of laws on organic products in Russia and the EAEU countries was a great victory for agricultural producers and consumers, there is still a lot of work to be done to bring the EAEU standards in line with international requirements. It is necessary to develop specific measures to support producers of organic agricultural products, to fine-tune control mechanisms, and to establish sanctions for violations in this area. In particular, it is advisable to establish legal liability for the use of organic labeling without proper grounds. As economic measures to support producers of organic products, it is necessary to establish a regulatory obligation of the state to purchase only organic products for military personnel, hospitals, orphanages, as well as other similar social needs. Producers engaged in environmentally friendly production should granted land tax benefits, as well as other taxes in Russia and the EAEU countries. The proposed directions of the state agrarian policy due to the fact, that production, and economic activities for the cultivation of organic products have certain features, carried out with minimal human intervention using the natural laws of the development of agricultural plants and animals. Given the relatively recent appearance of laws on the regulation of organic agricultural production in Russia and other EAEU countries, the number of topical issues will only increase.

Acknowledgements The reported study was funded by RFBR according to the research project No. 18-29-16132.

References

Belyakova, Z. Yu. (2018). Modern legal forms of ensuring the production and turnover of organic products. *Food Processing: Techniques and Technology, 48–3*, 140–151

Bydanova, Yu. N. (2020). Certification of organic products in Russia: state and prospects. In: *Management of Innovative Development of Agri-food Systems at the National and Regional Levels: Proceedings of the II International Scientific and Practical Conference*, Voronezh State Agrarian University, Voronezh.

Czarnezki, J. J. (2011a). Food, law and the environment: Informational and structural changes for a sustainable food system. *Utah Environmental Law Review, 31–2*, 263–290.

Czarnezki, J. J. (2011b). The future of food eco-labeling: Organic, carbon footprint, and environmental life-cycle analysis. *Stanford Environmental Law Journal, 30*, 3–49.

Czarnezki, J. (2015). Creating order amidst food eco-label chaos. *Duke Environmental Law and Policy Forum, 25*, 281–311.

Harrison, K. L. (2008). Organic plus: Regulating beyond the current organic standards. *Environmental Law Review, 25*, 211–234.

Hass, J. (2010). Don't take the bait: Why USDA organic certification is wrong for Salmon. *William & Mary Environmental Law and Policy Review, 34*, 589–613.

Isvaliev, S. D. (2017). The formation of agricultural clusters as a basis for the activation of investment activity in the regions (based on the materials of the Sughd and Khatlon regions of the Republic of Tajikistan). Ph.D. thesis. Tajik State University, Dushanbe.

Inshakova, A. O., Frolova, E. E., Galkina, M. V., & Rusakova, E. P. (2020). Development of social market economy under the influence of noneconomic factors: Modeling and regulation. *International Journal of Sociology and Social Policy*. https://doi.org/10.1108/IJSSP-03-2020-0067

Kurman, T. V. (2019). Theoretical and methodological foundations of legal support for the sustainable development of agricultural production. Abstract of the Doctoral thesis. Yaroslav the Wise National Law University, Kharkiv.

Matytsin, D. E., & Rusakova, E. P. (2021). Strategy of quality management in industry 4.0 and formation of a cognitive economy based on industrial and manufacturing engineering in the Russian Federation and countries of the EU. *International Journal for Quality Research*, 15–4. https://doi.org/10.24874/IJQR15.04-03

Matytsin, D. E. (2021). Internet-investing as a remote algorithm of the retail investment financing. In *Modern global economic system: Evolutional development vs. revolutionary leap. Springer book series "Lecture notes in networks and systems"*. Springer Nature.

Margera, E., Caro, K., & Durand, G. (2015). *Organic agriculture and Law*. Food and Agriculture Organization of the United Nations.

Meniv, A. I. (2016). Legal support for the use of GMOs in the cultivation of agricultural products of plant origin in Ukraine and the EU. Ph.D. thesis. Yaroslav the Wise National Law University, Kharkiv.

Morgera, E., Caro, C., & Duran, G. (2012). *Organic agriculture and the law*. Food and Agriculture Organization of the United Nations.

Navasardova, E. S., & Zakharin, A. N. (2019). Comparative analysis of the legislation of the EAEU member states in the field of organic agriculture. *Humanities and Legal Studies, 2*, 141–146.

Organization of Organic Agricultural Production in Russia: Information Publication. (2018). FGBNU "Rosinformagrotech", Moscow.

Ong, K. (2008). A new standard: Finding a way to go beyond organic. N.Y.U. *Journal of Environmental Law, 17*, 883–907.

Robert, P. (2012). The aftermath of Irene: Organic farming, consumer protection, and revising federal policies. *Journal of Environmental Law, 14*, 303–325.

Selivanova, K. A. (2016). Ecological and legal mechanism in the agricultural sector of the economy: Directions of implementation. Ph. D. of Legal Sciences thesis. Orenburg State Agrarian University, Orenburg.

Voronin, B. A. (2013). On the issue of legal regulation of the development of the market of organic agricultural products in the Russian Federation. *Agrar Land Law, 9*, 72–77.

Watnick, V. J. (2014). The organic foods production act, the process/product distinction, and a case for more end product regulation in the organic foods market. *UCLA Journal of Environmental Law and Policy, 32*(1), 40–79.

Yarandaikin, R. S. (1999). Organizational and legal problems of production and sale of environmentally friendly agricultural products. Doctoral thesis. Moscow State Law Academy, Moscow.

Zankovsky, S., Bezbakh, V., Inshakova, A., Rusakova, E. (2020). Social consequences of economic globalization: Experience of developed and developing countries and perspectives of optimization. *International Journal of Sociology and Social Policy*. https://www.scopus.com/inward/record.uri?eid=2-s2.0-85087079671&doi=10.1108%2fIJSSP-03-2020-0068&partnerID=40&md5=2c0385dcafc23892225bfca0c3a376fb. https://doi.org/10.1108/IJSSP-03-2020-0068

Legal Regulation of the Turnover of Biotechnologies in Russia, in Countries of the BRICS and of the EAEU: Certification and Labeling of GMO Products

Denis E. Matytsin, Aleksey P. Anisimov, and Agnessa O. Inshakova

Abstract This chapter justifies the conclusion that GMO technologies as a form of biotechnology are the greatest technological breakthrough of our time. The results of this technology are comparable in their significance to the appearance of metallurgy in the Neolithic period or the development of the first crops. With the help of GMOs, it is possible to increase the yield, which allows not only to solve the problems of world food security but also to preserve untouched lands from human impact (including through the creation of national parks or other types and forms of specially protected natural areas there). The use of GMO technologies reduces the use of pesticides and agrochemicals in agriculture. Undoubtedly, the introduction of such a complex technology into production inevitably creates certain risks for nature and human health, the study, and discussion of which continued. However, in conditions where the harm from GMO technologies and products has not been convincingly proven the search for an optimal model for their use should continue. The authors note that the concept of sustainable development implies a balance of environmental, economic, and social interests. Deviation from this balance in any direction entails many negative consequences for the entire society, both in terms of problems with the realization of the right to food, as well as in terms of environmental protection. In the case of Russia, this means that it is inappropriate to completely ban GMO technologies, which reduces the regulatory value of the law and reduces the potential for the competitiveness of Russian agriculture (although the establishment of separate bans on the consumption of such products, for example, for children, maybe justified). Considering the measures of state regulation of the use of GMO technologies and products, the authors focus on the prospects for the use of certification and labeling of GMO products, which has already had a positive effect in some BRICS countries and the EAEU.

Keywords GMO · Agriculture · Biotechnology · Business · Nature · Health

D. E. Matytsin (✉) · A. P. Anisimov · A. O. Inshakova
Volgograd State University, Volgograd, Russia
e-mail: Dmatytsin@ya.ru; gimchp@volsu.ru

A. O. Inshakova
e-mail: ainshakova@list.ru; gimchp@volsu.ru

E. G. Popkova and B. S. Sergi (eds.), *Geo-Economy of the Future*,
https://doi.org/10.1007/978-3-030-92303-7_44

JEL Code K 32

1 Introduction

Discussions about the possible impact of agricultural products grown using GMO technologies on the environment and human health have been at the center of public attention since the beginning of the XXI century. Periodically, new reasons for discussion are given by scientists' research, which does not always meet the standards of integrity and scientific character. Thus, one of such scandalous studies was the publication of Gilles-Eric Seralini, according to the results of which he linked the mortality of experimental groups of rats with their consumption of GM corn (Seralini, 2012). Biologists have subjected these results to reasonable criticism, but GMO technologies have suffered significant damage. Other studies of the potential dangers of GMOs are more objective, while the discussion between supporters and opponents of GMOs, who differently assess the degree of safety of products produced with their help, has no end. Despite the high public demand, there is still no accurate information about the presence of danger (or its absence) in the use of agricultural GMO products. It has not yet been possible to prove harm to human health from GMO technologies, although some types of products produced in violation of technologies have caused such harm, which has already been proven several times in the courts of some countries.

In addition to possible harm to health, a separate problem of handling GMOs is the presence of potential or real harm to the environment. The fact is that GMO technologies can not only improve the resistance to climate factors or weeds of cultivated agricultural plants but also lead to increased survival of the weeds themselves.

Based on the materials of such scientific discussions about the ambiguity of GMO technologies, indicating the pros and cons of this type of biotechnology, the Russian authorities concluded that it is advisable to ban these technologies almost completely. This led to the curtailment of scientific research and the cessation of cultivation of this variety of agricultural products, the country's loss from this most important sector of world agriculture. Meanwhile, a compromise between supporters and opponents of GMOs would be more appropriate, since the law, as a regulator of public relations, can build the necessary balance of interests between business, civil society, and public authorities through incentives, restrictions, permits, and prohibitions (Kamyshansky & Stanishevsky, 2018).

Traditionally, genetic engineering is considered as consisting of two sections-genetic and genomic engineering. The goal of genetic engineering is to introduce one or more (usually alien) genes into the recipient cell or to create new types of regulatory connections in the genome, without changing the species identity of the recipient organisms. Genetic engineering is faced with the task of deeper intervention in the genome, up to the creation of new species. The first GMO products (tomatoes Flavr Savr with delayed maturation of the company "Calgen" and herbicide-resistant soy of the company "Monsanto") appeared in the United States in 1994. Today, biotech firms supply the market with several transgenic plants: tomatoes, corn, potatoes, tobacco, soy, rapeseed, vegetable marrows.

Along with agriculture, GMO technologies can also be used in medicine, pharmaceuticals, etc. In terms of food, the number of genetically modified products can also be attributed to products containing individual ingredients obtained from GMOs (flour, oil, etc.) or additives (dyes, vitamins). At the same time, the scientific literature quite rightly notes that giving agricultural plants new qualities "will lead to an increase in the productivity of the agricultural sector, will improve the nutritional properties of agricultural products, will facilitate the process of processing raw materials, will reduce the amount of sprayed chemicals, will increase the profitability and productivity of agriculture" (Chuiko, 2011).

Taking into account the above circumstances it is necessary to study the problem of the use of GMO technologies in agriculture from the standpoint of legal science. Of course, such a study will not be able to give recommendations on the use of a certain type of GMO products (this question is rather from the field of biological or medical science). However, it is through legal means that the use of GMO technologies in individual countries or their regions can be recommended or prohibited, and a list of measures is defined to assess the quality of the product and give the results of such an expert assessment legal significance. Finally, it is the legal means that allow us to guarantee the information rights of consumers, to regulate the procedural features of the consideration of cases for compensation for harm caused by the use of GM products.

2 Methodology

In the course of the research, general scientific methods were used, including formal-logical, dialectical, system-structural, and critical cognition. Methods of synthesis, classification, and generalization were used to interpret the results of the study. Private scientific methods are also used in the work: formal-legal, the principle of assessing legal processes, the method of comparative analysis, etc.

3 Results

Today, more than 100 lines of transgenic plants are grown on a production scale in the world, the undisputed leader among which is soy, which accounts for 98% of the total number of genetically modified plants. Two-thirds of all genetically modified crops in the world are cultivated in the United States. The primary objectives of creating genetic modifications were to obtain high yields, achieve resistance to drought and cold, and some others. Now we are talking about the creation of vitamin-rich organisms, with a regulated fat content. The use of GMOs in the production of agricultural products and food products is of great importance in terms of food security. Despite this, the prevailing opinion about the use of GMOs is that there is a potential threat to human health and the gene pool of the nation as a whole, and there is no single definition of GMOs in legal science.

According to some authors, a genetically modified organism is an organism whose genetic material has been modified using genetic engineering methods (Bogatyreva, 2015). Others believe that a genetically modified organism used to grow agricultural products of plant origin is a plant organism (its parts, including seeds). In which the genetic material has received such changes that cannot be obtained in natural conditions in the process of natural recombination, and which is used to increase crop yields, change the quality indicators of agricultural products (Meniv, 2016). Agreeing with the latter definition it noted that GMO technologies should be considered in a broader context—as a kind of biotechnology that has long been used by humanity.

The term "biotechnology" was first proposed in 1917 by the Hungarian engineer K. Jereki (Bhardwaj, 2003). Trying to eliminate the gaps in national law, in the second half of the XX century, Sasson (1987) defined biotechnology as a technological process implemented using biological systems—living organisms and components of a living cell. At moment, there is no unified scientific opinion in the understanding of biotechnologies. They are interpreted as a set of methods and techniques for obtaining useful products and phenomena for humans using biological agents (Karavaeva & Kravtsov, 2005). As a special social activity aimed at the practical transformation of the surrounding world and oneself by a person through the use of biological processes and agents (Vorontsova, 2010a). As a science of ways to obtain target products using biosynthesis controlled by environmental parameters or genetic-engineering manipulations or a combination of these effects (Volova, 1999). The latter position seems to us the most successful.

So biotechnology is a field of applied biology, which involves the use of living organisms and biological processes in engineering, technology, medicine, and other areas in which biological products are needed. In the legal sense, "biotechnology" means any type of technology related to the use of biological systems, living organisms, or their derivatives for the manufacture or modification of products (processes) for their direct use. The Russian Federation has joined the Convention on Biological Diversity in 1992, but it has not signed the Protocol on Biosafety (the Cartagena Protocol) to this convention.

From the above, it should be concluded that GMOs should be understood as one organism or a group of them (and this may be a single-celled or multi-cellular formation), which are capable of transmitting hereditary information, and differ from natural organisms. Another, no less important, characteristic is that such organisms are obtained through the use of genetic engineering methods. Modification of the genome of an organism is one of the methods of biotechnology (although not the only one), and modern agricultural biotechnologies are almost identical to the use of genetic engineering and the use of GMOs. Taking into account the polarity of opinions on the usefulness/danger of GMO technologies and products for human health and the environment, we will try to systematize the positions of different sides of this discussion.

The introduction of GMO technologies in agriculture has made it possible to create some frost-resistant crops that are less susceptible to pesticides. Another important task of biologists was to obtain plants that were resistant to viruses, so the virus shells were introduced into the plant cells of the genes. As a result, transgenic plants obtained that can resist the effects of more than a dozen different viral infections.

The use of genetic engineering can reduce the use of insecticides by 40–60%. Genetic engineers have developed transgenic plants with extended fruit ripening periods (for example, some tomatoes can be removed from the bush red, without fear that they will ripen during transportation). Currently, the list of plants to which the methods of genetic engineering have successfully applied is about fifty species, including apple, plum, cabbage, eggplant, cucumber, wheat, soy, rice, rye, etc. In addition, the arguments in support of genetic engineering should include:

(1) the world is overpopulated, and for its existence, it is necessary to significantly (at least twice) increase food production;
(2) the growth of agricultural production is difficult to implement without plowing new land, which will inevitably lead to a deterioration of the state of nature;
(3) to protect crops from weeds and pests, farmers are forced to use pesticides and agrochemicals, and the volume of their use is growing every year, which also entails negative environmental consequences;
(4) the use of GMO technologies will reduce the burden on the environment, especially since officially recognized "natural products" are also a product of human-made breeding;
(5) critical comments on the prospects of GMO technologies are mainly expressed by representatives of developed countries where there is no food shortage. However, for many countries in Asia and Africa, the problem of food security remains very acute. Therefore, the use of GMO technologies in them will save the population of these countries from hunger (Vorontsova, 2010b).

In turn, opponents of GMOs report the first death of a person who ate food obtained using GMO technologies in a Spanish hospital in 2015. Medical examiners then determined the cause of death as a result of the consumption of tomatoes containing fish-related genes and antibiotic-resistant genes that prevented the development of white blood cells. There are also examples showing that concerns about the safety of GM foods have been justified concerning potential exposure and risk in some other countries. For example, a study conducted by the Department of Food Science and Technology at the University of Nebraska found that a gene produced from Brazil nuts when introduced into GM soy causes allergies in humans (Balashanmugam, 2018).

A separate problem is a danger of mixing products grown using GMOs with all the rest. In this sense, an example from the German judicial practice became almost "textbook", when the products of a Bavarian farmer were recognized as genetically modified, although he did not use GMO technologies. As it turned out, his bees inadvertently pollinated the GMO corn of the company "Monsanto", whose plantations were located nearby (Revenko, 2011).

In the context of such a pronounced polarization of opinions, we will try to formulate our position on this issue. In our view, the development and use of GMO technologies in agriculture have great potential in terms of the prospects for the realization of the human right to food. Solving the problem of food security by increasing the quantity and reducing the price of food will be of little relevance for developed countries, but it is very promising for countries where it was not possible to provide a high standard of living for the population (in particular, the countries that are members of the BRICS and especially the EAEU). Thus, GMO technologies are a great scientific discovery that can increase the yield of crops and, thereby, mitigate the world food problem. Along with this, it noted that in addition to agriculture, GMO technologies are increasingly used every year in the world of medicine and some other industries, in various sectors of science and technology (Matytsin & Rusakova, 2021).

The danger of GMO technologies is much more reasonable from the point of view of its impact on the environment. Opponents of GMOs are undoubtedly right in their concerns in the sense that ignoring the laws of evolution and population genetics and invading the genetic program of an organism is very dangerous, especially when it is caused by the short-term financial benefits of individuals, companies, or financial groups (Matytsin, 2021). Such risks are associated with the emergence of superweeds, the formation of new, poison-resistant insect populations, genetic contamination, and the permanent loss of traditional crop varieties (Scott, 2015).

On the other hand, all officially registered cases of harm to the health of citizens or nature occurred not because of the use of the achievements of genetic engineering as such, but because of the fact that specific companies-producers of agricultural GMO products incorrectly applied the appropriate technology. Accordingly, genetic engineering is a manifestation of scientific and technological progress, which cannot be prohibited (Tarakanov et al., 2020). Therefore, in our opinion, further discussions on the applicability of GMO technologies in agriculture conducted not from the standpoint of a complete ban or permission of GMOs, but in the direction of finding a balance of private and public interests, a compromise between the authorities, the public and agricultural producers. At the same time, it is necessary to look for new criteria of objectivity when evaluating the results of the use of genetic engineering methods, since this innovative industry has not yet received its development and acceptance by all countries of the world.

It seems that over time the results of the use of GMO technologies in agriculture in those countries where they allowed, would allow us to conduct convincing studies that can make an unambiguous conclusion about the danger or harmlessness of GMO products. At moment, the task of legal science is to offer an intermediate compromise, with the guarantee of the rights and legitimate interests of citizens (the development of labeling, expertise, judicial guarantees, etc.).

In the BRICS countries, the development of GMO technologies is as follows. The first GMO product was commercialized in Brazil in 1995. The National Technical Committee on Biosafety has approved more than 50 types of GMOs, most of which are plants, including cotton, soybeans, and corn. Brazil is a major exporter of agricultural goods and food (worth \$4.8 billion) to the United States and imports agricultural products, such as wheat and other products (worth $1.7 billion). In addition, Brazil is the largest exporter of GM cotton, corn, and soybeans.

In South Africa, GMO research and development has been very successful over the past three decades. This is because the African continent must adopt GMO technologies to cope with hunger and drought. South Africa recognizes the need to use GMO technologies to increase crop production and reduce grain imports from other countries. At the same time, the South African Government encourages various agricultural technologies, including GMO technologies. The first approval of field trials of GMO crops occurred in 1989, followed by commercial approval of insect-resistant GMO cotton and corn in 1997. Since then, there has been an increase in the commercialization of GMOs (Balashanmugam, 2018).

India has traditionally been among the countries where the production of agricultural products using GMO technologies is prohibited. In 2020, the Food Safety and Standards Authority of India approved a list of 24 product names that must be accompanied by the mandatory "GMO-free" label and have "GMO-free" certificates from January 2021. This comprehensive list includes some common items, including four of the world's leading crops: corn, wheat, rice, and soy, a variety of fruits (apples, pineapples, etc.), and vegetables (eggplant, potatoes). At the same time, the Indian expert community continues to debate the safety of GMOs and the need for their cultivation (and with the latter according to 30% of the public) (Neo, 2020). Note that although the government has banned the commercial cultivation of GM food crops in India, it has allowed the cultivation of non-human GMO cotton since 2002.

The GMO regulations established in China provide for the administration, evaluation, labeling, licensing of production, licensing of economic activities, examination, and approval procedures. Together, such laws regulate research, production, and processing of agro-GMOs; business operations; import and export; supervision and expertise. These measures and their associated procedures include detailed rules on safety assessment, importation, and labeling, which are designed to make laws more enforceable (Yu & Wang, 2012).

GMO technologies are widely used in China not only in agriculture but also in the pharmaceutical and food industries. The Chinese Ministry of Agriculture periodically issues new licenses for the commercial production of GMO food crops, including new varieties of cotton and tomatoes, sweet peppers, and petunias. The cultivation of GMO crops is successful—it involves 3.9 million hectares. Research and development in the field of GMO technologies identified by the Chinese government as one of the 16 main areas intended for breakthroughs. China is a major importer of cotton, papaya, soy, and corn from the United States. In addition, the country also imports millions of tons of soybean oil as farm animal feed and uses it as vegetable oil.

In Russia, by Article 21 of Federal Law No. 149-FZ of December 17, 1997 "On Seed Production", it is prohibited to import and use for sowing seeds of plants whose genetic program has been modified using genetic engineering methods and which contain genetically engineered material. The introduction of this material was not the result of natural (natural) processes, except for sowing (planting) of such seeds during examinations and research work. In addition, the Code of Administrative Offences of the Russian Federation No. 195-FZ of December 30, 2001, includes Article 6.3.1, according to which an administrative fine is imposed for violating legislation in the field of genetic engineering. Currently, in Russia, there are no varieties and hybrids of plants containing GMOs in the State Register of Breeding Achievements approved for use.

Although GMOs are banned in Russia, the situation is less clear in other EAEU countries. Thus, in the Republic of Belarus, the cultivation of genetically modified plants in the fields is not prohibited. But in practice, no one officially grows them, because first, you need to go through a fairly complex and strict control procedure: obtaining a permit from the Ministry of Natural Resources and Environmental Protection, testing such plants at the landfill, assessing all possible risks, then re-testing, medical evaluation. In the case of the production of such a plant, or the import of GMO products in the Republic of Belarus, it is subject to mandatory labeling.

The regulation of the turnover of GMO products in Kazakhstan is still limited to labeling. The republic has fixed requirements for the turnover of genetically modified organisms, their quantitative content in food products with a norm of no more than 0.9%, as established by the standards of the Codex Alimentarius. Everything above is subject to mandatory labeling "contains GMOs". Control over food products carried out monthly, monitoring at the stage of its implementation and appropriate laboratory examinations carried out. The identification of markers of genetically modified ingredients is a complex procedure that includes a qualitative analysis with a transition to a quantitative calculation and reaches a period of up to a month or more. Thus, in 2018, 1,706 units of food products from 27 countries were examined, including those of their production (Ivanilova, A. Mass production of GMO products opens up new prospects for Kazakhstan. https://mk-kz.kz/economics/2019/11/20/massovoe-proizvodstvo-produktov-s-gmo-otkryvayut-dlya-kazakhstana-novye-perspektivy.html. Accessed 23 April 2021).

In Armenia, no laws are prohibiting the import of products containing GMOs. The company "Monsanto", which produces GMO seeds, mainly supplies seeds of garden melons to Armenia. Such seeds undergo a preliminary examination their productivity analyzed only after that their use allowed. The marking of the presence of GMOs affixed to the products, the proportion of GMOs in which exceeds 0.9%.

It prohibited import both GM seeds and GMO products to the Republic of Kyrgyzstan (except for cases when the GMO standard in them is less than 0.9%).

In the BRICS and EEU countries, there are different national approaches to the cultivation and turnover of GMO products, ranging from its full permission (China) to a complete ban (Russia). This dynamic of decisions is determined by the economic, social, and environmental conditions in each country, as well as the position of the public and the country's leadership (Inshakova et al., 2020).

From the point of view of international law, there are two main approaches to the permissibility of the use of GMOs in agriculture:

(1) the principle of equivalence of products obtained using GMO technologies and "natural" products. This principle assumes the safety of GMO-containing

products, which is confirmed by the regulatory requirements for the quality and safety of food products.

This approach used in Canada, the United States, and some other countries. Accordingly, the main object of control is not the manufacturing process, but the product itself, which is subject to appropriate labeling.

(2) the precautionary principle of handling GMO products. Its main characteristic is the introduction of restrictions (or bans) on the production or import of GMO products into a particular country due to its potential danger to nature and the health of citizens. In particular, its principle was implemented in the Russian Federation.

The latter principle is closely related to one of the basic principles of environmental law, provided for by both international and national law. For example in Russia and other EAEU countries, the principle of prohibiting economic and other activities whose consequences are unpredictable for the environment is established (Article 3 of the Federal Law "On Environmental Protection" of January 7, 2002). The implementation of this principle is one of the additional guarantees of protecting the right of everyone to a favorable environment. For this guarantee to be able to work in full force, we consider it necessary to amend the civil legislation, recognizing GMO technologies as a type of source of increased danger with the resulting consequences.

It seems that even in countries where there is a complete ban on the use of GMO technologies in agriculture, it is necessary to develop research activities in this area. In our opinion, even if GMOs are banned in a particular country, research work in this sector should not be banned, since conclusions about the dangers of GMOs for nature and human health continue to be debatable. Moreover, the results of such studies in no case should be classified.

It is necessary to give a positive assessment of the experience of the European Union, which provides for strict requirements for the labeling of GMO products. This is especially important because it allows you to inform the buyer about the content of the product, allows the consumer to make a meaningful choice. Along with the legal regulation of the production of GMO products, EU legal acts pay much attention to such property of GMOs as "traceability" (traceability). This means the ability to detect GMOs and products made from GMOs at all stages of their placement on the market through, the use of production and sales chains (Prokudina, 2007).

In our opinion, it is necessary to recognize the justified bans established in some countries of the world on the cultivation of GMO products on land intended for the production of environmentally friendly (organic) products. Therefore, it seems appropriate to adopt special rules for the coexistence of genetically modified varieties of agricultural plants with traditional and organic crops. In addition, the issue of a single list of GMOs that do not pose a threat to human health or the environment also needs to discuss. A separate discussion by the scientific community requires the issue of drawing up a single (consolidated) list of types of GMO technologies that do not pose an environmental or other danger.

It is also necessary to fix in national legislation the obligation to create buffer zones for GMO producers when growing these agricultural products, separating the crops of GMO crops from traditional varieties of agricultural plants. At the same time, such zones were established on land plots. At the expense of their area of the commodity producer, that uses GMOs in the cultivation of agricultural products of plant origin (Meniv, 2016).

4 Conclusion

Currently, GMO crops cover more than 200 million hectares. This is 1.5% of the US territory or more than 10% of the world's crop. Since 1996, when GMOs began to be used commercially for the first time, their plantings have grown more than 100 times. At the moment, GMO technologies are rapidly developing, and they are among the most promising biotechnologies of the twenty-first century. They make it possible to increase crop yields and realize the human right to food for the inhabitants of developing countries, and preserve valuable natural ecosystems from plowing.

The use of GMO technologies reduces the use of pesticides and agrochemicals in agriculture. Undoubtedly, the introduction of such a complex technology into production inevitably creates certain risks for nature and human health, the study, and discussion of which continued. However, in conditions where the harm from GMO technologies and products is not convincingly proven, it is necessary to find a compromise between private and public interests. We believe that such technologies can also contribute to the achievement of the Sustainable Development Goals. Deviation from this balance in any direction entails negative consequences for the entire society, both in terms of problems with the realization of the right to food, as well as in terms of the state of the environment. In this regard, we are not in favor of banning GMOs, but of building a more complex balance of interests of society, business, and public authorities.

In this regard, it seems to be a more effective option to develop a system of restrictions, prohibitions, and permits in the use of GMO technologies with the help of law, which allows reducing the risks of side effects (including potential ones). For this purpose, it is necessary to develop the potential of contractual and administrative methods of regulating this sphere of relations, including through the wider use of labeling, certification, and quality expertise of agricultural GMO products.

Acknowledgements The reported study was funded by RFBR according to the research project No. 18-29-16132.

References

Balashanmugam, S. K. (2018). Safety assessment and liability regulations in the context of genetically modified food in the BRICS countries. *BRICS Law Journal, 5*(1), 27–55.

Bhardwaj, M. (2003). Global bioethics and international governance of biotechnology. *Asian Biotechnology Development Review, 6–1*, 39–53.
Bogatyreva, N. V. (2015). Issues of legislative definition of a genetically engineered modified organism. *Bulletin of the Saratov State Law Academy, 2*, 135–138.
Chuiko, N. A. (2011). Basic approaches to the regulation of genetically modified organisms in international practice. *Siberian Legal Bulletin, 1*, 160–165.
Inshakova, A. O., Goncharov, A. I., & Ershova, I. V. (2020). Digital blockchain registration of legally significant stages of complex good's export-import supplies by business entities of the EAEU and BRICS jurisdictions. In A. O. Inshakova, & E. I. Inshakova (Eds.), *Competitive Russia: Foresight Model of Economic and Legal Development in the Digital Age: Proceedings of the International Scientific Conference in Memory of Oleg Inshakov (1952–2018)*. Springer Nature.
Ivanilova, A. Mass production of GMO products opens up new prospects for Kazakhstan. https://mk-kz.kz/economics/2019/11/20/massovoe-proizvodstvo-produktov-s-gmo-otkryvayut-dlya-kazakhstana-novye-perspektivy.html. Accessed 23 April 2021.
Kamyshansky, V. P., & Stanishevsky, A. I. (2018). Genetic engineering in civil law. *The Power of Law, 4*, 33–44.
Karavaeva, E. I., & Kravtsov, R. V. (2005). Biomedical technologies: Issues of legal regulation and responsibility. *Siberian Legal Bulletin, 3*, 7–12.
Matytsin, D. E., & Rusakova E. P. (2021). Strategy of quality management in industry 4.0 and formation of a cognitive economy based on industrial and manufacturing engineering in the Russian Federation and countries of the EU. *International Journal for Quality Research*, 15–4.
Matytsin, D. E. (2021). Internet-investing as a remote algorithm of retail investment financing. In *Modern Global Economic System: Evolutional Development vs. Revolutionary Leap. Springer book series "Lecture Notes in Networks and Systems"*. Springer Nature.
Meniv, O. I. (2016). Legal support for the use of GMOs in the cultivation of agricultural products of plant origin in Ukraine and the EU. Ph.D. thesis. National Law University named after Y. Mudry, Kharkiv.
Neo, P. (2020). Economy vs science: GMO food tech causes clashes among Indian food sector. https://www.foodnavigator-asia.com/Article/2020/11/16/Economy-vs-science-GMO-food-tech-causes-clashes-among-Indian-food-sector#:~:text=India%20has%20traditionally%20been%20a,certificates%20starting%20from%20January%202021. Accessed 23 April 2021.
Prokudina, N. V. (2007). Main aspects of legal regulation of the use and production of genetically modified microorganisms in the European Union and the Russian Federation. *Agrar and Land Law, 5*, 74–83.
Revenko, L. S. (2011). Regulation of production and turnover of GMOs in the European Union: Current trends. *Russian Foreign Economic Bulletin, 4*, 3–8.
Sasson, A. (1987). *Biotechnology: Achievements and hopes*. Mir.
Scott, A. H. (2015). Genetically modified crop regulation: The fraying of American patchwork farm lands. *Villanova Environmental Law Journal, XXVI*, 145–167.
Seralini, G. E. (2012). Long term toxicity of a Roundup herbicide and a Roundup-tolerant genetically modified maize. *Food and Chemical Toxicol, 50*, 4221–4231.
Tarakanov, V. V., Inshakova, A. O., & Goncharov, A. I. (2020). Interactive demo on-line maintenance foreign trade deals of business entities of BRICS jurisdictions. In A. O. Inshakova, & E. I. Inshakova (Eds.), *Competitive Russia: Foresight Model of Economic and Legal Development in the Digital Age. Proceedings of the International Scientific Conference in Memory of Oleg Inshakov (1952–2018)*. Springer Nature.
Vorontsova, E. V. (2010a). On the possibility of a legal ban on the use of transgenic technologies in Russia: "pros" and "cons" in the light of the implementation of the constitutional right to health protection. *Izvestiya Kursk State Technical University, 3*, 113–120.

Volova, T. G. (1999). *Biotechnology*. SB RAS Publishing House.
Vorontsova, Z. I. (2010b). Initial principles of the activity approach to the analysis of biotechnologies as a form of scientific work. *Society and Law, 5*, 251–253.
Yu, W., & Wang, C. (2012). Agro-GMO biosafety legislation in China: Current situation, challenges, and solutions. *Vermont Journal of Environmental Law, 13*, 865–883.

Reproduction of Human Capital of Agricultural Production as the Basis of the National Food Security

Evgeny E. Shvakov, Natalia V. Boyko, and Evgeny A. An

Abstract At present, human capital is treated as one of the key factors of economic development—if it conforms—by its criteria—(quantitative and qualitative)—to the personnel needs of the national economy. Human capital of agricultural production is also a factor of achievement of national food security. The structural transformation of the national economy and change of production technologies require the reconsideration of approaches to formation and use of human capital. Agricultural production is not an exception. This leads to the necessity to study human capital of agricultural production from the positions of organization of its effective functioning and provision of national food security. In this paper, we consider the components of human capital, which formation is necessary in the modern conditions, and offer a methodology for its evaluation from the positions of sufficiency for organization of effective agricultural production and provision of national food security. We substantiate the measures of government regulation of the system of education and labor migration, which are to ensure the formation of human capital that is sufficient for effective organization of agricultural production and provision of national food security.

Keywords Human capital · Education · Food security · Russian economy

JEL Code Q5 · Q51 · Q59

E. E. Shvakov (✉)
Altai State University, Barnaul, Russia

N. V. Boyko
National Medical Research Center of Hematology of the Ministry of Health of the Russian Federation, Moscow, Russia

E. A. An
Al-Farabi Kazakh National University, Almaty, Kazakhstan

E. G. Popkova and B. S. Sergi (eds.), *Geo-Economy of the Future*,
https://doi.org/10.1007/978-3-030-92303-7_45

1 Introduction

For a country's economic system and development of its separate elements, labor resources—as a production factor—play the decisive role in formation of a competitive national economy (Cohen & Soto, 2007; Garza-Rodriguez et al., 2020; Goenka & Liu, 2020). At present, economic development is connected primarily to a human, more specifically human capital—as an economic category that reflects aggregate knowledge and skills of able-bodied population and its abilities to be involved in public production. From the positions of provision of national food security, reproduction of human capital for agricultural production of a country is necessary.

2 Methodology

A study of the retrospective and modern theories of human capital, given in the works of foreign and Russian scholars, confirms its importance for development of the national economic systems (Kapelyushnikov, R. Modern western concepts of labor force formation (p. 287). Moscow). A study by E. Denison has shown that education, as an element of human capital, is placed third in the system of factors of labor efficiency's growth (Denison, 1985). As for the Russian economy, this statement is also fair and has been confirmed by a range of studies of Russian scholars (Bozhechkova, 2013; Nureyev, 2000).

As for the issue of formation of human capital of agricultural production in the context of provision of national food security, it is necessary to consider two main aspects. The first aspect is quality of human capital, which is necessary for effective organization of national agricultural production. The second aspect is sufficiency of human capital of agricultural production from the positions of provision of national food security. These aspects of human capital have been studied with the use of the official statistics of the Russian Federation. The study has been performed with the use of general scientific and special methods of scientific analysis: analysis and synthesis, induction and deduction, and methods of statistical processing of information (grouping, comparison, and economic & mathematical modeling). The use of the above methods has ensured the objectivity of the performed analysis and correctness of the research results.

3 Results

As for the first aspect—quality of human capital, which is necessary for effective organization of national agricultural production—it should be considered through comparison of the characteristics of human capital and the modern conditions and technologies of agricultural production. In the modern understanding, human capital is a new form of production capital, which cannot be separated from its bearer—human—and which is formed based on innate abilities and in the process of life activities, by acquisition of knowledge, capabilities, and skills. The components of human capital from the positions of evaluation of its quality should conform to the current conditions of the national economy's functioning (Ershova et al., 2019). As for the modern conditions of the national economy from the positions of development of the existing theoretical view of human capital (Dyatlov, 2005; Ilyinsky, 2011), and in view of the conditions of functioning of agricultural production of the Russian Federation, the following components of human capital should be distinguished:

(1) health capital;
(2) education capital;
(3) mobility capital;
(4) digital capital.

While the first two components of human capital have been defined and studied within the modern theories of human capital (Ilyinsky, 2011), the distinction of mobility capital and digital capital reflects the modern tendencies of development of the world economy on the whole and Russia's agricultural production in particular. Distinction and formation of mobility capital within human capital have been predetermined by the modern dynamics of development of national economies and their separate components. Firstly, certain professions disappear, and new ones emerge. Secondly, certain production sub-systems of national economies are peculiar for large release of workforce, caused by the change of technological mode. This is a feature of the current agricultural production of the Russian Federation, in which, according to the Federal State Statistics Service, the number of the employed reduced in 2006–2019 by 40%—from 6,831,000 to 4,196,000 (Federal State Statistics Service, 2020a). In these conditions, an important role belongs to individual's abilities for professional and territorial mobility, which constitute mobility capital.

Distinction of the fourth component of human capital—digital capital—is a result of formation of the digital economy in the modern conditions. Digital capital—as a new element of human capital—individual's possessing knowledge, skills, and abilities to perform work in the digital environment with the use of IT tools and IT technologies.

Agreeing with the distinction of two types of capital (basic and acquired) in the structure of human capital, it is necessary to correlate its components with the types of capital. It is obvious that basic capital is determined by innate abilities of human, so it is formed by health capital, as a totality of hereditary physiological abilities of a person (Ilyinsky, 2011). Acquired capital forms in the process of formation of a human as a personality and within the participation in public production. The dominating role in creation of acquired capital belongs to the system of education (Gurban & Krutikova, 2011). Figure 1 shows participation of the national system of education in formation of the acquired components of human capital with regard to agricultural production.

As for sufficiency of human capital for organization of effective agricultural production, it should be assessed by correlating the agricultural companies' need for

Agricultural production's needs for workforce with the set professional and qualification characteristics

Elements of the national system of education	*Formed elements of human capital*
Pre-school education	Formation of basic knowledge and skills, which ensure the individual's inclusion in public production (education capital)
Secondary education	
Professional education	Formation of knowledge, skills, and capabilities that ensure the individual's execution in public production of specialized functions (education capital) in view of the modern tendencies of development of the national economic system (digital capital)
Further vocational education	Formation of knowledge, skills, and capabilities that ensure territorial and professional mobility (mobility capital) and the ability to participate in public production in the conditions of its digitalization (digital capital)

Formation of human capital with the set professional and qualification characteristics

Fig. 1 Formation of human capital of agricultural production in the national educational system. *Source* Compiled by the author

workforce with the number of students who are trained by the national educational system for agricultural production.

This measure is not completely new (Shvakov, 2020), but with regard to this research it is adapted to the specifics of agricultural production. The use of this method is predetermined by the fact that the system of education makes a decisive contribution to the formation of human capital (Pishnyak et al., 2020). Evaluation is performed in two stages—due to the necessity for various regulating measures at each stage. At the first stage, potential sufficiency of human capital is assessed. The assessment is built on the basis of correlation between the agricultural companies' need for personnel and the number of students of the national system of education on the specialties that are important for agricultural production. If the values of the obtained indicator—coefficient of potential sufficiency of human capital for agricultural production—is less than 1, this is a sign of insufficiency of personnel training for agricultural production by the national educational system. The measures of regulating influence should include the increase of personnel training for agricultural production within the national system of professional education (Table 1).

At the second stage, the factual sufficiency of human capital for agricultural production is evaluated. For this purpose, the obtained indicator is recalculated in

Table 1 Methodology of sufficiency of human capital for organization of effective national agricultural production

Stage of assessment	Calculated indicator	Methodology of calculation	Interpretation of the values of the calculated indicator
1	Coefficient of potential sufficiency of human capital for agricultural production	Kd = (Pi − Ci−n * K1)/Pi, where: Pi—the need for professional personnel with the education that is specialized in agricultural production; Ci−n—number of students accepted for the first year of study for specialties connected to agricultural production; n—duration of studies for obtaining professional education; K1—coefficient of preservation of the contingent of students (it is formed based on observations)	The value of the indicator that exceeds 1 shows sufficiency of personnel training for agricultural production by the national educational system. The value of the indicator that is below 1 shows the reverse and requires the increase of personnel training for agricultural production within the national system of professional education
2	Factual coefficient of sufficiency of human capital for agricultural production	Kdf = (Pi − Ci−n * K1 * K2)/Pi, where: K2—coefficient of employment of graduates with education that specializes in agricultural production (it is formed based on the data of the Pension Fund of the Russian Federation)	The value of the indicator that exceeds 1s shows that the national educational system covers the agricultural companies' need for personnel. The value of the indicator that is below 1 shows the absence of population's motivation for work in the sphere of agricultural production and requires the adoption of measures of state regulation within the national economic system

view of the factual employment of graduates with education that specializes in agricultural production. If the value of the obtained coefficient is below 1, and the coefficient of potential sufficiency of human capital exceeds 1, the working population does not have motivation for work in the sphere of agricultural production. Therefore, the state has to adopt measures within the national economic system that would ensure the presence of skilled personnel in rural areas and at companies that deal with processing of agricultural products.

With regard to the Russian Federation, comparison of the number of applicants who are accepted to specialties connected to agricultural production and the agricultural companies' need for personnel shows that the national system of education successfully performs its function on formation of human capital for agricultural production.

With the declared annual need for agricultural personnel of 3300–3700 (according to the Federal State Statistics Service, at least 5000 applicants are accepted by universities at specialties that are connected to agricultural production. The same situation is observed in the part of training of specialists of the middle level of qualification and skilled workers. According to the Federal State Statistics Service, agricultural companies have more than 30,000 vacancies each year (Federal State Statistics Service, 2020b), which shows the absence of motivation for employment in the national agricultural production and the priority of other spheres and types of economic activities during selection of employment.

4 Conclusion

The offered methodological approach to evaluation of human capital of agricultural production in the context of its effective organization and provision of national food security focuses on the following. Firstly, personnel training. For agricultural production—as a sector of the Russian economy that gradually decreases—it is necessary to implement not so much the strategy of personnel training (since the need for personnel is minimal) as the strategy of retraining of fired employees with their further employment. The dominating role should belong to further vocational education, as it is assigned a role of formation of mobility capital and digital capital—as the elements of aggregate human capital of population that is employed in agricultural production. Formation of these elements of human capital will ensure the increase of specialized mobility of population that works in agricultural production and its correspondence to the modern technologies of agriculture. Secondly, within the national economic system. From the positions of provision of human capital's sufficiency for organization of effective agricultural production, it is necessary to adopt measures on assignment skilled personnel in rural areas and increase of population's motivation for the employment in agricultural production. These measures could include the following:

- increase of interaction between the production sphere and agricultural production (Tsvetkov et al., 2019), and increase of personnel training by the national educational system by the orders from agricultural companies with further mandatory employment in these companies;
- creation of economic mechanisms that would stimulate the increase of population's interest in the employment in agricultural production (accessibility of accommodation, support for young specialists, etc.).

References

Bozhechkova, A. (2013). Econometric modeling of the impact of human capital on economic growth in the regions of Russia. *Audit and Financial Analysis, 1*, 90–99.

Cohen, D., & Soto, M. (2007). Growth and human capital: Good data, good results. *Journal of Economic Growth, 12*(1).

Denison, E. (1985). *Trends in American economic growth, 1929–1982* (p. 220). Brookings Institution.

Dyatlov, S. (2005). *Human capital of Russia: Problems of efficiency of use in the conditions of transition economy* (p. 280). St. Petersburg University of Economics and Finance.

Ershova, I., Belyaeva, O., & Obukhova, A. (2019). Investment in human capital education under the digital economy. *Ekonomichny Chasopis, XXI*(11–12), 69–77.

Federal State Statistics Service. (2020). Regions of Russia: socio-economic indicators—2019. Available at https://rosstat.gov.ru/folder/210/document/13204

Federal State Statistics Service. (2020). On the number and needs of organizations for employees by professional groups. Available at https://rosstat.gov.ru/compendium/document/13266

Garza-Rodriguez, J., Almeida-Velasco, N., Gonzalez-Morales, S., & Leal-Ornelas, A. (2020). The impact of human capital on economic growth: The case of Mexico. *Journal of the Knowledge Economy, 11*(2), 660–675.

Goenka, A., & Liu, L. (2020). Infectious diseases, human capital and economic growth. *Economic Theory, 70*(1).

Gurban, I., & Krutikova, M. (2011). The state of educational capital as a system-forming factor in the formation of human capital. *Bulletin of Ural Federal University Series Economics and Management, 4*, 136–148.

Ilyinsky, I. (2011). *Competitiveness of Russian human capital: State, problems, prospects* (p. 278). St. Petersburg University of Technology and Design.

Kapelyushnikov, R. Modern western concepts of labor force formation (p. 287). Moscow.

Nureyev, R. (2000). Development theories: New models of economic growth (contribution of human capital). *Economic Issues, 9*, 126–145.

Pishnyak, A., Goryainova, A., Nazarbayeva, E., & Khalina, N. (2020). Middle-class investment in human capital: Consumption of paid services in the field of education and medicine. *Questions of Economics, 11*, 69–85.

Shvakov, E. (2020). Content and current challenges of state policy in the field of training of highly qualified personnel (on the Example of the Altai Territory). In *Advances in Natural, Human-Made, and Coupled Human-Natural Systems Research. Lecture Notes in Networks and Systems*. Springer Nature.

Tsvetkov, V., Loginov, E., & Raikov, A. (2019). Integrated network interaction of education with science and production. *Educational Technologies and Society, 19*(1), 389–398.

Rural Families as the Key Potential for Developing the Agricultural Complex in the Republic of Bashkortostan

Liana N. Gabdrakhmanova, Eduard S. Gareev, Yuri N. Dorozhkin, Artem A. Cheremiskin, and Gulnara A. Yakupova

Abstract The paper substantiates and highlights the special importance of rural families in the development of the agricultural complex of the Republic of Bashkortostan (a subject of the Russian Federation). The authors reveal and analyze the problems of the formation of government policy on rural families as a key factor in developing the agricultural complex of the Republic of Bashkortostan. The authors disclose a factor determining the effective and sustainable development of the agro-industrial sector, namely the development of human potential. For developing, planning, and managing the agro-industrial sector, the authors investigated the sector's problems, tasks, and methods of their solution. The paper identifies the features and trends affecting the development of the agricultural sector of one of the Russian regions—the Republic of Bashkortostan. The authors highlight and structure the main factors in reducing young people employed in the agro-industrial complex. Additionally, they present an analysis of existing programs for the development of the agro-industrial complex and the main measures to support the labor potential of the agricultural sector of the Republic of Bashkortostan for further development of rural areas as a prerequisite for the retention of qualified personnel. Based on the research results, the authors identify the main directions for improving the agricultural and social policy (especially the work with rural families) to develop and improve the agricultural sector of the Republic of Bashkortostan.

Keywords Agricultural complex · Agriculture · Economy · Rural family · Republic of Bashkortostan

JEL Codes A14 · O1 · O4 · Q1

L. N. Gabdrakhmanova (✉) · E. S. Gareev · Y. N. Dorozhkin · A. A. Cheremiskin · G. A. Yakupova
Ufa State Petroleum Technological University, Ufa, Russia
e-mail: lianella2004@inbox.ru

E. G. Popkova and B. S. Sergi (eds.), *Geo-Economy of the Future*,
https://doi.org/10.1007/978-3-030-92303-7_46

1 Introduction

The agricultural sector plays a crucial role in the economy of the Republic of Bashkortostan. The Republic of Bashkortostan ranks 7th in Russia in terms of the volume of agricultural production. The Republic possesses 7.1 million hectares of agricultural land, which is 3% of the total area of agricultural land in Russia.

Rural areas of the Republic of Bashkortostan have certain potential, which, when used most productively and efficiently, can provide a high standard of living, quality of life, sustainable diversified development, and full employment of the rural population.

For effective development, planning, and management of the agro-industrial sector, it is necessary to investigate the strategic challenges that face the industry and can worsen soon. It is also necessary to identify key factors of sustainable development of the agricultural complex of the Republic of Bashkortostan. One of the factors determining the effective and stable development of the agro-industrial sector is the development of human potential.

In recent years, one of the main problems is the reduction of the rural population employed in the agro-industrial complex, which can negatively affect agricultural performance. The authors believe that the support of rural families should become a key policy of the Republic of Bashkortostan. One of the critical functions of the rural family is to provide food resources for the region and the country. This is especially relevant for the considered region, which has a considerable stock of land resources. Nevertheless, rural residents cannot effectively perform their role in current realities due to the demographic and socio-economic crisis.

Rural territories face the most significant outflow of rural families. The improvement of government policy in agriculture and increasing the rural population and rural families employed in the agricultural sector of the Republic of Bashkortostan will ensure the growth of indicators of the agricultural sector and determine the future of rural areas.

The employment dynamics of rural residents differ significantly from urban residents. Many rural residents work on private farms, which are based on manual labor, seasonal in nature, and have practically irregular working hours with no days off. Low income and the inability of rural residents to meet their needs and aspirations lead to a general deterioration in living standards: the motivation to work is reduced, and low self-esteem is formed, which negatively impacts the development of human potential.

Taking into account current realities, the rural family needs a comprehensive study of the features and trends of formation and development depending on the development specifics and ways to overcome socio-economic problems in the Republic of Bashkortostan.

2 Materials and Methods

In 2019, the production of agricultural products amounted to 167.1 billion rubles, which is 102% of the corresponding period last year (plan—101.6%). In 2020, there was an increase in production, which amounted to 177.1 billion rubles (6% higher than last year's level).

In the city, the functions of land are related exclusively to the location of production and social infrastructure facilities. In turn, in rural areas, land plays a special role as a production means and an object of labor. That is, the land is the primary source of rural production (Gaisina et al., 2019). In this regard, land contributes to the organization of everyday life and labor of rural households.

In connection with the ongoing transformation of the Russian agricultural complex (including the Republic of Bashkortostan), some processes are detrimental for the village and negatively impact agricultural performance.

Thus, the growth of rural population and rural families employed in the agricultural sector of the Republic of Bashkortostan will ensure the growth of indicators of the agricultural sector.

One of the factors of this transformation is the disintegration of the peasant class (Karimova et al., 2017). This term implies an increase in the share of the population not involved in agriculture, the emergence of agricultural holdings, and a decrease in the share of individual subsidiary farms. In 1995, the agricultural industry employed 10 million people (15.1% of all people employed in the economy). From 1995 to 2015, the number of people employed in the agricultural sector declined to 4.725 million (6.5%). As of 2019, the number of people employed in the agricultural sector was 4.196 million or 5.8% of all people employed in the economy.

Another feature directly affecting the development of agriculture is a significant drop in the standard of living of rural families. In rural areas, the population that lives below the subsistence level exceeds 50%. The level of well-being of rural residents of the Republic of Bashkortostan in terms of income remains low; the gap between urban and rural areas is increasing. According to 2018 data, the average wage per month in agriculture was 1953 rubles. Expenditures on food also significantly increased. This increase indicates poverty because the more a family spends on food, the poorer it is since it is forced to save on other needs. In 2017, rural families spent 38.9% of other expenses on food.

Reducing the number of ambulance stations, hospitals, post offices, kindergartens, schools, and other social institutions is also acute in rural areas. For example, from 2000 to 2020, the number of preschool educational organizations in rural areas of the Republic of Bashkortostan decreased by more than 65%, from 1284 to 433 units (Territorial body of the Federal State Statistics Service in the Republic of Bashkortostan, 2020, pp. 19–74). The same trend is observed in other socio-cultural facilities.

It is also necessary to mention the long-term agricultural crisis that started in Russia with the collapse of the Soviet Union and continues to this day. Over the past few years, the country's leadership has paid more attention to this issue; various state and regional programs to support young professionals in rural areas have been created. However, the system of import substitution is still quite undeveloped. The authors of this research believe that the reason for this is the consequences of liberal reforms, during which there was an abrupt and unprepared transition of agriculture to a system of market relations without consideration of the needs of the country's population. Since 1991, the area of agricultural crops in the country has almost halved (117,705 thousand ha to 74,759 thousand ha), with forage crops suffered the most (a decrease from 44,560 thousand ha to 16,974 thousand ha). Thus, we can conclude that agriculture is in a deep crisis.

The planned emergence of a system of farms did not happen either. Instead of state and collective farms, which completely ensured the food security of the USSR, people got private farms based on manual labor. Such transformations led to economic instability and mass unemployment, which force rural youth to migrate from rural territories to large cities, thus leading to the aging of the rural population and, consequently, a reduction in the proportion of women of reproductive age.

Nowadays, labor on the land is no longer the main and only source of earning money for villagers. Specific labor practices, such as logging, gathering, tourism, and so on, are developing. According to statistics, the percentage of rural employees leaving for seasonal work in large cities and district centers is growing (according to statistics, their number reaches up to 20 million people throughout the country). This trend leads to the destruction of the traditional way of life in the countryside.

3 Results

Based on these factors, we can conclude that the risk of falling into the group of poor and low-income rural families has significantly increased. The closure of social infrastructure facilities due to the optimization has caused a massive outflow of young people to large regional centers and cities, which leads to a reduction in the rural workforce employed in the agricultural sector (Yakupova et al., 2018). The demographic situation of the rural population is also marked with a decrease in the population due to a decrease in natural increase and migration losses. The main reasons for the decline in fertility are trends in current marital status, in particular:

- Decrease in the proportion of the married population;
- Increase in the divorce rate and decrease in the number of registered marriages;
- Increase in the age of starting a family and getting married;
- Increase in the share of single-parent families;
- Reduction of the number of children in families;
- Increase in the number of unregistered marriages.

According to the statistical collection "The Republic of Bashkortostan in Figures 2020," from 2005 to 2020, the number of the rural population in the Republic of Bashkortostan decreased from 40.21 to 37.77%.

There is a decline in the reproductive attitudes of young people in urban and rural areas.

Families predominantly have one or two children. Judging by demographers' estimates, families with more than three children are considered disadvantaged, which means an inability to provide a good education and difficulty in the subsequent realization of human capital.

Rural youth are more conservative than urban youth, but the number of unregistered marriages and out-of-wedlock births is also increasing. In 2019, the percentage of out-of-wedlock births in rural areas was 21.6% (Table 1).

Employment is an acute problem. Labor force survey statistics of the Federal State Statistics Service for 2018 showed that 12.4% of young rural residents are unemployed. This level of unemployment justifies the labor migration of rural youth to cities and other regions. In 2018, 148.1 thousand people from the Republic of Bashkortostan went to work in other places, primarily villagers. The results of a 2019 survey of rural youth conducted by the authors showed that more than half of those surveyed had some concerns about working in their region.

The ubiquitous policy of optimization, which has been the guideline of the managerial elite in recent years, leads to the destruction of rural infrastructure and contributes to the aggravation of unemployment. Over the past 15 years, there has been a reduction in preschool educational organizations from 1284 to 723. The same situation is observed for other objects of social infrastructure.

The authors believe that these factors can cause serious problems in the settlement of rural areas and the development of the agro-industrial complex of the Republic of Bashkortostan.

In recent years, the Republic of Bashkortostan shows interest in restoring and improving the potential of agricultural land. In 2019, considerable work was done in terms of technical modernization of production. The agrarians of the Republic purchased equipment for a total amount of 6.5 billion rubles (which is 2.3 billion more than in 2018). In terms of developing digital services in the agricultural sector, an agreement was signed with the Ministry of Agriculture of Russia on cooperation in developing the concept of the departmental project "Digital Agriculture." The Republic of Bashkortostan implemented an operational data monitoring project on the industry and geo-analytical field monitoring service. The Republic aims to achieve the growth of natural indicators, ensure the timeliness of agricultural activities, control work climate, and reduce the shortage of qualified personnel.

In 2019, systematic work began to support national products (horse breeding and beekeeping). Almost one billion rubles was allocated to support rural small businesses producing national products.

Despite the existing programs to improve the agro-industrial sector of the region and the allocation of cash subsidies ("New Farmer," "Family Livestock Farms"), the situation in rural areas is improving very slowly. Municipal authorities have no comprehensive approach to developing small businesses and increasing the share of young rural families employed in the agro-industrial sector of the Republic of Bashkortostan. Further development of agriculture is impossible without improving the quality of life of the rural population.

Table 1 The number and proportion of children born by marital status of the mother in the Republic of Bashkortostan, %

	2010		2011		2012		2014		2019	
Rural population										
Total	22,851	100.0	22,041	100.0	22,597	100.0	22,219	100.0	20,495	100.0
According to the joint statement	2175	9.5	2105	9.6	2128	9.4	2136	9.6	1878	9.2
According to the mother's statement	3294	14.4	3073	14.0	2974	13.2	2730	12.3	2555	12.5

Source Compiled by the authors based on (Territorial body of the Federal State Statistics Service in the Republic of Bashkortostan, 2020, pp. 17–38)

The research work "Studying the potential of young families in the municipalities of the Republic of Bashkortostan by focus-group method" revealed the most common problems of young families, which consist of ineffective assistance to young families. Such problems include housing conditions, insufficient incomes, the organization and modernization of the high-paying labor market, the lack of necessary and timely information about opportunities for participation in rural life, and receiving state support and additional sources of development (Lavreniuk & Chursina, 2019).

The authors of the research give the following recommendations for improving the agricultural and social policy of the Republic at all levels and increase the rural population:

- Increasing the level of employment of able-bodied rural population by opening new industries, increasing the number of jobs, increasing wages, and increasing the attractiveness of rural employment;
- Development of rural infrastructure, increasing the number of preschool and general education institutions;
- Organization of an effective institutional environment for the development of rural areas. The main components of the institutional environment include the development of agro-industrial production, reduction or abolition of taxes on small agricultural enterprises, the development of agricultural consumer cooperation as a measure of support and survival of small producers, getting rid of bureaucratic "red tape" unnecessarily burdening entrepreneurs. It is necessary to increase accessibility, transparency, and openness of the system of state support of agricultural producers, including information support. State support of the regional agricultural sector should be provided with the resources of the regional and federal budget;
- Allocation of targeted funds to solve pressing problems of rural residents, which should be determined through opinion polls at the local level;
- Measures to attract rural families to agricultural business, vocational training of unemployed citizens (including women on maternity leave), training of qualified managers of agricultural organizations of the Republic, competence-oriented training of specialists in the agricultural sector, taking into account the specifics of the regional economy;
- Growth of investment attractiveness of agriculture, industries, etc.;
- Development of rural tourism to increase employment of rural families and the share of profits of farms;
- Strengthening the institution of young families and the formation of family values among young people. Ensuring inter-agency interaction and coordination for the development of innovation and entrepreneurship of young rural families and the development of youth self-organization.

4 Discussion

One of the factors determining the sustainable development of the agricultural sector is the development of human potential (Ivanova, 2019). Many rural areas face an exodus of people, especially of those who have already formed families. The current rural economy and social infrastructure hinder the development of rural families. Due to an increase in the proportion of people not involved in agriculture, declining living standards, increasing risk factors for rural families contributing to poverty and low-income conditions, unemployment, low development of socio-cultural infrastructure, and reduced labor and professional socialization of the rural population, there is a threat of reducing the country's food security, which hinders the development of the agricultural complex of the Republic of Bashkortostan.

Most studies of rural families are conducted in terms of socio-economic and socio-demographic issues and deal with the problems of migration and unemployment, the decline in living standards and quality of life, and the content of value orientations. In works covering the problems of young rural families, the authors are interested in analyzing the problems of the rural population, youth (Gavrilyuk et al., 2016), and the family. Researchers see alcoholism and drug addiction, which require timely diagnosis and prevention, as an important problem in reducing the potential of young rural families to develop the Russian AIC. Researchers (Ustinova et al., 2016) demonstrate the ambiguous impact of globalization on the condition of young rural families. On the one hand, the process of digitalization facilitates education and the possibility of secondary income. On the other hand, it increases the migration of young people from rural areas to metropolitan areas. Researchers (Gaisina et al. 2017; Sekerin et al. 2018) point to the need for competence-oriented training of young professionals in the agricultural sector, taking into account the specifics of the regional economy.

5 Conclusion

Thus, the research revealed and analyzed the problems of the formation of public policy concerning rural families as a key factor in the development of the agricultural complex of the Republic of Bashkortostan. The authors identified and structured the main factors of the reduction of young people employed in the AIC. Additionally, they revealed the features and trends affecting the development of the agro-industrial sector of the Republic of Bashkortostan. The transformations occurring in rural families are civilizational and socio-economic factors. Due to historical circumstances, the civilizational factor cannot be affected significantly. The regional government must pay attention to socio-economic factors and help regulate them. The authors believe that the government's policy towards rural areas is contradictory—voiced statements about the revival of agriculture do not always correspond to reality. Nowadays, the Republic of Bashkortostan lacks a clear plan for the resettlement of rural areas.

Thus, the migration process is spontaneous. Rural settlements greatly vary in terms of population, social infrastructure, and economic development. It is necessary to consider the fact that rural families are heterogeneous in terms of social well-being. Therefore, public social support must be targeted and take into account the needs of families.

Our analysis identifies priority areas to improve the agricultural and social policy of the Republic of Bashkortostan and work with rural families to develop the agro-industrial sector of the Republic of Bashkortostan and Russia.

References

Gaisina, L. M., Barbakov, O. M., Koltunova, Y. I., Shakirova, E. V., & Kostyleva, E. G. (2017). Social management systems' modeling based on the synergetic approach: Methods and fundamentals of implementation. *Academy of Strategic Management Journal, 16*(S1), 83–95.

Gaisina, L. M., Shaikhislamov, R. B., Shayakhmetova, R. R., Kostyleva, E. G., Goremykina, L. I., & Gainanova, A. G. (2019). The essence and structural elements of a healthy lifestyle of students. *Espacios, 40*(21), 10.

Gavrilyuk, T. V., Garabazhii, V. A., Gaisina, L. M., Gabdrakhmanova, L. N., & Burenina, I. V. (2016). The dynamics of the global youth subcultures and movements: The Russian context. *Global Media Journal, 2016*, 1–9.

Ivanova, L. Kh. (2019). Republic of Bashkortostan: Development of human potential. *Analytical Bulletin of the Federation Council of the Federal Assembly of the Russian Federation, 7*(721), 23—35.

Karimova, R. M., Yakupova, G. A., & Yapparova, R. R. (2017). Structure and characteristics of the poor rural population of the Republic of Bashkortostan. In V. Sh. Guzairov, & I. V. Shavyrina, *Human resource management: Theory, practice, and perspectives* (pp. 315–321). Belgorod State Technological University named after V. G. Shukhov.

Lavreniuk, N. M., & Chursina, K. V. (2019). Interaction of young families and administrations of municipalities of the Republic of Bashkortostan. In N. K. Shamsutdinova, & E. V. Legotin (Eds.), *Republic of Bashkortostan. Demographic Report. Issue 3: Short version* (pp. 33–36). Lime.

Sekerin, V. D., Gaisina, L. M., Shutov, N. V., Abdrakhmanov, N. K., & Valitova, N. E. (2018). Improving the quality of competence-oriented training of personnel at industrial enterprises. *Quality—Access to Success, 19*(165), 68–72.

Territorial body of the Federal State Statistics Service in the Republic of Bashkortostan. (2020). Statistical collection "Republic of Bashkortostan in figures." Bashkortostanstat. Retrieved from https://bashstat.gks.ru/storage/mediabank/GHHcC2tr/komplexnyi-sbornik-2019.pdf. Accessed 1 February 2021.

Ustinova, O. V., Rudov, S. V., Kostyleva, E. G., Grogulenko, N. V., & Kulishova, N. D. (2016). The processes of globalization in the Russians' views. *Man in India, 96*(7), 2165–2177.

Yakupova, G. A., Valitova, N. E., & Gabdrakhmanova, L. N. (2018). The determining factors of functioning and development of rural youth families in Russia. *Vestnik of Economics and Management, 4*, 47–54.

Agro-Industrial Policy, Control, Supervision, and Management of Rural Territories for Sustainable Development of Geoeconomics

Gayk N. Aloyan, Albert T. Iskhakov, Anna N. Rida, and Olga A. Kuzmenko

Abstract The paper aims to identify the limitations of the contribution of agro-industrial policy, control, supervision, and management of rural territories to the sustainable development of agricultural economics and geoeconomics. Moreover, the paper aims to identify the prospects for improving agricultural economics and geoeconomics. The authors identify the advantages of institutions of agro-industrial policy, control, supervision, and management of rural territories associated, first, with the formation and high level of development of these institutions and, second, with a low disparity in the development of these institutions among countries with different levels of food security. The contribution of agro-industrial policy, control, supervision, and management of rural territories to the sustainable development of agricultural economics and geoeconomics is found low and even negative in most cases, which indicates the critical inefficiency of these institutions. The reason for the limited use of institutions is due to the focus on the interests of cities with insufficient attention to the interests of rural territories and the formal approach and low efficiency of agro-industrial policy, control, supervision, and management of rural territories. The prospects for improving these institutions are related to overcoming the identified limitations. The authors provide framework recommendations for overcoming the limitations identified.

Keywords Agro-industrial policy · Control · Supervision · Management of rural territories · Sustainable development · Agricultural economics · Geoeconomics

JEL Codes Q01 · Q15 · Q18 · O18 · R52

G. N. Aloyan (✉) · A. T. Iskhakov · A. N. Rida · O. A. Kuzmenko
Volga State University of Technology, Yoshkar-Ola, Russia
e-mail: kbua@inbox.ru

A. T. Iskhakov
e-mail: kbua@inbox.ru

A. N. Rida
e-mail: kbua@inbox.ru

O. A. Kuzmenko
e-mail: kbua@inbox.ru

E. G. Popkova and B. S. Sergi (eds.), *Geo-Economy of the Future*,
https://doi.org/10.1007/978-3-030-92303-7_47

1 Introduction

The agricultural economics and the related geoeconomics (land use and natural resource protection) occupy an important place among the Sustainable Development Goals (SDGs), as reflected in SDG 2 (zero hunger), SDG 10 (reduced inequalities), and SDG 11 (sustainable cities and communities). However, despite recognition at the UN level, agricultural economics and geoeconomics are developing at a slow pace compared to the implementation plan for the respective SDGs and other SDGs.

The problem lies in the fact that the existing scientific paradigm does not explain the logic (cause-effect relationships) of the development of agricultural economics and geoeconomics. Thus, during the COVID-19 pandemic and crisis, despite cuts in government funding, control, and oversight and the winding down of private environmental initiatives, there has been a reduction in pollution. Nevertheless, despite the continued work of institutions of agro-industrial policy, control, supervision, and management of rural territories, the gap between rural territories and urban areas (inequality of areas within countries and regions) has increased; the problem of global hunger and food security has worsened.

In this regard, it is necessary to create a new paradigm, the development of which is an urgent research area. This paper is intended to contribute to solving the problem (creating a new scientific paradigm). For this purpose, the authors put forward and test the hypothesis that agro-industrial policy, control, supervision, and management of rural territories are inconsistent, imperfect, and insufficiently contribute to the sustainable development of agricultural economics and geoeconomics. The paper aims to identify the limitations of the contribution of agro-industrial policy, control, supervision, and management of rural territories to the sustainable development of agricultural economics and geoeconomics. Moreover, the paper aims to identify the prospects for improving agricultural economics and geoeconomics.

2 Literature Review

The institutions of agro-industrial policy, control, supervision, and management of rural territories have been studied and described in sufficient detail in the existing literature, particularly in the works of Furmankiewicz et al. (2021), González-Corzo (2020), Krampe et al. (2021), Liu et al. (2021), Meyer et al. (2021), and Nurrochmat et al. (2020). The sustainable development of agricultural economics and geoeconomics concerning SDGs is studied in detail and presented in the works of Grajales (2020), Morozova et al. (2019), Morozova and Litvinova (2019), Popkova et al. (2021), Popkova and Sergi (2021), and Waeterloos (2020).

Nevertheless, the literature review revealed insufficient elaboration of causal relationships between the development of institutions of agro-industrial policy, control, supervision, and management of rural territories and sustainable development of agricultural economics and geoeconomics, which is a gap in the concept of sustainable development. This research aims to fill the identified gap.

3 Research Methodology

The authors apply econometric methods to test the hypothesis and determine the cause-effect relationships of the development of institutions of agro-industrial policy, control, supervision, and management of rural territories and sustainable development of agricultural economics and geoeconomics. The analysis of variation determines the degree of differentiation in the development of institutions of agro-industrial policy, control, supervision, and management of rural territories and sustainable development of agricultural economics and geoeconomics. The empirical basis of this research is presented in Table 1.

The method of correlation analysis is used to determine the relationships between the development of institutions of agro-industrial policy, control, supervision, and management of rural territories and the sustainability of the development of agricultural economics and geoeconomics. The research is based on a specially formed sample of countries with high (Finland, the Netherlands, and Japan), medium (Russia, Chile, and Brazil), and low (India, the Philippines, and Nepal) levels of food security in 2020 according to The Economist Intelligence Unit Limited (2021). The relevant statistics are presented in Table 2.

4 Findings

To determine the differences in the level of development of institutions of agro-industrial policy, control, supervision, and management of rural territories, the authors analyze the data variation from Table 1. The results of the analysis are presented in Fig. 1.

According to Fig. 1, the variation in the development of institutions of agro-industrial policy, control, supervision, and management of rural territories is generally low (e.g., 15% for nutrition standards) and moderate (ranging from 25.86% for food security programs to 42.38% for agricultural infrastructure). Only the political commitment to adapt agriculture to climate change is marked with relatively high variation (64.64%).

The authors conducted a correlation analysis to determine the impact of agro-industrial policy, control, supervision, and management of rural territories on the sustainable development of agricultural economics and geoeconomics. The detailed results of the conducted analysis are shown in Table 3.

Table 1 The level of development of institutions of agro-industrial policy, control, supervision, and management of rural territories in 2020

Country	Food safety programs	Agricultural research and development	Agricultural infrastructure	Food security and access policy commitments	Nutritional standards	Mechanisms to ensure food safety	Political commitment to adapting agriculture to climate change
	agr_1	agr_2	agr_3	agr_4	agr_5	agr_6	agr_7
Finland	100	76.3	72.5	100	100	100.0	95.4
Netherlands	100	33.8	82.8	50.0	73.5	80.0	95.4
Japan	100	30.7	87.5	50.0	100	100.0	83.8
Russia	100	47.9	45.8	50.0	73.5	100.0	50.8
Chile	75.0	40.8	40.1	50.0	73.5	100.0	57.0
Brazil	75.0	25.8	35.9	50.0	100.0	80.0	27.7
India	75.0	42.8	53.0	100.0	100.0	60.0	9.2
Philippines	50.0	33.5	21.9	50.0	73.5	40.0	20.0
Nepal	50.0	45.7	40.6	50.0	76.5	20.0	25.9

Source Compiled by the authors based on (The Economist Intelligence Unit Limited, 2021)

Table 2 Results in the sustainable geoeconomics in 2020

Country	Agriculture, forestry, and fishing, value added (% of GDP)	Food production index (2014–2016 = 100)	Rural population (% of the total population)	Irrigated agricultural land (% of total agricultural land)
	GEO_1	GEO_2	GEO_3	GEO_4
Finland	2.5	92.8	14.0	0.4
Netherlands	1.6	86.9	8.0	11.1
Japan	1.2	99.2	8.0	34.8
Russia	3.7	104.5	25.0	2.0
Chile	3.9	101.6	12.0	7.0
Brazil	5.9	107.7	13.0	2.9
India	18.3	111.8	65.0	38.1
Philippines	10.2	104.5	53.0	9.3
Nepal	23.1	106.6	79.0	29.7

Source Compiled by the authors based on (World Bank, 2021)

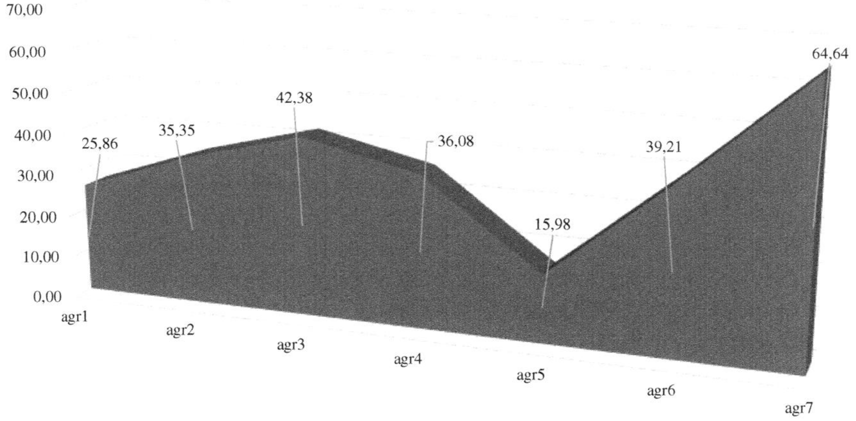

Fig. 1 Coefficients of variation in the development of institutions of agro-industrial policy, control, supervision, and management of rural territories, %. *Source* Calculated and compiled by the authors

Table 3 Detailed results of correlation analysis

Correlation	agr_1	agr_2	agr_3	agr_4	agr_5	agr_6	agr_7
GEO_1	−76.11	−0.05	−46.91	18.56	−5.10	−86.88	−76.17
GEO_2	− 60.22	−25.40	−69.10	4.14	15.96	−42.42	−91.91
GEO_3	− 75.82	4.99	−50.28	18.04	−15.59	−88.48	−75.48
GEO_4	− 20.87	−25.00	25.22	16.02	28.75	−40.44	−27.41

Source Calculated and compiled by the authors

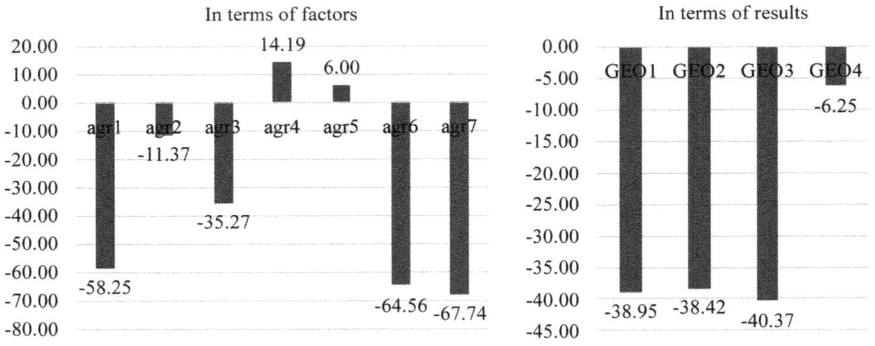

Fig. 2 Generalized results of the correlation analysis. *Source* Calculated and compiled by the authors

According to Table 3, most of the obtained correlation coefficients are negative. Therefore, agro-industrial policy, control, supervision, and management of rural territories negatively impact the sustainable development of agricultural economics and geoeconomics. The generalized results of the correlation analysis are shown in Fig. 2.

According to Fig. 2, only institutions such as food security and access policy commitments (correlation 14.19%) and nutrition standards positively contribute to the sustainable development of agricultural economics and geoeconomics (6%). The least negative cumulative contribution of the institutions of agro-industrial policy, control, supervision, and rural management is the development of irrigation (irrigated agricultural land: −6.25%). To clarify the economic meaning of the obtained results, let us conduct a qualitative (SWOT) analysis of the influence of agro-industrial policy, control, supervision, and management of rural territories on the sustainable development of the geoeconomics (Table 4).

The results of the SWOT analysis from Table 4 showed that the advantages of the institutions of agro-industrial policy, control, supervision, and management of rural territories are, first, the formation and high level of development of these institutions and, second, a low disparity in the development of these institutions among countries with different levels of food security. The shortcomings are related to the focus on the interests of cities with insufficient attention to the interests of rural territories and the formal approach and low efficiency of agro-industrial policy, control, supervision, and management of rural territories.

A certain threat is posed by further urbanization, the outflow of population from rural areas, and reduced emphasis on agricultural economics and geoeconomics amidst the escalation of other sustainable development issues. The opportunities, prospects, and recommendations of this research are related to the popularization of rural tourism as an incentive to increase the effectiveness of institutions and the development of AIC 4.0 and smart sustainable rural territories.

Table 4 SWOT analysis of the influence of agro-industrial policy, control, supervision, and management of rural territories on sustainable geoeconomics

Strong points (S)	Weak points (W)
– Formation and high level of development of institutions of agro-industrial policy, control, supervision, and management of rural territories; – Low disparity in the development of these institutions among countries with different levels of food security	– Orientation on the interests of cities with insufficient attention to the interests of rural territories; – Formal approach and low efficiency of agro-industrial policy, control, supervision, and management of rural territories
Opportunities (O)	**Threats (T)**
– Popularization of rural tourism as an incentive to increase the effectiveness of institutions; – Development of AIC 4.0 and smart sustainable rural territories	– Further urbanization and outflow of population from rural territories; – Diminished attention to agricultural economics and geoeconomics amidst the escalation of other sustainable development issues

Source Developed and compiled by the authors

5 Conclusions

The obtained results confirmed the posed hypothesis in an unexpected way. The contribution of the institutions of agro-industrial policy, control, supervision, and management of rural territories to the sustainable development of agricultural economics and geoeconomics was low and even negative in most cases, which indicates the critical inefficiency of these institutions. The reason for the limited use of institutions is due to the focus on the interests of cities with insufficient attention to the interests of rural territories and the formal approach and low efficiency of agro-industrial policy, control, supervision, and management of rural territories.

The prospects for improving these institutions are related to overcoming the identified limitations. It is recommended to popularize rural tourism as an incentive to increase the effectiveness of institutions and develop AIC 4.0 and smart sustainable rural territories. In future studies, it is advisable to approve the proposed recommendations and analyze the accumulated experience of improving the studied institutions.

References

Furmankiewicz, M., Hewitt, R. J., & Kazak, J. K. (2021). Can rural stakeholders drive the low-carbon transition? Analysis of climate-related activities planned in local development strategies in Poland. *Renewable and Sustainable Energy Reviews, 150*, 111419. https://doi.org/10.1016/j.rser.2021.111419

González-Corzo, M. A. (2020). Agrarian policy changes and the evolution of land tenure in Cuba. *Economic Papers, 39*(3), 239–258. https://doi.org/10.1111/1759-3441.12283

Grajales, J. (2020). A land full of opportunities? Agrarian frontiers, policy narratives and the political economy of peace in Colombia. *Third World Quarterly, 41*(7), 1141–1160. https://doi.org/10.1080/01436597.2020.1743173

Krampe, F., Smith, E. S., & Hamidi, M. D. (2021). Security implications of climate development in conflict-affected states implications of local-level effects of rural hydropower development on farmers in Herat. *Political Geography, 90*, 102454. https://doi.org/10.1016/j.polgeo.2021.102454

Liu, X., Liu, Z., Zhong, H., Jian, Y., & Shi, L. (2021). Multi-dimension evaluation of rural development degree and its uncertainties: A comparison analysis based on three different weighting assignment methods. *Ecological Indicators, 130*, 108096. https://doi.org/10.1016/j.ecolind.2021.108096

Meyer, E. L., Overen, O. K., Obileke, K., Botha, J. J., Anderson, J. J., Koatla, T. A. B., … Ngqeleni, V. D. (2021). Financial and economic feasibility of bio-digesters for rural residential demand-side management and sustainable development. *Energy Reports, 7*, 1728–1741.https://doi.org/10.1016/j.egyr.2021.03.013

Morozova, I., Litvinova, T., Mordvintsev, I. A., & Konina, O. V. (2019). The model of managing the development of agricultural machinery market infrastructure on the basis of public-private monitoring. In E. Popkova (Ed.), *The future of the global financial system: Downfall or harmony* (pp. 205–212). Springer. https://doi.org/10.1007/978-3-030-00102-5_21

Morozova, I. A., & Litvinova, T. N. (2019). The development of the agro-industrial complex in the cyber economy. In V. Filippov, A. Chursin, J. Ragulina, & E. Popkova (Eds.), *The cyber economy* (pp. 195–201). Springer. https://doi.org/10.1007/978-3-030-31566-5_21

Nurrochmat, D. R., Boer, R., Ardiansyah, M., Immanuel, G., & Purwawangsa, H. (2020). Policy forum: Reconciling palm oil targets and reduced deforestation: Landswap and agrarian reform in Indonesia. *Forest Policy and Economics, 119*, 102291. https://doi.org/10.1016/j.forpol.2020.102291

Popkova, E. G., Oudah, A.-M. M. Y., Ermolina, L. V., & Sergi, B. S. (2021). Financing sustainable development amid the crisis of 2020. A research note. In E. G. Popkova, & B. S. Sergi (Eds.), *Modern global economic system: Evolutional development vs. revolutionary leap.* (pp. 773–780). Springer. https://doi.org/10.1007/978-3-030-69415-9_88

Popkova, E. G., & Sergi, B. S. (2021). Energy efficiency in leading emerging and developed countries. *Energy, 221*, 119730. https://doi.org/10.1016/j.energy.2020.119730

The Economist Intelligence Unit Limited. (2021). Global Food Security Index. Retrieved from https://foodsecurityindex.eiu.com/Index. Accessed 8 September 2021.

Waeterloos, E. (2020). State-led agrarian reform in South Africa: Policy incoherencies and the concern for authoritarian populism. *Canadian Journal of Development Studies, 41*(3), 399–416. https://doi.org/10.1080/02255189.2020.1810000

World Bank. (2021). Agriculture and rural development. Retrieved from https://data.worldbank.org/indicator. Accessed 8 September 2021.

GPSR Compliance

The European Union's (EU) General Product Safety Regulation (GPSR) is a set of rules that requires consumer products to be safe and our obligations to ensure this.

If you have any concerns about our products, you can contact us on ProductSafety@springernature.com

In case Publisher is established outside the EU, the EU authorized representative is:

Springer Nature Customer Service Center GmbH
Europaplatz 3
69115 Heidelberg, Germany

Batch number: 10739612

Printed by Printforce, the Netherlands

Geo-Economy of the Future

Elena G. Popkova · Bruno S. Sergi
Editors

Geo-Economy of the Future

Sustainable Agriculture and Alternative Energy

Volume II. Transition to the Geo-economy of the Future based on Responsible Nature Use, Sustainable Communities and Alternative Energy

Springer

Editors
Elena G. Popkova
MGIMO University
Moscow, Russia

Bruno S. Sergi
University of Messina
Messina, Italy

ISBN 978-3-030-92302-0 ISBN 978-3-030-92303-7 (eBook)
https://doi.org/10.1007/978-3-030-92303-7

This Springer imprint is published by the registered company Springer Nature Switzerland AG
The registered company address is: Gewerbestrasse 11, 6330 Cham, Switzerland

Introduction to Volume I: Green Growth, Environmental Protection and Sustainable Agriculture in the Geo-economy of the Future

Sustainable Development of the Geo-economy and Environmental Management Best Practices

This book opens a new perspective on geo-economy, which for the first time means not economic geography and globalization (differences in the sociocultural and economic situation of countries—from the standpoint of the humanities), but land economics (from the standpoint of earth and planetary sciences). This makes it possible to systematize, in-depth study, and critically rethink current and future economic practices from the standpoint of consequences for the environment and sustainable development.

The uniqueness of this book lies also in the fact that, in contrast to the established environmental economy in science, which focuses on the problems of climate change and the reduction of production and consumption waste, the spotlight of this book is a geo-economy with a focus on responsible environmental management, sustainable agriculture, and alternative energy. Thus, this book fills a gap in the existing scientific knowledge at the intersection of economic geography and environmental economy, opening a new and wide field for scientific research on land economics.

The relevance of this book is explained by the fact that in the area of the new perspective of the geo-economy that has opened thanks to it, there is a wide range of acute, poorly studied, and unsolved issues. These issues include uncertainty about how to make the transition to a green economy and how to manage environmental management. The theoretical concept of the green economy is well developed and widely represented in the available scientific literature. However, it is not clear how to implement this concept in practice. The core of this concept is presented in the Sustainable Development Goals: SDG 12 (responsible consumption and production) and partly SDG 3 (in terms of the contribution of the green economy to health care).

This concept covers a wide range of areas of economic activity, because of which it has blurred lines, is not understood, and is not supported by many potentially interested economic entities in its implementation. For example, it is unclear whether the disposition of green investments is sufficient to qualify an enterprise as a subject of a green economy. The concept of a green economy does not give a clear and unambiguous answer to this question—the answer must be flexible and requires further in-depth scientific study from the standpoint of the geo-economy. For example, if an enterprise creates a large carbon footprint and causes serious damage to the environment, and its green investments are aimed at eliminating only superficial negative consequences of this damage, it is obvious that it cannot and should not belong to a green economy.

It also raises the question of what is meant by green investment. If the selection criteria are set too high, green investments will become unavailable for small and medium-sized enterprises (i.e., for mass implementation), and if the criteria are set too low, large businesses will formally approach green investments without actually contributing to the sustainable development of geo-economy. Consequently, the criteria should be flexible and take into account the specifics of different business practices and different business entities.

Among the problems under consideration is the ambiguity in the interpretation of the concept of "environmental quality," as well as the lack of a scientific and methodological approach to assessing and ensuring the environmental efficiency of economic practices. This concept is based on SDGs 13 (combating climate change), and 14 and 15 (biodiversity conservation). These concepts are based on SDG 12 (responsible consumption and production) and partially SDG 3 (in terms of the contribution of the "green" economy to health care). In particular, questions about whether and how to separate or combine environmental quality from the overall quality of goods and services need to be considered.

Likewise, it is worth considering whether environmental efficiency is viewed in isolation (and how to measure it in isolation) or in unity with the overall performance of an enterprise, country, and region. Although there are separate criteria for measuring and categorizing the given characteristics, coherent methodological support has not been developed yet. For practical application, the criteria and categories need clarification and should be supplemented with specified and detailed recommendations.

Separately, it is worth noting such an issue as the incompleteness of the concept of sustainable agriculture and the development of rural areas. This concept is based on SDG 2. In some cases, agricultural sustainability refers to the stability of the operation (normal, smooth functioning) of agricultural enterprises; in other cases, it refers to their contribution to the fight against hunger and ensuring food security.

Much less studied, but no less significant is the third area of sustainable development of agriculture and rural areas, namely the responsible land use. Although the responsible environmental management in the context of studying a new perspective of geo-economy applies to all sectors of the economy, it is most significant for agriculture, since, firstly, its conduct presupposes the closest naturality, and, secondly, agriculture has the greatest potential not only to reduce environmental damage but also to improve its condition based on restorative land use.

The described problems are comprehensively considered and studied in detail in the first volume of the book, and their promising solutions are also proposed. The purpose of the first volume of the book is to develop the scientific concept of the geo-economy of the future and to develop applied recommendations for its practical implementation while systematically considering the experience and prospects of sustainable green growth, environmental protection, and sustainable agriculture. The problems noted have determined the logic and structure of this volume of the book—they are sequentially explored in the first three parts (first, second, and third).

The first part is devoted to the green economy and environmental management with a detailed consideration of international experience (e.g., countries such as Russia and China) and industry practices. They also consider the health benefits of sustainable development of geo-economy and take into account the experience of sustainable development of geo-economy in the context of the COVID-19 pandemic and crisis. The second part examines the environmental quality and environmental efficiency of business practices. The third part explores sustainable agriculture and rural development.

The original perspective of the study of geo-economy provides a multidisciplinary approach to this book, due to which we hope it might be of interest and useful for representatives of various fields of science, including economic geography, environmental economics, agricultural sciences, business economics, regional economics (in terms of rural tourism and rural development), as well as the earth and planetary sciences.

Introduction to Volume II: Transition to the Geo-economy of the Future based on Responsible Nature Use, Sustainable Communities and Alternative Energy

A Scientific View of the Transition to the Geo-economy of the Future

The geo-economy of the future develops quickly but faces many problems on its path. One of the important problems is the lack of a clear idea of how the state regulation of sustainable development of geo-economy should be carried out, as well as what technologies should be used for responsible environmental management. This indicates a lack of scientific knowledge on how to build a system of state and corporate governance for the sustainable development of the geo-economy. This concept is based on SDG 16 (rule of law, justice, and strong institutions). As the practical experience of several countries and regions shows, often the state and corporate governance of geo-economy are in an imbalance—they are carried out separately and contradict each other, reducing overall efficiency.

We should also pay attention to such a problem as distancing from the practice of the existing concept of responsible industries, sustainable regions, cities, and communities. The basis of this concept is SDG 11. Issues such as empirical and engineering support—the development of guidelines for action on the organization of responsible environmental management by business entities and territories—deserve attention. Also, the accumulated experience of establishing responsible industries, sustainable regions, cities, and communities needs to be collected, generalized, and broadcasted among the academic community and beyond.

And finally, the last (but no less significant) of the considered issues is related to the incompleteness of the concept of sustainable and alternative energy. The core of this concept is SDG 7. In relation to this concept, the question of how to establish and institutionalize selected practices of using alternative energy and how to achieve a full-scale transition to sustainable energy deserves attention.

The existing concept offers only particular recommendations, most of which are academic. We need practice-oriented and universal recommendations, as well as ready-made frameworks. Also, the question of how to combine clean energy with other priorities for sustainable development of geo-economy, for example, environmentally responsible agriculture, needs to be considered.

The above problems are studied comprehensively and solved in Volume II of this book. The goal of Volume II is to form a systemic scientific view at the transition to the geo-economy of the future based on responsible nature use, sustainable communities, and alternative energy. This goal predetermined the structure of Volume II, which consists of three parts (fourth, fifth, and sixth). The fourth part focuses on responsible industries, sustainable regions, cities, and communities. The fifth part explores sustainable agriculture and rural development. The sixth (final) part of the book studies sustainable and alternative energetics.

Volume II (similarly to Volume I) is multidisciplinary—it contains studies that would be of high interest for representatives of economic geography, corporate economics, regional economics (in the aspect of sustainable development of cities and regions), sociology and social sciences (in the aspect of sustainable communities), management (in particular state and corporate governance), and earth and planetary sciences.

Contents

Environmental Quality and Ecological Efficiency of Economic Practices

State Legal Regulation and Technologies of Responsible Nature Use

Role of Government Policies in Improving Ecological Management of Industrial Economy

Gulzat K. Kantoroeva, Elena V. Kletskova, and Leila K. Bittieva

Abstract This study focuses on studying government policies, the importance of creating ties between politics, environment, and economy, and the policies of local authorities concerning industrial ecology. The study is based on the concept of political science—an approach that studies the government's understanding of government-citizen interactions. All competing definitions of public policies state that the policies are a result of decisions made by governments and that the government's inactions are equal to its actions. Modern analysts recognize the general need for a holistic and comprehensive approach to public policy.

Keywords Public policy · Environmental policy · Industrial economy · Industrial innovation · Industrial system

JEL Codes Q5 · Q52 · Q58

1 Introduction

Public policy is a process that includes several decisions. These decisions do not necessarily result from a conscious choice. The constraints imposed on governments can limit the range of possible options and alternatives, affecting their ability to implement decisions. Presenting public policy as a deliberate and target behavior of governments provides the standard for evaluating public policy.

Since this phenomenon is complex and consists of numerous decisions by numerous people and organizations, there are several strategies that only allow highlighting a limited range of factors. Research on politics mainly focuses on the impact and content of politics, concerning the nature of the policy tools at the disposal

G. K. Kantoroeva (✉)
Kyrgyz National University named after Zhusup Balasagyn, Bishkek, Kyrgyzstan

E. V. Kletskova
Altai State University, Barnaul, Russia

L. K. Bittieva
Kabardino-Balkarian State Agrarian University named after V.M. Kokova, Nalchik, Russia

E. G. Popkova and B. S. Sergi (eds.), *Geo-Economy of the Future*,
https://doi.org/10.1007/978-3-030-92303-7_48

of governments (Nekhvyadovich et al., 2020). However, it is important to examine the context of social and governmental institutions and the values that these institutions embody, as they determine both how a problem arises and the solutions to it (facilitating, hindering, or prohibiting certain decisions).

In this case, we assume that both individuals and institutions play a crucial role in policymaking, although in specific circumstances, one of them may be more dominant. Institutions are defined as the structures and organization of the state and society in the international system. The internal organization of the government partially determines its capabilities (Bogoviz et al., 2020). The potential of the government is also defined by its connection with society and the issues that it should solve using relevant policies. To implement policies effectively, the government needs the support of important social groups. The extent of support that these groups can offer depends on their own internal organization.

Unity within and among social groups creates a stable political environment that facilitates public policy development and implementation.

2 Methodology

Among social groups, business often plays the most important role in determining the political capabilities of the government, mainly because of the vital role of business in manufacturing. Manufacturing and production are fundamental activities in every society, the consequences of which extend far beyond the economy. The unmatched strength of businesses lies in the structural power of capital—the economic system where means of production are concentrated in enterprises and corporations. Globalization, caused by the rapid progress in communication and transport, as well as lower control over international economic operations, contributed enormously to the strengthening of the power of capital (Alekseev et al., 2018). Financial contributions to political parties and research provide additional leverage in policymaking. Ultimately, the power and influence of businesses can both undermine and advance social welfare (Glotko, Voronkova, et al., 2019).

In recent decades, governments at all levels introduced numerous environmental policies in response to the negative effects of economic development. Most of the current environmental policies have been implemented over the past 30 years and have been driven by human health and safety concerns. Now, governments consider new environmental policy issues that primarily aim to prevent pollution rather than control it (same as the industrial ecology). One of such issues is the way to promote ecological safety. Let us present the issue in the following way:

- Environmental policies present a prescriptive approach to integrating economic and ecological goals in the framework of a developed industrial economy. A developed industrial economy does not consider conservation and ecological safety to

be a limit to growth. Rather, it recognizes that new forms of environmental policies may positively impact economic growth. Similarly, the application of environmental policies should aim to leverage the power of businesses to generate environmental benefits (Costanza, 1996).
- New forms of political intervention that rearrange economic development (e.g., via spatial redistribution, sector balancing, changing technological composition, or readjusting the resource-intensity of socio-economic activities) can lower the environmental impact of businesses.
- The market failure theory offers several types of policy variants: regulation, incentives, disincentives, liability, market resolutions, and education and information for corporations and consumers. Such policies respond to insufficient or excessive use of environmental resources. Voluntary agreements and communal visions are two additional types of policy variants implemented by OECD countries. These recent policy innovations are particularly effective at the local and regional levels (Avkopashvili et al., 2019).

To apply prescriptive thinking to policies that stimulate industrial innovation, one must understand the logic behind the effectiveness of such policies. Efficiency depends on the extent to which the motives of enterprises and corporations are taken into account. Firms are assumed to have complex profit-maximizing motives, simplified to "lower costs" and "higher profits." Governments may choose to manage firms through traditional interventions via market failure policies or alternative innovative policies. The chosen measures must effectively motivate businesses to integrate economic and ecological issues in industrial production (Ragulina et al., 2019).

Local authorities (both municipal and regional ones) are often considered to be "closer to the people." More and more experts believe that the detailed development and implementation of policies that account for environmental and economic concerns are more likely to be achieved at the local government level. Moreover, local socio-economic and environmental conditions may require a specific blend of policies. Farthing described three current changes that led to higher interest in local environmental policies, which can additionally impact social and economic policies:

1. Despite the importance of global environmental issues, the costs and profits from environmental benefits available to the population, as well as its impact on the said population, are spatially limited.
2. It is increasingly recognized that the lifestyle of people within settlements can significantly affect the environment.
3. Local government policies that are not traditionally viewed as environmental may have significant environmental consequences (Glotko, Sycheva, et al., 2019).

In addition, local governments play an important role in existing international and national policy statements.

Several researchers have identified the stages of local environmental policy.

1. In the first stage, before the 1980s, environmental policies were fragmentary and largely unrecognized.

2. In the second stage, the term "sustainability" attained a higher status, and the environmental policies were more cohesive. Local authorities started developing "green" strategies, internal audits, and environmental reports; however, the response was largely passive.
3. In the third, current stage, local authorities implement more active environmental policies, such as the following:

   - Sustainable Development Goals and monitoring indices for progress in terms of sustainable development;
   - Key positions imposed on economic development by local environmental policies (encouraging the development, production, promotion, and use of clean technologies and processes, such as those that minimize resource use, waste production, and pollution; allow to reuse or recycle waste; amend the previous damage to the environment; measure and control the state of the environment).

It is also important to monitor and regulate the environmental performance of industrial production in ways that prevent environmental damage and improve the environment, create jobs, raise awareness and understanding of environmental issues in the business community, and encourage responsible attitudes.

Local governments increasingly incentivize rather than provide goods and services. Local governments take the role of catalysts, planners, coordinators, educators, and promoters of sustainable development. Adapting to these roles, combined with a holistic approach to policy that integrates economic development and environmental protection, is thought to contribute to a strong, sustainable local economy (Stroiteleva, Petrova, et al., 2020).

The development of existing eco-industrial parks (EIP) occurred in areas with strong sustainability trends; however, the role of local governments in these areas varied significantly from case to case. Factors influencing the type and level of involvement include: the internal organization of the local government, the strength of its ties to the business community, the level of public financial commitment, the planning processes in place, and whether a project is a private initiative or a public economic development project. In the case of public–private partnerships, local governments should play a central role in the development and management team.

Government codes and regulations aim to ensure public health and safety, promote responsible land use, and protect community values. However, they can hinder creative development, even if its goal is to achieve greater environmental protection and the health and safety of workers and communities. The environmental action initiative, which works with communities toward the industrial development that suits the local environment, warns that often the biggest challenge for communities interested in alternative economic strategies is working with existing rules and constraints that use conventional economic development strategies (Tashkulova & Kletskova, 2020).

Today, regulatory barriers are subdivided in two groups:

1. Those impeding the structural changes necessary for industrial parks to function more akin to ecosystems;
2. Those impeding activities necessary for the development of less energy-dependent EIP.

Structural design codes and regulations can become an obstacle when they do not account for the type of proposed development or are inflexible to other options. Regulations can act as a disincentive to environmentally beneficial creative designs (e.g., mandatory infrastructure solutions that can add unnecessary costs to a project and consume more resources and green space) (Vukovich et al., 2018). Examples of codes that often inhibit environmental innovation include zoning regulations, building codes, and parking requirements. In many cities, industrial districts are distant and unwanted, which increases infrastructure and transportation costs as well as commutes for workers. Colocation of buildings to take advantage of extra heat, energy, water, or materials; reduction of construction footprints; or the use of alternative methods for wastewater treatment often incur additional costs for developers since they require a lengthy process of obtaining permits. Parking requirements are often set based on the area of the facility to accommodate peak usage. This can result in underutilized areas that increase stormwater runoff and negatively impact the environment. However, overcoming normative obstacles to structural changes is easier than the obstacles to changes in processes and activities. For example, the conceptual transformation of waste into a resource requires appropriate government signals at all levels to close material loops.

3 Results

The key to operating EIPs as an "enlightened government agency that provides flexibility combined with voluntary guidelines" boils down to the Dutch model of local economic development as a template for public–private cooperation in achieving environmental goals. The model states that the main tasks of the government in incentivizing industrial ecosystems are designing and providing adequate infrastructure. Besides funding specific projects, there are four areas in which government agencies can support EIP development.

The first is improving regulatory measures aimed at discouraging pollution and encouraging greater resource efficiency. This requires higher regulatory flexibility and more clearly defined terminology to account for the byproduct exchanges, contributing to the overall state of the environment.

In addition, governments can encourage appropriate designs and activities via policy statements, guidelines, or codes of practice. Governments should have effective waste exchange programs and may regulate bans on certain materials in landfills (Tsvetkov et al., 2019). Therefore, municipal and regulatory land-use planning must consider the ecological functions of the landscape and plan accordingly.

The second is to develop appropriate market incentives, potentially including taxes and technology grants, to encourage the adoption of eco-industrial approaches. Financial disincentives with the aim of generating revenue and influencing behaviors that are effective at the EIP level include: business employment taxes, water use fees, sewer use fees, dumpsite fees, and single-use packaging taxes. Economic incentives include sales tax credits or rebates for the use of byproducts, used resources, and cleaner production technologies, as well as environmental credits and grants (Shenshinov, 2012). However, local economies are becoming increasingly integrated and dependent on global economic development, so the economic system cannot become more sustainable through local action alone. At a broader level, economic incentives for industrial ecosystems include: taxes on primary materials, energy taxes on non-renewable energy sources, special levies on non-recyclable materials, deposit refund fees, quantitative user fees, and quotas.

The third is to develop and provide guidelines for EIP development. Education, outreach, and research mechanisms aimed at fostering understanding and awareness of eco-industrial economic development business models can contribute to EIP development. The availability of easy-to-use information about recycling, reuse, and reclamation opportunities, material substitutes, and conservation practices is one of the most significant barriers to business change. In fact, local governments play an important role in information management (Merdesheva et al., 2020). Web-based information sources, publications, hotlines, and workshops are all effective ways to communicate information about waste exchange opportunities, relevant technologies, regulatory frameworks, incentive programs, and logistical information. The task of the local government is to keep itself and its representatives informed.

Policies have identified the importance of material flow data for public decision-making. The interpreted data improves natural resource decisions, takes sustainability into account, helps policymakers see the big picture, and shows where materials come from, how they are used, and where they end up. EIP is an open economic-environmental system that requires a systematic approach, involving an understanding of values as well as the physical and chemical characteristics of materials and energy flows. Easily accessible data on these flows at the regional and local levels could contribute to EIP growth.

The last one is to partner with public and private organizations to leverage public, financial, and informational resources. The industrial ecology paradigm is closely related to other innovations in environmental policies, planning, and management. Therefore, they should be used to strengthen the foundation of the industrial ecosystem (Stroiteleva, Kletskova, et al., 2020). Public sector experience has shown that partnering with the private sector and decentralizing R&D solutions have clear benefits, contributing to significant achievements.

4 Conclusion

Promoting eco-industrial parks will require significant changes in the nature and focus of local policies and regulatory frameworks. Links between the environment and economic policy are an important step. Long-term planning for the ecological restructuring of industrial production can replace compensatory environmental protection, which is becoming increasingly costly.

The success of green business and eco-industrial network planning demonstrates that economics and the environment do not conflict. Policy choices are clearly changing as government officials, industry, and lawmakers exchange ideas and experiences across international borders. A systemic and pragmatic approach of industrial ecology to analyzing the big picture can help improve sustainability-oriented policies.

References

Alekseev, A. N., Savenkov, D. L., Shvakov, E. E., Lebedev, N. A., & Ragulin, A. D. (2018). Financial problems of regional development and the ways of their solution in contemporary Russia. *Quality—Access to Success, 19*(S2), 71–75.

Avkopashvili, P. T., Polukhin, A. A., Shkodinsky, S. V., & Poltarykhin, A. L. (2019). The fundamental provisions of the concept of knowledge economy. In E. Popkova, Y. Ragulina, & A. Bogoviz (Eds.), *Industry 4.0: Industrial revolution of the 21st century* (pp. 57–64). Springer. https://doi.org/10.1007/978-3-319-94310-7_5

Bogoviz, A. V., Shvakov, E. E., Tretyakova, O. G., Zakharov, M. Y., & Abramov A. N. (2020). Globalization of education in the conditions of formation of the global knowledge economy: Regularities and tendencies. In E. Popkova (Ed.), *Growth poles of the global economy: Emergence, changes and future perspectives* (pp. 993–1000). Springer. https://doi.org/10.1007/978-3-030-15160-7_99

Costanza, R. (1996). Ecological economics: Reintegrating the study of humans and nature. *Ecological Applications, 6*(4), 978–990.

Glotko, A. V., Sycheva, I. N., Dunets, A. N., Poltarykhin, A. L., Zhuravlev, P. V., & Tubalets, A. A. (2019). Integration processes in the agro-industrial complex of the regions: Development and environmental problems. *Journal of Environmental Management and Tourism, 10*(3(35)), 613–621.

Glotko, A. V., Voronkova, O. Y., Miheeva, N. M., Skoblikova, T. V., Alekhina, N. A., & Zhminko A. E. (2019). The economic development of a transboundary region under the conditions of the Eurasian economic union. In *Proceedings of the IBIMA 2019: Education Excellence and Innovation Management Through Vision 2020* (pp. 7957–7968). Granada, Spain.

Merdesheva, E. V., Titova, O. V., & Avkopashvili, P. T. (2020). Methodological approach to the classification of digital economies by environmental efficiency and sustainability criterion. In E. Popkova, & B. Sergi (Eds.), *Scientific and technical revolution: Yesterday, today and tomorrow* (pp. 1360–1368). Springer. https://doi.org/10.1007/978-3-030-47945-9_148

Nekhvyadovich, L., Kostin, K. B., & Mironov, V. V. (2020). Government and public monitoring of entrepreneurship in a multicultural region: New perspectives in the conditions of the digital economy. In E. G. Popkova, & K. V. Vodenko (Eds.), *Public administration and regional management in Russia. Challenges and prospects in a multicultural region* (pp. 147–154). Springer. https://doi.org/10.1007/978-3-030-38497-5_17

Ragulina, Y. V., Semenova, E. I., Avkopashvili, P. T., Dmitrieva, E. A., & Cherepukhin, T. Y. (2019). Top-priority directions of implementing new internet technologies on the territories of rapid economic development. In E. Popkova, & V. Ostrovskaya (Eds.), *Perspectives on the use of new information and communication technology (ICT) in the modern economy* (pp. 182–188). Springer. https://doi.org/10.1007/978-3-319-90835-9_21

Shenshinov, Y. V. (2012). Labour quality management. *The Journal "North-East Asia academic Forum"*.

Stroiteleva, T. G., Kletskova, E. V., & Balaeva, S. I. (2020). Ontogenesis of green development model of digital economy: Essence, stages, and prospects. In E. Popkova, & B. Sergi (Eds.), *Scientific and technical revolution: Yesterday, today and tomorrow* (pp. 410–417). Springer. https://doi.org/10.1007/978-3-030-47945-9_46

Stroiteleva, T. G., Petrova, E. A., & Yungblyudt, S. V. (2020). Responsible information society as a social environment for creation of the "green" digital economy: Formation and monitoring. In E. Popkova, & B. Sergi (Eds.), *Scientific and technical revolution: Yesterday, today and tomorrow* (pp. 402–409). Springer. https://doi.org/10.1007/978-3-030-47945-9_45

Tashkulova, G. K., & Kletskova, E. V. (2020). Place and role of human in the system of circular reproduction in the digital regional economy. In E. Popkova, & B. Sergi (Eds.), *Scientific and technical revolution: Yesterday, today and tomorrow* (pp. 943–951). Springer. https://doi.org/10.1007/978-3-030-47945-9_101

Tsvetkov, V. A., Loginov, E. L., & Raikov, A. N. (2019). Integrated network interaction of education with science and production. *Educational Technologies and Society, 19*(1), 389–398.

Vukovich, G. G., Makuschenko, L. V., Bateykin, D. V., Titova, O. V., & Dobrosotskiy, V. I. (2018). Support of the territories of advanced economic development on human capital: Theory and practice. *Quality—Access to Success, 19*(S2), 157–160.

Environmental Investments as a Form of Social Responsibility of Business in a Mining Region

Evgeny E. Shvakov, Irina S. Trapeznikova, and Maria G. Sukhova

Abstract The need for sustainable economic development requires a rethinking of current economic practices, including approaches to the organization and implementation of the investment. Foreign and Russian companies widely implement the practice of financing projects within the framework of corporate or social responsibility of business. In Russian practice, these projects are often implemented to form a favorable public opinion on the company's activities. In this case, the adverse effects generated by the production activities of enterprises often remain outside the framework of the implemented projects. A striking example of such social initiatives is the implementation of social responsibility measures by mining enterprises. In mining regions, the social responsibility of business must be associated with overcoming the damage arising from production activities. The social responsibility of business in the mining regions should be considered an alternative direction of financing to preserve the environment. This circumstance makes it necessary to determine the damage caused by mining enterprises and correlate it with the measures implemented to overcome it. It is necessary to develop methodological approaches interconnecting these processes. Moreover, it is necessary to develop objective criteria and indicators of implementing the social responsibility of business, considering the goals and objectives of sustainable development of territories with the developed mining industry. The sufficiency of investment in the framework of social responsibility of business should be the key indicator of the developed methodological approach. This paper aims to justify the approaches to the development of this indicator.

Keywords Social responsibility of business · Sustainable development · Economy of a mining region · Environmental investments

JEL Codes Q5 · Q51 · Q59

E. E. Shvakov (✉)
Altai State University, Barnaul, Russia

I. S. Trapeznikova
Kemerovo State University, Kemerovo, Russia

M. G. Sukhova
Gorno-Altaisk State University, Gorno-Altaisk, Russia

E. G. Popkova and B. S. Sergi (eds.), *Geo-Economy of the Future*,
https://doi.org/10.1007/978-3-030-92303-7_49

1 Introduction

Traditionally, investments are considered long-term injections of funds to make profits. This notion of investment is well-established in current scientific research. At the same time, the practice of industrial enterprises contains a significant amount of funds invested in various projects not related to making profits and going beyond their production activities. The Soviet practice of economic organization recognized these forms of investment as a necessary element of the economic activity of any enterprise. A considerable part of social facilities (childcare, healthcare, cultural, and educational facilities) was created and maintained at the expense of enterprises. In a market economy, investments of this form proved unacceptable. Nowadays, projects of this kind are usually implemented as part of the social responsibility of business (Choi et al., 2018). These projects are often initiated and implemented to meet any needs of the territory in which the company carries out its production activities (territory of presence) (Dinius & Vergara, 2011; Littlewood, 2014; Syaparman et al., 2019). Investment activities within the framework of implementing the measures of social responsibility of business have become an influential factor in regional development (Dutta et al., 2020), including from the perspective of implementing the concept of sustainable development of territories. Consequently, this circumstance is also reflected in the transformation of the understanding of investment. According to this understanding, the injection of capital is recognized as an investment if any positive social effect is achieved.

2 Materials and Methods

The measures currently taken by the authorities to stabilize the environmental situation in the mining regions are insufficient and do not ensure the implementation of the principles of their sustainable development (Islamutdinov, 2016; Melnik et al., 2020). This necessitates the search for additional funding sources for environmental protection and environmental management at the state and regional levels and at the level of enterprises engaged in mining. In this regard, the Kemerovo Region is an excellent example since it is marked with a developed mono-industry economy specializing in coal mining and processing. This subject of the Russian Federation mines about 40% of all coal mined in the country. For the Kemerovo Region, the issues of investment in environmental protection, organization of rational nature management, and prevention of damage caused by the activities of coal mining companies are related to the need to find new sources of funding for environmental protection and environmental investment programs. The goal of these programs is to reduce environmental pollution and ensure that environmental policies and regional economic policies are complementary.

These additional sources of funding for environmental measures are the funds of coal mining companies, invested as part of the social responsibility of business.

Nowadays, the enterprises in the region minimally invest in environmental protection and take minimal measures within the framework of social responsibility of business. This situation with the financing of environmental measures in the framework of social responsibility of business is typical for all developing countries (Kim & Oh, 2019; Zheng et al., 2014). To assess the participation of coal mining business and authorities in the financing of environmental measures in the Kemerovo Region, let us calculate the relevant coefficients:

- Coefficient of participation in the financing of environmental measures by state and regional authorities (K_1);
- Coefficient of participation in the financing of environmental measures by coal mining enterprises (K_2).

The coefficients are calculated according to the following formulas:

$$K_1 = I_g/I_o, K_2 = I_p/I_o,$$

where:

I_g—financial costs for the implementation of environmental measures at the expense of budgetary funds;
I_{pg}—financial costs for the implementation of environmental protection measures at the expense of coal mining enterprises;
I_{og}—total financial costs for the implementation of environmental measures in the Kemerovo Region.

The calculations performed using the official data of the Federal State Statistics Service (Federal State Statistics Service, 2019) are presented in Table 1.

Table 1 shows that the coefficient characterizing the share of coal mining enterprises in costs for environmental protection ranges from 0.1 to 0.3.

Table 1 Evaluation of the participation of government and business in the financing of environmental protection in the Kemerovo Region

Year	Coefficient of participation in the financing of environmental measures by state and regional authorities (K_1)	Coefficient of participation in the financing of environmental measures by coal mining enterprises (K_2)
2012	0.737	0.263
2013	0.71	0.29
2014	0.84	0.14
2015	0.78	0.22
2016	0.81	0.19
2019	0.894	0.106

Source Calculated by the authors based on (Federal State Statistics Service, 2019)

Simultaneously, the vast majority of costs (70–90%) associated with the implementation of environmental protection measures are financed from the budget. Government activities do not harm the environment. The main economic actors whose activities negatively influence the main elements of the natural environment (atmosphere, air, and soil) are coal mining companies. The allocation of responsibility for financing environmental measures defies economic logic. Revenues from coal mining remain at the disposal of businesses, and financing of environmental protection measures is carried out at the expense of budgetary funds.

3 Results

In our opinion, all economic entities receiving economic rent from coal mining should participate in the financing of environmental measures:

- Coal mining companies (the form of their rent is the profit remaining at their disposal);
- Government (part of the natural rent, withdrawn in the form of taxation);
- Population (wages of workers at coal mining companies are also part of the natural rent).

According to independent consulting agencies, the government withdraws 25% of the natural rent from coal mining in the form of taxes. This value also includes the tax on the income of those working at coal mining companies. This tax, which is used to finance environmental protection measures, allows us to consider the population as a participant in environmental protection activities. Thus, the normal situation should be just the opposite of the current situation. The government should provide 25% of funding for environmental protection measures, and the remaining 75% should come from the coal mining companies. According to Table 1, if we take the value of the coefficient of participation in the financing of environmental measures by coal mining companies as a normative equal to 0.75, the annual financial investments of these companies in environmental protection should see at least 2.5-fold increase. If we take the 2019 figures, the increase will be more than seven times. At the same time, the distribution of costs for financing environmental measures is only the first aspect of this problem.

The second part of the problem lies in determining the amount of investment in environmental protection measures by coal mining companies as part of the implementation of their social responsibility. The application of the mechanism of social responsibility in the financing of environmental measures is due to their non-binding nature. However, it should be noted that environmental protection in the framework of social responsibility is carried out both in developing countries (Makarenko et al., 2019) and in countries with developed market economies (Wenk, 2006). Evidently, the calculated amount of investment in environmental protection as part of the implementation of social responsibility of businesses to the territory of their presence should be linked to the amount of damage caused by the activities of coal mining enterprises (Druzhinin et al., 2020; Dutta et al., 2020). When calculating the level of investment in environmental protection measures, it is necessary to consider the neutralization of damage in the following parts:

(1) Negative impact on the quality of atmospheric, air, and water resources, which accounts for at least 10–12% of the GRP;
(2) Negative impact caused by anthropogenic disturbance of land during open-pit coal mining, which reaches 3–4% of the GRP;
(3) Negative impact caused by the morbidity of the population in the region due to the environmental factor, which reaches 8% of the GRP.

To calculate the required amount of investment in environmental protection measures from the perspective of their fair distribution among the economic entities involved in the distribution of rents from coal mining, we propose to proceed from the calculation of the coefficient of the sufficiency of investment in environmental protection within the framework of social responsibility of business.

We propose to calculate this indicator by the following formula:

$$K_{di} = (I_t + N_t)/U_t,$$

where:

I_t—financial costs for the implementation of environmental protection measures as a part of the social responsibility of coal mining enterprises in the current calendar year in the territory of their presence (Kemerovo Region);
N_t—tax payments to the budget from coal mining enterprises in the current period (calendar year);
U_t—total damage caused by enterprises in the area of operation in the current year.

The benchmark value of the calculated indicator should be at least 0.75, since the government should carry out 25% of the investment in environmental protection measures as one of the recipients of the rent from coal mining. This simple transformation of the formula allows us to calculate the necessary level of investment in environmental protection measures within the framework of social responsibility of business, meeting the criterion of sustainable territorial development. A complete leveling of the damage caused to the natural and social sphere resulting from coal mining activities is taken as the criterion of sustainable territorial development.

$$I_t = 0.75\,U_t - N_t.$$

The proposed indicator is quite simplified, but it allows us to avoid complicated and lengthy mathematical calculations and determine whether the activity of business corresponds to the principles of social responsibility from the position of ensuring the sustainable development of the territory of presence. In turn, the total amount of investment by the government, region, and municipalities must be at least 25% of the level of business investment in environmental protection measures as part of its social responsibility following the criteria of sustainable territorial development.

4 Conclusion

According to expert estimates, the level of budgetary expenditures for implementing environmental protection in the Russian Federation will remain low and will amount to 0.2–0.3 of GDP by 2025 (Trapeznikova, 2019). Considering the measures implemented by enterprises and aimed at the disposal of accumulated industrial waste, reduction of harmful industrial emissions into the environment, and the restoration of the environment, the cost of environmental protection activities may increase and reach the value of 1–1.5% of GDP. According to expert estimates, the required level of environmental investment should be 6% of GDP, and the level of investment in environmental protection measures to improve the current state of the natural environment should be at least 8–10% of GDP to preserve the natural environment and ensure economic development in accordance with sustainable development criteria (Arkhipov et al., 2018). This circumstance necessitates the search for new sources of funding for environmental protection measures. The Kemerovo Region is no exception. According to the above data, the amount of funds allocated by the enterprises of the region for the implementation of environmental protection measures for the period 2012–2019 does not exceed 30% of their total amount. Taking into account budget expenditures, it does not cover the needs of the region in environmental investments. This circumstance indicates an insufficient level of social responsibility of coal mining companies in the region regarding the implementation of environmental measures.

In this regard, the implementation of the social responsibility of business at the regional level requires developing the investment mechanism to cover the costs for implementing environmental measures. This mechanism should link budget funding with the funding from coal mining enterprises in the Kemerovo Region. This mechanism should ensure: (1) distribution of responsibility for financing the implementation of environmental protection measures following the distribution of the natural rent from coal mining; (2) the formation of the necessary amount of investment in environmental protection activities.

The mechanism for investing in implementing environmental measures should not conflict with the regional economic mechanism. Therefore, it must have a profound theoretical and methodological economic and legal justification. To build this mechanism, we indicate the need to consider two leading economic indicators. The first economic indicator is the coefficient of participation in the financing of environmental measures by coal mining companies. The second economic indicator is the coefficient of the sufficiency of investment for environmental purposes within the framework of the social responsibility of business. The proposed indicators should become the basis for coordinated decisions between the coal mining business and the government to finance the measures of environmental protection that ensure the functioning of the regional economy in accordance with the criteria of sustainable development. This ensures (1) equitable distribution of the financial burden for the implementation of environmental measures between the authorities and coal mining enterprises of the Kemerovo Region and (2) the determination of the amount of financial costs for the implementation of environmental measures by the authorities and coal mining enterprises in the region.

The organization of effective interaction between the authorities and business entities in the financing of environmental protection measures is not only the most important condition for the implementation of social responsibility of business (Newey, 2019), but also a factor of sustainable development of the regional economy.

References

Arkhipov, A. I., Kormishkina, L. A., & Koroleva, L. P. (2018). Assessment of the resource potential of environmental investments. *Bulletin of the Institute of Economics of the Russian Academy of Sciences, 3*, 118–133.

Choi, J. J., Jo, H., Kim, J., & Kim, M. S. (2018). Business groups and corporate social responsibility. *Journal of Business Ethics, 153*(4), 931–954.

Dinius, O., & Vergara, A. (2011). *Company towns in the Americas: Landscape, power, and working-class communities (geographies of justice and social transformation).* University of Georgia Press.

Druzhinin, P. V., Shkiperova, G. T., Potasheva, O. V., & Zimin, D. A. (2020). The assessment of the impact of the economy's development on air pollution. *Economic and Social Changes: Facts, Trends, Forecast, 13*(2), 125–142.

Dutta, A., Jana, R. K., & Das, D. (2020). Do green investments react to oil price shocks? Implications for sustainable development. *Journal of Cleaner Production, 266*, 121956.

Federal State Statistics Service. (2019). Regions of Russia: Socio-economic indicators—2019. Retrieved from http://www.gks.ru/wps/wcm/connect/rosstat_main/rosstat/ru/statistics/publications/catalog/doc_1138623506156. Accessed 11 January 2021.

Islamutdinov, V. F. (2016). Evolution and specificity of the economic institutions of Khanty-Mansi autonomous okrug—Yugra. *Economy of Region, 12*(2), 463–470.

Kim, W. S., & Oh, S. (2019). Corporate social responsibility, business groups and financial performance: A study of listed Indian firms. *Ekonomska Istrazivanja, 32*(1), 1777–1793.

Littlewood, D. (2014). 'Cursed' communities? Corporate social responsibility (CSR), company towns and the mining industry in Namibia. *Journal of Business Ethics, 120*(1), 39–63.

Makarenko, I. O., Smolennikov, D. O., & Makarenko, S. M. (2019). Ukrainian national strategy for corporate social and environmental responsibility as a framework of responsible business conduct. *Espacios, 40*(22).

Melnik, M. V., Voronkova, O. Yu., Nikitochkina, Y. V., Tchuykova, N. M., Davidyants, A. A., & Titova, S. V. (2020). Corporate social responsibility of business as a factor of regional development. *Entrepreneurship and Sustainability Issues*, *7*(3), 2170–2180.

Newey, L. R. (2019). Well-being as a staged social responsibility process for business and society. *Social Responsibility Journal, 15*(1), 75–89.

Syaparman, S., Sri, S., & Etty, P. H. (2019). Supporting and restricting factors of corporate social responsibility (CSR) program of policy implementation of the at the coal mining company in Lahat district. *European Journal of Economics and Management Sciences, 3*, 39–48.

Trapeznikova, I. S. (2019). *Social responsibility of coal industry enterprises.* Moscow.

Wenk, M. S. (2006). Book review: Environmental protection and the social responsibility of firms: Perspectives for law, economics, and business. *Environmental Practice, 8*(3), 197–198.

Zheng, Q., Luo, Y., & Wang, S. L. (2014). Moral degradation, business ethics, and corporate social responsibility in a transitional economy. *Journal of Business Ethics, 120*(3), 405–421.

Improving Customs Services in the Context of Developing the Goods Traceability System

Petr N. Afonin, Viktoriya A. Ivanenko, Valentina A. Kondrashova, Anastasia Yu. Lebedeva, and German E. Myutte

Abstract The paper focuses on the optimization of implementing end-to-end customs control of goods moved across the customs border of the Eurasian Economic Union, which is being introduced into the system of public services. It proposes developing a multifactor system for assessing the effectiveness and efficiency of customs services in terms of comprehensive traceability of goods at the documentary and physical level at any stage of the trade turnover. According to the authors, the proposed system for assessing the effectiveness and efficiency of control and supervisory activities in applying the national goods traceability system will identify weaknesses in the implemented innovation. Moreover, it will modernize its mechanisms and algorithms to improve its functioning. The authors analyze the most significant positive effects on the country's economic processes. Additionally, the authors assess the risks of implementation and subsequent modernization of this innovation in the context of providing state customs services. Moreover, the authors provide and analyze expert opinions and assessments and propose measures to minimize the identified risks at the national and supranational levels. The author's optimization model for determining the optimal volume of products subject to mandatory labeling at the national level is proposed. A set of studies contributing to the solution of the proposed optimization problem is determined.

Keywords Traceability of goods · Customs control · Customs services · Information technology · Customs Identification · Labeling of goods · Risk management

---

P. N. Afonin (✉)
St. Petersburg Electrotechnical University "LETI", Saint Petersburg, Russia

V. A. Ivanenko
Vyborg Branch of RANEPA, Vyborg, Russia
e-mail: ivanenko-va@ranepa.ru

V. A. Kondrashova · G. E. Myutte
Saint Petersburg State University of Economics, Saint Petersburg, Russia

A. Yu. Lebedeva
Russian Customs Academy, Lyubertsy, Russia

E. G. Popkova and B. S. Sergi (eds.), *Geo-Economy of the Future*,
https://doi.org/10.1007/978-3-030-92303-7_50

JEL Codes O38 · A13

1 Introduction

Modern approaches to improving customs services imply the widest possible use of information technology (IT) that automates the functions of customs control at every stage of customs operations. On the one hand, this leads to a reduction in the time to perform customs operations. On the other hand, this increases certainty about private parameters of the location of goods and, accordingly, reduces the risk of customs offenses. One of the key areas ensuring the increase in the efficiency of customs control in recent years is the introduction and expansion of the goods traceability system. This system provides control over goods imported into the territory of Russia and other countries of the Eurasian Economic Union (EAEU) based on quantitative data automatically obtained using specialized customs identification technologies (Lipatova & Lubik, 2019; Tultseva & Vorotintseva, 2018). The introduced traceability technology is interdepartmental in nature. It is based on the use of high-tech digital information exchange involving the Federal Customs Service of Russia (FCS), the Federal Tax Service of Russia (FTS), participants in foreign economic activity, and other organizations (e.g., manufacturers of goods) (Tsymbalova, 2019). However, the expansion of the traceability system for other goods and the formation of a comprehensive traceability system for EAEU member countries is linked to solving the problems of optimizing the resources of customs and other government agencies and choosing a central control and supervisory body. The purpose of this research is to formalize the solution to this problem.

2 Materials and Methods

The development of the mechanism of the goods traceability system is discussed in the works of N. G. Lipatova (Lipatova & Lubik, 2019), T. M. Vorotintseva (2020), T. A. Zakharenko (2020), S. V. Zubarev and M. Yu. Shcherba (2020), G. V. Berlova (Samchenko & Berlova, 2020), and other authors. These works allow us to form a considerable categorical and analytical apparatus to increase the effectiveness of its application steadily.

The methodological basis for improving the effectiveness of the system of goods traceability includes system analysis, optimization theory, and decision-making methods in conditions of certainty. The combination of these methods allows us to form a conceptually complete mechanism for selecting objects of customs control for their inclusion in the loop of physical traceability of goods.

The legal basis for implementing the traceability system in Russia as a factor in improving customs services and the mechanism of interdepartmental interaction between the FCS and the FTS is Article 222 of the Federal law "On Customs Regulation in the Russian Federation and on Amendments to Certain Legislative Acts of the Russian Federation" (August 3, 2018 No. 289-FZ). This law determines the need for the interaction and cooperation of FCS and FTS on issues within their competencies based on using information technology for risk management purposes, implementation of traceability, and end-to-end control mechanisms (Russian Federation, 2018). According to EAEU regulations, traceability means the organization of accounting of goods subject to traceability and operations related to the turnover of such goods using national traceability systems. The traceability mechanism is a system functioning based on information exchange from national traceability systems and ensuring traceability to control the turnover of goods (Eurasian Economic Union, 2019).

3 Results

From the position of government control, the implementation of the traceability system allows preventing illicit trafficking of goods on the customs territory of the EAEU members. For participants in foreign economic activity, the application of this system provides the reduction of time for the release of goods into free circulation and the minimization of risks of additional costs on customs control, which arise in the absence of the necessary level of certainty about the origin of goods and the volume of their import and sale on the EAEU territory. For the end consumer, the introduction of the traceability system is a guarantee of purchasing quality goods produced legally outside the space of potential law violations.

The consideration of the category apparatus of the goods traceability system implies the definition of two directions of the system's development—physical and documentary traceability. According to the "Concept for the creation and functioning in the Russian Federation of the system of marking of goods by means of identification and traceability of the movement of goods" approved by Order No. 2963-r (December 28, 2018), the concept of physical traceability implies a type of traceability provided by the use of marking mechanisms. In turn, documentary traceability is provided by transferring information about the movement of goods in the accompanying commercial documents (Government of the Russian Federation, 2018a). This approach allows customs authorities to carry out comprehensive control of goods moving across the customs of the EAEU to prevent counterfeit products from entering the national market.

The goods traceability system has several mandatory elements (Fig. 1).

Figure 1 indicates the mandatory availability of information and technical base that meets the requirements of the current stage of scientific and technological progress and corresponds to the paradigm of comprehensive digitalization of public services and the concept of the digital economy. Public information systems are expected to be based on the functioning of identification tools to ensure physical traceability. The

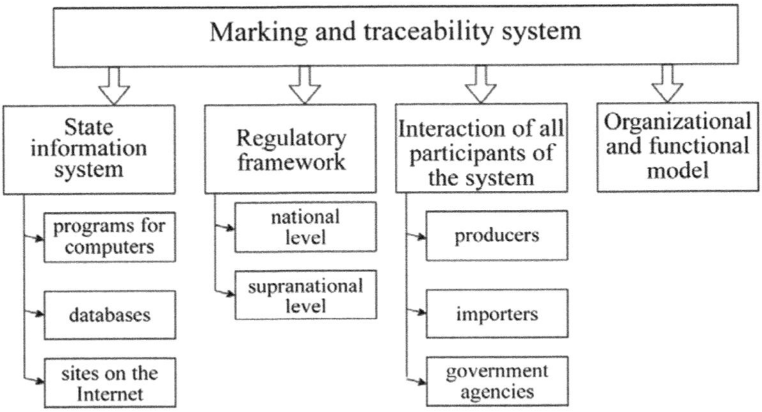

Fig. 1 Elements of the system of traceaability and marking of goods. *Source* Compiled by the authors based on (Eurasian Economic Union, 2019)

system will be integrated with document traceability tools to ensure the consistency of data on the movement of goods (Government of the Russian Federation, 2018a).

Following the objectives set for implementing the system from the perspective of the regulatory authorities, it is necessary to develop a multifactor system to assess the performance and efficiency of customs services in terms of comprehensive traceability of goods at any stage of the trade turnover. In this regard, it is proposed to introduce additional indicators of control and oversight activities into the list established by Order "On approval of performance and efficiency indicators of the Federal Customs Service of Russia, territorial customs authorities, and the central apparatus of the Federal Customs Service of Russia" (October 30, 2017 No. 1720) (Federal Customs Service of the Russian Federation, 2017). These indicators are as follows:

1. The amount of probable damage prevented by customs officials using the traceability system as part of actual control, which could have been caused by the introduction of counterfeit goods into the trade turnover across the EAEU customs border. It is proposed to calculate this amount using the following formula (1):

$$Q = \sum_{1}^{N} K_i \times P_i - S_{CR} - S_C \tag{1}$$

where:

K_i—the number (in units, kg) of detected counterfeit goods for which the marking and traceability system was used;
P_i (in mln. RUB)—market price of authentic goods, the analogs of which were transported across the EAEU customs border;

S_{CR} (in mln. RUB)—the value of goods—subjects of the copyright at the time of initiation of the case on the copyright, in case of a decision to terminate the case on the copyright;
S_C (in mln. RUB)—the value of goods in respect of which there was no administrative offense established by a court decision that has entered into legal force;
N—the number of cases of movement of counterfeit goods across the EAEU customs border detected using the traceability system for a given reporting period.

2. Share of counterfeit goods detected by customs officials as part of the actual control with the traceability system in the total number of detected counterfeit goods (2):

$$D_i = \frac{K_i}{K_{TOT}} \times 100\% \tag{2}$$

where K_{TOT}—the total amount (in units, kg) of detected counterfeit goods.

3. Share of economic damage prevented in the framework of current customs control with the use of the traceability system, which could have been caused by the introduction of counterfeit goods into the trade turnover in the total amount of prevented damage from the sale of counterfeit goods (3):

$$D_u = \frac{Q}{Q_{TOT}} \times 100\% \tag{3}$$

where Q_{TOT}—the total amount of damage prevented from the sale of counterfeit goods.

4. The amount of probable damage prevented by customs officials through the use of traceability mechanisms as part of decisions on the classification of goods moved across the customs border in accordance with the EAEU Commodity Nomenclature for Foreign Economic Activity (EAEU CNFEA). In order to calculate this indicator, the following formula is proposed (4):

$$Q_K = \sum_{i=1}^{k} S_i \tag{4}$$

where:

k—the number of valid customs decisions on the classification of goods in accordance with the EAEU CNFEA;
S_i (in mln. RUB)—the damage prevented due to additional charges of customs and other payments based on the ith decision on the classification of goods under the EAEU CNFEA made using the goods traceability system.

5. Share of triggering of the System of Risk Management (SRM) for goods subject to mandatory labeling (physical traceability mechanism) in the total number of triggering of the SRM for goods transported across the EAEU customs border (5):

$$D_R = \frac{R}{R_{TOT}} \times 100\% \tag{5}$$

where:

R—the number of SRM triggers for goods subject to mandatory labeling (physical traceability mechanism);
R_{TOT}—the total number of triggers of the SRM in relation to goods transported across the EAEU customs border.

It is assumed that the source of the initial data will be the database IAS "Tariffs-1" and the Central Database of the Unified Automated Information System of Technical Inspection.

The system of evaluation of indicators of effectiveness and efficiency of control and oversight activities within the framework of the national system of traceability of goods will allow to identify weaknesses in the implemented innovation and modernize its mechanisms and algorithms to improve its functioning.

Additionally, an important issue of optimizing the system of goods traceability is determining the optimal volume of products to be labeled (physical traceability). In this regard, we propose a mathematical model. The operation of this model uses the following designations:

B—a multitude of all products;
B_1—a multitude of products moved across the customs border;
B_2—a multitude of products in relation to which the risks of violation of customs legislation have been identified;
B_3—a multitude of products for which it is possible to apply a physical traceability system;
B_4—a multitude of products, the need for physical traceability of which is determined based on the results of the documentary traceability system.

In this case, the set of goods for which it is required to apply the physical traceability system will be determined by the following formula (6):

$$V_{pt} = V_1 \wedge V_2 \wedge V_3 \wedge V_4 \tag{6}$$

If we assume that the elementary operation of reading the marking element has a time duration of t_i and the number of individual items to be labeled and read is some value N in a given period (e.g., one day), then the total duration of all read operations will be $t_i \cdot N$.

Thus, we can determine the required number of customs officials D_n, whose tasks may include reading the marking elements, given, for example, that the shift duration is 12 h (7):

$$D_n = K_l \cdot t_i \cdot N \tag{7}$$

where K_l—the coefficient establishing the correspondence between the norm of the workload of officials and his or her ability to read the marking elements.

The solution to the problem of selecting the optimal volume of b_i goods for which it is possible to apply physical traceability can generally be described by the following system of constraints (8):

$$\begin{aligned} & b_i \in \mathrm{B}_{\mathrm{pt}} \\ & D_n \le D_{\mathrm{TOT}} \\ & f(b_i, D_n, \ldots) \to \min \end{aligned} \tag{8}$$

where $f\ (b_i,\ D_n,\ \ldots)$—a function determining the cost of the physical traceability system, the arguments of which are partial parameters of performance and cost of individual subsystems and elements of the physical traceability system.

The solution to this optimization problem requires the following set of studies:

1. Analysis of the current state of the provision of customs services with the use of traceability technology, in particular:
   - Analysis of the possibility of assessing the relationship between the development of customs services and the development of the goods traceability system;
   - Analysis of the current state of regulatory and legal support for traceability of goods in the standards of the World Customs Organization and the regulatory and legal framework of Russia and the EAEU member countries;
   - Analysis of the current state of the system for assessing the effectiveness of the traceability of the goods in Russia.
2. Analysis of the current state of information and technical support for traceability of goods, in particular:
   - Analysis of the current state of information and technical means of implementation of the goods traceability system (physical and documentary);
   - Development of methodological recommendations on assessing the applicability of the information and technical means of the system (tracking overloads, seal numbers, GPS tracking, and customs clearance of tracking systems);
   - Development of methodological recommendations on the complementarity of physical and documentary subsystems of the goods traceability system (including assessing the possibility of comparing the documents submitted to the customs administrations of different countries).

3. Development of methods to assess the effectiveness of customs control based on the application of the goods traceability system:
   - Analysis of the current state of the methodological apparatus for evaluating the effectiveness of the goods traceability system;
   - Analysis of the current state of mechanisms for assessing the resource-intensiveness of the goods traceability system;
   - Development of a methodology for implementing mechanisms for the traceability of goods, considering the implementation of the intelligent border crossing concept.

From the perspective of the national business community, one of the most significant positive effects of the traceability system is the optimization of financial, time, and human resources. However, the introduction of innovation can be accompanied by a high level of costs at the initial stage, which will negatively impact small and medium enterprises in the country and consumers of the final product. According to expert estimates, the introduction of the mechanism of digital labeling (physical traceability) will lead to adverse economic effects through increased costs for participants in foreign economic activity and national producers and higher prices for consumers (Alta-soft, 2020).

Forecasts indicate that the risk of destroying a significant segment of small businesses due to the lack of financial resources to ensure the physical traceability of goods is maximized. This approach can ensure a steady decline in the share of small and medium-sized enterprises in the Russian economy. The strategic goal of increasing the share of small business to 32.5% of GDP by 2024, outlined in the national project "Small and medium entrepreneurship" (Government of the Russian Federation, 2018b), will not be achieved. Additionally, digital labeling of goods will not actually increase tax revenues to the federal treasury. According to experts, the short-lived positive budgetary effect of revenue growth will only be caused by an increase in the tax base due to the transfer of labeling costs to the final prices of goods ("tax on tax") (Alta-soft, 2020).

4 Conclusion

According to the objectives set for the goods traceability system from the perspective of public authorities, it is necessary to develop a system to assess the effectiveness and efficiency of customs services in full traceability at each stage of the trade turnover. In this regard, it is proposed to introduce additional indicators of control and supervisory activities at the legislative level and optimize the system of traceability of goods by determining the optimal volume of products subject to labeling (physical traceability).

References

Alta-soft. (2020, March 12). A comprehensive study of the effects of widespread implementation of control tags. Retrieved from https://www.alta.ru/external_news/73081/. Accessed 27 January 2021.

Eurasian Economic Union. (2019). *Agreement on the mechanism of traceability of goods imported into the customs territory of the Eurasian Economic Union (May 29, 2019)*. ConsultantPlus. Retrieved from http://www.consultant.ru/document/cons_doc_LAW_325804/. Accessed 23 January 2021.

Federal Customs Service of the Russian Federation. (2017). *Order "On approval of performance and efficiency indicators of the Federal Customs Service of Russia, Territorial Customs Bodies, and the central office of the Federal Customs Service (October 30, 2017 No. 1720)*. ConsultantPlus. Retrieved from http://www.consultant.ru/cons/cgi/online.cgi?req=doc&base=EXP&n=709515&dst=100001#oO9WIeSvMeAp3alU. Accessed 24 January 2021.

Government of the Russian Federation. (2018a). *Order "On approval of the Concept for the creation and functioning in the Russian Federation of the system of marking of goods by means of identification and traceability of the movement of goods" (December 28, 2018 No. 2963-r)*. ConsultantPlus. Retrieved from http://www.consultant.ru/document/cons_doc_LAW_314921/. Accessed 23 January 2021.

Government of the Russian Federation. (2018b). National project "Small and medium entrepreneurship and support of individual entrepreneurial initiative." Retrieved from http://government.ru/info/35563/. Accessed 30 January 2021.

Lipatova, N. G., & Lubik, A. F. (2019). Traceability of the movement of goods as a tool to improve the efficiency of customs control. *The Russian Customs Academy Messenger, 3*, 39–45.

Russian Federation. (2018). *Federal law "On Customs Regulation in the Russian Federation and on Amendments to Certain Legislative Acts of the Russian Federation" (August 3, 2018 No. 289-FZ)*. ConsultantPlus. Retrieved from http://www.consultant.ru/document/cons_doc_LAW_304093/. Accessed 28 January 2021.

Samchenko, O. N., & Berlova, N. V. (2020). Directions of development of goods traceability in the Eurasian Economic Union. *The Russian Customs Academy Messenger, 4*(53), 109–115.

Tsymbalova, A. A. (2019). Digitalization of customs and traceability of goods. In S. N. Gamidullaev (Ed.), *Customs readings—2019: Science and education in the emergence of an innovative economy* (Vol. 3, pp. 167–171). Saint Petersburg Branch of the Russian Customs Academy.

Tultseva, A. S., & Vorotintseva, T. M. (2018). Traceability of goods as a mechanism for regulating international trade. *Scientific Review, 4*(53), 42–46.

Vorotintseva, T. M. (2020). On the creation of a national system of traceability of goods by the Federal Customs Service of Russia. *Customs Regulation. Customs Control, 5*, 57–63.

Zakharenko, T. A. (2020). Traceability of goods as a way of struggle against falsification and counterfeit. *Scientific Letters of Russian Customs Academy St.-Petersburg Branch named after Vladimir Bobkov, 3*(75), 14–17.

Zubarev, S. V., & Shcherba, M. Yu. (2020). Peculiarities of customs and other types of state control over the circulation of goods within the framework of the system of documentary and physical traceability of goods. In P. N. Bashly (Ed.), *Features of state regulation of foreign economic activity in modern conditions: Proceedings of the VII All-Russian Scientific and Practical Conference* (pp. 32–39). Rostov branch of the Russian Customs Academy.

COVID-19's Impulse to Intensify Neo-industrial Digital Technologies 4.0 in the Legal Regulation of Foreign Economic Activity of Economic Entities of the EAEU and the BRICS Member States

Agnessa O. Inshakova, Alexander I. Goncharov, and Denis E. Matytsin

Abstract The chapter contains a study of the impact of the new coronavirus pandemic 2019–2020 on the business activity of economic entities of the EAEU and BRICS jurisdictions in the implementation of foreign trade activities. The authors note the need and highlight the opportunities for wider and more intensive use of digital technologies of Industry 4.0 in the legal regulation of export–import transactions between the parties-residents of the member states of integration associations with the participation of the Russian Federation. The authors substantiate that in modern realities, infectious epidemiological risks are considered not as a force majeure circumstance but as restrictions on the collective work of employees of companies, including foreign trade commercial organizations. The authors argue that this pandemic set of legal procedures for the wider and more intensive use of digital technologies in Industry 4.0 should be enshrined in the provisions of the international convention, which countries participating in integration associations will be able to join, thereby ensuring the entry into force of the relevant regulatory provisions in their jurisdictions. The authors propose several legal procedures for wider and more intensive use of digital technologies of Industry 4.0 in the implementation of foreign economic transactions, which launched and implemented after the announcement of the next pandemic by the World Health Organization.

Keywords COVID-19 · Neo-industrial · Digital technologies 4.0 · Greater Eurasian Partnership · Infection; pandemic · Risks · Employees · Foreign trade companies · Foreign economic activity

---

A. O. Inshakova (✉) · A. I. Goncharov · D. E. Matytsin
Volgograd State University, Volgograd, Russia
e-mail: ainshakova@list.ru; gimchp@volsu.ru

A. I. Goncharov
e-mail: gimchp@volsu.ru

D. E. Matytsin
e-mail: Dmatytsin@ya.ru; gimchp@volsu.ru

E. G. Popkova and B. S. Sergi (eds.), *Geo-Economy of the Future*,
https://doi.org/10.1007/978-3-030-92303-7_51

JEL Codes G18 · G24 · K12 · K15 · L14 · L24 · L86 · O16

1 Introduction

Computerization of business processes of commercial organizations, including modern foreign trade companies, has been going on for at least 40 years. In recent years, the term "digital technologies" has become especially common. Although we mean all the same effective computer programs, the use of which is a particular business should help to reduce costs, increase the volume and quality of goods produced, work performed, and services rendered. At the same time, the method of transmitting commands inside computer programs in digits (1-yes; 0-no) has nothing to do with automation, internetization, robotization, computerization of modern economic activity (Matytsin, 2021).

Neo-industrial digital technologies 4.0 are the most advanced and complex computer programs used remotely via the Internet information and communication network to transmit specific commands to a specific unit of goods, as well as to obtain data on the initiator's request about the current state and geographical location of the goods (Matytsin & Rusakova, 2021).

Such technologies are especially in demand in foreign economic activity, in this area, huge flows of not only raw exchange-traded goods but also very expensive complex goods are constantly moving.

In March 2020s year, The World Health Organization has declared a COVID-19 pandemic. Across the planet, states imposed quarantine measures: the closure of public places, the termination of social contacts, bans on leaving places of residence. Some companies have completely stopped working, but a large number of economic entities, whose production process allows them have transferred their work collectives to the remote version of working from home using information and communication technologies via the Internet. According to the All-Russian Center for the Study of Public Opinion, 19% of employees in Russia who switched to remote work had to acquire new computer skills to work with messengers and video conferencing programs.

As part of the regulatory framework studied, the following documents. The Treaty on the Eurasian Economic Union and its Annexes (http://base.garant.ru/70670880/#ixzz6gzcJ0u7L). The Statement on the Digital Agenda of the Eurasian Economic Union (http://pravo.eaeunion.org/documentid=71553074byPara=1sub=1). The Decision of the Supreme Eurasian Economic Council of 11.10.2017 No. 12 "On the main directions for implementing the Digital Agenda of the Eurasian Economic Union until 2025" (http://pravo.eaeunion.org/documentid=71708158byPara=1sub=1). The main directions of the implementation of the digital agenda of the Eurasian Economic Union until 2025 and its Annexes (http://pravo.eaeunion.org/documentid=71708158byPara=1sub=1). The Moscow Declaration of the XII BRICS Summit (http://kremlin.ru/supplement/5581). The BRICS Economic

Partnership Strategy until 2025 (https://brics-russia2020.ru/images/114/81/1148133.pdf).

The study of doctrinal sources covers the scientific works of Russian scientists, including I. V. Ershova, E. E. Frolova, A. I. Goncharov, A. O. Inshakova, D. E. Matytsin, E. P. Rusakova, V. V. Tarakanov. The scientific works of some foreign authors were also studied, including A. Szalavetz; J. Muller, K. Voigt; H. Kagermann; A. Gilchrist; S. Weyer, M. Schmitt, M. Ohmer, D. Goreck.

2 Methodology

The content of this chapter was developed based on the materialist worldview and the general scientific method of historical materialism. General scientific methods of cognition are applied: dialectical, hypothetical-deductive method, generalization, induction, and deduction, analysis and synthesis, empirical description. The research also uses private scientific methods: dogmatic, comparative-legal, hermeneutic, structural–functional, etc.

3 Results

The chain reaction of the spread of a new coronavirus infection—COVID-19-around the world with great speed has undoubtedly had a negative impact on business activity in the foreign economic sphere. The desire of most states to protect the population from infection naturally led to a sharp decrease in the physical participation of people in business processes, which, in turn, quite naturally led to an economic downturn. For example, the United States exported $90 billion worth of goods in April 2020, and $95 billion in May, before the pandemic, the average monthly export level exceeded $120–130 billion. In terms of imports, stocks of food and consumer goods formed: the United States imported goods in April and May amount $170 billion, while before the pandemic, the monthly level exceeded $200 billion. France, for example, imported $246 billion worth of goods from May to October 2020, which is less than 50% of the volume of imports for the first six months of 2019 ($643 billion). France's exports for the same six months amounted to $198 billion, which is also less than 50% of the first half of 2019 ($556 billion). Foreign trade turnover also fell in Germany, Italy, Great Britain, and India. According to a study by the international insurance company Euler Hermes, the decline in export–import indicators directly depends on the dynamics of morbidity and the reaction of states. Every fifth company in Europe and the United States reported failures in supply chains caused by the pandemic of a new coronavirus infection.

The closure of our borders, the disconnection or complication of logistics routes, quarantine measures, partial and complete shutdowns of production, and other prohibitions and restrictive measures have significantly affected the intensity of foreign economic activity. According to the estimates of the Federal Customs Service of Russia, foreign trade turnover in January–August 2020 fell by 16% (imports—by 7%, exports—by 23%. At the same time, for 2020 as a whole, forecasts of a fall in Russia were: exports—23.5%, imports—7.6%). Objectively, it is necessary to point out the positive dynamics of the development of foreign trade by small and medium-sized businesses. In 2020, 2712 companies signed export contracts. In monetary terms, the Russian Export Center's export support for 17,630 companies exceeded $906 million. Focus on traditional offline events (exhibitions and business missions) shifted to an online format—placement on electronic trading platforms and search for foreign partners; as a result, 27% of export contracts concluded remotely. Commercial offers of small and medium-sized businesses placed by the Russian Export Center mainly on electronic trading platforms: 414 companies made export sales amount of $65.3 million.

In our opinion, in any pandemic, the response of state authorities in all countries is aimed primarily at curbing any possibility of cross-border spread of infection, as well as at preventing a sharp reduction in the stocks of goods that are critically necessary for the population during the pandemic. For example, in the EAEU space, the EEC Board in 2020 established a ban on the export of certain categories of goods to prevent a shortage of food products, as well as medical devices and personal protective equipment during the aggravation of the sanitary and epidemiological situation. The full list of products for which the export ban was in effect from 02.03.2020 to 01.06.2020 was established by the Decree of the Government of the Russian Federation No. 223 of 02.03.2020 "On the introduction of a temporary ban on the export of certain types of products from Russia" (expired).

In general, an impressive number of government measures were taken in Russia during the pandemic. Concerning certain categories of goods, changes made that provide for exemption from import customs duties when they are placed under the procedure of release for domestic consumption. Changes have been made to the Rules for Determining the Origin of Goods from Developing and Least developed countries, providing for the possibility of submitting an electronic copy of the certificate for obtaining tariff preferences instead of the original certificate of origin of goods from these countries. It established that participants in foreign economic activity are not liable for non-compliance by non-residents with the terms of foreign trade agreements due to the coronavirus pandemic if certain requirements are met. The necessary actions of residents were determined in case of non-fulfillment by foreign counterparties of the terms of delivery or payment of goods under foreign trade agreements concluded between residents and non-residents, or in case of non-return by non-residents of advance payments previously paid to them by residents. Non-fulfillment of obligations by a non-resident under a foreign trade agreement concluded with him due to force majeure, due to measures taken by foreign states against the spread of coronavirus infection, confirms the absence of guilt. The non-resident is not guilty of committing administrative offenses in the field of violation of currency legislation, provided for in Part 4 and Part 5 of Article 15.25 of the Administrative Code of the Russian Federation. At the same time, for non-attraction to administrative responsibility for the specified actions on the basis of such force majeure, the resident has the right to present any evidence. Such evidence must indicate the absence of guilt in his actions, as well as evidence confirming the adoption of measures to comply with the requirements of the currency legislation of the Russian Federation. Along with these measures, certain medical products were free of tax. In the federal project "Digital Technologies", the focus was on domestic IT solutions for ensuring effective interaction in the conditions of social fragmentation: remote work, training, television medical diagnostics, online services.

At the doctrinal level, our foreign colleagues did not touch the development of digital technologies from the perspective we presented (Weyer et al., 2015; Muller & Voigt, 2018; Szalavetz, 2019).

At the theoretical and methodological level, we propose to outline and at least mentally consolidate the anti-crisis set of digital technologies, which is called the "Pandemic List". According to this "Pandemic List", activities carried out in foreign trade companies upon the announcement of the next pandemic by the World Health Organization, in particular, in the space of the Greater Eurasian Partnership (EAEU + BRICS). The first group of three critical digital technologies identified below applies to every citizen, regardless of his/her profession and employment.

The first mandatory technology of the "Pandemic List" is the "Personal QR Code" software service, installed on the smartphones of the maximum number of citizens, provides them with prompt information about the current epidemiological situation in a particular state, the restrictive measures taken, techniques, and personal protective equipment for the health of people and pets. The specified software service should provide for the formation of temporary QR codes that reflect the infection of this citizen in the event of a positive result of his testing for the presence of infection. This red "QR-code-infection" becomes mandatory for this citizen to present a pass to all public places upon the introduction of quarantine measures. "QR-code-infection" self-destructs upon the fact of a negative test for infection in this citizen after recovery. At the same time, the maximum number of healthy citizens receive a green registration "QR-code-registration". These QR codes, which are electronic passes, are also required for electronic recording of each visit of a citizen in all places of his appearance in this locality. Such digital traces allow you quickly identify the circle of contact persons if someone from the people who were present at a certain time in a given place will later found to have an infection. After identifying the specified circle of persons, each person who was in contact with an infected person receives a yellow "QR code-risk-infection".

Also, this digital technology should provide for the ability to generate a digital label "QR code-antibodies"; these QR codes are assigned to each person whose individual health characteristics have ensured the production of antibodies in his body for an infection, the pandemic of which occurs during this period. Along with this, this digital technology should provide for the ability to generate a digital label "QR code-immunity"; such a QR code is assigned to each person who has been ill, treated and recovered, acquired immunity, and is no longer at risk of contracting the same infection. The logical conclusion of this complex of digital technologies of the First mandatory group of digital technologies should be a technology that provides for the formation of a digital label "QR code-vaccine", such a QR code assigned to each person who has been vaccinated with the appropriate drug against an infection, the pandemic of which occurs during this period.

The second mandatory technology of the "Pandemic List" is a software product that is also installed on the smartphones of citizens in need "Volunteers-Help". This technology is used by elderly citizens with poor health, as well as those people who are constantly at home for health reasons or not move independently at all. The service allows you to call the volunteers home to help when the risk of infection is very high, and visiting such citizens with poor health in any public place is associated with a high risk of infection.

The third mandatory technology of the "Pandemic List", also installed on the smartphones of the maximum number of citizens, is the software product "Electronic commerce and delivery of goods". This service will allow you to make transactions for the purchase and sale of goods that are vital to citizens during the pandemic and will be delivered to their homes by specially trained and protected crews in the event of, particularly, strict quarantine restrictions.

In general, the First mandatory group of digital technologies, which provides for the creation of unique personal QR codes in a citizen's smartphone, will ensure that their rights are respected, including when traveling within their country and between jurisdictions abroad. Of course, all of the above applies to all employees of foreign trade companies.

Let us further consider the second group—these are digital technologies that are used specifically for foreign economic activity.

The fourth digital technology of the "Pandemic List" is a complex of computer programs "Home-office", which must be quickly installed on the appropriate computer, transferred to the home of an employee of a foreign trade company in the shortest reasonable time after the announcement of a pandemic. As part of the specified set of computer programs, software products must be installed that allow the employee to establish uninterrupted video communication via the Internet information and telecommunications network. First, you need the ability to connect via teleconference to your line manager, to the head of the division and to the top manager of the company in the management vertical. Secondly, the ability to connect via teleconference to colleagues in the company, communication with whom is necessary for the preparation of decisions taken on behalf of the company. Third, you need the ability to work with databases that are available to the company and are necessary for the implementation of the employee's work functions. Fourth, we need the ability to connect via teleconference and communicate with foreign colleagues in connection with the preparation of a particular foreign trade transaction. Of course, these digital technologies should provide for the maintenance and storage of video and audio recordings on the company's parent server of all employee contacts from this work computer. It is stipulated by law that these videos and sound recordings are recognized as facts that have legal significance in the work of this employee (Inshakova & Goncharov, 2020).

The fifth digital technology of the "Pandemic List", conventionally called by us "Disinfection", should be used by those plants and factories that are interested in non-stop production of goods and even after the announcement of a pandemic, strictly observing sanitary and epidemiological requirements and releasing goods, will carry out disinfection of each unit of manufactured goods and its external packaging. Many modern manufacturers, as the goods are manufactured within the factory conveyors, load the goods immediately into standard transport (sea, automobile, aviation, railway) containers. It would be quite rational to disinfect these transport containers as well.

The sixth digital technology of the "Pandemic List", which we call "Customs Monitoring", should allow the local Customs authority, in conjunction with the parent computer server of the plant or factory that manufactures goods for export under a specific foreign trade contract, to conduct a piece-by-piece electronic registration of each unit of goods. The digital image of the goods intended for export must be tracked by the Customs Authority from the moment of their exit from the conveyor until they are packed in a standard transport container for shipment to a customs warehouse port (airport, railway station) (Inshakova et al., 2020b).

The seventh digital technology of the "Pandem-List", conventionally called by us "Customs drone", should provide the possibility of retrofitting a vehicle (truck) with a special robotic control device. This device will ensure its trouble-free movement of a vehicle (truck) with a container of goods (which has been loaded and disinfected using the "Customs Monitoring" technology) from the manufacturer to the customs warehouse without a driver. The final algorithms of the digital technology "Customs Monitoring" carry out the accounting of goods in this warehouse, the calculation of duties for it, the issuance of an electronic invoice for the payment of duties to the payer, the necessary marks in the Passport of a foreign trade transaction. This completes the electronic registration of the corresponding customs regime for the exported goods.

With the correct and precise configuration of this digital technology, the customs warehouse can function quite effectively with a minimum number of people. Further, using the digital technology "Customs Drone", standard transport containers with goods that have passed the established customs procedures can be delivered to the port (airport, railway station), where they can be put under automatic loading onboard a ship (aircraft, railway cars). Loading is carried out mainly by robotic lifting and transport mechanisms with a minimum number of people.

It is noted that the algorithms of this computer program provide for the movement of goods both for export–import on import. In this mode, the digital technology "Customs Drone" will manage the transportation of customs-cleared goods without a driver from the customs warehouse to the warehouse of the importer or the warehouse of a third party. It also clarified that customs warehouses should be equipped with robotic complexes that could use the above-mentioned digital technology "Disinfection" for additional sanitary treatment of transport containers with goods that are sent for export, as well as for disinfection of containers with goods coming from imported foreign trade transactions.

The eighth digital technology of the "Pandemic List", which we have conditionally called "Electronic State Border", should record the crossing of the state border by the transported cargo in standard transport containers of the jurisdiction of their dispatch and the jurisdiction of their delivery along the routes of foreign trade transactions. At the same time, the movement of goods is to be additionally traced through intermediate "digital gates", which should be installed in the places of cargo movement—in sea (river) ports of destination, at railway junctions, major interchanges, and intersections of highways (Inshakova et al., 2020a).

The ninth digital technology of the "Pandemic List", which we call "Roaming with the carrier", allows the initiator at any time to send a request to a satellite communication station installed onboard the carrier (ship, plane, train, truck). The functionality of the technology should provide the ability to communicate with the microcomputer of a standard transport container in which the goods transported under the smart contract corresponding to this transaction, and the ability to receive a report on its movement up to the exact coordinates of its current location (Inshakova et al., 2020c). In addition, this digital technology should provide real-time data transmission to the Internet portal "Foreign Economic Activity of the Greater Eurasian Partnership online", reflecting in an interactive mode the execution of existing smart contracts between foreign trade companies of the partner states (Tarakanov et al., 2020).

4 Conclusion

Undoubtedly, with its avalanche-like spread, COVID-19 generated a powerful impetus for the accelerated improvement, development, expansion, and intensification of neo-industrial digital technologies that ensure an appropriate legal regime and increase the efficiency of foreign economic activities of economic entities of the partner states of the Greater Eurasian Partnership (EAEU + BRICS). The authors propose a set of nine digital technologies formed in two groups, which at the public legal level are combined into a special list (Pandemic List) and applied from the day of the declaration of a particular pandemic by the World Health Organization. Such technologies in the period of epidemiological danger allow removing the absolute majority of employees of foreign trade companies from the risk of infection, without stopping foreign economic activity. The first group of three mandatory digital technologies should be brought to the maximum number of citizens at the expense of centralized targeted state funding. The specified list of digital technologies (Pandemic List) is proposed to fix as part of the provisions of the International Convention "On Digital Technologies of Foreign Economic Activity Used by Residents of the Greater Eurasian Partnership Jurisdictions".

Acknowledgements The reported study was funded by RFBR according to the research project No. 18-29-16132.

References

BRICS Economic Partnership Strategy until 2025. Available at: https://brics-russia2020.ru/images/114/81/1148133.pdf. Accessed April 22, 2021.

Decision of the Supreme Eurasian Economic Council of 11.10.2017 No. 12 "On the main directions of implementation of the Digital Agenda of the Eurasian Economic Union until 2025". Available at: http://pravo.eaeunion.org/document?id=71708158&byPara=1&sub=1. Accessed April 12, 2021.

Inshakova, A. O., & Goncharov, A. I. (2020). Negotiation: Features of regulation and prospects for law enforcement. In A. O. Inshakova, & A. V. Bogoviz (Eds.), *Alternative methods of judging economic conflicts in the national positive and soft law*. E. G. Popkova (Series Ed.), *Advances in research on Russian business and management*. Information Age Publishing. https://www.infoagepub.com/products/Alternative-Methods-of-Judging-Economic-Conflicts-in-the-National-Positive-and-Soft-Law

Inshakova, A. O., Frolova, E. E., Rusakova, E. P., & Kovalev, S. I. (2020a). The model of distribution of human and machine labor at intellectual production in industry 4.0. *Journal of Intellectual Capital 21*(4), 601–622. https://doi.org/10.1108/JIC-11-2019-0257

Inshakova, A. O., Goncharov, A. I., & Ershova, I. V. (2020). Digital blockchain registration of legally significant stages of complex good's export-import supplies by business entities of the EAEU and BRICS jurisdictions. In *Lecture notes in networks and systems* (Vol. 110, pp. 328–336).

Inshakova, A. O., Goncharov, A. I., & Salikov, D. A. (2020). Electronic-digital smart contracts: Modernization of legal tools for foreign economic activity. In *Lecture notes in networks and systems* (Vol. 91, pp. 3–13).

Matytsin, D. E. (2021). Internet-investing as a remote algorithm of the retail investment financing. In *Modern global economic system: Evolutional development versus revolutionary leap. Springer book series lecture notes in networks and systems*. Springer Nature.

Matytsin, D. E., & Rusakova, E. P. (2021). Strategy of quality management in industry 4.0 and formation of a cognitive economy based on industrial and manufacturing engineering in the Russian Federation and countries of the EU. *International Journal for Quality Research, 15*(4). https://doi.org/10.24874/IJQR15.04-03

Moscow Declaration of the XII BRICS Summit. Available at: http://kremlin.ru/supplement/5581. Accessed April 02, 2021.

Muller, J., & Voigt, K. (2018). Sustainable industrial value creation in SMEs: A comparison between industry 4 and made in China 2025. *International Journal of Precision Engineering and Manufacturing-Green Technology, 5*(5), 659–670.

Statement on the Digital Agenda of the Eurasian Economic Union. Available at: http://pravo.eaeunion.org/document?id=71553074&byPara=1&sub=1. Accessed March 02, 2021.

Szalavetz, A. (2019). Industry 4.0 and capability development in manufacturing subsidiaries. *Technological Forecasting and Social Change, 145*, 384–395.

Tarakanov, V. V., Inshakova, A. O., & Goncharov, A. I. (2020). Interactive demo online maintenance foreign trade deals of business entities of BRICS jurisdictions. In *Lecture notes in networks and systems* (Vol. 110, pp. 319–327).

The main directions of the implementation of the digital agenda of the Eurasian Economic Union until 2025 and their Annexes. Available at: http://pravo.eaeunion.org/document?id=71708158&byPara=1&sub=1. Accessed April 12, 2021.

The Treaty on the Eurasian Economic Union and its Annexes. Available at: http://base.garant.ru/70670880/#ixzz6gzcJ0u7L. Accessed February 02, 2021.

Weyer, S., Schmitt, M., Ohmer, M., & Goreck, D. (2015). Towards industry 4.0—Standardization as the crucial challenge for highly modular, multi-vendor production systems. *IFAC-Papers On Line 48*(3), 579–584.

Concepts and Categories of Interstate Regulation in the Implementation and Legal Risks in the Application of Digital Technologies of Industry 4.0 at the Level of the EAEU and BRICS

Alexander I. Goncharov and Agnessa O. Inshakova

Abstract The chapter of the monograph contains a scientific development of some concepts and categories from the interstate regulation of the procedures for the introduction of digital technologies in Industry 4.0 for their use by economic entities of the EAEU and BRICS jurisdictions in their foreign economic activities. The definition of legal risks that may arise during the implementation and application of digital technologies of Industry 4.0 formulated at the conceptual level. A systematic approach to the study of the problem is implemented by the authors by using several methods of scientific research in combination. The authors substantiate the scientific and methodological principle of inter-jurisdictional risk management of changes in legislation, as well as their neutralization and avoidance in the implementation of foreign economic activity by economic entities of the EAEU and BRICS jurisdictions. This approach of inter-jurisdictional management of legal risks proposed to be enshrined in the provisions of the international convention, to which the countries participating in integration associations will be able to join, ensuring the entry into force of the relevant regulatory provisions in their jurisdictions. A complex of information and communication technologies for continuous tracking of the movement of individual goods, as well as vehicles that transport goods under foreign trade transactions between the jurisdictions of the EAEU and BRICS associations, has been formed in five categories. This set of digital technologies applied on the basis of a new scientific and methodological approach to inter-jurisdictional risk management of collisions with legal provisions, as well as the neutralization and avoidance of these risks in the implementation of foreign economic activities by economic entities of the jurisdictions of these associations. The new approach provides for the coordination of the regulatory documents adopted by the member states of the integration associations in the mechanism of the inter-jurisdictional polysubject blockchain. The documents relate to the introduction and application of digital technologies of foreign economic activity, to the level of a single (identical) regulatory document for

A. I. Goncharov (✉) · A. O. Inshakova
Volgograd State University, Volgograd, Russia
e-mail: gimchp@volsu.ru

A. O. Inshakova
e-mail: ainshakova@list.ru; gimchp@volsu.ru

E. G. Popkova and B. S. Sergi (eds.), *Geo-Economy of the Future*,
https://doi.org/10.1007/978-3-030-92303-7_52

all participants by simultaneously registering it on the server of each of the foreign economic departments of the EAEU and BRICS as subjects of a distributed register.

Keywords Foreign economic activity · Digital technologies · Integration associations · Right risks · Industry 4.0 · Inter-jurisdictional polysubject blockchain

JEL Codes G18 · G24 · K12 · K15 · L14 · L24 · L86 · O16

1 Introduction

Economic interaction of economic entities—residents of the member states of integration associations with the participation of the Russian Federation, such as the EAEU and BRICS, has been currently improved on the basis of intensive implementation of computer digital technologies of industry 4.0. However, the legal regulation of this process at the interstate level between the members of integration associations is practically not developed and, in some cases, contradictory regulatory decisions are allowed. Such legal risks, if they arise, negatively affect the dynamics of foreign trade activities of economic entities that are residents of the EAEU and BRICS member States (Inshakova et al., 2020b). This problem has a complex causal basis; according to the authors, one of the key factors is the lack of scientific development of concepts and categories of interstate regulation in the implementation of digital technologies of industry 4.0, as well as legal risks in the application of these technologies by economic entities-residents of the EAEU and BRICS jurisdictions.

The following documents were examined as part of the regulatory framework: the Treaty on the Eurasian Economic Union and its Annexes, The Statement on the Digital Agenda of the Eurasian Economic Union, the Decision of the Supreme Eurasian Economic Council of 11.10.2017 No.12 "On the main directions of implementation of the Digital Agenda of the Eurasian Economic Union until 2025", the Main Directions of the Implementation of the Digital Agenda of the Eurasian Economic Union until 2025 and its Annexes, the Moscow Declaration of the XII BRICS Summit, and the BRICS Economic Partnership Strategy until 2025. The study of doctrinal sources covers the scientific works of some Russian scientists, including the following authors: A. O. Inshakova, E. I. Inshakova, A. I. Goncharov, I. V. Ershova, A. E. Kalinina, V. V. Tarakanov, K. A. Tolpekin, S. Yu. Glazyev, V. E. Ponomarenko, E. A. Chistyakova, O. Yu. Sokolova, S. V. Zakharova, E. S. Yankovskaya.

2 Methodology

The content of this chapter of the monograph was developed on the basis of the materialist worldview and the general scientific method of historical materialism.

General scientific methods of cognition were applied: dialectical, hypothetical-deductive method, generalization, induction, and deduction, analysis and synthesis, empirical description. The research also uses private scientific methods: dogmatic, comparative-legal, hermeneutic, structural–functional, etc.

3 Results

If we turn to the existing regulatory framework of the EEU, then, for example, the Statement on the Digital Agenda of the Eurasian Economic Union notes need to develop the digital economy of the member states and form the digital agenda of the EEU. One may be confident that the formation of the digital agenda of the EAEU will promote open, broad, and equal cooperation between the Member states, economic entities, and citizens, increase the efficiency and volume of the economy of each of the Member states, as well as the transition of the EAEU to a new level of economic, technological, and social development (http://pravo.eaeunion.org/document?id=71553074&byPara=1&sub=1). Currently, information support for integration processes in all areas affecting the functioning of the Union is defined by article 23 of the Treaty on the Eurasian Economic Union (Astana, 29.11.2014) (with amendments and additions) (http://base.garant.ru/70670880/#ixzz6gzcJ0u7L), and the Protocol on Information and Communication Technologies and Information Interaction within the Framework of the Eurasian Economic Union (Annex No. 3 to the Treaty).

The Decision of the Supreme Eurasian Economic Council of 11.10.2017 No.12 "On the main directions of implementation of the Digital Agenda of the Eurasian Economic Union until 2025" states the following—the member states independently develop, form, and implement national policies in the areas of digitalization of the economy, communications, and informatization, ensuring the sustainable functioning and security of the unified information space and communication infrastructure, including implementing national measures to develop the digital agenda. At the same time, it indicated that the lack of a coordinated policy of the member states in the digital sphere becomes an obstacle to achieving synergistic effects in the development of the digital economy of the member states and the digital space of the Union. The implementation of the digital agenda involves the use of new business processes, digital models, and the creation of digital assets (Matytsin, 2021). For the successful implementation of projects, it is possible to create an environment that ensures the development of digital transformation projects in the format of a regulatory "sandbox". Here we predict a quite noticeable financial and economic effect. First, knowledge, skills, and abilities will be formed. Secondly, competencies will accumulate. Thirdly, various promising projects will receive support both at the stage of approbation and at the stage of transformation into a commodity. At the same time, projects will also receive support at the initial start-up stage, in addition, at the testing stage, then at the stage of designing a template for solutions. Fourthly, the accumulation of young talented scientists and their teams in the projects being developed will be very useful. Fifth, an archive of process models will be gradually created. Sixth, digital asset business models will be tested at high speed. Seventh, the participants will make joint decisions in understandable conditions and algorithms. Eighth, all of the above components of the beneficial effect will ultimately contribute to a noticeable reduction in risks. The digital transformation of the Union's markets will lead to the development of a digital market, within the framework of which it is necessary to ensure the free movement of goods, services, capital, and labor. To achieve this result, it will be necessary to align the models of regulatory changes in the internal markets of the Union with the model of digital transformation of integration processes within the digital agenda.

The digital transformation of the markets of goods and services will lead to a significant simplification of trade procedures through the transition to digital form, the active use of electronic commerce, as well as to the effective implementation and use of "single window" mechanisms in the economy. This will require further harmonization in the field of rules for conducting cross-border electronic commerce, stimulating business in digital form, and coordinating actions in the field of protection of intellectual property rights. This will also include the rights of consumers of the digital market, as well as the inclusion of issues of regulating cooperation in the field of the digital economy in the agenda of the EAEU trade negotiations with third countries (http://pravo.eaeunion.org/document?id=71708158&byPara=1&sub=1).

It is important to note that in the Annex to the Main Directions of the implementation of the Digital Agenda of the Eurasian Economic Union until 2025, the priority of the development of initiatives within the framework of the implementation of the digital agenda of the Eurasian Economic Union until 2025 was the digital traceability of the movement of products, goods, services, and digital assets in the Eurasian Economic Union (http://pravo.eaeunion.org/document?id=71708158&byPara=1&sub=1).

Drawing attention to the BRICS regulatory documents it is clarified that the Moscow Declaration of the XII BRICS Summit also recognizes the role of the digital economy as an important tool for modernization and transformation, and for stimulating inclusive economic growth, as well as maintaining smooth global trade and commercial activity, and thus helping the BRICS economies achieve the Sustainable Development Goals. At the same time, the challenges associated with the unprecedented spread of digital technologies and e-commerce are highlighted, and the need for targeted measures to reduce the digital divide and support developing countries in overcoming its socio-economic consequences is emphasized. Taking into account the accelerated development of the e-commerce sector and the growth of online transactions worldwide, the BRICS member States will deepen cooperation within the framework of the BRICS Working Group on E-Commerce. The BRICS member States also aim to study the experience of the member States, other countries, and international associations in the field of consumer protection in the field of electronic commerce and to develop a practical framework for ensuring consumer protection in the "five" countries, including pilot projects and initiatives (Moscow Declaration of the XII BRICS Summit, 2020).

The BRICS Economic Partnership Strategy until 2025 states that in the context of the Fourth Industrial Revolution, accompanied by end-to-end digitalization of all sectors of the economy, the development and implementation of digital technologies is becoming a determining factor for sustainable economic growth of the association. To ensure a synergistic effect, the BRICS countries aimed at unlocking the potential of digital technologies and opportunities for the population of the BRICS countries to acquire advanced technological solutions as promising tools for improving competitiveness, as well as labor productivity and the quality of life of the population, ensuring economic growth, expanding social rights, and inclusion in digital processes (BRICS, 2020).

Regarding the interstate regulation of the introduction of digital technologies of industry 4.0 at the BRICS level, as well as regarding the legal risks in the application of digital technologies within the framework of this integration association, it was noted that there are no scientific developments on these issues (Matytsin & Rusakova, 2021). The doctrine contains scientific research on certain aspects of digitalization at the EAEU level.

For example, K. A. Tolpekin examines the implementation of the "Digital Agenda of the EAEU until 2025", but only in fragments in the light of ensuring the effectiveness of digital criminology (Tolpekin, 2017).

In the field of modernization of money and credit, the position of the Academician of the Russian Academy of Sciences S. Yu. Glazyev deserves support. The honored colleague Sergey Yu. Glazyev, the Minister for Integration and Macroeconomics of the Eurasian Economic Commission, correctly points out that individual countries that are integrated into the EAEU union, as well as the BRICS union, have small market spaces. The trade turnover between them is also small. Financial servicing of the turnover between the member states of integration associations should be carried out in the EAEU and BRICS areas simultaneously. This space can be considered a new idea of the Greater Eurasian Partnership. The Honored Academician proposes to build a new monetary and financial system of the Greater Eurasian Partnership based on "Eurascoin". It is necessary to establish the rules for its circulation, the same for all member states of the Partnership; such a system of monetary circulation should have free access for the member states of the Partnership. The "Eurascoin" basket will be backed up by the national currency of each member state of the Partnership, which will voluntarily participate in the new system of monetary circulation. Each member state of the Greater Eurasian Partnership and the new system of monetary circulation will be able to receive loans in the digital currency "Eurascoin" in proportion to its participation in this currency basket. The Honored Academician correctly proposes to issue "Eurascoin", as well as to carry out the circulation of this digital currency using blockchain computer technology. The distributed register of "eurazcoin" allows tracking financial transactions with each single "eurazcoin" from the moment of its creation and then for each participant in the settlement without exception. The new monetary and financial system of the Greater Eurasian Partnership, based on "Eurascoin", proposed by S. Yu. Glazyev, will allow you to control all transfers with this digital currency, exchange data on all digital money transfers online. This monetary and financial system of the Greater Eurasian Partnership is characterized by high reliability, while there is no need to use SWIFT technology (Glazyev, 2018).

We support the distinguished Academician Sergey Glazyev in the formation of a Large Eurasian Partnership in the format of the EAEU + BRICS.

Ponomarenko (2017) explores the possibilities and directions of digitalization of cross-border payments in the Eurasian Economic Union as a prerequisite for the formation of a single digital payment space of the EAEU. He concludes that there is a significant potential for the development of distributed registry technology in the payment sphere, and this author formulates proposals for the use of distributed registry technology in the payment space of the EAEU. The Interstate Bank may become a key subject in the implementation of the project of the unified digital payment space of the NPP.

Chistyakova et al. (2020) explore the creation of common digital ecosystems of employment, logistics, and industrial cooperation. The authors identify the areas of formation of a competitive digital trade ecosystem: accumulation of a mass of competitive products and immersion of the EAEU business in the digital environment; saturation of cooperation in the ecosystem and increasing its maturity. It also highlights the use of the latest technologies that improve the quality of decision-making; attracting a critical mass of consumers of the EAEU countries; promoting and increasing loyalty; creating favorable regulatory conditions. We should agree with these colleagues that in order to realize the strategic interests of the EAEU member states in the field of the digital agenda of the Eurasian integration, it is necessary to harmonize the regulatory framework in the field of the digital economy. It is also necessary to ensure the interoperability of digital systems, to make changes to technical regulations.

On the issues that we are considering in this scientific development, almost none of the scientists have yet spoken at the scientific level. Only my colleague Yankovskaya (2018) comparing three aspects of the development of foreign economic activity in the context of digitalization of the economic space of the EAEU-state policy, business processes, and their scientific understanding, makes the following conclusions. First, public policy measures are developed and implemented at a faster pace in comparison with the course of economic processes and the activation of business processes. Secondly, scientific research on this problem is developing at an insufficient pace and not able to form a scientific and methodological basis in advance, which is used in practical activities to improve the sphere of foreign economic activity in the EAEU. We agree with the above-mentioned author regarding the low intensity of scientific research, especially the issues of interstate regulation of implementation and legal risks in the application of digital technologies of industry 4.0 at the level of integration associations with the participation of Russia.

Considering the legal risks in the implementation and application of digital technologies of industry 4.0 at the level of the EAEU and BRICS, it is necessary, firstly, to formulate the very concept of these risks in the field of foreign economic activity. The legal risk of the impossibility of the introduction and application or improper use of digital technologies of industry 4.0 by economic entities-residents of the EAEU and BRICS member states is the probability of a special event for them. Such an event entails more than 1/3 of the planned increase in costs (losses) or receiving less than 1/3 of the planned income (lost profits) due to the collision of national norms of the EAEU and BRICS member states regulating the introduction and (or) use of digital technologies in the field of foreign economic activity.

The regulatory guidelines that we analyzed above can generally be considered quite appropriate, but there is no concentration of regulation in specific areas, in particular, on digital technologies of foreign economic activity of companies-residents of the member states of the EEC, and BRICS. The priority of digital traceability of the movement of products, goods, services, and digital assets is not clearly defined and regulated. At the same time, any foreign trade transaction has the same discrete stages, which are always repeated, having minor differences in terms of execution, due to one or another type of cargo transport. In general, the execution of a foreign trade transaction covers 5 stages: (1) production of export goods; (2) preparation of goods for shipment; (3) insurance, payments, and settlements for goods; (4) forwarding and international transportation of goods; (5) customs clearance, the crossing of state borders by cargo.

In a categorical perspective, we propose to the group and differentiate the advanced computer (digital) technologies of industry 4.0 used by companies involved in foreign economic activity in five areas of use in foreign trade transactions, based on discrete repetitive stages of transaction execution.

As part of the primary stage of the production of export goods, the "Digital marking of goods" category should include technologies for piece-by-piece chipping of each individual unit of goods that moved through a cross-border transaction. Modern complex products manufactured with identification numbers in the form of unique pin codes, as well as with non-separable remote Internet communication devices-microchips, which work by analogy with SIM cards of modern mobile phones. This product can communicate with the command center of communication when it is necessary for the owner (buyer) of the product. In particular, this microchip can perform the function of the digital marking of this product, turn on its "beacon" at the time of shipment from the factory conveyor, and carry out its digital traceability to the final point of receipt of this unit of goods by the buyer (Inshakova et al., 2020a).

At the second stage of the transaction, when preparing the goods for shipment, it is necessary to combine the technologies of equipment with digital markers of the standard sea, railway, aviation, and automobile transport containers in the "Digital marking of standard cargo containers" category. An Internet communication device corresponding to the communication loads—a similar large microchip (possibly a minicomputer)—must be attached inseparably to a standard transport container. Such a container minicomputer provides continuous tracking of the location of each unit of transported goods inside this container and supports tracking the movement of the container itself along the route of the transaction (Inshakova & Goncharov, 2019).

At the third stage of the cross-border transaction, insurance, payments, and settlements for the goods carried out. At this stage, the "Foreign Trade smart contracts" category should include technologies for automatic execution of payments for goods, other payments, and shipments of goods under transactions that are carried out on the basis of digital smart contract technologies. It is quite logical to connect the work of digital markers-microchips from each unit of goods, minicomputers, which are equipped with standard transport containers that move along the route of the transaction on a particular vehicle, to the functionality of the execution of smart contracts via the Internet connection. There are no technical problems to link the beginning of the movement or the refusal to start the movement of goods along the route of the transaction with the financial execution using remote banking technology: payments for insurance of risks in connection with the transportation of goods, payment of the price for the goods. All this in the end and create an algorithm for a smart contract for a specific foreign trade transaction (Inshakova et al., 2020c).

For the fourth stage, which is conditionally delimited as forwarding and international cargo transportation, it proposed to combine digital technologies of continuous satellite bearing from space in the "Roaming with carrier" category, as well as technologies for transmitting signals to the command center of communication for tracking standard cargo containers that are transported by ship, train, plane or car along the route of the transaction. Such transport containers, equipped with minicomputers that are on a continuous Internet connection, also connected via the Internet to the interactive portals "Foreign economic activity in the BRICS on-line association", "Foreign economic activity in the EAEU on-line association". These interstate Internet portals can function as independent Internet sites, can function as special tabs of the sites of the integration associations of the EAEU and BRICS themselves (Tarakanov et al., 2020).

At the fifth final stage of the execution of a foreign trade transaction, there are the technologies of forming an electronic cargo passport, monitoring and registering the passage of customs procedures by goods integrated into the "Electronic State Border" and "Customs Monitoring" categories. As well as crossing the state border of the jurisdiction of their dispatch and the jurisdiction of their delivery along the routes of foreign trade transactions by standard commodity containers. In addition, the movement of goods can be traced through intermediate "digital gates", which must be installed in intermediate sea (river) ports of destination, at railway junctions, major interchanges, and intersections of highways.

4 Conclusion

The complex of information and communication technologies for continuous tracking of the movement of individual goods and vehicles that transport goods under foreign trade transactions between the jurisdictions of the EAEU and BRICS associations are differentiated into five categories. This set of technologies applied on basis of a new scientific and methodological approach to inter-jurisdictional risk management of collisions with legal provisions. This also concerns an approach to neutralizing and avoiding these risks when carrying out economic activities by economic entities of the jurisdictions of these associations. The key feature of the new approach is the coordination of the regulatory documents adopted by the member states of the integration associations in the mechanism of the inter-jurisdictional polysubject blockchain. Such documents contain regulatory provisions on the introduction and application of digital technologies of foreign economic activity, up to the level of a single (identical) regulatory document for all participants by simultaneously registering it on the server of each of the foreign economic departments of the EAEU and BRICS (Kalinina et al., 2019).

This approach of inter-jurisdictional management of legal risks in the implementation and application of digital technologies of industry 4.0 in the EAEU and BRICS associations is proposed to be enshrined in the provisions of the International Convention "On Digital Technologies of Foreign Economic Activity used by Residents of the Greater Eurasian Partnership Jurisdictions". The member states of the EAEU and BRICS integration associations will be able to join this convention in the mechanism of the inter-jurisdictional polysubject blockchain, thereby ensuring the entry into force of the relevant regulatory provisions in their national territories.

Acknowledgements The reported study was funded by the Russian Science Foundation (project No. 20-18-00314).

References

BRICS Economic Partnership Strategy until 2025. (2020). https://brics-russia2020.ru/images/114/81/1148133.pdf. (date accessed: May 10, 21).

Chistyakovam, E. A., Sokolovam, O. Y., & Zakharovam, S. V. (2020). Common digital ecosystems of the EAEU countries. *Bulletin of the Samara State Economics University, 2*(81), 38–40.

Glazyev, S. Y. (2018). On the use of digital technologies to create the EAEU Capital Market. *Eurasian Integration: Economics, Law, Politics, 1*, 7–8.

Inshakova, A. O., & Goncharov, A. I. (2019). Innovation as an integral condition for the development of modern foreign trade turnover of energy resources and its legal regulation. In *Lecture notes in networks and systems* (Vol. 44, pp. 67–78).

Inshakova, A. O., Goncharov, A. I., & Ershova, I. V. (2020a). Digital blockchain registration of legally significant stages of complex goods export-import supplies by business entities of the EAEU and BRICS jurisdictions. In *Lecture notes in networks and systems* (Vol. 110, pp. 328–336).

Inshakova, A. O., Goncharov, A. I., & Salikov, D. A. (2020b). Electronic-digital smart contracts: Modernization of legal tools for foreign economic activity. In *Lecture notes in networks and systems* (Vol. 91, pp. 3–13).

Inshakova, A. O., Goncharov, A. I., Inshakova, E. I., & Tumchuk, Yu. A. (2020c). Digital technologies for alternative methods of resolving conflicts: The prospects of application in Russia and other BRICS countries. In A. O. Inshakova, & A. V. Bogoviz (Eds.), *Alternative methods of judging economic conflicts in the national positive and soft law*. E. G. Popkova (Series Ed.), *Advances in research on Russian business and management*. Information Age Publishing. https://www.infoagepub.com/products/Alternative-Methods-of-Judging-Economic-Conflicts-in-the-National-Positive-and-Soft-Law

Kalinina, A. E., Inshakova, A. O., & Goncharov, A. I. (2019). Polysubject jurisdictional blockchain: Electronic registration of facts to reduce economic conflicts. *Studies in Computational Intelligence, 826*, 205–213.

Matytsin, D. E. (2021). Internet-investing as a remote algorithm of retail investment financing. In *Modern global economic system: Evolutional development versus revolutionary leap. Springer book series lecture notes in networks and systems*. Springer Nature.

Matytsin, D. E., & Rusakova, E. P. (2021). The strategy of quality management in industry 4.0 and formation of the cognitive economy based on industrial and manufacturing engineering in the Russian Federation and countries of the EU. *International Journal for Quality Research, 15*(4). https://doi.org/10.24874/IJQR15.04-03

Moscow Declaration of the XII BRICS Summit. (2020). http://kremlin.ru/supplement/5581

Ponamorenko, V. E. (2017). On the grounds and prospects for the formation of a unified digital payment space of the EAEU. *Legal Science, 6*, 24–28.

Statement on the digital agenda of the Eurasian Economic Union "On The transition of the EAEU to a new level of economic, technological, and social development". http://pravo.eaeunion.org/document?id=71553074&byPara=1&sub=1. (date accessed: May 08, 2021).

Tarakanov, V. V., Inshakova, A. O., & Goncharov, A. I. (2020). Interactive demo online maintenance foreign trade deals of business entities of BRICS jurisdictions. In *Lecture notes in networks and systems* (Vol. 110, pp. 319–327).

The agenda of the EAEU trade negotiations with third countries (2020). http://pravo.eaeunion.org/document?id=71708158&byPara=1&sub=1. (date accessed: May 08, 2021).

The Treaty on the Eurasian Economic Union (Astana, 29.11.2014) (with amendments and additions). http://base.garant.ru/70670880/#ixzz6gzcJ0u7L (date accessed: May 10, 2021).

Tolpekin, K. A. (2017). International, national and integration norms of the EAEU on the use of information techniques of digital criminalistics in countering cybercrime. *International Cooperation of the Eurasian States: Politics, Economics, Law, 4*, 87–96.

Yankovskayam, E. S. (2018). Prospects for the development of foreign economic activity in the conditions of digital transformation of the Eurasian Economic Union. In *Scientific notes of the St. Petersburg branch of the Russian customs academy named after V.B. Bobkov* (Vol. 3, No. 67, pp. 78–82).

The Concept of Neo-industrial Modernization of Legal Regulation Strategy for the Intensification of Digital Technologies in the EAEU and the BRICS

Agnessa O. Inshakova, Alexander I. Goncharov, and Denis E. Matytsin

Abstract The chapter contains a scientific and methodological justification for updating the legal tools of the strategy of expanded and more active use of digital technologies, including economic entities of the EAEU and BRICS jurisdictions in the implementation of foreign trade activities, at the conceptual level. The concept of neo-industrial modernization of the strategy for the intensification of digital technologies is a scientific synthesis of three methodological segments that combine the theory, methodology, and legal mechanisms for the formation and implementation of this strategy in integration associations with the participation of the Russian Federation—the EAEU and the BRICS. The concept provides eight main directions for the formation of a single digital space for foreign economic activity. It proposes to consolidate the modernized legal tools as part of the provisions of the international convention, to which the countries participating in integration associations will be able to join, thereby ensuring the entry into force of the relevant regulatory provisions in their jurisdictions. It proved that it is methodologically mandatory for the member states of the EAEU and the BRICS to work out the newly introduced digital technologies for foreign trade activities in the mechanism of the inter-jurisdictional poly subject blockchain.

Keywords Foreign trade · Neo-industrial · Digital technologies 4.0 · Inter-jurisdictional poly subject blockchain · Greater Eurasian Partnership · Concept · Common digital space

JEL Codes G18 · G24 · K12 · K15 · L14 · L24 · L86 · O16

A. O. Inshakova (✉) · A. I. Goncharov · D. E. Matytsin
Volgograd State University, Volgograd, Russia
e-mail: ainshakova@list.ru; gimchp@volsu.ru

A. I. Goncharov
e-mail: gimchp@volsu.ru

D. E. Matytsin
e-mail: Dmatytsin@ya.ru; gimchp@volsu.ru

E. G. Popkova and B. S. Sergi (eds.), *Geo-Economy of the Future*,
https://doi.org/10.1007/978-3-030-92303-7_53

1 Introduction

The EAEU and BRICS member states are trying to solve the problem of broader and more intensive use of digital technologies in economic activity, and documents of these integration associations adopted at the interstate level on this issue. We can assume this trend of technological development its strategy and has received a certain legal regulation. However, the latest neo-industrial stage of the development of public relations naturally requires scientific understanding. Further, this will entail a modernization adjustment of the efforts of the member states of the Greater Eurasian Partnership in the development of the digital sphere of interaction.

The documents adopted at the beginning of 2021 at the interstate level do not contain an updated legal mechanism for the use of neo-industrial digital technologies 4.0 in foreign trade by economic entities of the EAEU and BRICS member states that meet the requirements of the time. Unfortunately, most of the basic documents should be included in these regulations. These are the Moscow Declaration of the XII BRICS Summit (http://kremlin.ru/supplement/5581) and BRICS Economic Partnership Strategy until 2025 (https://brics-russia2020.ru/images/114/81/1148133.pdf), for the integration association of the EAEU: the Treaty on the Eurasian Economic Union and its Annexes (http://base.garant.ru/70670880/#ixzz6gzcJ0u7L), the Statement on the Digital Agenda of the Eurasian Economic Union (http://pravo.eaeunion.org/document?id=71553074&byPara=1&sub=1), the Decision of the Supreme Eurasian Economic Council of 11.10.2017 No. 12 "On the main directions of implementation of the digital agenda of the Eurasian Economic Union until 2025" (http://pravo.eaeunion.org/document?id=71708158&byPara=1&sub=1), the main directions of the implementation of the digital agenda of the Eurasian Economic Union until 2025 and their Annexes (http://pravo.eaeunion.org/document?id=71708158&byPara=1&sub=1).

Foreign researchers, analyzing the order and possibilities of using modern digital technologies of Industry 4.0 (Muller & Voigt, 2018; Szalavetz, 2019; Weyer et al., 2015), first, practically do not concern the specifics of foreign economic activity. Secondly, the problems of digital technological support for foreign trade in the integration associations of the BRICS and the EAEU are not dealt with at all.

As part of the regulatory framework, the Treaty on the Eurasian Economic Union and its Annexes, the Statement on the Digital Agenda of the Eurasian Economic Union, Decision of the Supreme Eurasian Economic Council of 11.10.2017 No. 12 "On the main directions of Implementation of the Digital Agenda of the Eurasian Economic Union until 2025", the main directions of the implementation of the digital agenda of the Eurasian Economic Union until 2025 and their Annexes, Moscow Declaration of the XII BRICS Summit, and also, BRICS Economic Partnership Strategy until 2025 are studied.

The study of doctrinal sources covers the scientific works of many Russian scientists, including these authors: Goncharov A. I., Frolova E. E., Inshakova A. O., Kazachenok S. Yu., Kovalev S. I., Matytsin D. E., Rusakova E. P., Sevost'yanov M.

V. Also scientific works of a number of foreign authors, namely: Szalavetz, Muller, Voigt, Kagermann, Gilchrist, Weyer, Schmitt, Ohmer, Goreck.

2 Methodology

The content of this chapter of the monograph was developed on the basis of the materialist worldview and the general scientific method of historical materialism. General scientific methods of cognition were applied: dialectical, hypothetical-deductive method, generalization, induction, and deduction, analysis and synthesis, empirical description. Also, the following private scientific methods were used: dogmatic, comparative-legal, hermeneutic, structural–functional, etc.

3 Results

In our opinion, digital technologies of Industry 4.0 in the context of foreign economic activity in the integration associations of the EAEU and BRICS include, firstly, digital technologies of remote Internet support of the goods themselves (Matytsin, 2021). Such a materialized, modern, and relatively complex smart product must be equipped with an information and communication device for remote Internet communication that is inseparable from the entire service life of the product. The specified microcomputer should allow uniquely identify this unit of goods, track the location and technical condition of this product, as well as remotely receive commands to perform certain functions of the product, ensure the execution of the received commands by the smart product. Products with this functionality should be classified as neo-industrial products.

Secondly, we believe that digital neo-industrial technologies 4.0 should include digital technologies of remote purchase and sale of goods via the Internet and end-to-end tracking of the movement of goods from the initial shipment of the manufacturer to the receipt of the goods at the final logistics point designated by the buyer (Matytsin & Rusakova, 2021). Such remote information and communication technologies should provide both the buyer and the supplier with the opportunity to obtain information about the location of the goods transported along the route of the transaction at any time. In addition, the remote purchase and sale of goods via the Internet and the subsequent movement of smart goods in the transaction must be combined with the use of digital smart contract technology, which ensures the automatic execution of shipments and payments within this transaction. Such purchase and sale must initially be carried out according to the algorithm of the specified smart contract. Smart contracts for complex products are immanent and inherent in Industry 4.0 (Inshakova et al., 2020a).

Third, the participating states should agree as soon as possible on the formation and development of a single digital space for both the EAEU and the BRICS-for the countries of the Greater Eurasian Partnership (Inshakova et al., 2020b).

Fourth, we consider it mandatory to attribute the latest technology of the inter-jurisdictional poly subject blockchain to the digital technologies of Industry 4.0, which used to simultaneously build and legally regulate a single digital space for foreign economic activity of the Greater Eurasian Partnership. The formation and development of a single digital space is the red thread of the entire Concept of Neo-industrial Modernization of legal regulation of the strategy of intensification and unification of digital technologies for the foreign trade activities of economic entities of the EAEU and BRICS jurisdictions (Inshakova et al., 2020c).

Conceptually and terminologically, the single digital space for foreign economic activity of the Greater Eurasian Partnership covers and fully includes digital technologies, which we deductively relate to the above-mentioned first and second provisions of our theory; as general provisions of the Concept of a single digital space for Foreign Economic Activity of the Greater Eurasian Partnership. We understand how a set of legally binding documents on the implementation and application of digital technologies is openly prepared, coordinated, and archived on a digital platform independent of all participating states through the blockchain. The digital platform provides instant, simple, and non-discriminatory access for each member of the integration association to the inter-jurisdictional poly subject blockchain system—access to a specific root directory (archive folder) corresponding to a particular digital technology for foreign economic activity. All initiatives, objections, and changes in the documents of this root directory related to the implementation and application of a particular digital technology are necessarily duplicated after it is made. It duplicated on the server of each digital ministry and foreign economic department of the member state of the integration association, after which they can no longer be changed and supplemented without the consent of all partners.

In our opinion, it is methodologically mandatory for the member states of the Greater Eurasian Partnership to work out the newly introduced digital technologies for foreign trade activities in the mechanism of the inter-jurisdictional poly subject blockchain. This blockchain technology provides a step-by-step recording of every intention, every agreement, every legally significant fact, which will not need to be proved later in the event of a potential conflict (dispute) during its settlement. This provision is the target of building a single digital space: speed and simplicity of interaction of the partner states for coordination at the legal level of introduction and application of digital technologies of foreign trade activity at low risks of conflicts between participants.

Legal support for the provisions of the strategy for the expansion of digital technologies in the Greater Eurasian Partnership area is based on a system of normative legal acts regulating the construction and functioning of a single digital space. The elements of this system are the legislative acts of the member states of integration associations, the Treaty on the EAEU and its Annexes, acts of the EAEU organs,

Agreements within the framework of BRICS, international treaties within the framework of associations. In our opinion, the issue of nominating one national IT developer from each member country of the Greater Eurasian Partnership for the period 2021–2024 requires the most urgent coordination and solution, each of them, working in a team with other national IT developers, will create, implement and apply digital technologies for foreign trade activities. At the same time, the receipt of technical tasks for a particular digital technology by an international team of IT developers, reporting on the phased execution of work, and documents on the acceptance of the completed volumes of work recorded on the digital platform in the mechanism of the inter-jurisdictional poly subject blockchain.

In the period after 2025, the member States of the Greater Eurasian Partnership should continue to work on the final formation of a single digital space, focusing on the unification of neo-industrial digital technologies, on bridging the digital divide between countries, deepening digital integration, and harmonizing the legislation of the member states.

Within the framework of our Concept of Neoindustrial Modernization of the legal tools of the strategy for expanding digital technologies of the Greater Eurasian Partnership, we will outline eight main directions for the formation of a single digital space for foreign economic activity.

The first direction, of course, is the regulation of a single digital space. The main task of this regulation is to ensure the effective functioning of the single digital space, subject to the continuous prevention of regulatory arbitration, as well as the risks of destabilization of the single digital space. Here, when we mention regulatory arbitration, we mean unjustified competitive advantages of the use of neo-industrial digital technologies by economic entities of a particular jurisdiction, when these advantages arise due to differences in legal regulation. The regulation of the single digital space is based on the following three principles. First—the accessibility and clarity of both regulatory and supervisory rules. Second—the transparency of the activities of both participants and regulators of the single digital space. Third—the harmonization of both regulation and supervision based on advanced international recommendations and treaties. The system of regulation of the single digital space of foreign economic activity should integrate the interrelated activities of three groups of subjects. Digital and foreign economic ministries of the member states; a specialized body of the Greater Eurasian Partnership (such a body should be formed based on a special interstate agreement); the inter-jurisdictional regulatory body of the single digital space (with the competence established also based on an interstate agreement).

It is clarified that the provision of the interstate agreement on the powers of the inter-jurisdictional regulatory organ of the single digital space of foreign economic activity does not cancel. Or replace the obligations of each participating state (national digital ministry) to delegate on time to work for the benefit of the single digital space the above-mentioned national IT developer supplier of digital technologies.

The second direction of the formation of a single digital space for foreign economic activity is the harmonization of the legislation of the participating states. The main task of such harmonization is to create similar rules for the functioning of the digital space, provided that certain differences in the national regulatory framework do not hinder the development and create barriers to the digital space. Specific areas for the harmonization of legislation should be identified by the participating states also on the basis of an interstate agreement, which should include a corresponding plan. This plan should include time stages and differentiation of harmonization rates for different participating states. Here, the speed is determined by the model and features of the regulation of the digital space in a particular state; the degree of development of advanced digital technologies in it and the implementation of international digital standards; prevention of the risks of destabilization of the national digital space and, as a result, the unified digital space. The specified harmonization plan should also reflect the qualitative criterion of its sufficiency. In our opinion, the level of harmonization of legislation between states considered sufficient, first, provided each state accept the obligation to allow economic entities—residents of the partner's state to its internal national digital space; secondly, reducing to a minimum the regulatory interstate arbitration for economic entities—participants of the single digital space; third, the availability of unified tools for removing inefficient digital technologies from a single digital space. Such principles are important for ensuring the stability of both the internal national digital space and the unified digital space. As well as the principles that the partner states consider important for maintaining trust in digital technologies of economic entities and end end-users.

The third direction of the formation of a single digital space for foreign economic activity is the procedural provision of mutual admission of economic entities using digital technologies to the internal national digital spaces of the partner states. Such mutual access is necessary to provide an opportunity to apply digital technologies in foreign economic activity by residents of one partner state in the territories of other partner states. There is no doubt that their conclusion of an international treaty in this regard is presumptive. In our opinion, three forms of this mutual admission will be useful: first, a subsidiary organization can be established; second, a cross-border branch can be opened (according to the obligations of the partner states under the WTO); third, remote cross-border (without a presence in a foreign territory) provision of digital technologies is possible. It should be clarified here that the creation of a subsidiary IT developer and digital technology provider may be accompanied by its cross-border IT licensing. Such special licensing carried out by the digital ministry of the state of which the parent IT organization is a resident. The requirements for a subsidiary licensed entity must include all the maximum quantitative and maximum reputational licensing requirements established by the laws of the partner states.

The fourth direction of the formation of a single digital space for foreign economic activity is the regime of supervision over the activities of all its participants. It is obvious that, for example, the supervision of the activities of the parent IT company carried out by the digital ministry of the country of which it is a resident. If we consider subsidiary IT-company, respectively, the supervision implemented by the digital ministry of the country of its residence. The activities of cross-border branches are also controlled following the laws of their host countries. Supervision of remote cross-border (without a presence in a foreign territory) provision of digital technologies is carried out by the digital ministry of the country of origin of the IT supplier of a particular digital technology. In our opinion, mutual assistance is needed between the regulatory authorities of the country of origin and the host country. At the same time, the legal mechanisms and forms of supervision should also be determined on basis of an interstate agreement. In particular, the supervisory board, which has advisory and advisory competencies, contributes to an intensive and continuous dialogue between the supervisory authorities—the digital ministries of the partner states, seems to be quite efficient. Within the framework of the aforementioned fourth direction of the formation of a single digital space for foreign economic activity, the above-mentioned harmonization of the legislation of the partner states in the field of supervision was also carried out.

The fifth direction of the formation of a single digital space for foreign economic activity is the creation of its infrastructure. In this context, infrastructure is understood as a system of interconnected institutions that ensure the functioning of a single digital space. The purpose of this infrastructure is to create a business environment that would guarantee the availability, smooth functioning of a single digital space, respect for intellectual property rights in digital technologies, and control over risks for technology users.

The sixth direction of the formation of a single digital space for foreign economic activity is the protection of the rights and legitimate interests of IT developers and foreign trade companies using digital technologies. In our opinion, an adequate protection mechanism created taking into account the following conditions:

- the partner states should identify a priority set of digital technologies for which the partners will jointly coordinate the regime for protecting the rights and legitimate interests of IT developers and foreign trade companies;
- with the help of the inter-jurisdictional poly subject blockchain, IT developers and foreign trade companies continuously informed about the priority digital technologies for the partner states in the current time;
- the specifics of national legal regulation of digital technologies, including their creation and use in the partner states, is taken into account;
- a regime of interaction between national digital ministries and foreign economic agencies, representatives of the foreign trade business community of the partner states should be formed through information exchange on the practice of using digital technologies and the practice of preventing and resolving disputes (conflicts). It should be clarified here that this mode of interaction should be supplemented by the creation of an inter-judicial body for the pre-trial settlement of potential disputes (conflicts) over the use of digital technologies and the functioning of a single digital space;
- the development by the partner states of common approaches to countering digital fraud both within national territories and within the framework of a single digital space of foreign economic activity;
- the creation by the partner states of common approaches to the disclosure of information about digital technologies for foreign economic activity using the inter-jurisdictional poly subject blockchain.

In general, the main functions of the mechanism for protecting the rights and legitimate interests of IT developers and foreign trade companies using digital technologies should provide the following positive practices: stable transparency and accessibility within the national digital spaces of the partner states; speed and adequacy of the supervisory response of states to violations of the rights and interests of the subjects of the digital space; the predominance of the pre-trial method of settling disputes (conflicts) over digital technologies and their use by subjects within the single digital space; a steady trend towards harmonization within the national legislation on digital technologies of the partner states.

The seventh direction of the formation of a single digital space for foreign economic activity is to ensure cybersecurity. In our opinion, cybersecurity is achieved if some basic principles are followed. The first is the unity of standards (including the reception of advanced international standards) for the recognition of cybersecurity of digital technologies used in foreign economic activity. The second is the unity of approaches (rules) of the digital ministries of the partner states and subjects of the digital space to the prevention of cyber risks and cyber threats, as well as to ensuring the cyber stability of digital technologies. The third is the presence of a mechanism for responding to cyber-attacks that have taken place, which allows protecting the rights of the subjects of the single digital space. Fourth—continuous coordination of efforts and exchange of data on cyber incidents of digital ministries of partner states and subjects of the digital space, improvement of preventive measures. In general, in the field of cybersecurity, the partner states solve the following main tasks: confidentiality of commercial information about digital technologies; smooth functioning of the single digital space; effectiveness of the mechanism for protecting rights, including remote information and communication technologies for its use; trust of subjects in commercial information transmitted within the single digital space of foreign economic activity.

In this regard, we identify four main areas for the efforts of partner States to ensure cybersecurity:

- the creation of conditions for safe remote Internet interaction of the subjects of the single digital space in the implementation of foreign economic activity;
- the creation of a permanent platform for the analysis of cyber threats and cyber incidents, which should function with the help of an inter-jurisdictional poly subject blockchain;
- the creation of common standards for cybersecurity and cyber-resilience, and their control;
- coordination of policy in the field of protection of the rights and legitimate interests of the subjects of the single digital space.

The eighth direction of the formation of a single digital space for foreign economic activity is the creation of a single digital unit of account, which should be used by participants in foreign trade transactions instead of national monetary currencies and (or) foreign currencies of third countries.

4 Conclusion

The concept of neo-industrial modernization of legal regulation of the strategy of expansion and intensification of digital technologies in the EAEU and BRICS revealed as the theory, methodology, and legal mechanisms for the formation of a single digital space. From the standpoint of legal regulation, eight main directions for the formation of a single digital space for foreign economic activity of the Greater Eurasian Partnership developed. These main directions proposed to be fixed as part of the provisions of the International Convention “On Digital Technologies of Foreign Economic Activity Used by Residents of the Greater Eurasian Partnership Jurisdictions”. The member states of the EAEU and the BRICS integration associations will be able to join this convention in the system of inter-jurisdictional poly subject blockchain, thereby ensuring the entry into force of the relevant regulatory provisions in their national territories.

Acknowledgements The reported study was funded by the Russian Science Foundation (project No. 20-18-00314).

References

BRICS Economic Partnership Strategy until 2025. Available at: https://brics-russia2020.ru/images/114/81/1148133.pdf Accessed April 22, 2021.

Decision of the Supreme Eurasian Economic Council of 11.10.2017 No. 12 “On the main directions of Implementation of the Digital Agenda of the Eurasian Economic Union until 2025”. Available at: http://pravo.eaeunion.org/document?id=71708158&byPara=1&sub=1. Accessed April 12, 2021.

Inshakova, A. O., Goncharov, A. I., & Salikov, D. A. (2020a). Electronic-digital smart contracts: Modernization of legal tools for foreign economic activity. In *Lecture notes in networks and systems* (Vol. 91, pp. 3–13).

Inshakova, A. O., Goncharov, A. I., Kazachenok, S. Yu., & Sevost’yanov, M. V. (2020b). Med-arb: Concept, features of regulation, and prospects for law enforcement. In A. O. Inshakova & A. V. Bogoviz (Eds.), *Alternative methods of judging economic conflicts in the national positive and soft law*. E. G. Popkova (Series Ed.), *Advances in research on Russian business and management*. Information Age Publishing. https://www.infoagepub.com/products/Alternative-Methods-of-Jud-ging-Economic-Conflicts-in-the-National-Positive-and-Soft-Law

Inshakova, A. O., Frolova, E. E., Rusakova, E. P., & Kovalev, S. I. (2020c). The model of distribution of human and machine labor at intellectual production in industry 4.0. *Journal of Intellectual Capital, 21*(4), 601–622. https://doi.org/10.1108/JIC-11-2019-0257

Matytsin, D. E. (2021). Internet-investing as a remote algorithm of the retail investment financing. In *Modern global economic system: Evolutional development versus revolutionary leap. Springer book series lecture notes in networks and systems*. Springer Nature.

Matytsin, D. E., & Rusakova, E. P. (2021). Strategy of quality management in industry 4.0 and formation of a cognitive economy based on industrial and manufacturing engineering in the Russian Federation and countries of the EU. *International Journal for Quality Research, 15*(4). https://doi.org/10.24874/IJQR15.04-03

Moscow Declaration of the XII BRICS Summit. Available at: http://kremlin.ru/supplement/5581 Accessed April 02, 2021.

Muller, J., & Voigt, K. (2018). Sustainable industrial value creation in SMEs: A comparison between industry 4 and made in China 2025. *International Journal of Precision Engineering and Manufacturing-Green Technology, 5*(5), 659–670.
Statement on the Digital Agenda of the Eurasian Economic Union. Available at: http://pravo.eaeunion.org/document?id=71553074&byPara=1&sub=1 Accessed March 02, 2021.
Szalavetz, A. (2019). Industry 4.0 and capability development in manufacturing subsidiaries. *Technological Forecasting and Social Change, 145*, 384–395.
The main directions of the implementation of the digital agenda of the Eurasian Economic Union until 2025 and their Annexes. Available at: http://pravo.eaeunion.org/document?id=71708158&byPara=1&sub=1 Accessed April 12, 2021.
The Treaty on the Eurasian Economic Union and its Annexes. Available at: http://base.garant.ru/70670880/#ixzz6gzcJ0u7L Accessed February 02, 2021.
Weyer, S., Schmitt, M., Ohmer, M., & Goreck, D. (2015). Towards industry 4.0—Standardization as the crucial challenge for highly modular, multi-vendor production systems. *IFAC-PapersOnLine 48*(3), 579–584.

Features of Corporate Social Responsibility in Russia Within the Framework of Sustainable Development

Irina V. Degtyareva, Inna V. Bagautdinova, Guzel F. Tokareva, and Olga I. Shalina

Abstract Transforming business for sustainable development is the most important challenge facing companies around the world today. The study and dissemination of best practices allow all economic agents from different industries to implement a strategic approach to the management of corporate social activities, as well as to integrate the principles of responsible business management into the system of basic management functions. The technologies of the Fourth Industrial Revolution are rapidly changing the structure of markets and the strategies of companies, but they cannot be considered outside the context of the interaction of business, society, and nature. To implement more sustainable business models and address the main issues of the current decade related to social and environmental risks, an open dialogue between the state, science, and business is necessary. The article shows the features of corporate social responsibility (CSR) in Russia at the present stage, identifies the problems, and prospects for its development. The analysis of trends in the adaptation of Russian realities to the UN concept of sustainable development is carried out. The methodology used in the study contained a set of methods and approaches. In addition to general scientific methods of analysis and synthesis, induction and deduction, system, situational, process, and dynamic approaches were used. As a result of the research, several conclusions were drawn and practical recommendations were proposed to the state and the private sector for organizing and improving the effectiveness of the CSR system following current global trends. Also, the reaction of Russian companies to the Covid-19 pandemic is shown.

Keywords Corporate social activities · Responsible business · Social investment · Socially responsible companies · Sustainable development

JEL Codes L20 · L25 · M21

I. V. Degtyareva · I. V. Bagautdinova (✉) · G. F. Tokareva · O. I. Shalina
Ufa State Aviation Technical University (USATU), Ufa, Russia

O. I. Shalina
e-mail: olenkash@bk.ru

E. G. Popkova and B. S. Sergi (eds.), *Geo-Economy of the Future*,
https://doi.org/10.1007/978-3-030-92303-7_54

1 Introduction

Today, there is a steady increase in corporate social responsibility (CSR) in Russian business, which corresponds to global trends. Leading companies are gradually integrating the UN Sustainable Development Goals into the strategic management system. However, often joint initiatives of responsible business are limited to charitable projects. According to the Report on Social Investment in Russia—2019, the main indicators of social investment of Russian business in quantitative and qualitative terms from 2003 to 2018 do not show significant changes. The situation with the new coronavirus pandemic has also had a direct impact on corporate social responsibility. Since active production activities have a huge impact on the environment, including negative ones, the problem of corporate participation in the process of sustainable development comes to the fore. In this regard, any research on the theory and practice of corporate social responsibility, aimed at identifying problems and trends in responsible business behavior, is relevant and timely.

The main hypothesis of the study is the assumption that the formation of a National concept of sustainable development and social responsibility, including measures of state incentives for CSR, is one of the primary tasks in the post-pandemic economy (Degtyareva et al., 2016). It is assumed that this work can be carried out most effectively with the joint efforts and active participation of the state, the professional, and business community.

The purpose of the study is to identify the features of CSR development in Russia and to find the most effective approach to the implementation of corporate social responsibility within the framework of the sustainable development strategy.

The main issues of the work are:

1. Analysis of the peculiarities of CSR development in Russian business
2. Research of the current state of CSR of leading Russian companies
3. Identification of problems and prospects of social investment in Russia. **Suggestions and recommendations** The subject of the study is the features and nature of corporate social responsibility of large Russian companies at the present stage. The object of the study is large Russian companies that position themselves as socially responsible or strive for such positioning.

2 Methodology

The research methodology consists of the iterative application of some interrelated research methods. First of all, these are general scientific methods of analysis and synthesis, induction, and deduction. International CSR practice: existing corporate governance systems, regulatory framework (legal norms in force in different countries), methodological approaches (primarily those enshrined in the UN standards and recommendations, and Sustainable Development Goals). In addition, when writing the paper, the materials of the "Report on Social Investment in Russia—2019" were used, the data of which were prepared using a standardized survey by correspondence questionnaire of the largest Russian companies that position themselves as socially responsible, or strive for such positioning. The data analysis was based on general information about the company's corporate social activities: questions about the company's terminology, interaction with stakeholders, international standards used, the code of ethics and principles of conduct, participation in business associations, financial and non-financial reporting. The consolidated characteristics of the strategy in the field of corporate social responsibility and sustainable development were also taken into account: the goals of implementing the strategy, integrating the UN Sustainable Development Goals, corporate social investments, and evaluating their effectiveness, and charitable activities.

3 Results

The development of corporate social responsibility (CSR) in Russian business has been affected in recent years by some complex global and national factors related to the business environment, the development of the theory and practice of corporate social activity (Blagov & Sobolev, 2014). In particular, the concept of sustainable development, reflected in the UN Sustainable Development Goals (UN SDGs), has undergone fundamental changes—from focusing on saving resources for the benefit of future generations to urgent measures to preserve the main systems of global life support. In addition, the strategic trend of CSR development focused on creating interconnected value for business and society has shown its limitations in the context of the implementation of the UN SDGs. New approaches to the development of intra-and intersectional cooperation, ensuring more active involvement of business, are being updated. There is a noticeable trend of increasing institutional support for responsible business conduct at the national level. This is expressed both in the most important legislative initiatives, such as the introduction of the Draft Federal Law "On Public Non-financial Reporting", the adoption of amendments to the laws "On Charitable Activities and Volunteerism" and "On the Development of Small and Medium-sized Businesses in the Russian Federation", as well as in the adoption of National Projects for socio-economic Development. However, a noticeable increase in responsible business conduct, in general, was not accompanied by an adequate increase in interest in CSR and sustainable development issues on the part of Russian business (Degtyareva, 2010). Today, the adaptation of the UN SDGs to the activities of Russian commercial enterprises and their integration directly into the processes of strategic management occurs at different speeds and with different degrees of possible influence on the transformation processes within the enterprises' business model. This is also confirmed by the fact that, according to the study, CSR and sustainable development strategies in the extractive sector are mainly aimed at "reducing risks" (twice as often as in the processing industry, and 4 times more often than in the service sector) and creating sustainable competitive advantages, and are more often guided in setting priorities by the global Sustainable Development Goals (Blagov &

Petrova-Savchenko, 2020). Companies in the raw materials segment define their labor collectives as the target audience for such social investment programs, whereas for processing and service enterprises, such a target audience is traditionally customers and buyers. More energetic actions on the part of the state are also important to create a favorable environment: business expectations of greater state involvement and stimulating social investment remain high.

It is also necessary to note the speed with which Russian companies have changed course and declare their commitment to the UN Sustainable Development Goals until 2030. Of particular interest today are the prospects for CSR and corporate sustainability in Russia in comparison with international practice. A noticeable trend in the world is the desire for greater transparency and demonstration by companies of compliance with international standards, which implies their implementation and reporting (Degtyareva et al., 2016). Today, there is an active implementation of international standards in the company's management system and the rapid development of new standards. At the same time, the concepts themselves change faster than the standards. The transition to a new qualitative level is hindered by the lack of focus on partnership both within the corporate sector and on intersectional interaction with the state and NGOs. There is still a lack of information about how modern ideas of sustainable development affect the core business of Russian companies, and what innovative solutions they offer. Unfortunately, not all Russian companies have realized the importance of this issue. It becomes especially relevant in the context of global competition when responsible business behavior and its business reputation are the most important factors of competitiveness. At the same time, despite the difficult socio-economic conditions, 43% of the companies-respondents kept the same amount of social investment, and 40% even increased it. The events of the 2020 pandemic have strengthened the understanding of taking into account the interests of society and building open and transparent relations with the external environment. Using CSR and sustainable development as a competitive advantage requires their integration with the corporate governance system at the level of values, responsibility, and risk-sharing. The main obstacle to this is the limited scope of corporate governance and corporate social activities (limited number of public companies, insufficient development of the financial market, the weak interest of the state in supporting the competitive environment, etc.). The inclusion of CSR in the overall strategy and corporate governance system must be systematic.

Effective transformation of large Russian companies in the direction of sustainable development is impossible without updating the interpretations of CSR shared by the business community. The ever-increasing speed of change requires not only the adoption of a standardized approach to CSR but also the willingness of companies to regularly update their content. Thus, the voluntary international standard ISO 26000: 2010 "Guidelines for Social Responsibility" and other relevant standards in management practice acquire a significant role (Table 1). Since compliance with these standards ensures the functioning of an integrated corporate social activity management system in companies as a system of CSR principles, responsible business processes, and corresponding measurable results (Blagov, 2010).

Table 1 Implementation of international standards in the company's management system

Voluntary international standards	2014, %	2019, %
ISO 26000:2010	17	29
Other ISO standards	51	67
ILO occupational safety and health standards (OSH)	22	29
Standards for stakeholder engagement and social reporting Audit AA 1000 SES and AA 1000AS	15	22
No, but we plan to implement it shortly	17	2
No	17	18
Other	20	27

Source Compiled by the authors based on Blagov et al. (2008)

There is an obvious correlation between the degree of implementation of the international standards under consideration and the industry affiliation of the respondent companies (Table 2), as well as experience in conducting corporate social activities. First, the respondent companies related to the raw materials sector generally demonstrate a greater commitment to the standards under consideration, primarily specific standards for quality management, building effective environmental management systems, interacting with stakeholders, and auditing social reporting. It is these standards that are directly aimed at reducing social and environmental risks, largely due to industry specifics. In turn, the respondent companies related to the service sector are relatively more likely (41% versus 14% and 31% in the processing and raw materials sectors, respectively) to implement the system-forming ISO 26000:2010 standard, although this sector shows the largest share of companies (29%) that do not

Table 2 Implementation of international standards in the company's management system by economic sector

Voluntary international standards	Economic sector		
	Raw materials sector	Processing sector	Service sector
	%	%	%
ISO 26000:2010	31	14	41
Other ISO standards	92	71	47
ILO occupational safety and health standards (OSH)	62	29	6
Standards for stakeholder engagement and social reporting audit AA 1000 SES and AA 1000AS	31	14	24
No, but we plan to implement it shortly	0	7	0
No	8	7	29

Source Compiled by the authors based on Blagov and Petrova-Savchenko (2020)

use the standards in question at all. Secondly, the experience of conducting corporate social activities significantly affects the implementation of voluntary international standards by companies. For example, companies that have been conducting corporate social activities for 1 to 3 years (9% of the respondents) had only two cases of implementing the standards. In turn, the vast majority of companies-respondents (12 out of 13, i.e. 92%), following the ISO 26000:2010 standard; more than 2/3 of companies implementing 22 other ISO series standards; and 77% of companies implementing OSH standards, have been implementing corporate social activities for more than 10 years (Blagov et al., 2008).

Engaging companies in achieving the UN Sustainable Development Goals (SDGs) creates new opportunities for transforming companies' CSR and sustainable development strategies. This not only clarifies the goal setting, but also the search for and implementation of new business opportunities related to "corporate sustainability 3.0". The transformation of corporate strategies in the interests of sustainable development is impossible without balanced changes in all elements of the corporate social activity system as a set of CSR principles, responsible business processes, and measurable results of corporate behavior. The research in the framework of the Reports on Social Investment in Russia provides an opportunity to analyze the institutionalization of these elements in the relevant organizational structures, management practices, and tools. Over the past five years, the "strategic approach to CSR" itself has transformed, the generation of "social" value has acquired a new meaning in the context of the UN Sustainable Development Goals, and the role of the departments directly responsible for the implementation of relevant corporate strategies and involved in this process has been filled with more relevant content.

The analysis of the results of the responsible implementation of business processes in the "Reports on Social Investment in Russia" traditionally includes issues of evaluating the effectiveness of social investment of the respondent companies, as well as the preparation of public non-financial or integrated reporting. The overall indicators of the use of the social investment performance assessment by the respondent companies for the period 2014–2019 remained virtually unchanged. For example, 71% of companies (73% in 2014) said they would conduct such an assessment, and another 16% plan to do so in the future (10% in 2014). The changes, however, affected the qualitative content of the assessment: if in 2014 the share of respondent companies that track performance in the short and long term (immediate and deferred effects) was 84% of the total sample, then in 2019 this share was already 91%. At the same time revealing that quality content evaluation of projects in the field of corporate philanthropy has changed insignificantly: if the assessment itself holds 78% of the respondents (73% in 2014), the assessment of short-term and long-term periods—58% (59% in 2014). These data support the assumption that the historical identification of CSR with charitable activities is becoming obsolete, and the management of corporate social activities is increasingly moving beyond the management of corporate charity (Blagov et al., 2008).

The degree of involvement of respondent companies in the preparation of public non-financial and integrated reporting is steadily increasing, although the rate of this growth is relatively low.

Thus, it can be assumed that the respondent companies have made significant progress in the use of relevant international standards. Today, companies are successfully learning to shift their focus from the basic concepts of CSR to modern interpretations of sustainable development and corporate sustainability. Approaches to corporate social investment are increasingly focused on the development of partnerships—the most important factor in modern business transformation within the framework of the concept of sustainable development (Dyllick & Muff, 2016). Nevertheless, the interpretations of the interaction of business with the state and society remain largely contradictory, and the role of the criterion of innovation in the context of product, process, and marketing innovations remains underestimated. The company is consistently integrating CSR principles in the system of strategic management at all levels. However, the integration of the UN Sustainable Development Goals is developing slowly, only slightly contributing to the change in the structure of the entire system of corporate social activities for sustainable development. The new goal-setting mainly captures in new terms the already established system of interaction between companies and society, including those carried out in the traditional areas of corporate charity.

As a result, we can conclude that the analyzed companies are transforming the system of corporate social activity in the direction of current models of corporate sustainability. The analysis of changes occurring in the "principles", "processes" and "results" allows us to judge the formation of a group of leading companies that generally correspond to the "corporate sustainability 2.0" model and demonstrate the potential for transition to the "corporate sustainability 3.0" model. In turn, the broader coverage of companies by the ongoing transformation, as well as its acceleration, requires additional efforts at all levels of intra-and intersectoral interactions (He & Harris, 2020).

Separately, it is worth noting the impact of the Covid-19 pandemic on the social responsibility of business. The crisis has pushed some Russian companies (as well as foreign ones) to make short-term speculative profits from the pandemic by inflating prices or by misleading customers about the properties of products. The pharmaceutical industry companies reacted the most "predatory". For example, many pharmacies have increased the prices of critical medicines (antibiotics, antivirals) by 2–3 times. Companies that produce protective equipment have increased their prices tenfold. Some companies have conducted emergency marketing repositioning of products, identifying properties that are important during the pandemic, but not previously declared. Some firms have reduced their long-term investments in CSR, probably due to a lack of sufficient resources and growing pressure to survive. At the same time, at a time when resources are limited and survival is at risk, some firms have become more ethical and socially responsible. The main areas of social participation of Russian companies in the fight against coronavirus can be identified:

- charitable assistance by hospitals to volunteer organizations (for example, Gazprom NEFT served ambulances free of charge at branded service stations);

- restructuring of loans for small and medium-sized businesses by commercial banks (for example, Sberbank restructured about 20,000 loans for small and large businesses in the sectors most affected by COVID-19);
- large enterprises have provided installment payments to their small contractors;
- many departmental enterprises and large companies donated their infrastructure for activities for the fight against Covid-19. Thus, Russian railways provided their hospitals for the treatment of patients;
- providing discounts on services and goods of social significance. For example, Russian Railways introduced a reduction factor to the tariff for the transportation of socially significant goods;
- many enterprises tested for Covid-19 for free and vaccinated employees, provided financial assistance to those who fell ill.

And, finally, many companies did not lay off employees at the most difficult moment, attracting funds from the founders and using all available reserves (Bavel et al., 2020).

Thus, a critical situation allows you to recognize "who is who" and identify truly socially responsible companies. Many companies have become more human-oriented, improved feedback services, and become more flexible in responding to customer complaints and requests. The transition to remote work allowed for a 6% reduction in environmental emissions.

In this regard, the following *recommendations are made to the private sector:*

1. To develop a strategic approach to managing corporate social activities, integrating the principles of CSR and sustainable development in corporate governance, strategic planning, and the implementation of main management functions, conducting a comprehensive assessment of the effects of corporate social performance and reflecting the results in the non-financial and integrated reporting.
2. Consider the current change in corporate sustainability models as an objective global process that determines the survival and competitiveness of business, requiring the development of intra-and intersectional cooperation, trilateral and multilateral partnerships.
3. Conduct training in the management of corporate social activities focused on sustainable development, including training in effective interaction with internal and external stakeholders, at all levels of the management hierarchy. Support and develop the process of sharing experience and disseminating best practices in the field of CSR.
4. Consider the long-term reputational effects of socially responsible behavior: support your customers, contractors, and employees because, in the long run, this will affect the value of the business for the better (Tokareva & Shalina, 2015).

Recommendations to the State

1. Actively develop the institutional prerequisites for socially responsible business by improving and implementing relevant legislation (including in the field of

regulation of non-financial reporting and social entrepreneurship), strengthening the fight against corruption, and developing civil society institutions.
2. To more widely implement and fully support the mechanisms of trilateral (business—state—NGO) and multilateral partnership for sustainable development as the most important factor in the development of an up-to-date model of corporate sustainability.
3. Together with the business and professional communities, develop and approve a National Concept of Sustainable Development and Social Responsibility, including measures for joint indicative forecasting and planning, as well as state incentives for corporate social activities.
4. To develop a legal framework and infrastructure for the implementation of socially responsible investment. To date, key institutional investors in Russia do not offer financial instruments (investment units of funds, shares, etc.) that meet the criteria for socially responsible investment.
5. To carry out the promotion of the principles of social responsibility and social investment. Since the current Covid-19 pandemic has clearly illustrated the role of trust in the state, political polarization, and "information wars" in dealing with cataclysms.

4 Conclusion

Based on the study, the following conclusions were obtained:

1. The development of CSR in Russian business generally corresponds to the global trend of business transformation in the interests of sustainable development with a corresponding modification of the corporate sustainability model. However, the transition to the current model requires not only improving the management of corporate social activities of individual companies, but also qualitative changes in the system of intra- and intersectional interactions.
2. The current state of CSR of leading Russian companies is characterized by the presence of a stable group of leading companies that generally meet the best global standards of corporate social activity, with a relatively slow process of integrating the principles of CSR and sustainable development into corporate strategy.
3. The transition to the current model of corporate sustainability is largely constrained by the orientation of companies to create value for business and society, which usually does not imply active cooperation in the interests of sustainable development, and joint initiatives of responsible business are mainly limited to charitable projects, including corporate volunteering projects.
4. Quantitative and qualitative indices of social investment of Russian business in the period from 2004 to 2019 generally did not show positive dynamics. The structure of social investments is beginning to focus on supporting the local community and environmental protection activities.

5. The Covid-19 pandemic and the economic crisis clearly showed consumers and employees which manufacturers and employers adhere to social responsibility in practice, and which only declare it. The degree of trust in responsible companies will increase; their sales volumes and market value will increase.

References

Bavel, J. J. V., Baicker, K., Boggio, P. S. et al. (2020). Using social and behavioral science to support COVID-19 pandemic response. *Nature Human Behaviour, 4*, 460–471. https://doi.org/10.1038/s41562-020-0884-z (data accessed: February 20, 2021)

Blagov, Y. E. (2010). *Corporate social responsibility: The evolution of the concept*. Higher School of Management Publishing House.

Blagov, Y. E., Litovchenko, S., & Ivanova, E. A. (2008). *The report on social investments in Russia in 2008. Integration of CSR into corporate strategy*. Association of Managers.

Blagov, Y. E., & Petrova-Savchenko, A. A. (2020). *The report on social investments in Russia—2019: The transformation of business in sustainable development*. Association of Managers.

Blagov, Y. E., & Sobolev, I. S. (2014). *The report on social investments in Russia—2014: Creating value for business and society*. ATM Book.

Degtyareva, I. V. (2010). Social responsibility of business: Theory, models, public participation. *Social Policy and Sociology, 8*, 10–18.

Degtyareva, I., Tokareva, G., Shalina, O., & Fedorenko, O. (2016). Excellence of innovative activity management at the Russian enterprises. *International Journal for Quality Research, 10*(4), 785–798.

Dyllick, T., & Muff, K. (2016). Clarifying the meaning of sustainable business: Introducing a typology from business-as-usual to true business sustainability. *Organization & Environment, 29*(2), 156–174.

He, H., & Harris, L. (2020). The impact of Covid-19 pandemic on corporate social responsibility and marketing philosophy. *Journal of Business Research, 116*, 176–182. https://doi.org/10.1016/j.jbusres.2020.05.030

Tokareva, G. F., & Shalina, O. I. (2015). The main directions of the impact of the activities of institutional investors on social development in Russia. In *The collection: The vital potential of the region: socio-demographic problems of modern society (Aitov readings) Collection of materials of the international scientific and practical conference* (pp. 362–365). http://elibrary.ru/item.asp?id=25040463

The Economic Security Threats of the Region in Terms of Digitalization: Assessment and Development of Leveling Tools

Evgeniya K. Karpunina, Inessa G. Nazarova, Louiza I. Iljina, Irina N. Shvetsova, and Eleonora M. Chernenko

Abstract The authors substantiate the assumption that in the conditions of intensive digital development, the set of traditional threats to the economic security of the region is supplemented by threats from the digital environment. The authors attempt to systematize traditional threats to the economic security of the region and threats from the digital environment, and also described the peculiarities of their manifestation. The authors propose a methodology for assessing threats to the economic security of the region in the context of digitalization. The chapter proves that the reduction of threats to economic security created by the digital environment requires the formation of an institutional framework for cybersecurity of the digital space; the maintenance of digital hygiene and the introduction of cognitive goal-setting systems; the implementation of regional programs to improve the digital literacy of the population; the development of regional education systems.

Keywords Region · Economic security · Threats · Digital development · Threat leveling · Real economy · Financial sector · Social sphere

JEL Codes O11 · O33 · R11 · R58

E. K. Karpunina (✉)
G.R. Derzhavin Tambov State University, Tambov, Russia

I. G. Nazarova
Ukhta State Technical University, Republic of Komi, Ukhta, Russia

L. I. Iljina · I. N. Shvetsova
Pitirim Sorokin Syktyvkar State University, Syktyvkar, Russia

E. M. Chernenko
Kuban State Technological University, Krasnodar, Russia

E. G. Popkova and B. S. Sergi (eds.), *Geo-Economy of the Future*,
https://doi.org/10.1007/978-3-030-92303-7_55

1 Introduction

Currently, all economic entities of the regional economic system (individuals, households, enterprises, public administration bodies) are somehow included in the information space. They use a whole range of digital technologies to organize interpersonal interaction, work with contractors, develop internal communications, automate production processes, as well as manage various subsystems. Digital technologies make it possible to make their activities faster and more mobile, to get the maximum amount of required information in a short time, to take into account all the many factors of the external and internal environment to improve the quality of decisions made (Akatkin et al., 2017). For example, in 2020, more than 4.54 billion people worldwide had access to the Internet, which is 7% more than in 2019 (Web-Canape, 2020), 89% of companies have embarked on a digital business strategy (Forbes, 2018). E-government systems in many countries have long passed from the stage of formation to the stage of transformation of relations between government and citizens, enterprises and non-governmental structures using digital technologies, and further contextualization (Pavlyutenkova, 2019). As a result, the usual threats to the economic security of the region are supplemented by new ones created by the digital environment (Kapustina et al., 2020; Karpunina et al., 2020; Molchan et al., 2019).

In the context of the COVID-19 pandemic, the importance of the digital sector and information technology has increased significantly (COVID19, 2020). This made the problem of maintaining the security of the regional economy and society even more urgent.

It is necessary to develop effective tools for their identification at the time of occurrence and timely leveling to prevent emerging threats to economic security and reduce the amount of damage from their implementation. However, the solution to this problem is complicated by the lack of statistical information and the limitations of quantifying the probability of threats.

2 Materials and Methods

The purpose of the study is to reveal the factors of threats to the economic security of the region at the stage of digital development, as well as to find effective tools for assessing emerging threats and leveling them.

Research objectives: (1) systematize the economic security threats of the region at the stage of digital development; (2) propose a methodology for assessing threats to the economic security of the region in the context of intensive digital development; (3) offer tools for leveling threats to the economic security of Russian regions in the context of digitalization.

Research methods: analysis of scientific literature, comparative analysis method, graphic method, systematization method, method of economic and statistical analysis, system approach.

3 Theoretical Basis of the Study

The theoretical basis of the research is the publications of scientists on the issues of ensuring the economic security of the regional economic system, as well as the systematization of threats at the digital stage of development. Huber et al. (2010) argue that economic security is a characteristic of the state of an economic system with a certain level of stability and the ability to prevent emerging threats to the well-being of society The achievement of a state of economic security is influenced by many factors of a financial, political, economic, and social nature (Edelev, 2007; Kalinina, 2010; Kremlev et al., 2007).

In terms of intensive digital development, the range of traditional threats to economic security is expanding due to the emergence of new threats. And these are not always threats of an informational nature associated with data leakage and the development of cybercrime schemes (Karpunina et al., 2020). These are also social threats associated with increased social inequality due to different access to information technologies, rising unemployment due to automation of production activities and management processes (Kapustina et al., 2020). Digital technologies contribute to the development of the "on-demand economy", which leads to the reduction of small businesses and brings real economic damage to the economic system of the region (Chebotarev et al., 2016; Gorulev, 2018; Voronkov, 2019).

4 Results

The author's approach to structuring the economic security of the region in the context of digitalization involves the identification of two types of threats traditional economic security threats of the region and threats generated by the digital environment (Fig. 1).

The traditional economic security threats of the region can be attributed to the threats that arise in the real sector of the region's economy, in the monetary and financial sphere, as well as in the social sphere (Kupreschenko & Fedotova, 2010; Lomachenko, 2016; Orlova et al., 2016; Shatrovskaya, 2016) (Fig. 2).

In our opinion, the assessment of traditional threats to the economic security of the region can be based on a generally accepted methodology, which includes the following stages:

1. formation of a system of indicators of economic security (the real sector of the region, the monetary and financial sphere of the region, the social sphere of the region);
2. determination of their threshold values;
3. reduction of indicators to a dimensionless form (rationing);
4. calculation of the integral index of economic security of the region and comparison of its value with the integral threshold equal to one (Khasanov & Korableva, 2019).

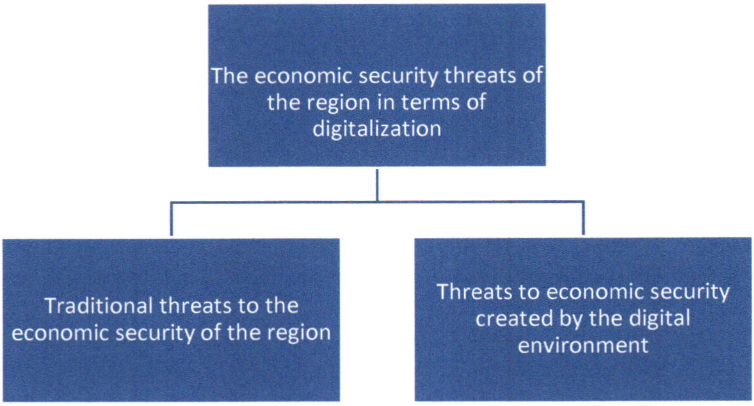

Fig. 1 Systematization of the economic security threats of the region in terms of digitalization. *Source* Compiled by the authors

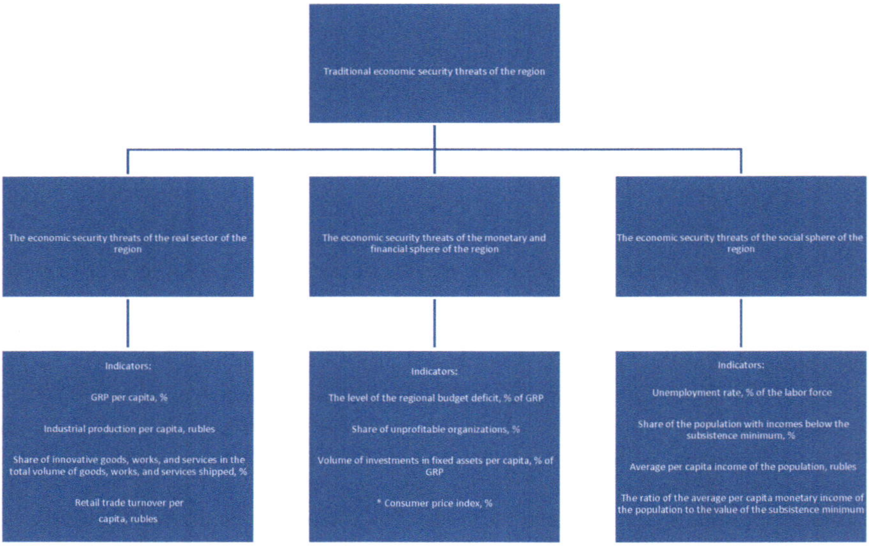

Fig. 2 Indicators for identifying traditional economic security threats of the region. *Source* Compiled by the authors

As part of the implementation of this methodology, we will analyze some indicators of the economic security of Russian regions and their threshold values (Table 1).

The evaluation of the results of the analysis is based on the comparison of the achieved values of the indicators of economic security of the regions with their threshold values (Krivorotov & Kalina, 2010; Mityakov et al., 2013; Tatarkin &

Table 1 Regional economic security indicators and their thresholds, 2019

Indicator	Threshold value	Actual value for 2019
Economic security of the real sector of the region		
Gross regional product per capita, rubles	Not less than the average for the Russian Federation	507,794
The volume of industrial production per capita, rubles	Not less than the average for the Russian Federation	402,142
The share of innovative goods, works, services in the total volume of shipped goods, works, services, %	Not less than the average for the Russian Federation	7.2
The degree of depreciation of fixed assets (%)	No more than 60	60
Retail trade turnover per capita, rubles	Not less than the average for the Russian Federation	203,029
Economic security of the monetary and financial sphere of the region		
The level of the regional budget deficit, % GRP	No more than 3	3
The share of unprofitable organizations, %	No more than the average for the Russian Federation	31.9
The volume of investments in fixed assets per capita, % of GRP	No more than 25	25
Consumer price index, December to December last year, %	No more than the average for the Russian Federation	102.5
Economic security of the social sphere of the region		
Unemployment rate, % of the labor force	No more than 4	4
The share of the population with incomes below the subsistence minimum, %	No more than 7	7
Average per capita monetary income of the population, rubles	Not less than the average for the Russian Federation	31,422
The ratio of the per capita monetary income of the population to the subsistence minimum	Not less than 3.5	3.5

Source compiled by the authors based on Federal State Statistics Service of the Russian Federation, (2020), Khasanov & Korableva, (2019)

Kuklin, 2007), as well as country averages. Exceeding the threshold values for positive indicators and finding the values of negative indicators below the threshold indicates the appearance of signs of a decrease in the economic security of the region. The choice of the normalization method determines the dynamic range of the results visualization. The most adequate method of normalization is proposed in the work of Mityakov et al. (2013), where the power dependence allows to ignore insignificant details in the case of a significant excess of the threshold values by the indicators. A

Table 2 Information for interpreting the results of a comprehensive (integrated) assessment of economic security in certain areas of the region

The value of the integral index of economic security	Conclusion
I < 1	Unstable (crisis, pre-crisis) state of the sphere under the study
I ≥ 1	Safe state of the sphere under the study

Source compiled by the authors based on Khasanov & Korableva, (2019); Mityakov et al., (2013)

comprehensive (integral) assessment of security in certain areas of the region is based on the calculation of the integral index as a weighted sum of normalized indicators:

$$I = \sum_{i=1}^{n} x_i * w_i \quad (1)$$

where x_i—normalized indicator value X_i, w_i—a weighting factor that reflects the degree of significance of indicator X_i, i = 1, ..., n. The conclusion about the state of the studied area of the region is formed based on the data presented in Table 2.

The intensification of digitalization becomes an objective factor in the development of the regional economic system, which brings unconditional benefits for the territory and expands the available opportunities. Numerous studies point to the benefits of digitalization for the development of various sectors of the regional economic system. For example, Molchan et al. (2019) attribute to the benefits of digitalization an increase in labor productivity in the real sector of the region's economy due to the introduction of more efficient technologies and systems for automating production activities. The digitalization of financial services creates conditions for the expansion of the consumer segment and the number of financial service providers, as well as a multiple acceleration of operations and an increase in the quality of services provided. However, the unconditional benefits of digitalization hide some threats to the safe functioning of the region (Fig. 3).

Threats of an information and technological nature in the region arise due to the increase in the number of enterprises operating in digital form and using digital technologies to optimize production and sales processes, as well as the strengthening of the information openness of the regional space caused by the expansion of access to the Internet. In terms of digitalization, the probability of information leakage and the development of various forms of cybercrime are growing (Karpunina et al., 2020). The high level of dependence of enterprises on digital technologies creates conditions for the emergence of information failures in the economy of the region (Table 3).

The data presented in Table 2 indicate an increase in access to the Internet and the use of digital technologies in the regions of Russia, and, therefore, a growing likelihood of threats of an information and technological nature. The leading regions

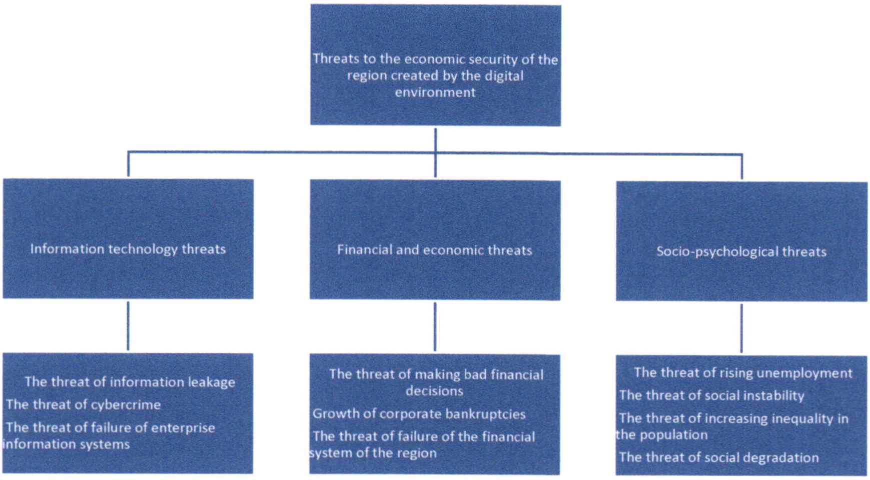

Fig. 3 Systematization of threats to the economic security of the region created by the digital environment. *Source* Compiled by the authors

of Russia in terms of the dynamics of digital development indicators are the Central Federal District, the North-Western Federal District, the Volga Federal District, and the Ural Federal District. According to the National Coordination Center for Computer Incidents, in 2018, there were more than 4.3 billion information attacks on the critical information infrastructure of the country (Rg, 2019). The targets of the attacks were objects of the financial sector—38% of the total number of attacks, public authorities—35%, the defense industry—7%. The regions of the Russian Federation with the highest growth rates of registered crimes committed using ICT technologies are St. Petersburg (462.7%), the Jewish Autonomous District (192.9%), Moscow (181.3%), the Republic of Ingushetia (180%), the Republic of Buryatia (167.5%) (Cybercrime & cyber conflicts: Russia., 2021). In general, regional-wide information technology threats can cause the under-production of GRP (Karpunina et al., 2019).

Financial and economic threats are realized in the form of increasing factors that lead to the bankruptcy of enterprises, failure in various elements of the financial system of the region. Such threats arise due to the unjustified use of digital tools to the detriment of traditional ones, and the uneven distribution of market power between financial service providers and consumers, leading to distortions in the structure of the regional financial system. The uncertainty of environmental factors and the reduction of business income of the population (as a result of the lack of adaptation to digital conditions in business structures) can lead to an increase in the number of unprofitable enterprises and a decrease in budget revenues in the region (Salikov et al., 2019). Ackerman et al. (2015), Brynjolfsson and McAfee (2014) note the trend of job polarization and the emergence of the threat of ousting only low-and medium-skilled professionals from the labor market, their skills are becoming insufficient to

Table 3 Indicators of access and use of digital technologies in the regions of Russia, 2013–2019

Region	The number of broadband Internet subscribers per 100 people of the population		The population using the Internet, as a percentage of the total population		The share of organizations using the Internet, % of the total number of surveyed organizations of the relevant subject of the Russian Federation		Organizations using cloud services, % of the total number of organizations in the business sector	
Years	2013	2019	2013	2019	2013	2019	2013	2019
Russian Federation	16.53	21.7	60.7	88.6	88.1	91.2	18	27.1
Central Federal District	18.73	26.0	61.6	90.9	87.9	93.1	20	31.6
North-Western Federal District	20.08	23.9	71.7	89.5	92.3	92.1	20	27.7
Southern Federal District	12.97	17.7	57.7	87.2	85.9	91.8	17	26.1
North Caucasus Federal District	5.17	8.6	38.8	88.6	89.2	77.7	20	18.6
Volga Federal District	18.2	22.0	60.3	88.5	88.7	92.8	16	26.3
Ural Federal District	18.66	24.6	65.4	87.9	91.2	91.7	20	28.6
Siberian Federal District	14.83	20.3	60.5	85.1	84.7	89.4	18	22.7
Far Eastern Federal District	13.47	17.5	59.4	87.0	86.0	91.1	16	23.2

Source compiled by the authors based on Federal State Statistics Service of the Russian Federation, (2020), HSE, (2020)

compete with digital technologies, artificial intelligence, and automated information processing and management decision-making systems. However, not all researchers recognize this threat to economic security. For example, Fossen and Sorgner (2019) conclude that the ongoing digital changes do not cause a decrease in employment in all professions, and the creation of digital platforms can contribute to the growth of non-standard employment, an increase in the number of short-term, part-time or low-paid jobs. This type of threat leads to a reduction in the company's income and aggregate demand, and as a result, a deterioration in the well-being of the region's population. In turn, this leads to an increase in emotional experiences associated with the fear of future job loss, loss of social status, and personal degradation, and it becomes the cause of another risk of economic security—the risk of social instability as a result of intensive digital consumption (Molchan et al., 2019). This problem was even more urgent for the Russian regions during the COVID-19 pandemic when the pace of development of digital technologies and the expansion of online services showed accelerated dynamics (COVID19, 2020). In Russia, as of April 2020, 82% of residents were involved in the digital environment, and 71% of them used the Internet daily, mainly to communicate with their loved ones and get news about the country and the world. The role of online services in such industries as trade and finance has significantly increased (Malysheva, 2020). Russians began to spend significantly more time (up to 19%) in social Internet services, and among young people, this figure was noticeably higher and reached 27% (WCIOM, 2020).

The assessment of the described threats to economic security created by the digital environment is currently difficult due to the lack of regional statistics reflecting these incidents. Nevertheless, at the state level, the assessment of such threats to economic security is possible using statistical methods of variance (formula 2) and mean square deviation (formula 3).

$$\sigma^2 = \left(\sum_{i=1}^{n} \left(x_i - x_{(mid)i}\right)^2 /n \qquad (2)$$

$$\sqrt{\sigma^2} = \sqrt{\left[\left(\sum_{i=1}^{n} \left(x_i - x_{(mid)i}\right)^2 /n\right]} \qquad (3)$$

where σ^2—variance; x_i—value of the ith indicator, $x_{(mid)i}$—the arithmetic mean of this sample; n—volume of collected empirical values; $(x_i - x_{(mid)i})^2$—the square of the residuals; $\sqrt{\sigma^2}$—mean square deviation.

The indicators of international ratings can serve as a basis for assessing threats to economic security from the digital environment. For example, the Business Digitalization Index (HSE), the Cybersecurity Index, the Human Development Index (HDI).

5 Conclusion

The most effective ways of leveling the traditional economic security threats of the region are:

1. implementation of measures to improve the efficiency of the use of natural, material, financial, and intellectual factors in the region;
2. development of a system of financial incentives for production growth and increasing the income of regional enterprises;
3. creating an enabling environment for human development for the benefit of the region;
4. development of interregional cooperation to meet the production and social needs of the region.

Reducing the threats to economic security created by the digital environment requires the implementation of the following priority areas of state policy:

1. formation of the institutional base of the cybersecurity digital space;
2. maintaining digital hygiene and implementing cognitive goal-setting systems implemented at the level of large companies in the region and the regional management system, related to the implementation of protection systems against external intrusions and the distribution of access rights to information (Gorulev, 2018);
3. implementation of regional programs to improve the digital literacy of the population;
4. development of regional educational systems to create opportunities for retraining specialists to prevent their mass displacement from the labor market.

References

Akatkin, Y., Karpov, O., Konyavsky, V., & Yasinovskaya, E. (2017). Digital economy: The conceptual architecture of the digital industry ecosystem. *Business Informatics, 4*(42), 17–28.
Akerman, A., Gaarder, I., & Mogstad, M. (2015). The skill complementarity of broadband internet. *The Quarterly Journal of Economics, 130*(4), 1781–1824.
Brynjolfsson, E., & McAfee, A. (2014). *The second machine age: Work, progress, and prosperity in a time of brilliant technologies* (1st ed.). W W Norton & Company.
Chebotarev, S., Chebotarev, V., & Zhurenkov, D. (2016). The initiative "Industry 4.0": A new industrial policy. *Society and Economy, 6*, 43–57.
COVID19. (2020). Three areas that can go completely online. Available at: https://covid19.fom.ru/post/tri-sfery-kotorye-mogut-polnostyu-perejti-v-onlajn. Accessed April 01, 2021.
Edelev, A. (2007). *Strategic stability and economic security of the subjects of the Russian Federation: Monograph*. Economic Security of the Ministry of Internal Affairs of Russia.
Federal State Statistics Service of the Russian Federation. (2020). Information Society. Available at: https://rosstat.gov.ru/folder/14478?print=1. Accessed April 01, 2021.
Forbes. (2018). The IDG's 2018 state of digital business transformation. Available at: https://www.forbes.com/sites/louiscolumbus/2018/04/22/the-state-of-digital-business-transformation-2018/#463a83de5883. Accessed April 01, 2021.
Fossen, F., & Sorgner, A. (2019). Mapping the future of occupations: Transformative and destructive effects of new digital technologies on jobs. *Foresight and STI Govern, 13*(2), 10–18.
Gorulev, D. (2018). Economic security in a digital economy. *Techn and Technology Problem of the Service, 1*(43), 77–84.
HSE. (2020). *Trends in the development of the Internet in Russia and foreign countries: Analytical report*. Higher School of Economics.
Huber, G., Rehm, P., Schlesinger, M., & Valletta, R. (2010). Economic security at risk: Findings from the economic security index. 32ГФ Шурчкова 23.04.2021. http://economicsecurityindex.org/assets/Economic%20Security%20Index%20Full%20Report.pdf. Accessed April 01, 2021.
Kalinina, N. (2010). Economic security of the region: The essence, threats and measures to ensure. *Siberian Trade and Journal of Economics, 10*, 1–21.
Kapustina, N., Strelnikova, T., Ioda, E., Furman, E., & Karpunina, E. (2020). Social risks of digital transformation. In *Proceeding of the 35th IBIMA conference, Seville, Spain*, (pp. 7682–7692).

Karpunina, E., Gorbunova, O., Moiseev, S., & Cheremisina, T. (2019). Resistance is not hopeless...regarding the policy of countering information threats of economic security. In *Proceedings of the 33nd IBIMA conference, Granada, Spain*, (pp. 2679–2687).
Karpunina, E., Shurchkova, J., Kochetkova, E., Ponomarev, S., & Tretyak, V. (2020). Cybercrime in the system of economic security threats. In *Proceeding of the 35th IBIMA conference, Seville, Spain*, (pp. 2679–2690).
Khasanov, R., & Korableva, A. (2019). Economic security index as a tool for assessing the development of agro-industrial regions. *Economic Policy, 14*(6), 82–101.
Kremlev, N., Fedorov, V., & Sergeev, M. (2007). Questions of economic security of the region. *Questions of Statist, 2*, 42–48.
Krivorotov, V., & Kalina, A. (2010). *Economic security of the state and Regions*. USTU-UPI.
Kupreschenko, N., & Fedotova, E. (2010). Analysis of real and potential threats to the social sphere of the regions of Russia and their impact on economic security. *Bulletin of the Academy of Economic Security of the Ministry of Internal Affairs of Russia, 8*, 86–93.
Lomachenko, T. (2016). Financial security of the region as the basis of national economic security of the state. *Economics and Management 2, 68*(4):70–79.
Malysheva, G. (2020). Socio-political aspects of the pandemic in the digital network society: The Russian experience. *Bulletin of Moscow Region State University (e-journal), 3*, 60–74.
Mityakov, S., Mityakov, E., & Romanova, N. (2013). Economic security of the regions of the Volga Federal District. *The Economy of the Region, 3*(35), 81–91.
Molchan, A., Karpunina, E., Kochyan, G., Petrov, I., & Velikanova, L. (2019). Effects of digitalization: New challenges for economic security systems. In *Proceedings of the 34rd IBIMA conference, Madrid, Spain*, (pp. 6631–6639).
Orlova, A., Lyshchikova, J., Nikulina, Y., & Anokhin, Y. (2016). Assessment of the level of economic security in the conditions of uncertainty. *International Journal of Economics and Finance, 6*(4), 1702–1706.
Pavlyutenkova, M. (2019). Electronic government versus digital government in the context of digital transformation. Monitoring of Public Opinion. *Economics and Social Challenges, 5*, 120–135.
Rg. (2019). Virtual and real wars. Available at: https://rg.ru/2019/08/14/chislo-opasnyh-kiberatak-na-obekty-v-rf-vyroslo-v-11-raz-za-tri-goda.html. Accessed April 01, 2021.
Salikov, Y., Karpunina, E., Kuzmenko, N., Chaikovskaya, L., & Roshchektayeva, U. (2019). Is it safe to live in the world of robots: Current trends in the processes of robotics, challenges to future development and threats to economic security. In *Proceedings of the 34nd IBIMA conference, Madrid, Spain*, (pp. 5685–5695).
Shatrovskaya, A. (2016). Security of the real sector of the economy in the system of national economic security. Theory and practice of the service. *Economics, Society Sphere, and Technology, 4*(30), 19–21.
TAdvisor. (2021). Cybercrime and cyber conflicts: Russia. https://www.tadviser.ru/index.php/Статья:Киберпреступность_и_киберконфликты_:_Россия. Accessed April 01, 2021.
Tatarkin, A., & Kuklin, A. (2007). *Complex methodology of diagnostics of socio-demographic security of the region*. Institute of Economics of the Ural Branch of the Russian Academy of Sciences.
Voronkov, A. (2019). Risks of economic security from digital economy technologies. Available at: http://www.e-rej.ru/Articles/2019/Voronkov.pdf. Accessed April 01, 2021.
WCIOM. (2020). The network as a rescue. Available at: https://wciom.ru/index.php?id=236&uid=10221. Accessed April 01, 2021.
Web-Canape. (2020). All Internet statistics for 2020—figures and trends in the world and Russia. Available at: https://www.web-canape.ru/business/internet-2020-globalnaya-statistika-i-trendy/. Accessed April 01, 2021.

The Modern Social Contract: Prerequisites for Renewal

Ekaterina I. Shumskaia and Natalia N. Chubaeva

Abstract The modern social contract requires revision and updating on some points. The reasons can be broadly divided into three groups: economic, political, and security issues. The authors investigate economic reasons, which include the lack of an equitable distribution of the benefits of economic growth over the past 30 years, ever-growing inequality both within and between countries, an incomparable increase in the return on capital compared with an increase in labor income, the negative impact of automation processes and robotization to the labor market. The authors conclude that the fundamental principles of the modern economic model have changed significantly over the past decades. As the Fourth Industrial Revolution emerges, national governments need to implement a series of institutional and governance reforms needed to ensure the positive impact of disruptive technologies: strengthening social protection policies and streamlining public services through greater use of automation.

Keywords Fourth industrial revolution · Social contract · Digital social contract · Inequality · Employment · Labor profitability · Middle class

JEL Codes O10 · O38 · O52 · P16 · P51

1 Introduction

A social contract can be described as an agreement reached by citizens on the rules and principles of government. Based on this agreement, people give up some of their freedoms in exchange for rules, customary practices, and public services (Rousseau, 1968). In the classical definition of T. Hobbes, the social contract allows people to move from a natural state to a social one, in which radical human freedoms are limited in exchange for public goods of various kinds (Hobbes, 1982). The sustainability of such a contract depends on a broad public consensus on how well the rules and regulations meet current needs.

E. I. Shumskaia (✉) · N. N. Chubaeva
MGIMO University, Moscow, Russia

E. G. Popkova and B. S. Sergi (eds.), *Geo-Economy of the Future*,
https://doi.org/10.1007/978-3-030-92303-7_56

The modern social contract is increasingly failing to meet these demands, thereby requiring revision and updating on several points. The active introduction of automation and robotization technologies poses the threat of a significant reduction in the existing number of jobs. Thus, at the Global Summit on Technology Governance in April 2021, Secretary-General of the International Trade Union Confederation S. Burrow emphasized that advances in technology threaten to exacerbate the already unstable global situation in which the world finds itself due to the COVID-19 pandemic (Burrow, 2021).

2 Methodology

The reasons for the failure of the modern social contract can be broadly divided into three groups: economic, political, and security issues. The authors investigate economic reasons, which include the lack of an equitable distribution of the benefits of economic growth over the past 30 years, ever-growing inequality both within and between countries, an incomparable increase in the return on capital compared with an increase in labor income, the negative impact of automation processes and robotization to the labor market.

To confirm the hypothesis put forward, the work uses comparative and statistical analysis. The latest data was sourced from the International Monetary Fund, OECD database, PwC, and McKinsey Research Institute.

3 Results

In modern society, the idea that everyone can receive worthy benefits from their efforts and costs has a strong position. At the same time, growing income inequality can undermine the foundations of a market economy and lead to inequality of opportunity. This suppresses social mobility and weakens incentives to invest in knowledge. All of this can finally undermine economic efficiency and growth potential in developed countries.

3.1 High Growth and Income Inequality

Against the background of modern economic development, the problem of uneven distribution of income is relevant at all levels, including international. Real disposable income among the general population is declining. Incomes are growing only among people who, according to statistics, are classified as wealthy. Against the background of the developing negative situation of lower incomes, part of the population from the "middle" strata is gradually moving to the poor.

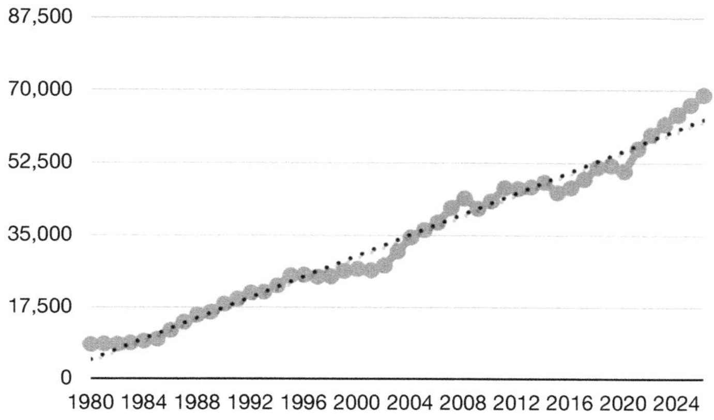

Fig. 1 GDP growth in developed countries, billion dollars. *Source* IMF, 2021

Thus, over the past decades, most OECD countries have experienced very high rates of economic growth (Fig. 1). US GDP grew from $2.8 trillion in 1980 to over $20 trillion by 2020. UK GDP, in turn, grew from $560 billion to over $2.6 trillion over the same period.

However, this cumulative growth has not affected the middle class in these two countries: the growth in disposable income has lagged behind GDP growth. The case of the USA is most acute: about 70% of households have not seen real income growth over the past thirty years. These countries show some of the highest Ginny ratios among the OECD countries (https://data.oecd.org/inequality/income-inequality.htm).

Over the past decade, research papers published by the OECD (OECD Report, 2008; Frankel, 2018; International Monetary Fund, 2017; OECD Report, 2011; Wilkinson & Pickett, 2009) have traced the evolution of income inequality. The main focus of the reports is the possibility of social mobility, which is increasingly elusive for low- and middle-income families. More than three decades ago, the total income of all middle-income households was almost 4 times the total income of high-income households. Today this ratio is less than 3 (Fig. 2).

The statistics shows astounding data: in the United States, the 1% of people with the highest income today receive more than 20% of total national income (before tax), which is twice as much as in the 1960s and more than the total income received 50% of the poorest (Jones, 2016).

A large return on capital in relation to labor is also demonstrated by the drop in the share of labor income in the national income of developed countries from about 54% in 1980 to 50.5% by 2014 (OECD Report, 2019). During this period, the decline in the share of the labor force took place everywhere in developed countries, but at the same time unevenly (Table 1). In general, the rise in the ratio of national wealth and income in recent decades is due solely to the growth of private wealth.

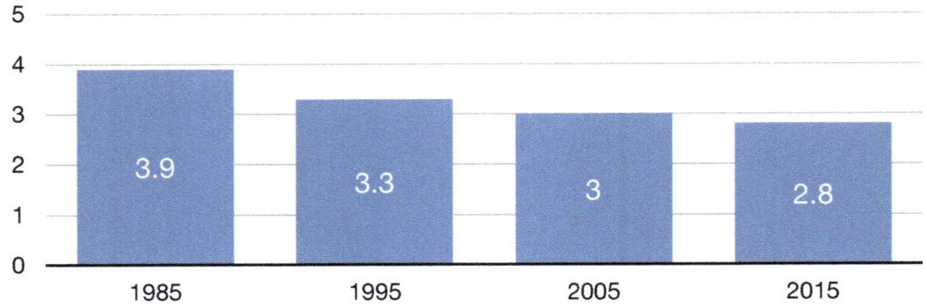

Fig. 2 Aggregate income of households with middle income to high income. *Source* Built by the authors based on data (https://data.oecd.org/inequality/income-inequality.htm)

Table 1 Total share of national income attributable to labor, %

	2017 to 1980	2017 to 2000
France	−7.7	2.2
Great Britain	−1.9	1.7
USA	−5.8	−5.3
Germany	−7.3	−2.5
Spain	−12.5	−4.5
Sweden	−6.2	1.1

Source Compiled by the authors based on data from McKinsey Global Institute (2019)

Research shows that globalization is also a driving force behind rising income inequality, as it is negatively correlated with labor force participation (Doan & Wan, 2017; Stiglitz, 2015).

Income inequality leads to inequality of opportunity. A higher level of financial support at the starting point of growth opens up access to more expanded and qualitative components of development. A person, possessing a high level of well-being, claims to have a higher quality level of education, higher qualifications, which in turn, together with the available "start-up capital", are the guarantor of more substantial income.

Today, the most effective regulator of income inequality, according to the American economist J. Stiglitz, should be the state, using tax and financial instruments (Stiglitz, 2015). Analyzing various factors (for example, the value of land, the amount of accumulated savings, the asymmetry of available information) in the economic model, R. Solow, J. Stiglitz cite the need to control the number of workers and entrepreneurs by distributing taxes as a tool to reduce income inequality.

I would like to point out separately that there is another way to reduce inequality. B. Gates and W. Buffett launched "The Giving Pledge" in 2010 to encourage extremely wealthy people to donate most of their wealth to charitable causes. Thus, as of August 2020, 211 people from 23 countries signed this commitment, compared with 40 participants in 2010.

Numerous studies of the problem of inequality are not accidental: it turned out to be extremely difficult to determine the reasons for such a transformation of society. Some scholars continue to insist on the negative effects of globalization and the pressure from low-paid workers in developing countries on the wages of the western middle class. However, an increasing number of scientists agree that it is necessary to take into account the impact of technology (Frankel, 2018).

3.2 *Automation and Robotization*

On the one hand, the introduction and diffusion of technologies of the Fourth Industrial Revolution can lead to a significant increase in the welfare of society. The global economy is expected to double by 2042 (OECD Report, 2019). But in the absence of adequate government intervention, unemployment will continue to rise and inequality will rise even more. The participation of national governments in the modern process of transforming the manufacturing economy into a service economy is necessary since it plays a central role in many rapidly growing sectors (such as education or health care).

In contrast to previous breakthrough technologies that have forced low-skilled labor out of the market, automation and robotization technologies blur the middle of the job distribution curve by reducing the demand for routine jobs of average skill (Autor & Dorn, 2013). We are seeing new jobs being created at the low-skilled and low-wage end of the spectrum, largely because it's not economically viable to automate tasks. Also, a small number of highly paid jobs are created with the requirements of high labor qualifications.

Modern technological transformations are happening faster, automating more jobs than in the preceding industrial revolutions. The changing nature of work can undermine the protection of labor rights, lead to long-term unemployment and job polarization in terms of skill levels and generate social grievances.

Technologies such as Big Data and Artificial Intelligence are already dramatically changing the corporate landscape. The emerging network effect is changing the way markets work, revealing winners and losers in the digital race. Firms that can collect and process information about their activities, customers, and others can increase their productivity and be more competitive. Thus, the OECD study shows that in the corporate sector, only a small group of companies can be distinguished, whose productivity has been growing over the past thirty years (Andrews et al., 2015).

This stratification of the private sector is a problem, as surplus profits are concentrated in a very limited number of firms, which then successfully develop sophisticated but legal schemes to reduce their tax burden.

The market power of tech giants deserves special attention from antitrust laws, which should take into account the growing role of innovative methods to increase the influence and degree of market presence of these companies. Previously, major companies such as Standard Oil or American Tobacco could influence the prices of goods and services. Modern giants manage information by creating and changing algorithms that give it out in the order and quantity they need. This can lead to misinformation, price discrimination, and undermining economic efficiency.

Data control leads to the loss of privacy in cybersecurity, where there is no longer a clear line between public and private property: both have a massive pool of data about every member of society. The lack of security leads to serious social, economic, and political problems.

4 Conclusion

The controversy over the factors that destabilize the middle class and rise in inequality will continue. However, it is clear that some of the fundamental principles of the modern economic model have changed significantly over the past decades: the number of jobs with average skills is declining, and only the owners of capital receive high incomes.

Technologies by themselves do not have predetermined results for society. An important question is to what extent government, business; trade union organizations, and other economic agents participate in and shape the innovation process for social purposes.

The modern social contract requires revision and updating on some points, including the protection of human rights while solving the problem of technology management. National governments need to undertake a series of institutional and governance reforms to ensure a positive impact during a period of active technology development. These should include strengthening social protection policies and streamlining government performance through increased use of disruptive technologies.

References

A new look at the declining labor share of income in the United States, McKinsey Global Institute, Discussion paper, May 2019.

Andrews, D., Criscuolo, C., & Gal, P. (2015). Frontier firms, technology diffusion, and public policy: Micro evidence from OECD countries, OECD Report. https://www.oecd.org/economy/growth/Frontier-Firms-Technology-Diffusion-and-Public-Policy-Micro-Evidence-from-OECD-Countries.pdf. Accessed: April 03, 2021.

Autor, D., & Dorn, D. (2013). The growth of low-skill service jobs and the polarization of the U.S. labor market. *American Economic Review, 103*(5), 1553–1597.

Burrow, S. (2021). Workers need a new social contract and technology has a key role to play, Global Technology Governance Summit, WEF.

Divided We Stand: Why Inequality Keeps Rising, OECD Report, 2011. https://www.oecd.org/els/soc/dividedwestandwhyinequalitykeepsrising.html. Accessed: April 07, 2021.

Doan, T., & Wan, G. (2017). Globalization and the labor share in national income, ADBI Working Paper, 639, Asian Development Bank Institute (ADBI), Tokyo.

Frankel, J. (2018). Do globalization and world trade fuel inequality? The Guardian, January 2, 2018. https://www.theguardian.com/business/2018/jan/02/do-globalisation-and-world-trade-fuel-inequality. Accessed: April 03, 2021.

Growing Unequal? Income Distribution and Poverty in OECD Countries, OECD Report, 2008. https://www.oecd.org/els/soc/growingunequalincomedistributionandpovertyinoecdcountries.htm. Accessed: December 18, 2020.

Hobbes, T. L. (1982). Penguin, 1651.

Income inequality. OECD data. https://data.oecd.org/inequality/income-inequality.htm

Jones, C. I. (2016). The facts of economic growth. In *Handbook of macroeconomics*, 2A. Elsevier B.V. ISSN 1574-0048. https://doi.org/10.1016/bs.hesmac.2016.03.002. (data accessed: February 01, 2021).

Rousseau, J. (1968). The social contract (1792). Penguin.

Stiglitz, J. (2015). New theoretical perspectives on the distribution of income and wealth among individuals. In *Part 4: Land and credit. National bureau of economic research*. https://www.nber.org/system/files/working_papers/w21192/w21192.pdf. (data accessed: April 01, 2021).

Under pressure: The squeezed middle class, OECD Report, 2019. https://www.oecd.org/social/under-pressure-the-squeezed-middle-class-689afed1-en.htm

Understanding the downward trend in labor income shares, World Economic Outlook: Gaining Momentum? International Monetary Fund, April 2017.

Wilkinson, R., & Pickett, K. (2009). *The spirit level: Why more equal societies almost always do better*. Bloomsbury Publishing.

Impact of Globalization on Internal Migration of Population

Natalya V. Vorontsova, Alexander V. Merzlov, Rafail R. Mukhametzyanov, Tatiana V. Ostapchuk, and Natalya N. Ivantsova

Abstract Since the emergence of the first civilizations, urbanization has been one of the features of the development of human society. This trend continues to this day. The increasing role of cities in society, the growth of the urban population, and the improvement of the quality of life in cities have long dominated the world. Opportunities for employment and career growth, higher income, and developed transport, engineering, educational, medical, and recreational infrastructure determine the attractiveness of urban life (especially in the metropolis) in the eyes of the rural population. However, recently, the developed countries of the world, especially the EU countries, witnessed a reverse flow of internal migration from the city to the countryside. An important role in this process is played by the features of globalization, namely global trends and challenges, advanced technology, and opportunities influencing political decisions and changes in the population's lifestyles and needs. The increased interest in rural areas among the urban population is due to several reasons. This paper aims to determine current global trends affecting rural development. The authors identify how these trends affect rural development and the extent to which they impact internal migration. As a result, the authors outline and systematize flows of internal migration in the urban–rural direction occurring in the current world.

Keywords Globalization · Internal migration · Urbanization · Urban population · Rural area

JEL code F69

N. V. Vorontsova (✉) · A. V. Merzlov · R. R. Mukhametzyanov · T. V. Ostapchuk · N. N. Ivantsova
Russian State Agrarian University—Moscow Timiryazev Agricultural Academy, Moscow, Russia

T. V. Ostapchuk
e-mail: tostapchuk@rgau-msha.ru

N. N. Ivantsova
e-mail: kozuch75@rgau-msha.ru

E. G. Popkova and B. S. Sergi (eds.), *Geo-Economy of the Future*,
https://doi.org/10.1007/978-3-030-92303-7_57

1 Introduction

Nowadays, the flow of internal migration in most countries is concentrated in the direction of the city. This trend has been forming for quite some time and has considerably increased over the last century. According to the World Bank, the urban population was only 3.0% in 1800 and 13.6% in 1900. In turn, about 34% of the world's population lived in cities in 1960, 47.4% in 2000, and 55.7% in 2019 (The World Bank, n.d.). Nevertheless, current realities suggest that the urban population, especially in developed countries, has a pronounced desire to live in the countryside. These phenomena have become apparent in the context of globalization.

Globalization is usually interpreted as the strengthening of interaction, interconnection, and complementarity of national economies and their integration into the global economic space (Korolkov & Priemko, 2017). However, globalization is also inherent in the great opportunities for movement due to the development of tourism and migration on various grounds. Additionally, contemporary trends and technologies provide new opportunities for the development of civilization. This ultimately affects the processes and directions of internal migration.

The research goal is to identify the relationship between globalization and internal migration from urban to rural areas.

The research tasks are as follows:

1. To identify factors influencing changes in migration priorities in the modern world;
2. To assess migration flows from urban to rural areas;
3. To identify current technological opportunities and global challenges affecting prospects for rural development.

The research hypothesis suggests that current technological opportunities and their development, as well as several negative aspects of emerging globalization, contribute to the desire of a part of the population to move from urban to rural areas.

The research novelty lies in identifying and systematizing the advantages of the countryside compared to the city under conditions of globalization.

2 Materials and Methods

The research uses comparative and systematic analysis, which analyzed and identified the factors influencing internal migration in the world. The statistical method allowed the authors to quantify internal migration flows. The impact of global challenges and technological opportunities on rural development was estimated using the induction method.

3 Results

During the COVID-19 pandemic, there were public perceptions that a mass exodus from cities to the countryside was inevitable (Weeden, 2020). This view is related to the changing lifestyles, perceptions, and priorities of many people worldwide due to the restrictions and epidemiological environment caused by COVID-19. Due to these trends, the opinions of scholars (particularly Peter F. Drucker) that the only way to preserve human life and health under conditions of total urbanization is to create urban communities (Drucker, 2007) are no longer accepted as truth in all instances.

According to the research results, we can say that the demand for rural areas among the urban population was present even before the COVID-19 restrictions, which is evidenced by the trend of public demand for relocation from cities. However, different countries have their own specifics and timeframes.

For example, in Germany, young families in the 1960s and 1980s mostly moved from rural areas to cities. The same happened in Switzerland, where people moved from the cantons to the city of Zurich. As a result, the birth rate and the number of young families increased in urban areas compared to rural areas. The demand was mainly for large cities. Since 2010, together with urbanization, Germany has seen a trend of population growth in the suburbs of large cities (Catella Group, 2020). In the Swiss cantons, the share of the population has not decreased since 2000.

The same trend is observed in other European countries (Hofer, 2018). For example, the French no longer consider large cities attractive to live in. There is a trend and a demand for the countryside. In Western Europe, there is summer and recreational migration to rural areas. For example, many rural settlements of Greece exist economically only at the expense of summer residents and homeowners from Germany.

In Canada and the USA, there is widespread demand for summer homes, second homes, single-story cottages, suburbanization, and the development of metropolitan suburbs (Pokrovsky, 2018).

In Russia, most of the urban population has a second home in the countryside (dacha (summer house) or house in a village). Nowadays, many residents of metropolitan cities actually live in rural areas but are officially listed as urban residents and remain employed in the city. This fact significantly distorts statistical data on the relocation of urban residents to the countryside (Vorontsova et al., 2021). Despite this fact, according to Federal State Statistics Service (Rosstat), the number of people leaving cities and arriving in rural areas has doubled in Russia since 2010, reaching 1.3 million people in 2019 (Rosstat, n.d.).

The validity of the statistics is also affected by the definition of the countryside itself. Rural areas do not necessarily coincide with administrative boundaries or the zone of influence of a particular economic sector (Novikov et al., 2013). As noted by many scholars, the term "rural" has a territorial emphasis regardless of the modes of land use, the degree of economic development, and the predominance of any economic sector. In practice, it is divided by the administrative boundaries of the subjects, municipalities, rural administrations, etc. (Merzlov et al., 2020).

Some specialists understand rural areas as a specific type of territory, including purely agricultural and separate, closely related to them, settlements of urban-type. In these settlements, employment in the sectors of material production is dominated by employment in the agro-industrial complex. Moreover, there are stable and active pendular labor movements in the urban–rural system (Novikov & Streltsov, 2014).

Given these aspects, we can conclude that there is internal migration from the city to the countryside. Nevertheless, it is statistically underestimated. During the study, the authors systematized urban–rural population movements in the following directions:

1. Expansion of the radius of suburbanization and the city's geographical boundaries;
2. Shuttle migration, short and long-distance movements to work;
3. Residential multilocality (living in two houses);
4. Seasonal migration, renting housing for a long summer or vacation period in rural or recreational areas;
5. Relocation to rural areas and small towns for permanent residence.

Rural areas attract city dwellers for various reasons, such as personal motivation, economic benefits, ecology, and many others (Vorontsova et al., 2021). The public policy being implemented plays an important role. Nevertheless, despite these or other motivations, the authors put forward the hypothesis that globalization currently impacts the decision of urban residents to relocate. Moreover, it influences the development of rural areas. The authors of the article "In Search of Community in Rural China" also note that globalization affects rural areas of China. This is especially evident in the shuttle migration when rural residents leave home to work in cities and towns. It is noted that many contemporary studies of migration from rural areas of China highlight its negative effects on rural life (Hill et al., 2020).

To confirm the posed hypothesis, the research considers global challenges and trends directly or indirectly affecting internal migration and rural development:

- *Environmental pollution.* The population's concentration, consumption patterns, modern lifestyles and economic activities in cities, garbage and greenhouse gas emissions only exacerbate the problem. Beginning in the 1950s, the United Nations began to adopt a series of measures and regulations to protect the environment. One of them is the strategy of transition to sustainable development—Agenda 21, signed by 179 countries (United Nations, 1992). This allowed to follow the directions of sustainable development in the planning and implementing public policies and strategies, including the development of rural areas (Vorontsova, 2021).
- *Poverty, Hunger, Unequal Opportunity.* With the world's technological advances, much of the world's population is below the poverty line, unable to eat adequately, and lacking access to basic household services. At the same time, another part of the population has a manifestation of luxury and affluence. The first group lives mostly in rural areas—globally, over 70% of the world's poorest people live in rural areas (FAO, 2015). In 2015, the UN Summit with Heads of State and

Government adopted a new agenda: "Transforming our World: The 2030 Agenda for Sustainable Development" (UN General Assembly, 2015). The 2030 Agenda aims to eradicate poverty and achieve sustainable worldwide development by 2030. Many countries around the world also follow the 2030 Agenda in supporting rural areas.

- *Climate change.* The change of weather conditions becomes more extreme, and sea levels rise. It destroys national economies and affects people's lives. Countries are taking a set of international (UNFCCC, 2015) and national technological and financial measures to combat the impacts associated with climate change.

About two-thirds of Russia's population live in areas designated as areas with harsh climatic conditions (Traywish, 2003). A similar situation is observed in Canada and Sweden. In recent decades, there has been a spontaneous migration of the Russian population to the most favorable areas of residence—the Black Earth Region and the European south of Russia. Simultaneously, Russia is still on a course for the development and arrangement of extreme territories. In turn, global warming increases the demand among the world's population to move to more northern areas. There is an emerging fashion for arctic tourism.

o *Security.* One of the current global trends is an increase in population migration in all regions of the world. The centers of migration tend to be large cities, scientific, business, and cultural centers. Migration flows quantitatively and qualitatively affect the demographic, social, and economic security of all countries (Pencea & Curteanu, 2020). Mass migration is often accompanied by the formation of marginalized groups (unemployed, homeless, beggars, and refugees), which create their subculture. The growth of such groups poses a danger to society because marginalized people are often hostile to the fundamental values of a particular society.

The world community also faces one of the most acute problems—terrorism. From the regions of traditional international conflicts, terrorism has moved to the developed and most prosperous states, namely to their centers, major cities, and places of gathering of people. Despite the measures taken by countries to combat emerging threats, these challenges undermine the ability to live safely in large cities and metropolitan areas.

o *Epidemics.* The modern world today is faced with a new viral disease—COVID-19. As recommended by the WHO, many countries apply various measures to control its spread. These measures include the closure of international borders, restrictions on travel, quarantine measures, sanctions for violations, isolation, distance work and study, vaccinations, etc. Urban residents, with the existing prosperous living conditions, were trapped in apartments with no way out. In turn, the rural population, less restricted in its freedom, was in danger of being left without qualified and affordable medical care. However, the risk of infection in large cities and crowded places is much higher than in rural areas.

The COVID-19 pandemic has radically changed the way people live and behave. There has been a worldwide trend of migration from cities to the countryside. Many tourists started to think about avoiding large crowds of people (cruise ships, large tourist complexes, museums, and cities) in favor of more dispersed distances—rural territories.

- o *Urbanization.* Throughout the world, there is an increase in the role of cities in society, which is accompanied by their growth, an increase in the proportion of the urban population, and the predominance of urban life over rural life. On the one hand, life in the city is more economically profitable than in the countryside. On the other hand, the concentration of industry, transport, and population in a small space worsens other conditions of quality of life. Thus, the rapid growth of urbanization in China contributed to the country's prosperity but caused several problems in the countryside. In turn, the problem of rural decline further threatens the sustainability of urbanization in China. Chinese scholars speak of the need for measures to revitalize the countryside for the effective long-term development of the country (Li et al., 2018).
- o *Advanced technology* connected with automation, decentralized production, and nanotechnology is opening up new production possibilities. These possibilities include new professions, uncrewed devices for the delivery of goods and services, modern methods of communication, electronic education and medicine, the ability to live and work at a distance, etc. However, new opportunities may cause new challenges for the global community that could have irreversible consequences. It is the violation of personal space and the protection of personal interests associated with the ability to access personal data, the displacement of traditional ways of life and communication, the reduction of employment and robotization, which may result in a decline in the value of and satisfaction with human life.

Measures to combat global challenges (e.g., pollution, inequality, poverty, and hunger) are reflected in public policies for rural development worldwide. Climate change, security threats, and epidemiological conditions influence decisions about the place of residency. Urbanization, modern opportunities, and digital technology contribute to internal migration from the city to the countryside.

4 Discussion

There is also a scientific view that globalization promotes the development of cities and migration from rural areas. Cities are seen as centers of power, wealth, and political and administrative governance. Some authors see urban centers as "gateways to the global world" (Makhnovsky, 2016). Nowadays, much attention in different countries is paid to creating smart cities aimed at the use of information and communication technologies to solve urban problems. For example, the National Strategic Program "Smart City," implemented in Korea, is marked with the use of 5G telecommunications technology and data processing (Yang et al., 2021). Many political leaders speak of the inefficiency of rural development and the high quality of life in cities (Vorontsova et al., 2021).

Undoubtedly, life is the most comfortable in cities. Moreover, the financial component plays an important role. However, cities have several disadvantages compared to rural areas. In this research, the authors attempt to identify the advantages of rural areas compared to urban areas in the age of globalization. These advantages are as follows:

1. *Ecology and natural foods.* Ecologically clean and green areas, the presence of forested areas, clean water bodies, and the availability of clean drinking water are becoming increasingly valuable. There is an increasing demand for organic food, clothing, and natural materials. Despite creating green recreational areas and stores with organic food in urban areas, rural areas win on all these indicators, with significant potential in terms of living conditions (clean air, water, and the absence of noise) and the possibility to produce and consume organic products.
2. *Space.* The crowdedness of cities and the limited living space of apartment buildings were most acutely felt by all urban citizens during the pandemic. The demand for space is the highest among families with children and older people. The opportunity for safe and affordable nature walks every day is an advantage of rural areas.
3. *Unique culture.* Rural areas have historically been a repository of traditions, language, customs, traditional forms and methods of farming, material culture (architecture and household items), and intangible culture (songs, dances, stories, and rituals), all of which are the basis of the national identity.
4. *Absence of information garbage.* Billboards, signs, and other information garbage of cities in the countryside are replaced by a diverse landscape and the beauty of nature.
5. *Advanced technology and innovation* contribute to the creation of autonomous comfort in rural areas.

Currently, society starts to think about more favorable conditions for life, with intangible values such as clean air, water, natural foods, lack of information garbage, availability of space, and distinctive culture playing an important role. These values can be found in rural areas. Competent use of these values will create conditions for quality of life in rural areas and contribute to the country's development.

5 Conclusion

There is a global trend toward internal migration to metropolitan residents to small towns and the countryside, the expansion of the suburbanization radius, various forms of mobility, and shuttle migration. Migration from cities cannot be considered equal to migration from rural areas. However, it is undeniable that the growth of people's needs and desire to live outside cities can be traced over the past decades in many developed countries. The COVID-19 pandemic spurred decisive action by city dwellers who had previously considered moving to the suburbs. Cities, especially metropolitan areas, are becoming business centers, while the closeness to nature gives a sense of freedom and security.

Global challenges are forcing humanity to think about directions to solve them. Governments of many countries are taking policy measures to combat global challenges. In this context, sustainable development and the reduction of disparities in social stratification are being pursued, which is also reflected in rural development policies.

Even though urbanization is still dominant worldwide, rural areas have all the potential that will help to attract the population if the right conditions are created.

References

Catella Group. (2020). Zwischen Urbanisierung und Suburbanisierung—wo wohnen die Deutschen zukünftig? Retrieved from https://www.catella.com/globalassets/global/mix-germany-corporate-finance/catella_market_tracker_urbanisierung-suburbanisierung.pdf

Drucker, P. F. (2007). Managing in the next society (E. V. Tribushna Trans. from English). Williams (Original work published 2002).

FAO. (2015, July 14). Investing in a future free from hunger: FAO Director-General José Graziano da Silva called for additional investments in social protection and rural development. In *Third International Conference on Financing for Development in Addis Ababa, Ethiopia*. Retrieved from www.fao.org/sustainable-development-goals/news/detail-news/ru/c/308018/

Federal State Statistics Service (Rosstat). (n.d.). The number and migration of the population of the Russian Federation. Retrieved from https://rosstat.gov.ru/compendium/document/13283?print=1

Hill, A. M., Wei, J., & Rose, S. (2020). In search of community in rural China. *Asianetwork Exchange, 27*(1), 127–149. https://doi.org/10.16995/ane.321

Hofer, B. (2018, August 5). Irgendwann wurde uns klar: Wir gehen nicht zurück in die Stadt. Kundendienst SRF. Retrieved from https://www.srf.ch/kultur/gesellschaft-religion/wochenende-gesellschaft/stadt-vs-land-irgendwann-wurde-uns-klar-wir-gehen-nicht-zurueck-in-die-stadt

Korolkov, A. F., & Priemko, V. V. (2017). *Agrarian policy: Textbook*. Russian State Agrarian University—Moscow Timiryazev Agricultural Academy.

Li, Y., Jia, L., Wu, W., Yan, J., & Liu, Y. (2018). Urbanization for rural sustainability—Rethinking China's urbanization strategy. *Journal of Cleaner Production, 178*, 580–586. https://doi.org/10.1016/j.jclepro.2017.12.273

Makhnovsky, D. E. (2016). Globalization and the development of a network of world cities. *Age of Globalization, 3*(19), 57–70.

Merzlov, A. V., Vorontsova, N. V., & Korolkov, A. F. (2020). Rural development in the countries of the world: Textbook. Russian State Agrarian University—Moscow Timiryazev Agricultural Academy.

Novikov, V., & Streltsov, V. (2014). Development of resources of work in rural territories: Problems, tendencies, prospects. *Economics of Agriculture in Russia, 2*, 66–72.

Novikov, V., Zhubarkin, S., & Chaly, V. (2013). Features of reproduction and research of labor capacity of rural territories of Russia: Branch, demographic and territorial aspects. *Economics of Agriculture in Russia, 7–8*, 137–148.

Pencea, G. C., & Curteanu, A. B. (2020). Internal and external migration. *Postmodern Openings, 11*(1), 57–70. https://doi.org/10.18662/po/108

Pokrovsky, N. (2018, April 18). Index card: de-urbanization. Is there life after the city? Scientific and educational portal IQ. Retrieved from https://iq.hse.ru/news/218311677.html
The World Bank. (n.d.). Official website. Retrieved from https://www.worldbank.org
Traywish, A. (2003). *Russia: Population and space*. Demoscope Weekly. Retrieved from http://www.demoscope.ru/weekly/2003/095/tema01.php
UN General Assembly. (2015). Resolution A/RES/70/1 "Transforming our world: The 2030 agenda for sustainable development" (September 25, 2015). New York, NY. Retrieved from https://www.un.org/ga/search/view_doc.asp?symbol=A/RES/70/1&Lang=E
United Nations. (1992). Agenda 21. Rio de Janerio, Brazil. Retrieved from https://sustainabledevelopment.un.org/content/documents/Agenda21.pdf
United Nations Framework Convention on Climate Change (UNFCCC). (2015). The Paris agreement (December 12, 2015). Paris, France. Retrieved from https://unfccc.int/sites/default/files/english_paris_agreement.pdf
Vorontsova, N. V. (2021). The relationship between global trends and rural development policies (Based on international experience). *Journal of International Economic Affairs, 11*(1). https://doi.org/10.18334/eo.11.1.111784
Vorontsova, N. V., Merzlov, A. V., Mukhametzyanov, R. R., & Platonovskiy, N. G. (2021). Assessment of the attractiveness of rural areas in terms of internal migration in Russia and the Eu countries. *Economy of Agricultural and Processing Enterprises, 6*, 40–47. https://doi.org/10.31442/0235-2494-2021-0-6-40-47
Weeden, S. A. (2020, September 2). The coronavirus pandemic is pushing Canadians out of cities and into the countryside. The Conversation. Retrieved from https://theconversation.com/the-coronavirus-pandemic-is-pushing-canadians-out-of-cities-and-into-the-countryside-144479
Yang, J., Kwon, Y., & Kim, D. (2021). Regional smart city development focus: The South Korean national strategic smart city program. *IEEE Access, 9*, 7193–7210. https://doi.org/10.1109/ACCESS.2020.3047139

ESG Banking: A Fundamental Component of Sustainable Development, or Development Without Prejudice to the Future of Russia

Elena P. Ermakova and Marina M. Vildanova

Abstract The authors investigate the features of the formation and development of ESG banking in Russia. It was revealed that: (1) the business model of ESG banking is the conduct of banking business under the UN's sustainable development standards (environmental, social, and governance (ESG)); a detailed description of these standards is called taxonomy; (2) Legal regulation of ESG banking is carried out in each country by the primary financial regulator and other specially authorized bodies; in Russia, the Bank of Russia acts as the leading financial regulator and VEB.RF has been a specially authorized body since November 18, 2020; (3) The primary sources of legal regulation of ESG banking in Russia are the documents of the Bank of Russia, namely: (a) "Recommendations for the implementation of the principles of responsible investment" dated July 15, 2020; (b) Regulation of the Bank of Russia dated December 19, 2019, No. 706-P "On the standards for the issue of securities"; (4) In July 2020, VEB.RF published "Methodological recommendations for the development of investment activities in the field of green finance in the Russian Federation", which contained a taxonomy of "green" finance. However, the adoption of the Federal Law "On Environmental Audit and Environmental Auditing and Amendments to Certain Legislative Acts of the Russian Federation" developed by the Ministry of Natural Resources and Environment of the Russian Federation would contribute to improving the implementation of "green" projects; (5) Research by Russian experts shows that Russian banks are ready to implement the ESG banking business model, but more than 60% of banks consider it necessary to establish regulatory standards for risk assessment, information disclosure, and accounting for ESG factors; (6) Russia has a positive experience in ESG banking: in February 2021, the RAEX-Europe rating agency published the ESG ranking of Russian companies, which for the first time included Russian banks.

E. P. Ermakova (✉)
Peoples' Friendship University of Russia (RUDN University), Moscow, Russia
e-mail: ermakova_ep@pfur.ru

M. M. Vildanova
The Institute of Legislation and Comparative Law Under the Government of the Russian Federation, Moscow, Russia

E. G. Popkova and B. S. Sergi (eds.), *Geo-Economy of the Future*,
https://doi.org/10.1007/978-3-030-92303-7_58

Keywords ESG banking · Bank of Russia · VEB.RF · Sustainable development · Taxonomy · Green finance

JEL Codes F36 · G15 · G21 · K24 · O16

1 Introduction

The popularity of ESG banking and green finance is increasing worldwide. People have become more aware of nature and their choice because they want to breathe clean air and save the planet for future generations. If a couple of decades ago, young people had material benefits in the first place and the speed and convenience of obtaining them, today's young people are ready to abandon fast-food goods, sort waste, go to stores with reusable bags. Currently, each person's social life, the unconditional observance of rights, and inclusion are the main priority. The same is happening in the corporate and financial world, where reputation is one of the most critical factors, and many organizations have already felt the benefits of introducing sustainable development principles into their processes (Mordyushenko, 2020).

In the financial world, interest in sustainable growth has long been limited to a small group of specialist operators, but that has changed in recent years. The 2015 Paris Agreement recognized the vital role of the financial system in promoting sustainable development. Since 2015, Environmental, Social, Governance (ESG) investment funds have increased the total assets they manage by over 170%. Between January and October 2020, net inflows to this category of funds in Europe amounted to more than 150 billion Euros, which is almost 80% more than in the same period last year, said Panetta (2021), Member of the Executive Board of the European Central Bank. According to him, this trend will continue.

Over the past year, the global economy has faced an unprecedented chain of escalating factors affecting ESG, which has affected the way companies do business (Vysokov, 2020). The most notable of these factors was the outbreak of the COVID-19 pandemic and the subsequent containment and vaccine development efforts. A stream of climatic disasters accompanied these events. Another year of record temperature reminded the population, tired of the restrictions caused by COVID, that the climate crisis is becoming more urgent without an obvious solution (Bendersky & Ferguson, 2021). Besides, numerous high-profile public disruptions have prompted a renewed focus on ethics among the general public, which will gain importance as global economies seek to equitably distribute stimulus funds to help overcome the economic downturn caused by COVID-19. The shocks caused by the Covid-19 have changed public opinion about what is crucial as it moves towards a sustainable recovery. The value of ESG factors has increased during the pandemic, which was confirmed by a study conducted by a group of specialists from Deutsche Bank. Analyzing investor sentiment and board data, the group found that climate change remains a top priority for corporations as companies discuss how to ensure that their economic recovery investments are as green as possible (ESG on the ground, 2020).

According to BloombergNEF (BNEF), the total sustainable investment in 2020 was $732 billion, up 29% over 2019 (Henze, 2021). Social Bonds and Sustainable Development Bonds showed impressive growth throughout the year, while green bonds rose sharply at the end of the year. The increase was driven by the category of social bonds issued to raise funds for social purposes such as employment, health care, and education. The issuance of these instruments in 2020 jumped seven times to $147.7 billion. The total volume of green bonds in 2020 was $305.3 billion, which is 13% higher than in 2019. The cumulative green bond issuance from 2007 to 2020 surpassed $1 trillion (Gardett, 2020).

The European Commission estimates that an additional €260 billion in annual investment is required to achieve the 2030 climate and energy targets set by the Green Deal (Schuller, 2020). The private sector is a key to financing the green transition. That is why regulatory issues of ESG business models should be a priority for the European Commission.

2 Methodology

The scientific development of the content of this chapter of the monograph is carried out based on the general scientific method of historical materialism. General scientific methods of cognition are used: dialectical, hypothetical-deductive method, generalization, induction and deduction, analysis and synthesis, empirical description. The study also used private science methods: juridical-dogmatic, statistical method, method of comparative legal analysis, other.

3 Results

3.1 ESG Banking Business Model

The ESG area in the investment risk assessment system is the most dynamically progressing aspect in the theory and practice of investment analysis, as it was noted in the 2020 Report of the Russian-British Working Group on Corporate Governance "International experience in the application of ESG (Environmental, Social, Governance) standards and the possibility of its use in Russia" (2020). In recent years, there has been a rethinking of approaches to investment and standards of doing business worldwide at all levels: civil society, business owners, investors, managers, politicians. However, there is currently no established international practice in regulating ESG issues. It is not only heterogeneous but also in constant development. The critical source of expertise for this practice now is the experience of voluntary disclosure standards related to sustainable development issues, generally recognized by both issuers and experts, but most importantly, by investors (Matytsin, 2021).

In December 2020, the Deputy Chairman of the Bank of Russia Sergey Shvetsov noted that "both at the national level until 2030 and in terms of the regulator, the development of the so-called social and "green" bonds is a new direction, when an investor does not just take on the function of a creditor, but he is also involved in changing the economy, turning it towards a lower environmental impact or social projects that are implemented by both businesses and regions. This market has potential. The infrastructure of "green" and social financing is being formed worldwide, and Russia is not on the sidelines. With market participants, Vnesheconombank, the Ministry of Economy, the Ministry of Finance, a "taxonomy and verifiers are being created". Sergei Shvetsov (2020) believes that by the end of 2021, this market should be working "to its full extent". Thus, the concept and internal content of ESG banking in Russia are just emerging, and it is expected that this process will be completed by the end of 2021.

As the authors have mentioned earlier, "green" finance refers to financial transactions that support the transition to a low-carbon economy and the fight against climate change (Ermakova, 2020). However, the term "green" finance is still understood in different ways, with many investors criticizing the investment industry and companies for advertising themselves as "green" or "sustainable" without an agreed definition of what that means (Inshakova et al., 2019). Lack of reliability and standardization has led to the need for normative consolidation of the concept and criteria of "green finance". This fact was emphasized by many Russian and foreign experts: "There are no generally accepted methodological approaches to regulating ESG issues and even a unified terminology in the world. This circumstance sometimes leads to suboptimal use of financial and human resources. Therefore, among the key participants in the process, there is a consensus on the need for the earliest possible formation of common approaches at the global level" (Report of the Russian-British Working Group on Corporate Governance "International experience in the application of ESG (Environmental, Social, Governance) standards and the possibility of its use in Russia", 2020).

It is the first time that such a definition has been formed in the European Union: the European Commission's website has defined "green finance" as financing that takes environmental and social considerations into account when making investment decisions. It was clarified that ecological concerns relate to climate change mitigation and adaptation. Social problems can relate to inequality, inclusion, labor relations, investment in human capital and communities, and human rights issues (Inshakova et al., 2020); governance of public and private institutions plays a crucial role in ensuring that social and environmental considerations are incorporated into decision-making. These three components—ecological, social, and governance—have come to be referred to as "ESG" for short (Frolova, 2020).

But the mere notion of "green finance" was not enough: different economic activity categories required a detailed listing. This way, the term "taxonomy" appeared for the first time—a kind of glossary that defines the performance criteria agreed with the Paris Agreement of 2015 for several economic activities (Piticchio, 2020). Taxonomy is a classification framework against which actions can be assessed to determine if they are environmentally sustainable.

At the end of 2019, the Council and the European Parliament agreed on a normative act—the "Taxonomy Regulation". On March 9, 2020, the Expert Group published its "Final Report on EU Taxonomy", which contained recommendations for a comprehensive EU taxonomy framework (Practical recommendations of the banking community on the implementation of ESG banking in Russia, 2020). The "Taxonomy Regulation" was adopted on June 18, 2020, and entered into force on July 12, 2020. Three weeks later—on July 13, 2020—the state corporation VEB.RF, together with ministries, departments, and the Central Bank of the Russian Federation, published the National Green Finance Standard of Russia, the first taxonomy of the Russian Federation on green finance (Sberbank, 2020). Thus, at the beginning of summer 2020, both the European Union and Russia adopted acts that outlined real prospects for the development of "green" finance and, ultimately, determined the opportunities for the financial sector to demonstrate its commitment to green finance and sustainable infrastructure, which is the essence of the 2015 Paris Agreement.

Experts note that the topic of ESG banking is relatively new for the Russian market, but the demand for it from customers is growing (ESG on the ground, 2020). Studies by Russian experts show that Russian banks are ready to implement the ESG banking business model, but more than 60% of banks consider it necessary to establish regulatory standards for risk assessment, information disclosure, and accounting for ESG factors (Vysokov, 2020).

Today it can be argued that there is a general understanding that ESG banking covers the practical actions of banks to implement the principles of Environmental, Social, and Corporate Governance (ESG). The ESG banking business model is the conduct of banking business following the sustainable development standards enshrined in the UN. The taxonomy provides for the identification and classification of phenomena and processes to assess ESG banking results.

3.2 Regulatory Issues of ESG Banking

In 2015, work was completed on various international agreements such as the Paris Climate Agreement ("COP 21") and the "UN Agenda and the 2030 Sustainable Development Goals" ("UN SDGs"). Russia signed the Paris Agreement in 2015 but only ratified it on September 23, 2019.

Russia is still only at the beginning of the path of "green" financing. On December 24, 2018, by the Presidium of the Council's decision under the President of the Russian Federation for Strategic Development and National Projects, the Passport of the National Project "Ecology" was approved. In May 2019, by its Resolution, the Russian government approved the "Rules for Subsidizing "Green Bonds"—Rules for Granting Subsidies to Compensate for the Costs of Payment of Coupon Yields on Bonds that enterprises will issue after January 1, 2019, as part of the implementation of BAT (Best Available Technologies)". Pilot "green bond" projects were launched by the Moscow Stock Exchange and the Bank of Russia.

The end of 2018, 2019, and 2020 became years of active awareness by the financial market and the Russian government to implement the principles of sustainable development and responsible investment. The green topic sounded on government and expert platforms, and new issuers of green bonds appeared. However, in 2019—early 2020 in Russia, there was a complete lack of government guidelines in "green" financing. In February 2020, Russian analysts noted an acute problem in the country with a lack of structured and complete information regarding "green" financing, one of the main factors of market deceleration (Investinfra, 2020). Full development of any market is impossible without its participants' access to reliable, confirmed, and verified information.

The primary financial regulator of Russia, the Bank of Russia, responded to the modern economy's challenges by adopting the "Regulation on securities issue standards" No. 706-P dated December 19, 2019, registered by the Ministry of Justice of the Russian Federation on April 20, 2020. The document provides for the possibility of issuing "green" and social bonds.

In July 2020, the Bank of Russia published its "Recommendations for the Implementation of Responsible Investment Principles", dated July 15, 2020.

According to the Order of the Government of the Russian Federation of November 18, 2020, No. 3024-r, the state development corporation "VEB.RF" is assigned the function of a methodological center in the development of investment activities in the field of sustainability, including green growth, and attracting extra-budgetary funds to the implementation of development projects in the Russian Federation. Currently, VEB.RF is developing a national "green" financing system, facilitating green projects on more favorable terms. VEB.RF plans to launch the Russian national "green" financing system by July 2021. The basis of this system will be the Methodological Recommendations on Green Finance, published in July 2020, which took into account the experience of the largest international organizations and the experience of the European Union and China.

VEB.RF, as a development institution, is systematically working to introduce standards of responsible behavior in Russia. Such work aims to reduce both the risks posed by companies to society and the environment, increase the sustainability of companies' business processes, and increase their competitiveness. With the expert support of the Russia-OECD Center, RANEPA, VEB.RF promotes standards both domestically and participates in international organizations' work to improve international approaches to companies' responsible behavior, including financial institutions.

In 2020, VEB.RF, together with the Russia-OECD Center RANEPA, prepared the Principles for Responsible Financing, which are planned to be signed by the heads of the BRICS development institutions. The principles will form the basis for the further work of VEB.RF to implement responsible financing standards in its work, promote responsible behavior among the bank's clients, and additional work in BRICS.

Along with the functions of the methodological center of VEB.RF, environmental projects are financed, and social impact bond (SIB—social impact bond), which are financial instruments implying a multilateral contract between the state, an investor, and a social service provider, where payment is made only after the achievement of social effect. Thus, at the international economic forum in St. Petersburg in 2019, agreements were signed on implementing the first pilot project of social impact in the Republic of Sakha (Yakutia). The project should be implemented in the period from 2019 to 2022, and its goal is to improve students' educational results in 27 schools of the municipal district "Khangalassky ulus".

To systematize the work of credit institutions in the field of ESG banking, the Association of Russian Banks (ARB) has prepared "Recommendations for Credit Institutions on the Implementation of ESG—the Principles of Environmental, Social, and Management Responsibility" (Association of Russian Banks, 2020). The taxonomy provides for the identification and classification of phenomena and processes to evaluate ESG banking results. The recommendations note the need to implement comprehensive ESG measures, including new risk criteria, adjustments to counterparties' actions based on current factors, for example, the "hydrocarbon footprint" or long-term reputational risks.

Within the framework of the ARB meetings, the participants discussed the taxonomies of projects developed by VEB.RF, the requirements for project verifiers, a model verification methodology, and methodological recommendations, regulating the process of obtaining a "green" or "transitional" status by a financial instrument. Based on other countries' practice, it was decided to develop a system of "transitional" projects, in which some deviation from international standards and the creation of its unique classification is possible. In May 2021, it is planned to launch state incentives for "green" finance, and the start of a fully functional National System of "green" finance is scheduled for July. Government incentives can push those who do not yet see the real sense in the ESG approach.

The factual implementation of ESG banking requires changes in the organization, planning, risk management, and reporting on the work done by each bank. Of course, special attention should be paid to the analysis of environmental risks. As Kichigin (2018) notes, "environmental legislation presumes that any planned economic or other activity carries a potential hazard or threat to the environment". For the measurement of the likelihood of negative consequences, it is necessary to assess the risks and determine their acceptability. Unfortunately, with rare exceptions, there are no methods to evaluate environmental hazards.

Improving the implementation of "green" projects would be facilitated by the adoption of the Federal Law "On Environmental Audit and Environmental Auditing and Amendments to Certain Legislative Acts of the Russian Federation", developed by the Ministry of Natural Resources and Environment of the Russian Federation; Federal Law "On Amendments to Certain Legislative Acts of the Russian Federation (regarding the creation of liquidation funds by subsoil users) and other regulatory legal acts, including GOSTs".

3.3 Russia Has a Positive Experience in ESG Banking

In February 2021, the independent European rating agency RAEX-Europe published the ESG ranking of Russian companies, which for the first time included Russian banks (New ESG-rating from RAEX Europe, 2021). PJSC Credit Bank of Moscow (MCB) became the leader among the largest banks, according to ESG assessment. The second place is taken by PJSC Sberbank, the third—by VTB Bank (PJSC), the fourth—by JSC Rosselkhozbank, the fifth—by JSC Gazprombank.

The banks' current positions depend on how much information about ESG practices the banks disclose in their public reports. However, currently, there is an active introduction of sustainable development principles among the Russian banking sector leaders. The rating compilers assessed whether the bank has policies on "green" and "social" financing, whether ESG risks are considered in credit policies, whether there are "green" and "social" projects on the banks' balance sheets, and others. Due to the lack of a clearly defined concept of "green" and "social" projects, the assessment was primarily qualitative (New ESG-rating from RAEX Europe, 2021). However, soon, due to the accumulation of a critical mass of practices and reporting in banks and the emergence of several regulatory documents (first of all, the "green taxonomy" of VEB.RF), it will become easier to "digitize" the green and social portfolio of banks according to a single standard (Inshakova et al., 2018).

Credit Bank of Moscow (MCB), with the support of international financial institutions such as the EBRD and the IFC, has developed its own environmental and social management policy, which allows all the Bank's borrowers to be classified according to their ESG risks. Also, in the MCB, there is a so-called "exclusion list". It is a list of projects and industries, which the bank refuses to finance. The bank implements projects in the field of "green" finance, has its ESG strategy. In addition to this, MCB is actively working on its impact on the environment and society, mainly through introducing the "green bank" concept.

In 2020, the Supervisory Board of **PJSC Sberbank** (hereinafter referred to as Sberbank) approved the ESG strategy, which has become an integral part of its development strategy until 2023. The following key areas of work for ESG were identified:

- in the environmental field: minimizing its impact on the environment, promoting the spread of greener business models and technologies, the transition to a low-carbon economy;
- in the social sphere: decent working conditions, inclusion, deterring to the manifestation of any types of discrimination, the development of new models of education, and an increase in financial literacy;
- in the management area: improving the quality and transparency of management systems, building an ESG risk management system, integrating ESG aspects and principles into management at all levels.

Sberbank is transforming itself and advising corporate clients on environmental and social risk management, and launching new investment products that follow sustainable development practices. Significant for Sberbank, as a financial institution, is the development of ESG—banking. First, we talk about lending to "green" projects (for example, in the construction of solar and wind power plants). In this way, in 2018, the bank financed the construction of the Samara solar power plant, and in 2019 supported the construction of the Staromaryevskaya solar power plant in the

Stavropol Territory and provided a credit line to PJSC Enel Russia to finance the construction of the Kola wind farm with a capacity of 201 MW (Russian National Taxonomy for Green Projects, n/a). Sberbank actively finances renewable energy projects to reduce CO_2 emissions into the atmosphere, municipal waste processing projects, and the Clean Air project, which is part of the Ecology national project. The project's goal is to reduce harmful emissions into the atmosphere in Russia's 12 most polluted industrial centers.

Simultaneously, it is necessary to note the development in the ESG bank risk management system and improvement of stress testing elements. The Bank begins to apply the practice of differentiating relations with clients depending on the level of ESG risks: the client base will be classified according to the level of environmental risk, and depending on the fulfillment of environmental requirements, the conditions for the provision of services can be adjusted. In 2019, Sberbank took part in the first syndicated lending, in which the borrower's progress in environmental performance was taken into account. The borrower was Rusal. The lending rate for this transaction varies depending on how the company is meeting its environmental performance targets. Sberbank recently opened a line of credit to AFK Sistema, in which the rate is pegged to ESG indicators.

Sberbank Private Banking has developed two strategies in responsible investment: (1) Solactive SPB Foodtech Index, which will allow clients to participate in the development of new projects in the field of fast-growing segments of food technologies that develop their business following the principles of ESG; (2) SPB Women Impact EUROPE Index, formed from the shares of 30 public European companies with a high indicator of compliance with gender balance.

In 2020, Sberbank became the organizer of the placement of "green bonds" of JSC Russian Railways in 100 billion rubles. The Expert RA rating agency recognized the "green bonds" of Russian Railways as complying with both the international ICMA methodology and the Russian national methodology developed under the leadership of VEB.RF. The VEB.RF Committee on Green Financing has officially confirmed the Expert RA's conclusion on the instrument's compliance with the VEB.RF Methodological Recommendations on Green Financing. It was the first time a financial instrument was verified under the new methodology. Bonds without maturity were issued by Russian Railways "for financing green projects, as well as refinancing the incurred costs for green projects" (Sberbank, 2020).

Thus, Sberbank's ESG strategy is aimed at customers, employees, shareholders, investors, society, and the state and includes completely different areas, primarily those related to business. The implementation of the "green" principles has already begun in Sberbank. One of the topical issues now is the rejection of "plastic"—the bank's digital services help in this direction. For example, in addition to classic banking products, Sberbank now offers digital cards to its private clients, for which the issue of "plastic" is not expected. Simultaneously, the set of services with such digital cards practically does not differ from those provided for ordinary credit cards.

Gazprombank (Joint Stock Company), one of the leaders in "green" financing, has formed a council to implement sustainable development principles to systematize the work on the bank's projects and is developing a methodology for evaluating

ESG projects. As noted by the Deputy Chairman of the Board of Gazprombank A. Matveev, “Companies adhering to ESG principles have shown great resilience to crises of any nature”. He also emphasized the importance of maintaining a balance between economic efficiency and social orientation when investing (Mordyushenko, 2020). Following these principal approaches, Gazprombank finances the production of solar modules and electricity at solar power plants and participates in creating an infrastructure for the gaseous motor fuel market. Thus, in 2019, 28 infrastructure facilities were built, contributing to the transition to gaseous motor fuel and decreased vehicle emissions into the air.

Some projects are aimed at Gazprombank itself to reduce the impact on the environment. Even before the pandemic, pilot projects were launched to separate the collection of recyclable materials and transfer them for recycling. After returning to office life, it is planned to create a “green” office to consume natural resources within the bank efficiently. In 2019, Gazprombank ranked third in the Forbes ranking of the best employers in Russia. The Bank carried out and continues to provide targeted assistance to people during the pandemic.

At the level of regional banks, **Center-invest Bank,** a large private bank in the South of Russia, can be recognized as the leader in developing “green” financing. Vysokov (2020) emphasizes the fundamental difference between ESG—banking and speculative banking: “The momentary profit of a banker-speculator is based on the difference in volatile risks and the hope that clients will not have defaults. ESG bank takes all measures so that clients do not have defaults either today or tomorrow, and the bank has no losses either to cover these defaults”.

4 Conclusion

According to the authors, the business model of ESG banking is the conduct of banking business under the UN’s sustainable development standards (environmental, social, and governance (ESG)); a detailed description of these standards is called taxonomy. Legal regulation of ESG banking is carried out in each country by the primary financial regulator and other specially authorized bodies; in Russia, the Bank of Russia acts like the leading financial regulator and VEB.RF has been a specially authorized body since November 18, 2020. The primary sources of legal regulation of ESG banking in Russia are the documents of the Bank of Russia, namely: (a) “Recommendations for the implementation of the principles of responsible investment” dated July 15, 2020; (b) Regulation of the Bank of Russia dated December 19, 2019, No. 706-P “On the standards for the issue of securities”. In July 2020, VEB.RF published “Methodological recommendations for the development of investment activities in the field of green finance in the Russian Federation”, which contained a taxonomy of “green” finance. However, the adoption of the Federal Law “On Environmental Audit and Environmental Auditing and Amendments to Certain Legislative Acts of the Russian Federation” developed by the Ministry of Natural Resources and

Environment of the Russian Federation would contribute to improving the implementation of "green" projects. Research conducted by Russian experts shows that Russian banks are ready to implement the ESG banking business model, but more than 60% of banks consider it necessary to establish regulatory standards for risk assessment, information disclosure, and accounting for ESG factors. Russia has a positive experience in ESG banking: in February 2021, the RAEX-Europe rating agency published the ESG ranking of Russian companies, which for the first time included Russian banks.

Acknowledgements The reported study was funded by RFBR, project number 20-011-00270 "a".

References

A conference on responsible financial and investment practice in Russia was held in Moscow. Investinfra. February 20, 2020. https://investinfra.ru/novosti/v-moskve-proshla-konferencziya-po-otvetstvennoj-finansovo-investiczionnoj-praktike-v-rossii.html. Accessed: 03.20.21 (In Russ.).

Bendersky, C., & Ferguson, M. (2021). *Sustainability in 2021: A bird's-eye view of the top five ESG topics*. S&P Global. https://www.spglobal.com/ratings/en/research/articles/210128-sustainability-in-2021-a-bird-s-eye-view-of-the-top-five-esg-topics-11818407 Accessed: 20.03.21.

Ermakova, E. P. (2020). Development of the legal framework for "green" financing in Russia, the EU and China: A comparative legal analysis. *Bulletin of the Peoples' Friendship University of Russia. Series: Legal Sciences, 24*(2), 335–352. https://doi.org/10.22363/2313-2337-2020-24-2-335-352 (In Russ.).

ESG on the ground. (2020). Deutsche Bank AG, Frankfurt am Main. October 23, 2020. https://flow.db.com/more/macro-and-markets/esg-on-the-ground. Accessed: 03.20.21.

Frolova, E. E. (2020). New ecosystem of the European Union financial market: Digitalization and sustainability. *Bulletin of the Peoples' Friendship University of Russia. Series: Legal Sciences, 24*(3), 673–694. https://doi.org/10.22363/2313-2337-2020-24-3-673-694 (In Russ.).

Gardett, P. (2020). Green bond issuance surges in 2020. Argus media. January 19, 2021. https://www.argusmedia.com/en/news/2178514-green-bond-issuance-surges-in-2020. Accessed: 20.03.21.

Henze, V. (2021). *Sustainable debt breaks annual record despite Covid-19 challenges*. Bloomberg BNEF. January 11, 2021. https://about.bnef.com/blog/sustainable-debt-breaks-annual-record-despite-covid-19-challenges/. Accessed: 20.03.21.

Inshakova, A. O., Frolova, E. E., Rusakova, E. P., & Kovalev, S. I. (2020). The model of distribution of human and machine labor at intellectual production industry 4.0. *Journal of Intellectual Capital, 21*(4), 601–622.

Inshakova, A. O., Goncharov, A. I., & Kazachenok, S. Y. (2018). Legal regulation of the syndicated lending: Banking innovations in contemporary Russia. In E. Popkova, & V. Ostrovskaya (Eds.), *Perspectives on the use of new information and communication technology (ICT) in the modern economy* (pp. 368–377). Springer. https://doi.org/10.1007/978-3-319-90835-9_43

Inshakova, A. O., Goncharov, A. I., & Kazachenok, O. (2019). Legal regulation of the syndicated lending: Banking innovations in contemporary Russia. *Revista Espacios, (July), 39*(28), 7.

Kichigin, N. (2018). Minimization and prevention of environmental and legal risks. *Journal of Russian Law, 8*, 144–154. (In Russ.).

Matytsin, D. E. (2021). Internet-investing as a remote algorithm of the retail investment financing. In *Modern global economic system: Evolutional development versus revolutionary leap. Springer book series "Lecture notes in networks and systems"*. Springer Nature.

Mordyushenko, O. (2020). Companies adhering to the ESG principles have shown great resilience to crises. Newspaper "Kommersant", "Money". Appendix, 34, 20. https://www.kommersant.ru/doc/4566604. Accessed: 03.20.21 (In Russ.).

New ESG-rating from RAEX Europe. (2021). RAEX-Analytica LLC.16.02.2021. https://raex-rr.com/country/RAEX-600/ESG_rating. Accessed: 20.03.21 (In Russ.).

Panetta, F. (2021). *Sustainable finance: Transforming finance to finance the transformation.* European Central Bank. https://www.ecb.europa.eu/press/key/date/2021/html/ecb.sp210125_1~2d98c11cf8.en.html. Accessed: 20.03.21.

Piticchio, T. (2020). *The integration and disclosure obligations on sustainability risks and factors in the UCITS and AIFs regulatory frameworks.* European Law Blog. https://europeanlawblog.eu/2020/01/31/the-integration-and-disclosure-obligations-on-sustainability-risks-and-factors-in-the-ucits-and-aifs-regulatory-frameworks/. Accessed: 02.02.2021.

Practical recommendations of the banking community on the implementation of ESG banking in Russia. (2020). Association of Russian Banks. February 2021. https://asros.ru/upload/iblock/160/PRAKTICHESKIE-REKOMENDATSII-BANKOVSKOGO-SOOBSHCHESTVA-PO-VNEDRENIYU-ESG_BANKINGA-V-ROSSII.pdf. Accessed: 03.20.21 (In Russ.).

Report of the Russian-British Working Group on Corporate Governance "International experience in the application of ESG (Environmental, Social, Governance) standards and the possibility of its use in Russia". (2020). Analytical Center "Forum". http://www.mfc-moscow.com/assets/files/analytics/doklad_ESG_june_2020.pdf. Accessed: 03.20.21. (In Russ.).

Russian National Taxonomy for Green Projects. (n/a). VEB.RF. https://veb.ru/files/?file=8b510cd8f05e94eee2a85d7e6c2495d9.pdf. Accessed: 03.20.21.

Sberbank. (2020). Sberbank is the organizer of the placement of "green" bonds of the Russian Railways for 100 billion rubles. September 25, 2020. https://www.sberbank.ru/ru/press_center/all/article?newsID=84587e90-92fb-482b-bb37-c49f69f62fa8&blockID=1303&regionID=77&lang=ru&type=NEWS. Accessed: 20.03.21.

Schuller, M. (2020). Banks and sustainability in 2021: The next steps on the bumpy road to climate neutrality. October 30, 2020. https://think.ing.com/articles/banks-and-sustainability-in-2021-taking-the-next-steps-on-the-bumpy-road-to-climate-neutrality. Accessed: 20.03.21.

Sergei Shvetsov's speech at a meeting of the Council for the Development of the Financial Market under the Federation Council. Bank of Russia. 08.12.2020. https://www.cbr.ru/press/event/?id=8349. Accessed: 03.20.21 (In Russ.).

Tarakanov, V. V., Inshakova, A. O., & Dolinskaya, V. V. (2019). Information society, digital economy and law. *Studies in Computational Intelligence, 826*, 3–15.

Vysokov, V. (2020). *ESG banking: Made in Russia: A scientific and practical guide.* Russian State University of Economics (RINH), 44P. ISBN 978-5-7972-2759-5. https://www.centrinvest.ru/files/smi/pdf/ESG.pdf. Accessed: 03.20.21 (In Russ.).

Socio-political and International Legal Basis for a Paradigm Shift in Russian Criminal Law

Elena N. Senina

Abstract It is a well-known general social law that the more flexible, diversified, and complex the system of social management is, the more effective its results ("the law of necessary diversity"). This law is especially pronounced since the object of criminal-legal action, crime, becomes increasingly complex and more diverse. Assessing the criminal policy pursued in Russia from the standpoint of the marked law and the results of this policy embodied in the criminal legislation and practice of its application, the author recognizes the lack of the necessary diversity in their system. This lack exists due to the dominance of punitive elements and insufficient development of restorative justice to resolve criminal conflicts. Simultaneously, mediation mechanisms and other restorative justice programs in criminal proceedings are well-proven, actively used in many countries, and strongly recommended by international legal acts. In this regard, the paper aims to evaluate the current Russian criminal legislation in terms of the extent to which it embodies the normative foundations of "punitive" and "restorative justice" and develop a strategy for its improvement based on the analysis of international legal instruments. This study allowed the author to justify the need to transform the paradigm of modern criminal-legal impact to combine "punitive" and "compensatory-remedial" measures organically. Not denying the need to achieve the traditional prohibitive and punitive objectives of criminal justice, this paradigm shifts the center of gravity to reducing harm from crime through the introduction and active use of mediation and other programs of restorative justice.

Keywords Criminal policy · Criminal law · Alternatives to repression · Restorative justice · Mediation

JEL code K14

E. N. Senina (✉)
Attorney at Law, Institute of Mediation of the Russian Academy of Advocacy and Notaries, Moscow, Russia

E. G. Popkova and B. S. Sergi (eds.), *Geo-Economy of the Future*,
https://doi.org/10.1007/978-3-030-92303-7_59

1 Introduction

In developed legal democracies, criminal law organically combines the elements of punitive and restorative justice, favoring the latter approach.

Traditional punitive justice considers crime as a conflict between the perpetrator and the country. Punishment is seen as the primary means of resolving this conflict and restoring the violated relationship. In contrast, the concept of restorative justice places the victim of the crime at the center, viewing as such not the state whose law was violated by the crime, but primarily the individual whose right or legitimate interest was violated by the perpetrator. Restorative justice is not limited to measures to compensate the property, physical, and moral harm caused to the victim, it also includes a wide range of measures of social support for victims of crime (from assistance in accessing justice to the prevention of revictimization).

The traditional punitive approach is based on ideas of intimidation and retribution. At first glance, it is more understandable in terms of achieving the utilitarian goals of punishment (e.g., prevention and correction). However, in the longer term, it is ultimately inferior in its effectiveness to the restorative approach since it entails certain social costs and losses and imposes high costs on society (e.g., maintenance of a considerable number of prisoners, the cost of the prosecution and justice system, penitentiary system, etc.). Restorative justice is based on a "biblical model of justice" and focuses on repairing the harm caused by the crime and correcting the perpetrator to avoid future misconduct (Zer, 2002, p. 164).

Another advantage of a restorative approach to justice is the fact that it considers a "victim" not only the person who has been harmed by the crime but also, in a sense, the offender himself. Hence, it places importance on restorative tasks related to the correction and resocialization of the offender.

The extent to which the national legal system embodies elements of punitive and restorative concepts allows us to judge its legal content, nature, and democratic orientation.

2 Methodology

To measure the repressiveness of Russian criminal policy and determine the proportion of "punitive" and "restorative" manifestations in the current criminal legislation, the author considers it appropriate to use the following research methods:

- Formal-logical (allowing the author to analyze the content of the Criminal Code of the Russian Federation regarding the degree of implementing these concepts in it);
- Sociological (allowing the author to diagnose the mental and socio-psychological attitudes of the authorities and the population regarding the past, current, and future means of resolving criminal conflicts);

- Comparative-legal (allowing the author to study international legal sources of mediation and other programs of restorative justice in terms of their content and prospects for implementation in the criminal legal system of Russia).

3 Results

From this perspective, the analysis of contemporary Russian criminal policy allows the author to conclude that it has a predominantly repressive orientation. The following factors evidence the dominance of the punitive concept in Russian criminal policy:

- Reduction of the age of criminal responsibility in many corpus delicti (Federal laws No. 73-FZ (April 21, 2004), No. 153-FZ (July 27, 2006), No. 370-FZ (November 24, 2014), and No. 375-FZ (July 6, 2006));
- Increased penalties for crimes committed out of negligence and their classification as serious crimes (Federal law No. 146-FZ (June 17, 2019));
- Increased upper limits of terms and penalties for several types of punishments (Federal laws No. 134-FZ (June 28, 2013), No. 215-FZ (July 27, 2009), No. 420-FZ (December 7, 2011), No. 130-FZ (May 5, 2014), No. 375-FZ (July 6, 2016), and No. 46-FZ (April 1, 2019));
- Reduced possibility of imposing the sentence below the lower limit, conditional sentencing (Federal Laws No. 130-FZ (May 5, 2014), No. 375-FZ (July 6, 2016), and No. 46-FZ (April 1, 2019)), conditional early release from serving a sentence (Federal Laws No. 25-FZ (March 9, 2001), No. 245-FZ (November 3, 2009), No. 352-FZ (December 9, 2010), No. 18-FZ (November 2, 2013), and No. 375-FZ (July 6, 2016)), delay of serving punishment (Federal Laws No. 130-FZ (May 5, 2014), No. 375-FZ (July 6, 2016)), and release from punishment of minors (Federal Law No. 162-FZ (December 8, 2003));
- Criminalization of organizing, instigating, aiding, abetting, and preparation for several types of crimes as separate offenses (e.g., Federal Laws No. 352-FZ (December 9, 2010), No. 130-FZ (May 5, 2014), and No. 97-FZ (May 4, 2011));
- Significant predominance of criminalization over decriminalization and penalization over de-penalization in lawmaking activities (Korobeev, 2019, p. 246).

These are just some of the facts indicating that legislators are currently relying on punitive methods. The prevalence of a punitive orientation of Russian criminal policy at the level of law enforcement is also evidenced by the presence of an accusatory bias in the activities of investigators and courts, including an annual decrease in the number of acquittals, a significantly limited list of cases tried by a jury, a relatively high number of prisoners, a high percentage of sentences of imprisonment, a significant prevalence of law enforcement errors related to increasing criminal responsibility, and torturous conditions of imprisonment (Andrianov & Pudovochkin, 2020, p. 97).

These facts give reason to agree with the conclusion of reputable scholars that "Russia implements a repressive criminal policy" (Babaev & Pudovochkin, 2020, p. 83).

The prevalence of the punitive concept is also evidenced by the fact that the proper criminal-legislative measures to compensate the harm caused by a crime, provided for in the norms of the Criminal Code of the Russian Federation (CC RF), are mainly aimed at compensating the country's costs related to criminal prosecution and justice (e.g., measures of pecuniary punishment included in the content of criminal punishment (Articles 46, 49, 50, and 53.1), court fines (Article 76.2), and confiscation of property (Articles 104.1 and 104.2) as measures of criminal law), and to a lesser extent on covering the damage caused by a crime to the victim (e.g., reparations for harm as a criminal law measure for exemption from liability and punishment (Articles 90 and 92 of the CC RF) and compensation for harm to the victim (Article 104.3 of the CC RF)).

Historically, the emergence and development of the concept of restorative justice is a natural process due to the shortcomings and costs of the traditional justice model with its typical "crisis of criminal punishment," which prompted the search and implementation of more flexible and effective alternative ways to resolve the conflict caused by crime and means to influence crime (Gilinsky, 2008).

One of the most important concepts in this direction is restorative justice.

The author shares the opinion of scholars who believe that current criminal policy should be reoriented from the resolution of criminal conflicts through primarily punitive mechanisms and measures of state coercion to its settlement in a maximally peaceful way with compensation for the harm caused by the crime to the victim and society (Babaev & Pudovochkin, 2014, p. 197).

However, in Russian reality, the promotion of the restorative concept and mediation is complicated by several national mental attitudes and patterns, including the following:

- "Stereotypes of repressive thinking and clearly hypertrophied ideas about the role and possibilities of criminal punishment in solving emerging social problems" (Kapinus, 2018, p. 37);
- "Criminal-legal fetishism—a sincere and unshakable belief of legislators in the omnipotence of criminal repression as the universal and only possible means of combating all forms and types of deviant behavior" (Korobeev, 2019, p. 246).

It must be recognized that this objective factor is difficult to overcome since the prohibitive-punitive concept is supported by most members of Russian society and is perceived in the public consciousness as the most understandable and justifiable form of resolution of criminal conflicts. Unfortunately, ordinary people in Russia are still characterized by punitive reflections and the prevalence of the ideology of a "strong" country, which is undoubtedly realized and used by the authorities to "ensure security and impose order," which inevitably leads to excessive criminalization and the prevalence of a punitive approach over a restorative one (Zhalinsky, 2015, p. 345).

The international community actively encourages the development of alternative means of resolving criminal conflict, introducing restorative justice, and saving punitive means. Several international legal acts recommend the use of effective alternatives, including restorative and conciliatory procedures, in resolving criminal law conflicts. Let us dwell on their key provisions.

The United Nations Standard Minimum Rules for the Administration of Juvenile Justice (Beijing Rules) (UN General Assembly, 1985) calls for the avoidance of formal juvenile proceedings by the competent authorities that result in sentencing and a criminal record (paragraphs 11.1 and 11.2). The document calls for limiting the use of punitive sanctions against juveniles, particularly those involving placement in correctional institutions (paragraphs 5 and 19.1), and increasing the role of social institutions (community and family), measures of community education (temporary supervision, guidance, and community service), and victim compensation (paragraph 11.4) in resolving criminal conflicts.

The Convention on the Rights of the Child (UN General Assembly, 1989) and the UN Guidelines for the Prevention of Juvenile Delinquency (Riyadh Guidelines) (UN General Assembly, 1990a) also contains the idea that juvenile delinquency should not be brought to trial and that the preference should be given to institutions, programs, and procedures considering the social and psychological features of children. The UN Standard Minimum Rules for Non-Custodial Measures (Tokyo Rules) (UN General Assembly, 1990b) also include several important provisions to develop alternatives to the traditional punitive concept of resolving criminal conflicts. This document outlines the reduction in the use of imprisonment (paragraph 1.5) and creating and using a more flexible and differentiated system of pre-trial and post-trial non-custodial measures (paragraphs 2.3, 2.4, and 21.1) based on increasing the role and participation of the public in this system and reducing the percentage of cases which fall into the judicial system (paragraphs 2.5, 2.6, 17.1, and 18.1) considering not only the characteristics of the offense and the perpetrator but also the interest of the victim, when choosing a measure (paragraph 3.2), to restore rights through restitution of property and other compensatory measures (paragraph 8.2).

The main international legal document recommending mediation in criminal proceedings is Recommendation No. R (99) 19 "On Mediation in Criminal Cases," adopted by the Committee of Ministers of the Council of Europe on September 15, 1999 (Committee of Ministers of the Council of Europe, 1999).

The Committee of Ministers notes the prospects and effectiveness of the use of mediation in addition to the classical criminal process since it makes the process of resolving criminal conflicts more flexible, humane, democratic, and focused on the full interests of the parties involved.

The Committee of Experts on Mediation in Criminal Cases, which prepared the Recommendations, proposes to increase the subjectivity, legal status, and guarantees of participation in the case of the victim and the offender, as well as the extent of their influence on the course and process of resolution of the criminal case, considering the interest of the victim to receive compensation and the interest of the offender to minimize the criminal-law consequences of the deed. Similar to previous documents, the Recommendations refer to increasing the role of the local community in the criminal process, which is a necessary condition for mediation and can improve the legal culture and literacy of a significant part of the population, and, consequently, the effectiveness of general crime prevention.

The principles formulated in the Recommendations are voluntariness, confidentiality, public availability of mediation, accessibility at any stage of the criminal process, and independence of the mediation service. These principles are valuable for the introduction of mediation in criminal cases into the national legal system.

Another important document promoting the introduction of restorative justice and mediation in criminal cases is the "Basic Principles on the Use of Restorative Justice Programs in Criminal Matters" (UN Economic & Social Council, 2002).

This document contains basic concepts for the development of the institution of mediation. In particular, the concept of "restorative process" is a procedure in which the victim and the offender are not passive observers but have sufficient power to resolve the criminal legal conflict between them and actively participate in the resolution of related issues with the assistance of a mediator, seeking to achieve a "restorative result."

"Restorative result" is an agreement between the victim and the offender, involving the satisfaction of interests violated by the crime and the restoration (reintegration) of social relations through compensation, restitution, and community service.

In addition to the "Parties," the "Basic Principles on the Use of Restorative Justice Programs in Criminal Matters" highlight the figure of the "Mediator," interpreting him or her as an actor whose role is to facilitate a fair and impartial resolution of the parties to a criminal legal conflict.

Unlike the traditional criminal process, the restorative approach to the settlement of a criminal-legal conflict has evident advantages for each party. It allows the victim of a criminal to obtain compensation for the damage caused by the crime and thereby satisfy his or her interests. It allows the guilty person to prove his or her correction by eliminating the consequences of the crime committed. Additionally, this approach, directed "face-to-face" with the person (the victim or the perpetrator) and viewing them not as objects of external power but as active participants in conflict resolution, allows community members to better understand the causes of crime and counteract it.

The declaration prepared as part of the work of the XII UN Congress on Crime Prevention and Criminal Justice (UN General Assembly, 2010) emphasized the advisability of increasing the proportion of non-custodial measures in national criminal law, in particular, community service, restitution, and the development of programs aimed at correcting offending behavior patterns and returning offenders to normal life (rehabilitation and socialization programs).

Thus, the aforementioned international legal instruments unanimously urge countries to introduce restorative justice programs (including mediation) in domestic criminal proceedings.

Russia follows up on such recommendations. The "Concept of long-term socio-economic development of the Russian Federation for the period up to 2020" recognizes the importance of introducing and developing programs of restorative justice and organization of conciliation technologies (Government of the Russian Federation, 2008).

Russia has taken the path of introducing restorative justice and mediation programs in the organization of criminal proceedings. The country takes the first steps in this direction in the area of criminal cases involving juveniles based on the experience of other countries that already use such programs.

The "Concept for the development of a network of mediation services" involves the introduction in Russia of mediation and restorative approach to resolving conflicts caused by offenses (Government of the Russian Federation, 2014). Similar to international documents, it provides a definition of "restorative justice," which implies the priority solution of restorative tasks to make amends for the property and moral damage caused to the victim, restores the violated social relations, and provides the perpetrator with the opportunity to prove the correction by his or her active actions.

The concept also defines the term "mediation" as a way to resolve disputes amicably by working out a mutually acceptable solution with the assistance of a neutral and independent person—the mediator.

4 Conclusion

Thus, the research shows that despite the numerous international legal sources of mediation and other programs of restorative justice that recommend countries to introduce appropriate methods in criminal proceedings for the optimal resolution of criminal legal conflicts, as well as the positive experience of other countries using such technologies, Russia is taking only the first steps in this direction with a significant lag.

The hope that Russia will continue to move in this direction is provided by the fact that countries using programs of restorative justice and mediation in the organization of criminal justice have generally begun to test them with cases involving juveniles and then extended the appropriate technology, with positive results, to adult offenders.

Nevertheless, it is possible that international standards and the positive experience of other countries in the use of mediation and other programs of restorative justice in criminal cases will not be fully demanded in Russian society for a considerable time due to the repressive trend in Russian criminal policy, which is gaining momentum from year to year, and the recent termination of integration of Russia into the international legal field, which complicates the development of cooperation with international and foreign organizations in this field.

Additionally, this concern has very substantial social reasons associated with the immaturity of public institutions and organizations, a high degree of distrust among citizens and the fragmentation of people, weak self-organization, and paternalistic attitudes in Russian civil society, which undoubtedly hinder the development of mediation and restorative justice.

References

Andrianov, V. K., & Pudovochkin, Y. E. (2020). Patterns and trends in the application of criminal law norms. *State and Law, 11*, 97–105. https://doi.org/10.31857/S102694520012520-3

Babaev, M. M., & Pudovochkin, Y. E. (2014). *Problems of the Russian criminal policy.* Prospekt.

Babaev, M. M., & Pudovochkin, Y. E. (2020). *Russian criminal policy of the XXI century.* Yurlitinform.

Committee of Ministers of the Council of Europe. (1999). *Recommendation "On mediation in criminal cases" (together with the "Explanatory Memorandum")* (September 15, 1999 No. R (99) 19). Council of Europe.

Gilinsky, Y. I. (2008, June 30). "The crisis of punishment" and its consequences. *Portal "Rodon."* Retrieved from http://www.rodon.org/society-080630145740. Accessed June 2, 2021.

Government of the Russian Federation. (2008). *Decree "On the concept of long-term socio-economic development of the Russian Federation for the period up to 2020"* (November 17, 2008 No. 1662-r). Moscow, Russia.

Government of the Russian Federation. (2014). *Decree "On approval of the Concept for the Development of a network of mediation services until 2020 in order to implement restorative justice for children, including those who have committed socially dangerous acts, but have not reached the age from which criminal responsibility begins in the Russian Federation"* (July 30, 2014 No. 1430-r). Moscow, Russia.

Kapinus, O. S. (2018). Criminalization and decriminalization of acts: The search for an optimal balance. *Social Sciences and Modernity, 4*, 37–46. https://doi.org/10.31857/S08690499000 0367-5

Korobeev, A. I. (2019). *Russian criminal law policy: From genesis to crisis.* Yurlitinform.

UN General Assembly. (1985). *United Nations standard minimum rules for the administration of juvenile justice (Beijing rules)* (adopted by resolution No. 40/33 of November 29, 1985). UN.

UN General Assembly. (1989). *Convention on the rights of the child* (adopted by resolution No. 44/25 of November 20, 1989; entry into force 2 September 1990). UN.

UN General Assembly. (1990a). *United Nations guidelines for the prevention of juvenile delinquency (Riyadh guidelines)* (adopted by resolution No. 45/112 of December 14, 1990). UN.

UN General Assembly. (1990b). *United Nations standard minimum rules for non-custodial measures (Tokyo rules)* (adopted by resolution No. 45/110 of December 14, 1990). UN.

UN Economic and Social Council. (2002). *Basic principles on the use of restorative justice programs in criminal matters* (adopted by resolution No. 2000/14 of July 27, 2000). UN.

UN General Assembly. (2010). *Salvador declaration on integrated strategies to meet global challenges: Crime prevention and criminal justice systems and their development in a changing world* (adopted by resolution No. 65/230 of December 21, 2010). UN.

Zer, H. (2002). Restorative justice: A new look at crime and punishment (L. M. Karnozova, & S. A. Pashina Trans. from English; L. M. Karnozova Ed.). NGO Center "Judicial and Legal Reform."

Zhalinsky, A. E. (2015). *Selected works: In 4 vols. Vol. 2: Criminal law.* Publishing house of the Higher School of Economics.

Book Review Inshakova, A. O., Bogoviz, A. V. (Eds.) "Alternative Methods of Judging the Economic Conflicts in the National Positive and Soft Law"

Agnessa O. Inshakova, Elena G. Popkova, and Denis E. Matytsin

Abstract This article is an overview of the collective monograph "Alternative Methods of Judging the Economic Conflicts in the National Positive and Soft Law", which was published by Information Age Publishing, Inc. in 2020 and has become an event for the modern academic community both in Russia and abroad. This chapter outlines the scientific message of this monograph, displays its main content, competitive advantages and scientific contribution, and also describes for which readership this monograph will be of interest.

Keywords Alternative methods · Judging the economic conflicts · National positive law · Soft law

JEL Codes D23 · D74 · F51 · J52 · J53 · Q34 · K12 · K13

1 Introduction

The release of the collective monograph "Alternative Methods of Judging the Economic Conflicts in the National Positive and Soft Law", published by Information Age Publishing, Inc. (Inshakova and Bogoviz, 2020) has become an event for the modern academic community both in Russia and abroad. The book is the result of the research of the team of professors, doctors of sciences, who have published hundreds of books, textbooks, methodological and practical manuals for both training students and those related to the proposals for improving practical economic activities in the

A. O. Inshakova (✉) · D. E. Matytsin
Volgograd State University, Volgograd, Russia
e-mail: gimchp@volsu.ru

D. E. Matytsin
e-mail: ppagp@volsu.ru

E. G. Popkova
MGIMO University, Moscow, Russia

D. E. Matytsin
Volga Branch of the International Law Institute, Volzhsky, Russia

E. G. Popkova and B. S. Sergi (eds.), *Geo-Economy of the Future*,
https://doi.org/10.1007/978-3-030-92303-7_60

field of economy and law, as well as a lot of scientific articles and monographs, indexed in the world science-based citation bases, including in the status of the editors.

2 Materials

The works of scholars such as Adhvaryu et al (2021), Asaoka (2021), Gready (2021), Heger and Neumayer (2021), Kalinina et al. (2019), Mawejje and McSharry (2021), Popkova et al. (2020) are devoted to the issues of economic conflicts and their resolution.

The perception of positive foreign experience in the application of various alternative methods of conflict resolution has led to their consolidation in the Russian legislation, as evidenced by the current legal framework (Federal law "On the Alternative Dispute Resolution with the Participation of an Intermediary (Mediation Procedure)" dated 27.07.2010 No. 193-FZ (State Duma of the Russian Federation, 2021a), Federal law No. 382-FZ dated 29.01.2015 "On Arbitration (Arbitration Proceedings) in the Russian Federation" Law of the Russian Federation dated 07.08.1993 (State Duma of the Russian Federation, 2021b), 5338-I "On International Commercial Arbitration") (President of the Russian Federation, 2021) and the development of various draft laws (the draft Federal law "On the Financial Commissioner for the Rights of Consumers of Financial Services" and others) (Presidium of the Supreme Court of the Russian Federation, 2021).

3 Results

The monograph examines the general provisions and theoretical and legal foundations of the alternative methods of conflict resolution. The conceptual apparatus of the studied methods is rich in various definitions that have been developed in the national legal doctrine.

The primary intentions of the publication are to present to the interested reader a comprehensive study of all currently available alternative methods of judging economic conflicts, with the help of which the economic agents can resolve mutual conflicts arising in the course of economic activity without resorting to state justice for this.

The secondary intentions of the publication are to disseminate the scientifically based knowledge and recommendations on the use of alternative methods of judging economic conflicts in economic practice, on their possible selective integration into national law. The monograph contains the first original studies, which have not been published anywhere before.

The content of the whole book is structured by the authors in six parts. *Part One* explores the alternative methods of judging economic conflicts arising from various forms of interaction between economic entities. *Part Two* explores the alternative methods of judging economic conflicts for participants in the integrated economic structures. *Part Three* reveals the actualization of the alternative methods for resolving international economic conflicts in the context of globalization. *Part Four* explores the evolution of the alternative dispute resolution methods in Russia and abroad, in particular, the factors of their development and the processes of modernization. *Part Five* discloses the alternative methods of judging economic conflicts governed by positive law. *Part Six* explores the alternative methods of judging economic conflicts, regulated by soft law. At the same time, the content of each part of the monograph is formed into separate chapters. In total, the book contains 22 chapters, whose presentation develops deductively—from more general questions to the special issues.

A new scientific result is the classification of the methods of judging economic conflicts developed by the authors. One of the unique advantages of the monograph is the novelty of an integrated approach to the study of alternative methods for judging economic conflicts. In previously published books this problem appeared to be uncoordinated and fragmented. Neither the works on arbitration proceedings nor the activities of international commercial arbitration or the specifics of mediation were published. In the book presented, the alternative methods of judging economic conflicts were first explored in their complete modern tools. In so doing, they are differentiated into ways attributable to positive law and soft law.

Another unique advantage of the book is the features of the current professional activity of its authors, which make it possible to effectively promote the monograph to the target readership audiences. The author's team includes, in particular, the first deputy rector of one of the largest Russian universities. The authors demonstrated their desire to prove the efficiency in protecting the violated rights of the economic turnover subjects by studying the essence and legal nature of the alternative methods.

The possibilities of using modern information and communication technologies to resolve disputes online (ODR), i.e. via the Internet in Russia and abroad, within the framework of the alternative dispute resolution procedures are considered by the authors. There is a large array (more than 60) of so-called ODR providers.

4 Conclusion

The audience of the book includes the primary audience—a professional community of businessmen engaged in entrepreneurial activities, including those who make export–import transactions with foreign partners. It also includes professional negotiators, mediators, auditors, judges, economists, and lawyers. The monograph contains some useful practical material for them, including the specific recommendations on the use of the alternative methods of judging economic conflicts, which they can directly use if such conflicts arise in their business. The secondary audience of the book is graduate students of universities, majoring in economics and law, students of the magistracy of the relevant qualification profiles, and postgraduate students whose training courses contain the compulsory subjects for which the book can be used directly.

References

Adhvaryu, A., Fenske, J., Khanna, G., & Nyshadham, A. (2021). Resources, conflict, and economic development in Africa. *Journal of Development Economics, 149*. https://doi.org/10.1016/j.jdeveco.2020.102598

Asaoka, D. (2021). A behavioural-economic perspective on conflicts of interest among shareholders, Debtholders, and directors. *Journal of Interdisciplinary Economics, 33*(1), 7–23. https://doi.org/10.1177/0260107920934204

Gready, S. (2021). The case for transformative reparations: In pursuit of structural socio-economic reform in post-conflict societies. *Journal of Intervention and Statebuilding*. https://doi.org/10.1080/17502977.2020.1852833

Heger, M. P., & Neumayer, E. (2021). Economic legacy effects of armed conflict: Insights from the civil war in Aceh, Indonesia. *Conflict Management and Peace Science*. https://doi.org/10.1177/0738894221994503

Inshakova, A. O., & Bogoviz, A. V. (Eds.). (2020). Alternative methods of judging the economic conflicts in the national positive and soft law. In E. G. Popkova, (Series Ed.), *Advances in research on Russian business and management*. Information Age Publishing. https://www.infoagepub.com/products/Alternative-Methods-of-Judging-Economic-Conflicts-in-the-National-Positive-and-Soft-Law

Kalinina, A. E., Inshakova, A. O., & Goncharov, A. I. (2019). Polysubject jurisdictional blockchain: Electronic registration of facts to reduce economic conflicts. *Studies in Computational Intelligence, 826*, 205–221. https://doi.org/10.1007/978-3-030-13397-9_24

Mawejje, J., & McSharry, P. (2021). The economic cost of conflict: Evidence from South Sudan. *Review of Development Economics*. https://doi.org/10.1111/rode.12792

Popkova, E. G., DeLo, O., & Sergi, B. S. (2020). Corporate social responsibility amid social distancing during the COVID-19 crisis: BRICS versus OECD countries. *Research in International Business and Finance, 55*(1). https://doi.org/10.1016/j.ribaf.2020.101315

Presidium of the Supreme Court of the Russian Federation. (2021). The draft Federal law "On the financial commissioner for the rights of consumers of financial services". https://www.consultant.ru/law/podborki/razyasneniya_po_voprosam_svyazannym_s_primeneniem_fz_ot_04.06.2018_%25E2%2584%2596_123-fz_ob_upolnomochennom_po_pravam_potrebitelej_finansovyh_uslug/

State Duma of the Russian Federation. (2021a). Federal law "On the alternative dispute resolution with the participation of an intermediary (mediation procedure)" dated 27.07.2010 No. 193-FZ. http://www.consultant.ru/document/cons_doc_LAW_103038/

State Duma of the Russian Federation. (2021b). Federal law No. 382-FZ dated 29.01.2015 "On arbitration (arbitration proceedings) in the Russian Federation, Law of the Russian Federation dated 07.08.1993. http://www.consultant.ru/document/cons_doc_LAW_191301/

The President of the Russian Federation. (2021). 5338-I "On international commercial arbitration". https://base.garant.ru/10101354/

Book Review Inshakova A. O., Frolova E. E. (Eds.) "The Transformation of Social Relationships in Industry 4.0: Economic Security and Legal Prevention"

Agnessa O. Inshakova, Elena G. Popkova, and Denis E. Matytsin

Abstract This article is an overview of the collective monograph "The transformation of social relationships in industry 4.0: economic security and legal prevention". This article substantiates the particular timeliness and relevance of the monograph at the present, reflects the contribution of this monograph to the growth of scientific knowledge, and discloses the content of the monograph.

Keywords Transformation of social relationships · Industry 4.0 · Economic security · Legal prevention

JEL Codes F52 · K23 · K31 · O31 · O32 · O33 · O38

1 Introduction

The dynamics of scientific and technological development of modern society is characterized by the high development rate, is accompanied by the digital economy algorithmization and, acts as a trigger of the appearance of new relations and the transformation of the existing public relations, in which the borders between the material, digital, biological worlds gradually disappear giving rise to the objective need for the complex social and economic and institutional transformations in the society requiring the respective legal basis. This determines the high timeliness and relevance of the monograph (Inshakova & Frolova, 2021) at the present.

A. O. Inshakova (✉) · D. E. Matytsin
Volgograd State University, Volgograd, Russia
e-mail: gimchp@volsu.ru

D. E. Matytsin
e-mail: ppagp@volsu.ru; dmatytsin@yandex.ru

E. G. Popkova
MGIMO University, Moscow, Russia

D. E. Matytsin
Volga Branch of the International Law Institute, Volzhsky, Russia

E. G. Popkova and B. S. Sergi (eds.), *Geo-Economy of the Future*,
https://doi.org/10.1007/978-3-030-92303-7_61

2 Materials

The modern communication and technical aspects of deep reconstruction of goods production, work implementation, service provision, which are, at the same time, the challenges for the investigators for various fields of science are called the fourth industrial revolution, or Industry 4.0. The modern scientific literature uses the term the "fourth industrial revolution" for the global processes of a radical change of the conventional methods and forms of the economic management related to the mainstreaming of innovative technologies, such as artificial intelligence, the Internet of things, unmanned transport vehicles, robotized systems, big data, etc.

In particular, these issues are addressed in the works of Abramova et al. (2019), Benzidia et al. (2021), Berges et al. (2021), Chaldaeva (2019), Fokina (2020), Guseva et al. (2019), Inshakova and Litvinov (2020), Inshakova et al. (2020), Karanina (2020), Mehta et al. (2021), Meindl et al., 2021), Narula et al. (2021), Popkova et al. (2021), Popkova and Sergi (2020).

This term is not new, although it gained popularity in a slightly modified form in 2011 when it was called "Industry 4.0" at Hannover Industrial Fair. According to the World Economic Forum, Industry 4.0 unites digital, physical, and biological systems. Some foreign countries (Germany, Japan, the USA, the UK, France, etc.), striving to start work aimed at solving the issues of the fourth industrial revolution, related to the wide commissioning of digital technologies and cyber-physical systems, took the respective strategic programs.

It is seen that the context of the revolutionary transformation includes the game-changing technologies whose study is the subject matter of this book in terms of modern economic and legal sciences.

3 Results

The conclusions reflecting the scientific novelty of the research based on a comprehensive economic and legal interdisciplinary analysis allow saying that the authors have developed proposals on the formation of an integral legal concept of neo-industrial modernization of contemporary Russia. Besides, they have proposed certain legal mechanisms for ensuring public–private law relations arising in the sphere of civil circulation as the main form of economic circulation in the context of Industry 4.0 and improving the legal infrastructure of key areas of the national technology initiative. The results of the research include:

- the determination of specificity, needs, and peculiarities of the 4th industrial revolution on the contemporary social development as well as the priority directions in the transformation of the economic and legal relations which are appropriate to the large-scale processes of digitalization and implementation of the newest neo-industrial technologies. Besides, they involve the clarification and addition of the categories in the legal regulation of the transition to the new economic development scenario in terms of the 4th industrial revolution;

- the formation of strategy and main directions of the economic and legal regulation of the technologies of the industry 4.0 development considering the priorities of the international integration including the participation of the Russian Federation (BRICS, EAEU);
- the evaluation of the traditional system of realization of rights and interests in compliance with the new requirements in terms of the 4th industrial revolution; the formation of a preventive legal mechanism, which includes the most effective means of protection of the rights of businesses in the context of industry 4.0.;
- the development of the methodology and legal mechanisms for the establishment and implementation of the legal concept of neo-industrial modernization.

4 Conclusion

The authors and editors of this book are convinced that their interdisciplinary collective study will make its contribution to the theoretical and practical development of the economic and legal regulation for the public relation transformations in terms of Industry 4.0 and will help to prepare an adaptive infrastructure of the fourth technological revolution.

References

Abramova, M. A., Dubova, S. E., Krivoruchko, S. V. (2019). Marxism and digital money as a new reality of the social and economic system. In M. L. Alpidovskaya, & E. G. Popkova (Eds.), *Marx and modernity: A political and economic analysis of social systems management* (pp. 411–422). A volume in the series E. G. Popkova (Ed.), Advances in Research on Russian Business and Management, Charlotte, NC, USA, Information Age Publishing. https://www.infoagepub.com/products/Marx-and-Modernity

Benzidia, S., Makaoui, N., & Subramanian, N. (2021). Impact of ambidexterity of blockchain technology and social factors on new product development: A supply chain and Industry 4.0 perspective. *Technological Forecasting and Social Change, 169*. https://doi.org/10.1016/j.techfore.2021.120819

Berges, I., Ramírez-Durán, V. J., & Illarramendi, A. (2021). A semantic approach for big data exploration in industry 4.0. *Big Data Research, 25*. https://doi.org/10.1016/j.bdr.2021.100222

Chaldaeva, L. A. (2019). Digital economy: A Marxist view of the present and future. In M. L. Alpidovskaya, & E. G. Popkova (Eds.), *Marx and modernity: A political and economic analysis of social systems management* (pp. 395–400). A volume in the series E. G. Popkova (Ed.,) Advances in Research on Russian Business and Management, Charlotte, NC, USA, Information Age Publishing. https://www.infoagepub.com/products/Marx-and-Modernity

Fokina, O. V. (2020). Marketing management of projects for the introduction of "smart" leaning technologies as a method of de monopolization of digital economy markets. *Portal "Scientific Narratives of Russia"*. https://iscconf.ru/маркетинговое-управление-проектами/

Guseva, I. A., Kulikova, E. I., & Rubtsov, B. B. (2019). Dialectics of the financial market category in the Russian economic science: from the Marx era to the digital economy. In M. L. Alpidovskaya,

E. G. Popkova (Eds.), *Marx and modernity: A political and economic analysis of social systems management* (pp. 401–410). A volume in the series E. G. Popkova (Ed.), Advances in Research on Russian Business and Management, Charlotte, NC, USA, Information Age Publishing. https://www.infoagepub.com/products/Marx-and-Modernity

Inshakova, A. O., & Litvinov, N. I. (2020). Digital institutions in the fight against the shadow economy in Russia. Portal "Scientific Narratives of Russia". https://iscconf.ru/цифровые-инс титуты-в-борьбе-с-теневой/

Inshakova, A. O., & Frolova, E. E. (Eds.). (2021). The transformation of social relationships in Industry 4.0: Economic security and legal prevention.

Inshakova, A. O., Frolova, E. E., Rusakova, E. P., & Kovalev, S. I. (2020). The model of distribution of human and machine labor at intellectual production in Industry 4.0. *Journal of Intellectual Capital, 21*(4), 601–622. https://doi.org/10.1108/JIC-11-2019-0257

Karanina, E. V. (2020). Digital approach to the fight against shadow economy in the region: Advantages for the economic security and investment attractiveness of the territory. Portal "Scientific Narratives of Russia". https://iscconf.ru/цифровой-подход-к-борьбе-с-теневой-эко/

Mehta, D., Tanwar, S., Bodkhe, U., Shukla, A., & Kumar, N. (2021). Blockchain-based royalty contract transactions scheme for Industry 4.0 supply-chain management. *Information Processing and Management, 58*(4). https://doi.org/10.1016/j.ipm.2021.102586

Meindl, B., Ayala, N. F., Mendonça, J., & Frank, A. G. (2021). The four smarts of Industry 4.0: Evolution of ten years of research and future perspectives. *Technological Forecasting and Social Change, 168*. https://doi.org/10.1016/j.techfore.2021.120784

Narula, S., Puppala, H., Kumar, A., ..., Prakash, S., & Talwar, V. (2021). Applicability of industry 4.0 technologies in the adoption of global reporting initiative standards for achieving sustainability. *Journal of Cleaner Production, 305*. https://doi.org/10.1016/j.jclepro.2021.127141

Popkova, E., Bogoviz, A. V., & Sergi, B. S. (2021). Towards digital society management and 'capitalism 4.0' in contemporary Russia. *Humanities in Social Sciences Communication, 8*(77). https://doi.org/10.1057/s41599-021-00743-8

Popkova, E. G., & Sergi, B. S. (2020). Human capital and AI in industry 4.0. Convergence and divergence in social entrepreneurship in Russia. *Journal of Intellectual Capital, 21*(4), 565–581. https://doi.org/10.1108/JIC-09-2019-0224

Mediation as an Alternative Dispute Resolution in Intellectual Property Disputes in Turkey and Russian Federation

Canan Kucukali, Liudmila A. Bukalerova, Saule M. Naurzaliyeva, and Alexandra S. Vasilenko

Abstract The mediation method, an alternative dispute resolution method, is becoming more attractive in Turkish law every day. With the article added to the Turkish Commercial Code of numbered 6102 compulsory mediation was introduced in some commercial cases, and therefore, due to the commercial nature of intellectual property disputes, compulsory mediation was introduced in pecuniary claims and compensation cases. Considering that intellectual property disputes are one of the most suitable disputes with the mediation method, it will be appropriate to include cases of infringement and determination within the scope of the compulsory mediation method in the field of intellectual property. Recently, with the decision of the Court of Cassation, mediation has been blocked as an alternative dispute resolution in the field of intellectual property. We hope that the Court of Cassation will abandon its case law that compulsory mediation should not be applied if the cases are piled up in intellectual property cases.

Keywords Alternative dispute resolution · Intellectual property · Mandatory mediation · Non-legal mediator · Court of cassation

JEL Code K41

C. Kucukali (✉)
Istanbul Ticaret Universitesi, Istanbul, Turkey
e-mail: ab28109@adalet.gov.tr

Ministry of Justice, Ankara, Turkey

L. A. Bukalerova · S. M. Naurzaliyeva · A. S. Vasilenko
Peoples' Friendship University of Russia (RUDN University), Moscow, Russia
e-mail: bukalerova-la@rudn.ru

S. M. Naurzaliyeva
e-mail: naurzalieva-sm@rudn.ru

A. S. Vasilenko
e-mail: vasilenko-as@rudn.ru

E. G. Popkova and B. S. Sergi (eds.), *Geo-Economy of the Future*,
https://doi.org/10.1007/978-3-030-92303-7_62

1 Introduction

The mediation method, which is an alternative dispute resolution (ADR) method, will greatly contribute to the reduction of the workload of the Intellectual Property (IP) courts. Alternative Dispute Resolution is widely used in countries with developed legal systems.

Alternative resolution of intellectual property disputes may be in the form of a mediation method, a consultation, conciliation, expert judgment, or a negotiation. Mediation is a practice in which an impartial facilitator intervenes to assist parties in resolving a conflict.

Since judicial proceedings in commercial cases are too costly and embarrassing for the parties, they tend to prefer private and confidential mediation or arbitration solutions.

2 Methodology

Mediation methods take several forms, but all usually are voluntary and non-binding. Mediation forms practiced today are facilitating, evaluative and transformative mediation.

Those who advocate facilitating mediation acknowledge that with the help of their legal advisors, the parties will be able to understand their situation better than the third party, and therefore, the parties can develop better solutions that the mediator can create (Stulberg, 1997).

Proponents of evaluative mediation counter that disputants often seek out the opinion of a neutral third party- an opinion often helps the construction of settlements. Turkey initially experimented with facilitating mediation practice. However, an attempt has been made to introduce evaluative mediation. In our opinion, the evaluative mediation method will increase the rate at which disputes are resolved and negotiations are concluded with an agreement.

This type of mediation is not legislated in the Russian Federation. Regulatory acts use the definition of mediation in a broad sense, giving the opportunity to independently choose the most effective type of mediation, as well as the technology of procedure in each case. Although at the doctrinal level, the term evaluative mediation is used (Shamlikashvili, 2013) and there are some proposals to legislate this form of mediation. Nowadays, mediation negotiations have become very easy with online mediation (Metha, 2009). The agreement reached at the end of the process becomes a court decision in Russia if it is written down and then approved by the competent judge or court.

3 Results

3.1 Mediation Practices in Turkey

With Law No. 6325 published in the Official Gazette dated 22.06.2012 the volunteer mediation method has been accepted in Turkey. In the preparation of the law, the UNCITRAL Model Law and the EU Directive, Green Book on ADR in Private Law, the Austrian, German, Hungarian, Bulgarian, and the Slovakian Mediation Laws have been considered.

Under Article 16/2 of the Mediation Law No. 6325, the time passed from the beginning of the mediation process until its end is not considered in the calculation of the foreclosures and the period of prescription. In the article, the provisions to be applied to the mediation process in case of resorting to mediation as a prerequisite for filing an action are specified.

Compulsory mediation was applied as a "prerequisite" as of 1 January 2018 in terms of employee and employer disputes in Turkey (Alkan, 2019). It has been started to be applied as of January 1, 2019 in commercial disputes regarding pecuniary claims and compensation. Finally, compulsory mediation for consumer disputes has been introduced by Law No. 7251 and Article "73 / A" of Law No. 6502 on Consumer Protection (O.G., 2020). With the amendment of the law pending in the Turkish Grand National Assembly, mediation will be implemented in family disputes soon.

In the Russian Federation in disputes about the violation of exclusive rights (exclusive rights as a property right are part of intellectual rights which also include personal non-property rights and other rights), the pre-trial procedure of dispute settlement is mandatory. But this pre-trial settlement procedure does not always involve mediation. So, according to the review of the Supreme Court, the pre-trial procedure is completed when sending a claim to the address specified in the contract, including electronic (Pavlova, 2020).

3.2 Mediation Practices in the Field of Intellectual Property in Turkey

Article 4/1 of the Turkish Commercial Code (TCC) provision covers the lawsuits stipulated in the Industrial Property Law No. 6769 (O.G: 10.01.2017, N: 29,944), the Law No. 5147 on the Protection of Integrated Circuit Topographies (O.G: 30.04.2004, N: 25,448) and the Law No. 5846 on Intellectual and Artistic Works (O.G: 30.04.2004, N: 25,448), provided that it concerns a commercial enterprise.

The most common examples where an intellectual property dispute arises are patents, know-how, brand licenses, franchising, computers, multimedia contracts, distribution contracts, common partnerships, R&D studies, technology-heavy business contracts, mergers and acquisitions where intellectual property values are important, sports marketing contracts, printing, music and film contracts (WIPO, 2020).

The mediation method is the preferred method in the field of intellectual property disputes because cases are complex and mediation is a confidential and secure process, and companies do not want their trade secrets to be exposed and gain publicity (WIPO, 2020).

In Turkish law, the dispute must be suitable for mediation to apply to the mediator in intellectual property disputes. Under the provision of Art.1 / 2 of the Mediation Law, the mediator can only be applied at the stage of resolving private law disputes arising from the work or transactions of which the parties can dispose without restrictions including those that have a foreign element.

In the Russian Federation intellectual property disputes it is often obvious that just resolution of conflict is risky because the outcome is unpredictable. At the same time, the parties are not entitled to a second chance since the courts do not reconsider disputes over claims that have already been considered by the courts. In contrast to the WIPO Arbitration and Mediation Center, where it is possible to file a complaint repeatedly, up to a positive decision of the WIPO Center. Several Russian scientists believe that this kind of dispute has quite a great potential for mediation (Sergo, 2017). According to the opposite point of view, the parties may face many difficulties: the agreement reached may contradict the law of one of the affected countries, not all of the affected jurisdiction's mediation agreements may be enforceable and not all jurisdictions disputes from the copyright and related rights are mediatable. Therefore, it is suggested that more emphasis be placed on the development of arbitration which has similar advantages but has established a simpler procedure for enforcing the award (Abrosimova, 2019).

3.3 Conditions of Being a Mediator in Turkey

In Turkey, it is necessary to be a graduate of a law faculty, to study in mediation institutions, and to be registered in the Ministry of Justice Mediation Department. In our opinion, following the spirit of mediation, having mediators from different fields on the intellectual property case will contribute to the resolution of disputes through mediation.

The additional Article 18 / A of the 5th Section of Law No. 6325: "Mediation as a Prerequisite for Action" was introduced with Article 23 of Law No. 7155 of 6/12/2018. In the article, the provisions to be applied to the mediation process in case of application to mediation as a prerequisite are specified. With Law No. 7155 (O.G. No. 30630, 2020) on the Procedure of Initiating the Proceedings for Money Receivables Arising from the Subscription Agreement, it was accepted that applying for mediation was accepted as a condition for filing a lawsuit in commercial cases, the subject of which is some amount of money and compensation claim in Turkey. According to Art. 5 / A of the Turkish Commercial Code, "*It is a prerequisite of the lawsuit to apply to the mediator before the lawsuit is filed against the commercial cases specified in Article 4 of the Law and other laws, the subject of which is the payment of a certain amount of money.*". Thus, the application to the mediator before the lawsuit has become a condition for some copyright and industrial property disputes (Karaca et al., 2019).

As per TCC Art. 4 and other laws, the obligation to apply to the mediator foreseen for commercial cases specified also includes cases arising from the intellectual property legislation. Because according to Article 4/1 of the TCC numbered 6102, the civil cases arising from the legislation on intellectual property law and the publication agreement of the Turkish Code of Obligations dated 11/1/2011 and numbered 6098 (Civil lawsuits arising from broadcasting agreements (Article 487 to 501) are considered commercial cases. Since broadcasters and producers are engaged in business related to the commercial enterprise, it is obligatory to apply for mediation before filing a lawsuit by or against them.

According to the provisions of Art. 4/1 of TCC; since the lawsuits arising from the rights regarding the intellectual and artistic works that do not concern any commercial enterprise are not commercial cases, they are not subject to mediation, which is a prerequisite of the lawsuit. Mediation can be applied if the lawsuits in the intellectual property legislation and the cases related to the intellectual property law in other legislations are commercial lawsuits.

It is understood that in commercial disputes in Turkey, in 57% of the disputes the parties have applied to the mediator in 2019 and it is observed that 43% of these applications ended with an agreement.

As a new provision by the Turkish Industrial Property Law numbered 6769, Article 19/4 states that the Turkish Patent and Trademark Office may encourage the parties to reconcile if deemed necessary during the trademark registration process and that the provisions of the Law on Mediation in Civil Disputes dated 7/6/2012 and numbered 6325 shall be applied in matters related to reconciliation.

In this regard, of interest the point of view of Kirillov, Honored Inventor of the Russian Federation, Senior researcher of the Military Institute (Engineering) of the Military Academy of MTO (Kirillov, 2021) who says that the advantage of mediation and as a method of protecting trademark rights is to satisfy the interests of both parties and find the best way out of the situation with minimal negative consequences and costs, without depriving them of the opportunity to use in the future as other non-judicial and judicial means of protecting exclusive rights to trademarks.

As per Article 81 of Turkish Law on Intellectual and Artistic Works, the label obligation is outside the scope of Law No. 6325 on Mediation in Legal Disputes. This article and penal provisions, which are regulated as criminal offenses, are not suitable for mediation. The mediation shall not be applied for violations such as non-label, irregular and false banderole, and the deterrence of penalties will continue effectively. The same rule applies to voluntary mediation. When an intellectual property dispute occurs, the parties prepare a report regarding the agreement between them.

Under Article 18/a-2 of the Law on Mediation in Civil Disputes No. 6325, the plaintiff must add to the petition, the original or a copy of the final minute, which was approved by the mediator.

In the event of non-compliance with this obligation, the court is going to send an invitation to the plaintiff with the warning that the final minute must be submitted to the court within one week; otherwise, the case shall be rejected. If the requirement of the warning is not fulfilled, the case shall be dismissed due to the procedural cause without notifying the other party. The essential expenses that must be incurred by the mediation office under this article shall be paid by the parties under the agreement if an agreement is reached at the end of the mediation activity, and in case of failure to reach an agreement, the expenses shall be collected from the budget of the Ministry of Justice to be collected from the future unfair party.

If a preliminary injunction has been issued before the lawsuit is filed, or if a precautionary attachment order has been issued, the filing period does not run until the date of the final minute from the application to the mediation office. Mediation negotiations are conducted within the jurisdiction of the justice committee of the first instance court to which the office that appointed the mediator is affiliated unless otherwise agreed by the parties.

In many intellectual property disputes, the mediation parties provide a more satisfactory solution than recourse to the court. However, it seems that the number of intellectual property disputes where the parties prefer to mediate remains low (Corbett, 2011).

The practice, which started in courts as a result of the regulation on the use of compulsory mediation methods in the field of intellectual property ended with the recent decision of the Court of Cassation. In the decision taken as a result of an appeal review of a decision of the 2. Istanbul Civil Court of Intellectual and Industrial Property Rights. In the case with payment of damages, infringement of the trademark rights, prevention of unfair competition, the first claim was filed without resorting to mediation. The Court of Cassation overturned the decision.

Mediation in the field of intellectual property has been blocked with the recent case law of the Court of Cassation. In a lawsuit brought before the Court of Cassation, it has been determined that if there are requests for infringement, prohibition, prevention, suspension, and pecuniary and non-pecuniary damages, there is a "backlog of cases". Accordingly, in the same lawsuit, although the receivable claim involving the payment of a certain amount of money is subject to mediation, the case regarding the determination of the absence of a valid partnership relationship is not subject to mediation; therefore, a claim filed with a non-mediation lawsuit shall not be subject to the mediation lawsuit requirement (Turkish Supreme Court, 2020). In intellectual property disputes, compensation claims are generally brought together with claims for the determination, cessation, and prevention of infringement. Therefore, it becomes impossible to recourse to a mandatory pre-trial mediator.

However, it is known that 95% of the disputes constituting the subject of intellectual property are between commercial companies which are interested in keeping their trade secrets. In our opinion, it will be beneficial to apply the compulsory mediation method in infringement cases filed together with the compensation or receivable claims within the scope of the lawsuit that will be subject to compulsory mediation.

It is more compliant with the law and equity to continue the trial in terms of requests not included in the scope of mediation, which is a prerequisite for the case, by giving a separation decision in terms of the case filed without applying to the mediator. Otherwise, while the plaintiff will be placed under an obligation that is not included in the legislation, the plaintiff's freedom to seek remedies will also be limited (Karaca, 2019). It is hoped that over time this practice will be expanded to all intellectual property disputes.

According to Article 18/5 of the Turkish Mediation Law in Civil Disputes, a lawsuit cannot be filed on the issues that are included in the settlement agreement document and agreed upon. In case of failure to comply with the settlement agreement concluded before the mediator, the right holder will suffer.

For this reason, it should be regulated that by adding this issue to the Law, a lawsuit can be filed for the agreement issues that do not obtain enforceability annotation.

In the Russian Federation, the choice of the most preferred way of dispute resolution in relations in the field of intellectual property depends on several factors in each particular case, such as the cost of dispute resolution, public awareness of their availability and advantages, reputational component, arbitrability/mediability, confidentiality, presence of a foreign element, as well as the legal force of the final document, as the decision made by arbitrators is comparable in legal force with the judicial (Abrosimova, 2019), and the settlement agreement will only be enforceable if additional conditions are met.

4 Conclusion

The mediation for resolving intellectual property disputes is a very convenient method whether in or out of court. This method is considered to be more effective, efficient, fast, and low cost, as well as benefit both parties in disputes (win–win solution). Mediation, which includes more flexible rules than arbitration, will be able to bring more appropriate solutions to the parties in the field of intellectual property. The technical aspect and complexity of the intellectual property field will allow the parties to create a wide range of bargaining and alternatives in the mediation process. Successful mediation provides the opportunity to look at the issues from a new perspective.

When the courts see that the dispute is appropriate for ADR, it encourages the parties to try mediation. The basic principles of mediation are confidentiality and impartiality. The mediating institution has been uniquely created in Turkey.

At the moment the mediation method is carried out only by mediators who are graduated from law faculty in Turkey. It is also very important that a person who mediates these disputes has knowledge and experience in this field. The complexity of the field of intellectual property necessitates the involvement of experts in different fields such as trademark and patent attorneys, mechanical engineers, electrical engineers, pharmacists, communication experts, as well as lawyers.

In the mediation agency, which entered into our law in 2013, compulsory mediation was applied as a "case condition" as of 1 January 2018 in terms of employee and employer disputes in Turkey. It has started to be applied as of January 1, 2019, in commercial disputes regarding trade receivables and compensation, including intellectual property disputes, and started to contribute to the reduction of the case files in the court process in most of the disputes. In intellectual property disputes, mediation is applied only on compensation claims. However, there is no obstacle to the resolution of the disputes with many demands such as infringement, ref, ban, with the mediator.

Making compulsory mediation a legal requirement in intellectual property disputes is a positive development. However, we believe that narrowing the boundaries of this with the case law of the Supreme Court will hinder the settlement of mediation in the country. In intellectual property disputes, going to the mediator to include all requests will help to resolve the dispute completely. Otherwise, splitting the requests and going to the mediator will make the dispute unsolvable.

An intellectual property product can be both a copyright issue and an industrial property issue. It is possible to easily resolve disputes regarding more than one intellectual property violation in front of the mediator.

The power of mediation does not arise from entitlement but focuses on the needs of the parties. Mediation training should be given to mediators regularly and frequently. Many issues that cause hesitation or are applied incorrectly will be resolved in this way. Such problems can be solved with the training of party representatives and mediators.

Currently, in Turkey facilitating mediation is pending the entry into force in the Parliament in March 2021 and evaluative mediation is expected to change with the Law on Mediation. In this case, it is aimed for the mediator to intervene in the events and to take a more active role, to offer solutions, and to be a legal guide. Same time with the law proposal in the TBMM commission, it is aimed to assign a lawyer by the bar association upon the request of the mediation office in case the parties request the assistance of a lawyer.

Another change to be brought by the law amendment to be issued, is that mediation centers will be put into operation. However, this situation is rightly criticized by currently active mediators for the reason that money can be earned thanks to the mediators and fair file distribution will not be made.

References

Abrosimova, E. A. (2019) Resolution of disputes in copy-right: is there an alternative to court and is it necessary. *Bulletin of Arbitration Practice, N 5.* https://mgimo.ru/upload/iblock/54a/razreshenie-sporov-v-avtorskom-prave-est-lialternativa-sudu-i-nuzhna-li-ona.pdf Data Accessed 20 June 2021

Corbett, Susan Felicity, Mediation of Intellectual Property Disputes: A Critical Analysis (January 27, 2011). New Zealand business law quarterly, March 2011 (Vol. 17, pp. 51–67). Available at SSRN https://ssrn.com/abstract=1802207 https://www.researchgate.net/publication/228160541_Mediation_of_Intellectual_Property_Disputes_A_Critical_Analysis Data Accessed 20 June 2021

Dogan Alkan, G. (2019). Mediation now mandatory for monetary re-lated IP disputes. *World Trademark Review*. https://www.worldtrademarkreview.com/enforcement-and-litigation/mediation-now-mandatory-monetary-related-ip-disputes Data Accessed 20 June 2021

Karaca, O. U. (2019). Industrial property disputes within the scope of mediation which is a condition of litigation. *Ankara Bar Association Intel-lectual Property and Competition Law Journal FMR, Y. 20*, 21(1) (47–55). https://iprgezgini.org/2020/07/02/dava-sarti-olan-arabuluculuk-sinai-mulkiyet-gundeminde-bana-yeniden-yer-acin/ Data Accessed 20 Nov 2020

Kirillov, N. G., & Kirillov, N.N. (2021) Modern ways of protection for trademark rights: a systematic approach IP. Indus-trial Property N 3. http://www.consultant.ru/law/podborki/sposoby_zaschity_tovarnogo_znaka/ Data Accessed 20 Nov 2020

Pavlova, Z. (2020). Supreme Court clarified the order of compulsory pre-trial regulation of arbitration disputes. https://www.advgazeta.ru/novosti/vs-razyasnil-poryadok-obyazatelnogo-dosudebnogo-regulirovaniya-arbitrazhnykh-sporov/ Data Accessed 20 June 2021

Sergo, A. G. (2017). Mediation in intellectual property disputes "IP. Copyright and Related Rights", N 1 http://www.consultant.ru/cons/cgi/online.cgi?req=doc&base=CJI&n=102606#08988956925660498 Data Accessed 20 June 2021

Shamlikashvili, T. S. (2013). The place of mediation in the system of al-ternative dispute resolution. When is the use of mediation preferable?. http://www.garant.ru/ia/opinion/shamlikashvili/464686/#ixzz6x05RJNBe Data Accessed 20 June 2021

Steve, M. (2009). Skype Ccalling—Mediation in the 21st century. https://www.mediate.com/articles/MehtaSbl20090511.cfm. Data Accessed 20 June 2021

Stulberg, J. B. (1997). Facilitative versus evaluative mediator orien-tations: Piercing the "Grid" lock. *Florida State University Law Review, 24*, (985–1005). https://ir.law.fsu.edu/cgi/viewcontent.cgi%3Farticle%3D1450%26context%3Dlr%26httpsredir%3D1%26referer Data Accessed 20 June 2021

The Turkish Supreme Court (2020). http://www.kazanci.com.tr

WIPO Mediation Arbitration Expedited Arbitration and Expert De-termination Rules and Clauses, (2020). WIPO ADR Center. https://www.wipo.int/edocs/pubdocs/en/wipo_pub_446_2020.pdf Data Accessed 20 June 2021

Interregional Differentiation of Income and Expenditure of the Russian Population: Assessment in the Long-Term Dynamics

Evgeniya K. Karpunina, Ekaterina A. Isaeva, Igor V. Petrov, Elena V. Korolyuk, and Saida B. Dzhimova

Abstract The research is aimed at identifying patterns of formation of interregional differentiation of monetary income and expenditure of the population using the method of statistical analysis. The authors investigated the prerequisites for the formation of interregional inequality in terms of per capita monetary income and expenditures of the population and assessed the impact of a complex of regional and all-Russian factors on it. The article offers a set of socio-economic indicators to analyze the degree of their impact on the regional average per capita income and expenditure, as well as their impact on the living standard in the region. The authors propose a method for constructing a weighted average composite rating of the degree of differentiation of income and expenses in federal districts. The study of regional and all-Russian factors made it possible to reveal problem areas of regional policy and identify promising areas for improving the quality of life in the Russian regions.

Keywords Federal district · Region · Income · Expenses · Population · Well-being

JEL Codes R11 · R58

E. K. Karpunina (✉)
G.R. Derzhavin Tambov State University, Tambov, Russia

E. A. Isaeva
Moscow University for Industry and Finance "Synergy", Moscow, Russia

I. V. Petrov · S. B. Dzhimova
Kuban State Technological University, Krasnodar, Russia

E. V. Korolyuk
Kuban State University, Krasnodar, Russia
e-mail: lenkor@bk.ru

E. G. Popkova and B. S. Sergi (eds.), *Geo-Economy of the Future*,
https://doi.org/10.1007/978-3-030-92303-7_63

1 Introduction

Currently, the functionality of the region in the development of a new economic mechanism of Russia is increasing. The dominance of the sectoral management principle in the economy, when differences between regions were leveled by direct regional management from the federal center, is a thing of the past. The independence of the regions in choosing the vector of development and the tools to ensure it contributed to an increase in the differences in the level and quality of life of the population (Glushanok, 2008; Korolyuk et al., 2021; Lisova et al., 2020; Ovcharova & Popova, 2013; Suvorov, 2001b).

This trend makes it necessary to develop new tools for regional development across the state based on creating a methodological basis for quantifying the interregional differentiation of income, as well as spending of the people in the long-term dynamics. In this context, it is necessary to choose the optimal tools for a comprehensive assessment of the level of the income of citizens, to study the structure of regional consumption, as well as to clarify the reasons for the stratification of the population (Averin et al., 2017; Ayvazyan, 2012; Galieva et al., 2020).

2 Theoretical Basis of the Study

The research is based on publications aimed at identifying the causes and factors of income inequality of the population, which, in turn, affect the quality of life of citizens (Arrow, 2000; Clements & Jiawei, 2018; Pigou, 1928; Sen, 2015; Shorrocks, 1978). A high level of differentiation of income and expenses of the population is characteristic of the Russian reality. This justifies the scientific interest in the study of this problem on the part of Russian scientists (Belyaevsky, 2012; Clements & Jiawei, 2018; Dankanich, 2011; Mayorova, 2013; Ovcharova & Popova, 2013). Statistical and econometric methods are often used to analyze the differentiation of the population's income and expenditure (Matveeva, 2008; Rakitskiy, 2019; Suvorov, 2001a, 2001b; Zharomsky et al., 2018; Zubarevich & Safronov, 2019). However, studies of interregional differentiation of the population's incomes and expenditures are needed for the reasonable adoption of measures aimed at interregional alignment.

3 Methodology

The purpose of the research is to form an author's method for assessing the interregional differentiation of income and expenses of the population in Russia, as well as to justify a set of effective state policy measures aimed at equalizing the Russian region's development.

Research objectives: (1) to assess the change in the structure of monetary income of the people by sources of income; (2) to clarify the structure of consumer spending of households by consumption goals; (3) to analyze the change in the ratio of the average per capita monetary income of the population to consumer spending per capita in regions; (4) to form a rating of Russian regions according to the interregional differentiation of income and expenditure of the population.

Research methods: the theoretical analysis, the comparative analysis, the systematization, the economic and statistical analysis, the system approach.

4 Results

The study of the differentiation of income and expenses in the regions from 2010 to 2019 will begin with an analysis of the real monetary income, the real accrued wages of employees, and the real size of assigned pensions, as well as an assessment of their weighted average changes (Table 1).

The weighted average change in the real monetary income, real accrued wages, and the real amount of assigned pensions for 2010–2019 allows us to segment the federal districts by the degree of change in the studied indicators and rank them by the amount of increase or decrease in the weighted average indicators. The spread between the extreme values of the weighted average change in real monetary incomes of the population ranges from 102.59% (Southern Federal District) to 99.95% (Ural Federal District). The changes in real accrued wages—from 103.4% (Far Eastern Federal District) to 102% (Ural Federal District), changes in the real size of assigned pensions—from 103.3% (Volga Federal District) to 102.51% (Southern Federal District).

We will analyze the degree of differentiation income in the regions by assessing the changes in the structure of the population's monetary income by the source of income in 2010–2019 (Fig. 1).

The income from property, income from the entrepreneurial activity, and another cash receipt in the overall income structure have decreased. This not only reduces tax revenues to the budget but also contributes to increasing the dependence of the population of the regions on wages and social benefits. Thus, we are talking about an increase in the dependence of the Russian population on state payments.

We will rank the regions according to the indicator of changes in income from entrepreneurial activities and income from property, they are most important for assessing the population well-being (Table 2).

The spread of values of changes in income from entrepreneurial activity in the total income structure is in the range from 1.3% (North-Western Federal District) to −9.1% (North-Caucasian Federal District), in the case of changes in property income—from 2.2% (North-Western Federal District) to -3.7% (Central Federal District).

We will evaluate the differentiation of Russian regions in the structure of consumer spending by household consumption goals (Fig. 2).

Table 1 The real monetary income of the population, the real accrued wages of employees, and the real amount of assigned pensions in the regions of Russia, changes for 2010–2019

Region	Weighted average change in monetary incomes of the population for the period from 2010 to 2019 (as a % of the previous year)	FD RATING based on the weighted average change in monetary income of the population	Weighted average change in the real accrued salary of employees of organizations	FD RATING based on the weighted average change in the real accrued salary of employees of organizations	Weighted average change in the real size of assigned pensions, for the period from 2011–2019 (as a % of the same period of the previous year)	FD RATING based on the weighted average change in the real size of assigned pensions
	2010–2019	2010–2019	2010–2019	2010–2019	2011–2019	2011–2019
Russian federation	101.09		103.0		102.8	
Central federal district	100.91	6	103.2	2	102.7	6
North-Western federal district	101.72	4	103.0	5	102.52	7
Southern federal district	102.59	1	102.6	7	102.51	8
North Caucasus federal district	102.05	2	103.04	3	102.8	5
Volga federal district	100.73	3	102.9	6	103.3	1
Ural federal district	99.95	8	102.0	8	103.3	2
Siberian federal district	100.60	7	103.01	4	103.1	3
Far Eastern federal district	101.67	5	103.4	1	102.9	4

Source Compiled by the authors according to (Rosstat 2020a, 2020b)

In the structure of consumer spending of households, expenses for social needs (purchase of food, alcohol, purchase of clothing and household items, payment for housing and communal services) and non-social expenses (payment for health and education services, payments for transport and communications, spending on recreation and cultural events) can be allocated. An increase in the social expenditures in the structure of household expenditures indirectly indicates a decline in household

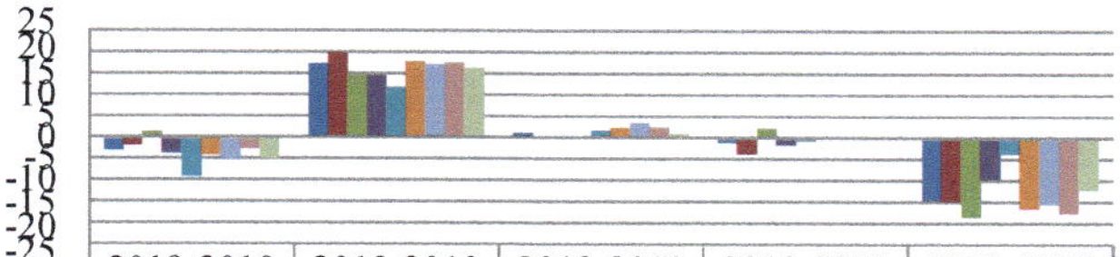

	2019-2010 Income from entrepreneurial activities	2019-2010 Payment of labor	2019-2010 Social payments	2019-2010 Property income	2019-2010 Other cash receipts
Russian Federation	-2.9	17.4	1.2	-1.1	-14.6
Central Federal District	-1.8	19.9	0.4	-3.7	-14.8
North-Western Federal District	1.3	14.7	0.3	2.2	-18.5
South federal district	-3.8	14.7	0.2	-1.6	-9.5
North Caucasus Federal District	-9.1	11.9	1.7	-0.7	-3.8
Volga Federal District	-4.1	17.9	2.4	0.4	-16.6
Ural Federal District	-5.2	17.3	3.5	-0.1	-15.5
Siberian Federal District	-2.6	17.5	2.5	0.2	-17.6
Far Eastern Federal District	-5.0	16.4	0.9	-0.1	-12.2

Fig. 1 Structure of the population's monetary income in regions of Russia by sources of receipt, 2010–2019. *Source* Compiled by the authors according to (Rosstat, 2020a, 2020b)

incomes, including those caused by the crisis phenomena. The increase in the non-social expenditures in the structure of the population's expenditures highlights the income growth since these expenditures are carried out by the population only if the initial needs of the person are fully met ("Engel's law") (Clements & Jiawei, 2018).

We will rank the regions by the weighted average change in non-social household expenditures in the overall structure of household expenditures (Table 3).

The table shows that the spread between the maximum and minimum values ranges from 0.82% (Central Federal District) to 0.02% (North-Western Federal District).

We will evaluate the changes in the structure of the use of monetary income of the population of the regions in 2010–2019 (Fig. 3).

During the study period, the structure of the use of monetary income of the population of Russia increased the share of expenses for the purchase of goods and services and the share of mandatory payments. These expenditures can be assessed as social, informing about the decline in the welfare and income of the country's population. Negative factors include a decrease in the cost of purchasing real estate, a growth (decrease) in financial assets and money from the population. These are "non-social" expenses, and their growth signals an increase in the welfare and income of the population.

We will rank the regions of Russia by the weighted average change in the share of non-social expenses (acquisition of real estate, increase (decrease) in financial

Table 2 The population's income from the entrepreneurial activity and property in the regions of Russia, changes for 2010–2019

Region	Change in share of income from entrepreneurial activities in the total income structure for 2010–2019, %	FD RATING on growth/decrease in the share of income from entrepreneurial activities in the total income structure of the population	Change in the share of property income in the total income structure for 2010–2019, %	FD RATING by growth/decrease in the share of property income in the total income structure of the population
Russian federation	−2.90	–	−1.1	–
Central federal district	−1.8	2	−3.70	8
North-Western federal district	1.30	1	2.20	1
Southern federal district	−3.80	4	−1.60	7
North Caucasus federal district	−9.10	8	−0.70	6
Volga federal district	−4.10	5	0.40	2
Ural federal district	−5.20	7	−0.10	4
Siberian federal district	−2.60	3	0.20	3
Far Eastern federal district	−5.00	6	−0.10	5

Source Compiled by the authors according to (Rosstat, 2020a , 2020b)

assets, and increase (decrease) in money from the population) in the overall structure of the use of monetary income of the population (Table 4).

The spread of the share of non-social expenses in the total use of monetary income of the population ranges from −3.6% in the Southern Federal District to −7.67% in the North Caucasus Federal District.

The ratio of the average per capita monetary income of the population and consumer spending per capita in the period 2010–2019 also changed (Table 5).

These tables allow us to determine the range of the spread of the indicator values: from −13.36% in the Southern Federal District to −28.8% in the Far Eastern Federal District. In 2010 the monthly income of the population was higher than the expenditure of the population by 43.7%. In 2019 the income exceeded the expenditure of the population per month by 23.8%.

We will present the data obtained earlier in summary Table 6.

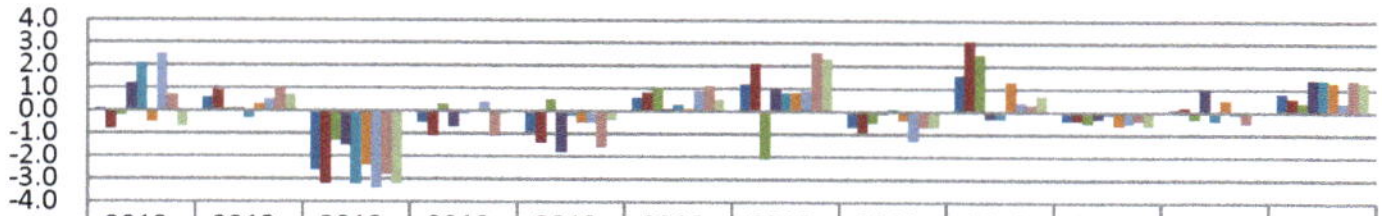

	2019-2010 Food and non-alcoholic beverages	2019-2010 Alcoholic beverages and tobacco products	2019-2010 Clothing and footwear	2019-2010 Housing and communal services and fuel	2019-2010 Household items, appliances and home care	2019-2010 Health care	2019-2010 Transport	2019-2010 Connection	2019-2010 Organization of recreation and cultural events	2019-2010 Education	2019-2010 Hotels, cafes and restaurants	2019-2010 Other products and services
Russian Federation	0.1	0.6	-2.6	-0.5	-0.9	0.6	1.2	-0.7	1.6	-0.4	0.1	0.8
Central Federal District	-0.8	0.9	-3.2	-1.1	-1.4	0.8	2.1	-0.9	3.1	-0.4	0.2	0.6
North-Western Federal District	-0.2	0.1	-1.3	0.3	0.5	1.0	-2.1	-0.5	2.5	-0.5	-0.3	0.4
Southern Federal District	1.2	0.1	-1.5	-0.7	-1.8	0.1	1.0	-0.1	-0.3	-0.3	1.0	1.4
North Caucasus Federal District	2.1	-0.3	-3.2	-0.1	-0.2	0.3	0.8	0.1	-0.3	-0.1	-0.4	1.4
Volga Federal District	-0.5	0.3	-2.4	0.0	-0.5	0.1	0.8	-0.4	1.3	-0.6	0.5	1.3
Ural Federal District	2.5	0.5	-3.4	0.4	-0.5	0.9	0.9	-1.3	0.4	-0.5	-0.1	0.4
Siberian Federal District	0.7	1.0	-2.8	-1.1	-1.6	1.1	2.6	-0.7	0.3	-0.4	-0.5	1.4
Far Eastern Federal District	-0.7	0.7	-3.2	-0.1	-0.4	0.5	2.3	-0.7	0.7	-0.6	0.0	1.3

Fig. 2 Structure of consumer spending in the regions of Russia by household consumption goals, %, 2010–2019. *Source* Compiled by the authors according to (Rosstat, 2020a, 2020b)

The data in Table 6 allow us to divide the regions into three groups depending on the level of differentiation of income and expenditure of the population (Fig. 4).

Thus, the results obtained refute the previously proposed hypothesis about a strong level of differentiation of the federal districts of Russia in terms of the population's income and expenditure.

The development of a set of state policy measures aimed at optimizing the population's income and expenses in the regions requires the identification of "bottlenecks" in the regional economies (Galieva et al., 2020) (Table 7).

5 Conclusion

Firstly, the authors used statistical methods to identify patterns of formation of interregional differentiation of monetary income and expenditures of the population. Secondly, the authors classified the main factors of differentiation of monetary incomes and expenditures of the population. Thirdly, the authors offered a set of statistical indicators to identify the degree of influence of socio-economic factors on the average per capita income and expenses of the population of the region and the standard of living of the population. Fourthly, the authors studied the relationship between statistical indicators of monetary income and expenditures of the population

Table 3 Weighted average change in "non-social" household expenditures in the total structure of household expenditures in the regions of Russia, %, 2010–2019

Region	Health care	Transport	Connection	Organization of recreation and cultural events	Education	Hotels, cafes and restaurants	Weighted average "non-social" change in household expenditures in the overall structure of household expenditures for 2010–2019	FD RATING based on the weighted average change in the share of "non-social" household expenditures in the total structure of household expenditures for 2010–2019
Russian federation	0.6	1.2	−0.7	1.6	−0.4	0.1	0.40	
Central federal district	0.8	2.1	−0.9	3.1	−0.4	0.2	0.82	1
North-Western federal district	1	−2.1	−0.5	2.5	−0.5	−0.3	0.02	8
Southern federal district	0.1	1	−0.1	−0.3	−0.3	1	0.23	5
North Caucasus federal district	0.3	0.8	0.1	-0.3	−0.1	−0.1	0.07	6
Volga federal district	0.1	0.8	−0.4	1.3	−0.6	0.5	0.28	4
Ural federal district	0.9	0.9	−1.3	0.4	−0.5	-0.1	0.05	7
Siberian federal district	1.1	2.6	−0.7	0.3	−0.4	-0.5	0.40	2
Far Eastern federal district	0.5	2.3	−0.7	0.7	−0.6	0	0.37	3

Source Compiled by the authors according to (Rosstat, 2020a, 2020b)

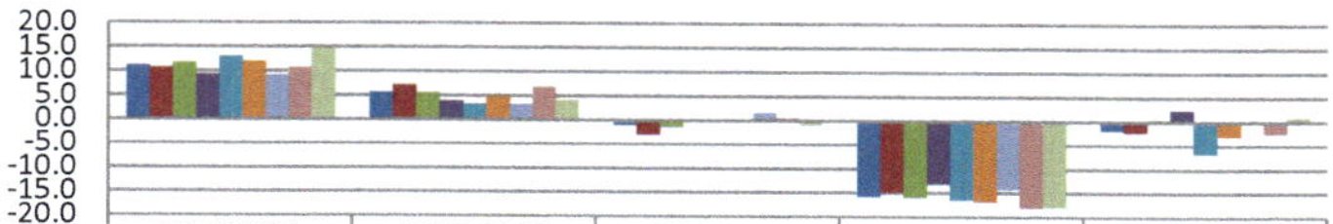

	2019-2010 Purchase of goods and payment for services	2019-2010 Mandatory payments and various contributions	2019-2010 Purchase of real estate	2019-2010 Increase (decrease) in financial assets	2019-2010 increase (decrease) in the population's money supply
Russian Federation	11.2	5.6	-1.0	-15.8	-1.8
Central Federal District	10.7	7.2	-3.0	-14.9	-2.2
North-Western Federal District	11.7	5.5	-1.3	-15.9	0.4
Southern Federal District	9.3	3.9	0.0	-13.2	2.4
North Caucasus Federal District	13.1	3.3	0.0	-16.4	-6.6
Volga Federal District	11.9	4.8	0.0	-16.7	-3.0
Ural Federal District	9.2	3.3	1.6	-14.1	-0.4
Siberian Federal District	10.7	6.8	0.5	-18.0	-2.4
Far Eastern Federal District	14.6	4.0	-0.8	-17.8	0.9

Fig. 3 Changes in the structure of the use of monetary income of the population, % of the total, 2010–2019. *Source* Compiled by the authors according to (Rosstat, 2020a, 2020b)

and determining factors by calculating the consolidated rating of regional differentiation for 2010–2019. The authors grouped the regions of Russia according to the differentiation of the population's income and expenses. Fifthly, the authors formulated recommendations for the implementation of state policy measures aimed at optimizing income and expenses in the regions of Russia.

Table 4 Weighted average change in the share of non-social expenditures in the total structure of the use of monetary income of the population, %, 2010–2019

Region	Purchase of real estate	Increase (decrease) in financial assets	Increase (decrease) in the population's money supply	Weighted average change in the share of acquisition of real estate, increase (decrease) in the financial assets and increase (decrease) in money held by population in the total structure of the use of the monetary income of the population for 2010–2019, as a % of the total volume	FD RATING based on the weighted average change in the share of real estate purchases, increase(decrease) in financial assets, and increase(decrease) in household money in the overall structure of cash income use
Russian federation	−1.0	−15.8	−1.8	−6.20	
Central federal district	−3.0	−14.9	−2.2	−6.70	7
North-Western federal district	−1.3	−15.9	0.4	−5.60	3
Southern federal district	0.0	−13.2	2.4	−3.60	1
North Caucasus federal district	0.0	−16.4	−6.6	−7.67	8
Volga federal district	0.0	−16.7	−3.0	−6.57	5
Ural federal district	1.6	−14.1	−0.4	−4.30	2
Siberian federal district	0.5	−18.0	−2.4	−6.63	6
Far Eastern federal district	−0.8	−17.8	0.9	−5.90	4

Source Compiled by the authors according to (Rosstat, 2020a , 2020b)

Table 5 The ratio of the average per capita monetary income of the population to consumer spending per capita changes for 2010-2019

Indicator/region		Russian federation	Central federal district	North-Western federal district	Southern federal district	North Caucasus federal district	Volga federal district	Ural federal district	Siberian federal district	Far Eastern federal district
The ratio of per capita monetary income of the population to consumer spending on average per capita (per month), %	Income 2010/Expenses 2010	143.71	145.77	145.01	126.72	142.29	140.38	147.31	148.13	155.52
The ratio of per capita monetary income of the population to consumer spending on average per capita (per month), %	Revenue 2019 / Expanses 2019	123.80	126.16	123.88	113.36	119.83	120.27	129.60	127.96	126.75

(continued)

Table 5 (continued)

Indicator/region		Russian federation	Central federal district	North-Western federal district	Southern federal district	North Caucasus federal district	Volga federal district	Ural federal district	Siberian federal district	Far Eastern federal district
The ratio of per capita monetary income of the population to consumer spending on average per capita (per month), %	(Revenue 2019 / Expenses 2019)–(Income 2010 / Expenses 2010)	−19.90	−19.61	−21.13	−13.36	−22.46	−20.10	−17.71	−20.17	−28.77
FD RATING of the ratio of the average per capita monetary income of the population to consumer spending on average per capita (per month)			3	6	1	7	4	2	5	8

Source Compiled by the authors according to (Rosstat, 2020a , 2020b)

Table 6 Summary rating of Russian regions by the level of interregional differentiation of income and expenditure of the population

Indicator	Year	Volga Federal District	Siberian Federal District	Southern Federal District	North-Western Federal District	Central Federal District	Far Eastern Federal District	Ural Federal District	North Caucasus Federal District
FD RATING based on the weighted average change in real monetary income of the population (Table 1)	2010-2019	3	7	1	4	6	5	8	2
FD RATING based on the weighted average change in the real accrued salary of employees of organizations (Table 1)	2010-2019	6	4	7	5	2	1	8	3
FD RATING based on the weighted average change in the real size of assigned pensions (Table 1)	2011-2019	1	3	8	7	6	4	2	5
FD RATING on growth/decrease in the share of income from entrepreneurial activities in the total income structure of the population (Table 2)	2010-2019	5	3	4	1	2	6	7	8
FD RATING by growth/decrease in the share of property income in the total income structure of the population (Table 2)	2010-2019	2	3	7	1	8	5	4	6
FD RATING based on the weighted average change in the share of "non-social" household expenditures in the total structure of household expenditures for (Table 3)	2010-2019	4	2	5	8	1	3	7	6
FD RATING based on the weighted average change in the share of real estate purchases, increase(decrease) in financial assets, and increase(decrease) in household money in the overall structure of cash income use (Table 4)	2010-2019	5	6	1	3	7	4	2	8
FD RATING of the ratio of the average per capita monetary income of the population to consumer spending on average per capita (per month) (Table 4)	2010-2019	4	5	1	6	3	8	2	7
WEIGHTED AVERAGE CONSOLIDATED RATING		**3.8**	**4.1**	**4.3**	**4.4**	**4.4**	**4.5**	**5.0**	**5.6**

Colour	Rating
Green	1-3
Yellow	4-6
Red	7-8

Source Compiled by authors

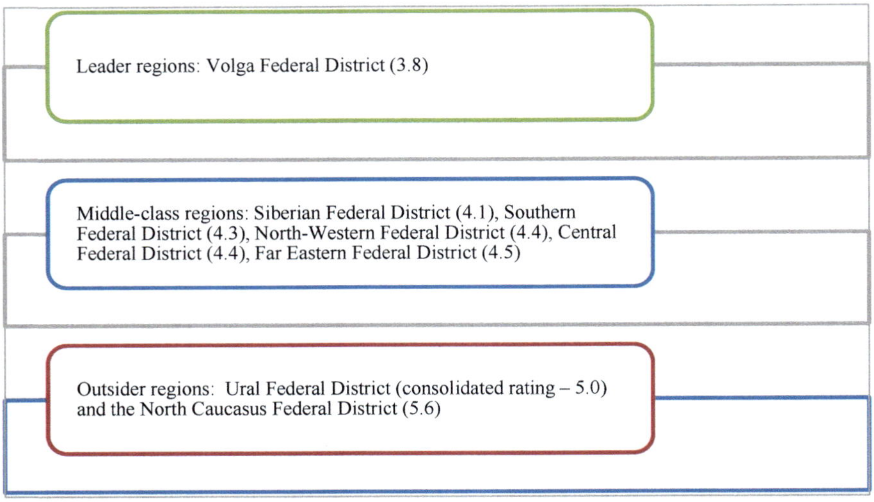

Fig. 4 Grouping of Russian regions by the level of differentiation of income and expenses of the population. *Source* Compiled by the authors

Table 7 State policy measures aimed at optimizing the income and expenditure of the population

Region/Measure	Increase in the real accrued wages of employees of organizations	Increase in the share of real estate purchases, the growth of financial assets and money from the population in the total structure of income use	Increase in real monetary income of the population	Increase in the actual size of the assigned pension	Growth of the share of property income in the total income structure of the population	The increase in the per capita monetary income of the population over the consumer spending of the population	Increase in the share of non-social household expenditure (health and education, recreation, etc.) in the total structure of household expenditures	Ensuring the growth of the share of income from business activities in the total income structure of the population
Volga federal district	+							
Siberian federal district		+	+					
Southern federal district	+			+	+			
North-Western federal district				+		+	+	
Central federal district		+	+	+	+			
Far Eastern federal district						+		+
Ural federal district	+		+				+	+
North Caucasus federal district		+			+	+	+	+

Source Compiled by authors

References

Arrow, K. (2000). Incomplete knowledge and economic analysis. Istoki. Issue 4. Higher School of Economics

Averin, A., Ponedelkov, A., Vorontsov, S., & Goloborodko, A. (2017). Monetary incomes of the population as an integral part of the social policy of the state. *State and Municipal Administration. Scientific Notes, 4*, 128–132.

Ayvazyan, S. (2012). Analysis of the quality and lifestyle of the population. Econometric approach. Nauka.

Belyaevsky, I. (2012). Consumer market as the most important factor of welfare. Bulletin of the UMO. *Economics, Statistics, and Informatics, 6*, 96–105.

Clements, K., & Jiawei, S. (2018). Engel's law, diet diversity, and the quality of food consumption. *American Journal of Agricultural Economics, 100*(1), 1–22.

Dankanich, S. (2011). The income inequality of the population: Types and consequences. *Problems of the Modern Economy, 3*(39), 59–63.

Galieva, G., Il'ina, L., Ponomarev, S., Khashir, B., & Karpunina, E. (2020). Consumption as a growth factor of the russian economy. In *Proceedings of the 36th IBIMA conference*, 4–5 November 2020, Granada, Spain, (pp. 3763–3772).

Glushanok, V. (2008). On determining the income of the population of the region. *Questions of Statistics, 8*, 19–21.

Korolyuk, E., Rustamova, I., Kuzmenko, N., Khashir, B., & Karpunina, E. (2021). Diagnostics of regional economic security problems during the 2020 crisis. In: *Proceeding of the 37th IBIMA Conference*, Cordoba, Spain, 1–2 April 2021, (pp. 5248–5257).

Lisova, E., Petrov, I., Koryakina, T., KalomboMulamba, V., & Karpunina, K. (2020). The well-being of russian regions under threat: Economic and social impact assessment of COVID-19. In *Proceeding of the 36th IBIMA Conference*, Granada, Spain, 4–5 November 2020, (pp. 7603–76160.

Matveeva, D. (2008). On improving the methodology of the balance of monetary income and expenditure of the population. *Questions of Statistics, 11*, 9–11.

Mayorova, L. (2013).Formation and investment potential of savings of the population. Taganrog" Publisher Stupin A.

Ovcharova, L., & Popova, D. (2013). Incomes and spendings of Russia's households: What has changed in the mass standard of consumption? *Russia's World, 3*, 3–34.

Pigou, A. (1928). *The Economics of Welfare*. King's College.

Rakitskiy, B. (2019). Methodological problems and guidelines for the study of social inequality. *The Standard of Living of the Russian Regions, 3*, 64–69.

Rosstat. (2020a). Report Socio-economic situation of Russia. URL https://rosstat.gov.ru/compendium/document/50801 Accessed 15 June 2021

Rosstat. (2020b). Statistical collection Russian regions, socio-economic indicators. URL https://rosstat.gov.ru/folder/210/document/13204 Accessed 15 June 2021

Sen, A. (2015). *Poverty and famines*. Oxford University Press

Shorrocks, A. (1978). Income inequality and income mobility. *Journal of Economic Theory, 19*(2), 376–393.

Suvorov, A. (2001a). Problems of analyzing income differentiation and building a differentiated balance of monetary income and expenditures of the population. *Forecasting Problems, 1*, 58–74.

Suvorov, A. (2001). Income and consumption of the population: Macroeconomic analysis and forecasting. MAKS-Press.

Zharomsky, V., Migranova, L., & Toksanbayeva, M. (2018). Socio-economic inequality in Russia: Dynamics and methods of assessment. *Population, 4*, 79–95.

Zubarevich, N., & Safronov, S. (2019). The share of food in total consumption of the population of Russian regions as an indicator of the standard of living and consumption modernization. Bulletin of the Moscow University. Series 5. *Geography, 2*, 61–68.

Monopolization of the Global Economy: On the Example of the American Multinational Enterprises

Yulia A. Konovalova and Monkisheme Verity-Alexia Liongo

Abstract The purpose of the chapter is to find out the current issues and tendencies of TNCs and to compare theoretical aspects with practical issues. The authors underline that the theoretical approaches to capitalism, its progress and transition to imperialism with determined characteristics, is actual today too. TNCs as the meaningful subjects of the global economy are the logical result of capitalist development. Monopolization of producers is the result of capital growth, output growth, competitiveness and free trade. It is noticed that the economic papers, and in particular, economic papers of Lenin are actual today too.

Keywords Monopolization · FDI · USA · Foreign affiliates · Policy of expansion · Imperialism

JEL Codes D42 · F21 · F23 · F54 · L12

1 Introduction

The topic of monopolization should be started from the statement of Lenin (1916), that capital monopolization or imperialism is the highest stage of the transitional process of the primary capital accumulation.

It should be noticed that the economic papers published by Lenin at the very beginning of the twentieth century, without too much modesty, are still actual, especially when we are talking about the issue and nature of capitalism, monopolism and imperialism.

One of the key features of capitalism is the concentration of the production process on the strongest and the biggest companies (corporations). At the same time,

Y. A. Konovalova (✉) · M. V.-A. Liongo
People's Friendship University of Russia (RUDN University), Moscow, Russia
e-mail: konovalova_yua@pfur.ru

M. V.-A. Liongo
e-mail: 1032165514@pfur.ru

E. G. Popkova and B. S. Sergi (eds.), *Geo-Economy of the Future*,
https://doi.org/10.1007/978-3-030-92303-7_64

the labour productivity is connected with and supported by the technical process, technical progress and the process of automation (automatic production line).

The concentration of production is implementing on the limited set of companies, so it can lead to monopolization. At the same time, specialization of monopolists on determined and selected types of economic activity could lead and promote the price collusion and governing of the market laws and rules.

2 Methodology

The methodology of the scientific topic is based on the traditional methods of economic analysis such as deduction and induction. But the main methods that have been implemented in this investigation are comparison and retrospective analysis due to the historical basis. At the same time, the authors used massive statistical data.

3 Results

One of the main characteristics of capitalism is a combination. According to Hilferding (1910), "combination" as the term (in the context of capitalism) is the opportunity to equal differences of conjuncture and to promote earnings.

Heymann (1904) underlined that mixed/combined companies are forming mixed/combined capital and chain of resources supply. At the same time, mixed/combined capital is connected with the banking system.

Hermann Levy (1909), the author of "Monopole, Kartelle und Trusts", marked that the size of big companies and the level of technical development are in the basis of forming monopoly; despite the high rate of costs, monopolies can ensure the high level of production and high rate of demand on goods.

According to the most famous economic paper of Marx "Das Kapital" (1867), monopolization is the logical result of production concentration and free competitiveness. The main difference of the global economy' subjects is the policy of implementation of protectionism or free trade policy, but the common point is the concentration of production in monopolies.

The very beginning of the twentieth century became the period of transition from the old capitalism stage to the new capitalism stage. Key characteristics of this transition are the destruction of households, appearing and setting up of cartels, big companies, organized producers; the growth of industrial production.

It needs to underline the role of international financial centres and the global leaders such as London and New York in the very beginning of the twentieth century: concentration of headquarters of TNC's (MNE's), advanced banking system and insurance system, advances market of stock exchanges, setting up of trusts and cartels, and so on. A technical advantage of the American MNE's and trusts and cartels—was one of the most serious and important positions on the global market. At the same time, MNEs developed a rather advanced and broad chain and net of foreign affiliates, developed scientific laboratories and started to buy patents and inventions to control the transition of knowledge and technologies on the national and global level.

When we are talking about TNC/MNE, we have to agree that the greatest and the biggest companies, their size, power, technical advantages, consecutiveness with the political circles, all of these play radical and vital on the countries-recipients.

According to Delyagin[1] (Delyagin, 2019), it is rather interesting and important to find out that TNC's interests are correlated logically and comfortable with the region/country of the host economy (recipient). TNCs promote the development of the new economic system under the governance of financial capital, technical advantages, broad markets of sales, assets located all over the world, aggressive investment policy and policy of expansion.

In the economic review, that has been published by UNCTAD in 2007 "The Universe of the Largest Transnational Corporations" (Heymann, 1904), TNCs from developed countries were the key drivers of the rapid growth in international trade and investment in all sectors of economic activity. Rapid expansion became the reason for the wave of cross-border mergers and acquisitions. UNCTAD gave the next one short meaning of transnationality: "Transnationality is a function of the extent to which a firm's activities are located abroad". So we can talk about the so-called "second economy".

At the very beginning of the 1990s, the approximate total number of TNCs was approximately 37,000 with 170,000 foreign affiliates; about 33,500 of 37,000 were parent companies from developed countries. In 2007 the number of TNCs was about 77,000 with more than 770,000 foreign affiliates.

Nowadays, following UNCTAD, TOP-100 non-financial TNC's by the value of foreign assets consists of 19 American corporations, 13 corporations from the United Kingdom, 11 corporations from Germany, 9 from Japan, 15 from France, 9 from China and so on.[2] Talking about the American companies, we have to say, that today the total value of all the American MNE's (Parents companies) made up about 43.4 trln. doll. (in 2018), the total value of liabilities made up about 32.1 trln. doll.

From this 43.4 trln. doll. of assets more than 8.9 trln. doll. are accumulated on Manufacturing (1.06 trln. doll.—Petroleum and coal products; 2.17 trln. doll.—Chemicals; 1.27 trln. doll. —Computers and electronic products; 1.46 trln. doll.—Transportation equipment), 1.93 trln. doll. are accumulated on Wholesale trade, 1.12 trln. doll. —on Retail trade, 3.04 trln. doll. —on Information, 23.93 trln. doll. —on Finance and insurance, and so on.[3]

---

[1] Delyagin M.G.—PhD, Doctor of Economics, Russian economist and political scientist, The head of the Institute of problems of globalization.

[2] World Investment Report—2020. UNCTAD. URL: https://unctad.org/webflyer/world-investment-report-2020 Accessed: 15.05.2021.

[3] Bureau of Economic Analysis of USA. U.S. Direct Investment Abroad (USDIA). Selected Financial and Operating Data of U.S. Parents by Industry of U.S. Parent 2018. URL: https://www.bea.gov/international/di1usdop Accessed: 15.05.2021.

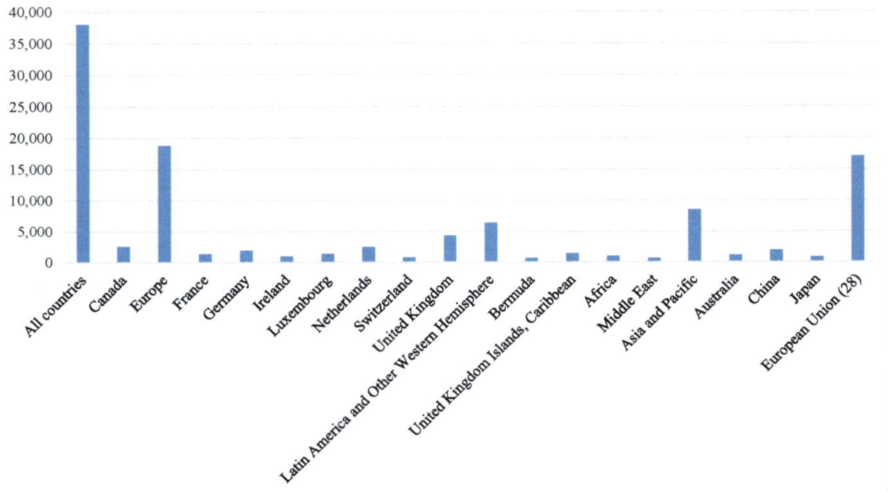

Fig. 1 Number of affiliates with assets, sales, or net income (±) greater than $25 million. *Source* Bureau of Economic Analysis of USA. U.S. Direct Investment Abroad (USDIA). URL: https://www.bea.gov/international/di1usdop

The broad and deep net of foreign affiliates is one of the features of the modern MNEs as the direct participants of modern capitalism/modern imperialism. Nowadays, there are about 38,083[4] the American foreign affiliates abroad, 2.5 thousand of them are located in Canada, 18.8 thousand—in Europe (1.4 thousand—in France, 1.9 thousand—in Germany, 1.3 thousand—in Luxembourg, 2.4 thousand—in the Netherlands, 4.3—in the United Kingdom), 1.2 thousand—in Mexico, 1.1 thousand—in Australia, 1.9 thousand—in China, 17 thousand—in European Union.

Despite the proportion of the American affiliates location by country, the greatest value of the assets of the foreign affiliate abroad is placed in regions/countries with certain low taxes regimes and investment climate (Figs. 1 and 2.).

Data, published by the Bureau of Economic Analysis of USA, showed that there were about 4,055 U.S. parent companies with certain conditions (in 2018). At the same time, we have to look at the data on total assets' distribution by industries. So, the biggest value of assets (by affiliates) was located in Finance and insurance, Manufacturing and Other industries.

The current process of monopolization can be proved by the dynamic of cross-border mergers and acquisitions (M&A). For example, on the period from 2010 to 2016 the American MNEs completed 55 deals worth over 3 billion dollars.[5]

The next point of monopolization—is the integration of the MNEs into the system of international foreign direct investment.

---

[4] The number of affiliates with assets, sales, or net income (±) greater than $25 million.

[5] World Investment Report—2017. UNCTAD. URL: https://unctad.org/system/files/official-document/wir2017_AnnexTables_en.pdf Accessed: 15.05.2021.

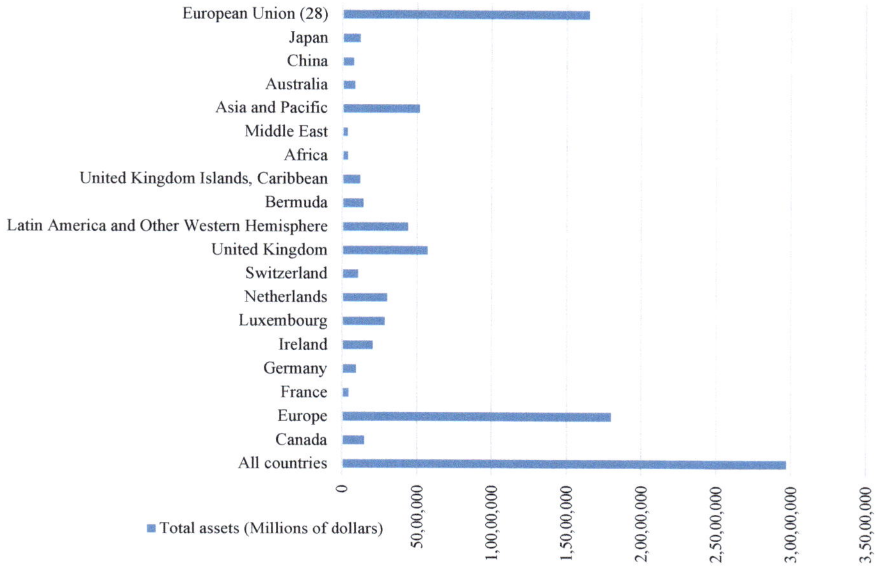

Fig. 2 Total assets (Millions of dollars). *Source* Bureau of Economic Analysis of USA (2021b)

In the scientific community, a significant number of studies have been dedicated to the problems of the U.S. in the global system of capital movement and FDI, especially. The issue of U.S. participation in the global flows of FDI is covered in works of the Russian and foreign scientists (UNCTAD, 2007; Aksenov, 2018; Arhangel'skij, 2019; Konovalova et al., 2021; Konovalova, 2020; Petrovskaja, 2018; Portanskiy, 2019; Sogrin, 2016; Ushanov et al., 2020; Ushanov & Konovalova, 2019; Ushanov, 2017; Volgina, 2015), who dedicated economic studies and investigations to the different aspects of the global economy, international economic relations, participation of certain countries in the international system of relations. The geography of the distribution of the American investments by countries and regions of the world in 2019 showed that 29.8 bln. doll. invested to Canada, European Union countries accumulated (−575 bln. doll.) of the American FDI at the end of 2019, of which 14.4 bln. doll. —in Germany, (−73.9 bln. doll.)—in Ireland, 12.2 bln. doll.—in Luxembourg, 32 bln. doll. —in the United Kingdom and so on.

It has been noticed that in 2019 the volume of repatriated U.S.' FDI was, mainly, from Ireland and Bermuda. In 2018 the volume of repatriated U.S.' FDI was, mainly, from Ireland, Netherlands, Bermuda, Singapore and European Union. At the same time, the main investors in the American economy are Canada (36.4 bln. doll.), European Countries in a whole (120 bln. doll.), France (7.7 bln. doll.), Germany (42 bln. doll.), Netherlands (13.1 bln. doll.), United Kingdom (23.2 bln. doll.), Bermuda (21.5 bln. doll.), Australia (16.2 bln. doll.), Japan (38.5 bln. doll.), EU (106.5 bln. doll.). Holding companies stay the main recipients of the U.S. direct investment

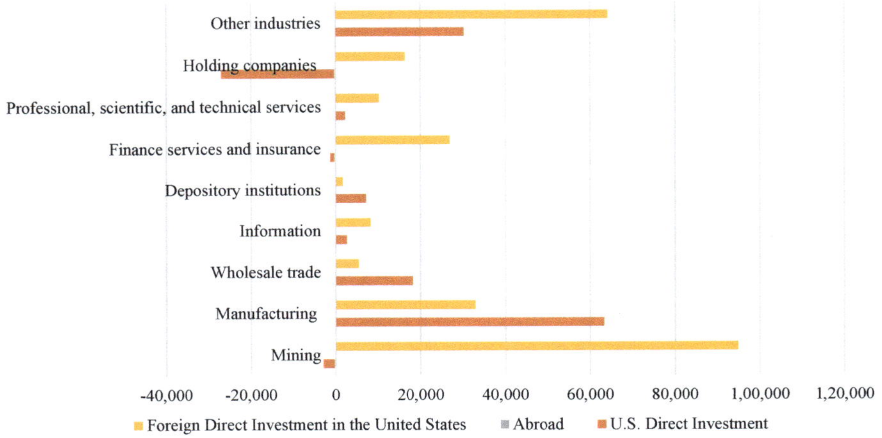

Fig. 3 Distribution of U.S. FDI' flows by industries in 2019 (mln. doll.). *Source* Bureau of Economic Analysis of USA (2021a)

abroad and the biggest volume of repatriation is coming from low tax jurisdictions (Fig. 3).

4 Conclusion

Theoretical aspects are proved by the practical issues and events: first of all, by the dynamic and size of M&A; the second one—is the size of MNEs, and the size (broadness) of their activity abroad (so-called second economy); the third one—is the FDI (directions, number of foreign affiliates, specialization), and places among TOP-100 non-financial TNCs by foreign assets. The dominance of MNEs is the result of capitalism transformation to imperialism. Despite the free trade and free competitiveness laws and nature, the tendency of capital accumulation led to the development of strong and significant subjects of the global economy, such as TNCs/MNEs.

Acknowledgements This paper has been supported by the RUDN University Strategic Academic Leadership Program and Grant of RUDN University No. 060119-0-000.

References

Aksenov, P. A. (2018). Trump's tax reform and American business. *International Trade and Trade Policy, 6*, 91–103.

Arhangel'skij, I. J. (2019). Antimonopoly regulation of foreign capital in the USA. *USA and Canada: Economics, Politics, Culture, 2*(590), 57–74.

Bureau of Economic Analysis of USA. (2021a). International trade and investment. Foreign Direct Investment in the U.S. and abroad. https://www.bea.gov/data/intl-trade-investment Accessed Data 15 May 2021
Bureau of Economic Analysis of USA (2021b). U.S. Direct Investment Abroad (USDIA). https://www.bea.gov/international/di1usdop
Delyagin, M. G. (2019). The end of an era. Be careful, the doors are opening. (Vol. 1. IPROG. pp. 832).
Heymann, H. G. (1904). Die gemischten Werke im deutschen Großeisengewerbe. Stuttgart. https://marxlib.ru/wp-content/uploads/2019/10/lenin-imperialism-the-highest-stage-of-capitalism.pdf Accessed 15 May 2021
Hilferding, R. (1910). Finance capital. https://www.marxists.org/archive/hilferding/1910/finkap/ Accessed 15 May 2021
Konovalova, Y. A. (2020). U.S. in the global FDI' flows Repatriation of foreign earning by the U.S. from EU members as the new global trend. *RUDN Journal of Economics, 28*(1), 172–183.
Konovalova, Yu. A., Ushanov, S. A., & Monkisheme, V.-A.L. (2021). Features of the American foreign direct investment: The role of the American holdings. *RUDN Journal of Economics, 29*(1), 238–252.
Lenin, V. I. (1916). Imperialism as the highest stage of capitalism. Public Domain (pp. 89).
Levy, H. (1909). Monopole, Kartelle und Trusts. https://archive.org/details/monopolekartelle00levyuoft/page/n15/mode/2up Accessed 15 May 2021
Marx, K. (1867). Capital. (Vol. 1–3). https://political-economy.com/capital-karl-marx/ Accessed 15 May 2021
Petrovskaja, N. E. (2018). Influence of foreign direct investment on employment (USA' experience). In *Megatrends of the global economy and international business*. International conference edition (pp. 144–146).
Portanskiy, A. P. (2019). US-Europe: Allied relations under threat. *Contemporary Europe, 6*(92), 30–40.
Sogrin, V. V. (2016). U.S. foreign policy at the beginning of 21st century. *USA & Canada: Economics, Politics, Culture, 5*(557), 53–63.
The Universe of the Largest Transnational Corporations. UNCTAD. (2007). https://unctad.org/system/files/official-document/iteiia20072_en.pdf Accessed 15 May 2021
Ushanov, S. A. (2017). The American transnational corporation and the questions of external economic safety. *The Eurasian Scientific Journal, 9*(3), 1–12.
Ushanov, S. A., & Konovalova, Yu. A. (2019). EU in the system of trade interests of USA. *RUDN Journal of Economics, 27*(2), 386–400.
Ushanov, S. A., Konovalova, Yu. A., & Zarubin, I. S. (2020). US-EU trade cooperation in the context of changing American foreign economic policy. *MGIMO Review of International Relations, 13*(5), 31–54.
Volgina, N. A. (2015). State TNCs in the world market of foreign direct investment: Modern trends. *RUDN Journal of Economics, 1*, 57–71.

The Role of Migration in the Contemporary Demographic Development of Russia

Roman V. Manshin and Evgeniya M. Moiseeva

Abstract The purpose of this research is to estimate what are the positive and negative implications of both internal and international migration for the contemporary demographic development of the country as a whole and its regions. The analysis is based upon official statistical data and standard statistical methods, including correlation and regression analysis, as well as demographic forecasts of the population change and migration till 2035 and 2050. The authors regard migration as a means of achieving the demographic sustainability of the Russian society and solving the related economic problems. The study provides an overview of the latest developments in the demographic and migration trends in Russia and its regions, which proves that migration has become the only source of population replacement. Scenario analysis is applied to estimate migration prospects in this regard. The authors also estimate the role of migration in the redistribution of the Russian population and its spatial disproportions. In conclusion, possible ways to further improve migration policy in the interests of the country's sustainable demographic development are suggested. The originality and significance of the study lie in the generalization of statistical data and a comprehensive approach to their analysis which makes it possible to embrace the complex interconnection between migration, population dynamics, spatial and socio-economic development.

Keywords Demographic development · Replacement migration · Population distribution · Population age structure · International and internal migration · Demographic forecasting · Migration policy

JEL Codes J11 · J61

R. V. Manshin (✉) · E. M. Moiseeva
Institute for Demographic Research FCTAS RAS, Moscow, Russia
e-mail: manshin@list.ru

R. V. Manshin
People's Friendship University (RUDN University), Moscow, Russia

E. G. Popkova and B. S. Sergi (eds.), *Geo-Economy of the Future*,
https://doi.org/10.1007/978-3-030-92303-7_65

1 Introduction

The demographic situation in Russia is worsening, despite the considerable efforts made by the Government during the past decades. Such strategically important to the national security of the country regions as Siberia and the Far East continue to lose their population especially fast, which is largely determined by migration outflow of their residents to the Central regions of the country, also known as the "Western drift." Over the past 10 years (2010–2019), the population of the Eastern part of Russia (the Far Eastern and Siberian Federal Districts) declined by 1% or 227.2 thousand people, while the population of the Western part of the country (the Central, Northwestern, Southern, North Caucasian, Volga, and Ural Federal Districts) increased by 3.5% or 4,053.6 thousand people (EMISS, n/a).

However, migration is a critical issue for Russia not only in terms of redistribution of the population but also in terms of its replacement. Since Russia entered stage two of the so-called demographic transition in the late 1980s, currently its working-age population is shrinking rapidly. Moreover, due to the infamous rough goings in the country's social and economic development spawned by the collapse of the Soviet Union, in the 1990s, mortality rates exceeded birth rates so dramatically that this unprecedented phenomenon got the name of the "Russian cross" (because of the shape which the corresponding curves had on the population movement charts) (Khalturina & Korotaev, 2006). Over the past 10 years (2010–2019), Russia has lost almost 1 million of its population due to natural decline, i.e., excess of the number of deaths over the number of births. At the same time, the country has hosted almost 2.5 million foreign citizens, and only thanks to migration its population has increased (EMISS, n/a). In light of this, immigration is regarded by some prominent Russian demographers as the only means to hamper the catastrophic population decline (Iontsev & Prokhorova, 2014).

In this research, we do not consider in detail the causes and consequences of the natural population decline and its components but analyze the role of migration in the population dynamics of Russia. That is, the goal of the study is to estimate what are the positive and negative implications of both internal and international population movement for the country as a whole and its regions. Since much research has already been made on this issue by many Russian scholars specializing in demography, economics, and migration studies, we concentrate on the most recent trends and developments that emerged over the past five years of 2015–2019. The research tasks include redefining the concept of "demographic development"; giving quantitative and qualitative assessment to the contemporary trends in the spatial population movement within Russia and across its borders, as well as their role in the population dynamics; providing some recommendations for further improvement in state migration policies in the interests of the demographic development of the country.

2 Materials and Methods

The very term "demographic development" seems to be intuitively comprehensible. Maybe that is the reason why it lacks any commonly agreed-upon definition either in official documents (e.g., Concept of the Demographic Policy of the Russian Federation through to 2025, 2007) or in academic literature. First of all, development can be both positive and negative, and in this sense, 'demographic development" might refer to any kind of change in the population and its structures over a certain period. However, in the academic discourse, we also use such notions as "population dynamics" or "demographic change". Making a brief content analysis via scientific citation databases or a specialized web search engine such as Google Scholar, we see that the former term is used mainly to describe mathematical models of the size and age composition of populations, while the latter term is used mostly to talk about demographic transition. At the same time "demographic development" has more to do with managing demographic processes (Rybakovsky & Khasaev, 2015) to achieve certain goals like the UN Sustainable Development Goals. That is why in this research we use the term "demographic development" implying sustainable demographic development. It also covers negative trends, such as the natural decline of the Russian population; however, by using this term we would like to emphasize that these processes should be managed as well to achieve social and economic growth for the society.

Second, population migration is a complex process that is linked not only to demographic trends but also to the economic situation, labour markets, living standards, infrastructure development, environmental change, culture, health, phycology, etc. But the only type of migration that has an impact on demographic processes is permanent or long-term resettlement (Iontsev, 2012). So, in this study, we analyze only long-term migration movements. According to the current statistical accounting in Russia, long-term migrants are those who registered at the place of stay for nine months or more.

The theoretical framework of our research is based upon the works of prominent Russian scholars specializing in population and migration studies; among them are Rybakovsky (2005), Zayonchkovskaya (2010), Vishnevskiy (2017), Ryazantsev (2019), Iontsev (2014), Topilin and Vorobyova (2020) and others.

The source of the empirical material we used was the official statistical data provided in open access by Rosstat (Russian Federal State Statistics Service). For their analysis, we applied standard statistical methods, including correlation and regression analysis using the least-squares method. Also, scenario analysis is applied to estimate the role of migration in the demographic development of the country under the forecasts provided by Rosstat, the United Nations Population Division, and the Institute for Demographic Research FCTAS RAS.

3 Results

The decline in the Russian population began in 1993, and since 1995, it became a stable tendency. The natural population decline was observed over a long period of 1992–2012. In 2013–2015, there was a short period of growth but since 2016, the number of deaths started to exceed the number of births again, and this tendency is unlikely to change in the foreseeable future due to the existing population age structure even if the total fertility rate increases. Net migration was positive ever since 1990, and over the past five years (2015–2019), it compensated 174% of the natural population decline giving the country over 1.13 million new residents (see Table 1).

However, according to the medium projections till 2035 produced by Rosstat, the migration inflow into the country is not going to increase but the natural population decline is going to continue; hence, the total population is going to shrink at a rapid pace. To cover the natural population losses, in 2035, the country would need to host over 113 thousand additional long-term migrants a year, compared to 2019.

But the total population decline by itself is not the only or the major problem of the demographic development of Russia. From the point of view of economics and social welfare, the decline in the workforce is even more problematic. Under the medium-case demographic scenario, the dependency ratio (of those not in the labour force and those in the labour force) is even likely to decrease from 0.77 in 2020 to 0.62 in 2035. But hereby, the share of the population under working age is going to decrease considerably from 18.7% in 2020 to 14.3% in 2023, which means that much fewer people will enter the labour force in the future (Table 2).

Since the majority of foreign citizens come to Russia either for work or study, the age distribution of migrants is quite favourable. For example, in 2019, 70% of them were of working age. Moreover, almost half of them (47% in 2019) are young people aged 15–35 (Fig. 1). Thus, we can conclude that the level of participation in the labour force among immigrants is high. That is why both scholars and officials

Table 1 Components of the population dynamics of Russia

Year	Total growth		Natural change		Net migration	
	People	‰	People	‰	People	‰
1990	608,665	4.1	332,865	2.2	456,062	1.9
1995	−168,299	−1.3	−840,005	−5.7	603,198	4.4
2000	−586,517	−4.1	−958,532	−6.6	241,755	2.5
2005	−564,464	−3.9	−846,559	−5.9	107,432	2.0
2010	31,931	0.2	−239,568	−1.7	158,078	1.9
2015	277,422	1.9	32,038	0.2	245,384	1.7
2019	−32,130	−0.2	−317,233	−2.2	285,103	1.9
2035*	−134,974	−0.9	−398,451	−2.8	263,477	1.9

Source Compiled by the authors based on the data by EMISS

Table 2 Russian population by major age groups as of January 1

Year	Under working age		Working-age		Overworking age	
	people	%	people	%	people	%
1990	36,100,900	24.4	83,943,392	56.8	27,620,789	18.7
1995	34,252,110	23.1	84,331,542	56.8	29,876,285	20.1
2000	29,579,756	20.1	87,172,272	59.3	30,138,100	20.5
2005	24,349,368	16.9	90,098,674	62.7	29,353,004	20.4
2010	23,086,254	16.2	88,561,160	62.0	31,186,088	21.8
2015	25,689,215	17.6	85,414,747	58.4	35,163,326	24.0
2020	27,442,371	18.7	82,677,671	56.3	36,628,548	25.0
2035	20,535,100	14.3	88,458,600	61.8	34,134,500	23.8
2050	22,490,884	15.4	66,888,473	45.8	56,519,299	38.7

Source Compiled by the authors based on the data by EMISS; National Demographic Report, 2020

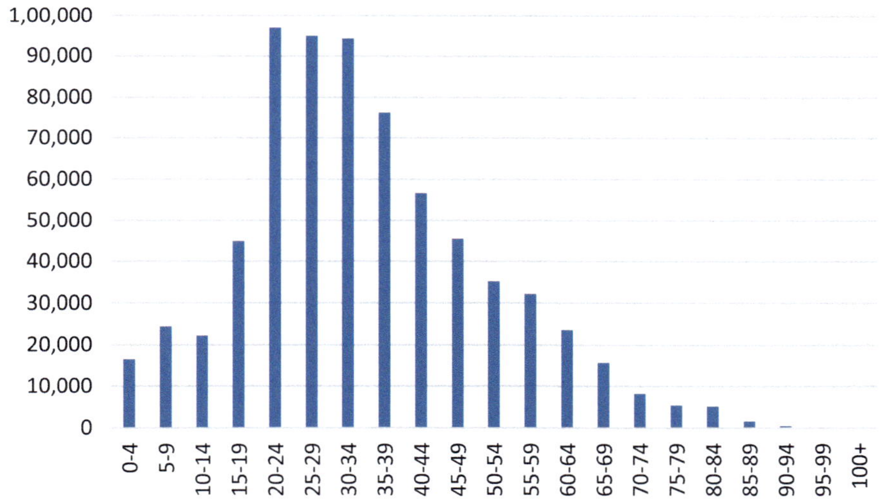

Fig. 1 Distribution of migrants who arrived in Russia in 2019 by age. *Source* Compiled by the authors based on the data by EMISS

tend to regard the issue of immigration in connection with the state of the labour market (Vorob'eva et al., 2020).

By our estimates, by 2050, to cover the workforce decline compared to nowadays, Russia will need to host more than 500 thousand long-term migrants each year, which is rough twice as many as in 2019. This number seems quite reliable since it corresponds with the estimates made earlier by other researchers (Vishnevsky et al., 2017; Ryazantsev et al., 2020). Meanwhile, according to a forecast provided by a research team from the Institute for Demographic Research FCTAS RAS, net migration in 2050 is most likely to reach only 313 thousand people, which will be enough to compensate for the natural population decline, but not the decline in the workforce. Nevertheless, if more measures are taken and the migration policies, as well as the economic situation in the country, are improved, net migration might reach 815 thousand people by 2050 (Ryazantsev et al., 2020). That would be enough to keep the size of the working-age population at its maximum reached in 2000. Meanwhile, according to a forecast by the UNPD, under this scenario, over ¼ of the Russian population would consist of post-1995 immigrants or their descendants (United Nations, 2001). However, the adaptation capacities of the society are a question yet, since many research works warn that a rapid increase in the share of migrants in a population might provoke ethnic conflicts and spur social tensions (Gadzhimuradova, 2020).

Another important effect of migration is population redistribution in which internal movements play a major role. Bringing people to the labour-deficient Eastern regions of the country has been an issue of the state demographic policy since the Soviet times of the mid-1970s when Komsomol was challenged to join the construction of the Baikal-Amur Mainline. However, after the collapse of the USSR, when the market economy mechanisms came into play, a massive population outflow from the Far East and Siberia started. Despite numerous development programs adopted and carried out by both federal and regional governmental bodies, this tendency persists and even worsens since 2005 (Manshin & Lukyanets, 2016). On the whole, the population of Russia drifts westwards, and only three Federal Districts (Central, Northwestern, and Southern) had a positive migration balance in 2015–2019 (Fig. 2).

Moreover, the migration gain is not distributed evenly across the territories of those regions, but there are centres of attraction within them where the newcomers tend to concentrate. Thus, in 2019, net internal migration to Moscow and the Moscow Region made up 132% of net internal migration to the Central Federal District, while St. Petersburg and the Leningrad Region obtained 139% of net internal migration to the Northwestern Federal District and the Krasnodar Territory concentrated 118% of net internal migration to the Southern Federal District (EMISS, n/a). These centres attract the population both from regions within and outside their Federal Districts and persist for so long that we cannot expect this tendency to change significantly in foreseeable future.

At the same time, in 2015–2019, all the Federal Districts of Russia had a positive migration balance in the population exchange with foreign countries. The largest migration gain was observed in the economically active Central Federal District; the Southern Federal District which is geographically closer to the states of Central Asia and the Caucasus, and also cheaper for migrants; and the industrial Volga, Ural, and Siberian Federal Districts (Fig. 3).

On the whole, the direction of migration flows is largely determined by economic reasons. According to the data provided by Rosstat, work became the reason for 10.4% of all resettlements in Russia in 2019; another 7.6% changed their place of residence to study; 10% migrated because of buying or inheriting housing; 13.8% resettled for unspecified personal reasons. Due to specifics of statistical accounting in Russia, the reason of 29.3% of all people's movements in 2019 was indicated as "a

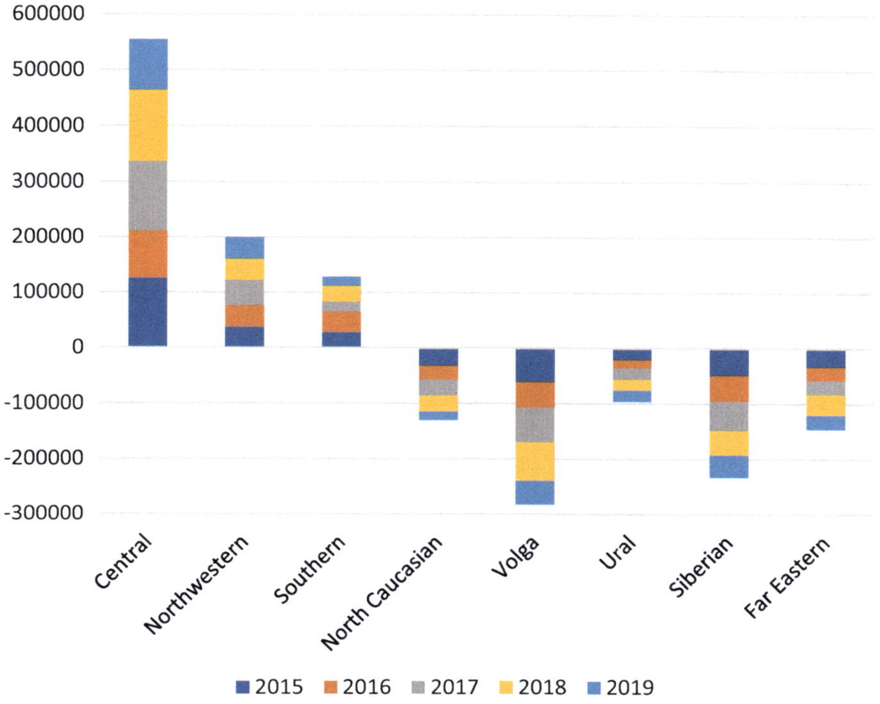

Fig. 2 Net internal migration in federal districts of Russia in 2015–2019. *Source* Compiled by the authors based on the data by EMISS

return after temporary absence"; however, in practice, it means just that the term of these people's registration at their place of stay expired, while it does not necessarily mean that they did return home.

This observation is supported by the results of a correlation-regression analysis, in which we used migration gain as a dependent variable and logarithmic values of gross regional product for all 85 subjects of the Russian Federation of 2019 as a regressor (Table 3). P-value (<0.0001) shows that there is a statistically significant correlation between net migration value and gross regional product (significance level of 0.01), i.e., the more economically developed the region is, the more people are willing to move there.

The data from Russian labour force sample surveys indirectly confirm the same tendency. More people come for work to the most economically developed regions: the "metropolitan" Central and Northwestern Federal Districts, as well as the "rich" oil and gas Ural Federal District (Fig. 4). Likewise, workers move within these regions: 96% of those who come to the Central Federal District go to Moscow and the Moscow Region; 82% of migrant workers in the Northwestern Federal District settle in St. Petersburg and the Leningrad Region; the Krasnodar Territory hosts 66% of those who moved for work to the Southern Federal District; and the oil and gas

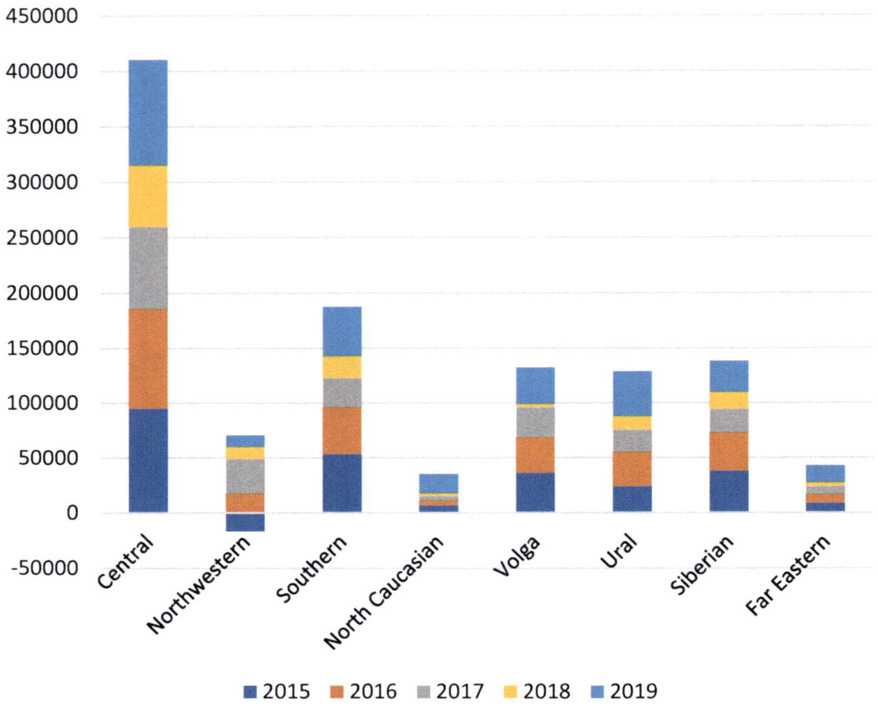

Fig. 3 Net international migration in federal districts of Russia in 2015–2019. *Source* Compiled by the authors based on the data by EMISS

Table 3 Regression analysis of net migration value and gross regional product

Model 1: least squares, observations used 1–85 dependent variable: net migration				
	Coefficient	Standard error	t-statistics	P-value
Const	– 108,997	27,464.6	– 3.969	0.0002
l_GRP	5,575.23	1,360.91	4.097	< 0.0001

Source Calculated by the authors based on the data by EMISS

Tumen Region gets 90% of the migrant labour force in the Ural Federal District (Labor Force Surveys, 2020). The large number of those who leave the subjects of the Central Federal District is explained by the workers' resettlement to Moscow. On the whole, their distribution is very much reminiscent of the spatial distribution of the total migration gain which was described above.

Although the majority of both internal and international migrants are of working age (in 2019 working-age population made up 70% of all migrants in the country), from a purely statistical point of view, there is no correlation between the share of the

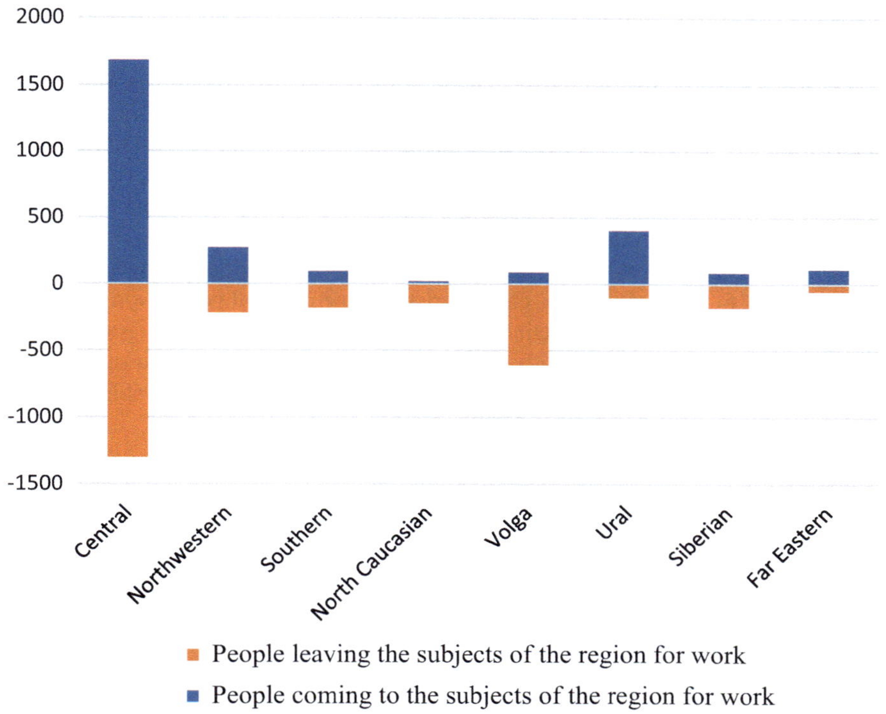

Fig. 4 Interregional labour migration of the employed population aged 15 and over by Russian regions in 2020 (thousand people) *Source* Compiled by the authors based on the data of the Labor Force Survey 2020

working-age population and the net migration rate (Table 4). For example, the share of the working-age population is relatively high in the Yamal-Nenets Autonomous District (63.6%), the Kamchatka Territory (60.2%), or the Republic of Daghestan (59.8%), which generally experienced a population outflow for the period under review. At the same time, it is below the national average in such regions as the Krasnodar Territory (54.7%), the Republic of Crimea (53.3%), or the Republic of

Table 4 Regression analysis of net migration rate and the share of the working-age population

Model 2: least squares, observations used 1–85
dependent variable: share of the working-age population

	Coefficient	Standard error	t-statistics	P-value
Const	0.553238	0.00234162	236.3	< 0.0001
Average net migration rate 2015–2019	– 2.15905e^{-05}	4.28418e^{-05}	– 0.5040	0.6156

Source Calculated by the authors based on the data by EMISS

Adygeya (54.3%,) which had a positive migration balance for the period of 2015–2019 (EMISS, n/a).

On the one hand, this might be explained by the fact that not all migrants stay in the regions where they came long enough to make up a significant stock and change the age structure of the local population. On the other hand, both Russian and international experience proves that the dependents (children, elderly relatives, etc.) often come with migrants, especially those who stay in the country or region permanently (Karachurina, 2007). Moreover, the population of migrants ages as well, while some research works also prove that the fertility rates in the migrant population tend to adjust to the fertility rates characteristic of the resident population (Kazenin, 2019). In addition, there are certain peculiarities in spatial trends of migration flows from the Russian north to the south: both statistical data and empirical research show that people often migrate to the labour surplus but climatically favourable southern regions as they retire (Lukyanets, 2020).

4 Conclusions and Recommendations

Migration is easy to manage compared to natural population movement, and it allows achieving the necessary goals of demographic development faster than, for example, fertility policies, which would bring economic results only in 20–25 years. The goals of migration policy might be easily adjusted to any changes in the demographic situation and labour market. However, migration cannot solve every problem of population ageing, decline, and redistribution. As rough estimates showed, to keep the share of the population aged 15–64 at its 2000's maximum Russia might have to host more than 800 thousand immigrants a year so that by 2050, post-1995 newcomers and their descendants might make up over ¼ of its population. Even if we leave aside the question of whether these migrants would be welcome in the society, it is difficult for Russia to attract such an unprecedented number of foreign citizens, since more and more people from traditional migrant-sending countries are now moving to work and live into more economically developed states. For example, migrants from Central Asia go to the Gulf countries or China (Ryazantsev & Rakhmonov, 2020), migrants from Eastern Europe go to the EU (Ryazantsev et al., 2020). The weakening Russian ruble makes the situation even less favourable.

While the trends in international migration flows are changing, internal migration flows in Russia demonstrate no considerable spatial reorientation. Moscow and the Moscow Region, St. Petersburg and the Leningrad Region, as well as the Krasnodar Territory, remain the overwhelming migrant attraction centres. On the whole, the population of the country continues to drift westwards. Both empirical data and the results of their statistical analysis show that the major reasons for that are economic ones. That is, the more developed a region is, the more migrants it attracts. This brings us to the conclusion that the problem of migration outflow from the strategically important regions of Siberia and the Far East cannot be solved by applying selective measures and short-term strategies. Comprehensive development programs aimed at improving the economic situation, infrastructural development, labour market conditions, and living standards are required. If the level of socio-economic development of different regions of the country becomes more even fewer people are likely to move for opportunities elsewhere.

And finally, we can conclude that so-called replacement migration might slightly hamper the population ageing, but in the long run, it is unable to stop the population change completely. First of all, the demographic transition works with the foreign-born population as well: permanent migrants are ageing too, and their fertility rates tend to decrease. To compensate for that, the country would need an ever-increasing number of immigrants, while both the adaptation capacity of the host society and the migration potential of sending countries are not unlimited, and the competition for migrants between hosting countries is growing. Moreover, there are also ethnic and cultural issues that were not regarded in this research. As for internal migration, it does not result in a considerable redistribution of the working-age population as well. Statistical analysis proved that there is no correlation between net migration rates and the shares of the working-age population in the regions of Russia.

Thus, to build a migration policy aimed at solving demographic problems and related economic issues, the most efficient ways to further improve such a policy appear to be: (a) improving legal and organizational mechanisms regulating immigration; (b) improving mechanisms of migrant adaptation and integration, which is an especially urgent task taking into account that the country needs more migrants to hamper the population decline; (c) reducing the disproportions in regional development and creating favourable conditions for internal migration. It should be noted that all of these objectives are declared in the current Concept of the State Migration Policy of the Russian Federation for 2019–2025 (2018). But unfortunately, the practice often goes in the opposite direction to the theory, as Russian specialists in migration studies have already noted many times (Ryazantsev, 2019).

Acknowledgements This paper has been supported by the RUDN University Strategic Academic Leadership Program and Grant of RUDN University No. 060119-0-000

References

Concept of the Demographic Policy of the Russian Federation through to 202.5 (2007) Approved by the decree of the president of the Russian Federation on 9 October 2007 No. 1351. http://docs.cntd.ru/document/436744671 Accessed 28 April 2021

Concept of the State Migration Policy of the Russian Federation for 2019–2025. (2018). Approved by the President of the Russian Federation on 31 October 2018. http://kremlin.ru/events/president/news/58986 Accessed 28 April 2021

EMISS (n/a). Unified interdepartmental information and statistical system. https://www.fedstat.ru Accessed 28 April 2021

Gadzhimuradova, G. I. (2020). Muslims in Europe: Identity crisis between european and islamic values. *Balkan Journal of Philosophy, 12*(2), 133–138. https://doi.org/10.5840/bjp202012216

Iontsev, V. A. (2012). On the formation and development of an independent academic discipline of migration studies. In I. A. Danilova et al. (Eds.), *Population Distribution*, Memoriam B. S. Khorev (pp. 312–313).

Iontsev, V. A., & Prokhorova, Y. A. (2014). International migration of population in the context of the demographic theory. *Living Standards of the Population in the Regions of Russia, 3*, 83–91. https://doi.org/10.12737/5647

Karachurina, L. B. (2007). The role of migration in demographic processes. In J. A. Zayonchkovskaya, I. N. Molodikova, & V. I. Mukomel (Eds.), *Methodology and methods of studying migration processes* (pp. 370). ISBN: 5–86103–046–4

Kazenin, K. I. (2019). Demographic processes among foreign labor migrants in the Russian Federation. SSRN 3361019. https://ssrn.com/abstract=3361019 Accessed 15 March 2021. https://doi.org/10.2139/ssrn.3361019

Khalturina, D. A., & Korotaev, A. V. (2006). Russian cross: Factors, mechanisms and ways of overcoming the demographic crisis in Russia (pp. 56). ISBN 978–5–397–03414–2

Labor Force Surveys (2020). Federal state statistics service. https://rosstat.gov.ru/storage/mediabank/oOqckdXs/ORS-2020-all.rar Accessed 28 April 2021

Lukyanets, A. S. (2020). Push factors of migration from the far north of Russia. In Migration processes in the Asia-Pacific region: Political. In *Socio-Economic and Demographic Effects Proceedings of the International Scientific and Practical Conference in Moscow*, Russia, 05–06 November 2020, (pp. 78–93). https://doi.org/10.38006/907345-84-3.2020.78.93

Manshin, R. V., & Lukyanets, A. S. (2016). Problems of assessment of possible sources of compensation of labor resources in the border regions of Siberia and the far east. *Journal of Economy and Entrepreneurship, 12–2*, 205–210.

Ryazantsev, S. V. & Rakhmonov, A. K. (2020). Labor force recruitment to the OECD and middle eastern Countries in The Republic of Tajikistan: Trends, mechanisms, consequences. *Central Asia and the Caucasus, 14046091*(4), 96–111. https://doi.org/10.37178/ca-c.20.4.10

Ryazantsev, S. V. (2019). Modern migration policy of Russia: Challenges and approaches to improvement. *Sotsiologicheskie Issledovaniya, 9*, 117–126. https://doi.org/10.31857/S013216250006666-5

Ryazantsev, S. V., Arkhangelsky, V. N., Vorobyova, O. D. et al. (2020). Demographic development of Russia: Trends, forecasts, and measures. national demographic report (pp. 156). https://doi.org/10.25629/HC.2020.13.01

Ryazantsev, S. V., Rusu, S. V. & Medved, V. A. (2020). Factors of migration to the EU Countries during the 2015–2016 crisis. *Scientific Review, Series 1: Economics and Law, 4*, 19–28. https://doi.org/10.26653/2076-4650-2020-4-02

Rybakovsky, L. L., & Khasaev, G. R. (2015). The strategy of the demographic development of Russia: concept and content. *Population, 2*(68), 52–63.

Rybakovsky, L. L. (2005). The demographic future of russia and migration processes. *Sotsiologicheskie Issledovaniya [social Surveys], 3*, 71–81.

United Nations. (2001). Replacement migration: Is it a solution to declining and ageing populations?. UN Population Division, Department of Economic and Social Affairs, ESA/P/WP.160, March 21.

Vishnevsky, A. G., Denisenko, M. B., Zayonchkovskaya, J. A. & Mkrtchyan, N. V. (2017). Internal migration challenges. *Demoscope Weekly*, 753–754. http://demoscope.ru/weekly/2017/0753/tema01.php Accessed 17 April 2021

Vorob'eva, O. D., Topilin, A. V., & Alikova, A. S. (2020). Sociodemographic consequences of internal migration of labor resources. *Herald of the Russian Academy of Sciences, 90*(6), 765–771. https://doi.org/10.1134/S1019331620060337

Zayonchkovskaya, J. A. et al. (2010). Migration and demographic crisis in Russia. In Series: Migration barometer in the Russian Federation (pp. 112).

Russian Regions: Assessment of Factors of Growth in the Living Standard and Well-Being of the Population

Natalia I. Kuzmenko, Galina K. Lapushinskaya, Ekaterina V. Lisova, Murat A. Tleptserukov, and Evgeniya K. Karpunina

Abstract The research is aimed at developing the author's approach to conducting a comprehensive analysis of long-term changes in the standard of living and the population well-being in the Russian regions, as well as determining the directions of socio-economic policy to overcome existing regional imbalances. The authors formed the rating of the Federal Districts of Russia based on the research of the dynamics of the indicator "change in the gross regional product per capita" in 2010–2019. The analysis of the changes in the population, as well as the indicators of fertility and marriage, allowed the authors to rank the Russian regions by the population status. The authors assessed the changes in the diet of the population of various regions and revealed positive long-term dynamics of changes in the structure of food consumption in Russia. The authors analyzed the improvement of housing conditions and ownership of long-term consumer goods by the population. The author's approach made it possible to present in the article a rating of Russian regions on changes in the standard of living in 2010–2019. The authors identified contradictions between the factors of population welfare growth and identified the directions of state regulation to ensure positive changes.

Keywords Federal district · Region · Welfare · Standard of living · Population · Consumption · Birth rate

N. I. Kuzmenko
Voronezh State Pedagogical University, Voronezh, Russia

G. K. Lapushinskaya
Tver State University, Tver, Russia
e-mail: laboratory-region@yandex.ru

E. V. Lisova
Autonomous Non-profit Organization of Higher Education "Institute of Business Career", Moscow, Russia

M. A. Tleptserukov
Kuban State Technological University, Krasnodar, Russia
e-mail: mat4791@yandex.ru

E. K. Karpunina (✉)
G.R. Derzhavin Tambov State University, Tambov, Russia

E. G. Popkova and B. S. Sergi (eds.), *Geo-Economy of the Future*,
https://doi.org/10.1007/978-3-030-92303-7_66

JEL Codes · R13 · R58

1 Introduction

The problem of significant differentiation of the population in terms of well-being and standard of living in Russia remains relevant for a long time (Belov & Stepanova, 2017), (Karpunina et al., 2020), (Suslova & Alyoshina, 2015). First of all, this concerns the income of citizens, property and consumption (Buneeva, 2012), (Galieva et al., 2020). Inaction in solving this problem can cause a decrease in the economic activity of the population, the under consumption and underinvestment, deterioration of the socio-psychological health of members of society (Belekhova & Rossoshansky, 2015).

Meanwhile, the change in the standard of living of the population and its well-being at the regional level is a very dynamic process that is determined by many factors. Identifying patterns of changes in the well-being of the population through the prism of identifying new growth factors, expands the possibilities of managing the process of changing the standard of living of the population of the regions. First of all, this is important for establishing "bottlenecks" in the implemented regional socio-economic policy and making informed management decisions aimed at overcoming existing problems.

2 Theoretical and Methodological Foundations of the Study

The standard of living and the population's well-being are socio-economic categories that reflect the satisfaction of the material and spiritual needs of citizens (Belyaeva, 2009), (Bobkov & Gulyugina, 2012), (Ovcharova, 2014), (Zherebin & Romanov, 2002). Numerous indicators and methods have been developed to assess the standard of living and well-being of the population. However, in practice, experts most often use two indicators—real per capita income and average life expectancy (Belov & Stepanova, 2017), (Gorshenina, 2016), (Talalushkina, 2013). Recently, new tools have been appearing to improve the comfort of a person's life. For example, digital technologies and devices serve this purpose. This further actualizes the need for a comprehensive analysis of long-term regional trends and the research of the relationship between factors of life status and the well-being of a person.

The purpose of the research is to develop an author's approach to conducting a comprehensive analysis of long-term changes in the standard of living and the population well-being in the Russian regions, to identify the factors that determine them, as well as to define the directions of state policy aimed at eliminating existing imbalances.

Research objectives: (1) to determine the nature of the mutual influence of living standards and GRP per capita; (2) to analyze the changes in the population, fertility, and marriage rates in the regions; (3) to assess changes in the diet of the people and the dynamics of the consumption of "proper" food; (4) to analyze changes in the housing conditions of the population and ownership of long-term consumption items; (5) to form a rating of Russian regions reflecting changes in the standard of living of the population; (6) to determine prospects of state regulation for improving the standard of living and the population well-being in the Russian regions.

Research methods: the theoretical analysis, the comparative analysis, the systematization, the economic and statistical analysis, a systematic approach.

3 Results

Let's start the research by analyzing the dynamics of the gross regional product per capita in 2010–2019. This indicator is most often used to assess the standard of living and the population's well-being (Table 1).

Table 1 shows the rating of the federal districts of Russia according to the indicator "change in the gross regional product per capita" in 2019 compared to 2010. In particular, in the North-Western Federal District, GRP per capita increased 1.6 times

Table 1 Dynamics of the gross regional product per capita, thousand rubles, 2010–2019

Federal district/indicator	Gross regional product per capita in 2010, thousand rubles	Gross regional product per capita in 2019, thousand rubles	Change in gross regional product per capita, % (2019-2010)/ 2010 * 100	FD rating by changes in the gross regional product per capita, 2010–2019
North-Western federal district	289,61	752,85	160,0	1
Far Eastern federal district	287,69	730,11	153,8	2
Ural federal district	423.5	1070,60	152.8	3
Volga federal district	190,72	480,46	151.9	4
North Caucasus federal district	94.92	232,01	144.4	5
Siberian federal district	222,85	535,32	140.2	6
Central federal district	350.2	835,86	138.7	7
Southern federal district	168,77	399,44	136.7	8

Source Compiled by the authors based on (3DPRO, 2020)

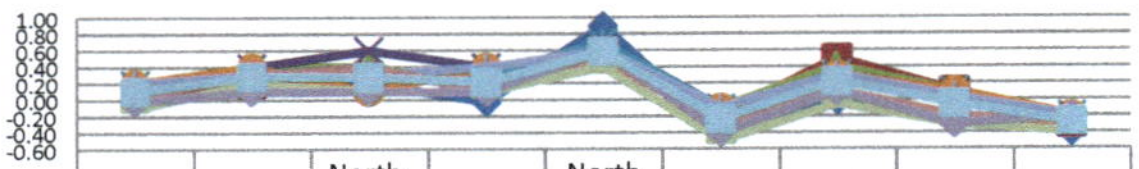

	Russian Federation	Central Federal District	North-Western Federal District	South federal district	North Caucasus Federal District	Volga Federal District	Uralsky District federal district	Siberian Federal District	Far Eastern Federal District
2010	0.02	0.3	0.2	-0.02	0.9	-0.4	-0.01	-0.2	-0.4
2011	0.1	0.2	0.3	0.2	0.6	-0.2	0.5	0.1	-0.3
2012	0.2	0.4	0.4	0.2	0.5	-0.1	0.4	0.1	-0.2
2013	0.2	0.4	0.6	0.4	0.5	-0.1	0.3	0.1	-0.3
2014	0.2	0.3	0.3	0.3	0.7	-0.1	0.3	0.1	-0.2
2015	0.2	0.4	0.1	0.4	0.6	-0.1	0.3	0.1	-0.2
2016	0.2	0.3	0.3	0.4	0.6	-0.1	0.3	0.03	-0.2
2017	0.05	0.3	0.4	0.1	0.5	-0.3	0.1	-0.2	-0.3
2018	-0.1	0.2	0.1	0.1	0.4	-0.5	-0.05	-0.3	-0.4
2019	-0.02	0.1	0.1	0.1	0.6	-0.4	0.1	-0.3	-0.2
Weighted average change in the population size (annual growth, in percent) for the period from 2010 to 2019	0.11	0.29	0.28	0.22	0.59	-0.23	0.22	-0.05	-0.27

Fig. 1 Population change in the Russian regions (annual growth; as a percentage), 2010–2019. *Source* Compiled by the authors based on (Rosstat (2020b). Statistical collection "Regions of Russia, socio-economic indicators, 2020"), (Rosstat (2020c). The report "The socio-economic situation of Russia")

(1st place in the rating). In the Southern Federal District, there was the smallest growth (8th place in the rating).

Assessing the standard of living and well-being requires studying the dynamics of population change, which is the starting point for the implementation of state social policy. The analysis of the change in the population of Russian regions in 2010–2019 is shown in Fig. 1.

The presented data allow us to identify the region with the highest weighted average population growth per year—the North Caucasus Federal District (0.59%), which occupies 5th place in the rating of GRP growth per capita in the period 2010–2019 (Table 1). It is followed by the Central Federal District, which is on the 7th position in the rating in terms of GRP growth per capita and has a weighted average increase of 0.29%. The North-Western Federal District during 2010–2019 had a weighted average population growth of 0.28% per year but ranked 1st in terms of GRP growth per capita. The Southern Federal District, which has an average weighted average population growth of 0.22%, is the second among the federal districts of Russia in the rating for GRP growth per capita. The Far Eastern Federal District, which has the maximum weighted average population decline (-0.27%), nevertheless, is on the 2nd place in the rating of regions in terms of GRP growth per capita.

Next, we will analyze the birth rate and growth rates of people entering into marriage. The dynamics of the total birth rate of the population in federal districts

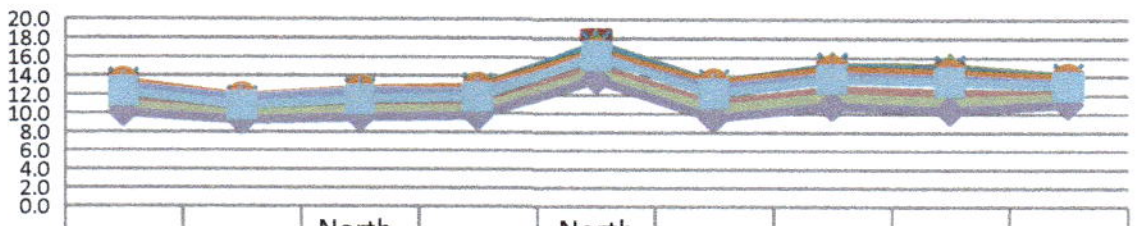

	Russian Federation	Central Federal District	North-Western Federal District	South federal district	North Caucasus Federal District	Volga Federal District	Ural Federal District	Siberian Federal District	Far Eastern Federal District
2010	12.5	10.7	11.4	11.8	17.3	12.4	14.1	14.1	13.2
2011	12.6	10.7	11.4	11.8	17.5	12.4	14.2	14.1	13.2
2012	13.3	11.4	12.2	12.6	17.4	13.3	15.1	15.0	14.0
2013	13.2	11.3	12.2	12.5	17.2	13.3	15.0	14.8	13.9
2014	13.3	11.4	12.3	12.8	17.3	13.3	15.1	14.7	14.0
2015	13.3	11.7	12.5	12.8	16.6	13.3	14.8	14.4	13.9
2016	12.9	11.6	12.4	12.3	15.9	12.8	14.1	13.8	13.3
2017	11.5	10.4	11.1	11.1	15.0	11.1	12.6	12.3	12.1
2018	10.9	9.9	10.4	10.5	14.4	10.6	11.9	11.4	11.9
2019	10.1	9.3	9.6	9.8	13.7	9.6	10.9	10.4	11.1
Weighted average change in the total birth rate per 1000 people of the population for the period from 2010 to 2019	12.4	10.8	11.6	11.8	16.2	12.2	13.8	13.5	13.1

Fig. 2 The birth rate of the population in the regions of Russia, 2010–2019. *Source* Compiled by the authors based on (Rosstat (2020b). Statistical collection "Regions of Russia, socio-economic indicators, 2020"), (Rosstat (2020c). The report "The socio-economic situation of Russia")

(the number of births per 1000 people of the population) for the period from 2010 to 2019 is shown in Fig. 2.

The North Caucasus Federal District, which ranks 5th in the rating for GRP growth per capita, in the period 2010–2019 has the maximum weighted average change in the birth rate—16.2%. The minimum weighted average change in the birth rate (10.8%) has the Central Federal District, this region occupies the 7th position in the rating of GRP growth per capita.

An analysis of the overall level of marriage in the period from 2010 to 2019 (that is, the ratio of the number of marriages concluded during a calendar year to the average annual population of the region) is shown in Fig. 3.

The North Caucasus Federal District with the highest birth rate has a minimum weighted average change in the total marriage rate per 1000 people: 6.4%. The Far Eastern Federal District provided the maximum weighted average change in the marriage rate of 8.5%, while it occupies the 2nd place in the rating in terms of GRP growth per capita and has a high birth rate (15.3%) (Table 2).

The aggregate weighted average rating by the status of the population is calculated as the weighted average sum of changes in the population, changes in the birth rate and marriage rate for the period 2010–2019. The calculation allowed us to rank the federal districts. The 1st place in the rating for this indicator was taken by the North Caucasus Federal District, and the last—the Ural Federal District.

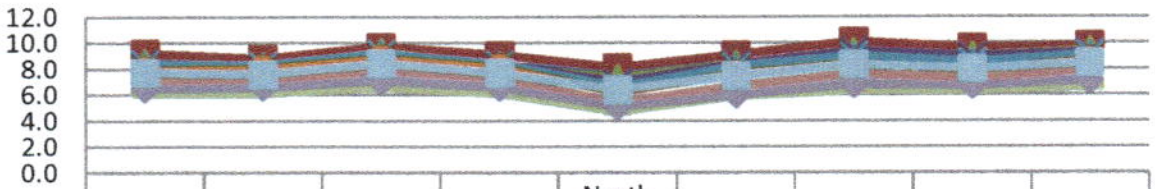

	Russian Federation	Central Federal District	North-Western Federal District	Southern Federal District	North Caucasus Federal District	Volga Federal District	Ural Federal District	Siberian Federal District	Far Eastern Federal District
2010	8.5	8.1	9.0	8.2	8.1	8.3	9.4	8.9	9.3
2011	9.2	8.8	9.7	9.1	8.1	9.1	10.1	9.6	9.8
2012	8.5	8.0	9.3	8.2	7.3	8.4	9.3	9.0	9.4
2013	8.5	8.2	9.2	8.4	7.0	8.4	9.4	9.0	9.3
2014	8.4	8.3	9.2	8.4	6.6	8.2	9.0	8.5	9.1
2015	7.9	8.0	8.8	8.1	6.2	7.6	8.4	7.9	8.6
2016	6.7	6.7	7.6	6.7	5.6	6.3	7.1	6.8	7.6
2017	7.1	7.2	7.9	7.3	5.4	6.7	7.6	7.3	7.9
2018	6.1	6.1	6.6	6.1	4.7	5.9	6.3	6.3	6.6
2019	6.5	6.6	7.1	6.6	5.1	6.0	6.8	6.7	7.1
Weighted average change in the total marriage rate per 1000 people of the population for the period from 2010 to 2019	7.7	7.6	8.4	7.7	6.4	7.5	8.3	8.0	8.5

Fig. 3 The total marriage rate in the regions of Russia, 2010–2019. *Source* Compiled by the authors based on (Rosstat (2020b). Statistical collection "Regions of Russia, socio-economic indicators, 2020"), (Rosstat (2020c). The report "The socio-economic situation of Russia")

The next aspect of research, the standard of living and the population well-being is the problem of improving the diet of citizens. The increase in the consumption of "proper" food is evidence of increased attention to the issues of a healthy lifestyle and the well-being of a person. Certainly, there are significant regional differences in the daily diet and the energy value of the consumed products, the volume and structure of consumption of basic food products, the share of food purchase costs in the total consumption expenditures of households in the region. The lower socially acceptable limit of food consumption is determined by the indicators of the consumer basket. However, the standards of the consumer basket are significantly lower than the rational standards recommended by the Ministry of Health of Russia (except for fish, potatoes, bread, and bakery products) (Buneeva, 2015), (Rosstat (2020a). Statistical bulletin "Food consumption in households-2019"), (Suslova & Alyoshina, 2015).

Let's analyze the changes in the structure of consumption of basic food products per capita in the Russian regions in 2010–2019 (Rosstat (2020a). Statistical bulletin "Food consumption in households-2019") (Fig. 4).

In Russia, for the period 2010–2019, the consumption of meat and semi-finished meat products increased by 10.1%, vegetables by 10.2%, eggs by 5.6%, vegetable oil by 4.5%, only the consumption of sugar has not changed. The consumption of bread products also decreased by 3.3%, milk by 4.5%, potatoes by 6.3%. In other words, the structure of food consumption in the Russian Federation improved during the study period. It is possible to notice a decrease in the consumption of milk and dairy

Table 2 Rating of Russian regions by population status, 2010–2019

Indicator/federal district	Russian federation	Central federal district	North-Western federal district	Southern federal district	North Caucasus federal district	Volga federal district	Ural federal district	Siberian federal district	Far Eastern federal district
Weighted average change in the population size (annual growth, in percent) for the period from 2010 to 2019	0.105	0.29	0.28	0.218	0.59	−0.23	0.224	−0.047	−0.27
FD rating based on the weighted average change in population (annual growth, in percent) for the period from 2010 to 2019		2	3	4	1	7	5	6	8
Weighted average change in the total birth rate per 1000 people of the population for the period from 2010 to 2019	12.36	10.84	11.55	11.8	16.23	12.21	13.78	13.5	13.06

(continued)

Table 2 (continued)

Indicator/federal district	Russian federation	Central federal district	North-Western federal district	Southern federal district	North Caucasus federal district	Volga federal district	Ural federal district	Siberian federal district	Far Eastern federal district
FD rating based on the weighted average change in the total birth rate per 1000 people of the population for the period from 2010 to 2019		8	7	6	1	5	2	3	4
Weighted average change in the total marriage rate per 1000 people of the population for the period from 2010 to 2019	7.74	7.6	8.44	7.71	6.41	7.49	8.34	8	8.47
FD rating based on the weighted average change in the total marriage rate per 1000 people of the population for the period from 2010 to 2019		6	2	5	8	7	3	4	1
Weighted average rating by papulation (population change, birth rate and marriage rate)		5.33	4.00	5.00	3.33	6.33	3.33	4.33	4.33

Table 2 (continued)

Indicator/federal district	Russian federation	Central federal district	North-Western federal district	Southern federal district	North Caucasus federal district	Volga federal district	Ural federal district	Siberian federal district	Far Eastern federal district
Aggregate weighted average rating by population (population change, birth rate and marriage rate)		6	3	7	1	8	2	4	5

Source Author's calculations

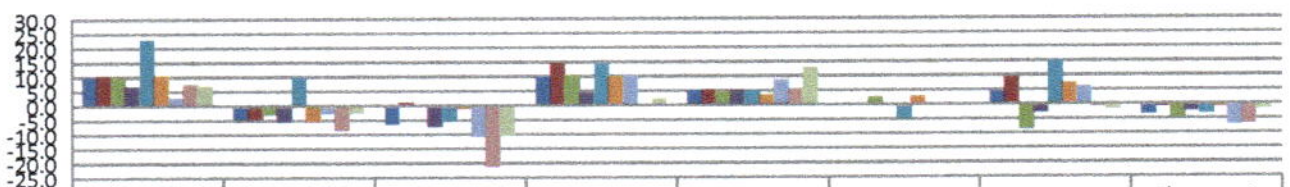

	Change in consumption of meat and meat products per year per capita, in kg, in 2019 to 2010,%	Change in consumption of milk and dairy products per capita, per year, in kg, in 2019 to 2010,%	Change in potato consumption per capita, per year, in kg, in 2019 to 2010,%	Change in consumption of vegetables and food melons per capita, per year, in kg, in 2019 to 2010,%	Change in egg consumption per capita, per year, in units, in 2019 to 2010, %	Change in sugar consumption per capita, per year, kg, in 2019 to 2010,%	Change in vegetable oil consumption per capita, per year, kg, in 2019 to 2010, %	Change in the consumption of bread products per capita, per year, kg, in 2019 to 2010,%
Russian Federation	10.1	-4.5	-6.3	10.2	5.6	0.0	4.5	-3.3
Central Federal District	10.5	-4.4	1.1	14.6	5.5	0.0	9.2	-0.8
North-Western Federal District	10.1	-3.0	0.0	10.6	4.6	2.6	-8.5	-4.8
Southern Federal District	6.9	-5.6	-7.2	4.4	5.5	0.0	-2.6	-2.4
North Caucasus Federal District	23.1	10.1	-5.3	14.8	5.5	-4.5	15.3	-3.1
Volga Federal District	10.6	-5.7	-1.0	10.0	3.6	2.8	7.2	-0.9
Ural Federal District	3.0	-2.4	-10.7	10.7	8.9	0.0	6.3	-7.1
Siberian Federal District	7.5	-8.5	-21.2	0.0	5.7	0.0	-0.8	-6.9
Far Eastern Federal District	7.0	-2.5	-10.5	2.2	12.9	0.0	-1.6	-1.7

Fig. 4 The consumption of basic food products per capita in the regions of Russia, 2010–2019. *Source* Compiled by the authors based on (Rosstat (2020a). Statistical bulletin "Food consumption in households-2019"), (Rosstat (2020b). Statistical collection "Regions of Russia, socio-economic indicators, 2020")

products, as well as a high level of sugar consumption in the diet of the population in the regions. This reflects a negative trend in changing the structure of consumption.

We will analyze the changes in the consumption of "proper" products (meat, milk, and vegetables) by the population of the regions of Russia and calculate the overall balance of changes in their consumption (Table 3).

The assessment of the overall balance of consumption of "proper" products by the population of the Russian regions serves as an indicator of improving the lives of people in the regions and allows us to assert the growth of their well-being (Kostrova, 2018), (Kostrova et al., 2020). In the whole country, the consumption of the proper products increased by 15.9%. In 2010–2019, the maximum increase in the consumption of "proper" products occurred in the North Caucasus (+48%), Central (+21%), and North-Western (+18%) Federal Districts. By contrast, in the Siberian Federal District, the consumption of "proper" products decreased by 1%.

The increase in the standard of living and well-being of the population is evidenced by the improvement of the population's housing conditions and the possession of long-term consumer goods (in particular, cars and vehicles) (Galieva et al, 2020), (Karpunina et al., 2021). We will form a rating of Russian regions by the population's ownership of long-term maintenance items (housing and cars) for the period 2010–2019 (Table 4).

In 2010–2019 in Russia, the indicator of the total area of residential premises per inhabitant increased by an average of 16.4%. In the North Caucasus Federal District, the increase was the maximum (20.4%), and in the Far Eastern Federal

Table 3 Change in the consumption of "proper" products (meat, milk, and vegetables) by the population in the regions of Russia %, 2010–2019

Indicator/federal district	Russian federation	Central federal district	North-Western federal district	Southern federal district	North Caucasus federal district	Volga federal district	Ural federal district	Siberian federal district
Consumption of meat and meat products (including category II offal and raw fat) per capita in 2010, kg	69	76	69	72	52	66	66	67
Consumption of meat and meat products (including category II offal and raw fat) per capita in 2010, kg	76	84	76	77	64	73	63	72
Change in consumption of meat and meat products per year per capita, in kg, in 2019 to 2010,%	10.14	10.53	10.14	6.94	23.08	10.61	3.03	7.46
Consumption of milk and daily products per capita, in 2010, kg	245	229	270	231	217	283	211	260
Consumption of milk and dairy products per capita, in 2019, kg	234	219	262	218	239	267	206	238

(continued)

Table 3 (continued)

Indicator/federal district	Russian federation	Central federal district	North-Western federal district	Southern federal district	North Caucasus federal district	Volga federal district	Ural federal district	Siberian federal district
Change in consumption of milk and dairy products per capita, per year, in kg, in 2019 to 2010,%	-4.49	4.37	-2.95	5.63	10.14	5.65	2.37	8.46
Consumption of vegetables and food melons per capita in 2010, kg	98	89	85	137	149	90	84	92
Consumption of vegetables and food melons per capita in 2019, kg	108	102	94	143	171	99	93	92
Change in consumption of vegetables and food melons per capita, per year, in kg, in 2019 to 2010,%	10.20	14.61	10.59	4.38	14.77	10.00	10.71	0.00
Overall balance of changes in consumption of "proper" foods in 2019 to 2010,%	15.86	20.77	17.77	5.70	47.98	14.95	11.37	-1.00

(continued)

Table 3 (continued)

Indicator/federal district	Russian federation	Central federal district	North-Western federal district	Southern federal district	North Caucasus federal district	Volga federal district	Ural federal district	Siberian federal district
Summary rating based on the level of consumption of "proper" products		2	3	7	1	4	5	8

Source Compiled by the authors based on (Rosstat (2020a). Statistical bulletin "Food consumption in households-2019"), (Rosstat (2020b). Statistical collection "Regions of Russia, socio-economic indicators, 2020")

Table 4 Rating of Russian regions by population ownership of long-term maintenance items (housing and cars), 2010–2019

Indicator/federal district	Russian federation	Central federal district	North-Western federal district	Southern federal district	North Caucasus federal district	Volga federal district	Ural federal district	Siberian federal district	Far Eastem federal district
Total area of residential premises per inhabitant on average, at the end of the year, square meters, 2010	22.6	24	24.7	21.9	18.1	22.7	22.3	21.7	21.2
Total area of residential premises per inhabitant on average, at the end of the year, square meters, 2019	26.3	27.5	28.1	25.4	21.8	27.3	25.9	25.1	23.7
Change in the total area of residential premises per average resident at the end of the year, square meters, (2019–2010) /2010*100,%	16.37	14.58	13.77	15.98	20.44	20.26	16.14	15.67	11.79

(continued)

Table 4 (continued)

Indicator/federal district	Russian federation	Central federal district	North-Western federal district	Southern federal district	North Caucasus federal district	Volga federal district	Ural federal district	Siberian federal district	Far Eastem federal district
FD rating by changes in the total area of residential premises per inhabitant on average for the period from 2010 to 2019		6	7	4	1	2	3	5	8
Number of own passenger cars per 1000 population, at the end of the year, units, 2010	228.4	249.2	253.7	230.2	154.7	209.3	254.1	217.3	225.7
Number of own passenger cars per 1000 people of the population, at the end of the year, units, 2019	315.5	320.8	324.5	321.1	230.2	323.1	353.9	296.6	320.5

(continued)

Table 4 (continued)

Indicator/federal district	Russian federation	Central federal district	North-Western federal district	Southern federal district	North Caucasus federal district	Volga federal district	Ural federal district	Siberian federal district	Far Eastem federal district
Change in the number of own passenger cars per 1000 people of the population, at the end of the year, units, (2019–2010)/2010*100,%	38.13	28.73	27.91	39.49	48.80	54.37	39.28	36.49	42.00
FD rating on changes in the number of own passenger cars per 1000 population for the period from 2010 to 2019		7	8	4	2	1	5	6	3
Aggregated weighted average rating for possessing of household items (housing and cars) for 2010 to 2019		6.5	7.5	4	1.5	1.5	4	5.5	5.5

(continued)

Table 4 (continued)

Indicator/federal district	Russian federation	Central federal district	North-Western federal district	Southern federal district	North Caucasus federal district	Volga federal district	Ural federal district	Siberian federal district	Far Eastem federal district
FD rating on possession of items of collective maintenance (housing and cars) for the period from 2010 to 2019		7	8	4	2	1	3	5	6

Source Author's calculations

District, only 11.8%. The number of own passenger cars per 1000 people of the Russian population in 2010–2019 increased by 38.1%. The maximum increase in this indicator was recorded in the Volga Federal District (54.4%), the minimum increase by 27.9% occurred in the North-Western Federal District. The North Caucasus and Volga Federal Districts are also leading in the rating for the possession of durable goods (housing and cars). On the contrary, the minimum value of the studied indicator is typical for the North-Western Federal District.

We will form a total rating of the regions of Russia on changes in the standard of living in 2010–2019 based on the summation of the results of the above analysis (Table 5).

The maximum value of the indicator in the weighted average rating of Russian regions for changes in living standards in 2010–2019 is the North Caucasus Federal District, and the minimum value of the indicator is reached by the Southern Federal District.

Table 5 Rating of Russian regions on changes in living standards in 2010–2019

Federal district/ indicator	FD rating by population (population growth, birth rate, and m am age rate)	FD rating by the level of consumption of "proper" products	FD rating on possession of collective maintenance items (housing and cars) for 2010–2019	FD rating by changes in the gross regional pro duct per capita, 2010–2019	Weighted average FD rating on changes in the standard of living 2010–2019 (the sum of 2–5)
1	2	3	4	5	6
North Caucasus federal district	1	1	2	5	2.25
Ural federal district	2	5	3	3	3.25
North-Western federal district	3	3	8	1	3.75
Volga federal district	8	4	1	4	4.25
Far Eastern federal district	5	6	6	2	4.75
Central federal district	6	2	7	7	5.50
Siberian federal district	4	8	5	6	5.75
Southern federal district	7	7	4	8	6.50

Source Author's calculations

Table 6 Guidelines of the state policy on improving the standard of living and the population's well-being in the Russian regions

Federal district/ guidelines of the state policy	The region's place in the rating on changes in the standard of living and the population's well-being in 2010–2019	Increase in GRP per capita	Improving the structure of food consumption by the population	Improving the living conditions of the population and increasing the ownership of long-term maintenance items	Improving the demo graphic situation	Improving the diet of the population
North Caucasus federal district	1	+				
Ural federal district	2		+			
North-Western federal district	3			+		
Volga federal district	4				+	
Far Eastern federal district	5			+	+	+
Central federal district	6	+		+	+	
Siberian federal district	7	+		+		+
Southern federal district	8	+			+	+

Source Compiled by the authors

The proposed approach allows us to identify positive and negative aspects of changes in the standard of living and the population's well-being for taking measures to optimize the state and social impact (Table 6).

4 Conclusion

The authors studied the relationship between the standard of living and the population well-being with the value of the gross regional product per capita, the nature of consumption, including "proper" products, with the degree of readiness of people to marrying and having children, as well as with the purchase of long-term consumer goods and improvement of housing conditions. The proposed approach to the study of the standard of living and the population well-being in the regions allowed the authors to rank the Federal Districts of Russia by changing the "basic" indicators (GRP per capita), as well as to link this indicator with "new" indicators (population size, birth rate, marriage, food consumption and possession of long-term items). The author's approach to the study of factors affecting the standard of living and the population well-being made it possible to assess the degree of social maturity of state and public institutions in the federal districts of Russia in 2010–2019, as well as to determine the guidelines of optimization the state and social impact.

References

3DPRO. (2020). Macroeconomic indicators. https://3dpro.info/panel/75f3ea67-d71b-4994-a2a7-b452eb26df48 (Data Accessed 12 June 2021)

Belekhova, G., & Rossoshansky, A. (2015). The standard of living in the estimates of the population. *Problems of Territory Development, 5*(79), 77–96.

Belov, V. (2012). Factor analysis of Russia's economic growth at the present stage. *Scientific works of the North-Western Institute of Management, 3, 2*(6), 8–26.

Belov, V., & Stepanova, T. (2017). Situational analysis of the standard of living of the population of the Russian Federation in the crisis period (2014–2015). In *Proceedings of the IX International Conference. North-Western Institute of Management of the RANEPA under the President of the Russian Federation* (pp. 26–29).

Belyaeva, L. (2009). The level and quality of life. Problems of measurement and interpretation. *Sociological Research, 1*, 33–42.

Bobkov, V., & Gulyugina, A. (2012). Inequality of the quality and standard of living of the population of the regions. *The Economy of the Region, 2*, 170–178.

Buneeva, M. (2012). Socio-economic component of consumer society. In *Proceedings of the International Conference "Scientific and Practical Aspects of the Development of Cooperation, Economics, and Education* (pp. 93–96). LIK Publishing House, Lipetsk.

Buneeva, M. (2015). Consumption as one of the most important spheres of society's life activity. In *Proceedings of the VIII International Conference "Problems of Economics, Organization and Management in Russia and the World*" (pp. 14–18). World Press.

Galieva, G., Il'ina, L., Ponomarev, S., Khashir, B., & Karpunina, E. (2020). Consumption as a growth factor of the Russian economy. In *Proceedings of the 36th IBIMA Conference*, 4–5 November 2020 (pp. 3763–3772). Granada, Spain.

Gorshenina, E. (2016). System of indicators of the level and quality of life of the population in the region-subject of the Russian Federation. *Economic Research, 4*, 47–62.

Karpunina, E. Molchan, A., Lapushinskaya, G., Ryabova, T., & Jusufov, A. (2020). Mechanisms for managing changes in the socio-economic development of lagging regions. In *Proceedings of the 35th I IBIMA Conference*, 1–2 April 2020 (pp. 6538–6549). Seville, Spain.

Karpunina, E., Galieva, G., Andryiashka, M., Vorobyeva, A., & Bakulin, O. (2021). Country risk assessment as a tool for improving the quality of state economic security management (on the Example of Germany). *Quality–Access to Success, 22*(183), 136–142.

Kostrova, Y. (2018). Methodological approaches to the research of the food market. In *Proceedings of the 3rd International Conference "Economic Security: Legal, Economic, Environmental Aspects*", Kursk, South State University (pp. 91–93).

Kostrova, Y., Shibarshina, O., Tuarmenskiy, V., & Lyaschuk, Y.(2020). Ensuring regional food security in the Ryazan region. In *Proceedings of the IOP Conference "FarEastCon 2019* (pp. 620–622).

Ovcharova, L. (2014). Dynamics of monetary and non-monetary characteristics of the standard of living of Russian households over the years of post-Soviet development: An analytical report. Liberal Mission Foundation.
Rosstat. (2020a). Statistical bulletin “Food consumption in households-2019”. https://rosstat.gov.ru/bgd/regl/b20_101/Main.htm Data Accessed 14 June 2021
Rosstat (2020b). Statistical collection “Regions of Russia, socio-economic indicators (2020). https://rosstat.gov.ru/folder/210/document/13204 Data Accessed 14 June 2021
Rosstat. (2020c). The report “The socio-economic situation of Russia”. https://rosstat.gov.ru/compendium/document/50801 (Data Accessed 14 June 2021)
Suslova, Y., & Alyoshina, O. (2015). The impact of the region’s infrastructure on the quality of life of the population. *Economic Problems of Regions and Industrial Complexes, 2*(54), 225–228.
Talalushkina, Y. (2013). The UN experience in the study of the quality of life of the population. *Economic Journal, 3*(31), 46–53.
Zherebin, V., & Romanov, A. (2002). *The standard of living of the population: The main categories, characteristics, and methods of assessment: Monograph.* UNITY-DANA.

Responsible Production and Sustainable Regions, Cities and Communities

Eco-industrial Parks (EIP) as a Tool for Modern Innovative Industry

Marina N. Lyovkina, Olga V. Titova, and Alikhan M. Mukozhev

Abstract The appeal of eco-industrial parks (EIP) to industry lies in the potential for increased profitability and cost savings among tenant enterprises due to economies of scale and value-added products. The amount and type of technological products with market value increases since they can be brought back into the production cycle as raw materials for other companies. The opportunity to increase competitiveness by improving operating performance is manifested in return on assets by 30–50% above the industry average. Eco-industrial parks create opportunities for waste exchange, recycling and innovative technologies, and production processes to improve energy and material efficiency. As a result, there appear new opportunities for technological development, such as low-temperature manufacturing, new recycling and recovery technologies, and research and development of new materials consisting of recycled resources. New production opportunities can create profitable niches in the regional economy.

Keywords Eco-industrial park · Innovative industry · Sustainable design · Environmental justice · Industrial development

JEL codes Q5 · Q51 · Q56

1 Introduction

Looking at brownfield case studies where high-performance eco-industrial parks (EIP) can increase property values for public or private developers, we can indicate that values are also increased by green buildings that preserve working capital and

---

M. N. Lyovkina (✉)
Gorno-Altai State University, Gorno-Altaysk, Russia

O. V. Titova
Altai State University, Barnaul, Russia
e-mail: otitova82@icloud.com

A. M. Mukozhev
Kabardino-Balkarian State Agrarian University named after V.M. Kokova, Nalchik, Russia

E. G. Popkova and B. S. Sergi (eds.), *Geo-Economy of the Future*,
https://doi.org/10.1007/978-3-030-92303-7_67

increase productivity. Creating business networks can lower operating costs, reduce inventory and space needs, and provide market advantages. Co-located companies can share the cost burden for infrastructure and services such as:

- Waste management;
- Purchases;
- Training and recruitment;
- Transport;
- Recreation and child care.

Umbrella permits, system-wide measurements, and monitoring for environmental compliance reporting allow for additional savings potential. Two co-benefits for companies are as follows (Glotko, Voronkova, et al., 2019):

- The green and socially responsible image that improves market access and acceptability;
- The solid foundation provided by an invested, engaged, and satisfied workforce and community.

Saving production, disposal, and regulatory costs can make companies more viable. It can also provide positive implications for business in terms of job retention and growth. The development of an eco-industrial park can create conditions conducive to attracting new businesses and jobs that can significantly outpace previously undervalued areas. New opportunities must be identified to attract upstream suppliers and downstream customers to participate in business and exchange relationships. The cost savings that significantly benefit small and medium-sized businesses can support the development of new businesses and jobs or allow existing companies to invest their savings in retaining employees, hiring, or training new employees. Many of the EIP currently being developed include incentives for training and hiring minorities and women, wage increases, and family-friendly policies. The emphasis on sustainable design also improves the quality of indoor and outdoor workspaces, thereby providing an environment for a more productive workforce (Glotko, Sycheva, et al., 2019).

Currently, the development of EIP is also used as a community development strategy. The goal is to add value to the economic base of the municipality by strengthening its industrial, social, and supportive institutions. The basis for this process is the development that is more competitive, efficient, and cleaner than traditional approaches. The need to engage a wide range of stakeholders in identifying community assets, opportunities, and challenges to address environmental justice issues is emphasized. The host neighborhood for the EIP can be significantly enhanced by closer partnerships between citizens, businesses, and the government.

The accessibility of nearby citizens is encouraged. For example, one EIP may offer companies incentives to hire local workers; another may offer improved environmental indicators that benefit everyone. Benefits include the reduction in greenhouse gas and toxic emissions, improved energy use, efficiency and conservation of materials and water use, promotion and diffusion of green technology, improved land use planning, and the development of green spaces in industrial and commercial sites. Industrial ecology promotes environmental friendliness at the enterprise, park, and community levels (Stroiteleva et al., 2020). Companies are encouraged to implement innovative processes and technologies to minimize the use of virgin materials, prevent pollution, improve energy efficiency, reduce water consumption, and reduce the amount of waste requiring landfill disposal. At the park level, managers must base location, infrastructure, and staffing decisions on the carrying capacity of the environment. Creating networks outside the EIP to form material sharing relationships requires community-wide collaboration to identify waste reduction opportunities (Karataev et al., 2020). It is necessary to create a more regional resource economy that recycles resources *back to industries*, reduces dependence on virgin and imported materials, and reduces demand for landfills. The reduction in significant energy and material consumption on a system-level leads to greater efficiency gains than optimizing processes individually on the company level.

2 Materials and Methods

Eco-industrial parks are an approach to developing public–private partnerships (PPP). These partnerships aim to optimize resources and improve economic and environmental performance. The most evident benefits of EIP to the government are increased tax revenues, the creation of local jobs, retention of existing businesses and redevelopment of the existing ones, and an increased value of operations and real estate. Moreover, EIP can strengthen neighboring areas for easier management. Industrial ecology encourages businesses to voluntarily improve their environmental performance and reduces the need for regulation, monitoring, and enforcement (Kostin et al., 2020). This could give municipal governments the ability to adopt environmental regulations that set higher standards, offering flexible incentives that focus on results rather than rules. Enterprises in industrial ecology networks can also benefit local authorities, helping to achieve sustainable development mandates and the goals of the green procurement policy.

Eco-industrial development offers advantages to municipal infrastructure. Landfills are a major community investment because of the high cost of regulatory and technical requirements. A significant reduction in waste streams can extend the life of landfills and divert materials to developing recycling facilities. Similarly, the return on water and wastewater infrastructure can be increased through increased use efficiency, programs for pollution prevention, water cascading, wetlands for storm water treatment systems and environmental wastewater treatment systems, new opportunities for economic development, and resulting public works benefits that contribute to reducing environmental burdens to help protect the environment (Merdesheva et al., 2020).

The developments of EIP are relatively untested and are being in the early stages of development. This creates several real and perceived implementation risks due to the lack of proven successes. The risks identified are almost entirely generated by industry. However, the related problems are listed below for each stakeholder group.

A potentially longer payback period could make financiers reluctant to develop an EIP. Calculating costs and savings over a longer time frame can help mitigate this perceived risk. The success of the agreement on material swap depends on the cost of recovering byproducts being less than the cost of disposal or the price of comparable virgin materials (Karataev et al., 2020). The emphasis on cooperation, interaction, and interdependence among businesses causes many problems.

First, there is the potential for additional costs to work with the community and other businesses in terms of time, workforce, transportation, recovery and exchange infrastructure, communications, and monitoring. Additionally, proprietary information can become available to competitors as they provide information about the ins and outs of businesses.

In bilateral transactions, the risk will be reduced because businesses will have the discretion to disclose information. Additionally, the exchange of byproducts could also lead to businesses continuing to rely on toxic materials. To avoid toxic trade, the strategies for pollution prevention and the design of technological processes must be prioritized; businesses must commit to continuing to protect the environment (Dragunov & Shenshinov, 2020).

Second, the ultimate risk associated with the interdependence of enterprises exists in the relationship of exchange of byproducts when the quantity and quality of supply are not ensured. Production shifts can affect supply, and sudden changes in the quality of byproducts can damage material handling equipment. In an increasingly interdependent industry sector, its inability to be critical of itself can be detrimental to the entire network. Conversely, it is argued that the risks associated with resource-exchange strategies prevail. In normal producer-supplier relationships, backup suppliers are required (Vukovich et al., 2018). Offerings are considered interdependent in nature. This statement is evidenced by the advantages of geographic co-location in the regions where the stability of EIP can be improved at the recruitment stage by providing a mix of tenant enterprises that have both diversity and redundancy.

Third, several regulatory risks have also emerged in the planning of EIP development. Regulators may prohibit possible innovations in process design and intercompany communications. Businesses and EIP developers perceive the risks of liability and confusion over definitions of hazardous waste in existing regulatory structures (Avkopashvili et al., 2019). Eco-industrial development is now discussing the need for greater flexibility in regulation at all levels to support more innovative and holistic approaches that focus more on the ecological system as a whole rather than on individual ecological environments.

3 Results

The problems facing society are less serious because the risks are spread more widely. Most of the problems are related to the local potential for successful planning and implementation. Industrial ecology and recycling research institutes highlight the following problems for communities:

- Creating local support for EIP;
- Ensuring the right to vote in the development process;
- Reaching consensus when setting EIP performance goals;
- Identifying appropriate EIP ownership and financing strategies.

The benefits of industrial ecology far outweigh the problems for communities. Organized communities with residents regularly involved in managing growth and development can be most effective in providing the greatest benefits of EIP development (Shenshinov & Al-Ali, 2020).

Government is in a good position to say that it too wants to be involved in eco-industrial development. The challenges of government are not much different from those trying to implement something new. Policies and processes must be changed, which is also required from the personnel and resources involved. The task set by the EIP is a test of the progressiveness and adaptability of the municipality. The following problems for the government are highlighted:

- Pressure to streamline zoning, permitting and adapting development regulations and infrastructure projects that require financial and human resources;
- Pressure for greater flexibility in environmental regulations;
- Developing appropriate technology, promoting technology transfer, and providing technical training;
- Facilitating and ensuring effective information exchange between business and the community.

The way eco-industrial development manifests itself in the community results from local partnerships formed between government agencies, community members, businesses, and industrial developers. The combination of tools and strategies to optimize resource efficiency between the collective industries of each park depends on what is most appropriate and feasible for the area's existing industrial complex, workforce, markets, and resources (Borodin et al., 2020). Any EIP project requires a considerable amount of energy and resources, and the community must determine the most viable option that will help overcome the significant challenges. The community can develop programs to meet the goals of various stakeholders. Building community support is integral to the success of EIP. Highlighting the potential benefits of EIP to the community can make people more willing to support EIP development with their time, energy, and resources.

Let us introduce the stakeholders that are included in working projects (WP):

- Developers-leaders of the industrial and financial community;
- Representatives of local companies and potential future tenants in the EIP;

- Chamber of Commerce;
- Public sector stakeholders from municipal governments and regional, provincial, and federal agencies;
- Public and environmental organizations;
- Practitioners with the capabilities needed in the project (e.g., architects, engineers, environmentalists, environmental managers, educators, and trainers).

4 Conclusion

The core of eco-industrial development is the strengthening of "business-business" and "business-community" networks to improve resource efficiency in the context of constant interaction with the environment (Costanza, 1996).

Similar to conventional industrial zones, eco-industrial parks are contiguous facilities containing a number of tenants sharing joint management, infrastructure, services, and often a tenant association. The EIP concept is innovative since it combines multiple sustainability strategies, whereas, in other countries, such practices and concepts have only been proven in isolation.

Every successfully implemented strategy adds value to the park and becomes a useful recruitment incentive.

The combination of eco-efficiency measures is supported by the application of natural systems principles, further distinguishing EIP from traditional industrial development. Industrial ecosystems are cyclical, relatively efficient, and tend to rely on alternative energy sources rather than chemical energy supply (Tsvetkov et al., 2019). A systemic view of production and consumption activities replaces the focus on reducing plant waste. The presence of byproducts from one industrial process as inputs to another industrial process mimics the ecological concept of the food web (Adarina et al., 2019). Another applicable attribute of natural systems is that of "nobody's" species or industries. This is where the species, represented in this case by the enterprise, can evolve to fill a particular form or function. Niches created for business development may include businesses, environmental monitoring and information businesses, transportation services, environmental management services, and remediation businesses. Additionally, the dynamic nature of ecosystems is beneficial in industrial systems. Changes in market conditions lead to the appearance of new products and the cessation of the production of others. Therefore, recruiting businesses with specific products or byproducts is continuous and creates a natural life story for the industrial ecosystem.

References

Adarina, R., Gazukina, Y., & Yankovskaya, K. (2019). Indicators of the "green economy" as a tool for monitoring the regional economy. *IOP Conference Series: Earth and Environmental Science, 395*, 012107. https://doi.org/10.1088/1755-1315/395/1/012107

Avkopashvili, P. T., Polukhin, A. A., Shkodinsky, S. V., & Poltarykhin, A. L. (2019). The fundamental provisions of the concept of knowledge economy. In E. Popkova, Y. Ragulina, & A. Bogoviz (Eds.), Industry 4.0: Industrial revolution of the 21st century (pp. 57–64). Springer. https://doi.org/10.1007/978-3-319-94310-7_5

Costanza, R. (1996). Ecological economics: Reintegrating the study of humans and nature. *Ecol. Applicat, 6*(4), 978–990.

Dragunov, A., & Shenshinov, Y. (2020). Effective development of innovative technologies as a method for increasing labor productivity in industry. *E3S Web of Conferences, 208*, 03001. https://doi.org/10.1051/e3sconf/202020803001

Glotko, A. V., Sycheva, I. N., Dunets, A. N., Poltarykhin, A.L., Zhuravlev, P. V., & Tubalets, A. A. (2019). Integration processes in the agro-industrial complex of the regions: Development and environmental problems. *Journal of Environmental Management and Tourism, 10*(3(35)), 613–621.

Glotko, A. V., Voronkova, O. Y., Miheeva, N. M., Skoblikova, T. V., Alekhina, N. A., & Zhminko, A. E. (2019). The economic development of a transboundary region under the conditions of the Eurasian economic union. In *Proceedings of the IBIMA 2019: Education excellence and innovation management through vision 2020* (pp. 7957–7968).

Karataev, E. M., Merkuryev, V. V., & Titova, O. V. (2020). Integrating the "green economy" into the model of digital future of the modern socio-economic systems. In E. Popkova, & B. Sergi (Eds.), Scientific and technical revolution: Yesterday, today and tomorrow (pp. 1402–1410). Springer. https://doi.org/10.1007/978-3-030-47945-9_153

Merdesheva, E. V., Titova, O. V., & Avkopashvili, P. T. (2020). Methodological approach to the classification of digital economies by environmental efficiency and sustainability criterion. In E. Popkova, & B. Sergi (Eds.), Scientific and technical revolution: Yesterday, today and tomorrow (pp. 1360–1368). Springer. https://doi.org/10.1007/978-3-030-47945-9_148

Nekhvyadovich, L. I., Borodin, V. A., & Avkopashvili, P. T. (2020). A systemic approach to the study of the sustainability of the global digital economy: Economies of scale, ratchet effect and hysteresis effect. In E. Popkova, & B. Sergi (Eds.), Scientific and technical revolution: Yesterday, today and tomorrow (pp. 1369–1377). Springer. https://doi.org/10.1007/978-3-030-47945-9_149

Nekhvyadovich, L., Kostin, K. B., & Mironov, V. V. (2020). Government and public monitoring of entrepreneurship in a multicultural region: new perspectives in the conditions of the digital economy. In E. G. Popkova, K. V. Vodenko (Eds.), Public administration and regional management in Russia. Challenges and prospects in a multicultural region (pp. 147–154). Springer. https://doi.org/10.1007/978-3-030-38497-5_17

Shenshinov, Y., & Al-Ali, A. (2020). The tools of increasing efficiency of human resource in the lean production environment: Conceptual study. *International Journal of Core Engineering and Management, 6*(7), 1–18.

Stroiteleva, T. G., Kletskova, E. V., & Balaeva, S. I. (2020). Ontogenesis of green development model of digital economy: Essence, stages, and prospects. In E. Popkova, & B. Sergi (Eds.), Scientific and technical revolution: Yesterday, today and tomorrow (pp. 410–417). Springer. https://doi.org/10.1007/978-3-030-47945-9_46

Tsvetkov, V. A., Loginov, E. L., & Raikov, A. N. (2019). Integrated network interaction of education with science and production. *Educational Technologies and Society, 19*(1), 389–398.

Vukovich, G. G., Makuschenko, L. V., Bateykin, D. V., Titova, O. V., & Dobrosotskiy, V. I. (2018). Support of the territories of advanced economic development on human capital: Theory and practice. *Quality—Access to Success, 19*(S2), 157–160.

Economic Assessment of Complex Damage from Extractive Industries, Considering the Environmental Factor

Takhir K. Toguzaev, Irina S. Trapeznikova, and Olga V. Zhuravleva

Abstract The activities of business entities in the extractive industry are always associated with the extraction of natural resources and the related damage to the environment. The vast majority of studies in this area involve attempts to estimate damages in monetary terms and to quantify them. The economic assessment of damage associated with the negative impact of extractive industries is a critical indicator in reconciling the interests of business entities and the interests of the territories of their presence. Thus, all participants making decisions on this issue should have a coherent understanding of the economic damage with the account of the ecological factors and the recognized methodology of its assessment. This paper provides an overview of approaches to determining environmental and economic damage and types of its assessment. Additionally, the paper assesses the complex damage to the economy from the activities of coal mining companies, taking into account the environmental factor. The calculations are based on the example of the Kemerovo Region as a region with developed mining industry.

Keywords Economic damage · Environmental violations · Damage assessment · Environmental factor

JEL codes Q5 · Q55 · Q57

1 Introduction

The mining industry, given the specifics of the activities of its constituent enterprises, is recognized as one of the main sectors of national economies with a high level of negative impact on nature (Slawikowskaja et al., 2016). At the same time,

T. K. Toguzaev (✉)
Kabardino-Balkarian State Agrarian University named after V.M. Kokova, Nalchik, Russia

I. S. Trapeznikova
Kemerovo State University, Kemerovo, Russia

O. V. Zhuravleva
Gorno-Altai State University, Gorno-Altaysk, Russia

E. G. Popkova and B. S. Sergi (eds.), *Geo-Economy of the Future*,
https://doi.org/10.1007/978-3-030-92303-7_68

mining enterprises can be considered the most indicative in the formation of adverse environmental impacts since they are marked with the following:

- Considerable amount of solid waste generated by coal mining and coal preparation;
- Significant amounts of toxic emissions contained in mine dumps and released into the water and air due to natural influences.

World science has long been concerned with the issue of economic changes in the environmental aspects of the operation of mining industries, including coal mining. This work resulted in the economic justification of the category of "environmental and economic damage," recognized in modern theory and practice. So far, there is no recognized methodology for assessing environmental and economic damage, which would consider all factors of the impact of coal mining enterprises on the environment and its components (Ministry of Natural Resources of Russia, 2010; Potravny et al., 2018).

Traditionally, the environmental and economic damage is assessed by the results of the impact of coal mining enterprises on the elements of the natural environment (water, air, and soil). The traditional analysis does not consider the damage to the social environment, including losses from the decline in public health, the loss of attractiveness of coal mining regions as a territory of permanent residence, and infrastructure losses (Disturbed lands reclamation Restoration of biological diversity, 2017; Potravny et al., 2015). Separate methodologies carry out the cost estimate of the listed losses, but the extensive damage is not calculated.

Russia's transition to a "green" economy led to an increased interest in developing methodological approaches to a comprehensive assessment of economic damages, taking into account the environmental factors for their subsequent use in the national economy.

2 Materials and Methods

Khaldeev was one of the first to define the economic damage caused by the negative impact of industrial activity on nature. In 1973, he defined this concept by linking it to the additional costs that society needs to incur to eliminate negative socio-economic phenomena and air pollution (Korneev et al., 1973). Hoffman detailed the proposed approach based on the division of costs into separate groups. He singled out the following groups:

- Costs aimed at reducing the volume of pollutants entering the environment;
- Costs necessary to eliminate the negative social impact of production;
- Costs resulting from the compensation of resources and products lost due to production activities (Hoffmann, 1985).

The development of theoretical ideas about economic damage resulting from environmental disturbances was contributed by such famous Soviet and Russian scholars as A. Belashova, A. Golub, A. Gusev, S. N. Bobylev, V. I. Danilov-Danilyan, O. E. Medvedeva, N. Pakhomova, A. I. Oleinik, E. V. Ryumina, E. Strukova, and others. Their theoretical and methodological research resulted in the adoption of methods for determining the economic efficiency of implementing environmental measures and the assessment of economic damage caused to the national economy by environmental pollution. This methodology also deals with the prevention of environmental impact, including pollution of the natural environment (Gosplan of the USSR, 1983).

Summarizing the above views on the content of economic damage caused by environmental disturbances and approaches to its economic assessment, we can conditionally distinguish the following main theoretical and methodological approaches:

1. Assessment of the amount of damage caused and prevented for a year. The authors of such approaches propose to assess damage for each of the natural environments—water resources, land resources, and atmospheric air. Harmful substances are assigned to a mono-pollutant, and the reduced emission rate is calculated. Coefficients characterizing the significance of the pollutant for different environments are applied when calculating economic damage. In particular, the coefficients of the relative danger of the pollutant and the nature of dispersion of admixtures are used to assess the damage caused by atmospheric air pollution. The coefficients of ecological suitability and ecological importance of the territory are applied to assess the damage caused by soil pollution (Gusev, 2017).
2. Economic assessment of prevented damage is a universal measure of the economic efficiency of environmental measures implemented by measuring the costs and results. This method allows estimating the prevented damage by the activities that affect the source of pollution (Methodology for determining the prevented environmental damage, 1999).
3. One more approach is the economic assessment of past damage. The aforementioned types of assessment allow calculating the annual damage. However, it is not fully compensated, which leads to the need to calculate the accumulated environmental damage. The generally applied consolidated method considers two main factors—geophysical (the ability of the environment to self-repair) and economic (an increase in the value of the damage, taking into account inflationary changes) (Evaluation of the past ecological damage accumulated in places of arrangement of organizations, 2011).

The analysis of existing methodological approaches demonstrates the ambiguity of the definition of economic damage caused by the activities of business entities. Therefore, we can assert that several groups of indicators can be used to determine economic damage as a result of environmental violations, including indicators reflecting the following:

- Pollution prevention costs;
- Consequence management costs;

- Losses from economic losses;
- Costs of compensating for losses to the social sphere.

It is no longer just about environmental damage but about a set of damages for different recipients (population, environment, and infrastructure).

3 Results

According to the authors, the damage from the activities of coal mining companies is a complex value composed of individual types of damage:

1. Damage to the environment caused by industrial activities, including: damage from air and water pollution, damage from the retirement of arable land, and depletion of natural resources (Bobylev et al., 2010). Losses from the pollution of the natural environment are calculated according to the method of determining the prevented environmental damage developed by the State Committee of the Russian Federation on Environmental Protection (Danilov-Danilyan, 1999). The calculations are performed based on regional indicators of the specific cost estimate of damage per unit of reduced mass of pollutants indexed to current year prices.
2. Real losses of the region's economy associated with a decrease in the total economically active population of the region. These losses are manifested in the reduction of the period of labor activity and an increase in the duration of temporary disability in the region due to the appearance and development of occupational diseases and diseases caused by the deterioration of the natural environment.

The most significant losses to the economy from the environmental factor occur when the integrity of the environment is violated in the organization of the open method of mining. Open-pit mining requires allocating significant land areas for open-pit and stockpile management, loading and unloading access roads, and other production facilities. The average area of a coal mine varies from 1000 to 2000 hectares (Panfilov, 2008).

In 2019, the coal mining industry in the Kemerovo Region disturbed 4,863 thousand hectares (Korneev et al., 1973). The estimated economic damage from these activities is defined by the number of costs required for reclamation of land disturbed by coal mining, it equals 1.215 billion rubles. This value is derived from calculating that the reclamation of one hectare of land requires about 250 thousand rubles (Panfilov, 2008).

If we consider the natural and economic significance of soils and land resources, the amount of damage will significantly increase. Damage is estimated based on the costs of land reclamation and the need to develop uncultivated land to replace the land taken away for the needs of coal mining companies. If we take the cost of development of new lands at the level of 1020 thousand rubles/ha (Korneev et al., 1973), then the amount of economic damage from the activities of coal mining companies in the region will reach 4.96 billion rubles.

If we take into account the cost of leasing land that is under dumps or occupied by open-pit mines, we get another 0.9 billion rubles (Board of Administration of the Kemerovo Region, 2008).

The work of coal mines is accompanied by the formation of a considerable amount of overburden placed in rock dumps. The average stripping ratio in the Kemerovo Region is 6.5 m^3/t (Boyko, 2015). In 2019, 3765 million tons of waste were generated in the region due to the activities of coal mining companies, of which 50.3% was utilized and neutralized, and 1565 million tons were placed at the region's waste disposal facilities. Let us calculate the damage from their placement based on the payment rates established by the Government Decree NO. 758. The final figure equals 1.72 billion rubles (Government of the Russian Federation, 2018).

Coal mining is accompanied by methane emissions. The gas content of seams does not depend on the method of coal mining. In coal mines, methane is concentrated in the air space of the mine, and the volume of its emissions can be tracked. In turn, open-cast mining is marked with unrecorded methane emissions into the atmosphere. The Global High-Level Commission on Carbon Prices specifies the cost of environmental damage caused by CO_2 emissions each year. To keep the temperature rise within two degrees Celsius while maintaining economic growth, the bottom line for the 2019 carbon price was $40 per ton (Generalskaya, 2017). The coal seams of the Kemerovo Region are considered gas-rich—the extraction of one ton of coal is accompanied by the emission of 30–40 m^3 of methane. Even if we count at the lower limit of emissions, the total damage to the region's economy from methane emission from the coal mining industry equaled 12,231 billion rubles in 2019.

The Kemerovo Region conducted repeated studies on the loss of economic indicators from the morbidity of the resident population due to environmental factors. The region developed a sufficient number of methods for assessing this kind of economic damage. The most famous is the methodological approach of Revich-Sidorenko (2006). Its application convincingly proves that the economic damage from the morbidity of the population working at coal mines or living in the area of their impact is 3.4% of GRP. This type of damage can be considered equal to 42.6 billion rubles in the Kemerovo region in 2019 (Table 1).

4 Conclusion

The assessment of damage was carried out only on violations of one year (2019). However, the damages accumulated in previous years continue to harm the environment and public health. Moreover, this damage limits the aspirations of the region of the presence of coal mining companies to achieve sustainable development goals (Pitulko et al., 2019).

Meanwhile, no more than 8% of regional budget revenues are generated by coal mining companies, and this share tends to decrease. The owners have no desire

Table 1 Assessment of complex economic damage from the activities of economic entities of the coal mining industry in the region in 2019 (on the materials of the Kemerovo Region)

Type of damage	Amount of damage, bln. RUB
Damage caused by waste generation from coal mining and violation of requirements for its handling	1.72
Damage as a result of disturbed coal mining lands	5.86
Damage from morbidity caused by the work of coal mining enterprises	42.6
Damage from the emission of greenhouse gas	12,231
Total:	**12,281**

Source Compiled by the authors based on (Department of Natural Resources and Ecology of Kemerovo Region, 2020)

to spend money and bear responsibility for the damage caused to the areas of their presence. Russian lawmakers must provide for the social responsibility of coal mining companies. Otherwise, the profits and dividends of business entities will be paid by the population's health and losses to the regional economy.

References

Board of Administration of the Kemerovo Region. (2008). *Decree "On approval of the procedure for determining the amount of rent for land plots"* (February 27, 2008 No. 62, as amended December 16, 2020). Kemerovo, Russia.

Bobylev, S. N., Minakov, V. S., Solovyova, S. V., & Tretyakov, V. V. (2010). *Environmental and economic index of the regions of the Russian Federation. Methodology and indicators for calculations.* In A. Ya. Reznichenko, E. A. Schwartz, & A. I. Postnova (Eds.), WWF Russia, RIA Novosti. Retrieved from https://wwf.ru/upload/iblock/dc8/index.pdf

Boyko, D. V. (2015). The possibility of using overburden and coal-bearing rocks for technological roads at Kuzbass coal mines, taking into account their frost resistance. In *Proceedings of the III Youth Environmental Forum*. Kemerovo, Russia. Retrieved from https://www.elibrary.ru/download/elibrary_25073752_75783059.pdf

Danilov-Danilyan, V. I. (1999). *Sustainable development and environmental policy issues*. Moscow, Russia, ECOS-Inform. Retrieved from https://istina.msu.ru/publications/book/60616279/

Department of Natural Resources and Ecology of Kemerovo Region. (2020). *Report on the state of the environment of the Kemerovo Region in 2019.* Kemerovo, Russia. Retrieved from http://ecokem.ru/wp-content/uploads/2020/02/doclad_2019.pdf

Disturbed lands reclamation. Restoration of biological diversity. (2017). GOST R 57446–2017 (effective December 1, 2017). Standartinform. Retrieved from https://docs.cntd.ru/document/1200145085

Evaluation of the past ecological damage accumulated in places of arrangement of organizations. (2011). GOST R 54003–2010 (effective January 1, 2011). Standartinform. Retrieved from https://docs.cntd.ru/document/1200082700

Generalskaya, K. (2017). Secrets of coalbed methane production and processing in Russia. *Extractive Industries*, *4*, 50–57. Retrieved from https://dprom.online/wp-content/uploads/2017/11/DP_8-ilovepdf-compressed.pdf

Gosplan of the USSR, Gosstroy of the USSR, Presidium of the USSR Academy of Sciences. (1983). *Temporary standard methodology for determining the economic efficiency of environmental measures and assessing the economic damage caused to the national economy by environmental pollution* (approved October 21, 1983 No. 254/284/134). USSR: Economy.

Government of the Russian Federation. (2018). *Decree "On the rates of payment for negative impact on the environment and additional coefficients"* (June 29, 2018). Retrieved from http://www.consultant.ru/document/cons_doc_LAW_204671/24bedf438cb57123e864ba598a271912f8736b86/

Gusev, A. A. (2017). On the economic evaluation of natural resources and its use in the national economy. *Economics of Contemporary Russia, 2*(77), 19–28.

Hoffmann, K. G. (1985). *Economic problems of natural resources management*

Korneev, Y. E., Khaldeyev, V. T., & Chupis, A. V. (1973). In *Multifactorial correlation of the incidence of bronchitis in the population. General methodological and theoretical issues of atmospheric air hygiene.* USSR: IOCG.

Methodology for determining the prevented environmental damage. (1999). Russia, State Committee for Ecology of the Russian Federation. Retrieved from https://docs.cntd.ru/document/1200035561

Ministry of Natural Resources of Russia. (2010). *Order "On approval of the methodology for calculating the amount of damage caused to soils as an object of environmental protection"* (July 8, 2010 No. 238). Retrieved from https://www.garant.ru/products/ipo/prime/doc/2072837/

Panfilov, E. I. (2008). Assessment of the impact on the subsoil and the possible consequences in the development of mineral deposits. *Russian Mining Industry, 2*, 26–32.

Pitulko, V. M., Kodolova, A. V., & Kulibaba, V. V. (2019). Past (accumulated) environmental damage in the context of rational environmental management in the Russian Federation. *Regional Ecology, 2*(56), 7–15.

Potravny, I. M., Novoselov, A. L., & Alnykina, E. M. (2015). Models of natural resource depletion and assessment of past pollution damage. *Scientific Bulletin of the Plekhanov Russian University of Economics, 2*(8), 53–78.

Potravny, I. M., Novoselov, A. L., & Novoselova, I. Y. (2018). The development of economic assessment methods of damage from environmental pollution and their practical application. *Economics of Contemporary Russia, 3*(82), 35–48.

Revich, B. A., & Sidorenko, V. N. (2006). Methodology of assessment of economic damage to public health from atmospheric air pollution. Manual on regional environmental policy. In V. M. Zakharova, & S. N. Bobyleva (Eds.), Acropolis, CEPR.

Slawikowskaja, Y. O., Rudakova, L. V., & Rudakov, R. B. (2016). Assessment of anthropogenic impact of mining enterprises on the environment. *Journal of Economic Theory, 4*, 124–137.

Effects of Education Transformation in Pandemic: A Regional Perspective

Yuliya R. Rudneva, Olga G. Kantor, Evgeny A. Shamonin, Nelly N. Galeeva, and Anastasia M. Rogacheva

Abstract In the COVID-19 reality, educational organizations worldwide faced the urgent need to transform the educational process to prevent the uncontrolled spread of infection while ensuring that students' competencies were built at the desired level. The primary way to solve this problem was an accelerated transition to distance education. In order to switch to this type of learning, educational institutions are devoting significant resources to training their staff in new teaching methods and improving technology for interacting with students. However, fast is not always good. Moreover, radical changes that do not correspond with the perceptions of the participants in the educational process may not bring the expected results, even if they are implemented in the spirit of the latest trends. In this regard, it is advisable to conduct research to update the ideas about the advantages and disadvantages of distance education, the timely identification of which can help improve the quality and effectiveness of the educational process. The paper presents the results of research conducted among the students of one of the flagship universities of Russia. The conducted study shows that students negatively perceive some advantages of distance education. Therefore, it is necessary to reconsider the conceptual foundations of distance education.

Keywords Distance education · COVID-19 · Blended learning · Advantages of distance education · Survey

JEL codes A19 · H00 · I21 · I25

Y. R. Rudneva (✉) · O. G. Kantor · E. A. Shamonin · N. N. Galeeva · A. M. Rogacheva
Ufa State Petroleum Technological University, Ufa, Russia
e-mail: julrud1976@yandex.ru

E. A. Shamonin
e-mail: shamonin@bk.ru

A. M. Rogacheva
e-mail: izabo@bk.ru

E. G. Popkova and B. S. Sergi (eds.), *Geo-Economy of the Future*,
https://doi.org/10.1007/978-3-030-92303-7_69

1 Introduction

Under the impact of the pandemic, distance education became an objective necessity and required an accelerated transformation of the entire education system. This situation forced educational institutions to accelerate processes that could have taken many years had to be accelerated. Priorities are forced to shift toward meeting the most necessary requirements, including the following:

- Meeting the deadlines for providing educational services and their required volume;
- Ensuring the reliability of telecommunication technologies used in the organization of interaction between teachers and students;
- Developing methods and tools for delivering information and monitoring the knowledge of students.

The acquired social experience forms important areas of discussion about the future of education developing in the context of its digitalization, focusing on individual requests and needs, and changing priorities, formats, forms, and values (Allen et al., 2002; Dietrich et al., 2020; Singh & Wulf, 2020; Valverde-Berrocoso et al., 2020).

Despite the increased attention to the problem of distance education, no one can predict when EdTech will replace the traditional (face-to-face) form of education and whether it will happen at all. Simultaneously, in the Russian reality, it is becoming evident that higher education institutions, introducing EdTech technologies, should generally improve the forms and methods of blended learning and combine the advantages of face-to-face communications and digital educational space. In this regard, each educational institution should determine the optimal proportions between traditional and distant education forms. Evidently, their individual features must be taken into account in the following contexts:

- Industry sector;
- Territorial reference;
- Social and cultural traditions, etc.

The determination of such proportions must be based on qualitative and quantitative analysis, which makes it relevant to identify potentially acceptable methods. The analysis of the secondary literature (Saba, 2014) showed that the following methods are used:

- Quantitative (statistical, comparative, regression, and factor analysis);
- Qualitative (in the context of case studies);
- Theoretical and analytical (a combination of several methods belonging to the groups of qualitative and quantitative methods of analysis);
- Historical;
- Survey (in particular, questionnaires);
- Meta-analysis;
- Focus groups, etc.

Some authors suggest using more sophisticated approaches to study the impact of distance education, particularly system dynamics (Saba, 2014) or approaches based on probabilistic models (Bernard et al., 2004).

The application of mathematical and statistical methods of analysis requires the availability of appropriate information support. However, a survey is almost the only method available in conditions of insufficient statistical data or their complete absence, which is characteristic of the early stages of research. Following the basic principles of this type of research, it is necessary to choose a goal. In this regard, it is advisable to use the model of the studied educational organization, which will reflect the most essential structural elements for the ongoing study and the connections between them.

The research object is the Ufa State Petroleum Technological University (USPTU). It is one of the largest sectoral universities of the Russian Federation, which is a supporting university of Russia. For seven years, USPTU steadily enters the rating of the 100 best universities of Russia (Interfax News Agency, 2020). For more than 70 years, the university has developed a stable academic school and a recognized tradition of teaching and monitoring knowledge. The available scientific and technical base allowed the university to provide an accelerated implementation of distance technologies without affecting the quantitative indicators of the quality of the educational process (Vanchukhina et al., 2019). However, the problem of identifying the degree of student satisfaction with the new form of education requires a comprehensive study to determine the preferences of students. The preferences of students are important for justifying the proportions between traditional and distance education and determining the degree of justification of the high expectations for the declared benefits of distance education (Moore et al., 2011; Traxler, 2018), as well as whether these benefits are such in reality.

This paper focuses on the attitude of the students of the USPTU on distance education. The study is based on the model of the educational process (Fig. 1).

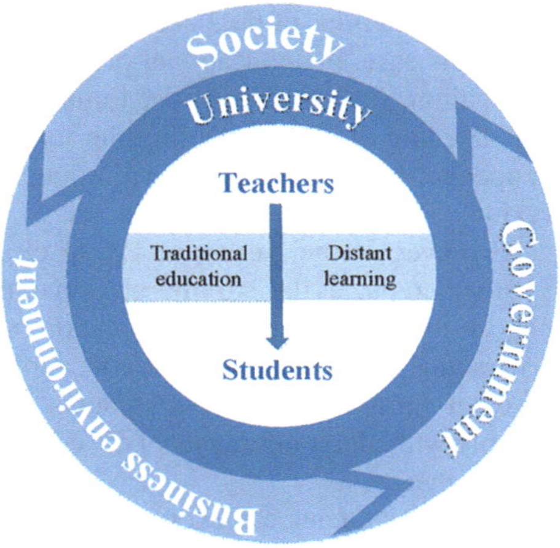

Fig. 1 Model of the educational process at the ufa state petroleum technological university. *Source* Compiled by the authors

According to it, the interaction between teachers and students is carried out in the system *University* through the involvement of tools and technologies of traditional and distance education.

2 Materials and Methods

Since the emergence of distance education, users noted its main advantage (Kuzmina, 2012; Oliveira et al., 2018)—*remoteness*—the absence of the need to attend offline classes in an educational institution during the learning process. This advantage allowed students to acquire knowledge from almost anywhere in the world. Accordingly, the next advantage is *convenience*—the learning process can be comfortably organized in place of residence, without the inconvenience caused by the presence of fellow students. In this way, *physical* and *social distance* is ensured. This leads to increased access to distance education for people with physical or psychological disabilities. Therefore, it allows for equalizing opportunities.

Additionally, as a rule, distance education is cheaper and gives more freedom in planning educational activities.

Simultaneously, the pandemic period showed that the listed advantages of distance education could turn into problems for students. To analyze the advantages and disadvantages of distance education (DE), we identified two types of it. They are as follows:

- DE as a form of education in secondary, higher, or additional education—regulated DE (RDE);
- DE as an opportunity to master individual courses independently—free DE (FDE).

In the case of RDE, the DE only transforms the interaction between the teacher and the learner by eliminating the distance between them through various means of communication—from regular mail (which was quite relevant 20–30 years ago) to all the possibilities of the Internet.

In this case, the educational process retains all its features: the presence of the curriculum, timing of training, timing of interim and final evaluations, and even the schedule of classes. During the forced transition to DE due to the COVID-19 pandemic, most educational institutions (higher and secondary) implemented this option.

However, in the current reality of electronic interaction, this form of learning gradually replaces the extramural form of training. In this case, students immediately choose the distance form of education, take the entrance exams, and enter into a contractual relationship with the educational institution. However, similar to students of other forms of education, students in extramural training are also required to comply with all requirements of the educational process and may be expelled if they violate them.

In RDE, students can choose to study any area of knowledge without being tied to the schedule of the educational process. These can range from courses from open sources (e.g., *Openedu* or *Coursera)* to fee-based courses offered by various educational organizations. It should be noted that most free courses will require the payment for the final certification and obtaining a certificate to confirm the qualifications.

In such courses, students have access to all necessary resources (text or video recordings of lectures, practical materials, and training and test assignments). At the same time, they have minimal time constraints. Although time constraints are outlined in the curriculum, they indicate general guidelines since the course remains open even after the formal deadline for its completion.

The COVID-19 pandemic arrived in Russia in early March 2020. By the end of March, most universities were forced to announce a transition to distance learning, which remained in place until the end of the 2019–2020 academic year. At the beginning of the new academic year, a significant slowdown in the coronavirus spread allowed universities to determine the preferred training format on their own.

In September 2020, the USPTU partially returned to contact learning, declaring the advisability of conducting all classes in a distance form. Thus, at the time of the study, students had been studying remotely for almost 9 months. Therefore, there is a need to assess the results and consequences of such a transformation.

According to the chosen research method, the authors conducted a survey using *Google Forms*. A total of 192 students participated in the survey. Of these, 64.2% studied economics and the humanities, while the rest studied technical subjects. The total number of surveyed students included six graduate students, but they were not considered in the further analysis due to the small number of respondents. According to the course of study, the remaining undergraduate students were distributed as follows:

- First year—32 people;
- Second year—23 people;
- Third year—58 people;
- Fourth year—56 people (The results of the conducted survey in Google Forms. (The results of the conducted survey in Google Forms, 2020).
- The study aims to determine the following:
- General feelings of students about distance education;
- Advantages and disadvantages of distance education students have experienced;
- Problems that students encountered during distance education;
- Changes for different participants of the educational process;
- What is necessary and what prevents the effectiveness of distance learning.

For some questions, students had the opportunity to give their own answers. This allowed us to expand the study's results.

3 Results

The survey began by exploring students' general understanding of distance education. More than half (53.4%) consider it "the same as contact education, but using a computer, phone, or tablet." At the same time, 29.5% identify it with extramural training. Another 15% consider it additional to full-time contact education since materials and tests are permanently available. This means that the vast majority do not initially have a negative attitude toward distance learning.

The next question (with multiple-choice) was "What are the advantages of distance education over traditional contact education?" The structure of the responses is presented in Fig. 2.

The main advantage indicated by the majority of students (80%) is the possibility of not attending the university. Notably, the priority of this characteristic among technical students is slightly higher than among economists. We should also note the trend of increasing significance in the students according to the year of study (from 67.7% in the first year to 91.2% in the fourth year). Evidently, senior students are more self-sufficient, and they often find a job, which confirms a similar increase in the importance of such a factor as the ability to combine attendance in class and work (from 25.8 to 42.1%).

The second most important factor was the possibility to return home during the study. The proportion of students who chose this advantage is almost independent of the education profile, but varies significantly from course to course. Another

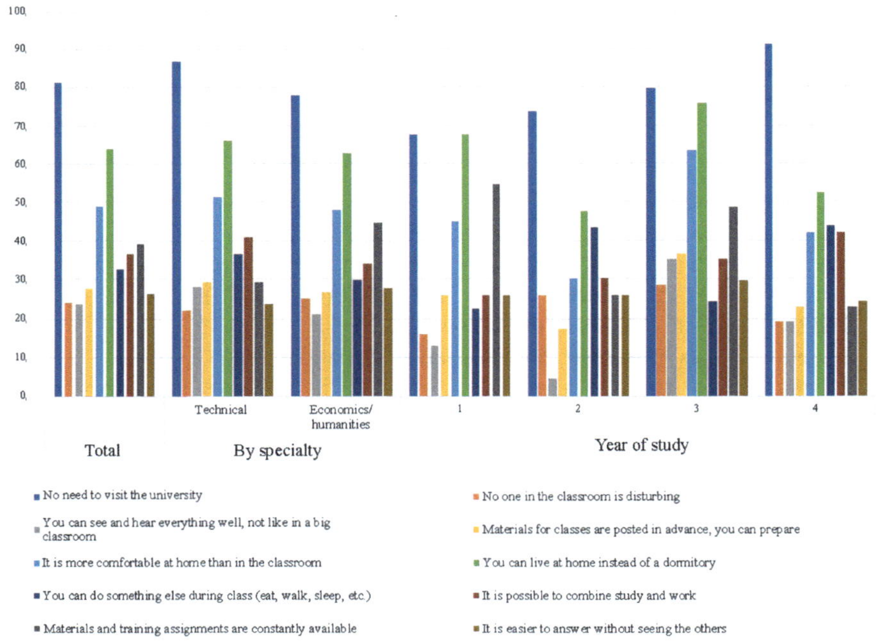

Fig. 2 Distribution of answers to the question "What are the advantages of distance education over traditional contact education?" in the context of specialties and years of study. *Source* Compiled by the authors based on the survey results

important factor related to the convenience of distance learning was that it was more comfortable at home than in a classroom.

It was surprising that students gave relatively low weight to such an answer as "You can see and hear everything well, not like in a big classroom." This is very important, especially in classes with many students or a lot of material presented on the board or screen. This can be explained by the fact that not all teachers had time to adapt and prepare visual material for online classes.

The following questions were aimed at identifying the negative aspects of distance education. Figure 3 shows the pattern of responses to the question (multiple choice) "What are the disadvantages of distance education compared to traditional contact education?".

The response "It is necessary to spend all time at the computer, phone, or tablet)" received the utmost significance—up to 100% of first-year students. However, this is the flip side of the classic advantage of DE, which is the ability to learn anywhere, which is currently provided only using gadgets.

The next most common answer for almost all categories of respondents was "No direct contact with the teacher." It is especially acute for first-year students who just finished school and experience all the problems associated with a lack of communication. They also have the highest percentage of answers "Studying alone, without groupmates" and "It is difficult to answer without seeing the teacher and other

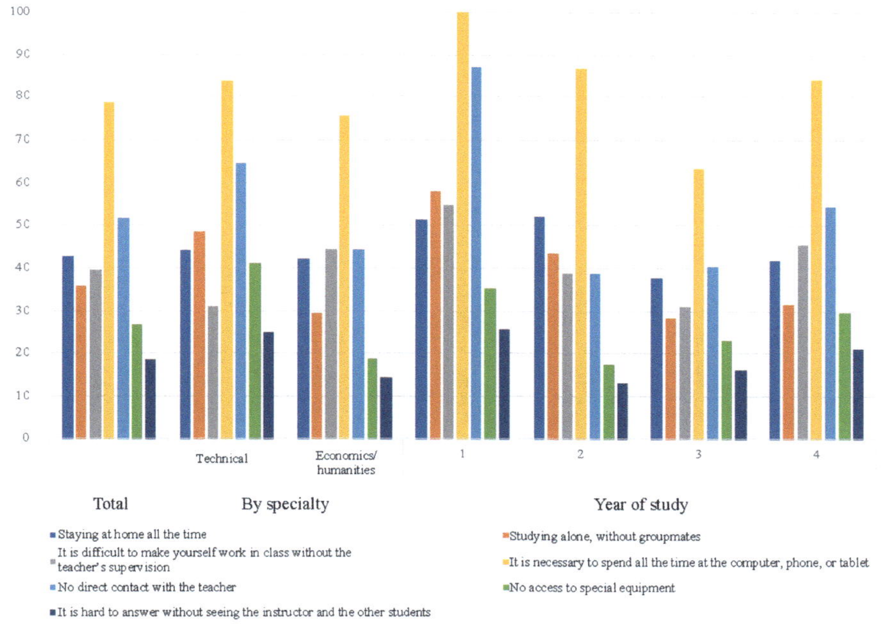

Fig. 3 Distribution of answers to the question "What are the disadvantages of distance education over traditional contact education?" in the context of specialties and years of study. *Source* Compiled by the authors based on the survey results

students." This problem is seen not only in Russian students (Kollalpitiya et al., 2020).

Although 70–90% of those surveyed chose not being able to attend university as an advantage, 40%–50% also had a negative view of being home all the time.

The difference in attitude to the lack of access to special equipment is also noteworthy. Students in economics/humanities do almost all laboratory work on computers. Therefore, it is possible to either do them at home or arrange remote access if a specialized program is required. In technical specialties, many laboratory works take place in chemical laboratories. Various computer simulators are offered as an alternative, but they cannot fully ensure the development and consolidation of the required professional skills.

Undoubtedly, the highlighted disadvantages of DE create specific problems in the lives of students. The structure of their responses to this question (with multiple-choice) is shown in Fig. 4.

The biggest problem lies in the connection to the Internet. First and foremost, this is due to the extremely dramatic transition of a considerable part of the population (including students and teachers) to working remotely via the Internet. The home networks of many people could not take the load. There may also be failures on

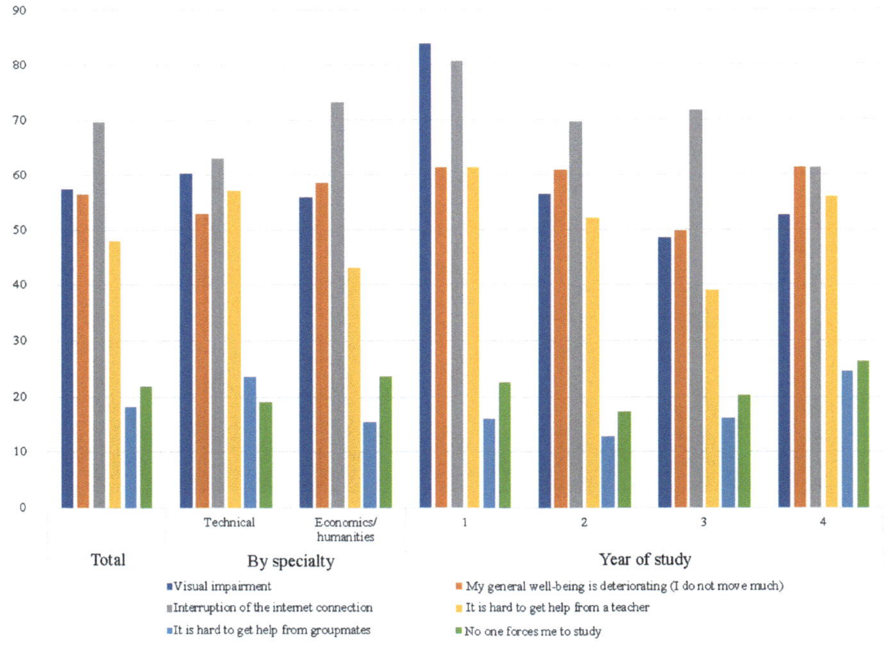

Fig. 4 Distribution of answers to the question "What are the problems in distance education?" in the context of specialties and courses of study. *Source* Compiled by the authors based on the survey results

the university's servers due to prolonged overloads. A similar problem was found in other countries in a study of eight weeks of forced DE (Almomani et al., 2020).

The next problems in almost all categories are the deterioration of vision and health (due to sedentary lifestyles). Thus, the widespread opinion that DE is convenient gets a downside – studying in comfortable conditions turns into a decrease in motor activity and, therefore, deterioration of the physical and psychological condition. Also, the possibility of getting advice from a teacher is essential to many students.

Psychologists highlight the following adverse effects of DE for students:

- Lack of socializing role of education—students do not learn how to build relationships with the outside world;
- Interaction and empathy disappear from the process of communication of the student with the teacher (Fadeev, 2017)—students do not get direct feedback,
- Unpreparedness of other family members for the constant presence of the child at home and the need for additional communication.
- The results of the transition to DE were also assessed as follows:
- 49.2% of students believe they are doing better in the university;
- 25.9% of students noted a decrease in the effectiveness of their work;
- 36.3% of students reported a decrease in teacher effectiveness.

In response to the open question about what qualities a student should have to learn effectively in distance learning, the most popular answers were responsibility (24.2%), discipline (19.8%), and assiduity (16.7%).

However, only 20.8% of the surveyed students express a desire to return to full-time contact education despite all the problems. This means that the opportunity to study in comfortable conditions and without being tied to the university plays a decisive role. Another 34.9% of students are willing to continue their education in the DE format. The rest choose various options for a combination of contact and distance education.

Distance education requires a higher level of computer literacy than the contact format. Therefore, students were also asked about the digital skills they needed (Fig. 5).

Almost all students find it necessary to master *Microsoft Excel* and *Microsoft Word*, due to the constant need to make calculations and draw up documents. The least attention has been paid to cloud technologies, which play a significant role in digital learning.

The disadvantages of this research include the fact that the survey mainly involved those who attend classes regularly. The students who miss the classes tend to have low motivation, but their opinions can help identify additional factors to increase the effectiveness of DE. In this regard, it seems that practical research aimed at identifying preferences among students may be fundamental in the context of understanding the impact of current circumstances and forming the information base for a comprehensive analysis using a wide range of methods, including mathematical and statistical.

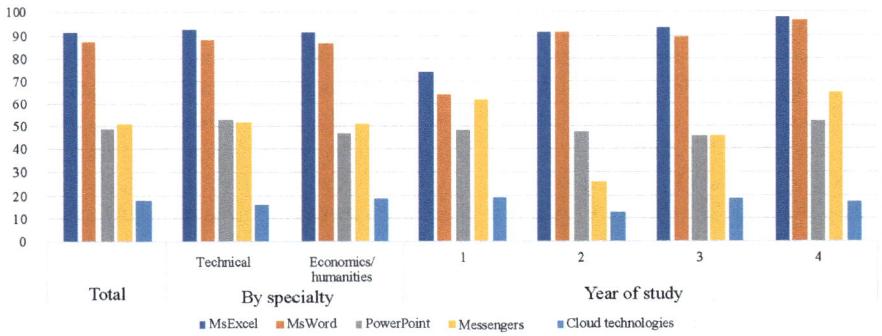

Fig. 5 Structure of responses to the question about the need for digital skills. *Source* Compiled by the authors based on the survey results

4 Conclusion

The presented results allowed us to question the indisputability of most of the classical advantages of DE. The obtained results show that in current conditions, the *flip side* of DE manifests itself in the form of disadvantages that are well-recognized in society.

For example, its convenience turned out to cause a decrease in motor activity and deterioration of health. Remoteness can only be achieved through the use of gadgets, which negatively affects the eyesight and increases Internet dependence. Physical and social distancing in the long term reveals an urgent need for them, turning from an advantage to a disadvantage. A free mode of working with study materials in the absence of a high level of responsibility and discipline leads to failure in meeting deadlines and expulsion from courses. Thus, for successful learning in a distance format, it is necessary to provide a sufficient material base and have a specific set of psychological and physical characteristics, which reduces the equality of opportunities.

Undoubtedly, some of the disadvantages of DE can be leveled with proper preparation of the teacher, students, and institution to work online. However, this requires increasing the computer literacy of all aspects of the learning process and personal development, as well as a significant investment of time and material resources. The urgent transition to DE took that opportunity away—everyone had to transform their participation in the learning process.

In a pandemic, it is challenging to solve the problems of mastering some of the practical skills involved in working with special equipment or executing and presenting the results of a collaborative project. The pandemic imposed additional restrictions on face-to-face student contact. Therefore, they had to do most of the projects either alone or with friends, using different means of communication, which made the work more difficult and reduced the quality of the result.

Nevertheless, it should be recognized that most students in the survey opted for the convenience of DE and acknowledged that the emerging problems with the organization of their time forced them to develop such qualities as responsibility and discipline.

References

Allen, M., Bourhis, J., Burrell, N., & Mabry, E. (2002). Comparing student satisfaction with distance education to traditional classrooms in higher education: A meta-analysis. *American Journal of Distance Education, 16*(2), 83–97. https://doi.org/10.1207/S15389286AJDE1602_3

Almomani, J. A., Alnasraween, M. S., & Almosa, N. A. (2020). Evaluating the distance university education experience after using the zoom application in Jordan from the students point of view. *Universal Journal of Educational Research, 8*(11B), 6239–6247. https://doi.org/10.13189/ujer.2020.082262

Bernard, R. M., Abrami, F. C., Lou, Y., & Borokhovski, E. (2004). How does distance education compare with classroom instruction? A meta-analysis of the empirical literature. *Review of Educational Research Fall, 74*(3), 379–439. https://doi.org/10.3102/00346543074003379

Dietrich, N., Kentheswaran, K., Ahmadi, A., Teychené, J., Bessière, Y., Alfenore, S., & Hebrard, G. (2020). Attempts, successes, and failures of distance learning in the time of COVID-19. *Journal of Chemical Education, 97*(9), 2448–2457. https://doi.org/10.1021/acs.jchemed.0c00717

Fadeev, E. V. (2017). Organizational and psychological problems of distance learning. *World of Science, Culture, and Education, 3*, 308–310.

Interfax News Agency. (2020). National university rankings. Retrieved from https://academia.interfax.ru/ru/ratings/?rating=1&year=2020&page=1 (Accessed 27 November 2020)

Kollalpitiya, K. Y., Partigianoni, C. M., & Adsmond, D. A. (2020). The role of communication in the success/failure of remote learning of chemistry during COVID-19. *Journal of Chemical Education, 97*(9), 3386–3390. https://doi.org/10.1021/acs.jchemed.0c00772

Kuzmina, L. V. (2012). Advantages and disadvantages of distance learning. *Bulletin of the Moscow University of the Ministry of Internal Affairs of Russia, 2*, 8–10.

Moore, J. L., Dickson-Deane, C., & Galyen, K. (2011). E-Learning, online learning, and distance learning environments: Are they the same? *Internet and Higher Education, 14*, 129–135. https://doi.org/10.1016/j.iheduc.2010.10.001

Oliveira, M., Penedo, A., & Pereira, V. (2018). Distance education: Advantages and disadvantages of the point of view of education and society. *Dialogia, 29*, 139–152. https://doi.org/10.5585/dialogia.N29.7661

Saba, F. (2014). Methods of study in distance education. In P. Anderson, & O. Zawacki-Richter (Eds.), *A critical review of selected recent literature* (pp. 151–171). AU Press. Retrieved from https://www.aupress.ca/app/uploads/120233_99Z_Zawacki-Richter_Anderson_2014-Online_Distance_Education.pdf

Singh, H., & Wulf, G. (2020). The distance effect and level of expertise: Is the optimal external focus different for low-skilled and high-skilled performers? *Human Movement Science, 73*, 102663. https://doi.org/10.1016/j.humov.2020.102663

The results of the conducted survey in Google Forms. (n.d.). Retrieved from https://docs.google.com/spreadsheets/d/1tuVW-hQOYc_N369MfOIEApDdJaeZ5NeOefoBU8Bw8G0/edit#gid=513372552 (Accessed 28 November 2020)

Traxler, J. (2018). Distance learning—Predictions and possibilities. *Education Sciences, 8*(1), 35. https://doi.org/10.3390/educsci8010035

Valverde-Berrocoso, J., Garrido-Arroyo, M., del C., Burgos-Videla, C., & Morales-Cevallos, M. B. (2020). Trends in educational research about e-Learning: A systematic literature review (2009–2018). Sustainability, 12, 5153. https://doi.org/10.3390/su12125153

Vanchukhina, L., Leybert, T., Rogacheva, A., Rudneva, Y., & Khalikova, E. (2019). New model of managerial education in technical university. *International Journal of Educational Management, 33*(3), 511–524. https://doi.org/10.1108/IJEM-08-2018-0270

Innovation Activity of Regions to Ensure Their Sustainable Development

Elena V. Petrukhina, Elena V. Dudina, Svetlana V. Derepasko, Irina A. Baranovskaia, and Nadezda P. Rudnikova

Abstract Regeneration of management processes and the obsolescence of interaction models between subjects of regional and municipal markets have led to a slowdown and, in some cases, complete incapacity of systems of innovative development of territories. First, this fact is associated with the lack of innovation activity of the regions at the proper level. Innovation activity is an element of innovation. Nevertheless, it is not considered a holistic tool for the sustainable development of the region. In accordance with this fact, the authors actualize the topic of innovation activity of regions to ensure their sustainable development. The paper aims to form a model of innovation activity of regions to ensure their sustainable development. The methodological basis of the study is implied by the following methods: comparison of scientific points of view, conceptual arrangement, demonstration of data, the implication of formulated proposals, and structurization. The research results are based on the study of the essential features and evaluation indicators of regional innovation activity, the formulation of the processes of implementation of the model of innovation activity of the regions, and the proposition of a model of regional innovation activity to ensure their sustainable development. This study will be further supplemented by a simulation model of innovative flows of the regional system of interaction of subjects of innovation and investment environment.

Keywords Innovation · Activity · Region · Investors · Municipal entity · Markets · Competitors · Model

JEL Codes Q01 · R11

1 Introduction

A decade of the innovation agenda for the development of the country and its regions has not formed a society of scientific knowledge focused on the technologization of the environment. The main factor in this problem was the reorientation of priorities

E. V. Petrukhina (✉) · E. V. Dudina · S. V. Derepasko · I. A. Baranovskaia · N. P. Rudnikova
Orel State University, Orel, Russia

E. G. Popkova and B. S. Sergi (eds.), *Geo-Economy of the Future*,
https://doi.org/10.1007/978-3-030-92303-7_70

and concepts of development of the country and regions. Nevertheless, the rudiments of innovative postulates, formed in the framework of regional development strategies, allowed to lay the foundations of innovative activity and start producing technological products for internal and external consumption of the Russian subjects. In this direction, it is quite important to study the issue related to the innovation activity of the regions to ensure their sustainable development. The following circumstances condition the statement of the highlighted condition.

The diversity of worldviews of scientists led to a different understanding of the category of innovation activity, which caused the formation of a considerable number of strategies aimed at segmenting sectors of the economy but not at the development of the whole region (Sliwka, 2003). Due to the lack of a common understanding of the innovation activity of the region, the differentiation of social and economic development of the region is formed within comprehensive technologization. Illarionova and Glekov consider the process of technologization of production as an innovative activity of the region (Illarionova & Glekov, 2019). Ushnurtsev and Shulepov understand this category as a set of innovation activities in the areas of the national economy (Ushnurtsev & Shulepov, 2016). Liu and White regulate the essence of this definition through research and development surveys forming the industrial potential of the territory (Liu & White, 2001). The fragmentation of these concepts does not allow us to formulate a general idea about the process of innovation activity of regions.

The above circumstances have led to the relevance of the research on the issue of innovation activity of the regions to ensure their sustainable development. Let us formulate the goal, objectives, and methods of our research. The paper aims to form a model of innovation activity of regions to ensure their sustainable development. The research goal assumes the following research tasks:

- To consider the essential features and evaluation indicator of innovation activity of the region;
- To formulate the main processes of implementation of the model of innovation activity of the regions;
- To propose a model of innovation activity of regions to ensure their sustainable development.

2 Materials and Methods

The methodological apparatus of this research is stated by the condition associated, on the one hand, with the identification of the features of the current model of innovative development and, on the other hand, with the author's proposals for the formation of a new system of relations aimed at ensuring sustainable development of the region. Thus, we used the following tools of scientific cognition: method of comparison of scientific points of view (Akaev & Hirooka, 2009), method of conceptual arrangement, method of data demonstration, method of the implication of formulated propositions (Barmuta, 2019), and structurization method.

3 Results

We consider the innovation activity of the region in the context of different conceptualizations of Russian and international authors.

The most generalized representation of the concept of innovation activity of the region is provided in the scientific article of Stiebale. He says that this process is a purposeful activity aimed at creating technological productions in the region (Stiebale, 2016). The highlighted definition considers the innovative activity of the region through the presence of technological industries in the system of the national economy. Stiebale indicates that the subjects of the realization of innovative activity are technological production. Technological production is a set of processes of technological readiness of enterprises to produce qualitative products different in their properties from the similar goods sold in the market.

A different point of view belongs to Nosov. He understands the essence of innovation activity of the region as a set of actions in the field of positive changes in the production of gross regional product (Nosov, 2020). The presented definition is quite generalized. Within the highlighted condition, the essence of innovative activity is highlighted through the presence of actions leading to changes. On the other hand, the definition proposed by Nosov focuses on the production component, the indicator of which is the gross regional product. However, the innovation activity of the region cannot be identical in its structure with the gross regional product. In our opinion, it is necessary to paraphrase the presented definition of Nosov and expand it from the position of adding the characteristic features of the innovative activity of the regions.

Innovation activity of the region is a comprehensive characteristic of innovation activity, which allows determining the intensity of implementation of innovative technologies in all spheres of the national economy (Czarnitzki & Kraft, 2004; Schnabel & Wagner, 1992). In the context of this definition, the innovation activity of the region became identified with innovation activity. We believe that this statement is not entirely true. Innovation activity is a process (Lapygin, 2019), and it is also the use of new forms and means to implement these processes (Lyapina et al., 2015; 2021; Stroeva & Kvak, 2014; Stroeva et al., 2021).

The most modern interpretation of the concept of "innovation activity of the region" is a definition that considers project management to be at the head of this category. Innovation activity of the region is a system of implementation of project activities in the field of innovation, allowing for significant sustainability of the region. As part of the formulation of this concept, A narrowly focused approach was chosen within the framework of the formulation of this concept. It regulates innovation activity only as a system of project activity. Innovation activity as a category is produced by technological processes. Limitations in the form of project management reflect only the sides of the strategic vision of innovation activity of the region but not its essence (Stroeva et al., 2021).

In accordance with the above definitions of innovation activity of the region, the following indices reflecting the essence of these definitions in the quantitative assessment were formed:

1. Index of technological productions i_p:

$$i_p \frac{y_t \times y_i}{y_c} \tag{1}$$

where
i_p—index of technological productions;
y_t—number of registered technological productions in the region;
y_i—number of regional industries engaged in the production of innovative products;
y_c—number of large, medium, and small businesses.

2. Index of the production of innovation activity i_a:

$$i_a = \frac{y_v}{\sum y_p}, \tag{2}$$

where
i_a—index of innovative production activity;
y_v—value of the gross regional product;
y_p—total value of sold innovative products produced in the region.

3. Index of innovation activity i_d:

$$i_d = \frac{y_k}{y_f}, \tag{3}$$

where
i_d—index of innovation activity;
y_k—value of innovative products, goods, and services produced in the region;
i_f—value of the total amount of products, goods, and services produced in the region.

4. Index of project management of innovation i_u:

$$i_u = \frac{y_r \times y_x}{y_w \times y_m}, \tag{4}$$

where
i_u—index of project management of innovation;
y_r—number of innovative projects implemented,
y_x—number of patents received as part of the innovative activities of enterprises,
y_w—number of innovative and intellectual products developed in scientific organizations;
y_m—number of publications of employees of scientific organizations on innovative topics.

The presented indices of innovation activity confirm the fact of differentiated understanding of this process from the positions of different scholars.

This affects the current model of innovation activity of regions, which is essentially focused on the conditions to support this process. Nevertheless, it does not involve strengthening the role of micro-level actors and the formation of mechanisms for their development.

The current model of innovation activity of regions functions based on the interaction of subjects, forming this process in the context of their powers. The main subjects of the model of innovative activity of the region are as follows:

- Subject of the Russian Federation (through the authorities) regulates the regulatory and programmatic function of creating innovative activity among other agents of the regional environment;
- Subjects of the innovative environment at the micro-level are the system of industries and other business structures, acting as the primary implementers of innovative ideas;
- Investors and partners are agents of an innovation activity forming additional profit within the framework of innovation activity of other subjects;
- Competitors and markets are the consumers of the innovation activity of the region.

The described model of innovation activity of the region is not aimed at ensuring its sustainability in the short or long term. First, this condition is formatted by the activities of the subjects of innovation activity of the region. The subject of the Russian Federation, acting through the authorities, carries out activities that contribute to the initial formation of innovative activity and the involvement of other agents in the framework of this process. The initial measures of innovative activity of the region by the subjects of the Russian Federation include the analysis of innovative activity, measures to support innovative development of industrial enterprises, implementation of innovative infrastructure, and creation of institutions of innovative development.

Subsequently, these activities are reclassified as a process of involving agents in the innovative activity of the region. These activities are based on the definition of methods and tools, creating conditions at the regional level, and the development of mechanisms to increase innovation activity.

The rest of the functionality of the subjects of innovative activity is more limited. This condition is manifested in accordance with the fact that the model of innovation activity is state-directed and aimed at the implementation of measures of a regional nature. Thus, the micro-subjects of innovative activity determine the directions of innovative activity, develop, produce, and implement innovative products, and form the strategy of innovative development. At the same time, the actions of subjects of innovative development at the micro-level are not independent.

The above characteristics of the model of innovation activity of the region determined the importance of the issue of modernizing internal processes associated with several circumstances. The first circumstance is the formation of different compositions and directions of the subjects of innovative activity of the region. In the current model of innovative development of the region, the final subjects were investors and partners, who received a profit from the sold products, and markets and competitors, who received an innovative product. The model of innovative activity of regions determines the agents of the micro-level as the final subject for their sustainable development.

The proposed model of innovation activity of regions is locked in the provision of micro-level subjects in the following relationships. The subject of the Russian Federation is the initiator of innovative activity of the region. The highlighted condition is determined by normative and control functions. Within the framework of the implementation of the powers of public authorities, the subject of the Russian Federation implements the following measures:

1. Development of the mechanism of commercialization of innovations—the allocated function is aimed at forming a system of commercialization of innovations at the national level. This system must function in the context of developing innovative activities to implement the needs of the region.
2. Development of provisions to increase innovation activity by differentiating the subjects-recipients of budgetary funds. The presented event has a dual meaning. On the one hand, the subject of the Russian Federation assigns to itself the criteria and postulates for assessing innovation activity. On the other hand, it strengthens the role of recipients of budgetary funds during innovation activity.
3. Decomposition of the old model—innovation infrastructure and problem assessment. The presented condition is reformatted from the current model of innovation activity of the region. The formation of innovative infrastructure at various levels promotes the activation of innovative activity. The faster the innovation infrastructure is modernized, the stronger are the effects of innovative activity of the subjects of the micro-environment. The assessment of existing problems allows accelerating this process by forming a plan to solve these problems.

Decomposition of crucial activities of the region is included in the system of organization of innovative activity at the micro-level. Under this condition, the subjects of the micro-environment act as the implementors of innovation policy, cooperators of innovative personnel, participants in projects in the development of the innovative activity, and managers of the innovative activity of the region. The formation of the presented roles of the micro-level subjects is carried out according to the implementation of national projects, based explicitly on the implication of federal projects in the subject of the Russian Federation. The number of roles of micro-level actors can vary within the main areas of federal projects, reflecting: productivity, digital economy, urban environment, science, and export activities. This list of areas of federal projects is chosen according to the research conducted by Matyugina and Klabukova. In their research, the authors identified key types of national projects correlated with an increase in innovative activity of the territory (Matyugina & Klabukova, 2020).

The highlighted specific characteristics of micro-level subjects depend on various factors of influence of other agents of the model of innovation activity of the region. It is proposed to use the municipal formation as an additional subject of innovative activity of the region. This condition is regulated by the fact that only the commonality of innovative activity of the municipality and the region will allow building a sustainable system of territorial development. The municipal formation is produced by the activities of local government, which, above all, are the subjects-evaluators of the innovative activity of agents of small territories. In simple terms, municipalities represent a rating subject forming an idea of the innovative activity of regions at the local government level. The main tools for the implementation of the evaluation function of municipalities are as follows:

- Rating of innovative activity. It allows to evaluate the level of development of innovative activity and the technologization of production of micro-subjects;
- Rating of innovation assistance. It evaluates the activity of local self-government bodies in terms of the implementation of support measures and other activities aimed at improving the innovation environment in the municipality;
- Development of directions of innovative activity. It is a tool that allows accumulating the main ideas about the levels of innovative activity. The analysis of the levels of innovative activity allows to carry out strategic planning of directions of development of a particular type of activity.

Let us note that the presented toolkit has different directions of implementation. Thus, ratings of innovative activity and assistance declare conditions of formation of these processes for micro-level subjects. The regional authorities finalize the development of directions of innovative activity.

The next subjects of innovative activity of the region are investors, competitors, and markets with a wider range of competencies. First, these subjects of innovative activity are the participants forming the demand for innovation. The subject of the Russian Federation acts as the leading participant in the current model of innovation activity of the region. Evidently, if we talk about the market for state innovation, the subject of the Russian Federation can shape the demand for this product or service. Simultaneously, the created system is closed, which contradicts market conditions and forms of realization of innovations. As noted earlier, the subject of the Russian Federation can be an agent of commercialization of innovations, but only the market can form the demand for these products or services through investors and competitors.

Despite the allocation of clear functions of investors, competitors, and markets for the management of innovation activity in the release of an innovative product or service, the formation of the innovation infrastructure is a fairly significant activity. Compared with the current model of innovation activity, in which the region is the main actor forming the innovation infrastructure, there is a differentiation of this function in another model. Through federal and regional programs, the subjects of the Russian Federation provide for the formation of innovative infrastructure at the territorial level. Nevertheless, the limited funds and the conditions for investors to make a profit allow us to conclude that the interests of the implementation of forms of innovative activity are shifting. According to Pavlov et al. the formation of regional innovation infrastructure should be based on the 70/30 proportion (Pavlov et al., 2020). Under this proportion, the region regulates the construction and implementation of 70% of the innovation infrastructure. The remaining 30% are assigned to agents making profits from innovation projects in the Russian subjects.

4 Discussion

In general, the presented model of innovation activity of regions is revised in the framework of the formation of sustainable development of the territory. In the context of this model, the role of micro-level actors characterizing the market of innovative goods and focused not only on profit but also on the development of conditions for innovation activity is strengthening. The region's role as a key subject of innovation activity is reduced to the formation of prerequisites for the organization of innovative activity and the creation of conditions for its implementation by all participants of the model of innovation activity. The addition of the subjects of the model shows their differentiation and regulates specific actions for the formation of innovative activity of the regions to ensure their sustainable development. This research allowed us to focus on the conventionality of the existing model of innovation activity of the region. In the future, this study will be supplemented by a simulation model of innovative flows of the regional system of interaction of subjects of innovation and investment environment.

5 Conclusion

The focus of our research on innovative activity of the regions to ensure their sustainable development allowed us to make the following conclusions:

1. Essential characteristic of the category of innovation activity of the region is manifested in the context of different worldviews of scientists, specifying this concept according to certain characteristics. First, the innovative activity of the region is associated with the study of aspects of the technologization of production, innovation, and project management. An estimated indicator of innovation activity of the region is presented to confirm these characteristics within each definition. Estimated indicators are regulated by indices of technological production, the innovative activity of production, innovative activity, and project management of innovations.
2. The model of innovation activity of regions currently being implemented is inefficient since it concentrates on implementing the main activities by the regional authorities. A market economy involves the formation and commercialization of innovations by market entities. The presented model of innovative development of the region is closed to the subject of the Russian Federation. It reduces the role of micro-level actors, investors, competitors, and the market. Thus, the postulates of the formation of innovative activity of the region were revised by changing the subjective function and their competencies.
3. The proposed model of innovation activity of regions aims to ensure their sustainable development according to the assignment of new powers to the subjects of this environment. Regional authorities act as agents who develop tools for promoting innovation. The subjects of the micro-level have a slightly

differentiated role, which is manifested in the framework of the implementors of innovation policy, cooperators of innovative personnel, participants in projects for the development of the innovative activity, and managers of the innovative activity in the region. A municipal entity, acting as a subject-evaluator of innovative activity of the agents of this environment, is proposed as a new element of the model. Investors, competitors, and markets implement market functions in the model. Interaction conditions between all subjects allow achieving sustainable development of the region in the short and medium term.

References

Akaev, A. A., & Hirooka, M. (2009). A mathematical model for long-term forecasting of the dynamics of innovative economic activity. *Doklady Mathematics, 79*(2), 275–279. https://doi.org/10.1134/S1064562409020331

Barmuta, K. A. (2019). Innovation in industrial clusters of the region in terms of spatial economics. *Scientific Notes of the Vernadsky Crimean Federal University. Economics and Management, 71*(2), 8–16.

Czarnitzki, D., & Kraft, K. (2004). Management control and innovative activity. *Review of Industrial Organization, 24*(1), 1–24. Retrieved from http://www.jstor.org/stable/41799182

Illarionova, E. A., & Glekov, P. M. (2019). Innovative activity of the region as a factor of social and economic development of the Russian Federation (On the example of the Belgorod Region). *The Eurasian Scientific Journal, 11*(6), 22.

Lapygin, Yu. N. (2019). Scientific and innovative policy of development of the region (On the example of the Vladimir Region). *Scientists Notes, S3*, 72–83.

Liu, X., & White, S. (2001). An exploration into regional variation in innovative activity in China. *International Journal of Technology Management, 21*(1–2), 114. https://doi.org/10.1504/IJTM.2001.002903

Lyapina, I. R., Torgachev, D. N., & Martov, S. N. (2015). Prospects of the enterprise in the market: The essence and features of the definition. *Central Russian Journal of Social Sciences, 10*(5), 190–196. https://doi.org/10.12737/14348

Lyapina, I. R., Zhuravleva, T. A., Skobliakova, I. V., Petrukhina, E. V., & Dudina, E. V. (2021). Conditions and factors of innovation cycle formation. In E. G. Popkova, & B. S. Sergi (Eds.), *Modern global economic system: Evolutional development vs. revolutionary leap* (pp. 963–973). Springer.

Matyugina, E. G., & Klabukova, A. A. (2020). The Competitiveness of the region in terms of the formation of the relationship "Innovative activity-investment attractiveness." *Economics: Yesterday, Today and Tomorrow, 10*(2–1), 300–306. https://doi.org/10.34670/AR.2020.32.57.030

Nosov, A. L. (2020). Innovative activity of the region as an element of an innovative national system. *Innovative Development of Economy, 1*(55), 54–61.

Pavlov, K., Nosova, O., & Nosova, T. (2020). Innovation policy as a condition for effective development of the region's economy. *Accounting and Analysis, 2*(278), 3–9.

Schnabel, C., & Wagner, J. (1992). Unions and innovative activity in Germany. *Journal of Labor Research, 13*(4), 393–406.

Sliwka, D. (2003). Organizational structure and innovative activity. *Economics of Governance, 4*(3), 187–214. https://doi.org/10.1007/s10101-002-0054-y

Stiebale, J. (2016). Cross-border M&As innovative activity of acquiring and target firms. *Journal of International Economics, 99*, 1–15. https://doi.org/10.1016/j.jinteco.2015.12.005

Stroeva, O. A., & Kvak, A. A. (2014). Tools for implementing a proactive regional innovation policy. *Central Russian Journal of Social Sciences, 6*(36), 102–109.

Stroeva, O. A., Lavrikova, N. I., Alekhina, T. A., Semenova, E. E., &Stepanova, M. A. (2021). Development of organizational diagrams and governance mechanisms of region's economy. In E. G. Popkova, & B. S. Sergi (Eds.), *Modern global economic system: Evolutional development vs. revolutionary leap* (pp. 1807–1816). Springer. https://doi.org/10.1007/978-3-030-69415-9_199
Ushnurtsev, D., & Shulepov, V. (2016). Innovative enterprise activity analysis. *Economics, Management, Information, Technology, 4*(4), 17–23.

Sustainable Development in the Context of Global Warming

Sergey D. Tsymbalov, Svetlana N. Kuznetsova, Elena V. Romanovskaya, Nataliya S. Andryashina, and Elena P. Kozlova

Abstract The purpose of the study is to assess a set of measures to counter global warming developed in the Russian Federation. The organizational structure of international interaction and aspects of the national strategy to ensure sustainable development of the Russian economy in the context of global climate change are considered. The research methodology is based on the analysis of information published in the open press. The article provides data on the causes and dynamics of global warming, as well as the possible consequences of it for the economy of the Russian Federation. The fundamentals of Russia's climatic doctrine and planning of measures to reduce greenhouse gas (GHG) emissions, the concept of forming a system for monitoring, reporting and checking the volume of GHG emissions, draft regulations in the field of economic regulation of GHG emissions and the principles of developing corporate strategies to reduce their emissions are considered. Measures to reduce the carbon footprint of products manufactured by the cement industry are proposed on the basis of the conducted research.

Keywords Global warming · Greenhouse effect · Greenhouse gases · Sustainable development · Climate doctrine · National strategy

JEL Code A 1 · O 1 · N 6

S. D. Tsymbalov (✉) · S. N. Kuznetsova · E. V. Romanovskaya · N. S. Andryashina · E. P. Kozlova
Minin Nizhny Novgorod State Pedagogical University, Nizhny Novgorod, Russia

S. N. Kuznetsova
e-mail: dens@52.ru

E. V. Romanovskaya
e-mail: alenarom@list.ru

N. S. Andryashina
e-mail: natali_andr@bk.ru

E. G. Popkova and B. S. Sergi (eds.), *Geo-Economy of the Future*,
https://doi.org/10.1007/978-3-030-92303-7_71

1 Introduction

The research is based on the analysis of the adopted international legal acts and regulations of the Russian Federation aimed at reducing the anthropogenic impact on the climate.

The Earth's climate system includes five components interacting with each other, while the bulk of the thermal energy of the climate system is concentrated in the waters of the World Ocean.

The temperature of the surface layer of the ocean has increased by more than 0.5 °C since the early 1970s, while the temperature of the near-surface layer of the atmosphere throughout the planet has increased by 1.0 °C by 2017 (IPCC AR5, 2013). Most scientists believe that this situation is due to an increase in the concentration of greenhouse gases (GHGs) in the atmosphere and is associated with anthropogenic activities. Numerous studies indicate that an increase in the temperature of the Earth's atmosphere by more than 2 °C from the level of pre-industrial development of mankind will lead to irreversible climatic consequences. In 2017, global GHG emissions amounted to 37 billion tonnes in CO_2-eq., of which 32.6 billion tons account for the combustion of fossil fuels, and about 4 billion tons account for the production of cement and other industries associated with CO_2 emissions into the atmosphere (International Energy Agency World Energy Outlook, 2018).

2 Methodology

The study analyzed a number of existing international agreements and regulations that relate to the problem of global warming. The Russian Federation has been participating in the process of consolidating the efforts of the international community aimed at reducing the anthropogenic impact on the climate for about the last 30 years. Our country signed and ratified the United Nations Framework Convention on Climate Change (UNFCCC) in 1994,[1] the Kyoto Protocol to the UNFCCC[2] in 2004 and the Paris Agreement of December 12, 2015[3] in 2019. In accordance with the Paris Agreement, Russia pledged to limit carbon dioxide emissions at the level of 70–75% of the level achieved in 1990, which will be about 3 billion tons in CO_2—equivalent in numerical terms.

Figure 1 shows the approximate structure of the most significant GHGs in the Earth's atmosphere. It is assumed that an international fund, which will accumulate about 100 billion US dollars annually for the period from 2020 to 2025 and which

[1] Federal Law No. 34-FZ of November 4, 1994 «On Ratification of the UN Framework Convention on Climate Change».

[2] Federal Law No. 128-FZ of November 4, 2004 «On Ratification of the Kyoto Protocol to the United Nations Framework Convention on Climate Change».

[3] Resolution of the Government of the Russian Federation dated 21.09.2019 No. 1228 «On the adoption of the Paris agreement».

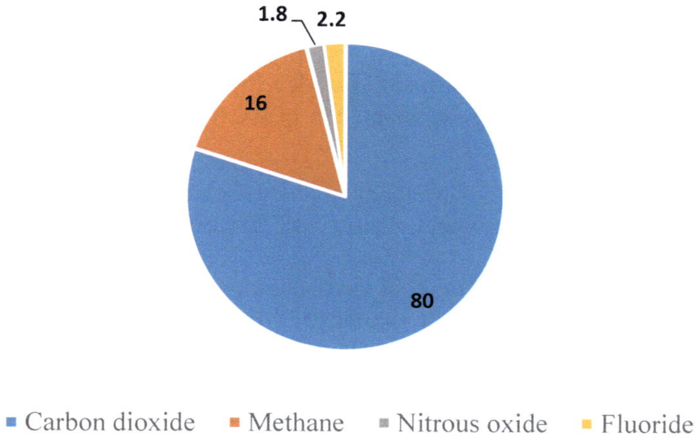

Fig. 1 GHG structure, %. *Source* Developed and compiled by the authors

will be sent to developing countries as support measures for sustainable development and the reduction of GHG emissions, will be created in accordance with the Paris Agreement. Basically, it is planned to reduce the volume of coal burned and increase the production and introduction of renewable energy sources.

Unfortunately, sometimes the carbon theme becomes a lever for the implementation of the mechanism of unfair competition. For example, EU countries recently initiated discussions on the introduction of cross-border carbon regulation (TUR), also known as the «Carbon Border Adjustment Mechanism» (CBAM). Russian experts and representatives of the business community note that it is necessary to ensure the implementation of the TUR solely as a mechanism for interaction between countries to reduce net GHG emissions. Otherwise, the TUR will be seen as an instrument of unfair competition, but not a climate effort within the framework of signed international agreements. Its introduction may lead to the formation of discriminatory conditions for goods imported into the territory of the European Union.

3 Results

The number of dangerous meteorological phenomena in Russia has approximately doubled over the past 15–20 years.[4] The negative consequences of the expected climatic changes for the Russian Federation may be: an increase in morbidity; intensification of droughts in some regions; precipitation, flooding and waterlogging; Forest fires; melting ice; violation of ecosystems; energy consumption.

[4] Report on Climate Features in the Russian Federation in 2017 (Roshydromet, Moscow, 2018).

The state system of measures[5] aimed at reducing the vulnerability of climate security was developed in our country to prevent negative challenges, a national plan for adaptation to climate transformation was developed.

The objectives of planning the adaptation of the Russian Federation to climate transformation are[6]:

- a scientific approach to the development and decision-making process aimed at increasing the degree of climate security and protecting the population from the effects of climate change;
- increasing the level of protection of the population and the state from the consequences of global warming and minimizing losses and damage to the country's economy from emergencies;
- obtaining additional benefits from global warming;
- updating the strategy for the development of economic activities and sectors of the economy, taking into account the impact of climate change on them and ensuring their implementation within the framework of state programs of public–private partnership and investment projects;
- implementation of measures to protect Russian enterprises engaged in foreign economic activity from unfair actions of foreign partners;
- fulfillment of the international obligations of the Russian Federation under the UN Framework Convention on Climate Change and other international treaties.

In accordance with the Concept, in order to solve problems related to greenhouse gas emissions at the level of organizations,[7] it is necessary to develop and adopt regulatory legal acts concerning:

- criteria assessment of organizations responsible for monitoring and providing information on greenhouse gas emissions;
- obligations of organizations to monitor and submit information (reports) on the volumes of emissions and on the reduction of GHG emissions to the authorized state executive authorities;
- guidelines and manuals for quantifying the volume of direct GHG emissions by organizations and indirect emissions during the production of heat and electricity;
- uniform requirements for the content of the report on GHG emissions;
- establishing the accreditation procedure for independent expert organizations that ensure the verification of reports on project GHG emission reductions;

[5] Order of the President of the Russian Federation. Dated 17 December 2009 No. 861-rp. On the Climate Doctrine of the Russian Federation.

[6] Order of the Government of the Russian Federation. Dated 25 December 2019 No. 3183-r. On Approval of the Action Plan for the First Stage of Adaptation to Climate Change for the Period Up to 2022.

[7] Order of the Government of the Russian Federation. Dated 22 April 2015 No. 716-r. On Approval of the Concept for the Development of a System for Monitoring, Reporting and Verification of Greenhouse Gas Emission Volumes in the Russian Federation.

- establishing the procedure for the collection, verification and registration of reports of organizations on GHG emissions by authorized state executive authorities.

Currently, a Draft Decree of the Government of the Russian Federation on the approval of the Concept of the system of accounting, registration, release into circulation, transfer and offset of the results of climate projects carried out on the territory of the Russian Federation has been prepared and is being discussed, which is necessary for the realization of joint implementation and carbon offset projects. The draft federal law «On State Regulation of GHG Emissions and on Amendments to Certain Legislative Acts of the Russian Federation» is also under discussion. It defines the legal framework for state regulation of GHG emissions in the Russian Federation in order to create conditions for sustainable development with low GHG emissions and increase the competitiveness and sustainability of the Russian economy in the context of the global transition to climate-sustainable development.

To date, methods for calculating direct[8] and indirect energy GHG emissions[9] have been developed and approved in Russia. The procedure for quantifying the mass of GHGs emitted by enterprises is established in the methodological guidelines[10] and the categories of emission sources that must be taken into account, including cement industry enterprises, are given. Direct emissions are calculated using conversion factors for the mass or volume of various fuels into conventional tonnes and GHG emission factors per tonne of fuel equivalent. At the first stage, control over Russian enterprises emitting more than 150 thousand tons of GHG per year in CO_2-eq. is established. In the future, organizations with an emission level of more than 50 thousand tons of CO_2-eq. will be required to submit reports on GHG to the authorized bodies. The largest companies and enterprises, such as PJSC Gazprom, UC Rusal, JSC Arkhangelsk PPM, etc., annually voluntarily submit reports on GHG emissions to the international non-profit organization CDP (Carbon Disclosure Project), whose head office located in UK, for publication. This information is used in the international market when purchasing products. At the same time, other things being equal, preference is given to those products that are produced with the lowest carbon footprint. There is an aspiration of market participants to make environmental reporting and environmental risk management a business norm. More than 8.4 thousand companies have publicly disclosed environmental information through CDP since 2002 (Carbon Project, 2021).

Companies are assigned climate ratings, which are published and posted on their portals by the world's leading news agencies (Bloomberg, Thomson Reuters, Google

[8] Order of the Ministry of Natural Resources and Environmental Protection of the Russian Federation. Dated 30 June, 2015 No. 300 On Approval of methodological recommendations and guidelines for quantifying the volume of GHG emissions by organizations carrying out economic and other activities in the Russian Federation.

[9] Order of the Ministry of Natural Resources and Environmental Protection of the Russian Federation. Dated 29 June, 2017 No. 330 "On the approval of guidelines for the quantitative determination of the volume of indirect energy emissions of greenhouse gases.

[10] See Footnote 8.

Finance), based on the results of an independent assessment of the submitted reports. Requirements for the disclosure of information on the GHG emissions by companies are fixed by law in many countries. Carbon reporting of companies becomes one of the main types of reporting that determine their investment attractiveness. When compiling reports, the companies apply principles established in the Greenhouse Gas Protocol (GHG Protocol), which allow quantitative assessment and management of GHG emissions (Ermakova, 2021). In Russia, the development of corporate strategies for low-carbon development is practiced using ISO 14090:2019 «Adaptation to climate change—Principles, requirements and guidelines». A fundamental element of any corporate carbon strategy is reducing the company's carbon footprint, and ideally achieving carbon neutrality. Zero carbon footprint can be achieved in a variety of ways, namely through:

- reduction of direct GHG emissions from sources that are directly controlled by the company;
- reducing indirect GHG emissions through energy and supply chain management;
- compensation (reimbursement) of unavoidable GHG emissions by reducing GHG emissions from third parties through participation in carbon projects.

Many Russian companies are developing strategies for the phased emission reduction for the period up to 2030 and beyond, taking into account the expected introduction of legislation on the economic regulation of GHG emissions. It is necessary to consistently solve the following tasks within the framework of the corporate strategy to reduce GHG emissions and increase the competitiveness of products:

- to carry out an inventory of GHG emissions and determine the carbon footprint of the company and the services or products produced;
- to develop a plan for specific technical and technological measures to reduce GHG emissions;
- to establish a system for monitoring GHG emissions during the company's economic and other activities;
- to improve energy efficiency of production;
- to establish a system of accounting and verification of GHG;
- to establish trade in carbon units in the domestic and foreign markets.

For example, let's consider the cement business, which is one of the largest GHG issuers. The amount of CO_2 generated during the combustion of process fuel during the burning of cement clinker and from the calcination of limestone is calculated in the production of cement. Also, GHG emissions from own boiler houses operating on fossil fuels or oil products, emissions from automobile and special vehicles equipped with internal combustion engines, emissions generated during gas welding, etc. are taken into account. These emissions are classified as direct ones and are calculated on the basis of data that can be obtained from the services of the enterprise (accounting, production and technical department, departments of the chief power engineer, supply, etc.). The plans to reduce GHG emissions annually include measures aimed at saving electric and thermal energy, reducing fuel and limestone consumption, using alternative fuels, strengthening the thermal insulation

Table 1 BATs directly or indirectly affecting the level of GHG emissions

BAT number	BAT name	Impact on GHG emissions
1a	Replacement of natural raw materials with production waste	Direct GHG emissions from calcination decrease with decreasing limestone consumption
1b	Reducing the content of the proportion of clinker in cement to the maximum allowable level	The share of material obtained with the release of GHGs in the final product of the enterprise decreases
2	Using carbonaceous waste as raw materials	Consumption of fossil carbonaceous fuels and thus GHG emissions are reduced
3	Reduction of heat losses during clinker burning due to the use of technical solutions	Direct proportional effect of heat losses on GHG emissions
4	Reducing the consumption of thermal energy by generating additional amounts of electricity or heat by combining factories with thermal power plants (TPPs) or heating plants that use heat from clinker kilns to recuperate air sent for fuel combustion in TPP boilers	Fuel consumption and GHG emissions are reduced by increasing the efficiency of fuel combustion in boilers of TPPs
5	Using waste as an alternative fuel	Direct impact on reducing fossil fuel consumption
6	Reducing the specific energy consumption for the production of 1 ton of Portland cement	Indirect proportional impact on reducing fossil fuel consumption
7	Continuous fulfillment of certain requirements of the energy management system	The introduction of technical solutions that provide energy savings for cement production entails a decrease in GHG emissions

Source Developed and compiled by the authors

of furnaces, introducing the best available technologies (BATs). In 2015–2017, 51 Information and Technical Reference Books on the BATs were published, including ITRB 6–2015 (cement production).[11] Table 1 shows BATs directly or indirectly affecting the level of GHG emissions in the cement industry.

Thus, 7 out of 16 BATs given in the handbook[12] directly or indirectly affect the level of GHG emissions. Analysis of the technical and technological state of the cement business shows that there are significant reserves to reduce GHG emissions. For example, a decrease in the specific consumption of limestone from 0.7941 to 0.7923 t/t of clinker due to optimization of the technology with a clinker production volume of 800 thousand t/year will ensure a decrease in CO_2 emissions by:

[11] Information and Technical Reference Book on the BATs 6-2015 (cement production), M., BAT Bureau, 2015, p. 305.

[12] See Footnote 11

Fig. 2 Biofuel supply unit. *Source* Developed and compiled by the authors

$$(0.7941 - 0.7923) \times 800000 \times 0.44 = 634\,\text{t/year CO}_2,$$

where 0.44—CO_2 emission factor during thermal decomposition of limestone. Using natural gas as a process fuel at a cement plant of this capacity instead of coal will reduce GHG emissions by about 15%. Many factories are beginning to use combustible waste (dried sludge from irrigation fields of treatment facilities, crushed car tires, RDF, etc.) as fuel for clinker burning, which makes it possible to reduce the consumption of fossil fuels and, accordingly, GHG emissions. For example, tests have been successfully carried out recently and the process of combustion of biofuel obtained from the digested sludge of domestic wastewater has been introduced into production at the factories of LLC «Group SMiKom» and LLC «Heidelberg cement Rus». For this purpose, special units for feeding biofuel up to 150 t/day to the calciner reactor of the clinker kiln were built at the enterprises (Fig. 2).

The calorific value of the alternative fuel is quite high and lies in the range of 2340–2400 kcal/kg, which is comparable to the calorific value of fuel peat and domestic waste (non-biological fraction). Taking the coefficient of conversion of sludge to equivalent fuel equal to 0.34 and the coefficient of CO_2 emission/tce equal to 3.11 from Table 1 of the methodological guidelines,[13] it is possible to roughly estimate the level of GHG emissions for 350 days of operation of a cement plant per year

$$150 \times 350 \times 0.34 \times 3.11 = 53513\text{t/yearCO}_2.$$

[13] See Footnote 8

Thus, in the general balance, this amount will replace GHG emissions from fossil fuels into the Earth's atmosphere.

4 Conclusion

There is no full-fledged emission control system in Russia yet. The responsible authority for the creation of the national system of economic regulation in Russia is the Ministry of Economic Development of the Russian Federation. It is planned to use the experience of European countries. In world practice, two main mechanisms of economic regulation are used: taxes on GHG emissions (carbon tax) and a quota and GHG emissions trading scheme (ETS).

There are also combined options: taxes and trade (tax and trade); quotas and taxes (cap and tax) (Tsymbalov & Sintsov, 2017; Tsymbalov et al., 2019). The first draft of the federal law on the introduction of charges for GHG emissions, discussed in 2019, was rejected due to serious disadvantages. To date, there is a regulatory framework for the formation of a GHG emissions register, however, the practice of verifying the reported data on GHG emissions has not yet become widespread.

References

Carbon Project: Wikipedia. Free encyclopedia. URL: https://en.wikipedia.org/wiki/Carbon_Disclosure_Project. Data accessed 29 Mar 2021.

Ermakova, M. S. (2021). Greenhouse gas emissions: Put it on the shelves. *Ecology of Production, 2*, 98–105.

International Energy Agency World Energy Outlook. (2018). www.iea.org.

IPCC AR5. (2013). AR5, Climate Change (2013): The Physical Science Basis. Cambridge University Press, Cambridge, United Kingdom and New York, NY, USA. SPM, in Russian (p. 6). www.ipcc.ch.

Tsymbalov, S. D., Sintsov, E. V. (2017). Development of a low-carbon development strategy for the company. In *Proceedings of the Conference «Modern Research of the Main Directions of the Humanities and Natural Sciences»*. Kazan.

Tsymbalov, S. D., Romanovskaya, E. V., Romanovskiy Yu.V., Bakulina N.A. (2019). Towards sustainable environmental development. Education and Law. M. **10**, 224–227.

Modern Approaches to the Sustainable Development of Adaptive Enterprise Systems Through the Prism of Geoeconomics (on the Example of Increasing the Efficiency of Air Purification Processes)

Victor P. Kuznetsov, Sergey D. Tsymbalov, Ekaterina P. Garina, Elena V. Romanovskaya, and Elena P. Kozlova

Abstract Today the economic framework conditions require efficient and high-quality work on planning the operating activities of industrial enterprises, this also including compliance with certain requirements for the implementation of technologies in polluting industries. These requirements largely determine costs. One of the operational costs of production management areas is the issue of environmental component, miscalculations which can lead to operating losses, and ultimately jeopardize the entire company. Mechanical engineering enterprises, solving the issue of cleaning up emissions, consider it as a factor of strategic success in the context of competitive development of production through increasing the efficiency of processes, as well as improving the company's image. The object of the research is the operating industrial and energy enterprises of the Nizhny Novgorod region. The subject of the research is the results of the operating activities of industrial enterprises (on the example of individual mechanical engineering industries) in terms of the environmental component. The purpose of the study is detailed engineering of adaptive systems for ensuring permissible emissions into the atmosphere.

Keywords Production systems · Increased efficiency · Effectiveness of implementation · Environment · Ecological economy

JEL Code Q5

V. P. Kuznetsov (✉) · S. D. Tsymbalov · E. P. Garina · E. V. Romanovskaya · E. P. Kozlova
Minin Nizhny Novgorod State Pedagogical University, Nizhny Novgorod, Russia

E. V. Romanovskaya
e-mail: alenarom@list.ru

E. G. Popkova and B. S. Sergi (eds.), *Geo-Economy of the Future*,
https://doi.org/10.1007/978-3-030-92303-7_72

1 Introduction

The development of large industrial centers has led to the rapid growth of negative externalities, such as the state of the air environment in large cities in which industrial and energy enterprises are located.

In current study examined a model of sustainable development through the prism of geo-economics. Sustainable development models describe various forms of transformations carried out by enterprises in the use of energy, material, human and other resources. They reflect on the search for new opportunities and their application (Alpidovskaya & Popkova, 2019; Inshakova Bogoviz, 2020; Popkova, 2017). These models represent the process of converting the base into an industrial production capacity (Popkova et al., 2020; Sedykh, 2019; Smirnova & Kochnova, 2019).

In the Nizhny Novgorod region, one of these is the metallurgical production of PJSC "GAZ", which produces up to 60 thousand tons of castings per year. This circumstance is caused by an in-depth analysis of the problem At the site of PJSC "GAZ", there is one of the largest operating foundries in Russia. Foundry and blacksmith shops do not have a sanitary protection zone in relation to other industries, and a zone to the residential area is also insufficient. Also, an applied technological process of making castings in one-time sandy-clay forms is characterized by a high level of formation. As a result, a release of various harmful substances into the atmosphere, and the level of dust emission is especially high. Dust is released during melting, molding, mixing, forming, cleaning, making rods. Since the classical technology of mass production of castings at the present time (more than 80% of the world production of castings from ferrous alloys, including in developed countries, falls on this method) is characterized by significant emissions into the atmosphere. The only method to reduce the harmful effect on the environment is to isolate the technological process and capture secretions (Garina et al., 2017, 2018).

2 Methodology

It was proposed to improve the efficiency of the process of cleaning emissions into the atmosphere through the introduction of devices in the first stage of cleaning. During development of a purification device, a scheme with internal water circulation was chosen, which allows to drastically reduce its consumption.

The issues of self-cleaning of droplet collection systems were also resolved. As a result, we managed to pick up such a structure and the device operation, which circulated in the droplet system significant amounts of water, which helped to prevent the overgrowth of deposits dust and virtually eliminate the need for forced cleaning. In devices of type PVM (wet ventilation dust collector), this problem is not solved. In the new devices, the cleaning cascade is made taking into account the use of several dust collection effects (shock, inertial, coagulation), which made it possible to collect dust quite effectively. The design of the apparatus called PBIV-30 (dust collector bubble-inertial fan with a nominal capacity of 30,000 m ^{3}/hour). This device is equipped with a system for monitoring and maintaining the water level, glazed windows to control its operation (Kozlova et al., 2020; Kuznetsov et al., 2018; Kuznetsova et al., 2018).

Several variants of devices of the first stage of cleaning have also been developed (cyclone of increased efficiency, louvered grill). In 2000, several samples of dry dust collectors of the inertial principle of operation (louvered grilles) were manufactured in production. This device on the dust of the 2nd group showed an efficiency of 80% with resistance at the level of a dry cyclone. The degree of cleaning when using the apparatus as the first stage is quite sufficient, but in comparison with cyclones, it is much more compact and easier to install. New ventilation systems in production are assembled on a modular basis. The ventilation system includes a standard fan and a standard set of cleaning devices, while the required performance is achieved by installing the required number of modules. For sources of the second and third group is set two-step purification: the first-stage cyclone or aerodynamic dry filter (louvre) or inertial trap ensuring purification rate of about 70–80% and in the second stage PBIV-30 apparatus. The total degree of purification at high input concentrations reaches 96–99%.

Practical experience allows us to assert that for a given period in foundries. It is possible to develop only wet dust cleaning systems because they are the cheapest, most compact, and easy to operate. Significant changes have also been made to the water supply system. The current design of the unit excludes the deposition of sludge in it and ensures guaranteed operation of the siphon unit by supplying makeup water to the drainpipe zone. In order to control the presence of water in the device, a reliable conductometric sensor is used, which blocks the operation of ventilation.

3 Results

The indicators obtained as a result of device operation when collecting dust from various sources of the foundry are shown in Table 1.

As seen from Table 1, the proposed treatment system apparatus ensures a sufficiently high degree of dust collection. In cases of application on dust sources of groups 2, 3, and 4 (which are shown in the table), not only the RF standards are fulfilled, but also rather strict standards adopted in the countries of Western Europe and the USA. On the basis of the above, it is possible to formulate the general principles of approaches to the modernization of ventilation equipment and the installation of newly installed equipment in foundries and forging shops (Sergi, 2018).

1. During major overhauls, gradually replace the fans and cleaning devices with the equipment of the GAZ PJSC system.

Table 1 Efficiency of cleaning the PBIV-30 apparatus from various sources of dust emission

No p/p	Dust source	Incoming concentration (mg/^{m3})	Output concentration (mg/^{m3})	Efficiency %
1	Passing the tumbler	416–1539	15–25	96,4–98,4
2	Shot-blasting chamber	5068–22,875	12,5–36,3	99.8
3	Knocking out molding lines	143–352	15,5–19,1	89,2–95
4	Stationary emery machines	610	8.3	98.6

Source The table was compiled based on data from the Committee for Nature Protection and Environmental Management of the Nizhny Novgorod Region

2. Ventilation equipment should be placed in convenient places accessible for repair and maintenance, seeking space for it due to the movement of technological equipment.
3. Place the cleaning device as close to the source of dust as possible, and exclude horizontal sections of the air ducts. In this case, it is more expedient to go for the complication of the waste disposal system than for the removal of cleaning devices from the source.
4. All ventilation exhaust systems must be equipped with cleaning devices, and low-power ventilation systems without cleaning must be eliminated.
5. Equipment projects must include equipment shelter and equipment for its sealing.

Based on the PBIV apparatus and the fan, a modular aspiration "Complex for air purification" is assembled. Our operating experience of the "Air purification complexes" shows that it is advisable to place them directly at the sources of the formation of emissions. This makes it possible to reduce the length of the air ducts, exclude their overgrowth, and reduce the labor intensity of maintenance. The completeness of the developed module makes it possible to implement this principle.

Six years of experience in operating such systems at PJSC GAZ allows us to conclude that they are simple, reliable, and economical. It is advisable to use such modules in foundries and forging shops to collect dust.

A wet cleaning apparatus, a bubble-inertial fan dust collector with a nominal capacity of 30,000 m^3/hour, is designed to capture dust particles contained in aspiration emissions from foundry technological units. It can be used for capturing gases soluble in water or in aqueous solutions of reagents. When the device is used to work with environments that are corrosive to carbon steel, the body, and other elements are made of corrosion-resistant materials.

The principle of operation is based on trapping dust at the water–air interface, formed when contaminated air is sucked in by means of a centrifugal fan through an apparatus partially filled with water. It uses a scheme of internal water circulation, which allows to significantly reduce its consumption. When moving dusty air through the apparatus, the following principles of wet cleaning are consistently implemented:

shock, inertial, bubbling, centrifugal. The droplet collection system of the apparatus is centrifugal-inertial.

Table 2 shows the technical characteristics of the PBIV-30 apparatus.

Figure 1 shows the design of the PBIV-30 apparatus.

This apparatus consists of a cylindrical body 1 with a conical bottom, a swirler 2, a precipitator 3, a drop catcher 4. It has an upper chamber, which forms an inlet cavity and an inlet pipe 5, which passes into the central pipe 6, an outlet cavity with an outlet pipe 7. There is a hatch 8 on the roof of the upper chamber, a water level control unit 9 is welded to the body, where the systems are mounted: maintaining a given water

Table 2 Technical characteristics of the PBIV-30 apparatus

1. Performance	Nominal maximum minimal	30,000 m^3/hour 35,000 m^3/hour not limited
2. Cleaning efficiency		85–99% depending on dispersion and concentration
3. Water consumption for make-up		0.15 m^3/h
4. Volume of water for one filling		2.6 m^3
5. Hydraulic resistance		290 kg/m^3

Source Developed and compiled by the authors

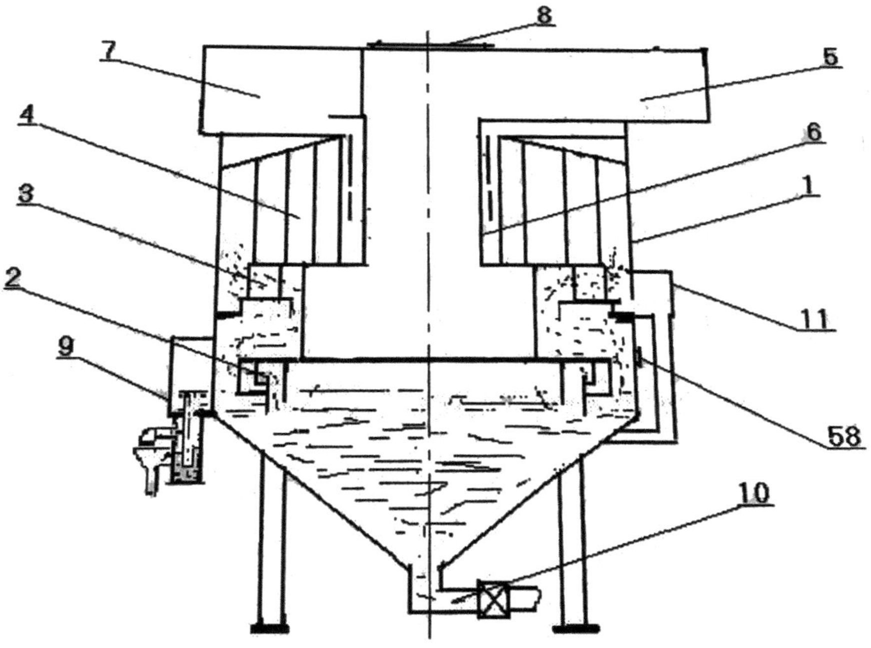

Fig. 1 Design of the PBIV-30 apparatus. *Source* Developed and compiled by the authors

level, refueling, and monitoring the presence of water. A sludge discharge unit 10 is attached to the flange of the conical bottom. On the body there are also welded pockets 11 with overflow pipes to drain water from the precipitator. Top pockets have threaded plugs for rinsing capability.

PBIV-30 device is easy to operate and requires practically no maintenance, except for changing the water. The frequency of water changes depends on the dust content in the emissions and can range from several hours to several days. It must be indicated in the passport of the ventilation system. During the operation of the device, it is necessary to periodically monitor the ammeter readings and the presence of water overflow into the funnel. A decrease in current against the nominal indicates an increase in the water level or clogging of the air ducts or apparatus with dust deposits. The water level may rise if the siphon unit is clogged. In the brief stop ventilation system, the device must be filled with water prior to its discharge into the hopper, it protects the machine from clogging of dust due to the presence of a natural draft ventilation system. When stopped for a long time, the air duct from the dust source to the device must be covered up. Compliance with the operating rules ensures that there is no need to clean a device. After a long period (more than a year), deposits may form on the louvre plates of the drip tray. It is recommended to inspect the plates once a year through the housing windows, after removing their covers, and, if necessary, remove deposits by flushing with water from a hose under pressure.

The body of the apparatus and the main units are designed for 15 years of operation in a non-aggressive environment and do not need repair during this period. If necessary, it is allowed to carry out welding work on the repair of the housing in the area above the swirler by filling the apparatus with water above the roof. In order to carry out this work, the device must be disassembled and the rubber removed. If the apparatus is made of corrosion-resistant materials, the service life is much higher, approximately 25 years. During this period, abrasive wear of the rubber lining is possible, which must be replaced with the disassembly of the device.

4 Conclusion

Under the target program of ensuring environmental safety of working conditions and reducing occupational morbidity in the foundry, 43 air purification complexes were installed and put into operation to replace the outdated ones. In 2007–2010, 12 air purification complexes were put into operation per year. The introduction of "Air purification complexes" allowed reducing occupational morbidity, reducing the release of suspended solids into the atmosphere, reducing water consumption and significantly improving the environmental situation at the working places of molding, cutting and smelting sections of foundries.

The construction of wastewater treatment plants in the blacksmith industry and the production of trucks (treatment facilities for zinc-containing wastewater) continues.

Due to the introduction of air protection measures, air emissions decreased by 68.1 tons per year, or by 0.4%.

The increase in the sale of production wastes was due to the in-depth processing of metal waste and a rise in prices.

Measures to optimize the use of production wastes made it possible to reduce the disposal volumes by 3.2 thousand tons.

In the field of occupational safety and health of employees, GAZ PJSC is guided by the main directions of state policy. In the production process, the priority is set to preserve the health of the company's employees. Also, the occupational health and safety management system was included in the unified management system of GAZ PJSC. Its aim was the creation and continual improvement of the conditions under which provided labor safety of workers, the quality and productivity of all employees. Thanks to this, health and well-being improved. The foundation is laid for a decent and stable future for the entire team.

Because of the work on the prevention of work-related injuries based on the OSH management system to reduce occupational injuries. The number of accidents decreased by 42%.

99.4% of workers employed in industries with hazardous working conditions have undergone preventive examination. The growth in the number of employees who underwent preventive treatment was more than 4%.

References

Alpidovskaya, M. L., & Popkova, E. G. (2019). Marx and modernity: A political and economic analysis of social systems management. In E. G. Popkova (Ed.), *Advances in Research on Russian Business and Management*. Information Age Publishing.

Garina, E. P., Kuznetsov, V. P., Egorova, A. O., Romanovskaya, E.V., & Garin, A. P. (2017). Practice in the application of the production system tools at the enterprise during mastering of new products. In *Overcoming uncertainty of institutional environment as a tool of global crisis management. "Contributions to Economics"* (pp. 105–112).

Garina, E. P., Kuznetsov, V. P., Romanovskaya, E. V., Andryashina, N. S., & Efremova, A. D. (2018). Research and generalization of design practice of industrial product development (by the example of domestic automotive industry). *Quality—Access to Success., 19*(S2), 135–140.

Inshakova A O., & Bogoviz A. V. (Ed.) (2020). Alternative methods of judging economic conflicts in the national positive and soft law. In E. G. Popkova (Ed.), *Advances in research on Russian business and management*. Information Age Publishing.

Kozlova, E. P., Kuznetsov, V. P., Garina, E. P., Romanovskaya, E. V., & Andryashina, N. S. (2020). Methodological bases of the assessment of sustainable development of industrial enterprises (technological approach). *Lecture Notes in Networks and Systems., 91*, 670–679.

Kuznetsov, V. P., Romanovskaya, E. V., Egorova, A. O., Andryashina, N. S., & Kozlova, E. P. (2018). Approaches to developing a new product in the car building industry. *Advances in Intelligent Systems and Computing., 622*, 494–501.

Kuznetsova, S. N., Romanovskaya, E. V., Artemyeva, M. V., Andryashina, N. S., & Egorova, A. O. (2018). Advantages of residents of industrial parks (by the example of AVTOVAZ). *Advances in Intelligent Systems and Computing., 622*, 502–509.

Popkova, E. G. (2017). Economic and legal foundations of modern Russian society. In E. G. Popkova (Ed.), *Advances in research on Russian business and management*. Information Age Publishing.

Popkova, E. G., Przhedetsky, Y. V., Przhedetskaya, N. V., & Borzenko, K. V. (Ed.). (2020). Marketing of healthcare organizations: technologies of public-private partnership. In E. G. Popkova (Ed.), *Advances in research on Russian business and management*. Information Age Publishing.

Sedykh, E. P. (2019). A system of regulatory support for project management in education. *Vestnik of Minin University., 7*(1(26)), 1.

Sergi, B. S. (Ed.). (2018). *Exploring the future of Russia's economy and markets: Towards sustainable economic development*. Emerald Publishing.

Smirnova, Z. V., & Kochnova, K. A. (2019). Training of employees of service enterprises using information technology. *Vestnik of Minin University., 7*(1), 5.

New Approaches to the Economic Security of Russian Regions in Terms of the COVID-19 Pandemic

Elena A. Yurina, Evgeniya S. Tishchenko, Elena V. Fedotova, Boris E. Markov, and Natalia I. Kuzmenko

Abstract The study aims to identify the problems of economic security of the Russian regions under the influence of force majeure caused by the pandemic COVID-19 and their assessment, as well as to determine ways to stabilize the socio-economic situation. The authors conducted a grouping of the economic security indicators of the federal districts and proposed a methodology for its assessment in terms of the coronavirus pandemic. The authors identified the problem areas of socio-economic development of the Russian regions and also proved the decline in their economic security during the pandemic. The authors evaluated the effectiveness of the internal state administration of the federal districts according to the anti-crisis measures adopted in 2020 (maintaining turnover in the main regional segments of the economy, avoiding tax debt on regional fees, increasing the profitability of the real business, raising the level of digital development), as well as measures to preserve employment and solvency of the population of each region. The authors justified the directions of normalization of the socio-economic development of the federal districts of Russia by eliminating the economic security problems.

Keywords Economic security · Indicator · Region · Federal district · Pandemic · Covid-19 · Crisis · Public administration

JEL Codes R11 · R58

E. A. Yurina (✉)
Tambov State University Named After G.R. Derzhavin, Tambov, Russian Federation

E. S. Tishchenko
Kuban State Technological University, Krasnodar, Russian Federation

E. V. Fedotova
Kaluga Branch of the Russian State Agrarian University, Moscow Agricultural Academy Named after K. A. Timiryazev, Kaluga, Russia

B. E. Markov
Krasnodar University of the Ministry of Internal Affairs of the Russian Federation, Krasnodar, Russian Federation

N. I. Kuzmenko
Voronezh State Pedagogical University, Voronezh, Russian Federation

E. G. Popkova and B. S. Sergi (eds.), *Geo-Economy of the Future*,
https://doi.org/10.1007/978-3-030-92303-7_73

1 Introduction

The global socio-economic crisis caused by the rapid spread of a new coronavirus infection requires a review of approaches to identifying problems of economic security of countries and their separate regions (Kulkova, 2020). This is so because this global challenge has had a strong impact on the production activities of enterprises, caused a reduction in employment and significant losses in the real sector of the economy and led to a decrease in budget revenues at all levels (Bank of Russia, 2020). The anti-crisis measures implemented in 2020 were aimed at preventing changes in the characteristics of consumer demand, reducing the supply of goods (Drobot et al., 2020), disrupting the supply chain (Salisu, 2020), destroying intersectoral ties, slowing economic activity, and falling incomes of citizens (Ashraf, 2020). All regions of Russia are facing an increase in threats to economic security that require timely detection and rapid leveling. It is necessary to focus on the internal results of the socio-economic policy of the federal districts in terms of the effectiveness of countering the sharp decline in business activity caused by the coronavirus pandemic. Today, it is relevant to assess the effectiveness of the internal state administration of the federal districts according to the anti-crisis measures adopted in 2020 (maintaining turnover in the main regional segments of the economy, avoiding tax debt on regional fees, increasing the profitability of the real business, raising the level of digital development), as well as analyzing measures to preserve employment and the solvency of the population of each region of Russia.

2 Literature Review

The analysis of legislative acts of the Russian Federation and scientific literature allows us to define the economic security of the state as a certain safe state of the national economy under the influence of different threats—internal and external. In this state, it is possible to ensure the unity of the economic space and implement the strategic national priorities of Russia (Government of the Russian Federation, 2014; State Duma of the Russian Federation, 2010). Thus, the main risk to the country's economic security is the uneven spatial development of the state and the likelihood of deepening the differentiation of regions by the level of socio-economic development (Bogomolov, 2012; Kalinina, 2010; Karavaeva et al., 2019; Shubina, 2017). The evaluation of the effectiveness of regional policies in response to the COVID-19 pandemic and other challenges of the global economy is the goal of the study. The choice of a regional agenda for discussing economic security risks during the COVID-19 pandemic is not accidental, since the possibility of stabilizing the socio-economic development of Russian regions and maintaining the well-being of the population is determined by the effectiveness and timeliness of measures taken by state authorities at all levels of government (Drobot, 2020; Karpunina et al., 2020; Lisova et al., 2020; Mejokh et al., 2020).

3 Methodology

The purpose of the study is to identify the problems of economic security of the Russian regions under the influence of force majeure caused by the pandemic COVID-19 and their assessment, as well as to determine ways to stabilize the socio-economic situation.

Research objectives: (1) to study the impact of the COVID-19 pandemic on the state of economic security of the Russian regions, as well as to assess the changes that have occurred; (2) to propose a methodology for assessing the internal economic security of the federal districts (without taking into account federal influence); (3) to determine the directions for stabilizing the socio-economic situation in the Russian regions by leveling the economic security problems.

Research methods: theoretical analysis, comparison, systematization, economic and statistical analysis, system approach, graphical method.

4 Results

The analysis of the economic security of the federal districts should be carried out by selecting the most significant factors for ensuring the sustainable growth of their economy. Increasing industrial production is a key factor in this sustainability. The index of industrial production in Russia as a whole in 2020 was 97.1% compared to the level of 2019. The situation in the federal districts is shown in Fig. 1 (Federal

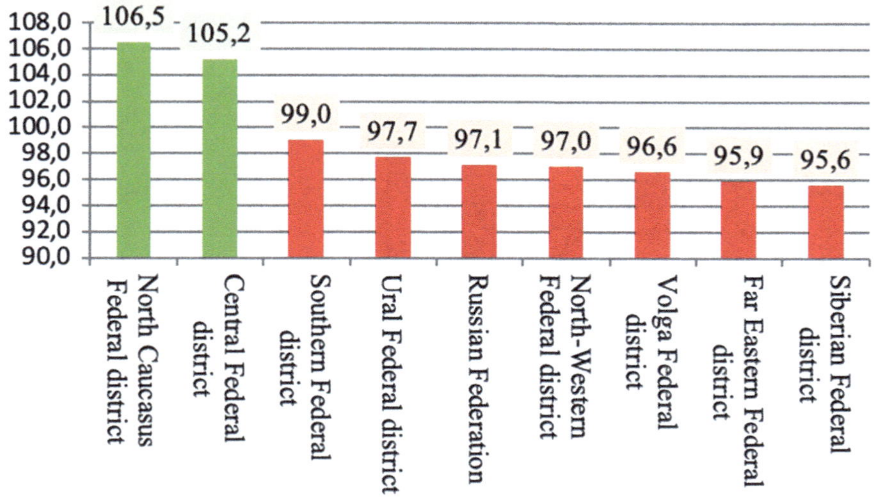

Fig. 1 Industrial production index for the federal districts, %, 2020 compared to 2019. *Source* Created by the authors based on (Federal State Statistics Service of the Russian Federation, 2021b)

State Statistics Service of the Russian Federation, 2021a).

Industrial production in 2020 compared to 2019 increased only in two districts—the North Caucasus and Central Federal Districts, and decreased in six federal districts. At the same time, in the North-Western, Volga, Far Eastern, and Siberian Federal districts, there is a decline in industrial production more than, on average, in Russia (97.1%). The coronavirus pandemic, restrictive measures, and a sharp decline in the business activity of organizations and citizens led to a drop in household incomes and an increase in unemployment. There is a decrease in the purchasing activity of citizens, which had an impact on the volume of retail trade, the market of paid services, and the decline in these sectors was more serious than in the regional industry. In 2020, the retail trade turnover in Russia amounted to 95.9%, the turnover of the paid services market decreased to 82.7% compared to 2019 (Federal State Statistics Service of the Russian Federation, 2021b). We can only note the growth of wholesale trade turnover to 101.3% from the level of 2019. We will analyze the retail market turnover for 2020 by federal districts, in comparison with 2019. This graph has a 3-dimensional axis for assessing changes in economic indicators: horizontally-retail trade turnover in 2020, in % by 2019; vertically-retail trade turnover for 2020, in billion rubles; circle size—the share of the federal district in the total retail market turnover in 2020 (Fig. 2).

Retail trade turnover decreased in all federal districts, but the smallest decrease was in the North-Western and Central Federal Districts, and the largest decline was typical for the Volga and North Caucasus Federal districts.

The analysis of the turnover of the paid services market by federal districts in 2020 is presented in Fig. 3 (horizontally—the turnover of the paid services market in 2020, as a % of the turnover in 2019; vertically—the turnover of the paid services

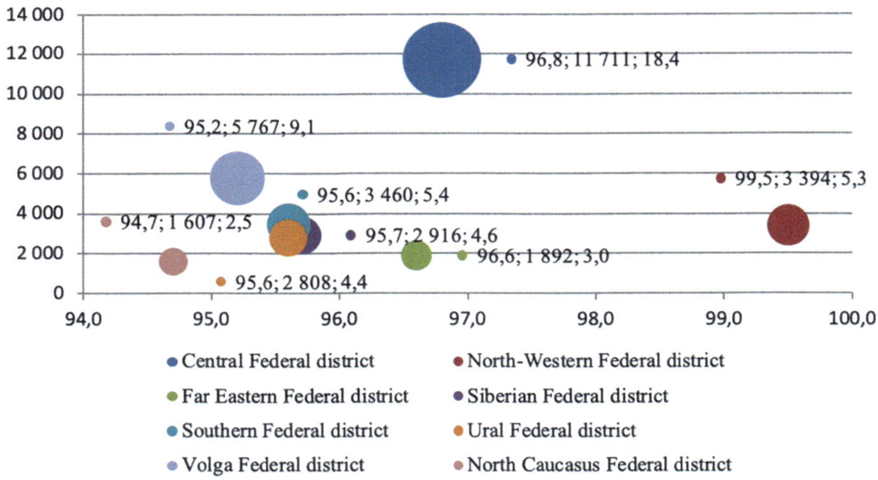

Fig. 2 Analysis of retail trade market turnover for 2020 by federal districts. *Source* Created by the authors based on (Federal State Statistics Service of the Russian Federation, 2021b)

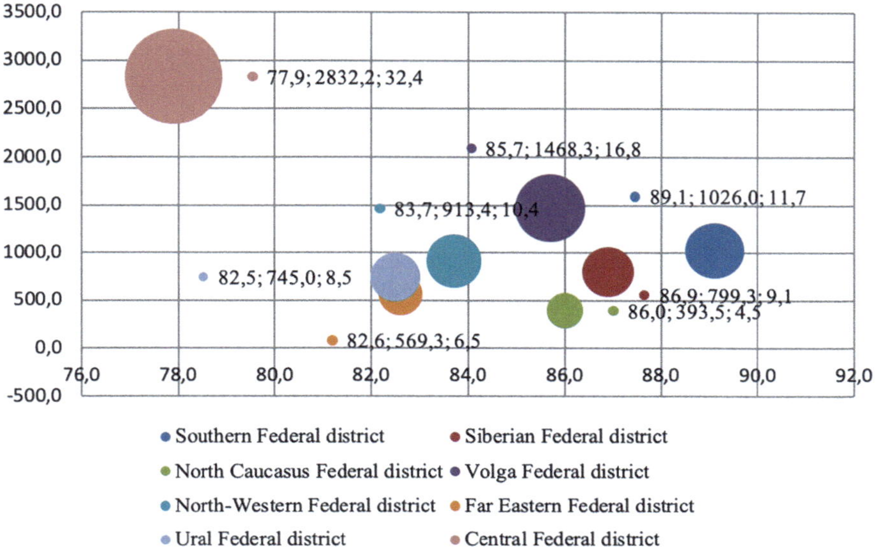

Fig. 3 Analysis of the paid services market turnover by federal districts, 2020. *Source* Created by the authors based on (Federal State Statistics Service of the Russian Federation, 2021b)

market for 2020, in billion rubles; the size of the circle—the share of the federal district in the total turnover of the paid services market in 2020).

The territories that are part of the Southern Federal district were the most effective in resisting the sharp decline in the paid services market, and the largest fall was achieved in the Central Federal district, the largest in terms of the turnover of the paid services market.

The wholesale market showed a positive trend in 2020 relative to the turnover in 2019. Figure 4 shows the analysis for the federal districts of Russia (horizontally-the turnover of the wholesale market in 2020, as a % of the turnover in 2019; vertically—the turnover of the wholesale market for 2020, in billion rubles; circle size—the share of the federal district in the total turnover of the wholesale market in 2020).

Wholesale market turnover increased in 2020 compared to 2019 in three federal districts (Southern, North-Western, and Siberian Federal districts), a slight decrease in turnover is typical for the Central Federal district, the turnover of the wholesale trade market declined in four federal districts (Volga, Far Eastern, North Caucasus, and Ural Federal districts). The assessment of the resilience of regional economic systems to the challenges of the pandemic can be made using the indicator "total debt on taxes and fees, insurance premiums, penalties and tax sanctions in the budget system", which reflects the effectiveness of tax policy (Fig. 5).

The largest amount of tax arrears is observed in the budget system of the Central Federal district since the turnover in various industries in this district is the largest relative to the turnover in other federal districts. There is also a significant tax debt to

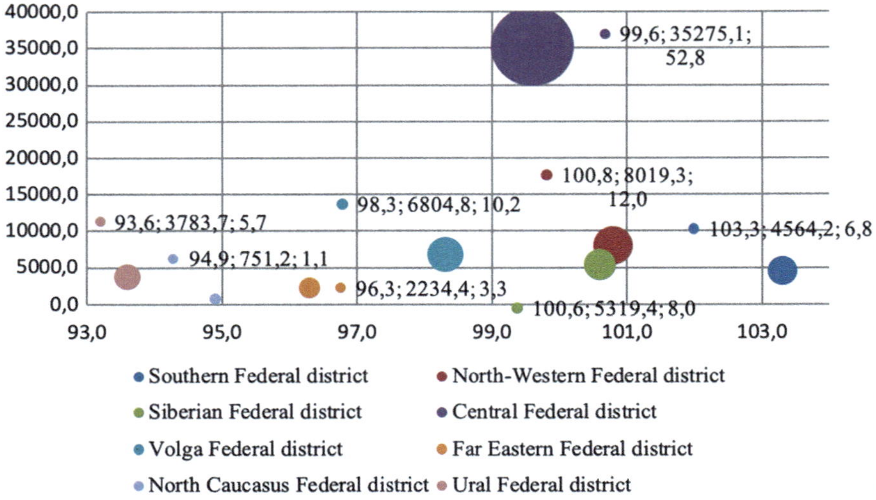

Fig. 4 Analysis of wholesale market dynamics by federal districts, 2020. *Source* Created by the authors based on (Federal State Statistics Service of the Russian Federation, 2021b)

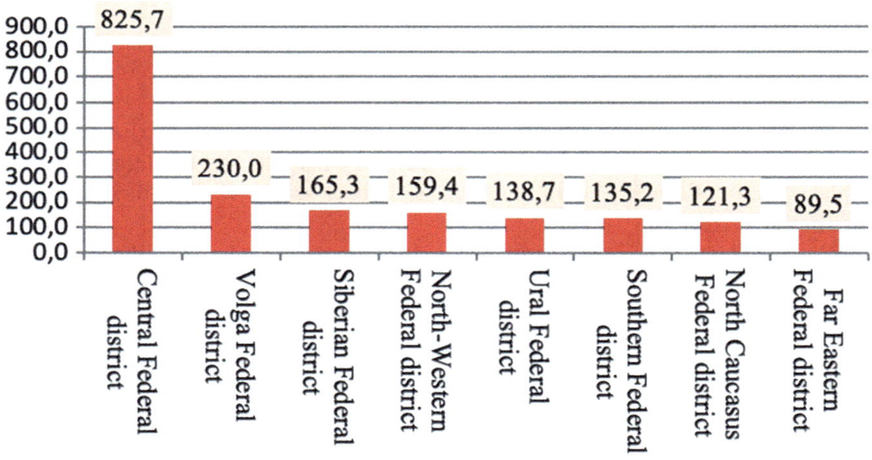

Fig. 5 Total debt on taxes and fees, insurance contracts, penalties, billion rubles, 2020. *Source* Created by the authors based on (Federal State Statistics Service of the Russian Federation, 2021b)

the budget system in the Volga Federal district, and other districts have a comparable level of tax debt.

The effectiveness of anti-crisis measures in 2020 and the level of economic security should be assessed using the indicator "the net financial result obtained by organizations without taking into account the financial result obtained by small businesses,

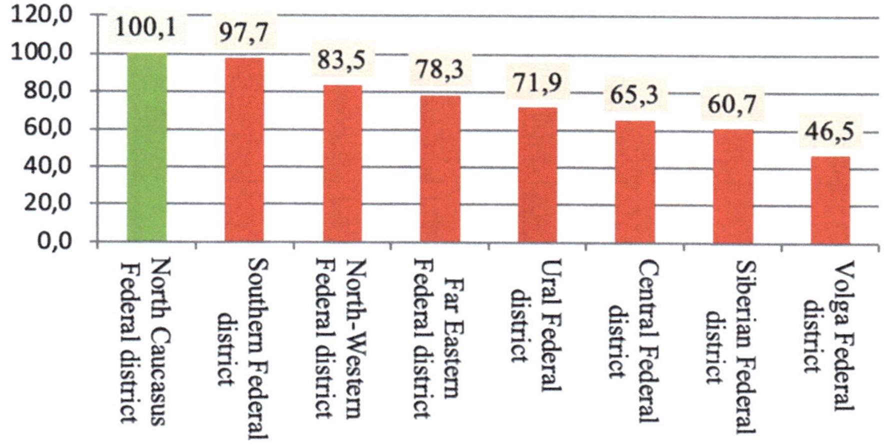

Fig. 6 Balance of profit and loss of organizations in January–November 2020, compared to January–November 2019, %. *Source* Created by the authors based on (Federal State Statistics Service of the Russian Federation, 2021b)

credit organizations, government agencies, and non-credit financial organizations". An analysis of the comparative dynamics of the financial result obtained by organizations in January-November 2020 compared to January-November 2019 is shown in Fig. 6.

Positive dynamics of profits and losses of organizations are noted only in organizations registered in the North Caucasus Federal District. Negative dynamics for 11 months of 2020 compared to the same period of 2019 is observed in seven federal districts, the largest decrease in the financial result was achieved in the Volga Federal District. We will conduct a situational analysis of changes in the consumer price index for goods and services from December 2020 to December 2019, reflecting the dynamics of income and purchasing power of the population (Fig. 7).

In general, in Russia, the consumer price index for goods and services from December 2020 to December 2019 was 104.9% (for food products-106.7%, for non-food products-104.8%, for services-102.7%). The consumer price index in December 2020 compared to December 2019 increased in all federal districts, the largest inflationary price growth occurred in the North Caucasus Federal district, and the minimum growth in the Ural Federal district. Another key indicator in assessing the economic security of the federal districts is the effectiveness of the implementation of employment support policies during the coronavirus pandemic. We propose to evaluate the effectiveness of measures to maintain employment at the pre-crisis level by analyzing the change in the registered unemployment rate in December 2020 relative to the unemployment rate in December 2019 (Fig. 8).

The graph shows that the least effective employment support policy is in the North Caucasus Federal District, and the most successful is in the Central, Far Eastern, and

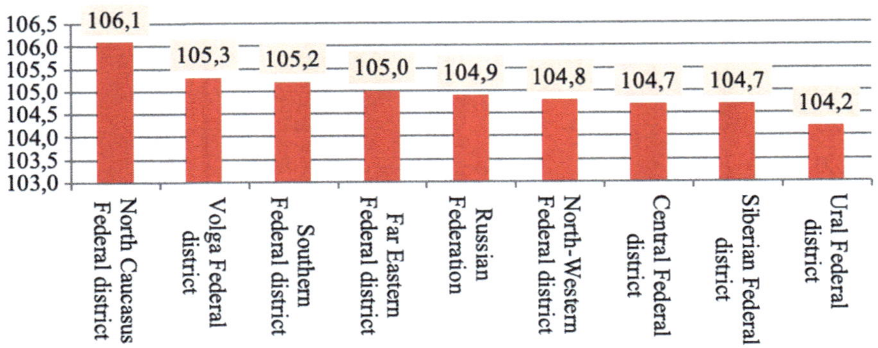

Fig. 7 Consumer price index for goods and services in December 2020 to December 2019, %. *Source* Created by the authors based on (Federal State Statistics Service of the Russian Federation, 2021b)

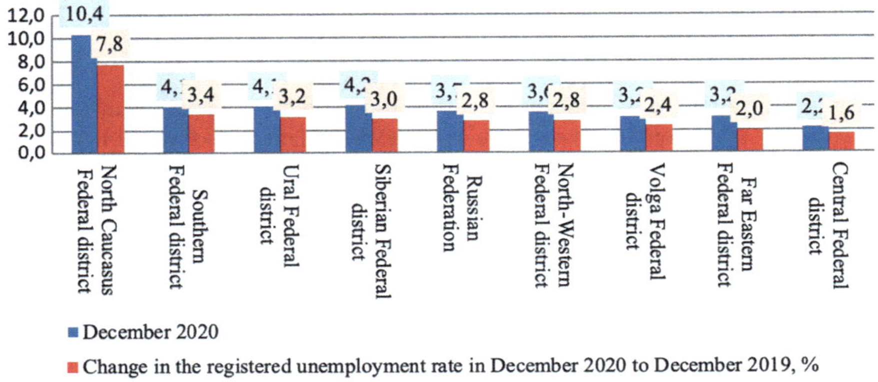

Fig. 8 The level of registered unemployment in December 2020 compared to December 2019. *Source* Created by the authors based on (Federal State Statistics Service of the Russian Federation, 2021b)

Volga Federal Districts, where the unemployment rate is lower than the national average.

The above indicators of economic security of the federal districts are situational, they reflect the "primary shock" from the pandemic. However, an objective assessment of the regional economic security should take into account the existing potential regional economy necessary for development (including the level of implementation of digital technologies, the level of scientific potential, and the quality of R&D) and the investment activity. The investment activity of the region can be estimated based on the analysis of the dynamics of fixed capital investment per capita (Federal State Statistics Service of the Russian Federation, 2021a). The available statistical data allow us to conclude that the dynamics of investment in fixed assets in Russia in 2019 in comparison with 2018 is positive and amounts to 108.7%. The same situation is typical for most federal districts: Siberian (114.1%), Central (113.12%), North Caucasus (111.6%), Far Eastern (109.5%), Volga (108.5%). In 2019, there was a decrease in investment potential in the Ural (98.1%), Southern (90.5%), and North-Western (87%) federal districts. The degree of depreciation of fixed assets, indicating the level of economic security, in 2019 in Russia amounted to 51.3%. In five federal districts, it is lower than the national average (Siberian – 49.6%, Central-49.6%, Southern-47.8%, North-Western-46.3%, and Far Eastern Federal District-44.2%) (Drobot et al., 2020). In terms of economic security, it is important to assess the level of innovative development of federal districts, which can be carried out based on an analysis of the level of costs for innovative activities of organizations (Federal State Statistics Service of the Russian Federation, 2021a). In 2019, the value of this indicator in the whole country reached 2.1%. The innovative development of five federal districts was lower than the national average: Siberian (1.8%), North-Western (1.8%), Southern (1.1%), Ural (0.7%), and North Caucasus (0.6%) federal districts. A significant indicator of the level of economic security of the region is its ability to spend on research and development, the introduction and use of digital technologies at the expense of internal resources without resorting to financial injections from the center, which is especially important in times of crisis. An analysis of the dynamics of the region's internal current expenditures on research and development in 2019 compared to 2018 showed that in the whole country they amounted to 110.4%, in five federal districts the value of this indicator is lower than the national average (in the Siberian (109.7%), Southern (108%), Ural (102.5%), North Caucasus (101.7%) and Far Eastern (95.1%) federal districts. In 2019, the share of domestic spending on the introduction of new digital technologies in the total amount of spending on digital development in four federal districts (Southern (76.7%), Siberian (76.1%), North-Western (75%), and Ural (72.4%) federal districts) is lower than in the whole country (78.9%) (Federal State Statistics Service of the Russian Federation, 2021a). Determining the place of each district in the overall economic security rating requires combining all previously analyzed indicators of the economic security of the regions (Table 1).

The combined analysis based on the weighted average assessment of all the analyzed indicators of the economic security of the federal districts allowed us to form a rating of the economic security and stability of the federal districts of Russia. According to the results of the author's analysis, the highest level of economic security currently has the Central (3.7), North-Western (3.8), Far Eastern (4.1), Southern (4.4), Siberian (4.6), North Caucasus (4.7), Volga Federal District (4.9%). The lowest level of economic security in the Ural Federal District (5.7).

5 Conclusion

The study concluded that "the primary shock" from the coronavirus pandemic in the short term caused damage to the economy of all federal districts. Even the Central

Table 1 Economic security indicators of the federal districts and the rating of the Russian regions

Indicator	Central federal district	North-Western federal district	Far Eastern federal district	Southern federal district	Siberian federal district	Volga federal district	North Caucasus us federal district	Ural federal district
Industrial production index in the district, in 2020, in % to 2019	105.2	97.0	95.9	99.0	95.6	96.6	106.5	97.7
Rating of the district by economic security indicator	2	5	7	3	8	6	1	4
Dynamics of retail trade turnover in the district, in 2020, in % to 2019	96.8	99.5	96.6	95.6	957	95.2	94.7	95.6
Rating of the district by economic security indicator	2	1	3	6	4	7	8	5
Dynamics of the paid services market in the district, in 2020, in % to 2019	77.9	83.7	82.6	89.1	86.9	85.7	86.0	82.5
Rating of the district by economic security indicator	8	5	6	1	2	4	3	7

(continued)

Table 1 (continued)

Indicator	Central federal district	North-Western federal district	Far Eastern federal district	Southern federal district	Siberian federal district	Volga federal district	North Caucasus us federal district	Ural federal district
Dynamics of wholesale trade turnover in the district, in 2020, in % to 2019	99.6	100.8	96.3	103.3	100.6	98.3	94.9	93.6
Rating of the district by economic security indicator	4	2	6	1	3	5	7	8
Consumer price index for goods and services in the district, December 2020 to December 2019, %	104.7	104.8	105	105.2	104.7	105.3	106.1	104.2
Rating of the district by economic security indicator	3	4	5	6	2	7	8	1
Total debt of the district on taxes and fees, insurance premiums, penalties, and tax sanctions to the budget system, as of December 1, 2020, billion rubles	825.7	159.4	89.5	135.2	165.3	230.0	121.3	138.7

(continued)

Table 1 (continued)

Indicator	Central federal district	North-Western federal district	Far Eastern federal district	Southern federal district	Siberian federal district	Volga federal district	North Caucasus us federal district	Ural federal district
Rating of the district by economic security indicator	8	5	1	3	6	7	2	4
Profit and loss balance of the district's organizations, in January-November 2020, in % as compared to January–November 2019	65.3	83.5	78.3	97.7	60.7	46.5	100.1	71.9
Rating of the district by economic security indicator	6	3	4	2	7	8	1	5
Change in the registered unemployment rate in the district, from December 2020 to December 2019, 2020–2019, %	1.6	2.8	2.0	3.4	3.0	2.4	7.8	3.2
Rating of the district by economic security indicator	1	4	2	7	5	3	8	6

(continued)

Table 1 (continued)

Indicator	Central federal district	North-Western federal district	Far Eastern federal district	Southern federal district	Siberian federal district	Volga federal district	North Caucasus us federal district	Ural federal district
Dynamics of investment in fixed assets per capita in the district, in actual operating prices, 2019/2018*100, %	113.1	87.1	109.5	90.5	114.1	108.5	111.6	98.1
Rating of the district by economic security indicator	2	8	4	7	1	5	3	6
Degree of depreciation of fixed assets of organizations in the district, in 2019, in %	49.6	46.3	44.2	47.8	49.6	58.3	53.6	61.7
Rating of the district by economic security indicator	5	2	1	3	4	7	6	8
Dynamics of internal operating expenses of organizations in the district for research and development, 2019/2018*100, %	110.7	115.5	95.1	108.0	109.7	110.7	101.7	102.5

(continued)

Table 1 (continued)

Indicator	Central federal district	North-Western federal district	Far Eastern federal district	Southern federal district	Siberian federal district	Volga federal district	North Caucasus us federal district	Ural federal district
Rating of the district by economic security indicator	2	1	8	5	4	3	7	6
Costs of innovative activities of organizations in the district, % of the total volume in 2019,%	3.0	1.8	2.5	1.1	1.8	2.9	0.6	0.7
Rating of the district by economic security indicator	1	4	3	6	5	2	8	7
The volume of innovative goods, works, and services in the district, as a percentage of the total volume of goods shipped, works, and services performed in 2019,%	5.0	5.6	3.0	2.7	2.6	11.3	5.3	3.3
Rating of the district by economic security indicator	4	2	6	7	8	1	3	5

(continued)

Table 1 (continued)

Indicator	Central federal district	North-Western federal district	Far Eastern federal district	Southern federal district	Siberian federal district	Volga federal district	North Caucasus us federal district	Ural federal district
The share of the district's internal expenditures on the introduction and use of digital technologies in the total amount of expenditures on digital technologies in 2019, %	79.5	75.0	82.2	76.7	76.1	80.1	89.9	72.4
Rating of the district by economic security indicator	4	7	2	5	6	3	1	8
Rating of the district by economic security indicator	*4*	*7*	*2*	*5*	*6*	*3*	*1*	*8*

Source Created by the authors based on (Federal State Statistics Service of the Russian Federation, 2021a, 2021b)

Federal District with the highest value in the rating of economic security, in the short term, could not effectively neutralize the negative trends in the wholesale trade sector, in the market of paid services, as well as timely eliminate high tax arrears to the budget. The high rating of economic security of the Central Federal District is provided by positive trends in the industry, including manufacturing, as well as good results of the retail segment of the region, an effective policy of preserving jobs, maintaining investment activity in the region, as well as a high level of internal costs for research and development, and a high share of financing for digital and innovative businesses. The quality and effectiveness of regional policy in terms of ensuring economic security should be based on the following principles: the preservation of jobs and the allocation of state subsidies, the promotion of investment activity to ensure the growth of consumption of goods and services, the adoption of operational measures to support the most vulnerable sectors of the regional economy in times of crisis. It is especially important to have a plan of operational assistance to the regional economy in the event of a sharp deterioration of the epidemiological situation in the country (the third wave of coronavirus), support for the system-forming sectors of the regional economy (manufacturing, mining, agriculture, etc.), and increase the reliability of the country's budget and tax system. The implementation of the costs of innovation and the development of advanced technologies is a priority for ensuring the competitiveness of the region and the country.

References

Ashraf, B. (2020). The economic impact of government interventions during the COVID-19 pandemic: International evidence from financial markets. *Journal of Behavioral and Experimental Finance, 27*, 100371. https://doi.org/10.1016/i.ibef.2020.100371

Bank of Russia. (2020). Assessment of the consequences of the epidemic for the Russian economy through the prism of the intersectoral balance. https://cbr.ru/Content/Document/File/112062/analytic_note_202008_dip.pdf. Data accessed: 12 Apr 2021

Bogomolov, V. (Ed.) (2012). Economic security: studies. 2nd ed., reprint. and add. UNITY DANA, Moscow.

Drobot, E. (2020). Global economy in the context of the COVID-19 pandemic: Results of 2020 and prospects for recovery. *Ekonomicheskie Otnosheniya, 10*(4), 937–960.

Drobot, E., Makarov, I., Nazarenko, V., & Manasyan, S. (2020). Impact of the COVID-19 pandemic on the real economy. *Journal of Economics, Entrepreneurship and Law, 8*, 2135–2150.

Federal State Statistics Service of the Russian Federation (2021a). Regions of Russia. Socio-economic indicators 2020. https://rosstat.gov.ru/storage/mediabank/fimgAF33/Region_Pokaz_2020.pdf. Data accessed: 14 Apr 2021.

Federal State Statistics Service of the Russian Federation (2021b). Report Socio-economic situation of Russia. https://rosstat.gov.ru/compendium/document/50801. Data accessed: 14 Apr 2021.

Government of the Russian Federation. (2014). Economic security strategy of the russian federation for the period up to 2030 No. 172-FZ (28 June 2014). http://government.ru/docs/all/111512/. Data accessed: 12 Apr 2021.

Kalinina, N. (2010). Economic security of the region: The essence, threats, and measures to ensure. *Siberian Trade and Economic Journal, 10*, 1–21.

Karavaeva, I., Buchvald, E., & Soboleva, I. (2019). Economic security of certain forecast parameters of socio-economic development and budget policy of the Russian Federation for the medium term. *Economic Security, 4*, 273–334.

Karpunina, E., Zabelina, O., Galieva, G., Melyakova, E., & Melnikova, Y. (2020). Epidemic Threats and their Impact on the Economic Security of the State. In *Proceeding of the 35th IBIMA Conference* (pp. 7671–7682). Seville, Spain

Kulkova, I. (2020). Adapting business behavior to changes caused by the coronavirus pandemic. *Journal of Economics, Entrepreneurship and Law, 1*, 2529–2540.

Lisova, E., Petrov, I., Koryakina, T., Kalombo Mulamba, V., & Karpunina, E. (2020). The well-being of Russian Regions under Threat: Economic and social impact assessment of COVID-19. In *Proceeding of the 36th IBIMA Conference in Granada* (pp. 7603–7616). Spain.

Mejokh, Z., Korolyuk, E., Sozaeva, D., Pilipchuk, N., & Karpunina, E. (2020). Economic security of Russian regions: Risk factors and consequences of the Covid-19 pandemic. In *Proceeding of the 36th IBIMA Conference in Granada* (pp. 8197–8205). Spain.

Salisu, A. (2020). Revisiting oil-stock nexus during COVID-19 pandemic: Some preliminary results. *International Review of Economics and Finance, 69*, 280–294.

Shubina, N. (2017). Conceptual approaches to the understanding of economic safety of region: Essence, structure, factors, and condition. *Bulletin of Ural Federal University. Series Economics and Management, 16*(2), 288–307.

State Duma of the Russian Federation. (2010). Federal Law of the Russian Federation No. 390-FZ (28 December 2010) "On Security". http://www.consultant.ru/document/cons_doc_LAW_108546/. Data accessed: 12 Apr 2021.

Sustainable Development of the Region: Environmental and Economic Aspects of Security

Julia V. Khodkovskaya, Ilvir I. Fazrakhmanov, Ekaterina E. Barkova, and Katsiaryna P. Korsak

Abstract The paper aims to explore the opportunities for sustainable development of the region, considering the environmental and economic aspects of security. The authors investigate the features of interaction between economic agents that shape regional development and regional security policy. The authors confirm the provisions of key theories and concepts on the importance of the determinants "resource endowment," "environmental friendliness," and "efficiency" in the sustainable development of regions and explain their differentiation. Moreover, the authors review the approaches to the environmental and economic component of regional security, as well as its manifestation and impact on the sustainable development of the region. The fundamental principles in an integrated approach to the content of sustainable development of the region are revealed—smart, sustainable, and inclusive economic growth. Using a systemic and dynamic approach, the authors identify threats to regional security. General scientific methods of analysis and synthesis allowed the authors to establish controversial aspects of the current practice of forming strategies for sustainable development of countries and regions (three-link spiral, four-link spiral, and five-link spiral). These methods also allowed for building a spiral model of sustainable development of the region, taking into account the criterion of resource-environment-efficiency.

Keywords Region · Security · Threats · Sustainable development · Economic growth · Resource endowment · Environmental friendliness · Efficiency

JEL codes O44 · Q01 · R11

J. V. Khodkovskaya (✉) · I. I. Fazrakhmanov · E. E. Barkova
Ufa State Petroleum Technological University, Ufa, Russia
e-mail: khodkovskiy@bk.ru

I. I. Fazrakhmanov
e-mail: ilvirus@inbox.ru

K. P. Korsak
Belarusian National Technological University, Minsk, Belarus

E. G. Popkova and B. S. Sergi (eds.), *Geo-Economy of the Future*,
https://doi.org/10.1007/978-3-030-92303-7_74

1 Introduction

The search for effective tools for ensuring sustainable development of the region in the context of growing competition in the markets of raw materials, capital, technology, and products increases the need for economic, environmental, and social development of regions. Moreover, it necessitates the need for security. The models and theories of management formed in the era of globalization require significant revision, improvement, and modernization of interaction between economic agents. Moreover, it is necessary to provide a constructive interaction between government, business, science, and civil society as strategic partners, as well as to form a territorially oriented innovation environment considering global and regional challenges affecting various aspects of regional security.

Current global and regional trends exacerbated the need for a systematic and science-based approach to ensuring stress resistance, security, self-sufficiency, and economic growth characterizing regional security. Therefore, the research objectives are as follows:

(1) To establish the impact of the environmental and economic component on the sustainable development of the region;
(2) To establish the key determinants of regional development that measure the current state of the region and the progressive dynamics of its growth;
(3) To address the problem of ensuring and improving security in the region, establish mechanisms and tools appropriate for the application, taking into account the interests of the government, business, science, and society.

The research subject is the forms and methods of sustainable development of regions and the impact of the environmental and economic component of security on regional development.

2 Materials and Methods

The introduction of social and environmental standards and models of sustainable development of territories and interaction with the market environment has recently become problem areas for many regions, limiting their growth and potential profitability. Disclosing social investments and inclusive green innovations in public reporting, introducing resource-saving technologies, and receiving public support, regions maintain their competitive position in the markets. Nevertheless, in practice, the export dependence of regions often remains, and the share of socially-oriented economic development of the region remains low. This indicates a decrease in the level of economic security and an increase in potential threats. As an integral part of national security, regional security is determined by the stability and efficiency of society, which can be seen as a manifestation of the stakeholder approach.

Analyzing the content and essence of stakeholder values (Apitz et al., 2017), taking into account economic, social, and environmental components, the authors studied the current concepts of sustainable development of countries and territories on the example of Scandinavian countries (Brandsen et al., 2016). The manifestations of corporate environmental responsibility and the nature of the impact on business performance in Canada have established a direct correlation between the activities of companies in the field of the environment and profit (Walker & Wan, 2012). Studies of regional dynamics in India have established structural features (geographic, demographic, resource, etc.) explaining the differentiation in the economic development of the Indian states.

Various scholars study the features of the formation of effective regional policy and the conditions for sustainable growth in the region:

- Smith (1962) and Ricardo (2016) studied the key theories and concepts of the regional economy;
- von Thünen studied agricultural location (Thünen, 1910);
- Weber studied the location of industrial production (Weber, 1922);
- T. Hagerstrand studied the diffusion of innovation (Hagerstrand, 1968);
- F. A. Perroux studied growth poles (Perroux et al., 1970);
- P. Baklanov studied the formation of economic space (Baklanov, 2007);
- L. Mints studied the influence of resource availability on production location (Mints 1967).

The focus on the sustainable development of the region in these studies proves the importance of economic equilibrium and specialization.

The analysis of modern concepts and strategies of national security of foreign countries (USA, China, and Venezuela) showed the importance of addressing economic issues and insufficient attention to environmental issues.

The study of the practices of environmental responsibility of market subjects was carried out using the standards "Environmental Management" (ISO 14000), "Information technology—Security techniques—Code of practice for information security management" (ISO 17799), "System of Environmental-Economic Accounting," Environmental Sustainability Index, etc. Several scholars point out that the application of international environmental standards reduces the negative impact of business on the environment (Dabelko & Barnett, 2019) and stimulates the development of environmentally effective business and environmental management combined with the implementation of social projects. Currently, there exist the following most common approaches to classifying regions by the level of development:

(1) Calculation of an aggregate indicator taking into account social, economic, and environmental criteria;
(2) Formation of a system of indicators assessing individual aspects of regional sustainability (institutional, economic, social, etc.).

Some indicators do not consider resource degradation in the regions and the depletion of natural capital. Moreover, the methodology for determining local indicators does not assess the self-development of regional infrastructure and the region's ecological reserves. The international practice of regional management confirms the inexpediency of implementing a single indicator in obtaining a reliable assessment of the region's state since many economic and social features are determined based on the GDP or GRP values (Brandsen et al., 2016). Thus, it is necessary to apply an adequate science-based approach considering the key determinants of regional development and measuring the current state of the region and the progressive dynamics of its growth.

The methodological basis of this study was the works of leading Russian and international researchers on the sustainable development of territories and countries (Lahsen & Turnhout, 2021; Linner & Wibeck, 2019). The authors summarized the approaches of modern researchers to the existing practice of forming strategies for sustainable development of countries and regions and the development of its key aspects, which allowed them to identify controversial aspects of adapting foreign models (three-link spiral, four-link spiral, and five-link spiral) to the Russian business environment (Manolov et al., 2020).

The authors studied the peculiarities of regional development, as well as the formation and use of the socio-economic potential of subjects of the Russian Federation, taking into account the criteria of dynamism, complexity, growth and diversification of potential, rationality and reduction of resource consumption, timeliness and regulation of the pace of innovation, etc. (Volkova & Golovinov, 2018). The works of Russian researchers allowed determining a direct correlation between the specialization of the region and the dynamics of its development (Grebenkin, 2020). Thus, the authors argue that the modern features of the specialization of the region are formed under the influence of digital technology and the capabilities of interregional communication networks (Balland et al., 2015, 2019). Simultaneously, independence, sustainability, and competitiveness depend on ensuring the economic security of the region (Sekiyama, 2020).

One of the key parameters ensuring the self-development of the region is the effective use of the resources of a particular territory (Belyakova & Proskurin, 2016; Peskova et al., 2019). Industry specifics of the regions, differentiation in socio-economic development, the level of investment and innovation activity, and the degree of digital transformation in the methodologies remain underestimated. Therefore, such estimates can only be used as baseline information to develop long-term socio-economic forecasts of regional development and the formation of sustainable development strategies. The adopted concept of long-term development of the Russian Federation "Strategy-2030" contains recommendations for forming regional strategies taking into account particular economic, resource, and other features of the regions to form a safe natural environment. In this regard, the application of horizontal forms of multiple governances is a sensible approach to regional specialization.

The concept of smart specialization traditionally has three dimensions (smart, sustainable, and inclusive economic growth), which corresponds to the promising directions of the Russian regional economy and the Europe-2030 strategy. Applying smart specialization in Russia should be based on the principle of forming a regional strategy, taking into account the competitive advantages and disadvantages, as well as the existing potential of a particular region (Manolov et al., 2020).

The authors propose a spiral model of sustainable regional development, taking into account the criteria of resource endowment, environmental friendliness, and efficiency.

Applying the method of system analysis, the authors used statistical data from the Federal State Statistical Service (Rosstat), information from analytical reviews of the economy, and the results of research by rating agencies.

3 Results and Discussion

Critical analysis of approaches to the environmental and economic component of regional security, as well as its manifestation and impact on the sustainable development of the region, showed that all studies focus on two aspects: the economic contribution of market actors in the region and their environmental responsibilities. The main problem is the uncertainty in assessing the effective use of resources in a particular territory since most methodologies use a deterministic approach that does not consider uncertainties and the impact of risk (Manolov et al., 2020). Research show that the formation of a competitive resource-efficient region with highly effective communication of the subjects (Manolov et al., 2020) is possible only by considering the environmental and economic components of security.

There are several principles of sustainable development of the region, taking into account the environmental and economic components:

(1) *Smart economic growth.* It assumes creating a regional information environment that generates innovative ideas, innovative solutions, and new knowledge in the business community. Moreover, the information environment forms a competitive regional economy contributing to the development of economic potential. The factor of environmental friendliness is one of the key criteria along with efficiency, informativeness, and dynamic indicators of the business community, initiating the development of monitoring and environmental safety and the growth of standardized environmentally friendly goods in GRP.

(2) *Sustainable economic growth* is focused on achieving a balanced development of the region, taking into account the environmental factor. Studies show that the sustainability of economic growth is achieved by applying political and economic instruments when the legal field of high-tech business environment is actively reformed, taking into account the priorities of digitalization, the features of business cooperation, etc. The environmental factor measures the quality of created goods and services. It is also built into the value chain, taking into account the region's specialization. The business community actively uses the practices of sustainable development that provide economic benefits to businesses and meet society's demands for a safe ecosystem and eco-products.

(3) *Inclusive economic growth* is associated with the environmental sustainability of territories. The importance of environmental friendliness is explained by the strengthening of integration processes in all spheres of human activity

and, therefore, by the threat of loss of natural sustainability, i.e., the ability of the biosphere and its constituent ecosystems to assimilate the consequences of various anthropogenic impacts on the natural environment (Gulam Mahiuddin, 2020; Vittenberg, 2010). . The environmental factor reinforces the importance of social values over commercial ones and contributes to creating a socially oriented environment at the regional level. This is evidenced by environmental legislation and the existence of national projects and programs to protect the environment, which form an active position of the government, society, and, in recent years, the business community to preserve the ecosystem. According to Liebig (1841) and Shelford (1912), ecology is the main phenomenon of sustainability of any living system. In the context of the region's sustainable development, the environmental and economic component is one of the main features of the region's development and competitiveness. However, the importance of the environmental and economic component is differentiated due to the different resource dependence of regions caused by the development of nature-intensive industries and their share in GRP. This statement is evidenced by the deterioration of the resource base structure of the Russian subjects, the growing share of hard-to-recover reserves, etc. According to the authors, the impact of the environmental and economic component depends on a range of mechanisms that ensure the effective use of regional resources through institutional and legal, market, industrial, technological, organizational, and managerial impacts.

Experts and researchers suggest various approaches to assessing the sustainable development of regions. However, taking into account the environmental and economic components (Dalby, 2020), the key criterion for differentiating regions can be considered "resource-environmental-efficiency" (resource endowment, environmental friendliness, and efficiency).

Taking into account resource endowment, all Russian regions can be divided into resource surplus (self-sufficient in resources, the current consumption of resources exceeds investment in the reproduction of the resource base in the region) and resource deficit regions (lack of resources, the current consumption of resources is not transformed into capital necessary for sustainable development of the region). This grouping is possible based on the indicative analysis. The indicators of the classification of regions are GRP growth rate, the share of extractive industries in GRP, the ratio of the share of manufacturing and extractive industries, assessment of the resource dependence of the region, resource intensity of business processes, etc.

Environmental friendliness as a determinant of sustainable development of the region affects the production, social, technological, and economic spheres of life. Exceeding environmental limits constrains the region's development prospects and increases the outflow of labor resources. Therefore, it seems possible to establish ecologically favorable regions (regional policy is aimed at maintaining the integrity of the ecosystem and ensuring the environmental security of the region) and ecologically unfortunate regions (there is an increase in anthropogenic environmental anomalies and the depletion of the ecological potential). The indicators of environmental friendliness of the region include environmental and economic efficiency of environmental protection activities in the region, indicators of environmental risk management, the level of implementation of environmental standards and technologies, the level of environmental education, environmental ethics and culture, corporate environmental responsibility, etc.

The determinant efficiency identifies the economic efficiency of the region. Taking into account this determinant, Russian regions can be grouped into effective and ineffective. Effective regions demonstrate positive dynamics and steady growth of economic indicators, stability of development, developed innovative environment, and availability of hybrid institutions. Additionally, net savings and net capital are invested or reinvested in the region's development, providing a relatively independent and sustainable position of the region. In effective regions, investment exceeds the consumption of economic goods. It creates a favorable economic environment for the reproduction of capital, the expansion of inter-regional relations, the development of export activities, the formation of prerequisites for progressive, sustainable development of the regional economy integrated into the national and international economy. Thus, effective regions are economically safe. In inefficient regions, economic growth is low or negative, indicating weak investment and low innovation capabilities not meeting the needs of the business environment, unproductive structural industrial policy, and the strong influence of external and internal threats to economic security. Indicators of efficiency of regions include economic growth rates, economic potential, proximity to markets, the intensity of use of new technologies, quality of regional management, dynamics of development of priority branches of the economy, state support of fast-growing market segments, etc. This criterion focuses on the most significant regional determinants that describe the global and cross-border nature of threats to regional security (Fig. 1).

Even though regional differences can be traced almost in all Russian territories and the crisis phenomena and socio-economic shocks (threats to the security of sustainable development of regions) is a feature of modern society, effective interaction between market actors is necessary, as shown in the spiral model of sustainable development of the region, taking into account the criteria of "resource-environmental-efficiency" (Fig. 2).

In the spiral model, the interaction of strategic partners and the resulting organizational solutions are focused on eliminating the problem of harmonizing the interests of all institutional forms (Manolov et al., 2020). Therefore, in addressing the problem of ensuring and enhancing security in the region, the trajectory of sustainable regional development must focus on the following most significant aspects:

- Promotion of effective competition policy ensuring stable reproduction and multiplication of resource, environmental, and economic potentials, differentiated state support of priority fast-growing sectors of the economy, and regulation of economic integration processes;
- Formation of an effective risk management system taking into account the dynamic development of the region, including monitoring of global and national

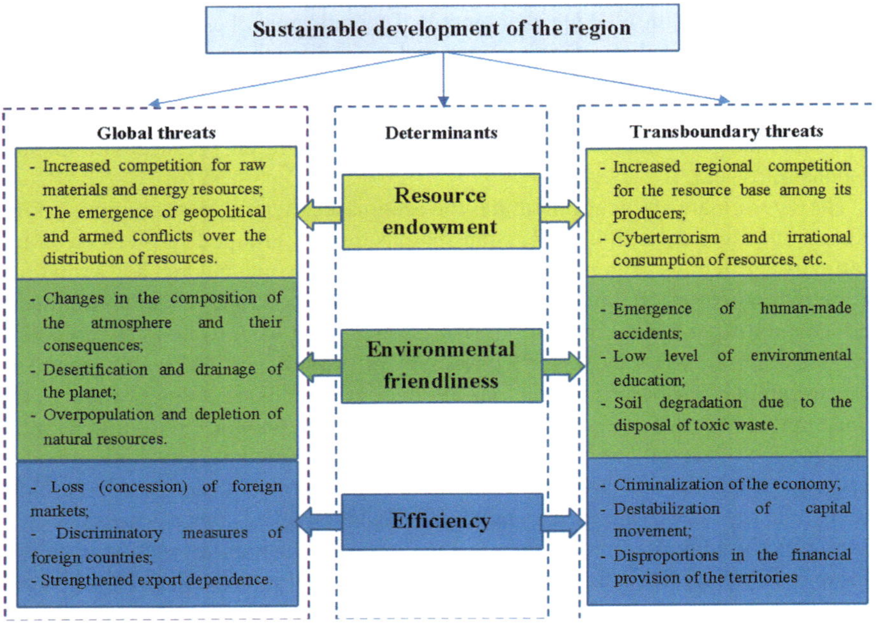

Fig. 1 Threats to the security of sustainable development in the region. *Source* Compiled by the authors

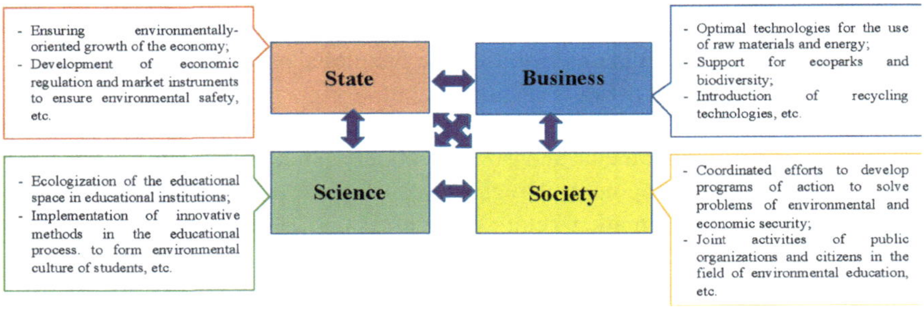

Fig. 2 The spiral model of the sustainable development of the region. *Source* Compiled by the authors

threats to the region's security, assessment of uncertainty and risks, and the use of effective mechanisms to improve security;

- Development of ecological and economic self-consciousness, the responsibility of the government, business, society, and science for the use and reproduction of the region's ecological potential. This creates conditions for openness and transparency in the activities of market actors, harmonization of their interests in preventing negative impacts on the ecological environment, achieving sustainable economic growth, and other priority factors affecting the safety of the region.

4 Conclusion

The research has established the importance of the environmental and economic component of regional security, the accompanying effects of which strengthen the activity/depression in the sectors of the economy involved in the creation and multiplication of the territory's capital. The approaches used to assess the environmental responsibility of market actors and the economic contribution to the region differ significantly, which is associated with the study of different aspects of regional security. The key criterion for differentiating regions is "resource-environment-efficiency," which focuses on the most significant determinants of sustainable regional development describing the global and transboundary nature of threats to regional security.

Applying a systematic approach to creating a rational mechanism of interaction between the government, business, society, and science, the authors propose the spiral model of sustainable regional development focused on the formation and use of effective instruments of security. The main components of institutional and infrastructural transformations in the region that determine the quality of life and level of security include transparency and openness of the business environment, free exchange and protection of information, reduction of risk and uncertainty, the concentration of resources to achieve sustainable growth, and additional capitalization of business using digital technology. Continuous dialogue between strategic partners, taking into account the regional interests, will form the conditions for sustainable regional development.

References

Apitz, S. E., Fitzpatrick, A. G., Mc Nally, A., Harrison, D., Coughlin, C., & Edwards, D. A. (2017). Stakeholder value-linked sustainability assessment: Evaluating remedial alternatives for the Portland Harbor Superfund Site, Portland, Oregon, USA. *Integrated Environmental Assessment Management, 14*(1), 43–62. https://doi.org/10.1002/ieam.1998

Baklanov, P. Y. (2007). *Territorial structures of the economy in regional management*. Nauka.

Balland, P.-A., Boschma, R., & Frenken, K. (2015). Proximity and innovation: From statics to dynamics. *Regional Studies, 49*(6), 907–920. https://doi.org/10.1080/00343404.2014.883598

Balland, P.-A., Boschma, R., & Ravet, J. (2019). Network dynamics in collaborative research in the EU, 2003–2017. *European Planning Studies, 27*(9), 1811–1837. https://doi.org/10.1080/09654313.2019.1641187

Belyakova, G. Y., & Proskurin, S. D. (2016). Smart specialization—A strategy for sustainable development of the regions. *European Social Science Journal, 10*, 30–36.

Brandsen, T., Ecchia, G., Eschweiler, J., Hulgård, L., & Nogales, R. (2016). Co-creating a Social Innovation Research Agenda for Europe. Social Innovation Europe/EMES Network.

Retrieved from https://emes.net/content/uploads/research/social-innovation-europe-sie/SIE-CoSIRA-report_July2016.pdf. Accessed 30 January 2021.
Dabelko, G., & Barnett, J. (2019). Environmental security. In A. Collins (Ed.), *Contemporary security studies* (5th ed., pp. 235–252). Oxford University Press.
Dalby, S. (2020). Environmental security and climate change. *Oxford Research Encyclopedia of International Studies*. https://doi.org/10.1093/acrefore/9780190846626.013.168(Accessed5January2021)
Grebenkin, I. V. (2020). Trends in industrial specialization and development dynamics in the Russian regions. *Economy of Region, 16*(1), 69–83. https://doi.org/10.17059/2020-1-6
Gulam Mahiuddin, A. (2020). *Environmental security and national security of Bangladesh (Dissertation of Post Graduate Diploma in International Relations)*. University of Dhaka.
Hagerstrand, T. (1968). *Innovation diffusion as a spatial process*. University of Chicago Press.
Lahsen, M., & Turnhout, E. (2021). How norms, needs, and power in science obstruct transformations towards sustainability. *Environmental Research Letters, 16*, 025008. https://doi.org/10.1088/1748-9326/abdcf0
Liebig, J. (1841). *Organic chemistry in its application to agriculture and physiology*. Johann Friedrich Vieweg.
Linner, B., & Wibeck, V. (2019). *Sustainability transformations: Agents and drivers across societies*. Cambridge University Press. https://doi.org/10.1017/9781108766975
Manolov, G., Orlova, D., Khodkovskaya, J., Barkova, E., & Nazarov, M. (2020). A Smart specialization strategy for sustainable development of regions. *E3S Web of Conferences 208*, 08009. https://doi.org/10.1051/e3sconf/202020808009.
Mints, L. E. (Ed.). (1967). *Intersectoral balance of the economic district*. Central Economics and Mathematics Institute of ASUSSR.
Perroux, F. A. (1970). Note on the concept of growth poles. In D. L. McKee, R. D. Dean, & W. H. Leahy (Eds.), *Regional economics: Theory and practice* (pp. 93–103). Free Press.
Peskova, D. R., Vasileva, J. P., Nazarova, U. A., & Sharifyanova, Z. F. (2019). World and Russian fuel and energy market development perspectives and drivers. *E3S Web of Conferences 124*, 04017. https://doi.org/10.1051/e3sconf/201912404017.
Ricardo, D. (2016). On the principles of political economy and taxation (P. Sraffa Transl. from English). Moscow, Russia. (Original work published 1817).
Sekiyama, T. (2020). Environmental security and Japan. Retrieved from http://ssdpaki.la.coocan.jp/en/proposals/47.html. Accessed 5 December 2020.
Shelford, V. E. (1912). Ecological succession IV. Vegetation and the control of land animal communities. *Biological Bulletin, 23*(2), 59–99. https://doi.org/10.2307/1536159
Smith, A. (1962). An inquiry into the nature and causes of the wealth of nations. Moscow, USSR: Sotsekgiz. (Original work published 1776).
Thünen, J. H. (1910). *Isolated state in relation to agriculture and the national economy*. Verlag von Gustav Fischer.
Vittenberg, E. Y. (2010). Ecological responsibility of the Russian business. *Bulletin of the Institute of Sociology, 2*(67), 31–48.
Volkova, Yu. A., & Golovinov, O. N. (2018). Innovation and innovation activities of the region. *Finance Accounting Banking, 3*, 27–32.
Walker, K., & Wan, F. (2012). The harm of symbolic actions and green-washing: Corporate actions and communications on environmental performance and their financial implications. *Journal of Business Ethics, 109*(2), 227–242. https://doi.org/10.1007
Weber, A. (1922). On the location of industries. Tubingen, Germany: Verlag von J.C.B. Mohr (Paul Siebeck).

Institution of Constitutional Supervision: The Experience of France and the Russian Federation

Marina F. Konstantinidi and Shamu Sh. Denilkhanov

Abstract The study analyzes the current state of the institution of constitutional supervision on the example of a comparative analysis of the experience of the functioning of such review bodies as the Constitutional Court of the Russian Federation and the Constitutional Council of the French Republic. Within the framework of this article, the authors consider the specifics of the constitutional judicial review carried out in the Fifth French Republic and the possibility of applying its model within the framework of the constitutional review carried out by the Constitutional Court in the Russian Federation. The specifics of this type of control is the presence of a quasi-judicial body of constitutional review—the Constitutional Council of France, which exercises control exclusively over the state Parliament, due to the deviation from the principles of parliamentarism in connection with the adoption of the French Constitution of 1958. The article analyzes its constitutional judicial powers, reveals the relationships and interactions that occur within the established model of government, taking into account the division of control functions with such a body as the Council of State of France. The result of the study is a proposal to apply the French experience in the Russian context in terms of the implementation of decisions of the Constitutional Council, which do not entail the need for the legislator to exclude a contradictory norm from the text of the law.

Keywords Constitutional practice · Constitutional judicial control · Constitutional review · Preliminary control · Law enforcement · Quasi-judicial body · Foreign experience · Constitutional council · Constitutional court · European model of governance

JEL Codes K 10

M. F. Konstantinidi (✉) · S. Sh. Denilkhanov
Pyatigorsk State University, Pyatigorsk, Russia

E. G. Popkova and B. S. Sergi (eds.), *Geo-Economy of the Future*,
https://doi.org/10.1007/978-3-030-92303-7_75

1 Introduction

As you know, the Russian Federation and the French Republic have a lot in common in the mechanism of constitutional and legal regulation. Both countries belong to the continental legal system and have a similar model of constitutional review. This is indicated by the institutions of the Constitutional Court and the Constitutional Council established in these states.

Despite the specific names of these bodies of constitutional review: the Constitutional Court and the Constitutional Council should be recognized as judicial bodies.

The Constitutional Council was created as a result of the adoption of the French Constitution of 1958, during the so-called the Fifth Republic.

At the beginning of our study, we should say what is understood by the Fifth Republic—this is a period of French modern history, which began in 1958 in connection with the adoption of a new Constitution of the State. This period is associated with the departure from the parliamentary state through the establishment of the post of president, who was elected by direct universal suffrage.

The adoption of the new Constitution of the French Republic also marked a new period of constitutional control in the State. With the adoption of the Constitution of 1958,[1] the Constitutional Council of France was also formed (Official website of the Constitutional Council of France, 2021), which was given the authority to exercise constitutional judicial control. This state body was endowed with specific powers, expanding from year to year through the adoption of laws; in addition it can be considered as a kind of quasi-judicial body.

The Council consists of 9 members; one third of the Council is replaced every three years. The members of the Council are appointed for a term of 9 years. The President of the Constitutional Council is appointed by the President of France. It is also interesting that the former French Presidents are by right life members of the Council.

Thus, the Council has the right to pre-review the constitutionality of laws adopted by the country's Parliament before their promulgation, pre-review international treaties before their ratification and other acts of an international nature, as well as to monitor the course of presidential elections and referendums. However, this council is not empowered to repeal laws that contradict constitutional norms, these powers belong to the State Council.[2]

"The execution of the decisions of the Constitutional Council is mandatory for all state bodies, including the Parliament, and the independence of the parliament is ensured by the fact that it has a discretion to implement new regulations aimed at

---

[1] The Constitution of the French Republic of 04.10.1958 (the Constitution of France) (with amendments and additions) from 23.07.2008) (2012). Constitutions of foreign states: Great Britain, France, Germany, Italy, the European Union, the United States of America, Japan: training manual. M.: Infotropik Media.

[2] See Footnote 1.

enacting Constitutional Court judgements, while taking into account the positions of the latter" (Kokotova, 2015, p. 55).

According to the current Constitution of France, the Constitutional Council is established as the body responsible for constitutional oversight. However, the Constitutional Council, when exercising its control powers, overlaps with the functions of another control body—the Council of State of France. Thus, the presence of two authorities with control and constitutional powers indicates that the system of constitutional control is dual. The model created in France is radically different from those created in other EU countries.

The specifics of the adopted model of constitutional control in France presupposes the functioning of two state authorities with similar powers: the Council of State and the Constitutional Council. The clear division between their responsibilities within the framework of constitutional control over the adopted acts could be achieved only through the definition of a list of types of normative acts under consideration for each of the specified bodies. This is the direction chosen by the French legislation, giving the Constitutional Council the right to control only normative acts adopted in the form of a law, and giving the authority to control the legality of normative acts, acts adopted at the level of executive authorities, to the State Council.

Thus, as we can see, the division of control powers between the State Council and the Constitutional Council is that only the Constitutional Council has the right to check the constitutionality of legislation.

The execution of decisions of the constitutional control body are largely explained by the type of control carried out, for example, a law adopted by the Parliament, prior to their signature by the President, is subject to preliminary review in the Constitutional Council (Official website of the Constitutional Council of France, 2021).

After such a review, the text of the law is published without those provisions that were deemed unconstitutional in accordance with the established procedure.

The implementation of the decisions of the Constitutional Council does not cause any special difficulties in terms of preliminary control.

However, constitutional control, which was only preliminary in nature, could be observed in France until 2008. Since 2008, the Constitutional Council has also been given the power to exercise subsequent control, which has led to the need to develop a mechanism for implementing such decisions.

The decision of the Constitutional Council, adopted during the subsequent control, does not entail the need for any changes to the law by the parliament, which confirms the supremacy of constitutional laws.

All this indicates the participation of the Constitutional Council in law-making activities, which in many ways distinguishes this quasi-judicial body from other bodies of constitutional judicial control. Therefore, the words of the French lawyer Mathieu seem very fair: "The Constitutional Council's activities cover all areas of law, including lawmaking" (Mathieu, 2010, p. 110).

2 Materials and Methods

Having conducted a comparative analysis of the model of constitutional judicial control used in France, the authors reveal in some detail all the features of its practical implementation and its impact on the political processes taking place in the State that has this model of governance.

Considering all aspects of the control powers in the French model of constitutional regulation, the authors make it possible to compare the existing bodies in the Russian Federation with similar powers and thereby determine the effectiveness and efficiency of the Russian model of constitutional control.

The analysis of scientific sources shows that at present there is an extensive theoretical and practical basis to reach objective and scientifically sound conclusions about the relationship between the Fifth Constitution and constitutional control. Despite the published results of scientific research on this topic, it should be noted that in general, the problem of a comprehensive study of the concept of constitutional control and its impact on the processes taking place in a state governed by the rule of law has not been sufficiently investigated.

3 Results

A distinctive feature of the Constitutional Council of France from other bodies of constitutional judicial review of the European model is that the Council exercises constitutional control over the activities of the Parliament. The Constitutional Council does not have the authority to exercise control over other state bodies of the country. It should be noted here that the French Government is authorized to apply to the Constitutional Council on issues related to trespassing on its prerogatives, in turn, the Parliament does not have the same rights due to the fact that the Constitutional Council exercises control only over the constitutionality of acts of Parliament.

It should also be emphasized that according to Article 34 of the French Constitution, Parliament has limited legislative competence, which indicates the abolishment of the principle of parliamentarism in the country.

In particular, according to article 6, the Constitutional Council, on a referral from the Government of France, may decide that the President of France will not be able to continue to perform his duties due to the circumstances specified in the Constitution.

As we noted earlier, the Constitutional Council controls the process of presidential elections, moreover, it plays a significant role in their conduct, in fact, the role of "the central election commission". In the event of electoral disputes, the Constitutional Council is authorized to conduct its own investigation.

According to article 59 of the French Constitution, the Constitutional Council has the right to decide on the constitutionality of the election of deputies and senators, in the event of a question of this nature.

Thus, in accordance with the decision taken by the Constitutional Council on the constitutionality of a normative act, the latter cannot be promulgated if its essential provisions are declared unconstitutional.

The form of decisions of the constitutional judicial review is also specific, which differs depending on the category of the constitutional dispute under consideration.

Thus, the form and content of such acts of the countries discussed above may vary depending on the type of constitutional proceedings, for example, in France, due to the specifics of the Constitutional Council, i.e. its quasi-judicial nature, it can make decisions in the field of the electoral process, namely on the constitutionality of the election of deputies. The council can also make a reasoned decision on the constitutionality of a legal act, on constitutional disputes of a public nature (Chirkin, 2013).

The Council can also make decisions concerning the approval and publication of a legal act, as well as a kind of conclusion on the existence of extraordinary circumstances in the country.

In the latter case, when the institutions of the Republic, the independence of the nation, the integrity of its territory or the fulfillment of its international obligations are threatened in a serious and immediate manner, and the normal functioning of the constitutional bodies of public power is interrupted, the President of the Republic takes all measures that are dictated by these circumstances, after formal consultation with the Prime Minister, with the Presidents of the Chambers, as well as with the Constitutional Council (Article 16 of the French Constitution).[3]

The authors consider it necessary to note that if the Constitutional Council reviews various international norms that are subject to ratification in France and reveals their inconsistency with the Constitution, the priority is given to the norms of international law, which means that the Constitution is subject to change. This process will require the involvement of the country's Parliament or it will be necessary to hold a referendum (Articles 54, 89 of the French Constitution).[4]

4 Conclusion

After analyzing the French experience in the implementation of constitutional judicial review, the authors believe that the creation of a quasi-judicial body of constitutional judicial control in our country is not only financially unprofitable, but not advisable at all. The Constitutional Court of the Russian Federation in the course of its activities has shown its competence and professionalism as a body of constitutional review. In addition, the current amendments to the Constitution of the Russian Federation[5] directly indicate the strengthening of the position of the Constitutional Court of the

[3] See Footnote 1.

[4] See Footnote 1.

[5] The Constitution of the Russian Federation (adopted by popular vote on 12.12.1993) (subject to amendments made by the Laws of the Russian Federation on Amendments to the Constitution of

Russian Federation and expand its powers and, as a result, strengthen its legal status and importance in the Russian judicial system. [6]

The Constitutional Court of the Russian Federation, in contrast to the Constitutional Council of France, is endowed, in addition to the right of constitutional control over the adopted federal laws, with other powers, and more extensive. So, in its practice, there are examples of constitutional verification of normative acts adopted by the Chambers of the Federal Assembly, the Government of the Russian Federation and even the President of the Russian Federation. However, the French experience of monitoring the implementation of decisions of the Constitutional Council is more remarkable. According to the authors, this approach should be applied in Russian practice as well. Thus, it should be noted that the recognition by the Constitutional Court of the Russian Federation of the norms of the law that do not correspond to the constitutional norms entails the need for the legislator to exclude this norm from the text of the law, but this action will be more formal, in view of the already adopted decision of the Constitutional Court of the Russian Federation. The implementation of such legislative procedures requires not only time, but also financial costs.

References

Chirkin V. E. (2013). *Constitutional law of foreign countries: textbook*. Norma, SIC Infra-M.

Kokotova, M. A. (2015). Decisions of the constitutional control bodies as a basis for changing legislation: The experience of France and Russia. *Constitutional and Municipal Law, 2*, 54–57.

Mathieu, B. (2010). Legal decisions of the constitutional council. *Bulletin of the Constitutional Court of the Republic of Belarus, 3*, 109–125.

Official website of the Constitutional Council of France. (2021). http://www.conseil-constitutionnel.fr. Data accessed: 11 Apr 2021.

the Russian Federation of 30.12.2008 No. 6-FKZ, of 30.12.2008 No. 7-FKZ, of 05.02.2014 No. 2-FKZ, of 21.07.2014 No. 11-FKZ).

[6] Federal Constitutional Law No. 2-FKZ of 05.02.2007 "On Amendments to the Federal Constitutional Law "On the Constitutional Court of the Russian Federation" (2007). Collection of Legislation of the Russian Federation, 7, St. 829.

Guidelines for the Region's Socio-economic Policies During the Pandemic

Natalja V. Polujanova

Abstract The study focuses on identifying the benefits of implementing the region's socio-economic policies according to the principles of lean production to overcome the negative effects of the COVID-19 pandemic. The author investigates the problems of socio-economic development of the regions of Russia during the COVID-19 pandemic, identifies the most significant threats to the regions (rising unemployment, falling average per capita incomes of the population, reduction of regional budgets and growth of regional public debt). The author proved that the principles of lean production can be successfully implemented as fundamental in the implementation of the socio-economic policy of the region. The focus of regional socio-economic policies during the pandemic should be the optimization of the cost of maintaining the region's public administration and increasing the quality of regional governance; improving regional policies to support people in the aftermath of a pandemic; developing the infrastructure and businesses implementing lean technologies. The author developed an algorithm for the formation and realization of regional socio-economic policy according to the principles of lean production during the pandemic, as well as describes the features of its implementation in the Belgorod Region.

Keywords Region · Regional socio-economic policy · Lean production · Regional governance · Pandemic

JEL Codes R11 · R58

1 Introduction

The well-being of regions, determined by the territory's available resource base and the quality of regional governance, is a prerequisite for the balanced development of the national economy.

The situation with ensuring stable socio-economic development of the regions of Russia, achieving their financial stability and maintaining a high standard of living

N. V. Polujanova (✉)
Belgorod National Research University, Belgorod, Russian Federation

E. G. Popkova and B. S. Sergi (eds.), *Geo-Economy of the Future*,
https://doi.org/10.1007/978-3-030-92303-7_76

of the population has always remained quite problematic. The pandemic has created new challenges associated with slowing economic growth due to the restraint of economic activity of economic entities and the increase in budgetary constraints (Milchakov, 2021; Zubarevich, 2021). This requires the search for new and effective regional governance tools.

One solution to this problem may be the implementation of regional socio-economic policies according to the principles of lean production. Indeed, thrifty technologies make work more efficient, reduce costs, improve the quality of services provided to the public and employee interest in work, as well as cope with budgetary constraints. Many countries demonstrate successful experience in implementing the principles of lean production in the activities of regional government bodies. For example, the United States effectively implemented this concept both regionally (Colorado, Iowa, Connecticut, Maryland, Meng, Ohio, etc.) and at the municipal level (Grand Rapids, Denver, King County) (Kaizen, 1986; Karpunina, 2019).

The study of mechanisms for implementing socio-economic policies according to the principles of lean production at the regional level is now becoming increasingly relevant, especially given the negative impact of the COVID-19 pandemic on the regional economic system.

2 Literature Review

The problem of spatial development is reflected in the studies of Zubarevich (2019), Atanov et al. (2014), Lyosh (2007). The issues of formation and implementation of the regional socio-economic policy are presented in the works of Filatov (2017), Karpunina et al., (2018, 2020a, 2020b), Friedmann (1973). The authors identified the tools of socio-economic policy that contribute to smoothing interregional differentiation.

The philosophy of lean manufacturing is based on the five principles of continuous improvement of the company's production process (Kaizen principles), which ensure a constant increase in productivity: neatness, order, cleanliness, standardization, discipline (1986).

Gradual improvement of each process, divided into stages, public discussion of problems allows managers to eliminate the situation with distortion or concealment of information, establish effective communications, develop a productive system of employee motivation, implement quality standards, as well as analyze all processes in the company. The concept of "lean manufacturing" is aimed at achieving a minimum of costs while maximizing efficiency. It proves its success on the example of Japanese companies Toyota, Canon, Philips, Nissan (Falaleeva, 2016). The concept of lean manufacturing can be implemented with a whole set of tools, including visual management, standardized operations, instructions, "KanBan", "poka-yoke", pulling production, equipment maintenance, multi-process processing, control schedules, etc. (Hobbs, 2007).

If the principles of lean production are adapted to the regional management system, the result of the implemented territorial socio-economic policy may be an increase in the efficiency of the regional economic system (Gorlov, 2016; Karpunina, 2019; Sychanina et al., 2018; Tyaglov & Takmasheva, 2019).

3 Methodology

The purpose of the study is to reveal the possibilities of the regional socio-economic policy implemented according to the principles of lean production to overcome the negative consequences of the COVID-19 pandemic.

Research objectives:

- systematization of the problems of socio-economic development of Russian regions during the COVID-19 pandemic;
- justification of the directions of implementation of the principles of lean production in the regional management system during the pandemic;
- formation of an algorithm for the implementation of regional socio-economic policy according to the principles of lean production during the pandemic.

Research methods: theoretical analysis, comparative analysis, economic and statistical analysis, graphical method, system approach.

Information base of the study: materials of the Federal State Statistics Service of the Russian Federation, information from the official portals of state authorities of the Russian Federation, OECD statistics.

4 Results

The lockdown introduced on the territory of Russia from March 30, 2020, led to a significant decrease in the economic activity of enterprises in the regions. There were serious concerns about falling incomes and maintaining overall employment. The total number of unemployed people in Russia aged 15 and older reached 4.3 million in 2020. The growth of this indicator compared to 2019 was 24.7%. The industrial production index for 2020 decreased by 2.9% compared to 2019. The volume of paid services to the population decreased in 2020 compared to the same period in 2019 by 17.3%. The drop in cargo turnover of all types of transport in 2020 compared to the same period last year was 4.9%. The drop in household income was one of the reasons for the 4.1% decline in retail trade turnover in 2020 (Korolyuk et al., 2021). In the same period, the consumer price index increased by 3.4% (Fig. 1).

The country-wide trends are visible in the context of the Russian regions. The pandemic has created prerequisites for the growth of unemployment in the regions of Russia. For example, the level of registered unemployment in December 2020 reached a critical level in several lagging regions of the country: the Republic of

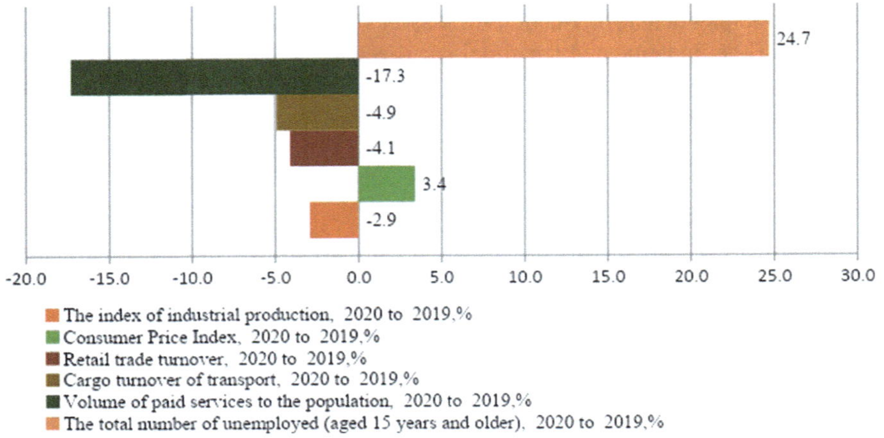

Fig. 1 Some indicators of the Russian economy development in 2020 compared to 2019, %. *Source* Compiled by the authors based on the Federal State Statistics Service of the Russian Federation (2020)

Tuva (19.3%), the Altay Territory (12%), the Kurgan Region (7.5%), the Republic of Kalmykia (6.8%), the Republic of Adygeya (5.9%). The average value of the registered unemployment rate in the country in December 2020 reached 4.2% (for comparison in December 2019—0.95).

The traditional problem of low per capita income in the Russian regions has only worsened in 2020. This problem was most pronounced in the lagging regions. For example, in the Republic of Tuva, the Republic of Kalmykia and the Altay Territory, the value of the average per capita monetary income of the population (the average value for 2017–2019, Russia = 100% (adjusted for the cost of living in the region)) it reached a critical level—54, 55 and 58%, respectively (Milchakov, 2021). The decline in the level of per capita monetary income of the population worsens the problem of increasing the dependence of the population on state payments and reducing the level of well-being in the regions.

In 2020, there was an increase in the burden on regional budgets due to a reduction in business income and tax revenues to the budget. From the federal budget, the Government of the Russian Federation allocated 290 billion rubles to maintain the stability of regional budgets. However, despite the stabilization measures, the number of regions with a budget deficit increased from 35 in 2019 to 57 in 2020. In general, the consolidated budgets of the regions in 2020 had a deficit of 676.6 billion rubles.

The level of debt burden (the amount of public debt relative to regional revenues without gratuitous receipts) on average for the regions in 2020 increased by 4.8 percentage points and amounted to 27.3% (Milchakov, 2021). The top 10 regions of the country with the highest increase in the level of debt burden on the regional budget were the Udmurtian Republic, the Republic of Khakassia, the Kemerovo Region-Kuzbass, the Tomsk Region, the Perm Territory, the Komi Republic, the

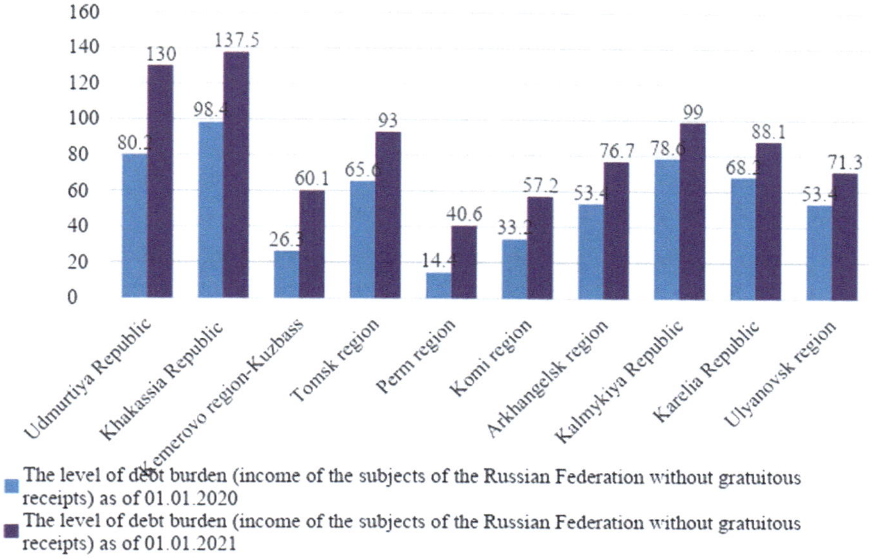

Fig. 2 Top 10 regions of Russia with the largest increase in the level of debt burden on the regional budget for 2020. *Source* Compiled by the authors based on the Federal State Statistics Service of the Russian Federation (2020)

Arkhangelsk Region, the Republic of Kalmykia, the Republic of Karelia, and the Ulyanovsk Region (Fig. 2).

Thus, the result of the pandemic was an increase in the number of regions with budget deficits and an increase in the total volume of debt obligations. This has become a threat to undermine the balanced development of Russia's regions.

Increasing the revenue base of regional budgets is possible if the regional economy stabilizes and investment activity rises. However, in 2020, many regions of Russia showed a sharp decline in the volume of investment per capita (the average value for 2017–2019, RF = 100%). This applies primarily to the Kurgan Region (30.2%), the Republic of Mari El (32%), the Republic of Tuva (35.6%), the Altay Republic (36.6%), the Chuvash Republic (39.4%).

At the same time, the Russian specifics of public governance also have a significant impact on the effectiveness of the implemented socio-economic policy. For example, Y. Adizes drew attention to the key problems in the system of public governance in Russia: lack of systematization and clarity of thinking, resulting from the specifics of Russian culture; absence of discipline, leading to abuse of power; predominantly authoritarian management as a reflection of the mental specifics of Russia and Russian authoritarianism concerning the power resource; widespread "focus on control", reducing the effectiveness of management decisions; fear generated by autocracy and control; "inefficient productivity", resulting from the centralization of services, excessive authoritarianism and lack of flexibility of management systems;

pronounced corruption component and the existence of bureaucratic barriers (Adizes, 2014).

Following the principles of lean production in the implementation of the territorial socio-economic policy can become a tool for overcoming the existing problems of the Russian regions during the pandemic.

The principles of lean manufacturing can be implemented in the regional governance system during a pandemic to (Karpunina, 2019):

(1) optimizing of expenses for the maintenance of public administration bodies in the region, as well as increasing the quality of regional management;
(2) improving the effectiveness of regional policies aimed at providing social support to the population during the pandemic;
(3) creating infrastructure and enterprises that are competent in lean manufacturing (Table 1).

Table 1 Goals and objectives of implementing the principles of lean manufacturing in the regional management system during the pandemic

Goals	Objectives
(1) Optimization of expenses for the maintenance of public administration bodies in the region, as well as enhancing the quality of regional management	• Reduction of the regional management apparatus; • Organization of effective working space for employees; • Elimination of losses of excessive information processing; • Reduction the time associated with the search and delivery of service information; • Implementation of the project management system and development of the project roadmap; • Levelling the corruption component, as well as digitalization of the activities of state authorities in the region (Repina, 2016)
(2) Improvement the effectiveness of regional policies aimed at providing social support to the population during the pandemic	• Expansion the range of support measures provided and their quality; • Improvement the quality of service to the population of the region; • Reduction turnaround time and waiting time; • Iimprovement of the comfort of the environment
(3) Creation of infrastructure and enterprises that are competent in lean manufacturing	• Popularization of the concept of lean production among managers and employees of enterprises in the region; • Motivation the improvement of the quality of services provided and goods sold by large, small and medium-sized enterprises in the region (Tyaglov & Takmasheva, 2019)

Source Compiled by the author

Table 2 Indicators of the thriftiness of the activities of regional authorities

Indicator group	Indicator name
Organizational and economic indicators of the "thrift" of the activities of regional authorities	The share of employees of the public authority who have been trained in the implementation of lean technologies (%)
	The share of implemented labor rationalization initiatives (%)
Indicators of the social effect of the introduction of lean production in the activities of regional authorities	Satisfaction of the population with the activities of the state authority (%)
	The share of the considered responses of business entities received during public consultations and taken into account in the development of the institutional framework of the region (%)
	Thrift index (the ratio of the number of implemented state support measures aimed at improving the quality of public services to the arithmetic average of the measures implemented by all executive authorities)

Source Compiled by the author based on (Tyaglov & Takmasheva, 2019)

Firstly, improve the quality of regional governance requires the development of organizational, economic and social indicators of the thriftiness of the activities of regional authorities (Table 2).

Secondly, during a pandemic, social support for the population becomes especially important, and measures to provide it are expanded many times. In particular, special measures are being implemented to support older people (Sinyavskaya et al., 2020). The expansion of measures to support vulnerable segments of the population in the region creates an additional burden on social institutions. In the social sphere, it is possible to use separate lean production tools to solve this problem:

- value stream (compilation of a list of processes, the definition of management processes, main and auxiliary processes; development of a model of interaction of these processes);
- mapping—development of a process passport and information maps for each process of the quality management system (planning, analysis of the quality control system by management, identification of recipients of social services, provision of social services, personnel management, infrastructure management, procurement, customer satisfaction analysis, monitoring and measurement of service results, etc.);
- standardized work (development of documents defining the order, time, frequency, and algorithm for providing services);
- solving the problem "one by one" (development of a process audit program, preparation of an internal audit plan, an internal audit report, a nonconformity

registration sheet and corrective actions; implementation of corrective actions) (Valiullina et al., 2019).

Thirdly, the implementation of the principles of lean production in the activities of enterprises during the pandemic allows management to:

- reduce the cost of maintaining office space by transferring the staff of the non-production unit to remote work;
- create personal training courses for the personnel of production departments and maintain working skills at the required level;
- implement a project for the stage-by-stage decommissioning of part of the unused equipment and premises of the enterprise, as well as reduce the cost of maintaining the infrastructure;
- to improve production and business processes (by searching for internal reserves to increase labor productivity, reduce costs and eliminate losses in the production of products and services) (Naugnova, 2014; NG, 2020).

The pandemic, on the one hand, has become a test for regional economic entities that implement lean technologies, has shown the unreliability of supply chains, the lack of transparency of relationships with suppliers, as well as has revealed the vulnerabilities of companies (Inproject, 2021). On the other hand, the positive impact of the pandemic on enterprises implementing the principles of lean production was manifested in the transition to an online mode and the emergence of the possibility of rapid decision-making without the need to move between production sites (LSSRussia, 2021).

The pandemic has created the conditions for the regional governance system to become as human-centered as possible. This applies to the work of public administration bodies in the region, the system of social support for the population, as well as the activities of enterprises in the region. Therefore, the key directions of the socio-economic policy of the region and the traditional algorithm for implementing the principles of lean production in the activities of regional economic entities should be adjusted depending on the degree of exposure to epidemiological risks. Thus, the algorithm for implementing regional socio-economic policy according to the principles of lean production during the pandemic has the following form (Fig. 3).

Currently, various approaches to the organization of lean production and the introduction of the concept of "Lean Region" are being implemented in the regions of Russia. The Belgorod Region is one of the most advanced regions implementing lean manufacturing tools.

The formation of a culture of "lean production" in the Belgorod region is aimed at creating conditions for all subjects of the regional economy to understand the practices of measuring values and losses, as well as the development of standards for the regional management system and the creation of a system for monitoring the results of implementation. In particular, in 2018–2020, the project "Formation of a culture of lean management in the executive authorities, state bodies of the Belgorod Region" was implemented. In 2021, the concepts of "lean management" and "lean

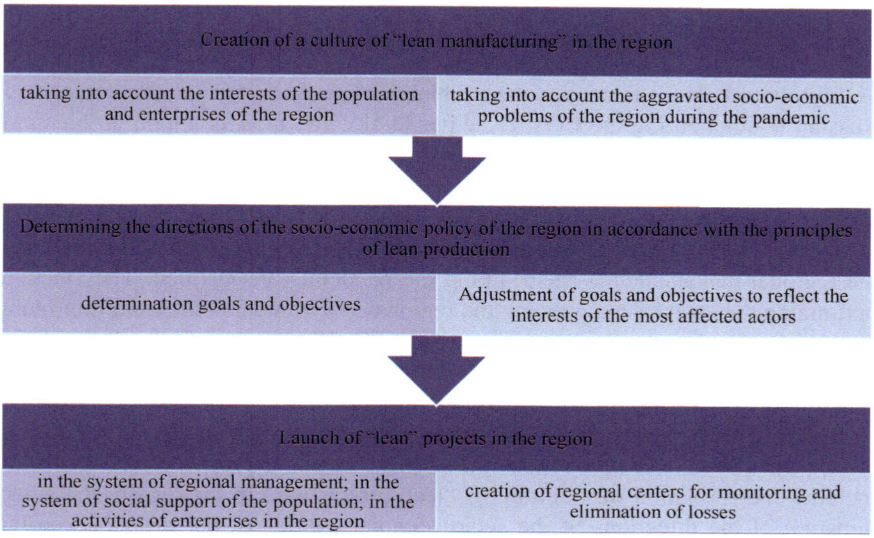

Fig. 3 Algorithm for the implementation of regional socio-economic policy according to the principles of lean production during the pandemic *Source* Compiled by the author

project" were fixed at the legislative level (Government of the Belgorod Region, 2018).

The next stage was to determine the directions of the socio-economic policy of the region according to the principles of lean production. For example, in the Belgorod region, the adoption of the regulation on the introduction of lean management has become an important step in determining the strategic guidelines for the development of the region: recognition of human resources as the main source of value creation; timely identification of changes in user requirements (to improve the quality of processes or services) and reduction of losses (Botvinyeva et al., 2021).

The launch of "lean projects" in the region is the result of the joint work of the regional lean and regional project offices. Employees of the regional lean office accompany the improvement team at the process mapping stage and help implement the appropriate tools. The regional project office is involved in the work at the stage of the development of project documentation. Employees of the regional project office monitor the timing and quality of the implementation of events. In the structure of the authorities of the Belgorod region, a register of processes has been created, which determines the order of their mapping. The region has implemented three waves of lean projects in three years of implementing lean technology tools. In 2018–2020, employees of the lean office conducted monthly training seminars for team leaders of regional and municipal authorities and employees of Competence Centers. Thus, more than 3000 people were trained. In the context of the pandemic, the implemented directions of the socio-economic policy of the region were adjusted according to the interests of the most affected subjects of the regional economy.

5 Conclusion

Firstly, the article systematizes the problems of socio-economic development of Russian regions during the COVID-19 pandemic. The most common problems of the Russian regions during the pandemic are the unemployment growth, the fall in the per capita income of the population, the reduction of regional budgets and the growth of regional public debt.

Secondly, the directions of implementing the principles of lean production in the regional management system during the pandemic are justified. They include optimizing the costs of maintaining the regional government bodies and improving the quality of regional management; improving the effectiveness of regional policies aimed at social support of the population during the pandemic; creating infrastructure and enterprises that are competent in lean production.

Thirdly, an algorithm for implementing regional socio-economic policy according to the principles of lean production during the pandemic is proposed. The algorithm includes the formation of a culture of "lean production" in the region; the determination of the directions of the socio-economic policy of the region according to the principles of lean production; the launch of "lean projects" with mandatory monitoring and elimination by a specialized regional center.

References

Adizes, I. (2014). *How to overcome management crises. Diagnostics and solutions of management problems* (English by N. Bragina Trans.). Mann, Ivanov and Ferber, Moscow.

Atanov, N., Munkodugarova, A., & Semenov, F. (2014). Strategic approaches of japan and the republic of Korea to regional development: Lessons useful for Russia. *Bulletin of the Buryat State University, 2*, 70–75.

Botvinyeva, T., Kirilova, I., & Orlova, A. (2021). 10 Belgorod region: how the lean principles moved from the executive to the legislative. https://lean.cdto.ranepa.ru/10-1-lean-v-belgorode-istoriya-vnedreniya. Accessed: 12 May 2021.

Falaleeva, E. (2016). The strategy of Kaizen for organizational changes. *Proceedings of the XIV International Student Scientific and Practical Conference, 7*(44), 94–106.

Federal State Statistics Service of the Russian Federation. (2020). Regional statistics. https://rosstat.gov.ru/regional_statistics. Accessed: 12 May 2021.

Filatov, S. (2017). On the state of socio-economic development of subsidized regions and options for managing changes. *Economics and Management, 2*(14), 15–21.

Friedmann, J. (1973). *Urbanisation, planning and national development*. Beverly Hills

Gorlov, S. (2016). Lean Region. *Accreditation in Education, 5*, 39–41.

Government of the Belgorod Region. (2018). Resolution No. 161-pp "On approval of the regulations on lean management in executive authorities and state bodies of the Belgorod region", 28 May 2018. https://docs.cntd.ru/document/550110788. Accessed: 12 May 2021.

Hobbs, D. (2007). *Hobbs the implementation of lean production*. Grevtsov Publisher.

Inproject. (2021). Too "lean" supply chains-a factor in the deterioration of the global response to new challenges. https://www.inprojects.ru/istonchenie-cepej-postavok-faktor-uhudsheniya-globalnogo-reagirovaniya-na-novye-vyzovy. Accessed: 12 May 2021.

Kaizen, M. (1986). *The key to the success of Japanese companies*. Alpina Publisher.

Karpunina, E. (2019). Principles of Kaizen in the service of regional management. In *Proceedings of the II All-Russian (national) Scientific and Practical Conference "Sayapin readings"* (pp. 316–325). Tambov State University named after G. R. Derzhavin.

Karpunina, E., Molchan, A., Lapushinskaya, G., Ryabova, T., & Jusufov, A. (2020a). Mechanisms for managing changes in the socio-economic development of lagging regions. In *Proceedings of the 35th IBIMA Conference* (pp. 6538–6549). Seville, Spain.

Karpunina, E., Egorov, S., & Radyukova, Y. (2018). Modernization of the mechanism of formation and implementation of the regional socio-economic policy: From subsidization to self-development. *Bulletin of the South-Western State University, 3*(28), 18–28.

Karpunina, E., Petrov, I., Klimentova, E., Sozaeva, J., & Korkishko, I. (2020b). Mechanisms of self-development of subsidized regions. In *Proceedings of the 35th IBIMA Conference* (pp. 2282–2293). Seville, Spain.

Korolyuk, E., Rustamova, I., Kuzmenko, N., Khashir, B., & Karpunina, E. (2021). Diagnostics of regional economic security problems during the 2020 crisis. In *Proceedings of the 37th IBIMA conference*. Cordoba, Spain.

LSSRussia. (2021). Examples of implementing lean manufacturing elements during a pandemic. https://lssrussia.ru/newslean/primery-vnedreniya-elementov-berezhlivogo-proizvodstva-v-period-pandemii/. Accessed: 12 May 2021.

Lyosh, A. (2007). *Spatial organization of the economy*. Science.

Milchakov, M. (2021). Implementation of regional development priorities in the context of a pandemic. *Financial Journal, 13*(2), 93–113.

Naugnova, I. (2014). *Domestic and foreign experience in the application of the lean production system at industrial enterprises*. Izvestiya RSPU named after A. I. Herzen (vol. 170, pp. 95–99).

NG (2020). The principles of lean manufacturing allowed the Sochi airport to survive during the pandemic. https://www.ng.ru/ideas/2020-10-05/7_7981_05102020.html. Accessed: 1 May 2021.

Repina, Yu. (2016). Lean housing and communal services (on the example of Kazan). In *Proceedings of the International Scientific and Practical Conference Innovations*, Novosibirsk: SibAK 3(52). Part I, 129–134.

Sinyavskaya, O., Andreeva, E., Gorvat, E., & Kareva, D. (2020). Covid-19 and social policy towards older citizens: a review of measures taken by foreign countries with different levels of development. https://www.hse.ru/data/2020/07/13/1594085462/HSE_Covid_09_2020_3_1.pdf. Accessed: 12 May 2021.

Sychanina, S., Mironchuk, V., & Sholin, Y. (2018). Lean management: Methods and tools of lean production in municipal management, prospects and practice of application. *Business. Education. Law, 4*, 198–203.

Tyaglov, S., & Takmasheva, I. (2019). The development of lean production in the conditions of transformation of the regional economy. *Journal of Economic Regulation, 10*(1), 107–119.

Valiullina, L., Bespalova, M., & Khadasevich, N. (2019). Integration of the principles of lean production and quality management system in the organization of labor of social service institutions. *Liderstvo i Menedzhment, 6*(2), 107–116.

Zubarevich, N. (2019). Spatial development strategy: priorities and tools. *Economic Issues, 1*, 135–145.

Zubarevich, N. (2021). The impact of the pandemic on socio-economic development and regional budgets. *Questions of Theoretical Economics, 1*(10), 48–60.

Chinese Pharmaceutical Companies Entering Global Market: The Case of Jiangsu Hengrui Medicine

Natalia A. Volgina and Na Li

Abstract Currently, the pharmaceutical industry is one of the fastest-growing knowledge-intensive industries in the world economy. Leading positions in the global pharmaceutical market are occupied by the largest companies from the USA, Japan and some European countries. However, in recent years, to reduce costs, links in pharmaceutical value chains have begun to be transferred to emerging market countries, which includes China. This created favourable opportunities for the development of national companies. Jiangsu Hengrui Medicine (JHM) is one of the most innovative pharmaceutical companies in China, founded in 1970 and listed on the Shanghai Stock Exchange in 2000. The purpose of this work is threefold: first, to trace the stages of formation of JHM; secondly, to identify the strategies that JHM used when entering foreign markets; thirdly, to highlight the factors of competitiveness that allowed the company to achieve success in the global pharmaceutical market. Several methods were used as research methods, in particular, comparative analysis, the method of analysis and synthesis, and the method of induction. As a result of the study, the authors came to the following conclusions. In its development, JHM went through a series of successive stages that were organically linked to each other: the production of basic pharmaceutical products (stage 1) laid the foundation for the production of generics (stage 2) and its innovative pharmaceutical products (stage 3). In parallel, the international development of JHM was going on, and the strategies for entering foreign markets were closely related to the increase in the degree of innovation of the manufactured pharmaceutical product. The main factors in the growth of the company's competitiveness in the international market were as follows: consistent investment in R&D, investment in the company's human capital, as well as cooperation with Chinese and foreign universities, innovation centres, etc. An important role in promoting Chinese companies abroad is played by government support for creating a favourable business macro environment.

Keywords Emerging markets · Pharmaceuticals · Jiangsu Hengrui Medicine (JHM) · Technological innovations · Internationalization

N. A. Volgina (✉) · N. Li
Peoples' Friendship University of Russia (RUDN University), Moscow, Russia

E. G. Popkova and B. S. Sergi (eds.), *Geo-Economy of the Future*,
https://doi.org/10.1007/978-3-030-92303-7_77

JEL Codes F15 · F23 · F63

1 Introduction

Economic globalization has intensified competition between enterprises, and the external environment of enterprises has changed significantly. In many aspects, such as resources, the scale of production, independent innovation, enterprises in most emerging market countries lag significantly behind enterprises in developed countries. According to the Report on Chinese Enterprises Globalization (2020) in 2018 in a difficult international environment, the global investment influence of Chinese enterprises continues to grow, accounting for more than 10% of global FDI flows for three consecutive years. At the same time, the level of foreign investment inflows in China remains the leading one in the world, ranking second in the world in terms of foreign investment value (Wang et al., 2020, p. 1). Some Chinese enterprises have achieved notable success on the way to the internationalization of production (Xie et al., 2019, p. 29). This includes the innovative pharmaceutical company Jiangsu Hengrui Medicine (JHM).

JHM is one of the most innovative large pharmaceutical companies in China. It was founded in 1970 and listed on the Shanghai Stock Exchange in 2000. By the end of 2019, JHM has more than 24,000 employees worldwide, including more than 3000 R&D employees and more than 2000 doctors, masters and scientists who returned from abroad.

The purpose of this work is threefold: first, to trace the founding stages of the JHM company; second, to identify the strategies that JHM used when entering foreign markets; third, to emphasize the factors of competitiveness that make the company achieve success in the global pharmaceutical market.

2 Methodology

The methodology for researching the formation of JHM and the stages of entry and consolidation in the global market includes a comparative analysis, which will enable a correlation of the positions of JHM in comparison with its main competitors. Research methods such as analysis and synthesis will provide an opportunity to highlight the key success factors of JHM in the global market, as well as provide estimates of future trends in its development, including in the Post—COVID period.

The main sources for writing this work were the official information of the JHM Company, as well as analytical papers of Chinese authors.

3 Results

3.1 Stages of Jiangsu Hengrui Medicine Company Development

At present, companies in Western Europe, the United States and Japan are leaders in the pharmaceutical market, both in terms of innovation and in terms of financial strength (Ni et al., 2007, pp. 1–2). These companies possess a significant share of the global pharmaceutical market.

However, in recent years, cost pressures and environmental controls in Europe and the United States have gradually increased. To reduce R&D costs, Western companies have begun to move mass production of raw materials, as well as R&D centres, to emerging market countries, including China. This gave Chinese enterprises room for development (Li, 2020, p. 22) and enabled them to gradually occupy the shares of the world pharmaceutical market.

According to PharmExec (Official Website, n/a) in 2020, four Chinese pharmaceutical companies entered the Global Pharmaceutical Top 50 List: Yunnan Baiyao Group (37th rank), Sino Biopharmaceutical (42nd rank), Jiangsu Hengrui Medicine (43rd rank), and Shanghai Pharmaceuticals Holding (48th rank).

In its development, JHM went through a series of successive stages that were organically linked to each other: the production of basic pharmaceutical products (stage 1) laid the foundation for the production of generics (stage 2) and its innovative pharmaceutical products (stage 3).

The first stage of the production of basic drugs took more than 20 years: 1970–1990s.

In 1970, Lianyungang Pharmaceutical Factory, the predecessor of Jiangsu Hengrui Medicine, was formally established, and in 1997 it was reorganized by the Government into JHM Co., Ltd. From 1970 till 1992 Lianyungang Pharmaceutical Factory was a small haemostatic disinfectant drug factory. There were many problems in the enterprise, especially the low level of pharmaceutical raw materials for tablet production, poor technical equipment, low profits, fierce market competition in this field and lack of own-brand products, which hinder the development of the enterprise (Ren & Liu, 2013, p. 98).

In the early 1990s, a company bought a patent for a new cancer drug—isophosphamide. The Chinese State Pharmaceutical Administration has approved drug use. This is JHM's first step towards generics production. In 1996, JHM sold nearly 100 million RMB anticancer drugs and carried out more than 20 new drug research and development activities, of which 5 were rated as "national key products".

To survive in a highly competitive Chinese pharmaceutical market and to strengthen the transition to the second stage of its development (which we date from the 1990s)—the production of generics—JHM began to use imitation innovations (Yang & Chen, 1997, p. 1). From 1994 till 2006 JHM has set up three research centres, the Clinical Medicine Department (Beijing Medical Department) and the

Enterprise Technology Center, and a post-doctoral research workstation. Gradually, the company launched a series of the first imitation anticancer drugs, which laid the foundation for the international transformation of the company's production activities.

In its activities, JHM focuses on two fast-growing pharmaceutical sectors: anticancer one and anaesthesia one, as well as on the production of some highly competitive products (Wang & Xiang, 2010, pp. 4–17). JHM is one of the most successful "import substitution" pharmaceutical companies in China. This allowed the company (along with the use of flexible marketing technologies) to occupy market shares of foreign companies in China.

In 2000, Jiangsu Hengrui Medicine was listed on the Shanghai Stock Exchange and issued 40 million shares. At the same time, a whole series of generics was launched into production. The creation of an R&D system and huge financial resources (Fig. 1) obtained from the sale of generics (in the domestic and world markets) paved the way for JHM to move to Stage 3 of development—the production of its innovative pharmaceutical products.

In 2011, the first innovative drug—Irecoxib, was approved for listing and became a new national first-class drug in China, marking the company's transition from a generics manufacturer to a manufacturer of original innovative drugs.

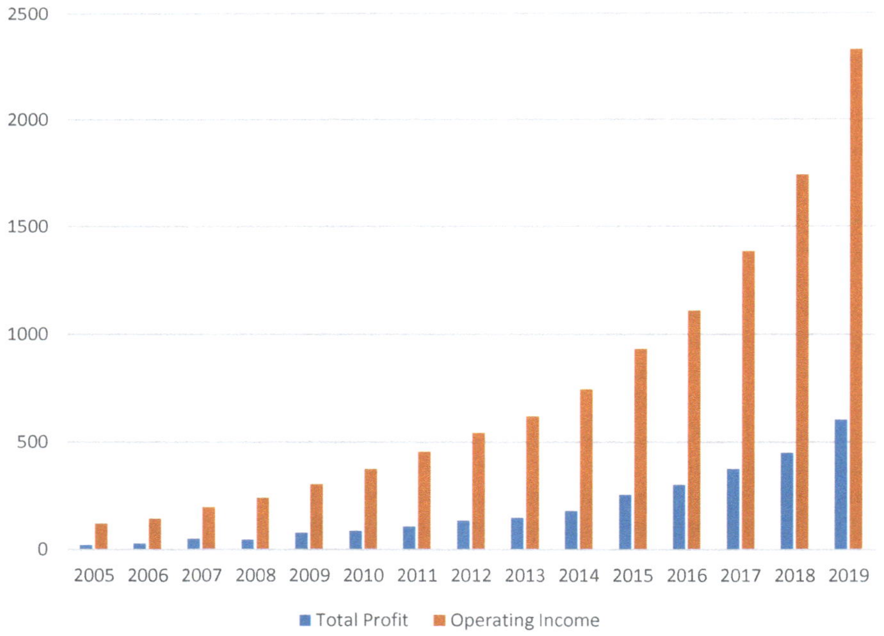

Fig. 1 JHM operating income and total profit, 2005–2019 (Unit of measurement: tens of millions of RMB). *Source* Compiled by the authors based on the Annual report JHM (2005–2019)

In parallel, the international promotion of JHM was going on, and the strategies for entering foreign markets were closely related to the increase in the degree of innovation of the manufactured pharmaceutical product. We classify these strategies as follows: simple imitation of a pharmaceutical product, in-house creation of an imitation, in-house innovation in pharmaceuticals. All of these products were offered to overseas buyers, thereby expanding JHM's presence in foreign markets.

3.2 *Factors of Company Competitiveness*

During this research, it was revealed that the main factors in the growth of the company's competitiveness in the international market were a consistent investment in R&D, investment in the company's human capital, as well as cooperation with Chinese and foreign universities, innovation centres, etc. Reliance on the domestic market sales played an important role in promoting the foreign operations of JHM.

In the process of internationalization of pharmaceutical companies, one of the main factors limiting the development of companies is the lack of innovation potential. And investment in R&D is a key factor in improving opportunities for technological innovation. The main sources of investment in R&D for JHM are operating income, as well as government subsidies and various grants (Chen, 2011, pp. 1762–1763). JHM is actively applying for various national, provincial and municipal science and technology competitions, seeking grants for national projects and support funds to obtain financial advantages and accumulate funds for R&D. JHM's R&D investment has continued to grow in recent years. The share of R&D investment in sales increased from 7.43% in 2005 to 16.7% in 2019.

In 2019, JHM invested 3.9 billion yuan in R&D funds, up 45.9% over the same period last year, while R&D investment accounted for 16.7% of operating income. This shows that Hengrui Medicine's investment in R&D has been steadily increasing in recent years. JHM's R&D investment and the percentage of R&D investment in sales from 2005 to 2019 are shown in Fig. 2.

Another important factor that contributed to the growth of JHM's competitiveness was the increased training of scientific and technological talents, as well as international scientific and technical exchange, that is, investment in human capital.

In the development process, JHM actively participates in the exchange, cooperation and training projects of medical personnel organized by the National Health Commission of China. At the same time, the company establishes close collaborative relationships with leading international medical institutions, which include, in particular: Mayo Clinic in the USA, Harvard Medical Institute and related hospitals, Anderson Cancer Center in the USA, University of Cambridge and its affiliated hospitals in the UK, French Institute of Public Health, University of Tokyo Hospital, Japan, etc.

In addition, JHM gives great importance to the training of scientific and technical talents. From 1997 to 2019 JHM's workforce has grown from over 10 people to over 24,000 people worldwide. As of 2019, JHM has an R&D team of more than 3400

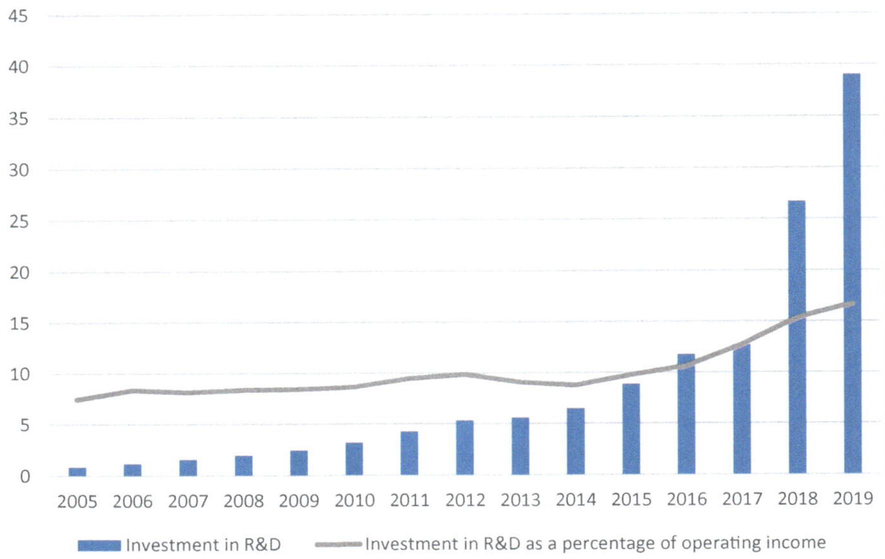

Fig. 2 JHM Investment in R&D and investment in R&D as a percentage of operating income 2005–2019 (Unit: tens of millions of RMB). *Source* Compiled by the authors based on the Annual report JHM (2005–2019)

people, including more than 2000 doctors, masters and more than 200 returnees with R&D talents around the world.

An important role in promoting Chinese companies abroad is played by government support for a favourable macro environment for enterprises in China, which also makes a tangible contribution to increasing the competitiveness of Chinese enterprises abroad. This is actively used by HJM. China's economy has a planned nature and develops in five-year terms. From 2006 to 2020 the Chinese government has consistently introduced some supportive measures to encourage pharmaceutical companies to innovate. In the long term, until 2035, the pharmaceutical industry will be included in the list of key developing strategic sectors of the country for development.

4 Conclusion

Throughout its development, JHM has gradually evolved from a small business into a system of scientific research centres engaged in the production and sale of both generics and its original medicines. In parallel, JHM entered foreign markets, squeezing out competitors from Western countries and occupying (albeit very slowly) their market shares. This was largely facilitated by the steady growth of R&D expenditures, the focus on the increasing role of human capital in the company's activities, as well as government assistance in creating a favourable business environment.

COVID-19 and its consequences will make significant changes in the situation in the global pharmaceutical market. The processes of internationalization in the pharmaceutical industry will continue, but the forms of these processes will change.

Acknowledgements This paper has been supported by the RUDN University Strategic Academic Leadership Program and Grant of RUDN University No. 060119-0-000

References

Chen, J. (2011). Collaborative innovation and the construction of national scientific research capacity. *Science of Science Research, 29*(12), 1762–1763.

JHM Annual Report 2005–2019 http://so.eastmoney.com/Ann/s?keyword=%E6%81%92%E7%91%9E%E5%8C%BB%E8%8D%AF%E5%85%AC%E5%8F%B8%E5%B9%B4%E5%BA%A6%E6%8A%A5%E5%91%8A&pageindex=1.pdf. Accessed: 30 May 2019.

Jiangsu Hengrui Medicine Official Website. http://www.hrs.com.cn/index.html.

Li, C. (2020). *Research on the key factors and formation mechanism of the international competitiveness of the pharmaceutical industry-based on the Empirical Analysis of the Pharmaceutical Industries in the United States* (pp. 1–2). China University of Political Science and Law.

Ni, Y., Jiang, B., Yan, Y., & Shi, L. (2007). An analysis of the current Indian pharmaceutical industry. *Chinese Pharmaceutical Technology Economics and Management, 1*(2), 22.

PharmExec Official Website. https://www.pharmexec.com

Ren, S., & Liu, Y. (2013). Enterprise transformation and upgrading based on technological innovation. *Business Management, 3*, 98.

Wang, H., Miao, L., & Yu, W. (2020). General report: status and characteristics of Chinese Enterprise Globalization in 2018–2019. In H. Wang & L. Miao (ed.), *Globalization Think Tank (CCG). Development Research Institute of Southwestern University of Finance and Economics (2020). Report on Chinese Enterprises Globalization (2020)* (p. 1). Beijing: Social Sciences Academic Press.

Wang, X., & Xiang, J. (2010). *China's Takeda, World's Henrui* (p. 4–17). Jiangsu Hengrui Medicine company research report.

Xie, P., Tai, X., & Cai, J. (2019). Stable, deep, proficient and the strategic core of reverse internationalization—Taking HengRui medicine as an example in the choice of Internationalization Strategic Path. *Tsinghua Management Review, 12*, 29.

Yang, D., & Chen, B. (1997). Imitation innovation, independent innovation and high-tech enterprise growth. *China Soft Science, 8*, 1.

Investment Activity of China in European Countries During the Pandemic

Inna V. Andronova and Daria S. Sokolan

Abstract The purpose of the article is to study the investment activity of China during the pandemic when the volume of investments around the world began to decline. The article provides a brief overview of China's investment activities in Europe over the past fifteen years. The main mergers and acquisitions in Europe during the pandemic were analyzed in detail. In the process of writing the article, general scientific research methods were used (analysis, synthesis, grouping method and graphical method). The statistical base was the data of the American Enterprise Institute, UNCTAD, Reuters, and BBC. Every year the role of China in the world economy is increasing; the Chinese government is skillfully using the main instruments of globalization to strengthen its influence in the world. Foreign direct investment is one of the main tools for China to achieve leading positions in the global economy, and in particular in the high-tech sector. During the pandemic, Chinese investors continued to invest in European countries. China chose to invest in the sectors, that most needed funding during the crisis.

Keywords Foreign direct investment (FDI) · Investment policy · M&A · European Union · Pandemic · Chinese investments

JEL Codes F20 · F21 · F23

1 Introduction

Today there we see the transformation of both the concept of "globalization" and the process of globalization itself. Twenty years ago the main drivers of globalization were developed countries led by their transnational corporations (TNCs), which have their branches throughout developing countries in Asia and Africa. Developed countries called on other countries to open markets, remove barriers to the movement of goods, services and capital. Today we are witnessing a completely new process: developing countries have become at the head of globalization with new

I. V. Andronova (✉) · D. S. Sokolan
Peoples' Friendship University of Russia (RUDN University), Moscow, Russia

E. G. Popkova and B. S. Sergi (eds.), *Geo-Economy of the Future*,
https://doi.org/10.1007/978-3-030-92303-7_78

levers of its advancement. To be more precise, China has become at the head of new globalization. Today, China's investment activities abroad are beginning to cause more and more fears from economists in developed countries. More and more often in the European press, there are headlines of economic articles with the following titles: "China Bought Europe", "How China Is Buying Its Way into Europe?" (Tartar, 2018). Nowadays China, using the already existing schemes of developed countries, is carrying out investment expansion in the countries and industries of interest to it.

2 Methodology

In the process of writing this article, an evolutionary approach was used to systematize Chinese investments over the past fifteen years. Grouping method, analysis, induction and graphical method were also used. The authors used data from articles from the leading world publishing houses Reuters, Financial Times, BBC. The statistical base was the data of the American Enterprise Institute, UNCTAD.

3 Results

Foreign direct investment is a direct instrument of the globalization process. Therefore, China decided to adopt FDI. Over the past ten years, the increase in China's foreign direct investment abroad has been at a record pace. Moreover, Chinese investments are growing in developed countries, especially in the EU countries. In general, China's investment activities in the EU can be divided into three stages:

Stage 1 (2005–2010)—investments in traditional sectors of the economy of developed countries,
Stage 2 (2011–2015)—investments in infrastructure projects of the country's most affected by the global financial crisis,
Stage 3 (2016-present)—investments in the high-tech sector and the financial sector of developed countries.

At various times, China has invested in certain countries and industries of interest. Chinese investors began to carry out the first major transactions in the EU countries in 2005. During this time, Chinese entrepreneurs were actively pursuing the goals of China's "Go Out" policy, which encouraged investment abroad. Investments in European countries gradually grew until the global financial crisis. In 2010 there was a slight decline, after which the growth of investments continued. In the period 2005–2010, China invested in the traditional sectors of the economy of the leading EU countries.

The second stage (2011–2015) can be characterized by the fact that Chinese investors slightly shifted the spectrum of their interests towards the EU countries most affected by the consequences of the global crisis and the Euro Zone crisis.

China has concentrated on investing in countries of southern Europe. During this period, the "One Belt—One Road" strategy (European Bank for Reconstruction and Development, 2021) was launched, pushing Chinese investors to become even more active, but more focused on infrastructure projects. In addition, the Chinese government decided to limit investments and highlight the most expedient and promising sectors for investment.

However, the third stage of Chinese investment abroad is of greatest interest. In 2016–2017 Chinese investments in the EU began to grow at a record pace. In 2017, the value of Chinese transactions in the EU amounted to 93 billion US dollars. Since 2016, China has been investing in the most developed countries of Europe (Germany, France, and the UK). Moreover, in 2015, another Chinese initiative, "Made in China—2025" (Institute for Security & Development Policy, 2018), was launched, the goal of which is to transform China into a global leader in technology export. Thus, with each stage, China's investments in the EU became more and more highly specialized. And to date, Chinese investors have focused on high-tech industries and alternative energy sources in the EU.

The pandemic has led to a massive economic crisis in all countries of the world. Considering China's tactics in the crisis and post-crisis years in the EU countries (when Chinese investors willingly and freely invested money in the most affected countries (Italy, Greece), gaining access to ports and logistics companies), an urgent question arises: "How will China take advantage of the crisis in this time?".

In 2020, Chinese investments around the world fell to 30 billion US dollars, almost half of the 2009 figures. However, Chinese interest in European countries remained the same: Europe accounted for about 10 billion US dollars (one-third of the total) of Chinese investments abroad. In total, 19 transactions were concluded in European countries, 11 of which were concluded in the EU countries. It is worth noting that in 2020 the most developed countries of Europe (Germany, France, and the UK) received most of the Chinese investments. However, there was a shift in the interests of Chinese investors towards France. Until 2020, the UK was the main recipient of Chinese FDI, and it is still the leader in the amount of accumulated investment. Nevertheless, as shown in Fig. 1, France accounted for 4.1 billion US dollars in 2020 (almost 42% of China's total investment in Europe). The UK ranks second in attracting Chinese FDI with 1.4 billion US dollars (15%) and the Netherlands closes the top three attractive countries for Chinese investors (9%). In Germany, deals were concluded at a more modest value—only 550 million US dollars. We also note that despite the crisis, Chinese investors were attracted by the Balkan countries (Bosnia and Herzegovina and Serbia), which were not previously included in the top countries of interest to China. In 2020, Chinese investors also channelled their capital to Spain, Ireland, Sweden, Switzerland, and Norway.

Let's move on to the sectoral structure of China's transactions in Europe in 2020. An interesting fact is that most of the investments fell on the entertainment sector, about 3.5 billion US dollars (39% of all Chinese investments in Europe). The technology sector attracted only 1.6 billion US dollars (16%), metallurgy—1.2 billion US dollars (12.24%), finance—1.1 billion US dollars (10.95%) (Fig. 2). Investments in the technology sector are explained by the already mentioned strategy "Made in

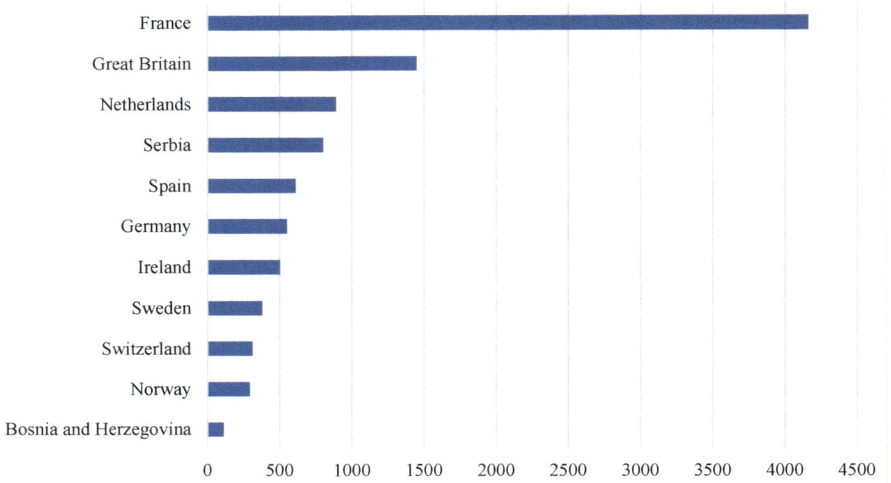

Fig. 1 Most Attractive European Countries for Chinese FDI in 2020. *Source* Compiled by the authors based on China Global Investment Tracker (2020)

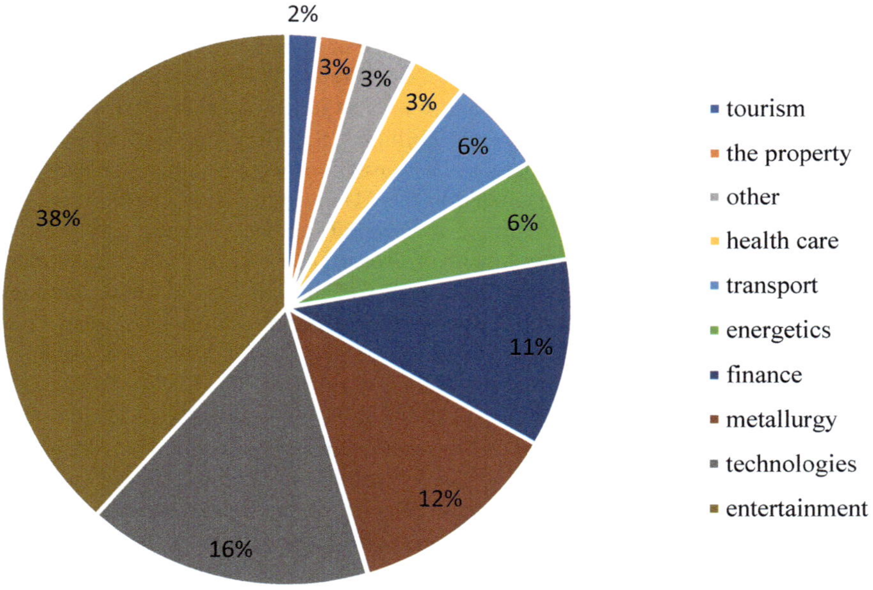

Fig. 2 Sectoral structure of Chinese FDI in Europe in 2020. *Source* Compiled by the authors based on China Global Investment Tracker (2020)

Table 1 China's largest deals in Europe in 2020

Investor	Quantity in millions	Share size (%)	Transaction party	Sector	Country
Tencent	3420	20	Vivendi	Entertainment	France
Wingtech	890	24	Nexperia	Technology	Netherlands
Zijin Mining	800	63	RTB Bor	Metals	Serbia
Three Gorges	610		X-Elio	Alternative Energy	Spain
Bytedance	500	100	Greenfield	Technology	Ireland
Evergrande	380	18	NEVS	Transport (Autos)	Sweden
Tencent	330	20	Voodoo	Entertainment	France
Huawei	240	100		Technology	France
Fosun	230	100	Bankhaus Lampe	Finance	Germany
Bank of China	180	13	Norwegian Air	Aviation	Norway

Source Compiled by the authors based on China Global Investment Tracker (2020)

China—2025", from which China did not deviate even during the pandemic. Most Chinese investors were attracted by the technology sector of the Netherlands, Ireland, and France.

Let's consider China's biggest deals in Europe in 2020. As the data in Table 1 show, the largest deal of 2020 was the purchase by the Chinese company Tencent of a 20% stake in the French media conglomerate UMG Vivendi.

The French company owns the world music leader Universal Music Group, which owns some record labels in the United States, video hosting Dailymotion, Video Game Company Gameloft and others. Note also that famous artists of Universal Music Group are Taylor Swift, The Beatles, Billie Eilish and Nirvana.

The Chinese holding company Tencent owns the WeChat mobile application, which ranks third in the world in terms of the number of users after Facebook and YouTube. This is not the first major deal of a Chinese company in Europe: in 2016, Tencent bought 84% of the shares of the Finnish company Supercell, which develops games for mobile devices (Supercell, 2021). Moreover, the Chinese corporation plans to buy out another 10% of UMG Vivendi in 2021 (Reuters, 2020a, b, c). An interesting fact is that the Chinese company is making such a large investment during the pandemic in the media sector (which is now experiencing not the best times). On the one hand, this deal can be viewed as a way for the Chinese side to gain the experience of a leading media corporation to promote their labels. Thus, it can be assumed that in the future, China is serious about promoting its performers and its music abroad (as the United States did in its time). This was not Tencent's only deal in France in 2020. Tencent invested 330 million US dollars in August 2020 to acquire 20% of French mobile game developer Voodoo (Hinata, 2020). Voodoo is consistently one of the twenty best mobile games on Gamer.biz. Thus, the Chinese company is trying to strengthen its position in the most popular areas.

The second-largest deal in 2020 was the deal between the Chinese company Wingtech and the Dutch company Nexperia. The Chinese company Wingtech accounts for about 27% of the global smartphone market (Counterpoint Global Smartphone ODM/IDH/EMS Tracker, 2020). In addition, the Chinese company has also held a leading position in the world in the production of smart devices. Before the deal with the Dutch company Nexperia, the Chinese company Wingtech did not produce only one element—semiconductors. Through a series of deals with Nexperia, the Chinese company has built its semiconductor value chain from design to electronics manufacturing (Wingtech, 2021). In turn, Nexperia is a world-famous semiconductor company, occupying 14% of the world semiconductor market (Nexperia, 2021). The Dutch company ranks first in the world in the production of diodes and transistors, it successfully competes with other European companies in the semiconductor field. Thus, by purchasing Nexperia, the Chinese company further strengthened its position in the smartphone market, becoming completely independent of overseas semiconductor supplies. Following the acquisition, Windtech management has invested even more in Nexperia's development to increase semiconductor shipments in Europe.

The third-largest deal in Europe was in Serbia: Chinese mining company Zijin Mining invested 800 million US dollars in Serbian copper smelter RTB Bor Group. The deal itself was concluded in 2019 when the Chinese company bought 63% of the shares of RTB Bor Group. It is worth noting that RTB Bor Group is the only copper producer in Serbia with four copper mines at its disposal. In turn, the Chinese company Zijin Mining is engaged in the exploration and development of deposits of gold, copper and other metals around the world. An important fact is also that the company has a relatively complete global value chain, the company has 12 investment projects abroad (Zijin Mining, 2021). This trend towards the formation of a complete value chain can be seen in many transactions of Chinese investors overseas.

In 2020, China has not forgotten about investments in alternative energy. China Three Gorges has invested 610 million US dollars in 13 X-Elio projects in Spain. X-Elio is a world leader in the design, financing, construction, and operation of photovoltaic installations. The Chinese company China Three Gorges is a world leader in the production of clean energy; it also develops and operates hydropower. The company has its branches in 47 countries of the world (China Three Gorges Corporation, 2021). Through this deal, China Three Gorges gained access to the Spanish energy market and strengthened its presence in Europe. Even before the pandemic, other Chinese companies showed interest in alternative energy in Europe. Given the current trends in the field of ecology, China's desire to strengthen its position in the global alternative energy market is understandable. China wants to take leading positions in all promising sectors of the world economy.

The operation of the Chinese company ByteDance in Ireland is also of interest. To begin with, ByteDance is a Chinese internet holding company that owns well-known video sharing platforms like Tik Tok and Xigua (ByteDance, 2021). It was the Tik Tok division that decided to open the first European data centre in Ireland. It is worth noting that Ireland is one of the largest data centres in Europe. Companies such as Facebook, Amazon, and Google Alphabet Inc are already operating there (Halpin, 2020). Tik Tok has grown rapidly in popularity and continues to attract new users. Today, more than 100 million people use Tik Tok in Europe. Thus, by opening a data centre, the Chinese company demonstrates its ambitions to expand its activities and increase the number of users.

In August 2020, the Chinese company Evergrande (a diversified investment holding in China) finally bought out the shares of the Swedish electric vehicle manufacturer NEVS. The purchase of the last 18% of the shares cost the Chinese side 380 million US dollars. The company itself, as a whole, cost the Chinese company almost 2 billion US dollars. The Swedish company is actively developing to create modern vehicles for smart and environmentally friendly cities (NEVS, 2021). Through the acquisition of the Swedish company, Evergrande is trying to strengthen its position as a leading manufacturer of electric vehicles in the future. The deal confirms China's commitment to strengthening its position in the green economy.

In December 2020, it is also worth noting the investments in France made by the world-famous Chinese company Huawei. Huawei has decided to invest 240 million dollars in the construction of a plant to manufacture equipment for mobile networks. China intends to increase its supply of 5G equipment to European customers (Reuters, 2020a, b, c). Huawei is one of the world's largest telecommunications companies. It owns 23 research centres around the world and collaborates with over 100 partner universities. More than 105,000 employees of the company work in the field of R&D. Over the past 10 years, the company has invested about 720 billion yuan in R&D. Despite the US claims against the company's activities, Huawei is focused on increasing its market and its influence in the world. An important fact is that France did not ban this kind of investment by Huawei, while the UK was wary of Huawei's 5G technologies.

A major deal worth 230 million US dollars was completed in Germany. Fosun, a large Chinese financial and industrial conglomerate, has acquired the German bank Bankhaus Lampe. Let's start with the fact that the Chinese conglomerate operates in the fields of insurance, finance, tourism, jewellery and every year it demonstrates the growth of assets abroad (Fosun, 2021). In turn, Bankhaus Lampe is an independent private bank in Germany, which was founded in 1852. It specializes in asset and capital management, corporate finance services (Reuters, 2020a, b, c). Through this deal, the Chinese conglomerate seeks to expand its presence in the European financial sector. An interesting fact is that Fosun carried out the deal through the German lender Hauck & Aufhaeuser, which he has owned since 2016. Thus, the interest of the Chinese investor in expanding their activities in Germany is visible.

The operation between the Bank of China and the Norwegian company Norwegian Air closes the top ten largest Chinese transactions in Europe. The Chinese investor acquired 13% of Norwegian Air and became the second-largest investor in the Norwegian company. Norwegian Air is the second-largest airline in Scandinavia and the third largest airline in Europe. It is worth noting that the Bank of China, a state-owned company, took advantage of Norwegian Air's financial difficulties (Nikel, 2020). In connection with the pandemic, the Norwegian carrier began to suffer losses in 2020, in connection with which the company's shares fell in price. After the restructuring, the Bank of China decided to buy the shares of the Norwegian company. Thus, China gained access to one of the largest air carriers in Europe.

4 Conclusion

The investment activity of China in Europe every year causes more and more concern on the part of European governments. In the context of the pandemic, which led to an acute shortage of capital in some sectors of the economy, a favourable environment has emerged for China's investment. Although overall Chinese investment around the world declined in 2020, Europe accounted for about 40% of all Chinese FDI. A study of China's ten largest deals in Europe in 2020 showed that Chinese investors are increasingly interested in the technology, alternative energy, and finance sectors. Most of the investments in the technology and finance sector were made in the developed countries of Europe—Great Britain, Germany, Ireland, and the Netherlands. In addition, Chinese investors have invested in metallurgical complexes in Serbia, Bosnia and Herzegovina. The continuing interest in the technology and alternative energy sector can be explained by the implementation of the "Made in China—2025" strategy and China's aspiration to become a leading technology power. In addition, most of the transactions are carried out by Chinese investors to create their global value chains and not depend on the supplies of other foreign companies. During the pandemic, China made major deals in the entertainment and aviation sectors, as these sectors were most affected by the pandemic and needed funding. China has even undertaken several greenfield projects to build its high-tech factories. Thus, we can conclude that China is confident in its strength and is not afraid to meet resistance from the European government. Moreover, China is skillfully taking advantage of the current crisis and is trying to strengthen its position in all promising sectors of the economy, as well as to strengthen its influence in Europe. If this continues, then China will seriously compete in advanced industries to developed countries.

Acknowledgements This paper has been supported by the RUDN University Strategic Academic Leadership Program and Grant of RUDN University No. 060119-0-000

References

ByteDance. (2021). Official website. https://www.bytedance.com/en/. Accessed: 10 March 2021.

China Three Gorges Corporation. (2021). Official website. https://www.hydropower.org/our-members/china-three-gorges-corporation. Accessed: 23 February 2021.

Counterpoint Global Smartphone ODM/IDH/EMS Tracker. (2020). https://www.counterpointresearch.com/smartphone-odms-see-12-yoy-growth-1h-2020-severe-market-headwinds/. Accessed: 5 March 2021.

European Bank for Reconstruction and Development (2021). Belt and Road Initiative. https://www.ebrd.com/what-we-do/belt-and-road/overview.html. Accessed: 2 April 2020.
Fosun (2021). Official website. https://www.fosun.com/about/. Accessed: 10 March 2021.
Halpin, P. (2020). TikTok to open the first European data centre in Ireland. Reuters, 6 August. https://www.reuters.com/article/us-usa-tiktok-ireland-idUSKCN2520P4. Accessed: 15 March 2021.
Hinata, Y. (2020). Tencent acquires stake in French. Nikkei Asia, 18 August 2020.
Institute for Security & Development Policy. (2018). Made in China 2025. https://isdp.eu/content/uploads/2018/06/Made-in-China-Backgrounder.pdf. Accessed: 12 April 2021.
NEVS. (2021). Official website. https://www.nevs.com/en/. Accessed: 22 March 2021.
Nexperia. (2021). Official website. https://www.nexperia.com/about/. Accessed: 8 March 2021.
Nikel, D. (2020). China Takes Stake In New-Look Norwegian Air. Forbes, 20 May. https://www.forbes.com/sites/davidnikel/2020/05/20/china-takes-stake-in-new-look-norwegian-air/?sh=686c7a007872. Accessed: 20 March 2021.
Reuters. (2020a). Fosun-owned Hauck & Aufhaeuser buys Bankhaus Lampe. 5 March. https://www.reuters.com/article/us-lampe-m-a-fosun-idUSKBN20S254. Accessed: 12 March 2021.
Reuters. (2020b). Huawei to build network equipment factory in France. 17 December. https://www.reuters.com/article/us-huawei-tech-france-idUSKBN28R2V7. Accessed: 21 March 2021.
Reuters. (2020). Tencent-led group to raise stake in Vivendi's UMG to 20%. 20 December 2020.
Supercell. (2021). Official website, available at: https://supercell.com/en/our-story/. Accessed: 5 March 2021.
Tartar, A. (2018). How China is buying Its Way into Europe. Bloomberg, 23 April 2018. https://www.bloomberg.com/graphics/2018-china-business-in-europe/.
The China Global Investment Tracker (2020). American Enterprise Institute https://www.aei.org/china-global-investment-tracker/. Accessed: 18 April 2021.
Wingtech. (2021). Official website. http://www.wingtech.com/en/WTJJ/9. Accessed: 5 March 2021.
Zijin Mining. (2021). Official website. http://www.zijinmining.com/about/About-Us.htm. Accessed: 10 March 2021.

Business Ecosystem: Opportunities and Threats

Nadezhda K. Savelyeva and Tatyana A. Timkina

Abstract The paper proposes the prerequisites for the formation of a new business model based on the ecosystem. The purpose of the study is to substantiate the positive and negative aspects in accordance with the principles of ecosystem functioning. The theoretical basis of the study is considering the essence of a new business trend by grouping the main elements, defining the main functions of each block. The practical basis of the study is the current state of information equipment of the country's population, the growth rate of the remote service level, as well as the study of the experience of the industry leader PJSC *Sberbank*. On the basis of the data obtained, positive aspects of activity on the ecosystem scale are formed; risks and barriers in terms of reducing the level of competition in the banking services market are identified.

Keywords Ecosystem · Internetization · Digitalization · Electronic platforms

JEL Codes F63 · G14 · O33 · F01

1 Introduction

Digitalization is an inherent part of modern life, the penetration of the Internet in all spheres of society is not the prospect of the coming years, but the reality. With the help of computers and mobile devices, it is possible to carry out many operations and transactions. This type of relationship between the consumer and the companies is supported by the possibility of remote payment for goods and services. The World Wide Web penetrates into remote areas; where previously there was only cellular communication, now it is possible to quickly and conveniently use the

N. K. Savelyeva (✉) · T. A. Timkina
Department of Economics/Institute of Economics and Finance, Vyatka State University, Kirov, Russia
e-mail: nk_savelyeva@vyatsu.ru

T. A. Timkina
e-mail: tatyana_timkina@inbox.ru

E. G. Popkova and B. S. Sergi (eds.), *Geo-Economy of the Future*,
https://doi.org/10.1007/978-3-030-92303-7_79

information search system, buy goods without leaving home, which leads to the growing number of customers, while the fee is still charged based on the digital platforms of banking organizations. Companies start providing remote service, which increases the competition between market participants, thereby giving advantages to consumers.

Internet coverage is gaining popularity; the growth rate has increased due to COVID-19. The pandemic resulted in completely reorganized work and guidelines. In order not to lose customers, to save the business, many companies have switched to the remote service, even those areas that could not exist in this format until recently, now work perfectly, e.g., consultation with a doctor, purchase of products, registration of loans, etc. The Internet has erased the territorial boundaries between the customer and the contractor; now, in order to provide a service or sell a product, you do not need to be in the same city or region with the consumer. "...The world is becoming more global, and information, goods, and services are becoming more accessible" (Report for public consultations "Ecosystems: Approaches to Regulation", 2021).

2 Methods

Business models based on the ecosystem are just evolving. The theoretical basis of the research is foreign and domestic developments. The modern world economy, influenced by globalization, the viral economic cycle, and post-pandemic changes in the disintegrated world economic systems, is characterized by a tendency to increase imbalances and does not fit into the theoretical models of the classical and traditional institutional economic school (Sozinova, 2021). Karpinskaya (2018) is one of the Russian scientists who consider the ecosystem as a new actor of economic development. In her work, Karpinskaya studies various approaches that determine the essence of the ecosystem as a unit of economic analysis, by analyzing the definitions of foreign and domestic scientists. The relevance of the prospects, positive and negative aspects of the ecosystem in the banking sector considered in the paper are confirmed by G. B. Kleiner "... social and economic ecosystems are now becoming a central element of the socio-economic landscape of the country..." (Karpinskaya, 2018). Goodwin (2015) justifies the relevance of modern Internet-based business models "... since the Industrial Revolution, the world has developed complex supply chains, from designers to manufacturers, from distributors to importers, wholesalers and retailers, this is what has allowed billions of products to be made, shipped, purchased, and used in all corners of the world. Recently, the power of the Internet, especially the mobile phone, has launched a movement that is rapidly eroding these layers and moving energy to new places..." In this case, territorial boundaries are being erased, the spheres of influence are being expanded, in this case, positions in certain markets are being strengthened, thus, the question of how such business conduct will affect the state of the entire industry remains open, provided that banking is still a business.

Prospects for improving the practice of making management decisions in conditions of economic crises are considered by Bogoviz et al. (2018), Fufacheva et al. (2017), Kataeva et al. (2017), Popkova et al. (2021a, 2021b), Savel'eva (2017), Soboleva et al. (2017), Sozinova (2020), Sozinova et al., (2017, 2019) and others.

Practical tools for applying ecosystems in the banking services market are described in the Bank of Russia report on approaches to ecosystem regulation (Report for public consultations "Ecosystems: Approaches to Regulation", 2021). Considering the prerequisites for the emerging and stable ecosystems, the level of the use of information technologies and information and telecommunications networks by the population should be taken into account, which will make it possible to use unified electronic platforms.

3 Results

The development of remote service increases the competition level. New players emerge in the market due to the fact that companies are not associated with some territory any longer. The growing number of companies creates a competitive struggle for the customer, thus, the relevance of the search for competitive advantages increases, and non-price competition in modern conditions is increasingly preferred, due to the additional services included, providing the customer with convenient services for performing operations in absolutely different areas. The main goal is to meet the needs in one place. The expanded list of services provided for each company prompted the formation of ecosystems. According to Ron Adner, "... the ecosystem is defined by the alignment structure of the multi-lateral set of partners that need to interact in order for a focal value proposition to materialize..." (Adner, 2016). This definition implies the companies' performance within a common structure, while preserving the integrity of its independent structural units.

Platforms are gaining popularity due to technological changes, Internet penetration, and most importantly, the increase in the level of public confidence in remote products. Practical skills, both for business and the consumer, were formed in the shortest possible time, and the restructuring of the usual business models in connection with the working conditions during the pandemic took the business to a new level. The transition or inclusion of remote services increased the number of potential customers. Let us define the term "ecosystem", as well as determine what the electronic platform is built from and what its functions are (Table 1).

The ecosystem is built on the work of several organizations united by a common software product. The main task, in this case, is to fully meet the needs of the customer. In modern conditions, it is convenient to receive services in one place, pay through one source, etc. Large ecosystems form a whole network of various services, if we talk about *Sberbank*, namely, this bank was one of the first to declare itself an ecosystem, including services for buying and selling real estate, food delivery, food,

Table 1 Elements of the ecosystem of financial organizations and their functions

Constituent part	Sphere of consumption	The list of services provided
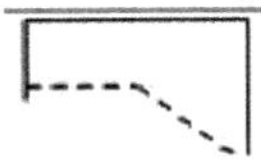	Financial basis	• Credit products: mortgages, consumer loans; • Investment products; • Deposits, current accounts; • Asset management; • Payment system
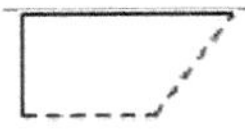	Electronic trading platform	Commodity marketplace, in this case, the bank's payment system is a tool for implementing commodity-money relations, which is why the electronic trading platform is a connecting part of the financial basis
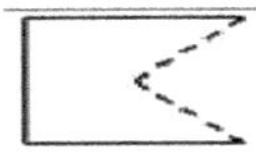	Needs	• Entertainment content (music, movies, games, social networks, etc.); • Booking and ordering (food, taxi); • Search for real estate, buying, selling
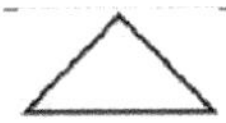	Information	• Information sources; • Operating System; • Television and the Internet
	Ecosystem	A set of all components on one electronic platform, while meeting all possible customer needs does not imply a monopoly position, in this case, the work of independent organizations united by one platform is concerned

Source Developed and compiled by the authors

entertainment content (movies, cash registers, music), pharmacy, delivery, buying and selling cars, taxis, etc.

It should be noted that this is considered to be an excellent marketing move, the impact on the subconscious, all the needs that the consumer faces, even the basic ones that are not related to lending or deposits, are associated with *Sber*—this name has already become household. But at the same time, the bank's credit products are not affected, you can assess the customer's solvency, its expenditure part, etc. In addition, the companies that are part of the ecosystem have their own independence.

Large companies tend to establish mutually beneficial cooperation with small firms rather than suppress them. This trend is driven by strong dynamics and rapidly changing development trajectories. Creating a product or providing services from scratch takes a lot of time, resources, and money, while the main deterrent is time due to the fact that while the product is being created, developing, and gaining momentum, the companies which initially operated in this area will already go ahead, thus, a less competitive product is created in advance. Therefore, it is more effective to attract

Table 2 Use of the Internet by the population aged 15–74 years in the Russian Federation for orders of goods and/or services by type of settlement and gender

	2017, thousands of people	2018, thousands of people	2019, thousands of people	2020, thousands of people	Growth rate 2020 verses 2021 (%)
Total	*32, 625.4*	*39,085.5*	*40,158.4*	*45,423.5*	16
Men	13,741.0	16,887.5	17,593.3	19,539.1	14
Women	18,884.4	22,198.0	22,565.1	25,884.4	18
City	*27,513.1*	*32,479.6*	*33,533.9*	*38,220.5*	17
Men	11,641.1	14,096.5	14,677.3	16,399.4	15
Women	15,872.1	18,383.1	18,856.6	21,821.1	19
Village	*5,112.2*	*6605.9*	*6624.5*	*7203.0*	11
Men	2099.9	2791.0	2916.0	3139.7	11
Women	3012.3	3814.9	3708.5	4063.3	12

Source Compiled by the authors based on Rosstat (2020)

existing companies to work competitively and minimize the cost of introducing a new product. These conditions encourage the development of ecosystems in different economic sectors. Commercial banks actively use ecosystem principles.

The purpose of the study is to analyze the current stage of development and the prerequisites for the emergence of ecosystems in the Russian banking services market.

The transition to remote services has allowed expanding the borders, the COVID-19 pandemic has created an unprecedented push and the restructuring of the usual areas of services provided and the sale of goods. The number of online trading users is increasing (Table 2).

According to the Statistics website, the number of consumers of remote purchases of goods and services increased by 5265.1 thousand people over the past year, which can be considered as the basis for the growth of customers and the expansion of demand for remote products. Electronic platforms for the sale of goods and the provision of services operate on remote payment, and, consequently, the Internet banking or a mobile application available. According to Table 2, the number of people in villages who use Internet to order goods increased by 578.5 thousand people. Residents of remote and sparsely populated territories, where it is not profitable for commercial banks to open their branches, face the problem of lack of financial accessibility. Since April 2020, ensuring the availability of financial services on the territory of the Russian Federation has been legally enshrined among the main functions of the Bank of Russia. The thing is that the lack of infrastructure deprives the population of meeting their needs since in such territories there are only stores for buying products (the number of items is reduced to a minimum) and post offices. Electronic platforms for sale will partially solve the problems of accessibility because to meet the demand for goods and services, offices and branches available are not a prerequisite.

4 Discussion

The optimal speed and banking services promoted are to be ensured from the banking institution to the end-user (customer), i.e., the speed at which the most favorable sales rates and money turnover, as well as the minimum cost of storing financial resources, on the one hand, are combined with the ability to fully and timely meet the demand, on the other hand (Sozinova, 2021). The increased number of market players, the rise in the level of accessibility due to the use of remote service increases the competition level, which leads to lower prices, a diverse choice of goods. Unified platforms, which are built on the cooperation of independent companies, make it possible to expand the market share by including new types of services. Another advantage of ecosystems from the point of view of the consumer is a single registration format, a simplified login procedure thanks to the single database. Besides, the customer gets simplified data exchange, reduced time lag between the search and the receipt of the desired product, possible recommendations for complementary products.

For business, modern business models also have positive aspects, such as expanding the customer base, the ability to work at a new level with more competitive firms, while not competing with them, but expanding and increasing the turnover, to apply new tools that were previously financially unavailable or did not matter when the occupied market share was concerned. These are the advantages of electronic systems for the consumer and supplier; now let us consider their disadvantages if large market players are involved.

According to the Central Bank, *Sberbank* holds a 40% market share in terms of revenue, which means that we have already had an undisputed leader, if *Sberbank's* services are even wider with the third-party services, subscriptions, and channels used and included, it will lead to even greater expansion. PJSC *VTB* occupies 19% of the market, and is preparing to launch the "Square Meter" ecosystem, the service is aimed at finding real estate, discounts from developers, design projects, repairs and secondary housing (PJSC VTB, n/a). From the list of services, we can conclude that the main goal is to attract mortgage customers, intermediary services in the purchase and sale of real estate. At the initial stage, the result is difficult to estimate, but the number of customers this service will attract can be predicted, especially if discounts from developers are provided. Thus, the share of the second largest bank will increase. If a significant market share is occupied by several leaders, these conditions are considered as barriers for other banks, which negatively affect the competition level.

Electronic platforms involve the additional services included, the main purpose of which is to make a certain set of services purchased collectively cheaper than individually, thus, imposing a number of products. The risk of foreign domination should also not be excluded if a significant market share is held by a foreign bank, which can negatively affect the national economy.

Transition and introduction of ecosystem principles occur rapidly, the question of practical application, the inclusion of new industries, and the exchange of data between related services involve a large amount of data, which does not exclude

Table 3 SWOT analysis of ecosystems in banking services

Strengths	Weaknesses
1. Increased availability of financial services; 2. Additional market players involved; 3. Price reduction while competing for the customer; 4. Expanded customer base; 5. Increase in services and products	1. Big players' dominance; 2. High level of ecosystem transition costs; 3. High implementation and transition costs; 4. Long-term payback period for investors and shareholders; 5. Disproportionate amount of costs and profits
Opportunities	*Threats*
1. Entering a new level by smaller companies; 2. Data exchange, confirmation of the customer's solvency, minimized debt risk; 3. Increased market share occupied; 4. Long-term profit growth	1. Cyclical nature of economic processes—crisis phenomena; 2. Exchange rate fluctuations; 3. Data leakage due to the progressive amount of information; 4. Consumer rejection of service packages; 5. Complexity of regional banks' activities, the decline in the level of competition; 6. Competition will be a struggle of "banking giants", with a reduction in the number of market players; 7. Regulatory and legal regulation; 8. Strict antitrust regulation; 9. Less developed companies suppressed; 10. Reduced concentration level

Source Developed and compiled by the authors

the possibility of personal data leakage. All the positive and negative aspects of the work of the national banking sector on the ecosystem scale can be expressed through SWOT analysis (Table 3).

The trend for expanding business in the ecosystem format is gaining momentum, and large banks are aiming to expand their services with third-party services that do not relate to the bank's direct activities. But what changes will this lead to for the whole country, what will happen to small banking organizations, e.g., regional banks because in the current situation they are significantly losing to federal competitors. On the other hand, the questions of the Government and the Central Bank about finding effective methods of regulating such systems are justified because the existing methods of evaluating a commercial bank do not give the proper result due to the fact that ecosystems combine many types of activities. "We are currently discussing the transition from regulation of a legal entity to regulation by type of activity," Nabiullina stressed . Of course, the market leaders, and even more so the "pioneer—*Sber*" has the financial capacity to restructure, but if we talk about small competitors, the pursuit of trends can provoke unprofitable financing of such a business model, and the subsequent collapse of a commercial bank. D. J. Teece (2018) correctly forms this problem, saying that "... Technological and innovative integrators present both coordination and market design challenges to the innovator, which tend to lead to market failure in the form of an excess of social over private returns...".

Now everything revolves around the customer, all the activities of the financial sector are aimed at attracting attention to the bank; in this case, the ecosystem is ideally created for this purpose, while meeting the basic needs of the consumer, a marketing technique is needed, i.e., a significant part of the basic operations is performed by one company: one account, payment system without constant confirmation of each financial transaction, all this will certainly have a positive impact on the demand for banking services. The analysis of the positive and negative aspects should be concluded with the phrase of Ron Adler: "… Ecosystems as mechanisms of interdependent business will only grow in importance and scale in the coming years. In the world, in practice, the concept of ecosystems is perceived with enthusiasm and at the same time chaotic. Ecosystems have created an opportunity for the world of research to shed light –both positive and normative—on a critical set of issues…" (Sozinova et al., 2019).

5 Conclusion

Due to the internetization of modern society, commercial banks tend to nationally replace the conventional business model with ecosystem principles. All business processes, meeting the needs of a particular individual are formed on the software. The penetration and scaling of the Internet into the life of society creates new needs; the expanded remote service provokes a change in the established business processes. The consumer wants to save time, minimize costs by not attending office centers. This is due to the conditions of the COVID-19 pandemic and a certain level of customer confidence in the online format. Previously, commercial banks acted as conduits in commodity-money transactions, being a kind of bridge between the consumer and the company. At the present stage, due to the constantly growing demand for remote services, the services provided, banks form their services, or attract existing companies to work together, while the scope of activity can be completely different, from subscriptions to watch movies to order food, grocery delivery, and taxi, united by one electronic platform—the ecosystem.

The question about the consequences of such business conduct, the risks to the country's economy, the role of the state in regulating and evaluating activities, and the impact on the competition of an entire industry remains open. The paper considers the positive and negative aspects of the modern business model. If we consider the impact on the level of competition and the concentration of participants in the banking sector, ecosystems will have a negative impact due to the indisputable advantages of price and non-price nature. At the same time, from the point of view of customer orientation, the ecosystem will undoubtedly simplify the ways of obtaining services, both of a financial nature, i.e. a direct function of the bank, and meeting the basic customer needs.

References

Adner, R. (2016). Ecosystem as structure: An actionable construct for strategy. *Journal of Management, 43*(1), 39–58. https://doi.org/10.1177/0149206316678451

Bogoviz, A. V., Sozinova, A. A., & Ostrovskaya, V.V. (2018). Approaches to Managing Economic Growth of Socio-economic Systems. In D. Endovitsky, E. Popkova (Eds). *Management of Changes in Socio-Economic Systems. Studies in Systems, Decision and Control* (vol. 135). Springer. https://doi.org/10.1007/978-3-319-72613-7_11.

Development of the financial market (n/a). Financial accessibility, Bank of Russia. https://cbr.ru/develop/development_affor/.

Fufacheva, L. A., Lepeshkin, S. M., Fokina, O. V., Sozinova, A. A., & Shchinova, R. A. (2017). Improvement of the mechanism of provision of food security of Russia within management of risk system of entrepreneurship. In E. Popkova (Eds.), *Overcoming Uncertainty of Institutional Environment as a Tool of Global Crisis Management. Contributions to Economics*. Springer. https://doi.org/10.1007/978-3-319-60696-5_2.

Goodwin, T. (2015). The Battle Is for the Customer Interface. TechCrunch, March, 2015. https://techcrunch.com/2015/03/03/in-the-age-of-disintermediation-the-battle-is-all-for-thecustomer-interface. Accessed 22 Apr 2021.

Karpinskaya, V. A. (2018). Ecosystem as a unit of economic analysis. In *System problems of Domestic Mesoeconomics, Microeconomics, and Enterprise Economics: Proceedings of the Second Conference of the Department of Modeling of Production Facilities and Complexes of the CEMI RAS* (vol. 2, pp. 125–141). CEMI RAS. https://doi.org/10.33276/978-5-8211-0769-5-125-141.

Kataeva, N. N., Marakulina, I. V., Sanovich, M. A., Sozinova, A. A., & Vasilyuk, N. (2017). Transformation of approach to market segmentation within crisis management of global entrepreneurship. In E. Popkova (Eds.), *Overcoming Uncertainty of Institutional Environment as a Tool of Global Crisis Management. Contributions to Economics*. Springer. https://doi.org/10.1007/978-3-319-60696-5_62.

PJSC VTB (n/a). Online services. Square meter. https://www.vtb.ru/personal/online-servisy/metr-kvadratniy/.

Popkova, E. G., Saveleva, N. K., & Sozinova, A. A. (2021a). A new quality of economic growth in "Smart" Economy: Advantages for developing countries. In E. G. Popkova, B. S. Sergi, (Eds.), *"Smart Technologies" for Society, State and Economy. ISC 2020. Lecture Notes in Networks and Systems* (vol. 155). Springer. https://doi.org/10.1007/978-3-030-59126-7_48.

Popkova, E. G., Savelyeva, N. K., & Sozinova, A. A. (2021b). Smart technologies in entrepreneurship: Launching a new business cycle or a countercyclical instrument for regulating the economic situation. In E. G. Popkova, B. S. Sergi, (Eds.), *"Smart Technologies" for Society, State and Economy. ISC 2020. Lecture Notes in Networks and Systems* (vol. 155). Springer. https://doi.org/10.1007/978-3-030-59126-7_188.

Report for public consultations "Ecosystems: Approaches to Regulation", Moscow (2021). consultation_paper_02042021. pdf (cbr.ru).

Rosstat. (2020). Selective federal statistical observation on the use of information technologies and information and telecommunications networks by the population. https://gks.ru/free_doc/new_site/business/it/ikt20/index.html.

Savel'eva, N. (2017). Integrated assessment technology in management of competition efficiency in business corporations. In I. Lapteva, O. Araslanova, Y. Matushkina, T. Koykova, & K. Graboyy, (Eds.), *MATEC Web of Conferences* (vol. 08086) SPbWOSCE-2016. https://doi.org/10.1051/matecconf/20171060.

Soboleva, O., Sozinova, A., Spengler, A., Fokina, O., & Savelyeva, N. (2017). Mechanisms of regulation of economic processes in a region. In E. Popkova, (Ed.), *Overcoming Uncertainty of Institutional Environment as a Tool of Global Crisis Management. Contributions to Economics*. Springer. https://doi.org/10.1007/978-3-319-60696-5_51.

Sozinova, A. A. (2021). Investigating the prospects of new institutional theory in the smart economy. In E. G. Popkova, B. S. Sergi, (Eds.), *Modern Global Economic System: Evolutional Development vs. Revolutionary Leap. ISC 2019. Lecture Notes in Networks and Systems* (vol. 198). Springer. https://doi.org/10.1007/978-3-030-69415-9_227.

Sozinova, A. A., Burtseva, T. A., Fokina, O.V ., Grabar, A. A., & Tufyakova, E. S. (2019). The concept of industrial enterprises' economic development amid the global financial and economic crisis. In E. Popkova, (Ed.), *The Future of the Global Financial System: Downfall or Harmony. ISC 2018. Lecture Notes in Networks and Systems* (vol. 57). Springer. https://doi.org/10.1007/978-3-030-00102-5_114.

Sozinova, A. A., Fokina, O. V., & Fufacheva, L. A. (2017). Reorganization of entrepreneurial structures within global crisis management: problems and perspectives. In E. Popkova, (Ed.), *Overcoming Uncertainty of Institutional Environment as a Tool of Global Crisis Management. Contributions to Economics*. Springer. https://doi.org/10.1007/978-3-319-60696-5_1.

Sozinova, A. A. (2020). "Dataset of fight against the viral threat based on management and marketing during the COVID-19 economic crisis", Mendeley Data, V1, https://doi.org/10.17632/g6p5kjz657.1 http://dx.doi.org/https://doi.org/10.17632/g6p5kjz657.1

Teece, D. J. (2018). Profiting from innovation in the digital economy: Enabling technologies, standards, and licensing models in the wireless world. *Research Policy, 47*(8). https://doi.org/10.1016/j.respol.2017.01.015.

WTO Crisis: Is There a Way Out?

Alexey P. Portanskiy

Abstract The author examines possible ways out of the WTO crisis, in which it found itself by the beginning of 2020. Recently, the global trading system was negatively affected by the US trade war with China and the WTO internal crisis associated with the suspension of the normal functioning of the Dispute Settlement Body. It has become clear that serious and urgent measures are needed to reform the WTO. The central problem of the reform should be the transformation of the consensus mechanism into a different way of decision-making. For its launch, it is critically important to achieving a convergence of the positions of China and the USA. Given the complexity of achieving multilateral agreements, the plurilateral format of negotiations among WTO members seems promising, followed by the signing of relevant agreements, the number of participants of which will grow over time.

Keywords World trade · WTO reform · Trade wars · Global crisis

JEL Codes B17 · F1 · F13

1 Introduction

In 2020, the World Trade Organization (WTO) found itself in a deep crisis. Two key causes of this crisis can be identified: first, the revealed lack of WTO instruments to end the trade war between the USA and China; second, the suspension of the normal functioning of the Dispute Settlement Body (DSB) at the WTO due to the non-constructive actions of the USA. Simultaneously, the crisis of the decision-making system remains the main problem previously accumulated in the WTO.

A. P. Portanskiy (✉)
National Research University Higher School of Economics, Moscow, Russia
e-mail: aportanskiy@hse.ru

IMEMO RAS, Moscow, Russia

E. G. Popkova and B. S. Sergi (eds.), *Geo-Economy of the Future*,
https://doi.org/10.1007/978-3-030-92303-7_80

2 Materials and Methods

The methodology of the scientific topic is based on the traditional methods of economic analysis, such as deduction and induction. Nevertheless, the main methods used in this research are comparison and retrospective analysis.

3 Results

3.1 The Consensus Mechanism is not Eternal

It is generally accepted that the WTO has become a victim of its success. Since signing the General Agreement on Tariffs and Trade (GATT 1947), the consensus mechanism has made the system reliable (World Trade Organization, 1947). However, it worked quite successfully only when the number of participants was 60–70 with a noticeable influence of developed countries. Later, when the number of members exceeded 150 (more than two-thirds of them were developing countries), it began to fail dramatically.

In recent years, several proposals have been made to transform the consensus mechanism into particular voting patterns (Elsig & Cottier, 2011). There are also comprehensive initiatives to reform the WTO, particularly the Concept of the EU (European Commission, 2018). Nevertheless, none of them has yet been implemented. The fact is that the creation of the WTO came at too high a price and required enormous effort and compromise, and, despite its well-known shortcomings, the system is functioning. Therefore, everyone understands the high risks of reform. On the other hand, the deep disagreement between the two leading players (the US and China) prevents WTO reform.

However, as already mentioned in previous works of the author, the need for new rules of international trade does not disappear, and if new rules are not implemented at the multilateral level, they can be developed outside the WTO. Mega-regional trade agreements (MRTAs) have brought significant advances and accomplishments to establishing the rules of international trade. The first large-scale examples were the Transatlantic Trade and Investment Partnership (TTIP) and the Trans-Pacific Partnership (TPP). Unfortunately, both of them were never implemented due to the fault of the Trump administration. However, the process has evolved. Several significant large-scale agreements were signed and entered into force in 2016–2019, including the following:

- Comprehensive and Progressive Agreement for Trans-Pacific Partnership (CPTPP);
- Comprehensive Economic and Trade Agreement between Canada and the EU (CETA);
- Japan-EU Free Trade Agreement (JEFTA).

The provisions that have been developed and formed the basis of the agreements listed above are now either ahead of the rules applied by the WTO or form them where multilateral rules do not yet exist.

It has also been noted in previous works that the emergence of MRTAs leads to a very different situation in international trade. In this regard, the risk of divergence and ambiguity with multilateral regulations is high enough. Agreements such as the former TPP, the CPTPP (which replaced the TPP), and the modern CETA contain rules that are more advanced than the rules of the WTO (this applies to provisions affecting the protection of intellectual property, the area of technical barriers to trade, trade in services, and trade facilitation) as well as the rules that do not yet exist in the WTO, particularly concerning competition, environment protection, investment, and labor standards.

The use of these rules by countries that are simultaneously parties to such agreements and the WTO generates risks of creating legal conflicts, especially in resolving trade disputes. The problem can only be solved by reforming the Multilateral trading system (MTS) (Portanskiy, 2019).

The failure of the decision-making system in the WTO has led to the fact that the organization has noticeably lost its most important function, the rules-making of international trade in the twenty-first century. In the twentieth century, the GATT-1947 coped with the tasks of regulating international trade exchange on principle "Produce goods here and sell them abroad" (World Trade Organization, 1947). Nevertheless, in the twenty-first century, trade became significantly more complicated when not only goods but also, industries, investments, know-how, and others began to cross borders. The need for new "rules of road" has arisen. Suffice it to mention such areas as e-commerce, labor standards, trade, environment, etc.

The stalemate in the Doha Round negotiations was a clear signal that multilateral agreements using a consensus mechanism were becoming less and less possible. Therefore, many experts believe that, in the foreseeable future, progress in trade negotiations can be achieved more likely in a plurilateral format (i.e., with a limited range of participants).

3.2 *Plurilateralism as the Salvation of the Multilateral System*

The WTO legal package contains two Plurilateral Agreements: Government Procurement Agreement (GPA) and Agreement on Trade in Civil Aircraft. Distinctive feature of these agreements is that WTO members are not obliged to join them.

Plurilateralism has been successfully developed for years within the MTS. Several multilateral trade agreements in the WTO legal framework (on anti-dumping, safeguard measures, subsidies, and technical barriers to trade) began as the Tokyo Round "Codes" or plurilateral GATT Codes, which were not accepted by all parties. These Codes became fully multilateral only with the creation of the WTO.

Similarly, the GPA agreement originally appeared as one of the Codes during the Tokyo Round in 1979. The GPA was adjusted in 1981. In 1994, it became a full part of the legal package of the Uruguay Round agreements. In 2012, the text of the GPA was revised, and a new version entered into force in 2014.

Forty-three countries that were already members of the WTO became members of the updated version of the GPA. Ten countries were in the process of negotiating accession, and five countries were considering accession.

The updated version of the GPA implied measures in the field of liberalization of the sector, namely—the extension of the GPA provisions to additional state contracts worth $80–100 billion (World Trade Organization, 2014). Russia does not exclude the possibility of negotiating the accession to the GPA.

The Information Technology Agreement (ITA) is also a right example. The document was signed in 1996 in Singapore (World Trade Organization, 1996). In the first phase, the ITA was signed by 29 countries. In 2013, 41 other countries joined the ITA. The Russian Federation also joined this agreement in 2013. The ITA was updated in 2015 (ITA-II). ITA-II is intended to further liberalize market access for information technology.

The agreements listed above show that the number of parties to plurilateral contracts increases over time, so does the weight of the market they cover. ITA, GPA, and Agreement on Trade in Civil Aircraft are examples of how plurilateral agreements can be successfully developed within the MTS. The 11th WTO Ministerial Conference held in December 2017 in Buenos Aires discussed several new plurilateral initiatives on e-commerce, investment facilitation, and micro-, small-, and medium enterprises (MSMEs), which may well become full plurilateral agreements in the future.

Thus, plurilateral agreements are not a stumbling block to MTS, as critics believed, but rather the opposite, a kind of building block for the multilateral system, contributing to its progress. As a result of the gradual involvement of new participants in existing plurilateral agreements, the rules and norms of the latter become the rules and norms of the MTS. This process is called *Multilateralising plurilateralism.*

So, in the current situation, the consensus mechanism has become the main brake on the conclusion of multilateral agreements in WTO, on the one hand. On the other hand, very important issues are often withdrawn from discussion in the WTO since they are of no interest to countries with a certain level of development and specific views on the MTS. This suggests that the future of the WTO will largely involve plurilateral arrangements. And the urgent task of the WTO is to engage in the institutionalization of plurilateral processes in order to allow them to develop normally (Medvedkov, 2020).

3.3 US-China Conflict—a Source of Problems on the Road to WTO Reform

The positions of the US and China are decisive for the present and future of the WTO, including its reform. The implementation of Trump's "America First" slogan since the beginning of his presidency has significantly impacted the MTS. After breaking the trade policies implemented by his predecessors over the past 80 years, Trump became the first US head of State to believe that it is possible to "make America great" through protectionism and that "Trade wars are good, and easy to win" (Davies, 2020).

The Trump administration has decided to respond to the "unfair trade practices" of other countries by refusing to follow WTO rules and principles and starting using a more "aggressive" trade policy to protect its economic interests. In 2017–2018, Washington's protectionist trade policies led to a series of serious clashes with critical trading partners (Kimberly, 2017).

The US used two legislative acts of the twentieth century to legally justify such actions: the Trade Act of 1974 (United States of America, 1975) and the Trade Expansion Act of 1962 (United States of America, 1962). The Trade Expansion Act of 1962 allows restricting the import of products if they constitute a "threat to national security." This Act underlies the increase in import duties on aluminum and steel implemented by the US in the summer of 2018 (U.S. Department of Commerce, 2017).

The imposition of duties on aluminum and steel based on the laws of the twentieth century greatly outraged the closest trading partners of the US—the EU, Canada, and Japan. These countries felt that these acts could not be applied in isolation from the rules of the WTO. The disregard for international trade rules by the Trump administration considerably damaged the WTO.

The trade war between the US and China initiated by Washington was the greatest test for the WTO as a rules-based system. In August 2017, despite Beijing's calls to avoid a trade war, president Trump has ordered an investigation into what Washington believes to be multiple episodes of intellectual property infringement against U.S. companies in China. As expected, the investigation led to the U.S. President's corresponding March 22, 2018 decision to impose increased tariffs on Chinese goods under the pretext of the need to reduce the U.S. trade deficit with China (BBC, 2018).

In fact, the trade war was chosen by Washington as a tool to counter China in its efforts to become the World's technological and economic leader. According to IMF estimates, this trade war decreased world GDP by 0.5% per year (Holland & Sam, 2019).

Judging by the first statements of President Biden and members of his team, tensions between the US and China are unlikely to decrease in the foreseeable future.

To understand the American approach to the issue of WTO reform, it is necessary to indicate that Washington has expressed serious dissatisfaction with the position in the WTO of a considerable group of countries, which joined the organization with the so-called "Status of developing country". To this day, the countries with this status enjoy the Special and Differential Treatment (SDT), which provides significant benefits for protecting the national market by import duties (Office of the U.S. Trade Representative, 2018). Despite impressive economic progress, such countries as the People's Republic of China, the Republic of India, and the Republic of Korea continue to use this status. From Washington's point of view, this gives China and several other WTO members unwarranted privileges.

The essence of Washington's approach to reforming the World Trade Organization is to eliminate the outdated privileges of a group of countries that now impede trade negotiations and the Multilateral Trade Agreements. This approach is understandable, but justifiable only in part. By insisting on its demands, the U.S. side, in fact, blocks from the very beginning any progress in the consultations on WTO reforms, which have barely begun, setting as a condition for the fulfillment of its demands.

The Biden administration will likely demonstrate a more constructive approach to the WTO than the previous administration, but the tough approach to China is unlikely to change.

Beijing's position on WTO reform, published in November 2018, includes three fundamental principles:

1. Protecting the fundamental values of multilateral trade;
2. Protecting the interests of developing countries in the WTO;
3. Respecting the practice of decision-making through the consensus mechanism (Zhu, 2019).

As the main defender of the interests of developing countries, China emphasizes the need to preserve their privileges, in particular the SDT regime in the WTO, which directly contradicts the position of the US and the EU.

In response to Washington's repeated criticism that China does not meet the criteria of a market economy and has a closed market, Beijing recognizes the importance of further reforms and greater openness in the economy. However, the country rejects the recommendations of the US on "three zeros in trade" (zero tariffs, zero market barriers, and zero subsidies), qualifying them as completely unfair to China as a developing country (Ghosal Singh, 2019).

At the same time, not wishing to remain in the position of justifying itself, China makes a number of claims against the United States. For example, the Trump administration's "America First" slogan undermines, in China's view, basic WTO principles of most-favored-nation treatment. Similar criticism is levied against Washington's well-known abuses of trade exemptions on national security grounds and unilateral measures to protect its market, which is blatant protectionism and undermines the established rules-based system, according to the Chinese government (Zhong & Ren, 2019).

The disruption of the normal functioning of the DSB due to the blocking of the Appellate Body (AB) on the eve of the 25th anniversary of the WTO is also largely related to Washington's discontent with Beijing's trade policy. The essence of this disruption was that the US prevented the appointment of new judges to the AB, despite the expiration of the mandates of two of the three acting judges, which made the operation of the AB impossible. The problem did not suddenly emerge; it was well known in 2018. Even then, the U.S. blocked the appointment of new judges to the AB, and it was clear to everyone what the final outcome would be by the end of 2019.

In 2018, as part of the WTO reform initiatives put forward by the European Union, the EU tried to address the American claims against AB as much as possible. However, the American side has consistently rejected all proposed options and compromises. President Trump believed that the WTO and its DSB acted in the interests of every country except the USA, which, by the way, was not true (Bown & Keynes, 2020).

However, the real irritant for Washington in this case, too, was China, which, having become economically on par with the United States, does not want to give up the said status of a developing country, and the corresponding benefits in the WTO, in particular, in the framework of litigation in the DSB. According to the US, the Chinese economy is not open to the extent required by the WTO (China became a WTO member in 2001), which is why the intransigence of the American government under President Trump regarding a possible compromise on the functioning of the AB. With Biden's team in the White House, Washington's approach to the AB problem is likely to become more constructive. Nevertheless, it will take time to restore the normal functioning of the DSB.

4 Conclusion

The sharp contradictions between the US and China, combined with the internal WTO crisis by early 2020, dictate the need for serious measures to support the MTS. The worst scenarios can potentially disrupt global trade and split the world into large trading blocs where trade relations will be primarily based on relative strength rather than rules. Such blocs or centers of power could be, for example, the US, the EU, and China. Participants in this new system will probably try to maintain a dispute settlement mechanism based on WTO principles. However, this mechanism is likely to lose its former reliability, resulting in unmanageable disputes that will dramatically reduce predictability in mutual trade, even for its most prominent players.

If the openness and predictability of the current GATT/WTO system encouraged exports, the new system could be marked with the displacement of imports, leading to a widespread escalation of protectionist measures.

Consequently, third countries run the risk of finding themselves in a tough position—many of them will be forced to make asymmetric deals with the US, China, or the EU. The interests of third countries and their negotiating potential would be seriously impaired (Akman et al., 2020).

The indicated facts describe the possible slide of the trading system to a *power-based system* formula. In such a scenario, the trading system would become less secure and less humane than the current MTS. It will become more challenging to solve global problems in a fragmented world economy, including climate change and pandemics. The materialization of such a scenario in the twenty-first century cannot be ruled out at all.

A counterbalance to the described risks is the worldwide consensus that the WTO must remain central to the regulation of world trade.

In the foreseeable future, the reformation of the WTO will have to begin one way or another. To launch this transformation, it is critical to get the US and China to converge on the reform. The central issue should be the transformation of the consensus mechanism into a different way of decision-making. Overcoming the consensus mechanism is necessary because, first, consensus in the WTO is now extremely difficult to achieve, and it takes very important issues out of the discussion just because they do not interest countries with a certain level of development or with certain views on the MTS as a whole; second, insurance is needed against blocking a decision by one member, which became especially obvious after the suspension of the Appellate body due to the US position.

The extreme difficulty of reaching consensus and the likelihood of overcoming this mechanism in the future brings to the fore the prospect of expanding plurilateral agreements, which do not undermine the multilateral system, but become its building blocks as the number of plurilateral formats (including MRTAs) grows—the norms of the latter will gradually become multilateral norms. This is the essence of the so-called "*multilateralization*" of plurilateral agreements.

Simultaneously, the plurilateral format allows reaching agreements in new areas (e.g., e-commerce, investment facilitation, environment, etc.). This format is also likely to become acceptable for international agreements in technology and digital giants.

At a time of global crisis, it is more important than ever to maintain the will and capacity for international cooperation, for multilateral negotiations, for the preservation of rules. The history of the twentieth century provides vivid examples of how countries have shown determination and the ability to strengthen institutions and rules in moments of crisis. The ability and capacity of the international community to deal with the most complex conflict and crises should allow it to cope with the current challenges to the global economy and trade, which leaves the chance to preserve the existing rules-based system.

References

Akman, M. S., Armstrong, S., Dadush, U., Gonzalez, A., Kimura, F., Nakagawa, J., & Braga, C. A. (2020). World trading system under stress: Scenarios for the future. *Global Policy, 11*(3), 360–366. https://doi.org/10.1111/1758-5899.12776

BBC. (2018). *Trump announces tariffs on $60bn in Chinese imports*. Retrieved from https://www.bbc.com/news/business-43494001.

Bown, C., & Keynes, S. (2020). *Why did Trump end the WTO's appellate body? Tariffs*. PIIE. Retrieved from https://www.piie.com/blogs/trade-and-investment-policy-watch/why-did-trump-end-wtos-appellate-body-tariffs.

Davies, G. (2020). *Donald Trump's trade wars have not been 'easy to win'*. Financial Times. Retrieved from https://www.ft.com/content/f2771098-36ed-11ea-a6d3-9a26f8c3cba4.

Elsig, M., & Cottier, T. (2011). Reforming the WTO: The decision-making triangle revisited. In M. Elsig, & T. Cottier (Eds.), *Governing the World Trade Organization: Past, present and beyond Doha. World Trade Forum* (pp. 289–312). Cambridge University Press.

European Commission. (2018). *Concept paper: WTO modernization*. Retrieved from http://trade.ec.europa.eu/doclib/docs/2018/september/tradoc_157331.pdf.

Ghosal Singh, A. (2019). *What does China want from WTO reforms? Within China, arguments rage about how to shape the future of the World Trade Organization. The Diplomat*. Retrieved from https://thediplomat.com/2019/05/what-does-china-want-from-wto-reforms/.

Holland, B., & Sam, C. (2019). *A $600 billion bill: Counting the global cost of the U.S.-China trade war. Bloomberg Economics*. Retrieved from https://www.bloomberg.com/graphics/2019-us-china-trade-war-economic-fallout/.

Kimberly, A. E. (2017). *The Trump administration's trade strategy is dangerously outdated. The Economist*. Retrieved from https://www.economist.com/finance-and-economics/2017/03/02/the-trump-administrations-trade-strategy-is-dangerously-outdated.

Medvedkov, M. (2020). Future rules of world trade-principles and limitations. *Trade Policy, 2*(22). https://doi.org/10.17323/2499-9415-2020-2-22-21-24.

Office of the U.S. Trade Representative. (2018). *USTR releases 2018: National trade estimate report*. Washington, DC: USTR. Retrieved from https://ustr.gov/about-us/policy-offices/press-office/fact-sheets/2018/march/ustr-releases-2018-national-trade.

Portanskiy, A. (2019). The imperative of WTO reform in an era of rising protectionism and trade wars. *International Organisations Research Journal, 14*(2), 238–251. https://doi.org/10.17323/1996-7845-2019-02-12

U.S. Department of Commerce. (2017). *Fact sheet: Section 232 investigations: The effect of imports on the national security*. Retrieved from https://www.commerce.gov/news/fact-sheets/2017/04/fact-sheet-section-232-investigations-effect-imports-national-security.

United States of America. (1962). Trade Expansion Act of 1962 (Pub. L. 87–794, October 11, 1962, 76 Stat. 872; codified at 19 U.S.C. ch. 7). Washington, DC.

United States of America. (1975). Trade Act of 1974 (Pub. L. 93–618, § 1, January 3, 1975, 88 Stat. 1978; codified at 19 U.S.C. ch. 12). Washington, DC. Retrieved from https://legcounsel.house.gov/Comps/93-618.pdf.

World Trade Organization. (1947). *The General Agreement on Tariffs and Trade (GATT)* (October 30, 1947). Retrieved from https://www.wto.org/english/docs_e/legal_e/gatt47_01_e.htm.

World Trade Organization. (1996). *Information Technology Agreement* (December 13, 1996). Singapore. Retrieved from http://www.wto.org/english/tratop_e/inftec_e/inftec_e.htm.

World Trade Organization. (2014, April 7). *Revised WTO Agreement on Government Procurement enters into force*. Retrieved from https://www.wto.org/english/news_e/news14_e/gpro_07apr14_e.htm.

Zhong, N., & Ren, X. (2019). *China submits proposal for WTO reform. China Daily*. Retrieved from http://www.chinadaily.com.cn/a/201905/15/WS5cdb46b5a3104842260bb9a0.html.

Zhu, V. (2019). *La Chine et la réforme de l'OMC: Qui veut le moins...veut le moins*. Newsletter from Institut Montaigne. Retrieved from https://www.institutmontaigne.org/blog/china-trends-1-la-chine-et-la-reforme-de-lomc-qui-veut-le-moinsveut-le-moins.

China's Foreign Trade Relations with the United States and the European Union in the Context of the PRC's Development Strategy Until 2050

Lora N. Fedyakina, Tian Hengwei, and Chen Hao

Abstract The paper focuses on China's foreign trade relations with its main partners, competitors, the USA, and the EU in 2010–2019, covering different periods in the development of the global economy and politics. The years 2010–2017 were the years of globalization and open markets. The year 2017 saw attacks on China, accusations and threats from President D. Trump. In 2018, D. Trump, appealing to national legislation, launched a trade war with China. In violation of international agreements and principles, he imposed additional duties against Chinese imports, to which China responded quickly with similar measures. By the end of 2019, the confrontation between the parties eased to a certain extent. On January 15, 2020, the USA and China concluded an agreement, which formally completed the first stage of the settlement of trade and economic contradictions between the parties. The analysis showed that China's economic cooperation with the leaders of the world economy developed quite evenly and steadily in 2010–2018, and only in 2019 there was a drop in exports and imports of goods in relations with the USA. During the research, the authors apply the methods of systematic logical, economic, and statistical analysis, induction and deduction, expert assessments, and empirical generalizations. The applied methods allow the authors to determine trends in the development of foreign trade relations of China with the main partners and identify the features of their development in changing foreign policy and economic conditions, which is the purpose and novelty of this research.

Keywords Trade deficit · Customs duties · Middle-income society · Modernization · Development strategy

JEL Codes F13 · F14 · F21 · F51 · F53 · 057

L. N. Fedyakina (✉) · T. Hengwei · C. Hao
Peoples' Friendship University of Russia (RUDN University), Moscow, Russia
e-mail: lnf76@yndex.ru

E. G. Popkova and B. S. Sergi (eds.), *Geo-Economy of the Future*,
https://doi.org/10.1007/978-3-030-92303-7_81

1 Introduction

The relevance of the research is determined by the geopolitical and economic position of China, the USA, and the EU in the world economy and politics, their influence on the development of the world economic system, and the great importance of China's trade and economic ties with the leaders of the world economy in increasing its international competitiveness and in implementing the Development Strategy of the People's Republic of China until 2050.The paper aims to determine the trends and prospects for the development of trade and economic relations of China with the USA and the EU and identify the features of the foreign economic policy of partners, problems, risks, as well as the methods of overcoming differences and conflicts and implementing agreements.

In this regard, the authors set the following tasks:

- To investigate changes in the balance of power of partner countries in world GDP and the world market for goods and services;
- To analyze China's foreign trade in goods and services with the USA and the EU;
- To identify trends in exports and imports of goods and services, the dynamics of the trade balance, and the commodity structure;
- To analyze China's relations with partners in the mutual exchange of foreign direct investment, which are directly related to the production and export of goods and services.

It is necessary to pay the closest attention to studying the foreign trade policies of partner countries, especially the USA, toward China and China's response. This necessity is caused by a sharp transition of the Trump administration from the policy of globalization to national protectionism by any, including illegal, actions under various pretexts and accusations. China's foreign economic policy requires consideration of its conceptual foundations, which are formed by the Development Strategy of the PRC's until 2050, adopted in October 2017. (Xinhua News Agency, 2017).

In the first stage of the Strategy (2020–2035), China will carry out mainly socialist modernization; it will rise to the level of innovative leader countries. After the second stage (2035–2050), China will reach the level of developed countries in per capita income, human development, industrialization, application of high technology, urbanization, and modernization. It will become one of the leading countries in total national power and international influence.

2 Materials and Methods

The authors used traditional and well-proven methods of systematization, comparison, statistical and logical analogy, analysis, synthesis, induction, deduction, and retrospection to develop the main provisions of the research.

The informational and statistical basis of the research includes the following:

- Xi Jinping's Report to the 19th CPC Congress (October 18, 2017) (Xinhua News Agency, 2017);
- Publications of international and national economic organizations (including World Bank (Data of The World Bank, n.d.), International Monetary Fund (IMF, 2013, 2021), World Trade Organization (WTO, 2020), Organization for Economic Cooperation and Development (OECD. Stat, n.d.), US Bureau of Economic Analysis, US Department of Commerce (n.d.; 2011–2020, 2010–2019), and Chinese Ministry of Commerce (Ministry of Commerce of People's Republic of China, 2020a, 2020b);
- Articles of prominent Russian scientists published in the journals World Economy and International Relations, Russian Foreign Economic Bulletin, as well as Internet resources (Adno et al., 2003; Maslov, 2018; Mikheev & Lukonin, 2020, 2021; Smbatyan, 2021).

3 Results

The research showed that by the beginning of the new millennium, China had made significant economic progress along the path of reform and opening up to the world, which led to a marked improvement in the well-being of citizens and placed it among the leading countries in several macroeconomic indicators in the world economy. Between 1980 and 2000, China's share of world GDP (purchasing power parity) increased from 2.5 to 10.7%, making it second in the world after the USA. The share of China in world industrial production increased from 2.1 to 11.1% (second place after the USA). The country's share in world agricultural production increased from 15.8 to 23.8% (first place in the world) (Adno et al., 2003, pp. 506–602).

During the 2007–2009 crisis, China's GDP (PPP) increased 1.4 times to \$5.1 trillion. China's GDP (PPP) was \$12.3 trillion in 2017 and \$14.3 trillion in 2019, equaling 19.2% of world GDP and giving the country first place in the world (Fig. 1).

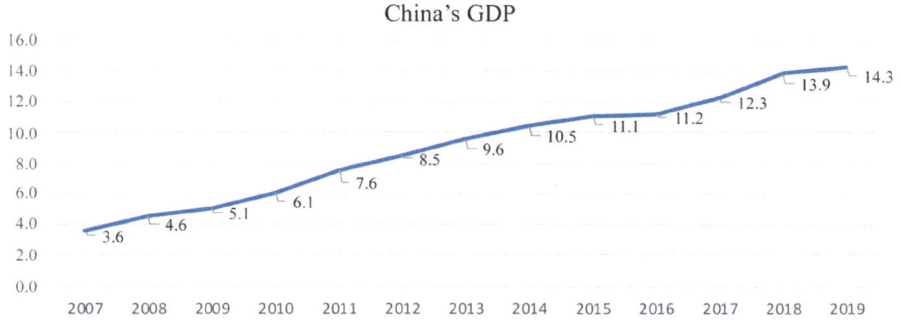

Fig. 1 China's GDP at current prices in 2007–2019, trillion dollars. *Source* Compiled by the authors based on (Data of The World Bank, n.d.)

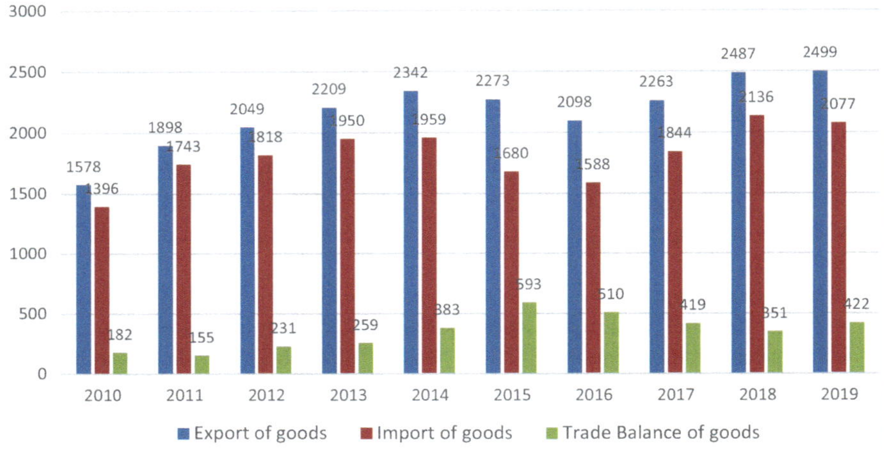

Fig. 2 China's foreign trade in goods in 2010–2019, $ billion. *Source* Calculated and compiled by the authors based on WTO, 2020

China's accession to the WTO (2001) opened up new opportunities and gave a powerful impetus to further development of the country. Foreign trade has become an essential factor in China's economic development and growth. In 2002, China was fourth (after the USA, Germany, and Japan) in world merchandise exports with a share of 4.3%. In 2009, it became the leader with a share of 9.7% and second in world merchandise imports (8.0%) (IMF, 2013). In 2010–2019, China's merchandise exports increased 1.6 times, imports increased 1.5 times, and trade surplus increased 2.3 times and accounted for 19.2% of exports. China's share of world merchandise exports has steadily increased, reaching 10.3% in 2010, 12.7% in 2017, and 13.2% in 2019, and 9.0%, 10.4%, and 10.8% in world imports, respectively (Fig. 2).

China–United States merchandise exports were growing steadily until 2018, with a 1.5-fold increase between 2010 and 2018, peaking at $539.4 billion. In 2019, at the height of the trade war, it was down 19.5% to $451.3 billion. China's merchandise imports from the USA peaked in 2017 at $131.7 billion, an increase of 1.4 times since 2010. By 2019, it had fallen to $107.9 billion (−22.0%). Meanwhile, in 2018–2019, Chinese export revenues were 4.2–4.4 times higher than import payments; China's surplus (to the extent of the National debt of the United States) was $417.3 billion and $365.8 billion (Fig. 3).

In 2019, the main commodities of Chinese exports to the USA were consumer goods (50%); capital goods (34%); processed goods and materials (9%); vehicles, parts, and engines (4%); food, feed, and beverages (1%). Chinese imports from the USA were capital goods (42%); manufactured goods and materials (27%); food, feed, and beverages (12%); vehicles, parts, and engines (9%); consumer goods (9%) (Bureau of Economic Analysis. U.S. Department of commerce, 2011–2020).

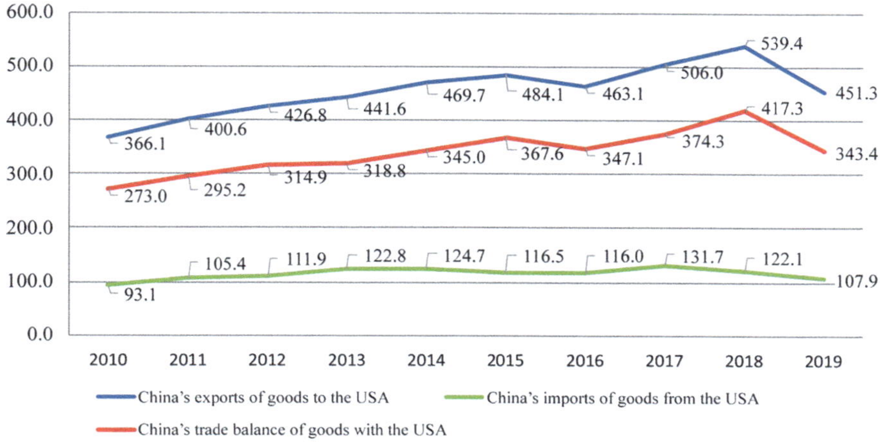

Fig. 3 China–United States merchandise trade in 2010–2019, $billion. *Source* Compiled by the authors based on (Bureau of Economic Analysis. U.S. Department of commerce, n.d.)

Between 2010 and 2019, Chinese exports to the EU increased by 37.7%; imports increased by 64.1%. China's export earnings exceeded its import payments by 55%, and China's surplus (in the EU deficit) was 35.4% to exports ($152 billion) (Fig. 4.).

China's main exports to the EU (2019) include non-agricultural consumer goods (49%), machinery and equipment (27%), chemicals (7%), and food and raw materials (5%).The main items of China's imports from the EU are industrial goods (46%), machinery, equipment, cars, and airplanes (27%) (Directorate-General for Trade of the European Commission, 2021).

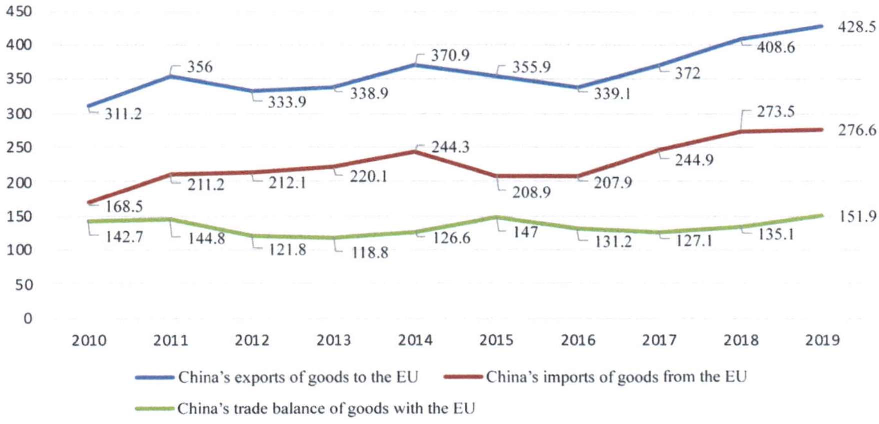

Fig. 4 China–European Union merchandise trade in 2010–2019. *Source* Compiled by the authors based on Ministry of Commerce of People's Republic of China, (2020a)

Table 1 China's trade in services with the USA and the EU in 2010–2019, $ billion

	China's exports of services to the USA	China's imports of services from the USA	China's exports of services to the EU	China's imports of services from the EU
2010	11.5	20.5	22.9	26.0
2011	12.6	25.3	24.7	30.3
2012	13.5	29.9	25.3	32.5
2013	14.6	35.2	27.8	36.4
2014	14.9	41.6	30.7	40.2
2015	15.6	46.8	31.7	48.5
2016	16.6	53.4	35.6	47.8
2017	18.0	56.1	35.8	52.0
2018	18.9	58.5	39.3	62.6
2019	19.8	59.4	39.6	65.5
Total for 2010–2019	156.0	426.7	313.4	441.8

Source Compiled by the authors based on (Bureau of Economic Analysis. U.S. Department of commerce, n.d.; OECD.Stat, n.d.)

China's trade in services with the USA and the EU was in favor of its partners. From 2010–2019, China's average annual export of services to the USA was $15.7 billion, imports were $42.1 billion, and the deficit was $26.5 billion. During the same period, China's average annual export of services to the EU was $31.3 billion, imports were $44.5 billion, and the deficit was $13.2 billion, which is not comparable to the American and European trade deficits with China (Table 1).

Mutual bilateral flows of foreign direct investment (FDI) between China and its main partners were developed in 2010–2019. Over the reviewed period, total exports of FDI from China to the USA were $39.7 billion, total imports of American FDI to China were $56.7 billion, with total imports of American FDI exceeding total exports of Chinese FDI by 1.4 times (Table 2). At the same time, the export of FDI by Chinese companies to the USA occurred with an upward trend. In 2010–2014, the average annual export was $1.9 billion; in 2015–2019—5.9 billion, it increased by 3.1 times. The export of FDI to China by American companies increased 1.7 times.

The total export of Chinese FDI to the EU was $79.4 billion; the total import of Chinese FDI from the EU was $70.3 billion; exports exceeded imports by 12.8%. China's investment ties with the EU developed more evenly and steadily with an increasing trend. By 2019, Chinese exports in aggregate exceeded FDI imports from the EU by 11.3%.

Between 2010 and 2019, the cumulative FDI exports of Chinese companies to the USA and the EU were $119.1 billion, with China receiving $127.0 billion in FDI from the American and European companies.

Before the presidency of D. Trump (January 20, 2017), China's foreign trade relations with the USA and the EU developed quite normally and according to the

Table 2 Mutual flows of foreign direct investment between China and its partners in 2010–2019, $ billion

	China's exports of FDI to the USA	China's imports of FDI from the USA	China's exports of FDI to the EU	China's import of FDI from the EU
2010	1.0	5.4	6.0	5.6
2011	1.1	−1.7	7.6	5.3
2012	3.4	−1.2	6.1	5.4
2013	1.9	7.4	4.5	6.5
2014	2.2	11.0	9.8	6.2
2015	5.1	6.0	5.5	6.5
2016	18.0	8.5	10.0	8.8
2017	2.2	7.9	10.3	8.3
2018	0.3	6.4	8.9	10.4
2019	4.3	7.0	10.7	7.3
Total for 2010–2019	39.7	56.7	79.4	70.3

Source Compiled by the authors based on (Bureau of Economic Analysis. U.S. Department of commerce, 2010–2019; Ministry of Commerce People's Republic of China, 2020b)

WTO principles and rules, despite the chronic and excessive imbalance in favor of China, especially in goods. By the presidency of D. Trump, China had already taken a strong position in the world economy. It firmly held first place in world GDP (at PPP) and world exports of goods and services. It was the world leader in the world market of intermediate goods and foreign exchange reserves. Moreover, the country was the largest holder of the foreign public debt of the USA.

With his desire to "make America great again" and under the slogan "America first," President Trump immediately started to attack his main rival and competitor. The reason for these attacks was the fact that the bilateral trade relations with China were developing not in favor of the USA. In 2017, the USA merchandise deficit with China increased nearly 10.9% from 2016; it was 3.8 times the United States–China exports and accounted for 43.4$% of the total USA merchandise deficit of $862 billion (WTO, 2020).

In 2018, Trump launched a real trade war against China, which lasted until the end of 2019. On March 22, 2018, he signed a "Presidential memorandum targeting China's economic aggression." In legal terms, D. Trump relied on the US Trade Act of 1974 and the results of an investigation against the PRC government for violations of intellectual property rights and unfair and discriminatory technology transfer policies, which allows the USA to establish barriers to imported goods unilaterally.

On July 6, 2018, the USA imposed additional customs duties of 25% on China's List No. 1 Chinese 818 tariff sub-items for Chinese goods, totaling about $200 billion. On September 24, 2018, the government imposed additional duties on List No. 2 of

5,745 tariff sub-lists totaling about $200 billion per year. On May 10, 2019, the American government increased the rate of additional duties on the goods from List No. 2 from 10 to 25% (Smbatyan, 2021).

China has responded swiftly and harshly to all unlawful actions of the USA by imposing similar duties on American goods. Some Russian experts suppose that in response to US sanctions, China has an opportunity of putting American supplies of crude oil and liquefied gas in a difficult position, for about 40% of all agricultural products supplied from the United States, for American states-suppliers of these products, for modern automotive products, etc. (Maslov, 2018). China also has the necessary instruments to make the life of American pharmaceutical and electronic companies in China as difficult as possible (e.g., by increasing the licensing period for their products, which will give a significant competitive advantage to European and other companies in the Chinese market).

China survived a trade war (2018–2019) in 2018. That year, China's merchandise exports to the USA increased by 10.7%, imports increased by 7.3%, export revenues exceeded import payments 4.4 times; the USA trade deficit with China was 43.9% of the National debt of the United States. In 2019, Chinese merchandise exports to the USA were down 19.5%, and imports were down 13.2%. However, exports exceeded imports 4.2 times, and the USA deficit was 37.2% of the National debt of the United States.

China's merchandise trade with the EU in 2018–2019 had a positive trend. Chinese exports increased by 9.9 and 4.6%. China's imports increased by 11.7 and 1.1%. Chinese exports exceeded imports by 1.5 times for each year. The surplus increased by 6.2 and 12.4%. The EU's trade deficit with China was over half the total EU trade deficit (54.6 and 53.1%). In 2018–2019, Chinese FDI exports to the USA averaged $2.3 billion per year, compared to a 2010–2019 average of $4 billion per year. China's imports were slightly above the ten-year annual average. Average annual (2018–2019) exports of Chinese FDI to the EU were $9.8 billion; Chinese FDI imports from the EU were $8.8 billion, exceeding the 2010–2019 annual average ($7.9 billion and $7.0 billion). In December 2019, the EU and China agreed to conclude an investment agreement. Nevertheless, this agreement was never signed under American pressure.

On January 15, 2020, the USA and China entered into the first stage of the agreement on the intention and willingness of the parties to resolve a wide range of bilateral trade issues. This agreement formally closed the first stage of resolving the trade and economic contradictions between the parties. The parties agreed to a gradual reciprocal return of payments in the amount of additional import duties from the level of the end of 2017. China pledged to purchase American products worth at least $200 billion over two years and agreed to further liberalize the access of American companies to the Chinese market and strengthen the protection of intellectual property rights (Mikheev & Lukonin, 2020). China has filed a complaint with the WTO Dispute Settlement Body, where an arbitral panel is considering the legal basis of the complaint and the dispute "United State—Countervailing Duty Measures on Certain Products from China."

4 Conclusion

The study of China's trade and economic relations with the USA and the EU showed that their long cooperation, until the election of D. Trump as president of the United States, developed rapidly and allowed China to achieve great success in partner markets and global markets for goods, services, and investment. During the difficult years of the Trump administration, China withstood a trade war with the USA with a slight loss in 2019.The country resisted during a pandemic.

In May 2020, the National People's Congress (NPC) held a session saying that China had coped with the pandemic, but the fight was not over. The country was facing serious challenges caused by a deep global economic downturn (Mikheev & Lukonin, 2021). Negative factors have led to a decline in domestic consumption in China, a drop in investment, and a decline in Chinese exports. The leadership of the Communist Party of China showed great professionalism in managing and monitoring the economy during the global economic crisis; China maintained and even strengthened its position in the global economy.

By the end of 2020, China has reached a positive growth rate in GDP (2.3%) and merchandise exports (4.0%) with negative values in most countries, including the USA, Germany, and other EU countries. China's balance on current estimates amounted to 2% of GDP, its share in the world market of goods and services increased by 1.6% and reached 12.4%, in the goods market—14.7%; in world GDP (at PPP)—18.3% (IMF, 2021).

The first meetings of Chinese representatives with the new US administration, which continues to talk to China from a position of strength, do not inspire much optimism, and no one has yet canceled Trump's initiatives. The interdependence of the world's economic leaders—China, the USA, and the EU–is so great that their foreign trade, investment, and other economic ties will develop and expand.

The USA and the EU need Chinese goods, investment, a marketplace, and exported commodities. China needs markets, capital, and technology. China is determined to continue to pursue an independent and autonomous foreign economic policy, make efforts to smooth differences through negotiations, and actively develop international cooperation based on mutual respect, equality, and mutual benefit.

Acknowledgements This paper has been supported by the RUDN University Strategic Academic Leadership Program and Grant of RUDN University No. 060119-0-000

References

Adno, Y.L., Alexandrova, I.I., Baykov, N.M., Bereznoy, A.V., Bolotin, B.M., Borisov, V.V. ... Elyanov, A.Ya. (2003). *The world economy, global trends over 100 years*. In I. S. Korolev (Ed.). Economist.

Bureau of Economic Analysis. U.S. Department of commerce. (2010–2019). Foreign direct investment in the U.S.: Balance of payments and direct investment position Data/Country by

industry detail (cross-classified by selected country and selected industry)/Financial transactions without current-cost adjustment, 2010–2019. Retrieved from https://www.bea.gov/international/di1fdibal. (Accessed April 15, 2021).
Bureau of Economic Analysis. U.S. Department of commerce. (2011–2020). U.S. international trade in goods and services China. Retrieved from https://apps.bea.gov/international/factsheet/factsheet.cfm?Area=650&UUID=9a761f65-1f76-473b-bbfc-56d0b4a250f7. (Accessed April 20, 2021).
Bureau of Economic Analysis. U.S. Department of commerce. (n.d.). U.S. trade in goods and services by selected countries and areas, 1999-present. Retrieved from https://www.bea.gov/data/intl-trade-investment/international-trade-goods-and-services. Accessed April 12, 2021.
Data of the World Bank. (n.d.). GDP (current US$)-China, 2007–2019. Retrieved from https://data.worldbank.org/indicator/NY.GDP.MKTP.CD?end=2019&locations=CN&start=2007&view=chart. Accessed April 10, 2021.
Directorate-General for Trade of the European Commission. (June 2, 2021). Overview China. Retrieved from https://webgate.ec.europa.eu/isdb_results/factsheets/country/overview_china_en.pdf. Accessed April 21, 2021.
IMF. (2013). International Financial Statistics Yearbook, 2013. Retrieved from https://www.imf.org/en/Publications/International-Financial-Statistics/Issues/2016/12/31/International-Financial-Statistics-November-2013-40658. Accessed April 4, 2021.
IMF. (2021). World Economic Outlook, 2021. Retrieved from https://www.imf.org/en/Publications/WEO/Issues/2021/03/23/world-economic-outlook-april-2021. Accessed April 20, 2021.
Maslov, A. (2018, July 10). Battle of the Titans: Who wins in the US-China trade war. Forbes. Retrieved from https://www.forbes.ru/biznes/364351-bitva-titanov-kto-vyigryvaet-v-torgovoy-voyne-ssha-i-kitaya. Accessed April 21, 2021.
Mikheev, V. V., & Lukonin, S. A. (2020). Beijing's pain points. *World Economy and International Relations, 64*(5), 5–15. https://doi.org/10.20542/0131-2227-2020-64-5-5-15
Mikheev, V. V., & Lukonin, S. A. (2021). Beijing's pain points—2 (glance from mid-2020). *World Economy and International Relations, 65*(1), 74–75. https://doi.org/10.20542/0131-2227-2021-65-1-70-81
Ministry of Commerce of People's Republic of China. (2020a). Tables 6–7 from the China Foreign Trade Report. Retrieved from http://images.mofcom.gov.cn/zhs/202012/20201209112029503.pdf. Accessed April 20, 2021.
Ministry of Commerce People's Republic of China. (2020b). Statistical Bulletin of FDI in China, 2020. Retrieved from http://images.mofcom.gov.cn/wzs/202012/20201230152644144.pdf. Accessed April 15, 2021.
OECD.Stat. (n.d.). EBOPS 2010—Trade in services by partner economy. Retrieved from https://stats.oecd.org/Index.aspx?DataSetCode=TISP_EBOPS2010. Accessed April 12, 2021.
Smbatyan, A. S. (2021). WTO decision on additional U.S. duties on Chinese imports. *Russian Foreign Economic Journal, 3*, 38–56. https://doi.org/10.24411/2072-8042-2021-3-38-56
WTO. (2020). World Trade Statistical Review, 2020. Retrieved from https://read.oecd-ilibrary.org/trade-monitoring/world-trade-statistical-review-2020_0a4fefd8-en#page1. Accessed April 11, 2021.
Xinhua News Agency. (November 3, 2017). Full text of Xi Jinping's report at 19th CPC National Congress. Retrieved from http://xinhuanet.com/english/special/2017-11/03/c_136725942.htm. Accessed February 12, 2021.

Prospects for the Development of the Transport Sector of the EAEU in the Context of Digitalization

Irina B. Teslenko, Aleksandr A. Vinarchik, and Elena N. Gorbatenko

Abstract The purpose of the chapter is to study the prospects for the development of a unified transport network of the EAEU states in modern conditions of digitalization of the economy. The authors conclude that the development of the transport network of the EAEU countries in the context of digitalization is impossible without modernizing approaches to managing the transport system and using new technologies, such as intelligent transport systems, digital logistics; systems of electronic exchange of information and document circulation; online services based on the "one-stop-shop" principle. According to the authors, building an ecosystem of digital transport corridors will reduce transport costs; increase the throughput of the EAEU ITC and their competitiveness, which in turn will contribute to the creation of new production of goods and services, attracting labor resources to the economic activities of the integration space.

Keywords Transport · Transport policy · Logistics · Transit traffic

JEL Codes L91

1 Introduction

The development of the transport system plays a huge role in the development of any state since it unites disparate areas of the territory into a single whole. That is why the transport system is rightfully compared to the country's circulatory system. Based on the development of the transport system, a territorial division of labour is formed, migration flows arise, and meets the needs of the population in transportation and tourist services. It is the level of development of the country's transport system that determines the state of its socio-economic development. The success of

I. B. Teslenko (✉) · A. A. Vinarchik
Vladimir State University named after Alexander and Nikolai Stoletovs, Vladimir, Russia
e-mail: iteslenko@inbox.ru

E. N. Gorbatenko
Financial University under the Government of the Russian Federation, Moscow, Russia

E. G. Popkova and B. S. Sergi (eds.), *Geo-Economy of the Future*,
https://doi.org/10.1007/978-3-030-92303-7_82

integration processes and the formation of a single economic space largely depend on the transport network and its effective functioning.

The Eurasian Economic Union is the youngest integration association in the post-Soviet space and includes Russia, Belarus, Kazakhstan, Kyrgyzstan and Armenia. The strategic goal of the association was the modernization of national economies, the growth of their competitiveness and the welfare of the population. All this is achieved by jointly combining the efforts of the countries of the union and achieving a synergistic effect from the implementation of joint projects.

The EAEU Treaty includes various aspects of the interaction between member states, including transport issues. Effectively built transport corridors are an important factor in the development of production, increasing competitiveness, attracting qualified labour and giving a powerful impetus to the socio-economic development of integration.

2 Materials and Method

To write this article, the authors used the methods of statistical analysis, studied the regulatory documents of the EAEU countries on the development of the transport system, articles in the open press, and conducted a content analysis of Internet sources.

3 Results

The efforts of the EAEU countries are aimed at creating institutional conditions for the development of a modern unified transport network that would contribute to the cheapness and speed of international transportation. The Department of Transport and Infrastructure was created within the framework of the European Economic Commission of the EAEU, the Advisory Committee on Transport and Infrastructure of the EEC operates, whose functions include the examination of projects in the field of transport infrastructure, the development of proposals for the development of road facilities, logistics, development of road infrastructure.

In December 2016, the Main Directions of the Transport Policy were approved, suggesting the removal of restrictions on transportation by all modes of transport until 2025. And in 2017, within the framework of these areas, "road maps" were approved containing specific measures to create a single transport space and a common market for transport services in the EAEU.

In 2018, the Council of Heads of Authorized Bodies in the Field of Transport of the Member States of the Eurasian Economic Union was created (Cooperation in the field of transport with the EAEU member states, n/a).

The development of the transport sector is of particular importance for the EAEU since apart from Russia, none of the Union countries (Belarus, Kazakhstan, Armenia and Kyrgyzstan) has an outlet to the sea. Calculations show that a landlocked country incurs transport costs for every 1000 km distance seven times more than for maritime powers. A doubling of transport costs, according to available estimates, reduces the country's foreign trade turnover by about 80% (Chibuchkhyan et al., 2018).

The development of the transport sector is influenced by many factors: the state of the world economy, political events (for example, the Nagorno-Karabakh conflict), the geographical remoteness of states from world markets, regional specificity and the level of development of the transport infrastructure of countries, uneven transit flows, country barriers, etc.

There are about sixty obstacles in trade between the EAEU countries alone. Eliminating at least some of them could reduce business transaction costs by $100–120 million, and the cumulative positive effect would be from 1.5 to 2 percent of the value of mutual trade (Russian Newspaper, (n/a)).

One of the serious barriers to the development of the Union is the various requirements for the overall dimensions and weight of the cargo (the maximum permissible weight of a loaded vehicle (News of Eurasia, n/a). So, if in Belarus the maximum permissible weight of a two-axle truck is 20 tons, and a five-axle one—41 tons, then in the Russian Federation these norms are 18 and 35, in the Republic of Kazakhstan—18 and 38 tons, respectively.

In Armenia, the maximum permissible vehicle weight is 36 tons (the lowest rate in the territory of the EAEU), and in the Russian Federation—44 tons, which leads to additional costs for Russian transport companies, and not only. Such discrepancies complicate the planning and implementation of road transportation within the EAEU.

Failure to recognize compulsory motor vehicle liability policies is a serious barrier in the field of international logistics. Only Russia and Belarus recognize the national MTPL policies "Green Card" (Sputnik, n/a).

The paper format of documents, differences in customs and transit procedures, etc., hinder the acceleration of the movement of goods.

Real freight traffic in 2020 was strongly influenced by the pandemic. The fact is that some of the projects included in the national project "Transport part of the comprehensive plan for the modernization and expansion of the main infrastructure for the period up to 2024" is off-budget; therefore, their implementation directly depends on the decision of private investors. Investor rejection risks may lead to a revision of the project parameters.

Despite all the difficulties, the EAEU has significant economic potential: a network of important transcontinental international transport corridors (ITCs) runs across the territory of the Union countries along the West–East and North–South lines, which can ensure the growth of the economy of the EAEU member states. However, the action of objective and subjective factors does not provide the required efficiency of freight traffic (Vysotskaya, 2019).

Addressing the issues of reducing transport costs, the EAEU countries take measures to obtain the possibility of access to seaports. So, in Armenia, a project is being implemented to build an Iranian-Armenian railway as a transit link of the ITC "Persian Gulf-Europe". Thus, the Republic of Armenia organizes direct rail links with the EAEU countries, which will solve the problem of the isolation of its territory from the countries of the integration association (Almetova et al., 2018).

The proposals of Uzbekistan on joint implementation of the Trans-Afghan Railway project with the EAEU, Kazakhstan's proposal to create within the EAEU a single network of wholesale distribution centers providing specialized services (storage and processing of goods) before their distribution and sale are discussed (Sputnik, n/a).

An important task of the EAEU is to ensure an increase in transit freight traffic between China and Europe. The transit land route through the territory of the EAEU is faster than the sea route and cheaper than the air route. Building relations between China and the EAEU according to the "One Belt—One Road" type is of great importance for the creation and development of the Eurasian partnership (Vysotskaya, 2019). The PRC presents this partnership as a kind of symbiotic system with a free structure, high inclusiveness, and the main principle—"partnership without forming a union" (Blog of Nikolai Kapustin, 2018).

Currently, the volume of trade between China and Western Europe, according to some estimates, is about 1200 billion dollars, and only about 4–5% of all traffic goes through the territory of the CIS countries. This is the potential that the EAEU needs to master for further economic development (Sputnik, n/a).

Improving the transport component of the EAEU countries in conditions of digitalization is impossible without modernization of approaches to the management of the EAEU transport system and the use of new digital technologies (intelligent transport systems, digital logistics; information exchange and document management systems; online services based on the "single window" principle (The Ministry of Transport of the Russian Federation, n/a).

In 2019, a project was developed to create an ecosystem of digital transport corridors (ECTC) of the EAEU with a horizon until 2025. According to estimates, the total cost of the formation of the ECTC will amount to 10 billion rubles, and the effect—154 billion rubles.

The ecosystem of digital transport corridors is an open system of information services based on a digital platform aimed at efficient interaction of transport companies and cargo owners in all member states EAEU, as well as from third countries.

The ecosystem of digital transport corridors allows for the development and implementation of many innovative projects. An example is a project for the implementation of digital navigation seals, which are essentially an onboard computer containing all the necessary information for transportation (Internet Portal CIS Space of Integration, n/a). In the future, trade flows will be controlled virtually.

The implementation of the project for the development of an ecosystem of digital transport categories involves the creation of common information resources, which will include information on all main routes of the countries of the integration group, roadside service facilities, checkpoints and customs control. The implementation of this project will significantly reduce the transportation time by building optimal transport routes. The advantage of this project is also that the number of services can be increased in the future, which significantly increases the efficiency of cross-border cooperation.

Recently, the EEC approved a program for the trial operation of a certification center of a trusted third party service of an integrated information system for the formation of cross-border electronic document circulation within the union.

The Russian Export Center and the Russian-Singapore Business Council launched a pilot project for the placement of Russian companies on a B2B electronic platform (the platform is entirely Russian development).

Since, according to analysts, by 2028, almost half of global consumption will come from Asian countries, this project is a real opportunity for high-tech Russian companies to enter the promising markets of Southeast Asia, Africa and Europe (Russian Newspaper, n/a). The project is one of the most important for the development of digital transport corridors "East–West" and "North–South" and the creation of a network of technology transfer, industrial cooperation and subcontracting.

For the implementation of the ECTC project, certain efforts will be required. They are related to the fact that although there are platforms and information systems for transport in the EAEU, countries do not carry out any exchange of information, there are no uniform standards and rules for exchange within countries and no sectoral mechanism that supports these rules. Without the creation of such a mechanism at the national level, the emergence of a supranational platform is impossible. In addition, the project will remain on paper if there is no access to the internal information of all participating countries.

At present, the countries have agreed to take an inventory of all technological resources, to open a joint technology competence center. The EEC proposed to create a system of cross-border space of trust, develop uniform standards for the provision of transport and logistics services and introduce an integrated supervision system (Internet Portal CIS Space of Integration, n/a).

It is predicted that from 2022, due to the implementation of the concept of an ecosystem of digital transport corridors in international transport and logistics transportation, the annual mileage will increase by about 20%, and the price of the transport component in the final product will decrease from 20 to 12–15% (Vysotskaya, 2019).

It is reasonable to consider the proposal to create transit transport hubs in transit regions, where it is possible to redistribute multidirectional freight traffic (Almetova et al., 2018). This proposal will reduce the cost of carrying out transport work and increase the efficiency of road transport in transit.

Some initiatives have already begun to be implemented, including the development of digital transport corridors in the EAEU (integration of information about vehicles, crews, cargo, permits and accompanying documents at all stages of transportation throughout the EAEU).

In 2021–2023, it is planned to carry out work on improving legislation in the field of multimodal and transit transportation, the use of unified documents for the transportation of goods and the rules for carrying out transportation; rational interaction of countries to ensure effective mobility of transit cargo flows in the Union.

4 Conclusion

Summing up, it should be noted that in the development of the transport sector of the EAEU many problems and factors leading to the slowdown in the development of international transport corridors. The most pressing are the issues of harmonization and unification of the regulatory framework and standards for transportation.

The introduction of innovative approaches to the international transport system will allow you to solve these problems quickly and efficiently. The introduction of digital platforms and the creation on their basis of an ecosystem of digital transport corridors will reduce transport costs, increase the throughput of international transport corridors of the Eurasian Economic Union, increase the competitiveness of the transport system, which in turn will create a multiplier effect and contribute to the development of new industries for the production of goods and services attracting qualified labour resources to the economic activities of the integration space.

In the future, concerted efforts to create a modern transport and logical infrastructure, apply progressive digital solutions and unify approaches to administration can make the entire Greater Eurasia region a single transport territory, create modern transport hubs, and significantly reduce prices for the provision of transport and logistics services.

References

Almetova, Z. V., Sheremet, A. A., Samartseva, A. V., & Dolgushina, N. Yu. (2018). Integration processes of the transport systems of the Eurasian Economic Union. *Bulletin of the South Ural State University. Economics and Management, 12*(3), 161–168 (166 Ser.).

Blog of Nikolai Kapustin. Conjugation of the Eurasian Economic Union and the Silk Road Economic Belt: An Integration Dilemma?. https://russiancouncil.ru/blogs/nikolay-kapustin/33975/

Chibuchkhyan, S., Chibukhchyan, G., & Gasparyan, J. (2018). The development of the logistics capabilities of the EAEU is the basis of the economic growth of the union. *Logistics, 5*(138), 52–55.

Internet Portal CIS Space of Integration (n/a). Digital transport corridors—the future of the EAEU. https://e-cis.info

News of Eurasia. Logistic and customs barriers within the EAEU. https://alta.ru

Russian Newspaper (n/a). By 2025, the EAEU will create a system of digital transport corridors. https://rg.ru

Sputnik (n/a). Development of transport systems and common markets. Where is the CIS heading 30 years later?. https://uz.sputniknews.ru

The Ministry of Transport of the Russian Federation. (n/a). Cooperation in the field of transport with the EAEU member states. https://mintrans.ru

Vysotskaya, V. (2019). Prospects for the development of the transit potential of the Eurasian Economic Union. https://cyberleninka.ru/article/n/perspektivy-razvitiya-tranzitnogo-potentsiala-evraziyskogo-ekonomicheskogo-soyuza

Prospects for the Formation of the Digital Space of the EAEU Countries

Nadezhda V. Muravyeva and Vera U. Rudakova

Abstract The purpose of the chapter is to study the prospects for the formation of a single digital space for the EAEU countries. According to the authors, the digital agenda of the EAEU provides exceptional opportunities for the implementation and scaling of digital initiatives, that is, it initially sets favorable conditions for the accumulation of internal potential and the development of a fairly large internal market. Combining the efforts of the EAEU states in building a single digital platform will avoid the absorption of the information space by large transnational players. The authors note that digital transformation is the driving force behind the innovative, sustainable growth of the Union countries, as well as the driver of integration processes.

Keywords Digital economy · Information space · Digital platform · Digital transformation

JEL Codes F15

1 Introduction

Digitalization has become an integral part of the life of most of the world's population. It changes the style and forms of everyday life, transforms the ways and models of economic activity.

The created Internet platforms allow manufacturers and sellers to sell goods all over the world, saving on transaction costs. State Internet portals provide fast and effective interaction between citizens, businesses and government services. The hard and monotonous work of workers is automated, replaced by robots; the joint work of people and robots is organized. New specialties are emerging, and the remote form of activity is spreading. Communication of people from different cities, regions and countries of the world becomes possible at any time, thanks to programs for organizing online video conferencing, etc.

N. V. Muravyeva (✉) · V. U. Rudakova
Vladimir State University named after Alexander and Nikolai Stoletovs, Vladimir, Russia

E. G. Popkova and B. S. Sergi (eds.), *Geo-Economy of the Future*,
https://doi.org/10.1007/978-3-030-92303-7_83

Digital transformation is a global phenomenon that is revolutionizing not only labour and human relations, but also interstate and, above all, relations within already established integration associations. Such associations are initially created to maintain stable economic growth of each country entering the integration, to increase the competitiveness of national economies and, most importantly, in the interests of improving the quality of life of the population of the participating countries.

These goals formed the basis for the creation of the Eurasian Economic Union (EAEU), which included the Republic of Armenia, the Republic of Belarus, the Republic of Kazakhstan, the Kyrgyz Republic, and the Russian Federation.

Since the inception of the Union, it has been of paramount importance to pursue a coordinated policy in the member states in almost all spheres of life. In the past few years, an increasing place in the formation of a coherent policy has been given to digital transformation, as the main factor in the future development of the Union.

2 Materials and Method

To prepare this article, the authors used the program and regulatory documents of the EAEU member countries, reports of the statistical services of the republics, content analysis of Internet sources. All these sources of information allowed the authors not only to assess the level of development of digital technologies in each country but also to determine the prospects for creating a single digital platform for the countries of the commonwealth.

3 Results

The document "Main directions for the implementation of the digital agenda of the Eurasian Economic Union until 2025" adopted in 2017 opens up new opportunities for all-around cooperation for the EAEU member states based on the creation of a common digital space.

The digital space of the Union is a zone that integrates digital processes, means of digital interaction, information resources, as well as a set of digital infrastructures, based on regulations, mechanisms of organization, management and use (On the Main Directions for the Implementation of the Digital Agenda of the Eurasian Economic Union until, 2025, 2017).

Experts predict that the creation of a digital space will eliminate barriers to the movement of goods, services, capital, labour, information, will provide an opportunity to develop new business models of interaction and even entire sectors of the digital economy, which will ultimately lead to the achievement of the set integration goals.

To create a digital space, first of all, significant funding will be required for the formation of institutional foundations, training the population in digital skills, creating a digital infrastructure, etc.

So far, the main obstacle to creating a digital space is the low level of citizens' trust in digital solutions, lack of understanding of their benefits and insufficient digital literacy.

The implementation of a joint EAEU policy is accompanied by the implementation by each country of its digitalization program. In Russia, a national project "Digital Economy" was adopted, which should be implemented until 2024. As a result of the project, it is planned to increase internal costs for the development of the digital economy from all sources at least three times compared to 2017; creation of stable and secure information and telecommunications infrastructure for high-speed transmission, processing, and storage of large amounts of data, available to all households; the use of predominantly domestic software by government agencies, local governments and organizations.

In order to obtain the intended results, it is necessary to create an ecosystem of the digital economy, which should include a system of legal regulation of digitalization, an infrastructure for processing and storing large amounts of information, a system for training highly qualified personnel, the development and implementation of digital platforms in all spheres of government, ensuring information security and implementation of a coordinated and balanced policy with the EAEU countries in the field of the digital economy, etc. (National project "Digital Economy", 2019).

In Belarus, in February 2021, the State Program "Digital Development of Belarus" for 2021–2025 was also adopted, which is aimed at introducing digital technologies in all sectors of the national economy, in public administration and all spheres of life of the population. The State Program of Belarus sets such important tasks as the introduction of information and digital technologies in education, improving the quality of medical care based on the use of digital platforms, the development and implementation of smart cities technology, the introduction of digital platforms in all spheres of the national economy, ensuring information security (About the State Program "Digital Development of Belarus" for 2021–2025, 2021).

Earlier (2017), a similar state program, the implementation period of which was 3 goals, was adopted in Kazakhstan. The global goal of this program is, using digital technologies, to create conditions for the transition to a fundamentally new model of development of the national economy of Kazakhstan—the economy of the future. The goal involves solving such problems as the introduction of digital technologies in all sectors of the economy, increasing the level of coverage of the territory with communication networks, ensuring the country's information security, supporting, and financing innovative entrepreneurship and increasing digital literacy of the population, etc. (State program "Digital Kazakhstan", 2020).

According to this program, the volume of investments in the digital economy of Kazakhstan will amount to about $1.2 billion (The rhythm of Eurasia (2021).

The strategic plan "Agenda for Digital Transformation of Armenia until 2030" was developed and adopted in Armenia. In this regard, the government has identified six key pillars for building the digital economy: digital government, digital skills, infrastructure, cybersecurity, the private sector, and infrastructure. In total, the strategic plan includes three sequential stages of implementation: infrastructure development, inventory and updating of available resources, large-scale investments in the digital sector and accelerating the development of the economy based on digitalization. By 2030, a new information space of the country should be built, the foundation of which will be digital technologies. Building a digital state in Armenia involves ensuring TOP-20 in the e-government development index and TOP-30 in the global competitiveness development index (Digital Agenda of the Republic of Armenia, 2018).

In Kyrgyzstan, the Concept of digital transformation "Digital Kyrgyzstan"—2019–2023 was adopted. As a result of the program, technological changes should take place in the country, the competitiveness of the economy, the standard of living of citizens and the efficiency of the state should increase (Concept of digital transformation "Digital Kyrgyzstan"—2019–2023, 2018).

In general, already in the first years of the implementation of the adopted digitalization agendas in the EAEU member states, significant progress has been recorded in many areas of digital development. Digitalization processes in the countries of the Union were accelerated by the coronavirus crisis.

The digital transformation of the EAEU member states is the key to creating a unified digital policy for the Union. The speed of progress in solving the tasks outlined in the EAEU Development Strategy until 2025 will depend on how effectively the interaction of countries will be built in "digital".

The EEC representatives believe that in the crisis and post-crisis periods they should become not so much a springboard as a trigger for the intensification of the EAEU countries' aspirations for digitalization.

In 2021, representatives of the EEC proposed to bet on integration in the field of high technologies. So, Kazakhstan owns the idea of creating and implementing a joint educational project in the field of artificial intelligence and digital technologies.

The basis for building a unified information and communication space of the EAEU should be an Integrated Information System, which greatly simplifies the exchange of data. The decision to create such a system was made before the coronavirus epidemic, but its implementation in the life of the EAEU is slow. Out of the planned 89 general processes of the Integrated Information System of the EAEU, only 19 have started working, and 48 processes are under testing.

Experts in the field of digital technologies note that the speed and efficiency of work to create a unified information system directly depends on the level of development of digital technologies and the digital segment of the economy of the EAEU member states.

They explain the slow pace of building a single digital space for the following reasons.

Firstly, in the countries participating in the integration association, the level of development of digital technologies differs significantly in content, focus, strategy and level of development of digitalization infrastructure.

For example, in Russia and Kazakhstan, emphasis is placed on the full digitalization of all sectors of the national economy using government regulation measures to create a favourable ecosystem for the development of information and communication technologies. In Armenia and Kyrgyzstan, the digital technology strategy is aimed at embedding the projects of these countries in more global and successful projects of the European Union or Asian countries. In Belarus, the development of infrastructure components and personnel training have been chosen as the central strategic line.

In terms of the accumulated experience in the implementation of national digital agendas. Russia and Kazakhstan are leading in the formation of national competencies, and Belarus has become an important participant in international chains of development of information and communication products (Digital sovereignty & digital agenda of the EAEU, 2021).

Secondly, there is a low level of mutual trust of all parties involved in terms of ensuring digital sovereignty (this is the independence of the state in managing digital transformation and forming a new ecosystem, which excludes the possibility of external influence on its functioning and sustainability) (Digital sovereignty & digital agenda of the EAEU, 2021).

Thirdly, there are differences in national document management standards (the definitions of the concepts of an electronic document and an electronic digital signature differ).

Nevertheless, the digital agenda of the EAEU provides exceptional opportunities for the implementation and scaling of digital initiatives, that is, it initially sets favourable conditions for the accumulation of internal potential and the development of a fairly large internal market.

According to the digital agenda of the EAEU until 2025, the priority areas for the countries are digital transformation of economic sectors and cross-sectoral transformation; the digital transformation of management and integration processes; the digital transformation of markets for goods, services, capital and labour resources; development of digital infrastructure and ensuring the security of digital processes.

The EAEU has an institutional framework for the implementation of joint and common digital projects: the Digital Initiatives Fund of the Eurasian Development Bank, which has already launched the first digital projects in the EAEU (Digital sovereignty & digital agenda of the EAEU, 2021).

Experts believe that the ongoing changes should be brought into a common digital strategy, in which the development of cross-border digital projects will be not only a tool for integration, but also a means of gaining national advantages in the domestic, regional, and international markets.

This requires:

- creation of a clear strategy for the development of the digital economy of the EAEU within the framework of the stated main directions,
- the formation of common interests in the field of digitalization in the context of roadmaps by industry and direction (energy, transport, logistics, industry, trade, agriculture, etc.),
- development of a plan for comprehensive digitalization with a transition to a new technological order based on strategies and roadmaps,

– a project-based approach to the implementation of the digital transformation plan.

4 Conclusion

The unity of actions of the EAEU countries gives each country more chances to achieve its goals. Experts believe that by implementing digitalization projects separately and without participating in cross-border projects, the EAEU countries will either have to build high barriers, while having high transaction costs, or they will be doomed to digital absorption by large transnational corporations (Digital sovereignty & digital agenda of the EAEU, 2021).

Despite the difficulties, there are already certain successes in the implementation of the joint agenda. In particular, in the digitalization of customs processes. In accordance with the 2017 agreement between Russian Railways and Chinese Railways, electronic data exchange of SNGS consignment notes (Agreement on International Rail Freight Traffic) is carried out, which is currently 95%. Border crossing can be accelerated by making a full transition to electronic legally significant document flow between carriers and regulatory authorities. Work continues on the implementation of electronic waybills for road freight transport (Digital interface between the EAEU and the Belt and Road: a view from Russia and China (2020)).

The leaders of the EAEU countries have approved the concept of a common financial market. It will allow the development of advanced financial technologies, making services in this area more accessible (Mishustin proposed to create a single digital platform for the EAEU (2020)).

World Bank experts calculated that the economic effect of the implementation of the digital agenda will increase the Union's GDP by 2025 by about 10.6% of the total expected growth of the aggregate GDP of the member states by that time. At the same time, the potential effect is almost twice the possible size of an increase in the GDP of the EAEU countries in the case of digital development without the implementation of a common digital agenda (Eurasian Economic Commission, 2017).

According to Mishustin, the EAEU countries need to create a unified digital identification system, which requires recognition of the electronic digital signature throughout the Union space.

The EAEU should conduct an inventory of all technological resources and open a joint center of competence in this area, since the EAEU still does not have its software, its processors, or its digital security system.

The Union needs to create its ecosystem and make digital transformation a Eurasian project.

Summing up the overall result, it should be noted that digital transformation is the driving force behind the innovative, sustainable growth of the Union countries, as well as the driver of integration processes.

Recognition of the need for a coordinated policy to achieve the set objectives will strengthen the EAEU's position in the world market, preserve a single cultural space in the interests of all countries participating in the integration, and make the EAEU governance flexible and mobile.

References

About the State Program "Digital Development of Belarus" for 2021–2025. (2021). https://pravo.by/upload/docs/op/C22100066_1612472400.pdf. Accessed March 20, 2021.

Concept of digital transformation "Digital Kyrgyzstan"—2019–2023. (2018). http://www.ibc.kg › ru › download › 1452. Accessed June 16, 2020.

Digital Agenda of the Republic of Armenia. (2018). https://d-russia.ru/tsifrovaya-povestka-respubliki-armeniya.html. Accessed June 07, 2018.

Digital interface between the EAEU and the Belt and Road: a view from Russia and China. (2020). https://news.rambler.ru/world/45105983-tsifrovoe-sopryazhenie-eaes-i-poyasa-i-puti-vzglyad-iz-rossii-i-kitaya/. Accessed October 07, 2020.

Digital sovereignty and digital agenda of the EAEU. (2021). https://globalaffairs.ru/articles/czifrovoj-suverenitet-eaes/. Accessed March 09, 2021.

Eurasian Economic Commission (2017). Digital reboot of the EAEU. A new look at the economy! http://www.eurasiancommission.org/ru/act/dmi/workgroup. Accessed February 02, 2018.

Mishustin proposed to create a single digital platform for the EAEU (2020). https://www.kommersant.ru/doc/4236294. Accessed January 31, 2020.

National project "Digital Economy". (2019). https://strategy24.ru/rf/management/projects/natsional-nyy-proyekt-tsifrova-ekonomika. Accessed January 23, 2019.

On the Main Directions for the Implementation of the Digital Agenda of the Eurasian Economic Union until 2025 (2017). https://www.garant.ru/products/ipo/prime/doc/71708158/#20. Accessed June 20, 2020.

State program "Digital Kazakhstan" (2020). https://primeminister.kz/ru/gosprogrammy/gosudarstvennaya-programma-cifrovoy-kazahstan-9111459. Accessed December 9, 2020.

The rhythm of Eurasia (2021). With respect to the "figure". Meanwhile, digitalization is still stalled https://www.ritmeurasia.org/news--2021-03-13. Accessed March 13, 2021

Development of the Institutional Environment of the Region's Ecosystem

Vera V. Borisova and Tamila S. Tasueva

Abstract The paper focuses on the design of the institutional environment of the region in an ecosystem format. The authors study the institutional environment of the region's ecosystem in the context of the problem-oriented paradigm through the prism of the problems facing the region. The paper explores the prospects of creating special regulatory regimes for participants of digital transactions in the regional economy. The authors consider the features of the development of internet transactions based on the replacement of paper documents with virtual messages. Moreover, the authors focus on regulating electronic signatures, including the use of cloud technology in paperless document management. It is shown that the electronic interaction of the government with all participants of the institutional environment of the ecosystem contributes to the accelerated development of the digital economy. The paper substantiates the need to develop the region's digital ecosystem, which plays an integrating and regulating role in a hybrid (virtual and real) world. The emergence of new institutions and the transformation of existing ones due to the digital transformation of the regional economy determine the relevance of this research. The authors determine that the digital transformation has covered all spheres of life of the region's economy—from document management to the possibility of processing all public services in an online format. It is proved that the digital transformation formed a favorable environment for launching new business models and startups. The digital transformation involves creating an adequate institutional environment of the region's ecosystem.

Keywords Institutes · Institutional environment · Region · Network-centrism · Ecosystem · Digitalization

JEL Code M110

V. V. Borisova
Saint Petersburg State University of Economics, Saint Petersburg, Russia

T. S. Tasueva (✉)
Grozny State Oil Technical University named after Academician M.D. Millionshchikov, Grozny, Russia

E. G. Popkova and B. S. Sergi (eds.), *Geo-Economy of the Future*,
https://doi.org/10.1007/978-3-030-92303-7_84

1 Introduction

The digital ecosystem of the region expands the opportunities for socio-economic development of the territory through the use of virtual assistants and the construction of communication platforms and information repositories. In this context, institutions become a kind of interface connecting business entities in the region's digital ecosystem (Borisova, 2020).

The analysis shows that new digital business models, including new services and platforms that integrate markets, are forming quite rapidly in the region's economy. Services and applications of an ecosystem operate in different sectors of the economy: e-commerce, transportation, tourism, financial, consulting, and other market segments. Members of the regional digital ecosystem retain their autonomy and can be partners of several ecosystems.

The evolution of digital ecosystem models indicates the expansion of the boundaries of a partnership between its participants: the differences between the system elements are gradually erased, the focus of digital services and services is shifting towards customer preferences, and the connections are acquiring a cross-functional nature. This expansion is manifested in the peculiarities of the management structure, the solution of priority tasks based on the interaction between partners, and the decentralization of the decision-making process (Trofimov, Zakharov & Frolov, 2019).

The prerequisites for implementing ecosystem modeling of the regional institutional environment form the cross-functional competencies of the partners. The success of the institutional environment of the region's ecosystem is determined by the expansion of digital services and applications, the ability of its participants to build cooperative relationships going beyond existing value chains, the convergence of innovations, and the integration and application of expertise from different fields (Borisova, 2018). This indicates that the regional institutional environment in the form of an ecosystem will be the most probable structural subsystem in the organization of the country's economy.

2 Materials and Methods

The ecosystem approach to the development of the institutional environment in the region is based on a new network-centric model of value chain management, which ensures the optimization of the regional system of goods exchange and its integration into inter-regional marketplaces. The network-centric nature of the ecosystem will expand the institutional capacity of the integration platform-coordinator of inter-regional goods flows as a basic element of the territory's institutional environment (Borisova, Yuan & Tang, 2020).

The key points of the digital concentration of regional institutions were identified using economic-statistical groupings, visual modeling, and expert evaluations. In particular, the analysis of transformational changes in the institutional environment allowed the authors to develop a foresight map of structuring the goals of the institutional design of the regional ecosystem. The authors developed a system of goals to create institutions, including their quantitative assessment and qualitative interpretation. Additionally, the authors analyzed the structure of institutions in terms of their compliance with the system of goals.

Generalization of the identified transformations in the regional economy was carried out based on project-oriented conceptual modeling of the processes of creation and development of new institutions (Borisova, 2020). The authors studied the institutional environment of the region with the help of an expert focus group. The focus group analysis allowed us to identify specific features, problems, and bottlenecks in the functioning of institutions operating in the region. The authors formulated recommendations for the transformation of institutions based on quantitative assessments of the effectiveness and rationality of their structure and construction principles.

3 Results

Global practice shows that the largest companies by market capitalization (Amazon, Alibaba, Apple, Facebook, Google, Tencent, and Microsoft) work in the format of digital ecosystems. At the beginning of the 21st century, European countries launched a project to support small- and medium-sized enterprises through Internet technology and digitalization. This project was based on the principle of a biological ecosystem, which connects living organisms by stable ties (Sokolov, 2021).

The term digital ecosystem is used figuratively as a metaphor to describe its participants as mixed communities and ecosystems where digital and human resources are combined (Borisova, 2020). This analogy underscores the interdependence of participants in the digital business environment. Increased interaction between participants in the digital ecosystem has been facilitated by producing and distributing smart products and services, increasing Internet coverage, increased demand for smartphones, and the introduction of technology for collecting and analyzing big data (Popkova & Sergi, 2020).

The three largest Russian companies (Yandex, Sber, and Mail.ru Group) demonstrate their experience in creating digital ecosystems. Their total revenue in the pandemic year of 2020 increased by 24% to 400 billion rubles. Digital ecosystems showed their effectiveness during the COVID-19 pandemic. Statistics show that the COVID-19 crisis has contributed to an increase in online sales and remote customer service. There began a competition to develop at the level of digital ecosystems.

We describe the region's digital ecosystem as a large, complex, socio-economic, socio-technical, open, and adaptive system with signs of self-organization, where the search and transfer of information on the movement of goods and services are carried out through the Internet channels. The institutional environment of the ecosystem integrates digital services and a set of partners' assets on a communication platform under a single brand. The participants of this environment acquire a single identification number (ID—a unique attribute of the object) and become users of all services, including operational support from a single center.

The analysis of the institutional transformations and infrastructure of the studied regions, including the assessment of the prospects for the emergence of new institutions in the basic sectors of the economy, has shown that the preconditions for the formation of inter-regional network communities in the form of digital platforms are emerging. From our point of view, within the institutional environment of the region's ecosystem, it is advisable to create an integration platform-coordinator of regional and inter-regional goods flows (Rogachev, Ostrovskaya, Natsubidze, Litvinova & Yakovleva, 2018).

"Economic theory applies the term 'institutional arrangement'—as the result of the effective functioning of infrastructure institutions aimed at the development of the system, reducing transaction costs, and minimizing market failures and interactions of economic entities with the institutional infrastructure" (North, 1997).

"Institutions are the "rules of the game" in society, man-made restrictive frameworks that organize relationships between people. Consequently, they set the structure of incentives for human interaction, whether in politics, social sphere, or economics" (North, 1997).

The prospects for the emergence of new institutions in the basic sectors of the region's economy based on the application of technology foresight, the construction of road maps, and a scenario method were assessed in conjunction with institutional design. This allowed the authors to identify promising inter-regional innovation relationships, including the consistent implementation of a system of methods of stimulating, forecasting, and programming the prerequisites for creating regional infrastructure facilities in a digital economy.

Targeted organization of the digital institutional environment is implemented on the principles of stage component completeness of the project, sufficient to stimulate the project and protect it from opportunistic influence. The analysis of the design of the regional ecosystem's institutional environment and the management of the development of digital institutions in the region occurs on the threshold of significant transformations. It is important to maintain the relevance of the key elements of the region's institutional environment: strategies, goals, structural components, formal and informal norms, and mechanisms for launching new institutions.

The totality of organizational and functional methods of creating the structure of the digital institute ensures the reproduction of norms and complementary links between them within the institute. This creates conditions for the sustainability, quality, and reliability of projected institutions. The improvement of the institute's organization and the alignment and realization of its potential reserves in the context of relative standardization, unification, and regulation of its elemental composition are associated with the principle of the rationalization of institutional design (Tambovtsev, 1997).

Systemic assessment of the institution in terms of the principle of rationality and sustainable development of the regional economy is based on the analysis of scenarios for developing the institutional environment and the prospects for achieving the stated objectives. The following actions are taken as a part of the scenario analysis and selection of the best institutional project:

- Setting the goal of the institutional project (the correct goal setting is based on the results of a focus-group study of the parameters affecting the projected digital transformation in the regional economy);
- Determination of resource constraints (including material, labor, financial, information and communication, power and administrative, and other constraints);
- Expertise of scenarios of development of the institutional project based on the given target criteria of option selection (minimization of costs, resource provision, consideration of participants' interests, on which the results of the institutional project will be directed, etc.);
- Agreement and approval of the project;
- Execution and correction (if necessary) of the project.

The institutional factor is recognized as one of the critical factors in the digital transformation of industries and regions of the country. Experts note that quantitative assessment of the qualitative features of institutions is a rather complex methodological task. "Institutions cannot be analyzed without considering their semantic content, which complicates their quantification" (Podshivalova, 2014).

The authors propose the concept of network-centric modeling of inter-regional network communities in the form of digital platforms and interpretation of the institutional environment of the region in the format of an ecosystem (Borisova, 2018, 2020). To develop our concept, we will represent the set of institutional links in the form of a set of institutions (formal and informal) forming the framework of the region's ecosystem. In this case, the set institutional environment of the regional ecosystem forms (provides) a set of alternatives for making institutional decisions by economic entities. Of the many options for network-centric linkages in the regional ecosystem, economic agents choose the best ones in terms of target settings.

Assessing the quality of the institutional environment of the region's ecosystem, we proceeded from the current institutional links and their relevance in the digital transformation of the region's economy. This circumstance is especially relevant for the national republics of Russia.

Modeling the institutional environment of the region's ecosystem is possible in matrix form (Kirdina, 2001) One of the possible options is a matrix that quantitatively unites institutions and possible connections of institutions. The matrix elements can have a zero value (if there is no institution in the nexus) or can be equal to one (if there is an institution in the nexus).

The sufficiency of the institutional environment depends on the goals of regional policy and the set strategic objectives. They were determined by conducting a focus-group study involving experts from civil servants, representatives of ministries, business, and science. The degree of the gap between the parameters of sufficient and legislated level of the institution (institutional environment) demonstrates their qualitative features. The measures of mismatch (parameter gaps) will be the corresponding coefficients used for particular institutions and the whole institutional environment.

Different variants of events are possible, including the complete absence of parameter gaps and inconsistencies. An increase in the coefficients indicates a decrease in the qualitative features of the institution (the institutional environment of the region). Not only the importance of the qualitative features of institutions is relevant but also the state of network-centric relations in the ecosystem of the region formed by them.

The following grading scale can be used for the coefficients obtained as a result of the calculations (Podshivalova, 2014):

- High quality of the institutional environment—the calculated parameters are in the range 0–1;
- Satisfactory quality of the institutional environment and the gap in ties does not exceed 30%—the calculated parameters are in the range 1–3;
- Low quality of the institutional environment and a significant gap in most bundles—the calculated values of the coefficients range from three and above.

The generalized model of the institutional environment of the regional ecosystem includes a set of norms characterizing a particular institution (norms of support, monitoring, and control, enforcement mechanisms, forms of preservation of institutions, value norms, etc.). The ecosystem approach to the development of the institutional environment of the region allows us to compare the quality of various institutions in time. Moreover, it allows us to assess the prospects for the emergence of new structures within the entire institutional environment of the region based on the analysis of the qualitative features of particular institutions corresponding to the key problems of economic development in the region.

4 Conclusion

The development of digital institutions is hampered by the contradiction between general and special approaches to the design of regional structures of the new type of ecosystem. The solution of local digital institutional tasks and the introduction of innovative pilot projects occur in the absence of a holistic view of the institutional environment of the region as an ecosystem. The prerequisites for the ecosystem approach to the design of regional institutions are already in place. They include sectoral and government support for institutional transformations in the Russian economy.

The objective of the regional ecosystem is to increase the well-being and satisfaction of citizens with digital services, supporting high-quality public services at all stages of interaction with the population. The pandemic and the COVID-19 crisis radically changed the speed of transformation of the institutional environment and the form of interaction between economic entities in the region. Many social institutions, the formation of which was expected in a few years, emerged in a matter of months, which allowed several Russian regions to adapt to the new normality and take advantage of the opened-up opportunities for growth. Transformational and structural changes in the economy of the Russian regions caused rapid restructuring and adaptation of institutions to the new situation.

Integration of partners in a single, safe, and seamless digital institutional environment allows for the following:

- Improving the choice of strategic and tactical objectives and, accordingly, rationalizing the forms and methods of achieving the objectives;
- Increasing the efficiency of developing alternative solutions and criteria for assessing management tasks to choose the best option;
- Applying methods of data analytics that provide more reliable and in-depth forecast assessments of market-forming factors;
- Enhancing the analysis and monitoring of key indicators.

Any of the services of the region's digital ecosystem is equally accessible to its members. The quality of its services is supported by the guarantee of the coordinating platform.

The basic requirements for building digital ecosystems are as follows (Borisova & Bilczak, 2020):

- Resource and technological openness;
- Shared formation and use of infrastructure (including networks, cloud storage, and data centers);
- Creation of systematic management of the data lifecycle and its archival storage;
- Management of critical infrastructure risks and security;
- Implementation of a comprehensive strategy to integrate partners and economies of scale.

The application of the tools proposed in this research to quantify the qualitative features of the institutional environment of the regional ecosystem will allow for the following:

1. Identifying the institutions that need to be promptly corrected due to their significant impact on the institutional environment of the entire ecosystem;
2. Identifying institutional traps (North, 1997);
3. Calculating quantitative assessments of qualitative features of the institutional environment of the region;
4. Forming analytics of data on the state of the institutional environment of the region for the territorial authorities;
5. Determining the quality of the region's institutional environment and applying timely measures to prevent destructive situations.

In the next few years (5–7 years), super complex inter-branch and inter-regional ecosystems will account for up to 30% of revenues and 40% of all global organizations in the world. This fact indicates the feasibility of further comprehensive analysis of the digital transformation of the institutional environment in the regions.

Acknowledgments The study was supported by RFBR project No. 20-010-00141 "Formation of the institutional framework of the region's infrastructure in the digital economy."

References

Borisova, V. V. (2018). Formation of a digital ecosystem of logistics. In *Digital revolution in logistics: Effects, conglomerates, and growth points* (pp. 47–51). Rostov-on-Don, Rostov State University of Economics (RINH).

Borisova, V. V. (2020). Digital institutional transformation of region's logistics infrastructure. *Bulletin of the Rostov State University of Economics (RINH), 2*(70), 45–53.

Borisova, V. V., Yuan, H., & Tang, L. (2020). Development strategies of Aliexpress electronic platform in Russia. *Bulletin of the Rostov State University of Economics (RINH), 4*(72), 110–115.

Borisova, V., & Bilczak, M. (2020). Methodological aspects of the typology of regions and territorial formations. *Olsztyn Economic Journal, 15*(3), 229–239. https://doi.org/10.31648/oej.6540

Kirdina, S. G. (2001). The theory of institutional matrices: Toward the new paradigm. *The Journal of Sociology and Social Anthropology, 4*(1), 101–115.

North, D. (1997). *Institutions, institutional change, and economic performance* (A. N. Nesterenko Transl. from English). The Foundation for Economic Books "Nachala." (Original work published 1990).

Podshivalova, M. V. (2014). Institutional environment of Russia: Current status and trends. *Bulletin of Ural Federal University Series Economics and Management, 1*, 17–26.

Popkova, E. G., & Sergi, B. S. (2020). A digital economy to develop policy related to transport and logistics. Predictive lessons from Russia. *Land Use Policy, 99*, 105083. https://doi.org/10.1016/j.landusepol.2020.105083

Rogachev, A. F., Ostrovskaya, V. N., Natsubidze, A. S., Litvinova, T. N., & Yakovleva, E. A. (2018). Tools for sustainability management of socio-ecological systems in the globalizing world. In E. Popkova (Ed.), *The impact of information on modern humans* (pp. 241–247). Springer. https://doi.org/10.1007/978-3-319-75383-6_31

Sokolov, A. (2021). Where ecosystems are headed. Vedomosti. Retrieved from https://www.vedomosti.ru/technology/articles/2021/02/23/858980-ekosistemi (Accessed February 20, 2021)

Tambovtsev, V. L. (1997). Theoretical issues of institutional design. *Economics Matters, 3*, 82–94.

Trofimov, O. V., Zakharov, V. Ya., & Frolov, V. G. (2019). Ecosystems as a method of organizing the interaction of the production and services sectors in the context of digitalization. *Vestnik of Lobachevsky State University of Nizhni Novgorod. Series: Social Sciences, 4*(56), 43–55.

Sustainable and Alternative Energy

Improvement of Methodology for Assessing Sustainable Development of Oil Companies as a Basis for Ensuring Economic Security

Alsu M. Akchurina and Mamduda A. Khalikova

Abstract The paper aims to improve the methodological approach to assessing the sustainable economic development of oil companies, taking into account the specifics of production processes, the structure of the business model, economic performance, and economic security. Based on the study of theoretical and methodological literature, the authors propose a methodical approach for assessing the sustainable economic growth of the oil company, which allows to manage economic growth based on planning and analyzing key performance indicators. This work is relevant since the management system of enterprises in the oil industry, forming a single technological chain, includes different objects and subjects of management in organizational terms; these objects and subjects need a balanced, comprehensive development program. A comprehensive program for the development of enterprise will only be balanced if it is based on a modern system of managing production and economic mechanism, which will provide a reasonable rate of sustainable economic growth and, consequently, the economic security of the enterprise. This paper is devoted to studying this problem.

Keywords Business model · Assessment of sustainable economic growth · Key success factors · Key performance indicators · Economic security · Rate of economic growth

JEL Codes P47 · L53 · J28

1 Introduction

The problem of creating a methodology for assessing the sustainable economic growth of the enterprise is of scientific and practical interest to a considerable number of specialists [2-4; 6; 9-11]. The theory of the use of sustainable economic growth

A. M. Akchurina · M. A. Khalikova (✉)
Ufa State Petroleum Technological University, Ufa, Russia

A. M. Akchurina
e-mail: alsu-akchurina@list.ru

E. G. Popkova and B. S. Sergi (eds.), *Geo-Economy of the Future*,
https://doi.org/10.1007/978-3-030-92303-7_85

rates in management has been actively researched in the literature since the late 1970s. The model developed by Robert K. Higgins [7; 13] shows the relationship between the level of the company's economic growth and financial performance.

Nowadays, there is no single comprehensive model for assessing the sustainable economic growth of oil and gas companies, which allows considering the specifics of production processes, the structure of the business model, and economic performance indicators (Amouzesh et al., 2011).

The vertically integrated business model of oil companies and the peculiarities of production, distribution, and sales of products affect the performance of companies, which provides opportunities for a wide range of areas to improve economic growth. To assess the rate of economic growth of an oil company and determine its sustainability, it is necessary to adapt classical methods of assessing sustainable growth to the specifics of oil companies. This paper considers two main business segments—"Exploration and production" and "Refining, commerce, and logistics." The research object is the economic growth of the oil company over a long period of activity.

The research aims to create an improved methodology for assessing the sustainable economic growth of an oil company, which will allow it to manage its level based on planning and analyzing key performance indicators.

The implementation of the research aim required the solution of the following tasks:

- To study the existing methods for assessing the level of economic growth of industrial enterprises;
- To analyze the peculiarities of the functioning of oil companies;
- To investigate the factors affecting the level of economic growth of the oil company;
- To form a comprehensive system for assessing economic growth based on the planning of key performance indicators (KPIs) to ensure economic security in an unstable economical environment.

2 Materials and Methods

The analysis of the actual level of the achievement of economic growth in the enterprise allows one to identify growth reserves in all areas of activity to increase the investment attractiveness and competitiveness of the enterprise, that is, the growth of its market value.

The economic growth of the enterprise can be considered in two aspects:

- In the microeconomic sense, many authors [12; 15] understand the economic growth as an increase in volume indicators over time;
- In the macroeconomic sense, economic growth is identified with the increasing rate of reproduction of goods and services in the dynamics (Tonkikh et al., 2012).

Many authors understand the economic category of sustainable economic growth of the enterprise as the progressive dynamics of change in economic indicators over a long period and the balance between the growth rate of volume indicators and basic resources of the company (Tonkikh et al., 2012).

In microeconomic terms, economic growth can be assessed through the dynamics of changes in economic indicators (e.g., revenue, the amount of equity, and net profit for a certain analytical period). Several indicators, such as labor productivity, indicators of production efficiency, etc., are used as private indicators.

Achieving sustainable economic growth for any company is possible through extensive and intensive development. The extensive way of development of oil companies assumes an increase in oil and gas production through the development of new fields, as well as the growth of oil and gas products through the expansion of oil refining capacity. Achieving sustainable economic growth through intensive development involves improving the technological processes of production based on the modernization of equipment and introducing new production and processing technologies.

Thus, the level of economic growth of the company can be measured through the integral index. The methodology for calculating the integral index should be based on analyzing the rates of change of economic indicators of extensive and intensive ways of development. The methodology should consider the company's external and internal conditions, the analysis of the dynamics of the structure of borrowed and own funds as sources of funding production. It is necessary to analyze the growth rate of borrowed and own funds and bring them into correlation with the growth rate of sales when planning the growth rate of sales and indicators of production efficiency.

The rate of economic growth of an oil company depends on the following factors:

- State of the external economic environment. If the economic environment is stable and favorable, the level of sustainability of economic growth can be predicted for the long term. If the economic environment is unstable, the level of planned performance indicators requires constant monitoring and adjustment;
- Level of state regulation of the company's activities. Active and passive regulation of companies through laws and regulations in the economic development of the country can have different effects on the output of its activities;
- Degree of development of the legislative framework in the field of subsoil use of the Russian Federation also affects the volume indicators of companies;
- Level of the company's production and resource potential. The level of economic growth of the oil company will be positively influenced by:

- Growth and structure of hydrocarbon reserves exceeding the volume of production;
- Quality of oil produced;
- Geography of mining and processing assets;
- Logistics;
- Modernization of production;
- Refining depth;

– Distribution.

- Degree of technological and technical improvement of the company's production processes. Sustainable economic growth of the company depends on the implementation of technological developments, which affects the level of costs, indicators of production efficiency, and the provision of own funding sources;
- Management quality. The quality of the company's management allows it to adapt quickly to the current manageable and unmanageable challenges in the country's economy and the world.
- Organizational structure of the business. An effective business model aimed at increasing the company's value will also increase the level of sustainability of the company's economic growth.
- Corporate culture. A high corporate culture manifests value orientation associated with sustainable development, contributing to self-learning in the company.

A comprehensive comparative analysis of the factors of economic growth of oil companies is a methodological and informational basis, the results of which reveal causal relationships between the rate of change of these factors and the rate of economic growth of the oil company.

Table 1 presents the main internal and external factors, which directly impact the dynamics of economic growth, taking into account the specifics of the enterprises in the oil and gas industry.

Qualitative and quantitative analysis of these factors will determine the level of economic growth of the oil company, its place among the competitors, and key success factors in the company's development. It should be noted that the list of key success factors may change due to the variability of the external and internal

Table 1 List of internal and external factors that directly impact the assessment of economic growth of the oil company

Internal factors	External factors
Climatic and geological factors of the development of hydrocarbon field	Unstable geopolitical business environment
Low refining depth	Volatility of prices on oil and oil products on foreign and Russian markets
Schemes of financing working capital High interest rates for credits	Variability in exchange rates and inflation rates
Utilization of production capacity in oil refining	Taxation system in the oil and gas industry
Physical and moral depreciation of fixed assets	Changes in tariffs of natural monopolies
Lack of labor resources of demanded professions and qualifications	Increase in electricity prices
Underdevelopment of production infrastructure and pipeline system	Reduced share of foreign investment in investment projects in the oil and gas industry

Source Compiled by the authors

environment of the oil company. Table 2 presents a general list of key success factors (KSF), which can influence the rate of the company's economic growth.

Table 2 List of key success factors for oil companies

Key success factors (KSF)	Deciphering the key success factors
KSF determining the manufacturability of production processes:	• Scientific and methodological support of the integrated structure of the company's production and technological processes; • Degree of innovation in production and technological processes; • Degree of application of existing high-tech industries
KSF determining the efficiency of production processes:	• Saving fixed costs in the production and marketing of petroleum products; • Saving fixed costs of exploration and production of hydrocarbons; • Increasing the share of high-quality refined products; • Increasing the utilization rate of production facilities in oil refining; • Geographical location of the company, which reduces logistics costs; • Access to skilled labor
KSF determining the effective sale of products:	• Creation of an effective wholesale and retail sales segment of the company; • Availability of oil and oil product marine terminals; • Development of exchange trade in oil and petroleum products
KSF leading to the growth of professional skills:	• Increasing the level of expertise by increasing the cost of staff training on advanced training courses; • Obtaining new qualifications based on increased costs for programs of additional vocational training
KSF determining marketing effectiveness:	• High qualification of employees of the company's sales segments; • Improvement of the incentive system for regular customers of the retail network; • Expanding the range of petroleum products;
KSF determining the effectiveness of corporate governance:	• Improving the internal communications system; • Improving the employee motivation system based on the introduction of modern methods of staff stimulation; • Increasing the effective management of the company based on a humanitarian and mechanistic model of diagnostics

(continued)

Table 2 (continued)

Key success factors (KSF)	Deciphering the key success factors
Other KSF:	• Creating or maintaining a high level of company image; • Increase in social and charitable projects; • Possibility of diversifying the sources of financing of the company's investment activities; • Availability of licensed activities

Source (Gershun & Gorsky, 2005)

The composition of key success factors is influenced by the achievement of strategic development goals of the company through the level of achievement of KPIs and the unstable external and internal environment of the oil company (Kaplan & Norton, 1996). Therefore, the methodology of evaluating sustainable economic growth should be developed, considering the achievement of strategic objectives of the oil company development, quantified through a system of KPIs. This task requires additional research.

The goals of long-term development, which ensure the sustainable economic growth of oil companies, are to maintain leadership in the main indicators of operating activities, reach the level of world companies on performance indicators, improve competitiveness, increase oil production, and increase the value of the company.

In this paper, the authors propose an improved approach to assessing the sustainable economic growth of an oil company for the business segments "Exploration and production" and "Refining, commerce, and logistics," taking into account the development strategy of the oil company.

This approach was based on the Higgins economic growth model (Higgins, 2007):

$$SGR = \frac{(Eq_0 + NewEq - Div) \times \left(1 + \frac{\Delta}{Eq}\right) \times \frac{S}{A} \times \frac{1}{S_0}}{1 - \left[\frac{Np}{S} \times \left(1 + \frac{\Delta}{Eq}\right) \times \frac{S}{A}\right]} - 1 \tag{1}$$

where

SGR—economic growth rate;

NewEq—the amount of equity;

Div—the amount of annual dividends;

$\frac{S}{A}$—the ratio of sales volume to total assets;

Eqo—initial equity;

S_0—initial sales.

$\frac{NP}{S}$—return on sales ratio (ratio of net profit to sales);

$\frac{\Delta}{Eq}$—debt to equity ratio.

SGR—we take this indicator as the level of economic growth, which can be calculated based on the analysis of KPIs. In our opinion, this model should be complemented by Van Horn's economic growth model (Van Horne et al., 2003). In this representation, the economic growth of the company corresponds to the maximum growth rate of sales, subject to the following inequalities (Horne et al., 2003):

$$100\% < Temp(A) < Temp(SR) < Temp(NI) \quad (2)$$

where
A—total assets;
SR—sales revenue;
NI—net income.

The authors propose to analyze the degree of implementation of key KPIs for each business segment ("Exploration and production" and "Refining, commerce, and logistics") when assessing the level of economic growth of an oil company (Fig. 1). This will allow comparing the correspondence between the rate of economic growth in the company and the rate of KPIs fulfillment by business segments and the company. In turn, this correspondence will allow to identify bottlenecks in the KPIs fulfillment and develop programs for achieving a balanced rate of economic and KPIs growth.

Thus, the achievement of sustainable economic growth of the oil company relies on the analysis of the fulfillment of KPIs by business segments and the whole company, which will allow for planning long-term management decisions for achieving consistency between the growth rate of volume indicators and financial resources (Wet, 2004).

The scheme for assessing the level of economic growth of an oil company consists of the following steps:

1. Study of the strategic goals of an oil company, which are measured through the KPIs system of an oil company;
2. Based on the analysis of the company's external and internal environment, the key success factors of the oil company are identified, which are the basis for planning KPIs for the business segment and the whole company;
3. Definition of quantitative key performance indicators of the oil company for the business segments ("Exploration and production" and "Refining, commerce, and logistics") and the whole company with further analysis of the real level of KPIs fulfillment;
4. Based on the results of the analysis of the achievement of the planned KPIs by business segments and the whole company, the rates of sales and assets growth, revenues, and profits are assessed for inequality correspondence for business segments and the whole company;
5. Calculation of the actual level of economic growth of the company;
6. Determination of reasons for non-fulfillment of KPIs and development of recommendations for KPIs adjustment separately for business segments and the whole company;
7. Proposals to adjust KPIs that will allow reconsidering inequalities in the growth of assets, revenues, and profits for segments and the whole company. Moreover,

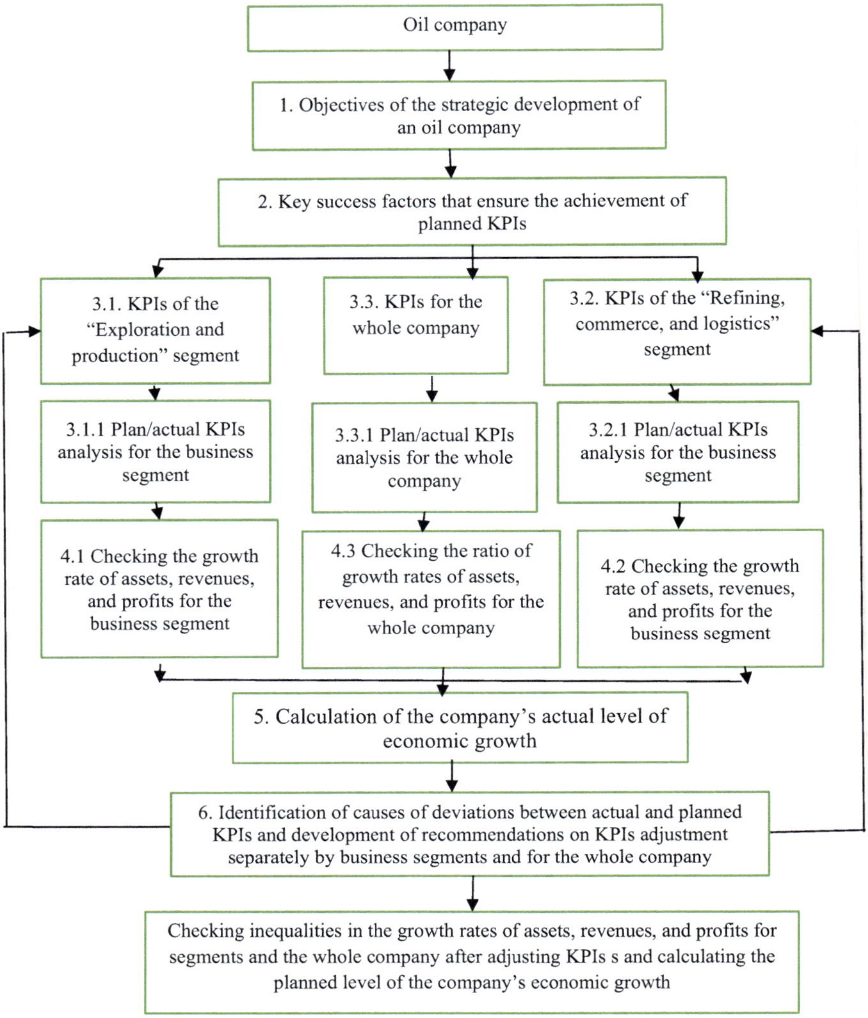

Fig. 1 Scheme for assessing the level of economic growth of an oil company. *Source* Compiled by the authors

they will allow calculating the corresponding planned level of economic growth of the company.

Thus, the proposed algorithm links the assessment of the level of economic growth of the company with the level of achievement of the planned KPIs. In turn, KPIs depend on the variability of the impact of KSF as a result of external and internal risks of the company. Therefore, the proposed algorithm allows for the development of measures to improve the level of economic growth and achieve a balanced ratio between economic growth and the growth of assets, revenues, and profits for segments and the whole company as a whole.

3 Results

In this paper, the authors assess the level of economic growth of PJSC "Rosneft" for 2018–2019. Rosneft's strategic goal is to become one of the top four global oil and gas companies in the field of industrial safety, occupational health, and environmental protection (PJSC "Rosneft", n.d).

The list of KPIs available on Rosneft's website includes the following (PJSC "Rosneft", n.d):

- Increase of the return on average capital employed (ROACE);
- Increase of the volume of hydrocarbon production;
- EBITDA;
- Leverage ratio (Net debt / EBITDA);
- The ratio of Rosneft's total shareholder return to the industry average total shareholder return for Russian companies;
- Decrease in costs of the reporting period relative to the previous period under comparable conditions;
- Indicators on the implementation of the innovative development program.

The authors analyzed the level of KPIs fulfillment and the proportions of asset, revenue, and profit growth in the business segments "Exploration and production" and "Refining, commerce, and logistics." The analysis of the main proportions calculated by formula 2 allowed the authors to conclude that to achieve KPIs, the company must increase oil production in the segment "Exploration and production" in 2019 by intensifying production and reducing production costs.

The analysis of proportions based on 2018–2019 results determined that the proportion ratio of asset, revenue, and profit growth in the segment "Refining, commerce, and logistics" are met. This is due to the fact that PJSC "Rosneft" increased the volume of sales of petroleum products for export in 2019. Moreover, the volume of sales in the Russian market also increased by 16% compared to 2018.

To maintain the proportions of sustainability for the segment "Refining, commerce, and logistics," the authors recommend the following:

- Increase the depth of oil refining and the yield of light petroleum products;
- Reduce the cost of refined products;
- Expand marketing channels.

For the whole company PJSC "Rosneft," it is proposed to implement technological projects in business segments aimed at reducing variable costs for oil production and refining.

The authors assess the level of sustainable economic growth for 2018–2019 according to the proposed algorithm. The results are shown in Table 3.

Table 3 Calculation of PJSC "Rosneft" economic growth for 2018–2019

Indicator	Actual		
	2018	2019	Growth rate, %
Initial volume of sales, billion rubles	6011	8238	37.04
Initial amount of equity capital, billion rubles	3619	4053	11.99
Amount of annual dividends, billion rubles	274	354	29.20
Sales volume, billion rubles	8238	8676	5.32
Total assets	13,163	12,950	−1.62
Amount of borrowed capital, billion rubles	8486	7798	−8.11
Amount of equity capital, billion rubles	4053	4516	11.44
Net profit, billion rubles	549	708	28.96
Level of economic growth	6.31	2.12	−66.40

Source PJSC "Rosneft"(PJSC "Rosneft", n.d.)

Thus, the actual economic growth for 2019 is down 66.4% compared to 2018. This is due to an excess of expenses in the "Corporate governance" business segment, expenses related to acquiring new production assets, the growth of tariffs of natural monopolies, and wage indexation.

Using the proposed methodology, we can assess the correlation between the growth rate of assets, revenues, and profits for the whole company and the level of economic growth:

$$100\% > 98.4\% < 105.32\% < 128.9\%$$

The ratio of the rates of asset growth decreased due to a decrease in the share of debt capital in the capital structure by 8.11%, which affected the growth rates of revenue and net profit.

4 Conclusion

The analysis revealed that the decrease in the amount of borrowed funds and large dividends does not correspond to assets and sales growth rate. To achieve the strategic goals, the company solved several problems simultaneously—growth in sales, reduction of the use of borrowed capital, and a high level of dividends. If we consider each task separately, it corresponds to the company's development goals. Nevertheless, the direction of their growth rates does not match each other. This is the consequence of not meeting many of the KPIs for the business segments and the company.

The calculations confirm the main conclusion that operational control over the implementation of the planned KPIs of the oil company and a timely program to adjust them will allow the oil company to maintain high rates of economic growth in the long term.

Thus, the proposed algorithm for calculating the level of economic growth of an oil company, based on the assessment of KPIs for business segments and the whole company, will determine the level of achievement of long-term goals of the company, identify the growth reserves of economic growth, and ensure economic security in a volatile economic environment.

References

Amouzesh, N., Moeinfar, Z., & Mousavi, Z. (2011). Sustainable growth rate and firm performance: Evidence from Iran stock exchange. *International Journal of Business and Social Science, 23*(2), 249–255.

De Wet, J. H. (2004). Chapter 6: Growth sales and value creation terms of the financial strategy matrix. Retrieved from https://repository.up.ac.za/bitstream/handle/2263/27371/06chapter6.pdf?sequence=7

Gershun, A., & Gorsky, M. (2005). Balanced management technology. Moscow, Russia: Olymp-Bussiness.

Higgins, R. C. (2007). Analysis for financial management: Tools for making business decisions (A. N. Svirid Transl. from English). "Williams" Publishing House. (Original work published 1983).

Van Horne, J. C., & Wachowicz J. M., Jr. (2003). *Fundamentals of financial management*. In Yu. M. Teslenko, S. A. Lukin, S. A. Dymov, M. L. Semanenko, O. V. Skachkova, E. A. Maksimov, … D. N. Fedotov Transl. from English; Ya. V. Sokolov (Eds.), Finance and Statistics. (Original work published 1971).

Kaplan, R. S., & Norton, D. P. (1996). *The balanced scorecard translating strategy into action* (1st ed.). Harvard Business School Press.

Kim, A. (n.d.). Management system. Retrieved from http://www.lessonslearned.ru/systema-upravleniya-predpriyatiem

PJSC "Rosneft" (n.d.). Official website. Retrieved from https://www.rosneft.ru/

Tonkikh, A. S., Ostaltsev, A. S., & Ostaltsev, I. S. (2012). The modeling of enterprise economic growth: The preconditions for alternative models elaboration. *Management of Economic Systems, 9*(45), 1–15.

Methodology for the Formation of Oil Company's Product Program in the Context of Sustainable Development

Vera V. Biryukova, Evgeniy V. Evtushenko, Marina M. Gayfullina, Gylnara Z. Nizamova, and Gulzira B. Estakova

Abstract The article proposes a methodology for the formation of a program for the development of a range of petrochemical products at an oil company. An algorithm for the formation of a program for the development of the range of petrochemical products of vertically integrated oil companies in the retail segment has been compiled. An analytical matrix of assortment strategies "production conditions—market attractiveness" has been developed. For the purposes of the study, petrochemical product groups were formed that are recommended for sale in the roving segment of the oil company: petrochemicals used for maintenance and repair of vehicles; petrochemicals used for auto cosmetic purposes; hygienic protection products based on petrochemicals; polymer products. Based on the example of the oil company Public Joint Stock Company Oil Company Rosneft, the attractiveness was assessed and the priority directions for the development of petrochemical products were selected for the company's own retail. It has been established that the following product groups are promising directions for the development of petrochemical industries for sale through grocery retail for the oil company PJSC Oil Company Rosneft: petrochemicals used for maintenance and repair of vehicles, as well as petrochemicals based on polymers. Proposals have been developed for the development of new and areas of improvement of existing petrochemical products for the oil company PJSC Oil Company Rosneft.

Keywords Auto chemistry · Assortment · Assortment matrix · Assortment strategy · Petro chemistry · Petrochemical industry · Oil company · Product program · Strategy · Sustainable development

V. V. Biryukova (✉) · E. V. Evtushenko · M. M. Gayfullina · G. Z. Nizamova
Ufa State Petroleum Technological University, Ufa, Russia

M. M. Gayfullina
e-mail: marina_makova@list.ru

G. Z. Nizamova
e-mail: gulya182004@list.ru

G. B. Estakova
Kazakh-British Technical University, Almaty, Kazakhstan

E. G. Popkova and B. S. Sergi (eds.), *Geo-Economy of the Future*,
https://doi.org/10.1007/978-3-030-92303-7_86

JEL Codes D20 · D41 · M31

1 Introduction

One of the components of the company's sustainable development is its relationship with consumers. At the same time, the company needs, on the one hand, to satisfy the needs of consumers as fully as possible, and on the other hand, to increase the efficiency of its activities. Important components in matters of relationship with consumers are a correctly defined target audience of consumers of the company's products and a well-chosen range of products. In these conditions, the urgent tasks for the company are the formation of an effective assortment policy and an effective product program of production.

The task of assortment development is an integral element of the company's assortment strategy. At the same time, high added value in the market often doesn't depend on the technologies used in production, on quality, cost, but depends on competent marketing, choice of a market niche, positioning, branding, communication channels with consumers, etc. As a rule, a highly profitable product isn't the one that is better, but the one for which consumers are willing to pay a higher price.

The development of petrochemical industries is becoming an important segment of the oil company's development in the context of unstable world prices and unstable demand for oil and petroleum products to ensure the sustainable development of an oil company. The current stage in the development of petro chemistry is characterized by the rapid expansion of the list of products of low-tonnage and high-tech chemistry (Kryukov & Shmat, 2021).

The main purpose of the article is to propose a methodology for the formation of an oil company's product program using the example of the petrochemical segment. At the same time, the priority is given to the sale of petrochemical products through the company's own retail.

The subject of the research is the problems of effective management of the range of products of an oil company in the development of the petrochemical segment.

The object of this study is the development of a range of petrochemical products under its own brand, sold through the oil company's own retail network (filling stations and shops).

The methodological basis of the study is the methods of analyzing the assortment of products, grouping, generalization, and expert assessments.

The most common methods for analyzing a company's assortment are:

1. McKinsey Model (McKinsey 7 s Model)—allows making informed decisions regarding the expansion of production volumes and also its modification, taking into account the seven key elements of the company (Klundert & Potters, 2021; McKinsey Global Institute, 1996).
2. M. Porter Model (Porter's Five Forces)—allows achieving competitive advantages in the market by developing an effective range of products, taking into account the company's position in the market [9; 17].

3. Matrix Boston Consulting Group (BCG matrix)—allows prioritizing products manufactured and determining the degree of demand for a particular product (Boston Consulting Group, 1970). The BCG matrix is used for product positioning and strategic planning (Stern & Deimler, 2009). There are various modifications of this method in the scientific literature, in particular, the using of the BCG matrix in combination with the Delphi approach (Myllylä & Kaivo-oja, 2015).
4. Methods of ABC-analysis and XYZ-analysis.

The method of ABC-analysis of the product range allows identifying the most significant in terms of sales volume (Pawelek et al., 2017) and also classifying products based on the principle of mutual similarity of products (Li et al., 2021).

There are a number of proven marketing tools that allow a company to maintain a leading position in a highly competitive environment, including through the right product strategy (Mushketova & Fedorova, 2021), taking into account consumer behavior (Frederiks et al., 2015), and using various planning methods (Armstrong & Brodie, 1994).

It should be borne in mind that the industry factors in which the company operates has a significant impact on the methodology for forming the product program of the enterprise.

Certain industry factors influencing the formation of the product program of an oil company are touched upon in scientific research. Among the aspects considered are the factors of competition in the oil industry (Gong, 2020), production conditions and the state of fixed assets of oil companies (Gaifullina et al., 2017), the values of operational efficiency indicators (Eller et al., 2021), etc.

The analysis showed that, despite a sufficient number of publications on assessing the attractiveness of the market and the competitiveness of products, the issues of taking into account industry specifics, in particular the specifics of the activities of oil companies, including from the standpoint of the attractiveness and competitiveness of petrochemical products, remain insufficiently studied.

2 Methodology

Within the framework of this study, a methodology for the formation of an oil company's product program in the petrochemical segment is proposed.

It will be necessary to perform a number of stages that provide scientific validity and a methodological sequence of actions and decisions to form a product program of an oil company in the petrochemical segment. The proposed algorithm for the formation of an oil company's product program in the petrochemical segment is shown in Fig. 1.

The formation of an oil company's product program in the petrochemical segment begins with a strategic analysis of the company's external environment.

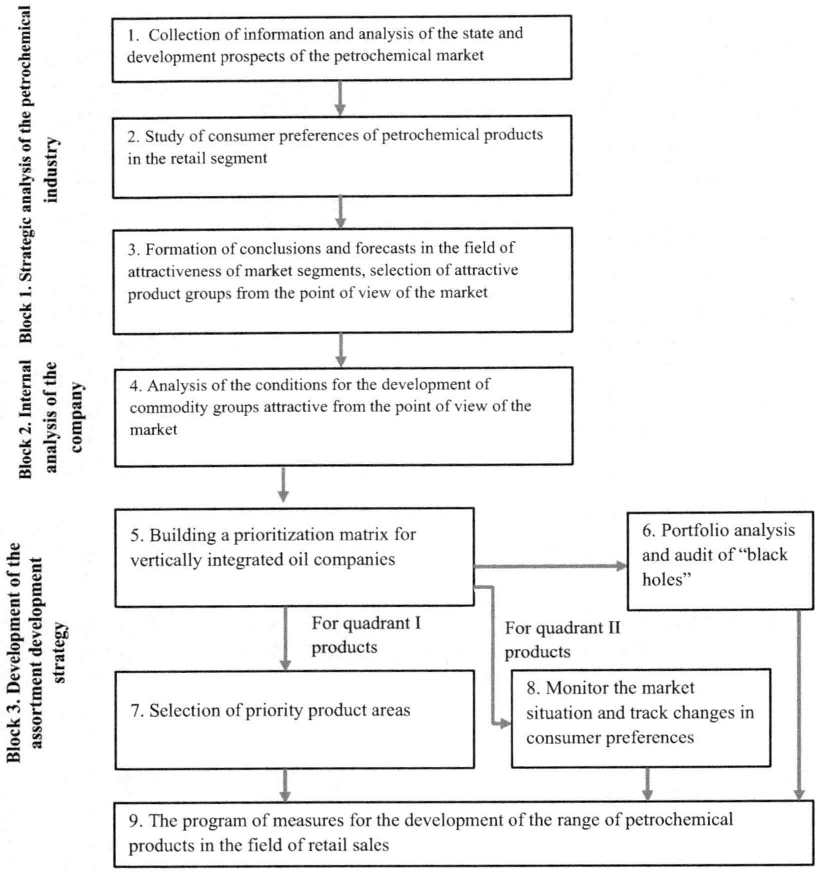

Fig. 1 Algorithm for the formation of the product program of an oil company in the petrochemical segment. *Source* compiled by the authors

At this stage, the analysis of the state and development prospects of the petrochemical market in the retail segment is carried out (the main method of collecting information is desk research).

Next, consumer preferences in the petrochemical retail segment are explored based on field research.

Within the framework of this stage, it's proposed to firstly carry out a primary test of research hypotheses and identify insights in a focus group, and then conduct a secondary (expanded and corrected) test of research hypotheses using a questionnaire).

At the stage of drawing conclusions and forecasts in the area of attractiveness of market segments, highlighting attractive product groups from the point of view of the market, it's necessary to determine the opportunities and threats from the far and near external environment in the petrochemical market in the retail segment (Burenina et al., 2018).

Further, the quality of conditions for the development of petrochemical areas in the company is analyzed and also a portfolio analysis of the company's petrochemical products in the retail segment is carried out.

Based on the results of the analysis, recommendations for the development of product areas based on a model of strategic priorities are worked out.

The proposed model for prioritizing product areas is formed depending on two factors: the resources of the enterprise and the attractiveness of the market segment (Fig. 2).

The proposed principle of prioritizing the product directions of an oil company is similar to the McKinsey matrix and the "Market share - market growth" model. But in contradistinction to the "Market Share - Market Growth" model, which was proposed by the Boston Consulting Group, new product profiles are superimposed on this matrix. A set of actions is proposed for each product group in the advisory quadrants.

The characteristics of the quadrants of the product directions prioritization model are shown in Table 1.

Based on the proposed model for prioritizing product areas, it's possible to form measures for the development of petrochemical products in the retail segment.

Let's consider how the assessment is formed along the axes of the product areas prioritization matrix.

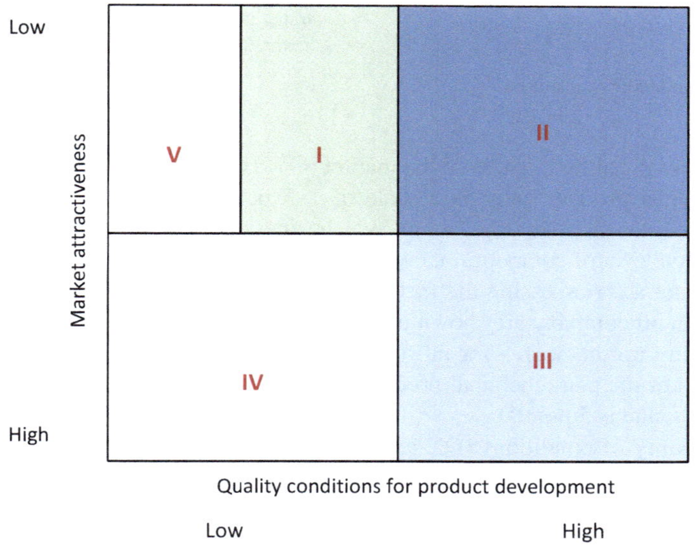

Fig. 2 Matrix of prioritization of product directions of an oil company. *Source* compiled by the authors

Table 1 Brief description of the product line prioritization model

Quadrant	Product line name	Explanation
I	Promising directions for the development of production	Products with the maximum potential for import substitution. It's necessary to identify the factors constraining the quality of production conditions and to develop measures to eliminate them, to include them in the assortment strategy. These may include: technical, technological limitations, low level of competence, etc
II	Developed directions	Products falling into this area have the most favorable environment for achieving success in the short term
III	Unpromising	Products for which it's necessary to monitor the market situation and track changes in consumer preferences. In case of an increase in market attractiveness, immediately introduce into the range
IV	Moderately promising for the development of production	Products for which it's necessary to monitor changes in market demand and should be included in the production development plan in the case of high marginality
V	Unpromising directions of production development	Products with a minimum potential for import substitution, or requiring significant investment development costs (the payback period is longer than the horizon of a strategic assessment of the attractiveness of the market)

Source compiled by the authors

To assess the attractiveness of the market (OY axis), it's proposed to evaluate the petrochemical product lines on a scale of 1–5 points according to the following criteria: market size (for the company's business), market growth rate, product profitability, level of competition, degree of seasonality/demand cyclicality.

The criteria for assessing the factors of attractiveness of petrochemical product lines for an oil company are shown in Table 2.

Summing up the scores for all 5 criteria, the authors determine the attractiveness index of the petrochemical product line (the maximum value is 25 points; the minimum value is 5 points).

The quality of conditions (OX axis) for the development of the petrochemical product line of an oil company is also assessed using a special index, which reflects such factors as: the company's relative market share, product quality, level of competence, technology availability, and production capacity. Factors are assessed on a scale from 1 to 5 points; the scales for assessing factors are shown in Table 3.

Table 2 Criteria for assessing the attractiveness of petrochemical product lines for an oil company

Factor	1 point	2 points	3 points	4 points	5 points
Market size (for the company's business)	Below the markets of the company's existing products, there is no growth potential	Below the markets of the company's existing products, the growth potential is doubtful	Matches the markets of the company's existing products, there is potential for growth	Corresponds to the markets of the company's existing products, there is potential for growth	Exceeds market share in the industry, there is potential for growth
Market growth rate	Decline in market volume	Stagnation	Below the industry average	Corresponds to the industry average	Above the industry average
Profitability	Significantly below the industry average	Slightly below the industry average	Corresponds to the industry average	Above the industry average	Significantly above the industry average
Competition level	Monopoly	Low level (oligopoly)	Tighter tendency	Average level	High level
Seasonality/cyclicality of demand	Explicitly expressed (once a year, duration less than a month)	Expressed (once a year, duration from 1 to 3 months)	Expressed (several times a year, or lasting at least 3 months)	There is an hourly cycle	Seasonality is absent

Source compiled by the authors

Summing up the scores for all 5 criteria, the authors determine the quality index of conditions for the development of the petrochemical product line (the maximum value is 25 points; the minimum value is 5 points).

The assessment of the factors that determine the attractiveness of petrochemical product lines and the quality of conditions for the development of the product line is carried out by experts on the basis of a strategic analysis of the industry and the company in which it's planned to develop petrochemical production, as a result of interviews with the company's specialists.

Thus, a total score is calculated for each product, the model accepts the equivalence of factors, in accordance with the company's development strategy, the weight values of the factors can be revised.

The proposed approach to the selection of petrochemical product lines based on the needs of target customers allows the company to determine a product position that differs from and surpasses the position of competitors' products.

Table 3 Criteria for assessing the conditions for the development of the petrochemical product line of an oil company

Factor	1 point	2 points	3 points	4 points	5 points
The relative share of the company in the market of the product group	Less than 0.3	0.3–1	Equal 1	More than 1	More than 2
The ability to provide the required quality of goods (consumer requirements)	Can't provide basic needs	Provides basic needs	Provides basic and non-essential needs	Provides basic, non-essential needs and "hidden requirements"	Provides basic, non-essential needs and satisfaction engines
Competence level	No specialists, no learning ability	No learning ability, but there is an opportunity to attract	There are no specialists, but there is the ability to learn	There are no specialists, but there is the ability to learn, there is the ability to attract	There are specialists
Availability of technologies	No, there is no opportunity to purchase	No, but there is an opportunity to purchase imported technology (not included in the list of prohibited)	No, but there is the ability of import substitution	Yes, but there is the ability to import substitution/acquisition (not included in the list of prohibited)	There are technologies
Availability of production facilities	No, construction of new facilities is required	Modernization required (provision in the medium time)	Minor modernization required (provision in the short time)	Existing capacities meet product requirements	Existing capacity exceeds requirements, there is potential for growth

Source compiled by the authors

3 Results

Approbation of the proposed approach was carried out on the example of the oil company PJSC Oil Company Rosneft.

Based on the analysis of the foreign and Russian experience of the retail market of oil companies in petrochemical products under their own trademark, a possible list of petrochemical products for sale through Rosneft's own retail company has been formed:

1. Petrochemicals used for the maintenance and repair of motor vehicles;
2. Petrochemicals used for auto cosmetic purposes;

3. Consumer petrochemical products;
4. Other petrochemical products.

The results of expert assessments of the factors of attractiveness of petrochemical product lines in the retail segment for various products and also factors that determine the quality of conditions for the development of various petrochemical product lines in Rosneft, are shown in Table 4.

Taking into account the conducted research, the authors formed a priority list of high-margin petrochemical products for sale at gas stations and in stores of PJSC Oil Company Rosneft. The prioritization matrix of product areas for Rosneft is shown in Fig. 3.

Product groups (quadrant I) are promising areas for the development of petrochemical industries for grocery retail in Rosneft:

(1) Petrochemicals used for auto cosmetic purposes;
(2) Hygienic protection products based on petrochemicals.

The main limiting factor for the development of these industries in Rosneft is technological limitations, lack of experience in these industries.

The developed areas for the development of petrochemical industries for grocery retail in Rosneft are the following product groups (quadrant II):

(1) Petrochemicals used for the maintenance and repair of motor vehicles;
(2) Petro chemistry based on polymers.

Prospective and developed areas within each of the petrochemical groups, as well as the recommended assortment strategies of petrochemical product areas for Rosneft, are shown in Table 5.

Work on goods of priority areas is that to identify factors that restrain the quality of production conditions and develop measures to eliminate them.

Products classified as developed areas have the potential for development, for this it's necessary to investigate consumer "pains" and consider the possibility of changing the sales technology, which may be the subject of further research.

4 Conclusion

In modern conditions, the petrochemical segment is becoming an important segment for oil companies from the standpoint of development. At the same time, the development of petrochemicals has a positive effect on the development of the country's economy as a whole. So, according to studies by M. Maitah and A. J. Bassam, there is a direct relationship between the export of petrochemical products and economic growth (Maitah & Bassam, 2021).

Undoubtedly, the price factor plays an important role in the choice of the product by the buyer, which is confirmed by the research of P. Verleger (Verleger, 2021). However, in the future, product marketing strategies will play a special role in the petrochemical segment of the market.

Table 4 Assessment of factors of attractiveness and conditions for the development of various product lines of petro chemistry in PJSC Oil Company Rosneft

Name of product	Cumulative factor score (maximum 25)	Cumulative condition score (maximum 25)
1 Group "Petrochemicals used for the maintenance and repair of vehicles":	19.6	14.7
1.1 Motor oils, for engines	21.5	24.0
1.2 Brake fluids for hydraulic transmissions, antifreezes, fluids for cooling systems (radiator)	20.0	22.5
1.3 Grease for the car	16.5	19.5
1.4 Adhesives and sealants	21.5	10.0
1.5 Anti-gravel coatings to prevent corrosion	20.5	10.5
1.6 Rust converters	20.5	10.0
1.7 Tire repair products (polymer based sealing materials)	20.0	10.0
1.8 Anti-freeze liquid for the car	18.0	15.5
1.9 Air conditioner and ventilation system cleaners	18.0	10.5
2 Group "Petro chemistry used for auto cosmetic purposes":	20.6	11.8
2.1 Tire care products	20.0	10.5
2.2 Glass care products	19.5	13.0
2.3 Polishing and cleaning products for the car	22.0	12.5
2.4 Petro chemistry for washing and detailing cars	21.0	12.5
2.5 Deodorizing and aromatizing agents	20.5	10.5
3 Group "Consumer products of petrochemicals":	22.2	10.3
3.1 Soaps and detergents for personal use	20.5	10.5
3.2 Hand antiseptic	24.0	10.5
3.3 Wet wipes for hands	23.5	10.0
3.4 Plasters and dressings	20.5	10.0
3.5 Oral and dental hygiene products	21.5	10.5
3.6 Household and rubber gloves	23.0	10.0

(continued)

Table 4 (continued)

Name of product	Cumulative factor score (maximum 25)	Cumulative condition score (maximum 25)
4 Group "Other petrochemical products" (petrochemicals based on polymers)	20.0	19.5

Source compiled by the authors

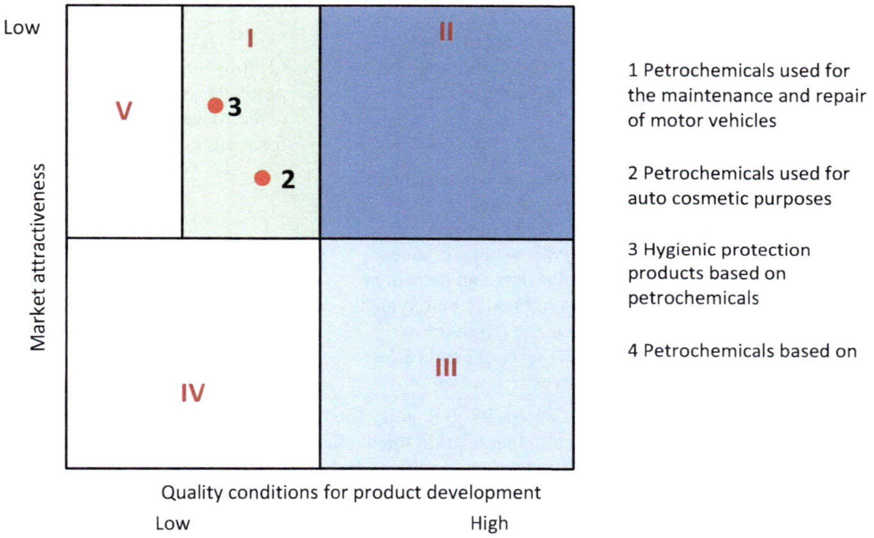

Fig. 3 Matrix of prioritization of product directions for PJSC Oil Company Rosneft. *Source* compiled by the authors

Today in modern economic science there is no unified, generally accepted approach to the formation of the company's product program. Various concepts, typologies, approaches are presented in the literature. The approach proposed in the article to prioritize the product lines of development of an oil company in the petrochemical segment is similar to the McKinsey matrix and the "Market share - market growth" model. But in contradistinction to the "Market Share - Market Growth" model proposed by the Boston Consulting Group, this matrix is superimposed on the profiles of new product solutions. A set of actions is proposed for each product group in the advisory quadrants.

The formation of the oil company's product program according to the proposed approach involves the construction of a matrix based on two attributes (factors) using a system of horizontal and vertical coordinates of the economic space, which express the quantitative or qualitative characteristics of the corresponding market parameters.

Table 5 Assortment strategy of petrochemical product lines of PJSC Oil Company Rosneft

Commodity group	Promising directions	Developed directions
Petrochemicals used for the maintenance and repair of motor vehicles	(1)Adhesives and sealants; (2)Anti-gravel coatings to prevent corrosion; (3)Rust converters; (4)Polymer-based sealing materials; (5)Cleaners for air conditioning and ventilation systems	(1)Motor oils, for engines and other systems; (2)Brake fluids for hydraulic transmissions, antifreezes, fluids for cooling systems (radiator); (3)Grease for the car; (4)Antifreeze liquid for a car;
Petrochemicals used for auto cosmetic purposes	(1)Tire care products; (2)Deodorizing and aromatizing agents	(1)Glass care products; (2)Polishing and cleaning products for the car; (3)Petrochemicals chemistry for washing and detailing cars
Hygienic protection products based on petrochemicals	(1)Soap and detergents for personal use; (2)Hand sanitizer; (3)Wet wipes for hands; (4)Plasters and dressings; (5)Oral and dental hygiene products (rinses); (6)Household and rubber gloves	
Strategy	It's necessary to identify the factors that restrain the quality of production conditions and develop measures to eliminate them and add them to the assortment strategy	To conduct an analysis of points of growth in sales of goods, including consumer requirements, changes in sales technology

Source compiled by the authors

As factors of market attractiveness (direction), it's proposed to use a special index determined on the basis of market size, market growth rates, profitability, degree of competition, seasonality and cyclical demand. Factors are assessed based on strategic industry analysis. The quality of conditions for the development of the product direction is also assessed using a special index, which reflects factors such as the relative share of a company in the market, quality of goods, level of competence, availability of technologies, availability of production facilities. Their intersection forms areas (quadrants, strategic sectors), which are reflecting the firm's position in the market.

The proposed methodology for the formation of a product program allows an oil company:

- To define target segments of consumers of petrochemical products with high added value of the company's own retail;
- To determine new directions of product development in the area of existing solutions from competitors and in the search for the new petrochemical products;

- To highlight the directions of development of existing product solutions in the area of eliminating negative user experience that will increase the attractiveness of solutions for the consumer.

  At the same time, you can't disagree with the research by Hassani et al. (Hassani et al., 2017) that the development of product lines in the petrochemical industry, diversification of the product portfolio are impossible without innovations and new technologies in the petrochemical industry.

  It's becoming obvious that due to the high growth potential of petrochemical products, oil companies can increase the profitability of this group through their own brand, which will allow them to differentiate their own products and increase their value in the eyes of the consumer.

  Based on the results of calculations, it was established that the development potential of the petrochemical segment of a vertically integrated oil company has two principal directions:
- Improving product properties to eliminate dissatisfaction with their current quality;
- The creation of new products to meet various consumer needs.

  The authors see promising directions for further research in this area in the study of:
- Attributes of petrochemical products sold through the company's own retail chain.

References

Armstrong, J. S., & Brodie, R. J. (1994). Effects of portfolio planning methods on decision making: Experimental results. *International Journal of Research in Marketing, 11*, 73–84.

Boston Consulting Group (1970). *The Product Portfolio.*

Burenina, I. V., Evtushenko, E. V., Kotov, D. V., Battalova, A. A., Gayfullina, M. M., & Gamilova, D. A. (2018). Integral assessment of the development of Russia's chemical industry. *Journal of Environmental Management and Tourism, 8*(5), 1075–1108.

Eller, S. L., Hartley, P. R., & Medlock, K. B. (2021). Empirical evidence on the operational efficiency of National Oil Companies. *Empirical Economics, 40*, 623–643. https://doi.org/10.1007/s00181-010-0349-8(Dataaccessed30January

Frederiks, E. R, Stenner, K., Hobman, E. V. (January 1, 2015). Household energy use: Applying behavioural economics to understand consumer decision-making and behavior. *Renewable and Sustainable Energy Reviews, 41*, 1385–94.

Gaifullina, M. M., Nizamova, G. Z., Musina, D. R., & Alexandrova, O. A. (2017). Formation of strategy of effective management of fixed production assets of oil company. *Advances in Economics, Business and Management Research, 38*, 185–190.

Gong, B. (2020). Total-factor spillovers, similarities and competitions in the petroleum industry. In *Shale Energy Revolution* (pp. 151–217). Springer. Data Accessed January 22, 2021. https://doi.org/10.1007/978-981-15-4855-0_8

Hassani, H., Silva, E. S., & Al Kaabi, A. M. (2017). The role of innovation and technology in sustaining the petroleum and petrochemical industry. *Technological Forecasting and Social Change, 119*, 1–17

Ketels, C. H. M. (2006). Michael Porter's competitiveness framework—recent learnings and new research priorities. *Journal of Industry. Competition and Trade*, *6*, 115–136. Data Accessed January 25, 2021. https://doi.org/10.1007/s10842-006-9474-7

Kryukov, V. A., Shmat, V. V. (2020). Russian petrochemical industry in space and time. *Studies on Russian Economic Development*, *31*, 629–635. Data Accessed January 30, 2021. https://doi.org/10.1134/S1075700720060088

Li, J., Moghaddam, M., & Nof, S. Y. (2016). Dynamic storage assignment with product affinity and ABC classification—a case study. *The International Journal of Advanced Manufacturing Technology, 84*, 2179–2194. Data Accessed January 25, 2021.https://doi.org/10.1007/s00170-015-7806-7

Maitah, M., & Bassam, A. J. (2015). The economic role of petrochemical industry in Iran. *Modern Applied Science, 9*(11), 101–108. Data Accessed January 22, 2021. https://doi.org/10.5539/mas.v9n11p101

McKinsey Global Institute (1996). *Capital Productivity*. McKinsey & Company.

Mushketova, N. S., Fedorova, S. V. (2019). Strategic marketing planning in the petrochemical market. In *IOP Conference Series Materials Science and Engineering*, (483(1), 012053). Data accessed 25 January 2021. https://doi.org/10.1088/1757-899X/483/1/012053

Myllylä, Y., Kaivo-oja, J. (2015). Integrating Delphi methodology to some classical concepts of the Boston consulting group framework: arctic maritime technology BCG Delphi foresight—a pilot study from Finland. *European Journal of Futures Research*, *3*, T. 2. https://doi.org/10.1007/s40309-014-0060-7

Pawelek, B., Pociecha J., & Baryla M. (2017). ABC Analysis in corporate bankruptcy prediction. In *Abstracts of the IFCS Conference* (pp. 17). Japan, Tokyo.

Porter, M. E. (1998). Clusters and competition: new agendas for governments, companies, and institutions. In M. Porter (Ed.), *On competition.* Harvard Business School.

Stern, C. W., & Deimler, M. S. (2009). *The Boston consulting group on strategy. Classic concepts and new perspectives* (2nd ed.). Wiley.

Van de Klundert, T. C. M. J., & Potters, J. (1997). McKinsey on capital productivity. *De Economist*, *145*, 101–109. Data Accessed January 22, 2021. https://doi.org/10.1023/A:1002911613975

Verleger, P. (2011). The margin, currency, and the price of oil. *Business Economics, 46*, 71–82. Data Accessed January 22, 2021. https://doi.org/10.1057/be.2011.3

Collective Motivation for Safe Labor as a Factor of Corporate Environmental Responsibility in Enterprises of the Fuel and Energy Complex

Nadezhda V. Grogulenko, Elena G. Kostyleva, Nika E. Valitova, and Eduard S. Gareev

Abstract The authors justify the need to form collective motivation of employees to safe behavior as an essential factor of corporate environmental responsibility of enterprises. The relevance of studying new approaches to the development of corporate social environmental responsibility (CSER) is due to the fact that the functioning of enterprises of the fuel and energy complex (FEC) is associated with great environmental risks. The authors systematize the existing approaches to studying CSER and highlight the need to consider environmental factors and labor protection in the organization of FEC enterprises. The authors agree that CSER should be considered not only in terms of the impact on the environmental state of natural resources and the company's competitiveness but also in terms of industrial safety, social policy, and the formation of corporate culture with environmental values and norms. The conclusions formulated in the research are based on the data of the author's sociological study, which was conducted among employees and representatives of the top management of oil and gas companies.

Keywords Collective motivation · Labor motivation · Economic security · Corporate social responsibility · Corporate environmental responsibility

JEL Codes D21 · D22 · M14 · O15

1 Introduction

The study of collective motivation for safe work is of great practical relevance in developing the company's strategy of corporate social environmental responsibility (CSER). Collective allows forming internal and external communications based on trust, which significantly reduces environmental and production risks.

The relevance of the research is confirmed by the fact that, over the past decades, there have been many international and Russian indices and ratings evaluating the

N. V. Grogulenko (✉) · E. G. Kostyleva · N. E. Valitova · E. S. Gareev
Ufa State Petroleum Technological University, Ufa, Russia

E. G. Popkova and B. S. Sergi (eds.), *Geo-Economy of the Future*,
https://doi.org/10.1007/978-3-030-92303-7_87

effectiveness of company development in terms of its environmental and social responsibility.

The processes of entering the information society have led to the fact that the business community understands the term "corporate social environmental responsibility" in connection with the need to make strategic decisions aimed not only at a profit but also at increasing the intangible assets of the company, particularly the elevation of the image and the formation of a positive reputation and consumer loyalty.

To increase the company's competitiveness, along with the introduction of innovations, increasing productivity, and stimulating human resources, it is necessary to implement CSER. It is impossible to improve the competitiveness of enterprises without CSER aimed at external and internal targets.

2 Methodology

The analysis of Russian and international literature showed that there exist many approaches considering the problems of CSER from the perspective of business ethics, corporate activity, etc. When defining CSER, most authors consider the totality of fulfilling the interests of the company and society (Belenkova et al., 2019; Vanchukhina et al., 2016).

Belgian researcher E. Davignon calls to avoid associating CSER with charity and promotions. He emphasizes the underlying interest of the entrepreneur to carry out systematic activities to improve the environment and society (Gusarov, 2000). P. Robbins, in his study of CSER, stresses the importance of the obligations of business entities to fulfill strategic socially significant goals, which they fulfill in addition to statutory norms and requirements (Robbins & Coulter, 2004, pp. 206–209).

When considering the concept of CSER, Russian researcher Yu. V. Blagov gives a central place to the company's fulfillment of obligations (to clients, partners, founders, and employees) that exceed the legal framework and aim to improve the quality of life of the members of labor collective (Blagov, 2006).

Determining the concept of CSER, O. V. Danilova focuses on the role of intellectual activity to stabilize the economy beyond the enterprise, aimed at the reputational capital of the company (Danilova, 2009, p. 25).

We agree with the position of N. Petrova and A. Titkova, in which they propose to consider the phenomenon of CSER in two ways. In terms of globalization, the phenomenon is based on meeting the needs of society (Ustinova et al., 2016). From a narrower perspective, the emphasis is made on the company's need to redistribute a portion of income to public consumption funds to take care of public needs. Yu. N. Popov complements this approach and emphasizes the importance of CSER as an instrument of "social audit" for the redistribution of income and interests between employers and employees (Petrova & Titkova, 2010, pp. 421–423).

L. Figlin and V. Moseyko write that CSER is based on the level allowing the company management to evaluate the impact of its activities on the natural environment and society and assess the compliance with the principles of democracy and equal rights (Figlin & Moseyko, 2003).

B. M. Shpotov proposes to evaluate CSER in terms of maximizing the benefits and advantages of the company and minimizing the negative effects of the company on society and the environment (Shpotov, 2003).

D. Grayson and A. Hodges raise the status of CSER to the level of a strategic concept. According to the researchers, this concept serves for building the company's activities and caring for all its partners, the general public, and nature (Grayson & Hodges, 2004, pp. 150–151).

Additionally, we agree with the approach of O. I. Sergienko and O. S. Pavlova (Sergienko & Pavlova, 2011). They propose a model of CSER based on a multi-purpose analysis of all stakeholders, which, in our opinion, stimulates the development of a collective identity (Gaisina et al., 2015).

3 Results

To analyze the content, primary directions, and tasks of the current policy on the CSER of enterprises in the fuel and energy complex (FEC), in the winter of 2021, the authors surveyed employees and top managers of subsidiary companies of "Rosneft," "Gazprom," and "Sibur." The sample size was 360 and 120 people. Employees were sampled using the cluster method. Experts were sampled using the "snowball" method. Additionally, as part of the study, the authors conducted an expert survey by interviewing 50 engineering and technical employees responsible for occupational safety and health to identify ways to encourage safe behavior.

The research results show that the implementation of the CSER policy at FEC enterprises is an integral and obligatory activity along with the innovation policy.

Let us look at the rating of respondents' answers to the question "What results can the company achieve through the implementation of the CSER policy?" (Table 1).

According to Table 1, most surveyed employees and managers of FEC companies chose the answer "increase the efficiency of the company's operations." However, their opinions diverged further. Employees were more interested in the possibility of improving the social aspects of their activities, welfare, and environmental issues through the CSER policy. Top managers intended to use an effective CSER policy to gain support from the business community and optimally implement the development program of their companies. The answers "improving the environmental situation" and "improving the living standards of employees" are at the bottom of the rating for representatives of top management.

The opinions of top managers and employees coincided regarding the possibility of increasing the loyalty of state authorities and the public and improving the company's image and reputation through the CSER policy.

Table 1 Rating of answers to the question "What results can the company achieve by implementing the CSER policy?"

No	Answer options	Rating of results on the CSER policy	
		Employees	Top management
1	Improving the efficiency of the company's operations	1	2
2	Improving the employees' standard of living	2	7
3	Improving the environmental situation	3	6
4	Loyal attitude to the company on the part of state authorities and the public	4	4
5	Creating a positive image and reputation of the company	5	5
6	Implementation of a program for the future development of the company	6	3
7	Gaining the support of influential partners and shareholders	7	2

Source Compiled by the authors

Answering the question "How can CSER of a FEC company contribute to solving environmental problems of the region and the country?" the majority of surveyed top managers (55.5%) noted that this program would allow delineating the areas of responsibility and obligations of all stakeholders (government agencies, the public, and the business community). This approach will allow for real action without shifting problems to other parties involved. Another quarter of respondents (26.5%) believe that the CSER policy will unite the efforts of all participants in the process of creating an environmentally safe product by oil and gas companies, from production to processing, receipt, and promotion of finished products on the market. Additionally, 18.0% of top managers indicated that the analysis of the needs of the company's stakeholders should be conducted at the initial stage of creating a CSER policy to develop the document on their collective expectations.

The answers of top management to the question "What are the main directions of CSER of FEC enterprises?" are of particular interest (Table 2).

As shown in Table 2, experts identify four main directions for developing CSER of FEC enterprises. First, it is to increase satisfaction with the quality of products (32.4%) and ensure the company's development (30.2%). The next directions include the improvement of the environmental situation (20.3%) and the development of socially responsible activities (14.3%). Respondents explained the answer "other direction" as improving occupational health and safety (2.3%).

During the research, the authors conducted an expert survey of engineering and technical employees of FEC enterprises on the formation of collective motivation for safe work.

In general, employees gave a positive assessment of motivation to work safely. Responding to the question "How do you assess the level of motivation for safe

Table 2 Answers of top managers of FEC enterprises to the question "What are the main directions of CSER of FEC enterprises?", %

No	Answer options	Top managers
1	Increasing satisfaction with product quality	32.4
2	Ensuring the development of the company	30.2
3	Improving the environmental situation	20.3
4	Development of socially responsible activities	14.3
5	Other direction	2.3
6	No answer	0.5

Source Compiled by the authors

production at your FEC enterprise?", 62.0% of experts chose the answer "high level." The answer "low level" was chosen only by 10.0% of employees. Nevertheless, 28.0% of employees rate the level of motivation as average. The assessment given by engineers shows that the work performed at FEC enterprises to improve labor safety must be based on the results of qualitative cuts and be systematic and progressive in nature, taking into account the existing errors.

To the question "Does your company work to increase motivation for safe work?", 80% of respondents answered in the affirmative.

To understand the type of motivation for safe behavior prevailing at FEC companies, respondents were asked to assess the effectiveness of incentives for safe production at their company (Table 3).

According to Table 3, engineering and technical employees believe that incentives based on approval and censure can be effective (60% and 20%, respectively). However, they still consider positive incentives to be more effective. Simultaneously, 20% of the surveyed employees assessed negative incentives as ineffective.

The authors analyzed responses to the question, "What types of positive incentives for safe behavior are most often used at your company?". It turned out that non-material incentives (gratitude, praise, promotion, additional bonuses, etc.) are used most often. Material incentives are used more often in the form of bonus payments, but their amounts are small.

However, according to our research, negative incentives for employees are more often practiced at FEC enterprises (Table 4).

Table 3 Assessment of the effectiveness of incentives for safe production by engineering and technical employees, %

Types of incentives	Effective	Rather effective	No answer	Rather ineffective	Ineffective
Positive incentives	60.0	26.0	4.0	10.0	–
Negative incentives	20.0	26.0	4.0	30.0	20.0

Source Compiled by the authors

Table 4 Answers of engineering and technical employees to the question "What measures are applied at your enterprise to employees for violation of labor safety rules?", %

Types of incentives	Always	Often	Rarely	Never	No answer
Dismissal	15.0	63.0	4.0	3.0	15.0
Reprimand	78.0	13.0	–	–	9.0
Forfeiture of bonus	78	10.0	7.0	–	5.0
Penalty and payment of forfeit or damages	–	8.0	5.0	81.0	6.0
Demotion	–	–	12.0	80.0	8.0
Additional training in occupational safety and health	25.0	70.0	–	–	5.0
Other	–	–	–	20.0	80.0

Source Compiled by the authors

As can be seen from the sum of the first two columns of the table ("always" and "often"), the most practiced incentives at FEC enterprises are additional training on labor protection (95.0%), reprimands (91.0%), deprivation of bonus (88.0%), and dismissal (78.0%). Only 8.0% of experts indicated a system of fines or penalties. According to 12.0% of engineers and technicians, demotion is practiced rarely.

It is indicative that 99.0% of the surveyed engineers answered affirmatively to the question, "Do you consider it necessary to modernize the existing model of incentives for safe work at your FEC enterprise?" The rationalizing ways were as follows:

- Correlation of wages based on the compliance with occupational safety rules (99.0%);
- Financial remuneration for compliance with all norms (98.0%);
- Team-building activities (78.0%);
- Opportunities for career and professional growth (79.0%);
- Administrative penalty for violations (79.0%).

Moreover, the significant potential is shown by improved working conditions (68.0%) and a well-designed insurance system (66.0%).

Experts see improved ways of getting employees to work, providing more authority to employees, and verbal praise with certificates as ineffective methods of encouraging safe behavior.

It is necessary to highlight the answers of experts to the question "How should the implementation of the CSER policy be monitored?". After analyzing the answers, the authors obtained the following control mechanisms:

- Coordination of decisions on the implementation of the objectives in the field of CSER among all stakeholders;
- Personal responsibility of all participants of the production process for the implementation of CSER (from top management to entry-level employees);
- Publication of the report on the implementation of the company's CSER policy on the official website of the company;

- Implementation of activities based on international standards of CSER;
- Coverage of all ongoing activities and social projects in the media and social networks;
- Strengthening the information support of all social projects from the moment of application to the final events;
- Introduction of open innovative forms of conducting socially-oriented companies (online brainstorming sessions, discussions in social networks, monitoring the opinions of all stakeholders, etc.).

4 Conclusion

Russian financial experts define CSER as a voluntary activity of a company related to its main profile to contribute to developing the economic, social, and environmental sphere. International financial experts consider CSER as a result of the globalization of the world economy, which cannot be implemented without the combination of a competent professional approach and mutual benefit of society and a company.

A competently structured CSER policy will enable the company's long-term development program by increasing its reputational capital, collective identity, and collective motivation to work safely.

When designing the development strategy for the CSER of FEC companies, it is necessary to consider the needs of state authorities, the public, public organizations, the scientific community, partners and shareholders, consumers of products, and employees of the company.

The approach of collective motivation to safe labor considered by the authors will increase the effectiveness of the company in implementing a policy of corporate social environmental responsibility and improve the system of corporate governance at FEC enterprises.

References

Belenkova, O. A., Vanchukhina, L. I., & Leibert, T. B. (2019). Creative potential of the human capital as the key resource of development of the techogenic civilization. *Espacios, 40*(10), 28. Retrieved from http://www.revistaespacios.com/a19v40n10/a19v40n10p28.pdf. Accessed February 25, 2021

Blagov, Yu. E. (2006). The genesis of the corporate social responsibility concept. *Vestnik of Saint Petersburg University. Management, 2*, 3–24.

Danilova, O. V. (2009). *Social Responsibility of Business in the Market Economy (Theoretical and Economic Aspect)* (Synopsis of Dissertation of Doctor of Economics), Moscow, Russia: Financial Academy under the Government of the Russian Federation.

Figlin, L., & Moseiko, V. (2003). Quality management model of social responsibility of the organization. *International Journal of Management Theory and Practice, 2*, 86–91.

Gaisina, L. M., Gareev, E. S., Valitova, N. E., Khairullina, N. G., & Ustinova, O. V. (2015). Corporate staff identity as a factor of increasing labor productivity. *Mediterranean Journal of Social Sciences, 6*(5), 274–285. https://doi.org/10.5901/mjss.2015.v6n5p274
Grayson, D., & Hodges, A. (2004). *Corporate social opportunity! seven steps to make corporate social responsibility work for your business*. Greenleaf Publishing.
Gusarov, Y. A., & Davignon, E. (2000). Perspectives for the twenty-first century. In *Ideas of Europeanism in the second half of the twentieth century.* Moscow, Russia: Institute of Scientific Information on Social Sciences of the Russian Academy of Sciences. Retrieved from https://cyberleninka.ru/article/n/e-davinon-perspektivy-na-xxi-stoletie/viewer. Accessed February 25, 2021
Petrova, N., & Titkova, A. (Eds.). (2010). *Power, business, and society in the regions: The wrong triangle*. ROSSPEN.
Robbins, S. P., & Coulter, M. (2004). *Management* (Transl. from English). Williams.
Sergienko, O. I., & Pavlova, A. S. (2011). Formation of corporate social environmental responsibility of business on the basis of stakeholder analysis. *Scientific Journal NRU ITMO. Economics and Environmental Management, 1*(8). Retrieved from http://economics.ihbt.ifmo.ru/file/article/6624.pdf. Accessed February 25, 2021
Shpotov, B. M. (2002). Business ethics and management: Contemporary approaches. *International Journal of Management Theory and Practice, 1*, 89–90.
Ustinova, O. V., Rudov, S. V., Kostyleva, E. G., Grogulenko, N. V., & Kulishova, N. D. (2016). The processes of globalization in the Russians' views. *Man in India, 96*(7), 2165–2177.
Vanchukhina, L. I., Leybert, T. B., & Khalikova, E. A. (2016). Methodological approaches to evaluation and analysis of labor efficiency in the spheres of fuel and energy complex. *Journal of Environmental Management and Tourism, 7*(4), 585–593. https://doi.org/10.14505//jemt.v7.4(16).04

Ensuring Economic Security by Modifying Renewable Energy Systems

Aigul Z. Kharisova, Irina V. Shvayko, Olga I. Shalina, Aida F. Ostryakova, and Ildus Yu. Singizov

Abstract The paper presents the possibility of ensuring economic security by modifying the energy system of renewable energy sources, including solar, wind, hydropower, biofuels, etc. This process is related to the transition of economies to a less carbon-intensive and more sustainable energy system. Recovery in fuel demand and stability in international markets are driving a recovery in production in 2021 and sustained growth through 2025. The paper explores economic and policy incentives focused on clean energy that can directly or indirectly support renewable energy. Experts predict that renewable energy sources should take the leading place in the world's electricity sector. The authors propose using a dynamic information model to refine the forecasts. This research tool will provide data and forecasts in all sectors employing renewable energy technologies. The authors provide up-to-date indicators, analysis, and information on energy security and sustainability on a global scale. Moreover, they quantify the effects of the widespread global recession caused by COVID-19 and consider measures in the clean energy sector to address them.

Keywords Economic security · COVID-19 · Crisis · Renewable energy sources · Energy markets

JEL Codes B41 · E6 · F01 · F62 · F64 · L20

A. Z. Kharisova (✉) · I. V. Shvayko
Ufa State Petroleum Technological University, Ufa, Russia
e-mail: harisovaaigul@bk.ru

O. I. Shalina
Ufa State Aviation Technological University (USATU), Ufa, Russia
e-mail: olenkash@bk.ru

A. F. Ostryakova
Ufa Law Institute of the Ministry of Internal Affairs of Russia, Ufa, Russia

I. Yu. Singizov
Bashkir State University, Ufa, Russia

E. G. Popkova and B. S. Sergi (eds.), *Geo-Economy of the Future*,
https://doi.org/10.1007/978-3-030-92303-7_88

1 Introduction

The COVID-19 crisis has caused enormous economic damage. The main indicators of the world economy have deteriorated. Nevertheless, there has been an increase in the use of renewable energy sources (RES). Renewable energy markets, especially power generation technologies, have proven to be resilient to the crisis.

At the heart of ensuring economic security by modifying renewable energy systems are the ideas of transition to a less carbon-intensive and more sustainable energy system, which implies renewable energy sources—solar, wind, hydropower, biofuels, and others.

The consumption of renewable energy has rapidly grown in recent years due to political support from governments and dramatic cost reductions for solar photovoltaics and wind power. The electricity sector has remained the main focus of renewable energy. Solar photovoltaics and wind power stand out. Electricity accounts for only one-fifth of global energy consumption. The expansion of renewable energy in the transportation and heating sectors is particularly important (Kurbanova & Shalina, 2019, p. 36).

Figure 1 shows the change in energy demand and renewable energy production in electricity, heat, and transportation for 2019–2020.

In 2019, global production of renewable electricity grew by 6%. Wind and solar photovoltaic technologies accounted for 64% of that growth. The share of renewable energy in global electricity production reached nearly 27% in 2019. The International Energy Agency (IEA) plans that renewable energy will account for half of the production by 2030 (Fig. 2). According to the IEA sustainability plan, renewable energy needs to expand significantly, which will require accelerating the pace of annual capacity additions (Dafnomilis et al., 2020).

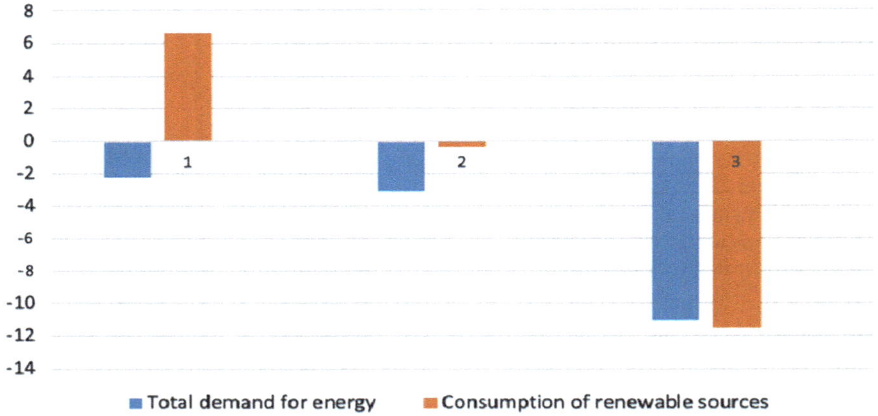

Fig. 1 Changes in the demand for energy and renewable energy in 2019–2020. *Source* Compiled by the authors based on the International Energy Agency (IEA) data Dafnomilis et al., 2020

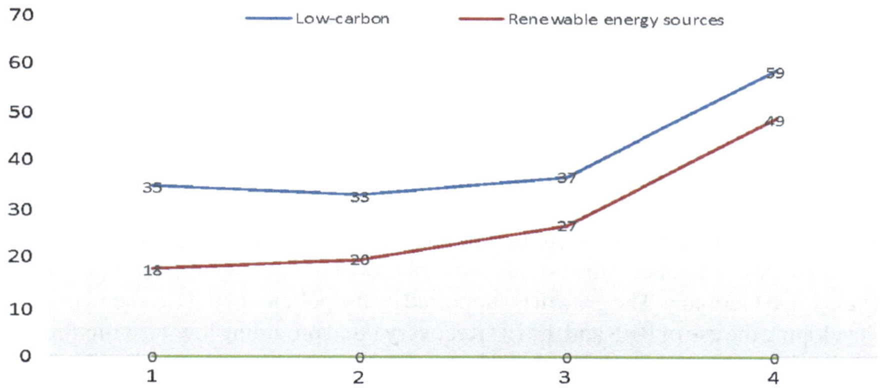

Fig. 2 Share of renewable energy sources in electricity generation in 2000, 2010, 2020, and 2030, (%). *Source* Compiled by the authors based on the IEA data Dafnomilis et al., 2020

The sustainability of renewable energy supply is determined by the specific energy sector.

Unlike all other fuels, renewable energy used to generate electricity is up nearly 7% in 2020. Global energy demand is down 5%. Nevertheless, long-term contracts, the continuous installation of new power plants, and free access to the power grid are ensuring the growth of renewable electricity. This compensates for the decline in the use of bioenergy in industry and biofuels in transport, resulting from the decline in economic activity. The total growth in demand for renewable energy in 2020 equaled 1%.

Despite impending economic uncertainty, investor interest in renewable energy remains strong. From January to October 2020, there was 15% more renewable capacity offered at auction than in the same period last year. At the same time, shares of the manufacturers of renewable equipment and project developers outperformed most major stock indexes and the overall energy sector due to expectations of economic growth in the medium term. In October 2020, shares of solar companies around the world more than doubled compared with December 2019. Thanks to China and the USA, clean renewable energy capacity saw a 4% increase globally in 2020 and amounted to 200 GW. Higher growth rates in wind and hydropower are driving global renewable capacity to a new record this year, accounting for nearly 90% of the increase in total capacity worldwide. The growth of solar energy is expected to remain steady (Kharisova, 2017, p. 85).

The increase in renewable capacity is at a record level (10% in 2021). Experts predict that two factors should lead to the fastest growth since 2015.

First, the commissioning of deferred projects in markets was disrupted. Prompt government action in key markets in the USA, India, and some European countries has allowed developers to complete projects months after the stipulated deadline, which was initially the end of 2020 (Farkhutdinova & Fakhrislamov, 2019, p. 211).

Second, some markets (e.g., the USA, the Middle East, and Latin America) will see growth in 2021 due to political support and lower costs.

Europe and India will lead the ranking of countries using renewable energy in 2021. India is expected to be the largest contributor to the growth of RES in 2021, with annual revenues nearly doubling from 2020. Most of the wind and solar photovoltaic projects up for auction will begin operations after delays caused not only by COVID-19 but also by contract negotiations and land acquisition issues. In the EU, capacity additions are projected to increase sharply in 2021. This increase is mainly due to previously auctioned utility-scale solar photovoltaic and wind power projects in France and Germany. The growth is supported by the policies of the EU member states developing the use of RES and the EU Recovery Fund providing low-cost funding and grants. In the Middle East, North Africa, and Latin America, renewable energy will come to the previous levels in 2021 with the commissioning of government-supported projects (Amirkhanova et al., 2017, p. 45).

Renewable energy sources are resilient to the COVID-19 crisis but not to political uncertainty. The expiration of incentives in key markets and the resulting political uncertainty could lead to a slight decline in the increase of renewable energy capacity in 2022. In China, subsidies for onshore wind power and solar photovoltaics expire this year, while support for offshore wind power ends in 2021. Political projects for 2021–2025 will be announced in late 2021, raising uncertainty about the pace of the expansion of renewable energy in China in 2022 and beyond. Technologies using renewable energy will also be delayed in 2022 due to the expiration of tax credits for onshore wind production in the USA, the ongoing financial struggle of distribution companies in India, and the delayed auctions in Latin America. Particularly, the growth of onshore wind power is expected to decline by 15% globally, while the expansion of offshore wind power continues globally (Izilyaeva, 2018, p. 32).

Cost reductions and sustained political support are expected to contribute to the strong growth of renewable energy starting in 2022. Despite the problems arising from the COVID-19 crisis, plans to expand the use of renewable energy have not changed. Solar photovoltaics (PV) and onshore wind are already the cheapest ways to add new power plants in most countries. In countries with good resources and cheap financing, wind and solar PV will challenge existing fossil fuel plants. Projects on solar energy now offer some of the cheapest electricity in history ("Vyvoz.org" Portal, 2020).

Overall, renewable energy sources should provide 95% of the world's net increase in energy capacity by 2025. The total installed capacity of wind and solar PV installations will exceed natural gas in 2023 and coal in 2024. Solar power alone accounts for 60% of all added renewable capacity by 2025, and wind power for another 30%. Further cost reductions will cause an increase in the annual revenue from wind farms, bringing it to one-fifth of total production (Petrov, 2021).

The COVID-19 crisis hit the biofuel industry the most. In 2020, global production of transportation biofuel was projected to be down 12% from 2019. This is the first decline in annual production in two decades. It is caused by the decline in demand for transportation fuels and the decline in fossil fuel prices, making biofuels less

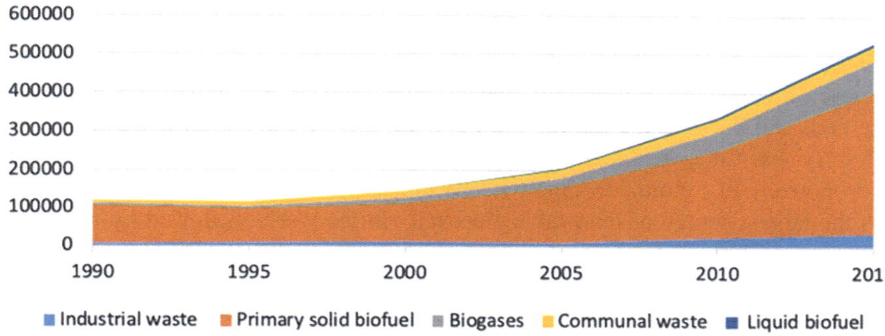

Fig. 3 Production of electricity from biofuels and waste by sources, 1990–2018. *Source* Compiled by the authors based on the IEA data Dafnomilis et al., 2020

economically attractive (Fig. 3). The biggest annual production drops are observed in the USA and Brazilian ethanol and European biodiesel.

2 Materials and Methods

The analysis of data on the use and modification of RES systems allows us to argue that this system, as an object of management and forecasting, is marked with the presence of various kinds of uncertainties, including uncertainties in mathematical models, uncontrolled changes in the parameters of internal subsystems, the effect on the system of random external factors, etc. Therefore, when solving management problems in information systems, it is preferred to use the means of modeling fuzzy data and knowledge, fuzzy logical allocation, methods of the theory of adaptive systems. The theory of identification takes one of the central places in solving the problems of synthesizing mathematical models. In this case, the dynamic information model is represented by diagrams reflecting RES indicators.

Simultaneously, the decision on the structure of the model of the studied system should be taken based on the principle of information completeness of the analyzed (used in the modeling) data. Thus, the factors affecting the construction of RES forecasts can be divided into two categories: economic and political. These categories require the definition of several new characteristics of the information model.

The proposed dynamic information model differs from the known ones by considering the observed structural-information set of data. This will improve the accuracy of structural identification of the system under the influence of various kinds of interference. The study revealed several new properties of the system complementing the dynamic set. In the future, this will reduce the level of uncertainty in the process of solving the problem of structural identification and making predictions (Semenov & Davydov, 2011).

3 Results

The use of a dynamic information model allows us to conclude that RES will overtake coal and become the largest source of electricity generation in the world by 2025. By that time, they are expected to supply a third of the world's electricity. Hydropower will continue to supply nearly half of the world's renewable electricity. It is the largest source of renewable electricity in the world, followed by wind and solar power. The continuing decline in the cost of renewable energy is changing the makeup of investors and the role of politics. The share of the growth of renewable energy attributable to purely market conditions outside of political programs (e.g., bidding and feed-in tariffs) will triple from 5 to 15% by 2025. This includes corporate agreements on purchasing power plants with higher prices on power or other contracts.

The most significant increases in ethanol production are observed in China and Brazil. The leading producers of biodiesel and hydrotreated vegetable oil are the USA and Southeast Asia.

The drop in the economic activity caused by the pandemic has affected industrial heat consumption. This has led to an increase in demand for renewable energy and especially the use of bioenergy in industry. COVID-19 had a limited direct impact on the short-term consumption of renewable thermal energy. Although global demand for heat in industry and buildings is falling, the consumption of heat-related renewable electricity in 2020 grew in both sectors.

The share of renewable heat consumption is expected to remain broadly unchanged over the next five years. Global renewable heat consumption in 2025 is projected to be 20% higher than in 2019, with the construction sector seeing a stronger uptick than industry. Despite this, renewable energy will account for only 12% of global heat consumption by 2025 since the overall market is expected to expand under the influence of industrial activity. As a result, total heat-related carbon dioxide emissions in 2025 are expected to be only 2% lower than in 2019. The magnitude of the change in energy consumption is shown in Fig. 4.

4 Conclusion

Political and economic incentives focused on renewable energy can support RES. Some countries have announced stimulus measures related to energy. Most of the \$470 billion is for short-term economical aid. About \$108 billion were allocated for providing economic growth with clean energy development. These measures can support renewable energy by providing additional financial support through buildings, grids, electric vehicles, and low-carbon hydrogen. The EU's economic recovery plan, which is expected to spend about \$310 billion on climate-related issues, also indicates clean energy development.

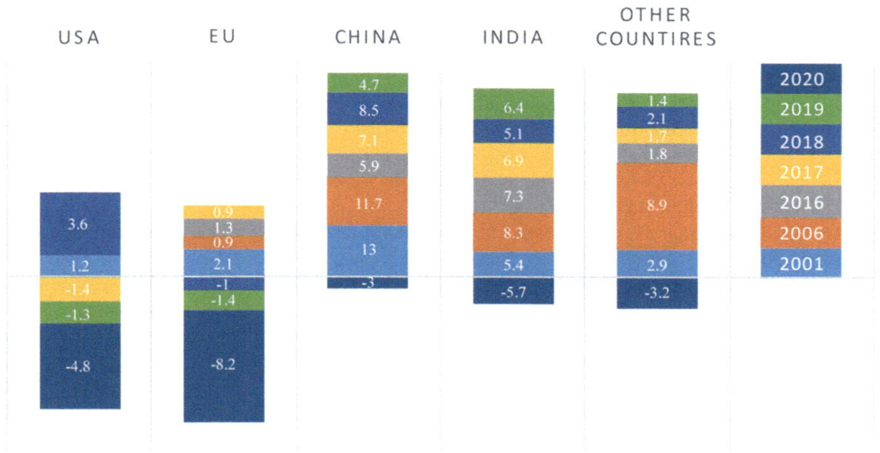

Fig. 4 Average annual growth rates of electricity demand in selected regions for 2001–2020, %. *Source* Compiled by the authors based on the IEA data Dafnomilis et al., 2020

The strategy of economic security in the Russian Federation is also related to the modification of renewable energy systems. The program of implementation of renewable energy sources in Russia foresees that the share of energy obtained from them will increase by 1–2% within ten years. This figure is not high, but experts believe that such a growth rate is quite acceptable for the domestic economy and the energy sector.

Renewable fuels for transportation are an area of particular potential support because this sector has been hit hard by the crisis. Net-zero emissions targets in key markets are expected to accelerate the adoption of renewable energy sources. Following the EU and several European countries, the three largest Asian economies recently announced zero-emissions goals: Japan and South Korea by 2050 and China by 2060. These announcements will accelerate the adoption of renewable energy in all sectors, which will significantly impact global markets.

References

Amirkhanova, L. R., Bikmetov, E. Y., Zinnurov, U. G., & Kharisova, A. Z. (2017). Management of university competitiveness based on image creating research. *Journal of Business and Retail Management Research (JBRMR), 11*, 40–48.

Dafnomilis, I., den Elzen, M., van Soest, H., Hans, F., Kuramochi, T., & Höhne, N. (2020). Exploring the impacts of the COVID-19 pandemic on global energy markets, energy resilience, and climate change. Retrieved from https://www.pbl.nl/sites/default/files/downloads/pbl-new-climate-institute-2020-exploring-the-impact-of-covid-19-pandemic-on-global-emission-projections_4231.pdf. Accessed February 22, 2021.

Farkhutdinova, I. R., & Fakhrislamov, T. R. (2019). Political internet communications in modern Russia. *Eurasian Law Journal, 7*(134), 359–361.

Izilyaeva, L. O. (2018). Russian-American relations through the prism of national security strategies states. *Economics and Management: Research and Practice Journal, 5*(143), 31–34.
Kharisova, A. Z. (2017). Improving the competitiveness of Russia: In a new age with new energy. *Bulletin of Kazan State Power Engineering University, 3*(35), 84–90.
Kurbanova, E. R., & Shalina, O. I. (2019). Social and economic development of Russia and the world. In I. V. Degtyareva, A. V. Maryina, & T. N. Yakovleva (Eds.), *Current issues of economic theory: Development and application in the practice of Russian transformation* (pp. 22–25). Ufa State Aviation Technical University.
Petrov, A. (2021, March 2). What is happening with renewable energy in Russia? Tinkoff Journal. Retrieved from https://journal.tinkoff.ru/russia-green-energy. Accessed 22 February 2021.
Semenov, S. G., & Davydov, V. V. (2011). Dynamic model of the information system based on the observed structural information portrait. *Bulletin of the National Technical University Kharkov Polytechnic Institute. Series: Informatics and Modeling, 36*, 156–163. Retrieved from https://cyberleninka.ru/article/n/dinamicheskaya-model-informatsionnoy-sistemy-na-osnove-nablyudaemogo-strukturno-informatsionnogo-portreta/viewer (Accessed February 22, 2021.
Vyvoz.org Portal. (2020, October 14). Renewable energy sources in Russia. Retrieved from https://vyvoz.org/blog/vozobnovlyaemye-istochniki-energii-v-rossii/. Accessed February 22, 2021.

Comparative Analysis of Changes in Prices for Fuel and Energy Commodities for Sustainable Company Development of the Power Sector

David A. Gertsekovich, Anastasia N. Kopeykina, Sergey N. Larin, Oleg L. Podlinyaev, and Nina M. Baranova

Abstract The article analyzes energy commodities at the producer and consumer prices. The characteristic features of the dynamics of energy commodities for 2000–2019 are considered: expected return, risk, return risk ratio, quantitative ratio of dynamics of producer, and purchaser prices. These indicators will be used to select the energy commodities for the development of decision-making systems. The choice is carried out based on the "risk-return" model, which is based on the fundamental propositions of the portfolio theory, the principle of constructing the "winner's portfolio", momentum strategies and the concept of "optimization of the learning sample length". The analysis made it possible to establish the leading energy commodities and draw up a rating table of preference patterns from the standpoint of the producer and the consumer. The advantage of investing in business related to the sale of the considered energy commodities is noted. This approach makes it possible to project obtaining a higher expected return than directly industrial production with a more favourable ratio of the expected return to the risk level. The producer price growth and the consumer price growth were compared based on synthesized empirical models. The reasons that form the demand for energy commodities at present and the increased need for them in the future are investigated.

Keywords Energy prices · Rate of return · Risk · Estimation of investment attractiveness · Portfolio performance · Markowitz model · Sustainable development

D. A. Gertsekovich · A. N. Kopeykina
International Institute of Economics and Linguistics of Irkutsk State University, Irkutsk, Russia

S. N. Larin
Central Economics and Mathematics Institute RAS, Moscow, Russia

O. L. Podlinyaev
Institute of Pedagogics of Irkutsk State University, Irkutsk, Russia
e-mail: podlinyaev@inbox.ru

N. M. Baranova (✉)
Peoples' Friendship University of Russia (RUDN University), Moscow, Russia
e-mail: baranova_nm@pfur.ru

E. G. Popkova and B. S. Sergi (eds.), *Geo-Economy of the Future*,
https://doi.org/10.1007/978-3-030-92303-7_89

JEL Codes C02 · D04 · E17 · O14

1 Introduction

Interest in the fuel and energy industry only increases every year as the global economy develops. Russia is among the largest leading countries in hydrocarbon production and reserves. The fuel and energy industry in the Russian economy is one of the most developed, demanded, and profitable industries.

To date, there are no generally accepted criteria and methods for the preference pattern formation when choosing energy carriers, both for producers and consumers. It is necessary to develop a mathematical model for an optimal investment policy, both for energy producers and energy consumers to solve some economic problems of sustainable development. The results obtained will serve as the basis for choosing the groups of the most promising fuel and raw materials resources and increasing the efficiency of investment in the production and consumption of energy resources.

2 Materials and Method

More recently, many scientists have been taking up the problems of energy development. Nazarova et al. (2020), Ivanov (2019) and others were engaged in mathematical modelling of the changes in price for energy commodities, production and consumption of electricity, fuel, etc. Studies of Bouchouev (2017), Mikhailova et al. (2019) and others are devoted to the analysis of the movement of oil prices. Investments in the company were carried out by Damodaran (2012), Gitman (2008), Sharpe et al. (1998) and others. The researches by Markowitz (1952), Red'kin (2019), Jegadeesh (1993) and others were devoted to the portfolio analysis formation and application in various industries; the company development strategy was formed based on these studies. Risk-Return Relationship and Portfolio Management were handled by Dhankar (2019), Gertsekovich and Babushkin (2019) Gibson (2008), Skoglund and Chen (2015) and others.

A method for selecting the most attractive types of energy commodities through the average producer and consumer prices was formulated for the first time in this paper. Previously, these prices were not compared through portfolio performance and were carried out based on "classical criteria": net revenue, gross profit, return, etc.

3 Results

3.1 Performance Attribution

The term return in this paper is understood to be the price growth rate (in percent) over the previous year. We will distinguish producer prices and consumer prices within the framework of the task. They are actual prices at the time of registration for produced and shipped energy commodities intended for sale on the domestic market (free of commodity tax—VAT, excise tax, etc.).

The consumer price includes, in addition to the producer price, VAT, excise tax, transport, selling, marketing, and intermediary and other costs (Russia in figures. Statistical handbook, 2020). To begin with, consider the expected return on goods at producer prices and the risk level (Table 1). Table 1 presents the producer returns (first-line) and the purchaser returns (second-line) for each energy commodity for 2000–2019 (Russia in figures. Statistical handbook, 2020).

The maximum and minimum values are highlighted in bold and underlined (Table 1). Statistical analysis of consumer prices shows that brown coal has minimum values, both in terms of return and risk. This provides a better return risk ratio for the consumer. If the return risk ratio is considered as a criterion for the effectiveness of the management structures of enterprises-consumers, then it can be assumed that they, in general, function much more efficiently compared to the management structures of producers for this energy carrier.

Considering steam and hot water, the risk level is minimal at both producer prices and consumer prices. This energy carrier is the most attractive target for risk-averse investors. Fuel oil is less attractive for investors since the return risk ratio is minimal. Further, the price behavior of some energy commodities will be analyzed (Table 1).

The prices for motor gasoline and diesel fuel (at producer prices) have been steadily growing (except for some years). The purchase price of crude oil decreased by $119.54 in 2014 compared to 2013. The return growth rate decreased by 31.79%. The consumer oil price in 2020 turned out to be lower than the producer price. There are several reasons for this price change.

(1) Lack of balance between supply and demand. The structure of the oil production market has changed, which was caused by the growth of shale oil in the United States and tar sands in Canada. This set in motion the entire global chain from a producer–consumer position.

(2) China, one of the largest oil consumers, reduced imports as the country's economic growth slowed down. In addition, the Chinese authorities made attempts to improve the energy efficiency of their production.

(3) Iran and Saudi Arabia have cut export contract prices. If Saudi Arabia managed to maintain high exports, then the Russian fuel supply began to decline. Russia acts on the world oil market as a "price-taker" and not as a "price-maker" (Bouchouev, 2017; Mikhailova et al., 2019; Skoglund & Chen, 2015). An important factor in the impairment of value was the lack of preparedness for the OPEC countries to agree on a reduction in oil production.

Table 1 The expected return and risk level of producers and consumers of the main types of energy commodities

No	*Energy commodities*	*Expected return (Rt),%*	*Risk level (Rs),%*	*Return/Risk ratio (Rt/Rs),%*
1	Coking coal (CC)	13.9	**41.2**	0.3
		6.2	27.4	**0.2**
2	Coal, less anthracite, coking and brown coal (C less A)	8.9	11.0	0.8
		6.4	6.9	0.9
3	Brown coal (lignite) (BC)	7.6	9.2	0.8
		5.7	4.5	1.3
4	Crude oil (CO)	12.7	16.7	0.8
		12.2	23.6	0.5
5	Combustible natural gas (CNG)	**14.9**	29.6	0.5
		6.5	7.2	0.9
6	Motor gasoline (MG)	4.7	9.2	0.5
		9.4	5.5	1.7
7	Diesel fuel (DF)	9.1	13.7	0.7
		10.4	10.2	1.0
8	Fuel oil (mazut) (HFO)	3.7	32.2	**0.1**
		6.4	**39.1**	**0.2**
9	Electricity (E)	4.8	8.0	0.6
		6.5	8.6	0.8
10	Steam and hot water (S&HW)	7.4	4.2	1.7
		6.2	4.5	1.4
Producer's Maximum		14.9	41.2	1.7
Consumer's Maximum		12.2	39.1	1.7
Producer's Minimum		3.7	4.2	0.1
Consumer's Minimum		5.7	4.5	0.2

Source Prepared by the authors based on Russia in figures. Statistical handbook (2020), U.S. Energy Information Administration. Independent Statistics and Analysis (n/a), International Energy Agency. Data and statistics (n/a)

The period 2000–2019 saw an upward trend in return on gasoline, steam, and hot water, at both producer prices and consumer prices (excluding the gasoline price in 2019). This is the result of the relatively low level of competition. The return of steam and hot water and crude oil at the producer price turned out to be higher than at the consumer price in 2014 and 2015 (Russia in figures. Statistical handbook (2020), U.S. Energy Information Administration. Independent Statistics and Analysis (n/a), International Energy Agency. Data and statistics (n/a).

3.2 Comparison of the Growth Rates of Producer and Consumer Prices

Not all resources in Table 1 under consideration show uniform change. This suggests that when the situation on the market changes in different periods, there may be so-called "temporary drawdowns" or price measures instability. The most "stable" energy commodities were coking coal, fuel oil, and diesel fuel.

The assessment of the strength of the relationship between the producer return and the consumer return showed that the close correlation is observed only for fuel oil ($R^2 = 0.98$), diesel fuel ($R^2 = 0.93$) and coking coal ($R^2 = 0.90$).

The regression equations were constructed to quantify the correlation between the producer and the consumer returns for each energy commodity. Let us write down only models that have the reliability of the 95% confidence level.

$$Y = 0.70 \cdot X + 4.11; \; R^2 = 0.87 \tag{1}$$

$$Y = 1.20 \cdot X; \; R^2 = 0.97 \tag{2}$$

$$Y = 0.58 \cdot X; \; R^2 = 0.81 \tag{3}$$

X is the producer return, Y is the consumer return, R^2 is the determination coefficient.

Model 1 is built for diesel fuel, model 2 is for fuel oil, and model 3 is for coking coal.

(1) The diesel fuel, fuel oil, and coking coal markets should be attributed (somewhat conventionally) to the established and more stable markets based on the results of assessing the quantitative correlation between producer and consumer prices.

(2) Let's examine the fuel oil (model 2). The inclination of the line (k = 1.2) shows that the consumer return will increase by about 1.2% with an increase in the

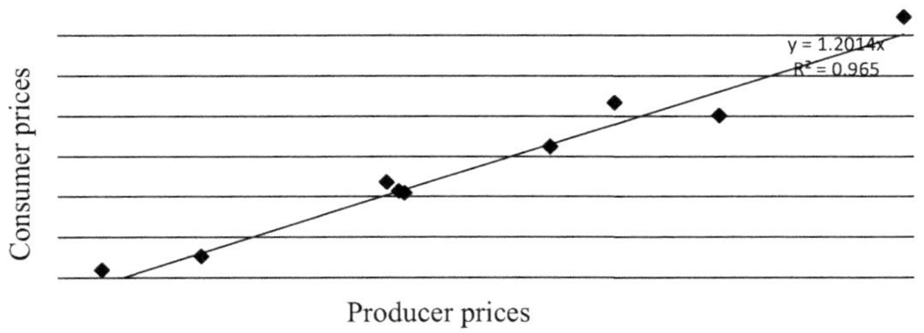

Fig. 1 The producer and the consumer return ratio (%). *Source* Prepared by the authors based on Russia in figures. Statistical handbook (2020), U.S. Energy Information Administration. Independent Statistics and Analysis (n/a), International Energy Agency. Data and statistics (n/a)

producer return by 1% (Fig. 1). It is only after that ratio that it is a truly attractive business for the consumer, which can generate higher profits.

(3) R^2 indicates that producer and consumer prices for diesel fuel, fuel oil and coking coal only bring return with a high degree of synchronicity.

Crude oil as a commodity provides a fairly good sales return. On the other hand, investing in this commodity is highly risky. Motor gasoline and coal less anthracite are significantly way below other energy commodities, both in the price growth rate and in the price measures stability.

Investment recommendations based solely on a comparative price analysis would have little effect (Markowitz, 1952; Dhankar, 2019). Price developments help to assess the price acceptability, determine the supply–demand situation on the market, see the future financial capacity. However, it is necessary not only to take into account the expected return but also to take an assessment of the expected risk level to select the most attractive group of leading energy carriers from the standpoint of future sales-boosting (Damodoran, 2012; Gitman, 2008; Markowitz, 1952; Dhankar, 2019; Gertsekovich & Babushkin, 2019; Gibson, 2008; Skoglund & Chen, 2015; Gertsekovich et al., 2020; Ferry, 2010).

3.3 Cost-benefit Evaluation

It is possible to draw up a preference table, calculating the average return of each of the considered energy commodities, and thereby highlight the most profitable pattern of the planned investment activity (Tables 1, 2 and 3).

Table 2 Return risk ratio at producer prices (by leaders)

No	*Energy commodity*	*Expected return (Pr),%*	*Risk (Rs),%*	*Return/Risk (Pr/Rs),%*
1	Steam and hot water	7.4	4.2	1.7
2	Brown coal	7.6	9.2	0.8
3	Coal, less anthracite, coking and brown coal	8.9	11.0	0.8
4	Crude oil	12.7	16.7	0.8
5	Diesel fuel	9.1	13.7	0.7
6	Electricity	4.8	8.0	0.6
7	Combustible natural gas	14.9	29.6	0.5
8	Coking coal	13.9	41.2	0.3
Maximum producer prices by leaders		14.9	41.2	1.7
Minimum producer prices by leaders		4.8	4.2	0.3

Source Prepared by the authors based on Russia in figures. Statistical handbook (2020), U.S. Energy Information Administration. Independent Statistics and Analysis (n/a), International Energy Agency. Data and statistics (n/a)

Table 3 Return risk ratio at consumer prices (by leaders)

No	*Energy commodity*	*Expected return (Pr),%*	*Risk (Rs),%*	*Return/Risk (Pr/Rs),%*
1	Diesel fuel	10.4	10.2	1.0
2	Motor gasoline	9.4	5.5	1.7
3	Coal, less anthracite, coking and brown coal	6.4	6.9	0.9
4	Steam and hot water	6.2	4.5	1.4
5	Brown coal	5.7	4.5	1.3
Maximum consumer prices by leaders		10.7	10.2	1.7
Minimum consumer prices by leaders		5.7	4.5	0.9

Source Prepared by the authors based on Russia in figures. Statistical handbook (2020), U.S. Energy Information Administration. Independent Statistics and Analysis (n/a), International Energy Agency. Data and statistics (n/a)

The greatest cost-effectiveness at producer prices was observed when producers sold combustible natural gas ($Rt = 14.9$), coking coal ($Rt = 13.9$), crude oil ($Rt = 12.70$), etc. (Tables 1and 2, column "Expected Return"). The leaders in terms of expected risk, i.e. the anti-leaders are coking coal ($Rs = 41.2$), fuel oil ($Rs = 32.2$) and combustible natural gas ($Rs = 29.6$).

The data in Tables 1 and 3 (column "Expected Return") on consumer prices show that the top-performing energy commodities are crude oil ($Rt = 12.2$), diesel fuel ($Rt = 10.4$) and motor gasoline ($Rt = 9.4$). The anti-leaders in terms of the risk level include fuel oil ($Rs = 39.1$), coking coal ($Rs = 27.4$) and crude oil ($Rs = 23.6$). In that way, a joint analysis of indicators for these energy commodities does not allow to form an objective preference table. A more accurate study is provided by an analysis based on the "Risk return" model. It is based on the fundamental propositions of portfolio performance, the winner's portfolio construction (De Bondt, 1987; Ferry, 2010; Graham, 2006; Jegadeesh, 1993; O'Shaughnessy, 2005; Sharpe et al., 1998), and the optimization of the learning sample length concept (Damodoran, 2012; Fama & French, 1988; Gershenhorn, 1977).

Let us take the risk and the expected return comparison first at producer prices and then at consumer prices. Such commodities as motor gasoline and fuel oil should be excluded from the further analysis for 2000–2019 (Fig. 2).

A synthesized group of energy leaders at producer prices is shown in Fig. 2.

Let us construct a linear regression for the formed energy leaders (model 4):

$$Pr = 0.24 \cdot Rs + 5.87;\ \mathrm{R}^2 = 0.72 \tag{4}$$

The synthesized "Risk return" model (for producers) allows us to assess the quantitative relationship between return and risk. Thus, an increase in risk by 100% will lead to an increase in return by 24% (4).

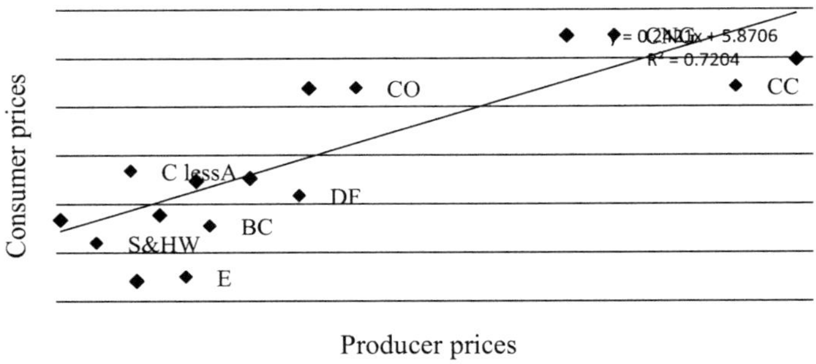

Fig. 2 Energy leaders at producer prices. *Source* Prepared by the authors based on Russia in figures. Statistical handbook (2020), U.S. Energy Information Administration. Independent Statistics and Analysis (n/a), International Energy Agency. Data and statistics (n/a)

The presented group of energy leaders (Fig. 2) should be divided, somewhat conventionally, into two subgroups.

(1) Electricity, steam and hot water, brown coal, coal less anthracite, diesel fuel, and crude oil fits into the low-profit and less risky commodities for investment. This subgroup of investment patterns is suited to "old school" and people who prefer stability (Dhankar, 2019; Markowitz, 1952). Moreover, crude oil and combustible natural gas are more preferable, as they are above the trend.

(2) Coking coal and combustible natural gas are placed into a subgroup focused on investors who are not afraid to take responsibility for great risk.

The investor addresses the issue of achieving the maximum return with the lowest risk level. This calls for setting priorities in the fuel and energy sector in such a way as to "protect" himself from the changes in the considered energy carriers cost and the losses (Table 3). Let us construct a scatter diagram to improve the efficiency of investment decisions. Then a group of energy leaders is formed in terms of consumer prices based on the analysis using the "Return risk" model (Fig. 3).

The "Return risk" model (5) is constructed for energy commodities at consumer prices and appears as follows:

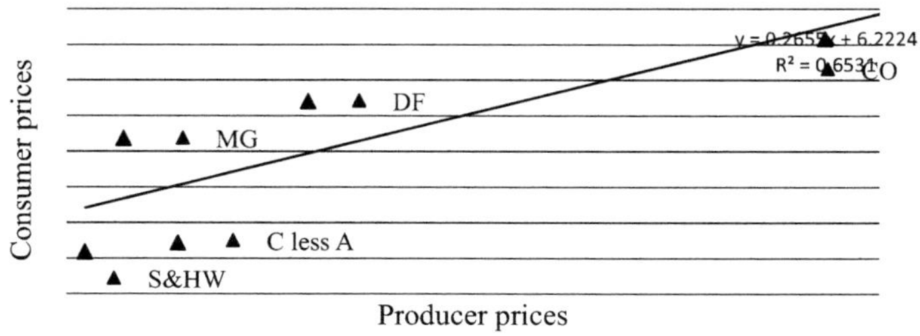

Fig. 3 Energy leaders at consumer prices. *Source* Prepared by the authors based on Russia in figures. Statistical handbook (2020), U.S. Energy Information Administration. Independent Statistics and Analysis (n/a), International Energy Agency. Data and statistics (n/a)

$$Pr = 0.28 \cdot Rs + 6.24;\ R^2 = 0.65 \quad (5)$$

“Risk return” model (5) allows assessing the quantitative relationship between return and risk. Thus, an increase in risk by 100% will lead to an increase in return by 28% (4). It follows from model (5) that the trend line has a larger slope than that of the model (4). The slope differences are significant and amount to approximately 15%. This means that an investor who «operates» with consumer prices, all other things being equal, can count on a significantly lower risk level.

Figure 3 shows the relationship between risk and return at consumer prices, which can be seen that gasoline, steam and hot water, diesel fuel, coal less anthracite, and crude oil are the optimal groups for prospective investors. But this energy commodity is not meant for all investors at that risk level.

Steam and hot water, diesel fuel, coal less anthracite and crude oil are the leaders, both at the producer and at consumer prices; therefore, it is recommended to make the final choice in favour of these energy commodities. Whereas fuel oil does not belong to the leaders, at both the producer and consumer prices.

The firm conclusion follows from the comparative analysis of extreme values in terms of return, risk and returns risk ratio for the leaders in producer and consumer prices: according to the data of the researched energy commodities, the business associated with their consumption is much preferable to the business of their production during 2000–2019. This conclusion is correct for all three criteria: return, risk, and return risk ratio (Tables 2 and 3).

The paper’s results were obtained using the fundamental principles of the portfolio theory through the statistical data. Therefore, one should not rely only on the computations when making investment decisions, because deviations and some discrepancies are possible in real life. Some scenarios are either impossible to predict, or their occurrence is unlikely (for example, COVID-19). In the long run, the trends following can help prevent mistakes when choosing the most preferred energy commodity by minimizing the risk.

4 Conclusion

The article analyzes the prices for the main energy commodities, gives objective assessments of their investment attractiveness according to the criteria: return, risk and return risk ratio. Investors should be guided by these criteria when choosing the most attractive investment pattern in the fuel and energy industry.

The synthesized groups of energy leaders are in relatively stable demand, which may be associated with the stable demand for them and the lack of alternatives to these energy commodities at present and shortly. Therefore, the energy leaders defined in the article are capable of turning into a big profit to both their producers and their consumers.

The choice of such energy commodities as motor gasoline, steam, and hot water, and diesel fuel as the most optimal group gives hope for high efficiency of investment strategies (Bagdasarov, 2017, p. 30).

Gasoline and diesel fuel will be in demand as long as transport exists, or until more environmentally friendly fuel (for example, biofuel) comes to replace it. Therefore, prices for these energy commodities will rise through the increased demand. Steam and hot water are widely used in heat exchange processes in production and oil and gas, oil refining and other industries.

However, one should not forget about the factors that can negatively affect demand and reduce prices. It is necessary not only to carry out research but also to monitor the political, economic, ecological, and other processes taking place in the world to reduce the potential losses risk.

5 Results and Discussion

This paper was the first to substantiate the possibility of using the portfolio analysis of statistical data on average prices of producers and consumers to select the most attractive types of energy commodities for investment in their production and consumption and to meet the goals of sustainable development. Previously, such a problem-solving approach was not used, and the problems were solved based on "classical criteria" such as net revenue, gross profit, return, etc. The proposed approach can be applied both in Russia and in other countries, subject to the availability of similar statistical data. On its basis, sustainable development of enterprises engaged in the production and consumption of certain types of energy commodities can be ensured.

References

Bagdasarov, L. N. (2017). *Popular oil refining*. Gubkin University.

Bouchouev, I. (2017). Measuring financial supply & demand for oil derivatives. In *Energy Risk Europe Conference, Koch Supply & Trading*.

Damodoran, A. (2012). *Investment Valuation: Tools and techniques for determining the value of any asset* (3rd ed.). Wiley.

De Bondt, W. F. M. (1987). Further evidence on investor overreaction and stock market seasonality. *Journal of Finance, 42*(3), 557–581.

Dhankar, R. S. (2019). Stock Market Overreaction. In: *Risk-Return Relationship and Portfolio Management*. India Studies in Business and Economics, Springer.

Fama, E. F., & French, K. R. (1988). Permanent and temporary components of stock prices. *Journal of Political Economy, 96*(2), 246–273.

Ferry, R. (2010). *All about asset allocation* (2nd ed.). McGraw-Hill Education.

Gershenhorn, G. I. (1977). A software package for constructing empirical differential equations. *Long-term natural phenomena forecasts* (pp. 133–137). Science publishing house.

Gertsekovich, D. A., Podlinyaev, O. L., & Larin, S. N. (2020). Vertical-aligned economic and mathematical model of the investment decision-making system in agriculture. *Journal of Economy and Entrepreneurship, 14*(9), 748–754.

Gertsekovich, D. A., & Babushkin, R. V. (2019). Dynamic portfolio analysis of world stock indices. *World of Economics and Management, 19*(4), 14–30.

Gibson, R.C. (2008). *Investment portfolio formation: Asset allocation: Balancing Financial Risk* (3rd ed.). McGraw-Hill.

Gitman, L. J. (2008). *Fundamentals of investing* (10th ed.). Pearson Addison-Wesley.

Graham, B. (2006). *The intelligent investor: The definitive book on value investing*. HarperCollins Publishers Ltd.

International Energy Agency. Data and statistics. https://www.iea.org/data-and-statistics/data-tables. Accessed December 25, 2020

Ivanov, N. A. (Ed.) (2019). *Shale revolution and energy transition, Nestor-History*. Moscow-St. Petersburg.

Jegadeesh, N. (1993). Returns to buying winners and selling losers: Implications for stock market efficiency. *Journal of Finance, 48*(1), 65–91.

Markowitz, H. M. (1952). Portfolio selection. *Journal of Finance, 7*(1), 77–91.

Mikhailova, A. Yu., Burakov, D. B., & Didenkoc, V. Yu. (2019). Relationship between oil price and macroeconomic indicators in Russia. *Finance: Theory and Practice, 23*(2), 105–116.

Nazarova, Z. M., Zabaykin, Yu. V., & Yakunin, M. A. (2020). Formation of a stable price for energy resources by solving a non-linear dynamic problem as an element of socio-economic stability of the economy. *News of the USMU, 57*(1), 170–181.

O'Shaughnessy, J. (2005). *What Works on Wall Street* (3rd ed.). McGraw-Hill.

Red'kin, N. M. (2019). Investment portfolio optimization on Russian stock market in context of behavioral theory. *Finance: Theory and Practice, 23*(4), 99–116.

Russia in figures. Statistical handbook (2020). https://rosstat.gov.ru/storage/mediabank/GOyirKPV/Rus_2020.pdf. Accessed December 25, 2020.

Sharpe, W. F., Gordon, A. J., & Bailey, J. W. (1998). *Investments* (6th ed.). Prentice Hall.

Skoglund, J., & Chen, W. (2015). *Financial risk management: Applications in market, credit, asset and liability management and advanced financial management*. Wiley Finance Series.

U.S. Energy Information Administration. Independent Statistics and Analysis. https://www.eia.gov/international/data/world. Accessed December 25, 2020

Sustainable Agriculture and Alternative Energy Based on High-Tech Entrepreneurship Development in the Agricultural Machinery Market

Tatiana N. Litvinova and Olga M. Zemskova

Abstract The paper aims to substantiate the need and develop recommendations for sustainable agriculture and alternative energy based on high-tech development of entrepreneurship in the agricultural machinery market. The authors use the method of regression analysis and the method of scenario analysis. The authors formed a sample of countries leading in exports of agricultural machinery in the world in 2020. As a result, the authors prove that sustainable agriculture and alternative energy based on high-tech development of entrepreneurship in the agricultural machinery market allows the systematic implementation of Sustainable Development Goal 2 and Sustainable Development Goal 7. Moreover, it unlocks the potential of the development of the geo-economy in the period up to 2030. AI and big data are the most promising vectors of developing high-tech businesses in the agricultural machinery market to achieve the geo-economy of the future. The authors recommend a 77.23% increase in the use of AI and big data for high-tech entrepreneurial development in the agricultural machinery market. This will increase agricultural sustainability by 9.70% and the share of alternative energy by 2.80%.

Keywords Sustainable agriculture · Alternative energy · High-tech development · Entrepreneurship · Agricultural machinery market

JEL Codes L26 · O13 · O14 · Q01 · Q13 · Q42

1 Introduction

The Sustainable Development Goals (SDGs) are a systematic view of the priorities of strategic development of society and the economy. The SDGs are comprehensively formulated since they are closely interrelated. However, the SDGs are often implemented in isolation in practice. In some cases, this leads to limited results for

T. N. Litvinova (✉) · O. M. Zemskova
Department of Management and Logistics in Agribusiness, Volgograd State Agrarian University, Volgograd, Russia

E. G. Popkova and B. S. Sergi (eds.), *Geo-Economy of the Future*,
https://doi.org/10.1007/978-3-030-92303-7_90

the implemented SDGs. In other cases, it causes negative externalities for related SDGs, preventing their achievement.

The priorities of the geo-economy of the future are combined in a common block that includes sustainable agriculture (SDG 2) and alternative energy (SDG 7). The separate implementation of two SDGs distorts their meaning and leads to contradictory results. On the one hand, alternative energy is evolving from fossil fuels to renewable but environmentally hazardous nuclear power. Accidents at nuclear power plants increase agricultural risks. On the other hand, the sustainability of agriculture is achieved through energy-intensive automation.

Thus, the current practice of implementing these two SDGs contradicts the essence of the geo-economy of the future. This essence consists of the systematic introduction of green innovation in energy and agriculture and responsible agricultural production based on clean (solar, wind, etc.) energy. This paper hypothesizes that the high-tech development of entrepreneurship in the agricultural machinery market will solve the problem and ensure the comprehensive implementation of SDG 2 and SDG 7 since agricultural machinery is the link between energy and agriculture and, therefore, serves the core of the future geo-economy.

The paper aims to substantiate the need and develop recommendations for sustainable agriculture and alternative energy based on the high-tech development of entrepreneurship in the agricultural machinery market.

2 Literature Review

The scientific literature studies the problem's components separately, which leads to the fragmented nature of findings. Thus, the issues of sustainable agriculture are considered by Sazanova and Ryazanova (Sazanova & Ryazanova, 2019), Sergi et al. (Sergi et al., 2019), and Sofiina (Sofiina, 2020). The issues of alternative energy are reflected in the publications Adedoyin et al. (Adedoyin et al., 2021), Kulasuriya et al. (Kulasuriya et al., 2021), Olvera-Gonzalez et al. (Olvera-Gonzalez et al., 2021), Pedersen et al. (Pedersen et al., 2021), Popkova and Sergi (Popkova & Sergi, 2020), and Popkova et al. (Popkova et al., 2019).

The issues of high-tech development of entrepreneurship in the agricultural machinery market are disclosed in the works of Litvinova (Litvinova, 2020, 2021) and Singh (Singh, 2018). To overcome the gaps at the intersection of the noted issues, this paper conducts a systematic study of sustainable agriculture and alternative energy based on the high-tech development of entrepreneurship in the agricultural machinery market.

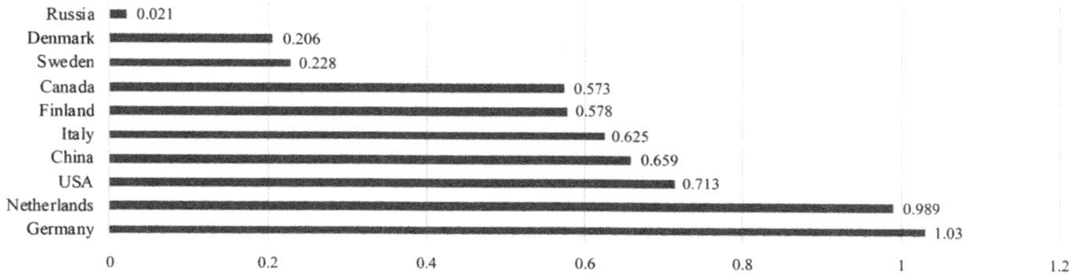

Fig. 1 Exports of agricultural machinery in the sample countries in 2020, $ billion. *Source* Compiled by the authors based on the materials of OEC (2021)

3 Research Methodology

To test the hypothesis, the authors selected the method of regression analysis as the most accurate method of economic statistics. The authors formed a sample of countries leading in exports of agricultural machinery in the world in 2020. Figure 1 presents the statistics of exports of agricultural machinery from these countries in 2020.

According to Fig. 1, the largest exporter of agricultural machinery in 2020 was Germany ($1.03 billion), with the Netherlands in second place ($0.989 billion) and the USA in third place ($0.713 billion). The samples included developing countries such as China ($0.659 billion, fourth place) and Russia ($0.021 billion).

To test the hypothesis, the authors determine the correlation between the vectors of high-tech development of entrepreneurship (linkage to the market of agricultural machinery is provided by a special sample of countries leading in exports of agricultural machinery)—robotization, artificial intelligence (AI), and big data (according to IMD), the share of alternative energy (according to the Economist Intelligence Unit Limited), and sustainability of agriculture (according to UNDP). The statistical basis of the research is presented in Table 1.

4 Findings

The contribution of high-tech entrepreneurial development in the agricultural machinery market to sustainable agriculture and alternative energy is reflected in the following multiple linear regression models derived from the analysis of data from Table 1:

- $y_1 = 12.22 + 0.47*x_1\text{-}0.10*x_2$, $R^2 = 42.57$;
- $y_2 = 53.30 + 0.39*x_1\text{-}0.10*x_2$, $R^2 = 70.83$.

According to the obtained regression models, robotization in the agricultural machinery market does not contribute to the sustainable development of agriculture and alternative energy. The most promising vector of high-tech entrepreneurship in

Table 1 Statistics on sustainable agriculture, alternative energy, and high-tech business development in countries with a developed agricultural export market

Country	Share of alternative energy in total primary energy supply, %	Natural resources and resilience, points 1–100	Robotization, place 1–63	Artificial intelligence and big data, place 1–63
	y_1	y_2	x_1	x_2
Germany	14.633	52.9	5	46
Netherlands	7.177	61.5	21	20
USA	7.915	51.4	4	9
China	–	51.2	1	8
Italy	18.236	50.7	6	59
Finland	34.126	73.2	33	15
Canada	16.374	54.5	13	4
Sweden	40.782	67.4	18	7
Denmark	36.931	57.6	30	12
Russia	–	55.0	32	33

Source Compiled by the authors based on the materials of IMD (IMD, 2021), the Economist Intelligence Unit Limited (The Economist Intelligence Unit Limited, 2021), and UNDP (UNDP, 2021)

the agricultural machinery market for the future geo-economy is AI and big data. Their promotion to first place contributes to a 0.10-points increase in the sustainability of agriculture. Moreover, the share of alternative energy increases by 0.10%. The importance of AI and big data for the geo-economy of the future is evidenced by the close correlation of the indicators: the correlation was 42.57% and 70.83%, respectively.

In order to determine the prospects of the geo-economy of the future, the authors conduct a scenario analysis of sustainable agriculture and alternative energy in countries with a developed export agricultural machinery market. The authors compiled histograms of the normal distribution of forecast y_1, y_2, and x_2 (Fig. 2) until 2030 based on the arithmetic mean and standard deviations of data.

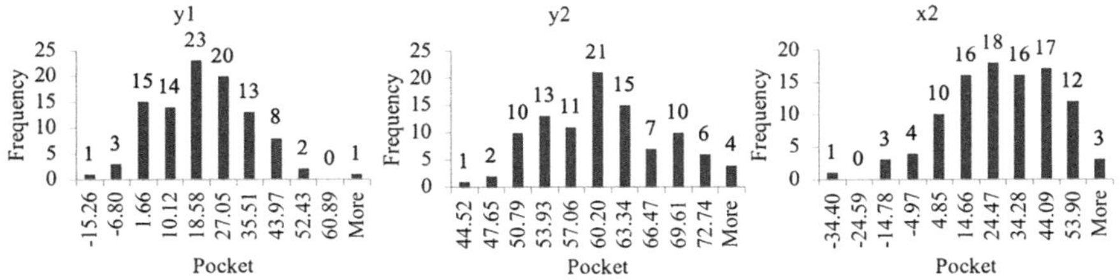

Fig. 2 Histograms of the normal forecast distribution y_1, y_2, and x_2. *Source* Calculated and compiled by the authors

Table 2 Scenario analysis of sustainable agriculture and alternative energy in the countries with the developed export markets of agricultural machinery up to 2030

Scenario	Sub-scenario	Share of alternative energy in, %		Agricultural sustainability, points 1–100		Artificial intelligence and big data, place 1–63	
		Value, %	Growth*, %	Value, points 1–100	Growth*, %	Value, place 1–63	Growth*, %
Baseline value in 2020		17.62	–	57.54	–	21.30	–
Realistic scenario	**	*18.58*	5.45	*60.20*	4.62	–	–
Pessimistic scenario	**	*1.66*	−90.58	*44.52*	−22.63	–	–
Optimistic scenario	***	19.33	9.70	59.15	2.80	*4.85*	−77.23
Optimizing	***	19.73	11.98	59.52	3.44	*1.00*	−95.31

Note * Compared to the base value in 2020; ** Regardless of the development of high-tech entrepreneurship in the market of agricultural machinery; *** Depending on the development of high-tech development of entrepreneurship in the market of agricultural machinery. *Source* Calculated and compiled by the authors

Based on Fig. 2, the authors determined probable scenarios in accordance with the obtained regression models (Table 2).

In Table 2, the authors highlighted the values of the indicators taken as the basis of the corresponding scenarios in bold and dark background. Scenario analysis in Table 2 shows that under the realistic (most probable) scenario without targeted development of high-tech entrepreneurship, the agricultural machinery market will see an increase in the sustainability of agriculture by 5.45% and alternative energy by 4.62%. In the pessimistic scenario, there would be a 90.58% decline in agricultural sustainability and a 22.63% decline in alternative energy by 2030. However, the measures necessary to implement this scenario are unknown, which makes its practical implementation difficult.

Supporting the already launched trend of high-tech business development in the market of agricultural machinery (77.23% increase in the use of AI and big data) will allow implementing the optimistic scenario, in which there will be a 9.70% increase in agricultural sustainability and a 2.80% increase in alternative energy. The maximum possible (by increasing the use of AI and big data to the first place) increase in the sustainability of agriculture and alternative energy through high-tech development of entrepreneurship in the market of agricultural machinery is 11.98 and 3.44%.

5 Conclusions

Thus, sustainable agriculture and alternative energy based on high-tech development of entrepreneurship in the agricultural machinery market allows for the systematic implementation of SDG 2 and SDG 7. Moreover, it unlocks the potential of the future geo-economy in the period up to 2030. This fact confirms the hypothesis put forward. The authors recommend increasing the use of AI and big data by 77.23% for high-tech entrepreneurial development in the agricultural machinery market. This will increase agricultural sustainability by 9.70% and the share of alternative energy by 2.80%.

References

Adedoyin, F. F., Alola, A. A., & Bekun, F. V. (2021). The alternative energy utilization and common regional trade outlook in EU-27: Evidence from common correlated effects. *Renewable and Sustainable Energy Reviews, 145*, 111092. https://doi.org/10.1016/j.rser.2021.111092

IMD. (2021). World Digital Competitiveness Ranking 2020. Retrieved from https://www.imd.org/centers/world-competitiveness-center/rankings/world-competitiveness/. Accessed July 6, 2021.

Kulasuriya, C., Vimonsatit, V., & Dias, W. P. S. (2021). Performance-based energy, ecological and financial costs of a sustainable alternative cement. *Journal of Cleaner Production, 287*, 125035. https://doi.org/10.1016/j.jclepro.2020.125035

Litvinova, T. N. (2020). Management of the development of infrastructural support for entrepreneurial activity in the Russian agricultural machinery market based on "smart" technologies of antimonopoly regulation. Retrieved from https://iscconf.ru/управление-развитием-инфраструктурн/. Accessed July 6, 2021.

Litvinova, T. N. (2021). Infrastructure maintenance of "Smart technologies" for entrepreneurship in the agricultural machinery market: Needs vs. opportunities. In E. G. Popkova, & B. S. Sergi, (Eds.), *"Smart technologies" for society, state and economy* (pp. 1670–1677). Springer. https://doi.org/10.1007/978-3-030-59126-7_182

OEC. (2021). Other agricultural machinery: Exporters and importers, trade by country. Retrieved from https://oec.world/en/profile/hs92/other-agricultural-machinery#exporters-importers. Accessed July 6, 2021.

Olvera-Gonzalez, E., Escalante-Garcia, N., Myers, D., Ampim, P., Obeng, E., Alaniz-Lumbreras, D., & Castaño, V. (2021). Pulsed led-lighting as an alternative energy savings technique for vertical farms and plant factories. *Energies, 14*(6). https://doi.org/10.3390/en14061603

Pedersen, T. T., Victoria, M., Rasmussen, M. G., & Andresen, G. B. (2021). Modeling all alternative solutions for highly renewable energy systems. *Energy, 234*, 121294. https://doi.org/10.1016/j.energy.2021.121294

Popkova, E. G., & Sergi, B. S. (2020). Energy efficiency in leading emerging and developed countries. *Energy, 221*(1), 119730. https://doi.org/10.1016/j.energy.2020.119730

Popkova, E. G., Inshakov, O. V., & Bogoviz, A. V. (2019). Motivation of energy saving within the corporate market responsibility of economic entities. In O. Inshakov, A. Inshakova, & E. Popkova (Eds.), *Energy sector: A systemic analysis of economy, foreign trade and legal regulations* (pp. 95–105). Springer. https://doi.org/10.1007/978-3-319-90966-0_7

Sazanova, S. L., & Ryazanova, G. N. (2019). Problems and opportunities of development of the agricultural industry of Russia from the point of view of Marxist theory. In M. L. Alpidovskaya, & E. G. Popkova (Eds.), *Marx and modernity: A political and economic analysis of social systems*

management (pp. 599–608). Information Age Publishing. Retrieved from https://www.infoagepub.com/products/Marx-and-Modernity. Accessed July 6, 2021.
Sergi, B. S., Popkova, E. G., Bogoviz, A. V., & Ragulina, Yu. V. (2019). The Agro-industrial Complex: Tendencies, scenarios, and policies. In B. S. Sergi (Ed.), *Modeling economic growth in contemporary Russia*. Emerald Publishing. https://doi.org/10.1108/978-1-78973-265-820191009
Singh, S. (2018). Strategy, current activities, and future prospect for advancing Indian agricultural machinery into the African market. *AMA, Agricultural Mechanization in Asia, Africa, and Latin America, 49*(2), 31–42.
Sofiina, E. V. (2020). Economic return from land use as a factor in the "smart" antitrust regulation of the agricultural market. Retrieved from https://iscconf.ru/экономическая-отдача-от-землепользо/. Accessed July 6, 2021.
The Economist Intelligence Unit Limited. (2021). Global Food Security Index 2020. Retrieved from https://foodsecurityindex.eiu.com/index. Accessed July 6, 2021.
UNDP. (2021). Sustainable development report 2020: Data explorer. Retrieved from https://dashboards.sdgindex.org/explorer/share-of-renewable-energy-in-total-primary-energy-supply/chart. Accessed July 6, 2021.

Model of Uninterruptible and Highly Efficient Operation of Vertical Farms Based on Alternative Energy in the Interests of Sustainable Agriculture

Elena G. Popkova

Abstract The paper aims to develop a model for the uninterruptible and highly efficient operation of vertical farms based on alternative energy for sustainable agricultural development. The author uses regression analysis to model the impact of the share of alternative and nuclear energy on the sustainability of agriculture. As a result, the author proves that alternative energy is already demanded in agriculture worldwide, as evidenced by the established high regression dependence of relevant statistical indicators. Nevertheless, smart vertical farms cannot completely switch to alternative energy since it is more suitable for horizontal farms with low energy consumption that can continue their operation even in the temporary absence of power supply. For large smart vertical farms with large production capacity and correspondingly high energy intensity, the "3E" model was developed. This model involves the combination and flexible use of solar (E1), wind (E2), and electric (E3) energy. A smart vertical farm prioritizes the use of alternative energy depending on its availability, monitored and controlled by AI through the IoT. If alternative energy is not available, the vertical farm switches to electric power. The proposed model ensures the uninterruptible and highly efficient operation of vertical farms based on alternative energy for sustainable agricultural development.

Keywords Vertical farm · Alternative energy · Sustainable agriculture · Smart agriculture · AI · IoT

JEL codes M11 · M14 · M21 · O13 · P28 · P48 · Q01 · Q12 · Q42

1 Introduction

The Fourth Industrial Revolution broadened the horizon of agricultural development. Throughout previous industrial revolutions, automation covered various spheres of economic activity but bypassed agriculture, limiting it to fragmented mechanization.

E. G. Popkova (✉)
Center for Applied Research of the Chair "Economic Policy and Public-Private Partnership" of MGIMO University, Moscow, Russia

E. G. Popkova and B. S. Sergi (eds.), *Geo-Economy of the Future*,
https://doi.org/10.1007/978-3-030-92303-7_91

For example, at the beginning of the twenty-first century, when the industry was reduced to the control of automated production, agriculture (crop production) was dominated by manual labor. All plants were planted manually, care and control of plant growth were also performed by humans, and only the harvest was performed using agricultural machinery.

Industry 4.0 has brought breakthrough scientific and technological advances to agriculture for the first time in the history of industrial revolutions. It allowed for the creation of smart autonomous vertical farms controlled by artificial intelligence (AI). The transformation of agricultural production into cyber-physical systems based on the Internet of Things (IoT) allows considerably reducing the labor intensity of agricultural production, achieving a revolutionary increase in labor productivity, guaranteeing the high quality of agricultural products, and reducing the risks of their production due to their independence from the environment. This allows for a fundamental change in the approach to food security and the full implementation of the Sustainable Development Goal 2 (Zero Hunger).

Nevertheless, the lack of power supply is a serious problem for the automation of agriculture. Rural areas experience a shortage of energy, while the energy intensity of smart vertical farms is very high. The prospects for solving the problem are related to the transition of agriculture to alternative energy. This research aims to develop a model for the uninterruptible and highly efficient operation of vertical farms based on alternative energy for sustainable agriculture.

2 Literature Review

Opportunities and prospects for the development of alternative energy were sufficiently studied in the works of such authors as Ashok et al. (2021), Askari Fard et al. (2021), Neves et al. (2021), Popkova and Sergi (2020), Popkova et al. (2019), Saraswat and Digalwar (2021), and Sillero et al. (2021). The specifics of vertical farms and their significant contribution to the sustainable development of agriculture are noted in the studies of such scholars as Litvinova (2020), Sazanova and Ryazanova (2019), Sergi et al. (2019), and Sofiina (2020).

Nevertheless, these issues are studied separately, which causes a gap at their junction, associated with the uncertainty of the prospects for using alternative energy for sustainable agriculture. This paper develops a model for the uninterruptible and highly efficient operation of vertical farms based on alternative energy to fill the identified gap.

3 Research Methodology

This study hypothesizes that alternative energy increases agricultural sustainability. To test the hypothesis posed, the author implements regression analysis to model

Table 1 Statistics of alternative energy and agricultural sustainability for the sample countries in 2021

Level of development of alternative energy	Country	Alternative and nuclear energy, % of total energy consumption	Agricultural sustainability, points 1–100
		Alternative and nuclear energy, % of total energy use	Natural Resources and Resilience, points 1–100
High	France	49.1	59.0
	Sweden	43.2	67.4
	Switzerland	39.9	64.2
	Tajikistan	37.5	37.0
Average	Germany	12.9	52.9
	USA	11.9	51.4
	Brazil	10.8	47.1
	Russia	8.2	55.0
Low	Malaysia	1.2	47.5
	Thailand	1.1	59.5
	Kazakhstan	0.6	83.7
	Singapore	0.2	82.3

Source Compiled by the author based on The Economist Intelligence Unit Limited (2021) and The World Bank (2021)

the impact of the share of alternative and nuclear energy (x) on the sustainability of agriculture (y):

$$y = a + b * x \qquad (1)$$

The hypothesis is considered proven if a positive value of the regression coefficient *b* is found. To obtain accurate and reliable results that can be extended to different countries, the author forms a sample, which includes countries with different levels of the development of alternative energy (Table 1).

4 Findings

According to Table 1, a regression curve is constructed reflecting the dependence of agricultural sustainability on alternative energy in 2021 (Fig. 1).

According to Fig. 1, a 1% increase in the share of alternative energy increases agriculture's sustainability by 3.1088 points (moderate correlation—33.04%). Consequently, the hypothesis is confirmed: in Industry 4.0, smart vertical farms are heavily

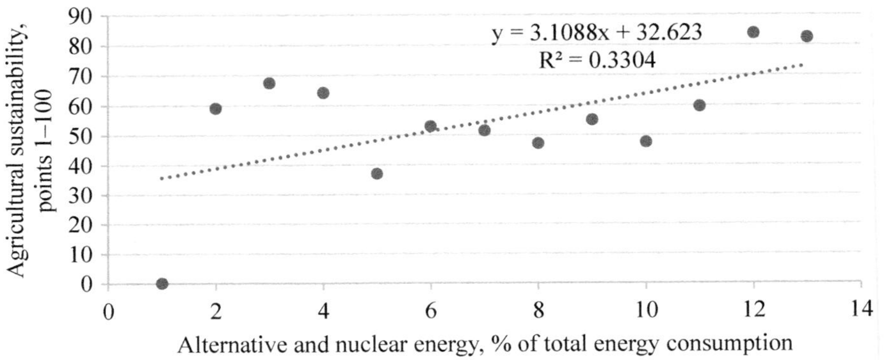

Fig. 1 Regression curve of the dependence of agricultural sustainability on alternative energy in 2021. *Source* Calculated and compiled by the author

dependent on alternative energy. However, with the current approach to using alternative energy in agriculture, smart vertical farms face energy shortages due to the insufficient capacity of alternative sources (solar panels and windmills).

Moreover, vertical farms depend on intermittent alternative energy production tied to the favorability of the environment (sunlight and wind strength). Thus, a shortage of alternative energy causes disruptions in vertical farms. This fact reduces their competitiveness compared to horizontal farms and prevents them from unlocking the potential contribution of vertical farms to food security.

Therefore, a new approach to using alternative energy in agriculture is needed, which will ensure high efficiency and the uninterruptible operation of smart vertical farms. The "3E" model is proposed as such an approach (Fig. 2). A special feature and an advantage of the new model is the simultaneous use of three energy sources.

The first source is solar energy generated by the vertical farm using solar panels. The second source is wind power. The vertical farm obtains it with its windmills. The third source is electricity obtained by the vertical farm through an external power supply.

The vertical farm prioritizes alternative (solar and wind) energy derived from the environment. Thus, it depends on interactions with it. If no alternative energy is available, which is detected by AI through the IoT, the vertical farm automatically switches to the use of pre-stored electricity in the amounts necessary (its reserve is created).

This makes the "3E" vertical farm environmentally friendly and highly flexible in combining various energy sources available to it. This allows to fully meet the energy needs of a smart vertical farm and does not require manual control of energy supply, and switching of energy flows from different sources. Moreover, the proposed model allows for the establishment of large smart vertical farms with great production capacity, which will completely solve the problem of food security in the long term.

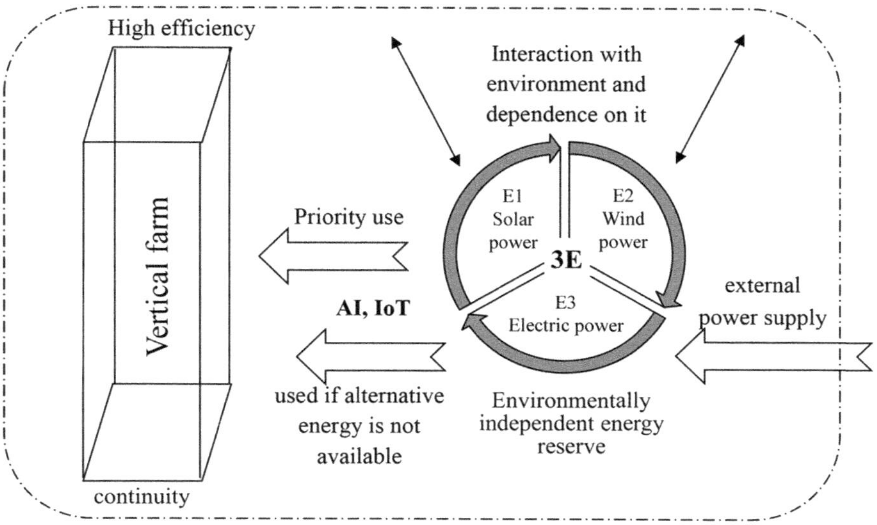

Fig. 2 Model "3E" for the uninterruptible and highly efficient operation of vertical farms based on alternative energy for sustainable agriculture. *Source* Calculated and compiled by the author

5 Conclusions

Alternative energy is already in demand in agriculture worldwide, as evidenced by the established high regression dependence of relevant statistical indicators. Nevertheless, smart vertical farms cannot switch entirely to alternative energy; it is more suitable for horizontal farms with low energy consumption that can continue working even in the temporary absence of power supply.

For large smart vertical farms with high production capacity and high energy consumption, the "3E" model was developed. This model involves the combination and flexible use of solar (E1), wind (E2), and electric (E3) energy. A smart vertical farm prioritizes the use of alternative energy depending on its availability, monitored and controlled by AI through the IoT. If alternative energy is not available, the vertical farm switches to electric power.

The proposed model ensures the uninterruptible and highly efficient operation of vertical farms based on alternative energy for sustainable agricultural development. However, it is necessary to note that the high cost of implementation and maintenance is a disadvantage of the "3E" model, making it inaccessible to the broad masses of small and medium agricultural businesses, which is a limitation of this model and the results of this study. The "3E" model is focused on large vertical farms. Future research should pay attention to the development of promising ways to transition small and medium-sized vertical farms to alternative energy to overcome the noted limitation.

References

Ashok, K., Natarajan, R., Kumar, P., Sharma, K., & Mathur, M. (2021). Sustainable alternative futures for agriculture in India—The energy, emissions, and resource implications. *Environmental Research Letters, 16*(6), 064001. https://doi.org/10.1088/1748-9326/abf0cd

Askari Fard, A., Hashemy Shahdany, S. M., & Javadi, S. (2021). Automatic surface water distribution systems: A reliable alternative for energy conservation in agricultural section. *Sustainable Energy Technologies and Assessments, 45*, 101216. https://doi.org/10.1016/j.seta.2021.101216

Litvinova, T. N. (2020). Management of the development of infrastructural support for entrepreneurial activity in the Russian agricultural machinery market based on "smart" technologies of antimonopoly regulation. Retrieved from https://iscconf.ru/управление-развитием-инфраструктурн/

Neves, D., Baptista, P., & Pires, J. M. (2021). Sustainable and inclusive energy solutions in refugee camps: Developing a modelling approach for energy demand and alternative renewable power supply. *Journal of Cleaner Production, 298*, 126745. https://doi.org/10.1016/j.jclepro.2021.126745

Popkova, E. G., & Sergi, B. S. (2020). Energy efficiency in leading emerging and developed countries. *Energy, 221*, 119730. https://doi.org/10.1016/j.energy.2020.119730

Popkova, E. G., Inshakov, O. V., & Bogoviz, A. V. (2019). Motivation of energy saving within the corporate market responsibility of economic entities. In O. Inshakov, A. Inshakova, & E. Popkova (Eds.), Energy sector: A systemic analysis of economy, foreign trade and legal regulations (pp. 95–105). Springer. https://doi.org/10.1007/978-3-319-90966-0_7

Saraswat, S. K., & Digalwar, A. K. (2021). Evaluation of energy alternatives for sustainable development of energy sector in India: An integrated Shannon's entropy fuzzy multi-criteria decision approach. *Renewable Energy, 171*, 58–74. https://doi.org/10.1016/j.renene.2021.02.068

Sazanova, S. L., & Ryazanova, G. N. (2019). Problems and opportunities of development of the agricultural industry of Russia from the point of view of Marxist theory. In M. L. Alpidovskaya, & E. G. Popkova (Eds.), Marx and modernity: A political and economic analysis of social systems management (pp. 599–608). Information Age Publishing. Retrieved from https://www.infoagepub.com/products/Marx-and-Modernity

Sergi, B. S., Popkova, E. G., Bogoviz, A. V., & Ragulina, Y. V. (2019). The Agro-industrial complex: Tendencies, scenarios, and policies. In B. S. Sergi (Ed.), *Modeling economic growth in contemporary Russia*. Emerald Publishing. https://doi.org/10.1108/978-1-78973-265-820191009

Sillero, L., Prado, R., Welton, T., & Labidi, J. (2021). Energy and environmental analysis of flavonoids extraction from bark using alternative solvents. *Journal of Cleaner Production, 308*https://doi.org/10.1016/j.jclepro.2021.127286

Sofiina, E. V. (2020). Economic return from land use as a factor in the "smart" antitrust regulation of the agricultural market. Retrieved from https://iscconf.ru/экономическая-отдача-от-землепользо/ Accessed 6 July 2021

The Economist Intelligence Unit Limited. (2021). Global Food Security Index 2020. Retrieved from https://foodsecurityindex.eiu.com/index Accessed 6 July 2021

The World Bank. (2021). Alternative and nuclear energy (% of total energy use). Retrieved from https://data.worldbank.org/indicator/EG.USE.COMM.CL.ZS?view=chart Accessed 6 July 2021

Fiscal Policy, Control, and Supervision to Ensure Food and Energy Security and Sustainable Development

Irina B. Mironova, Lilia R. Moiseeva, Elmira F. Nigmatulina, and Marina G. Mirgorodskaya

Abstract The paper aims to identify prospects and develop recommendations for improving fiscal policy, control, and supervision to ensure food and energy security and sustainable development. The authors apply the regression analysis method to prove that fiscal policy, control, and supervision can and should be carried out in the interests of food and energy security and sustainable development. To maximize food security (90.18 points), energy security (100 points), and sustainable development (87.38 points), the authors find a Pareto optimal. It recommends improving fiscal policy and supervision as follows: increasing insolvency resolution by 59.62%, increasing contract enforcement by 65.02%, improving taxation by 37.27%, and increasing protection of minority investors by 49.33%. The recommendations developed are derived from a representative sample of countries and are universal. The proposed recommendations open up the possibility of a new approach to food and energy security and sustainable development based on creating a favorable environment for business through fiscal policy, control, and supervision of the government. The advantages of the developed author's approach are, first, the systemic provision of food and energy security and sustainable development while obtaining synergies in the form of simultaneous maximization of their results. Second, the involvement of business in ensuring food and energy security and sustainable development while reducing government involvement and, particularly, funding.

Keywords Fiscal policy · Control and supervisory · Food security · Energy security · Sustainable development

JEL codes E62 · M21 · Q01 · Q18 · Q48

I. B. Mironova (✉) · L. R. Moiseeva · E. F. Nigmatulina · M. G. Mirgorodskaya
Volga State University of Technology, Yoshkar-Ola, Russia
e-mail: kbua@inbox.ru

E. G. Popkova and B. S. Sergi (eds.), *Geo-Economy of the Future*,
https://doi.org/10.1007/978-3-030-92303-7_92

1 Introduction

The issues of food and energy security and sustainable development have become particularly urgent against the backdrop of the changed context connected with the pandemic, the COVID-19 crisis (lack of funding), and the worsening of climate change (deteriorating conditions for agriculture and increased energy demand). The problem is that the current approach to resolving these issues has two significant shortcomings.

The first shortcoming is that food security, energy security, and sustainable development are considered and pursued separately. Therefore, the initiatives to achieve them are fragmented and lead to fragmented results. The second shortcoming is that the main role in ensuring food security, energy security, and sustainable development is given to the government with business being aloof, while these issues cannot be solved without the support of the business community.

To address this problem, this paper proposes a new approach to food and energy security and sustainable development based on fiscal policy and government control and supervision. The authors hypothesize that by creating a favorable environment for business, the government can engage and stimulate the participation of businesses in the systemic provision of food and energy security and sustainable development. The logical reasoning behind the hypothesis is that massive public and business support in lockdown (2020) has markedly reduced the economy's energy intensity, decreased carbon emissions, and improved the climate.

The paper aims to identify perspectives and develop recommendations to improve fiscal policy and control and supervision in the interests of food and energy security and sustainable development.

2 Literature Review

Fiscal policy, control, and supervision as components of government regulation of the economy and factors in the formation of the business environment (conditions for doing business) are considered by Gao et al. (2021), Götze and Hartmann (2021), Jiang and Cheng (2021), and Spyrakis and Kotsios (2021). The interconnectedness of food security, energy security, and sustainable development is substantiated and emphasized by Campiotti al. (2016), Ibitoye et al. (2021), Morozova et al. (2019), Popkova and Sergi (2020), Popkova et al. (2019), and Sergi et al. (2019).

The literature review shows that the experience and prospects of using the mechanisms of fiscal policy, control, and supervision in the interests of food and energy security and sustainable development are studied insufficiently (which is a gap in the system of available scientific knowledge). This research is devoted to its in-depth elaboration (filling the gap).

3 Research Methodology

To test the hypothesis, the authors have chosen one of the most reliable and accurate methods of economic statistics—the method of regression analysis. Using this method, the authors determine the contribution of fiscal policy, control, and supervision (World Bank, 2021) to the provision of food security (The Economist Intelligence Unit Limited, 2021), energy security (World Energy Council, 2021), and sustainable development (UNDP, 2021).

To obtain reliable data for the world economy, the authors form a representative sample, including countries from different parts of the world and with different levels of income (World Bank, 2021). Table 1 presents the statistics on the implementation of fiscal policy, control, and supervision in 2020.

Table 2 presents the statistics on food and energy security and sustainable development in 2020.

To determine the prospects for improving fiscal policy, control, and supervision in the interests of food and energy security and sustainable development, the authors search for a Pareto optimal based on the obtained regression models by the least-squares method. In the Pareto optimal, all three dependent variables (y_1–y_3) can be maximized by changing the factor variables (x_1–x_5).

Table 1 Statistics on the implementation of fiscal policy, control, and supervision in 2020, points 1–100

Level of income	Country	Business registration	Protection of minority investors	Taxation	Contract security	Resolution of insolvency
		x_1	x_2	x_3	x_4	x_5
High	New Zealand	100.0	86.0	91.0	71.5	69.5
	Singapore	98.2	86.0	91.6	84.5	74.3
	USA	91.6	71.6	86.8	73.4	90.5
Above average	Thailand	92.4	86.0	77.7	67.9	76.8
	Russia	93.1	60.0	80.5	72.2	59.1
	China	94.1	72.0	70.1	80.9	62.1
Below average	India	81.6	80.0	67.6	41.2	62.0
	Indonesia	81.2	70.0	75.8	49.1	68.1
	Philippines	71.3	60.0	72.6	46.0	55.1
Low	Niger	91.5	42.0	49.4	54.7	39.3
	Nepal	81.7	58.0	47.1	46.0	47.2
	Mozambique	69.3	32.0	64.0	39.8	47.8

Source Compiled by the authors based on (World Bank, 2021)

Table 2 Statistics on food and energy security and sustainable development in 2020, points 1–100

Level of income	Country	Global food security index	Energy trilemma index	Sustainable development index
		y_1	y_2	y_3
High	New Zealand	77.0	79.5	79.13
	Singapore	75.7	70.5	69.89
	USA	77.5	79.8	76.01
Above average	Thailand	64.0	65.2	74.19
	Russia	73.7	73.8	73.75
	China	69.3	67.0	72.06
Below average	India	56.2	56.2	60.07
	Indonesia	59.5	66.8	66.34
	Philippines	55.7	60.3	64.51
Low	Niger	47.6	28.8	49.53
	Nepal	53.0	43.0	66.52
	Mozambique	40.6	40.9	51.50

Source Compiled by the authors based on (The Economist Intelligence Unit Limited, 2021; UNDP, 2021; World Energy Council, 2021)

4 Findings

The contribution of fiscal policy, control, and supervision to food and energy security and sustainable development is reflected by the following regression models obtained based on statistics from Tables 1 and 2:

$$y_1 = -6.40 + 0.23x_1 + 0.10x_2 + 0.26x_3 + 0.27x_4 + 0.11x_5 \tag{1}$$

According to the obtained model, all factors have a positive impact on food security. The reliability of the obtained regression model is evidenced by the high coefficient of multiple correlations (94.92%) and the Fisher test—automatically obtained calculated F is 10.92, and the tabulated F for 12 observations and 5 variables at a significance level of 0.05 is 4.39 (the calculated F is greater than the tabulated F: 10.92 > 4.39).

$$y_2 = 1.09 - 0.28x_1 + 0.19x_2 + 0.63x_3 + 0.23x_4 + 0.19x_5 \tag{2}$$

According to the obtained model, all factors (except business registration) positively impact energy security (x_1). The reliability of the obtained regression model is evidenced by the high coefficient of multiple correlations (93.64%) and the Fisher test—automatically obtained calculated F was 8.54, and the tabulated F for 12 observations and 5 variables at a significance level of 0.05 is 4.39 (the calculated F is greater than the tabulated F: 8.54 > 4.39).

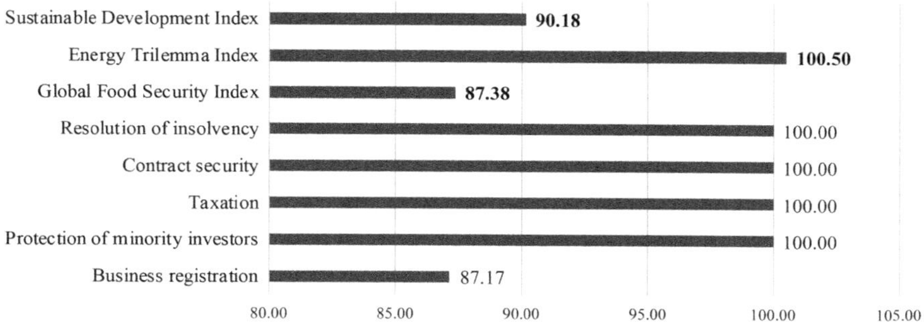

Fig. 1 Pareto optimal, in which all three dependent variables (y_1-y_3) can be maximized by changing the factor variables (x_1–x_5). *Source* Calculated and compiled by the authors

$$y_3 = 37.15 - 0.13x_1 + 0.21x_2 + 0.04x_3 + 0.26x_4 + 0.13x_5 \quad (3)$$

According to the obtained model, all factors (except business registration) positively impact energy security (x_1). The reliability of the obtained regression model is evidenced by the high coefficient of multiple correlations (85.12%) and the Fisher test—automatically obtained calculated *F* is 3.15, and the table *F* for 12 observations and 5 variables at a significance level of 0.10 is 3.11 (the calculated *F* is greater than the tabulated *F*: 3.15 > 3.11).

Using the method of least squares based on Eqs. (1–3), the authors found a Pareto optimal, in which all three dependent variables (y_1-y_3) are maximized by changing the four consistent factor variables (x_2–x_5) (Fig. 1).

According to Fig. 1, the maximum values of the Food Security Index (90.18 points), Energy Trilemma Index (100 points), and Sustainable Development Index (87.38 points) in the found Pareto optimal are achieved in case of the most favorable business climate (100 points) in terms of resolving insolvency, enforcement of contracts, taxation, and protection of minority investors. On this basis, the authors identify the prospects for improving fiscal policy, control, and supervisory to ensure food and energy security and sustainable development (Fig. 2).

According to Fig. 2, the authors propose the following recommendations to improve fiscal policy, control, and supervision in the interests of food security, energy security, and sustainable development:

- Increase the resolution of insolvency by 59.62%;
- Increase contract security by 65.02%;
- Improve taxation by 37.27%;
- Increase the protection of minority investors by 49.33%.

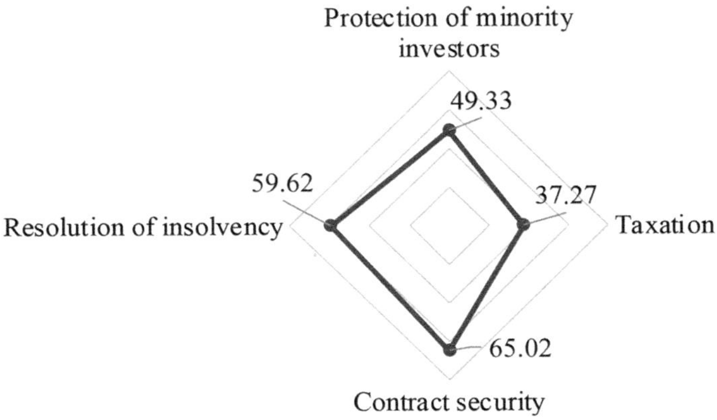

Fig. 2 Prospects for improving fiscal policy, control, and supervision to ensure food and energy security and sustainable development. *Source* Calculated and compiled by the authors

5 Conclusions

Thus, the authors confirmed the hypothesis that fiscal policy, control, and supervision can and should be carried out in the interests of food security, energy security, and sustainable development. To maximize food security (90.18 points), energy security (100 points), and sustainable development (87.38 points), the authors found a Pareto optimal, according to which it is recommended to improve fiscal policy, control, and supervision as follows:

- Increase the resolution of insolvency by 59.62%;
- Increase contract security by 65.02%;
- Improve taxation by 37.27%;
- Increase the protection of minority investors by 49.33%.

The developed recommendations are derived from a representative sample of countries and are universal.

The proposed recommendations open up the possibility of a new approach to food and energy security and sustainable development based on creating a favorable environment for business through fiscal policy, control, and supervision of the government. The advantages of the developed author's approach are, first, the systemic provision of food and energy security and sustainable development while obtaining synergies in the form of simultaneous maximization of their results. Second, the involvement of business in ensuring food and energy security and sustainable development while reducing government involvement and, particularly, funding.

Nevertheless, a limitation of the obtained results is that they do not consider the peculiarities of each particular country. It is suggested that future work be devoted to identifying and addressing these specificities with the help of case studies.

References

Campiotti, C. A., Bibbiani, C., Campiotti, A., Schettini, E., Viola, C., & Vox, G. (2016). Innovative sustainable strategies in agro-food systems and in buildings for energy efficiency. *Rivista Di Studi Sulla Sostenibilita, 2*, 79–96. https://doi.org/10.3280/RISS2016-002008

Gao, Y., Shu, Y., Cao, H., Zhou, S., & Shi, S. (2021). Fiscal policy dilemma in resolving agricultural risks: Evidence from china's agricultural insurance subsidy pilot. *International Journal of Environmental Research and Public Health, 18*(14), 7577. https://doi.org/10.3390/ijerph18147577

Götze, V., & Hartmann, T. (2021). Why municipalities grow: The influence of fiscal incentives on municipal land policies in Germany and the Netherlands. *Land Use Policy, 109*, 105681. https://doi.org/10.1016/j.landusepol.2021.105681

Ibitoye, S. E., Jen, T.-C., Mahamood, R. M., & Akinlabi, E. T. (2021). Densification of agro-residues for sustainable energy generation: An overview. *Bioresources and Bioprocessing, 8*(1), 75. https://doi.org/10.1186/s40643-021-00427-w

Jiang, Q., & Cheng, S. (2021). How the fiscal and monetary policy uncertainty of China respond to global oil price volatility: A multi-regime-on-scale approach. *Resources Policy, 72*, 102121. https://doi.org/10.1016/j.resourpol.2021.102121

Morozova, I. A., Dneprovskaya, I. V., & Kozhukhova, M. T. (2019). Sustainable development of entrepreneurship as the national economic security's mainframe condition. In E. Popkova (ed.), The future of the global financial system: Downfall or harmony (pp. 972–986). Springer. https://doi.org/10.1007/978-3-030-00102-5_104

Popkova, E. G., & Sergi, B. S. (2020). Energy efficiency in leading emerging and developed countries. *Energy, 221*(1), 119730. https://doi.org/10.1016/j.energy.2020.119730

Popkova, E. G., Inshakov, O. V., & Bogoviz, A. V. (2019). Motivation of energy saving within the corporate market responsibility of economic entities. In O. Inshakov, A. Inshakova, & E. Popkova (Eds.), Energy sector: A systemic analysis of economy, foreign trade and legal regulations (pp. 95–105). Springer. https://doi.org/10.1007/978-3-319-90966-0_7

Sergi, B. S., Popkova, E. G., Bogoviz, A. V., & Ragulina, Y. V. (2019). The agro-industrial complex: Tendencies, scenarios, and policies. In B. S. Sergi (Ed.), *Modeling economic growth in contemporary Russia*. Emerald Publishing.

Spyrakis, V., & Kotsios, S. (2021). Public debt dynamics: The interaction with national income and fiscal policy. *Journal of Economic Structures, 10*(1), 8. https://doi.org/10.1186/s40008-021-00238-4

The Economist Intelligence Unit Limited. (2021). Global Food Security Index. Retrieved from https://foodsecurityindex.eiu.com/Index Accessed 7 September 2021

UNDP. (2021). Sustainable Development Report. Retrieved from https://dashboards.sdgindex.org/rankings Accessed 7 September 2021

World Bank. (2021). Doing Business. Retrieved from https://russian.doingbusiness.org/ru/data/doing-business-score?topic=protecting-minority-investors Accessed 7 September 2021

World Energy Council. (2021). Energy Trilemma Index. Retrieved from https://trilemma.worldenergy.org/ Accessed 7 September 2021

Effect of Raw Material Moisture on Quality of Fuel Pellets from Common Reed

Vasilii E. Kostin, Vladimir G. Kochetkov, Natalya A. Sokolova, Aleksey A. Sukhov, Aleksandr I. Belyaev, and Anna M. Pugacheva

Abstract In the steppe regions of the South and South-East of Russia, there are practically no wood species that can be used to produce fuel pellets. At the same time, there are large reserves of non-wood vegetable feedstock that are quite suitable for these purposes. This feedstock can be a dry leaf-stalk mass of ordinary reed, which occupies a large area. The authors carried research to identify such areas and determine the operational reserves of common reed in the territories of the Volgograd and Astrakhan Regions. The technological scheme for obtaining fuel pellets from ordinary reed has been developed, the necessary equipment was selected, and optimal parameters were determined at various stages of production. At the crushed feedstock granulation stage, one of the most important parameters is the humidity of feedstock entering the granulator. The authors experimentally determined the optimal humidity parameters of both directly mown reed biomass and crushed cane chips. It was found that at the low humidity of wood chips, the granulation process does not occur, and at high humidity, the granules are deformed. The optimal moisture content of the feedstock is in the range of 8–12%. In addition to solving the problem of providing fuel pellets, mowing dry cane biomass can prevent landscape reed fires, which are very common in the study area.

Keywords Non-wood plant material · Common reed · Reed leaf-stem biomass · Fuel pellets · Moisture · Landscape fires · Feedstock granulation · Granulator

V. E. Kostin · V. G. Kochetkov · N. A. Sokolova
Volzhsky Polytechnic Institute, Volgograd, Russia
e-mail: vek@post.volpi.ru

N. A. Sokolova
e-mail: natalissa_72@list.ru

A. A. Sukhov (✉) · A. I. Belyaev · A. M. Pugacheva
Federal Scientific Centre of Agroecology, Volgograd, Russia
e-mail: suhov-a@vfanc.ru

A. I. Belyaev
e-mail: director@vfanc.ru

A. M. Pugacheva
e-mail: pugachevaa@vfanc.ru

E. G. Popkova and B. S. Sergi (eds.), *Geo-Economy of the Future*,
https://doi.org/10.1007/978-3-030-92303-7_93

JEL Codes 631.92 · 67.03 · 662.6

1 Introduction

The production of fuel pellets from wood raw materials has become widespread in many countries of the world. In 2018, almost 36 million tons of wood fuel pellets were produced in the world, and according to INFOBIO estimates, by 2028 experts expect this volume to increase 2.5 times (Barmin & Golub, 2000). The leaders in the production of fuel pellets are the United States; Russia is one of the top ten countries in the production of wood fuel pellets. The main production facilities in Russia for the production of wood fuel pellets are concentrated in the North-West and Siberian Federal District, in which logging and timber processing are also developed. Technologies for the production of wood fuel pellets are also quite well developed, especially considering that the production of wood fuel pellets in Russia is export-oriented (92%), which means that the pellets must meet modern quality standards. The European EN plus standard applies to wood fuel pellets and provides for various requirements for granules intended for household use (EN plus- A1 and EN plus- A2 certificate) and "industrial" used in industrial enterprises and utility boiler houses (EN-B certificate).

In work (Barz et al., 2006) it is noted that all indicators describing the process of fuel pellets production can be divided into three groups:

- Input (mode) controlled parameters,
- Unmanaged impacts,
- Output technical and economic indicators.

One of the most significant mode parameters is the moisture content of the feedstock. For wood pellets, the optimum moisture value of the processor chip is in the range of 10–12%. With a raw material humidity of less than 7%, the fuel pellets have low strength, with a raw material humidity of more than 12%, the wood pellets also have low strength and increased looseness, and with a raw material humidity of more than 16% due to internal moisture pressure, the pellets break when leaving the matrix (Information & Analytical Agency "INFOBIO", 2019; ISO, 2014).

Thus, it is apparent that control of the moisture content of the feedstock from the time of harvesting to the pelletizing operation is necessary to obtain quality fuel pellets. So the humidity of freshly cut wood is 50–100%, after long-term storage in air, the humidity decreases to 15–20%. That is, when using wood as a raw material for the manufacture of fuel pellets in the process, drying is necessary, this is one of the most energy-intensive technological operations in the process of producing fuel pellets and significantly affects their cost.

2 Methodology

The technological process of production of fuel pellets is based on the following scheme (Information & Analytical Agency "INFOBIO", 2019; ISO, 2014; Kostin et al., 2019):

- coarse crushing;
- drying;
- fine crushing;
- mixing, water treatment;
- pressing;
- cooling, drying;
- bagging.

However, in addition to wood, another feedstock of plant origin is widely used for the production of fuel pellets. The quality of pellets from non-wood feedstock is regulated by ISO 17225–6:2014 "Solid fuels from biomass (biofuels)—Technical characteristics and fuel classes—Part 6: Sorted non-wood fuel pellets" (Kostin et al., 2012a).

One of the types of such feedstock is the common reed, which is widely distributed throughout the Russian Federation, except for the Polar Regions. In many regions, in particular, the Volgograd and Astrakhan Regions, reeds pose a serious threat due to the systematic occurrence of uncontrolled landscape fires (Kostin et al., 2012b, 2017a).

Carried to determine the operational resources of reed in the Volgograd and Astrakhan Regions studies have shown that only the Northern regions of the Astrakhan Region and the southern regions of the Volgograd Region have the potential for winter harvesting of cane in the amount of more than 70 thousand tons. According to the Federal State Statistics Service of Russia, the fuel pellets production volume in Russia amounted in 2018 to 1,411,108.3 tons, thus, only the potential of the explored region is about 5% of the total wood pellets production in Russia (Kostin et al., 2017b).

In this regard, it is of practical interest to determine the optimal operating parameters for the fuel pellets production from ordinary reed, including the humidity influence assessment.

As shown by field research, the yield of reed winter harvest is on average about 5 t/ha in dry matter, and the feedstock humidity is usually in the range of 5 to 14% (the average according to research results is 8.1%) (Safonov, 2011).

Thus, compared to wood, reed feedstock does not require energy-intensive drying operations.

In the course of research on the influence of feedstock humidity on the quality of fuel pellets made from reed, the main technological operations were reproduced.

Large crushing of the original leaf-stem biomass of the reed was carried out using a knife grinder with a circular knife. The result of the coarse crushing operation is process chips with a particle size of more than 10 mm. The bulk density of reed leaf biomass technological chips does not exceed 100 kg/m^3.

The moisture content of the chips was determined by an analytical method: drying a control sample in a drying oven until a constant weight was established. For these studies, a SNOL 67/350 laboratory low-temperature electric cabinet was used. Analysis of the results of determining the moisture content of the technological chips of cane leaf-stem biomass showed that the residual moisture content in the technological chips meets the standard requirements and does not exceed 15%.

A rotary knife mill PM 120 was used for fine crushing of reed leaf-stem biomass in laboratory conditions. After fine crushing, reed chips have a characteristic linear particle size of no more than 2 mm and are suitable for granulation.

In the process of fine grinding of grinding, some decrease in the moisture content of the feedstock is possible due to heating of the chips particles during grinding, as well as a multiple increase in the contact area of the surface of the chips particles with air, however, it should be noted that finely ground chips are more hygroscopic. Therefore, before granulation, additional moisture control was carried out by two methods: analytical and express method using a portable moisture meter BIO MoistureWood. The analytical method gives higher accuracy, but requires a significant investment of time, as a result of which it is not applicable in industrial production. The express method based on the use of a portable moisture meter BIO MoistureWood takes a few minutes and allows you to determine the moisture content of technological chips immediately before the start of the pelletizing process. Analysis of the moisture measurement results showed that the discrepancy between the values obtained with the instrument and the analytical method did not exceed 5%.

As noted earlier, reed chips have a low residual moisture content and do not require additional drying, but if the moisture content of the chips is less than 8%, then moisture is required to form granules. The required amount of water to moisten the wood chips to optimal conditions was calculated analytically, using the formula:

$$\Delta \mathrm{m_w} = \left(\frac{w}{100} + 1\right) * \mathrm{m_{dm}} - \mathrm{m_{sm}}, \tag{1}$$

where Δm_w—the required amount of water to moisten the chips, w -reed moisture, m_{dm}—dry matter mass, m_{sm}—sample mass.

The granulation operation was performed using a granulator ZLSP120B (Fig. 1).

Pellets and non-granulated reed chips are fed through a tray into a collection container, where a sieve is installed to separate the pellets from the reed chips.

In the process of pelleting, the temperature of the chips being pressed increases significantly, therefore, the pellet at the exit from the die hole has an elevated temperature and has low physical and mechanical characteristics. When the granule is cooled to ambient temperature, its strength significantly increases.

Fig. 1 Granulator ZLSP120B. *Source* Developed and compiled by the authors

3 Results

Research on the influence of the moisture content of reed chips on the possibility of its granulation and the quality of the resulting granules made it possible to draw the following conclusions:

- if the moisture content of the technological chips of leaf-stem biomass of reed is less than 8%, then the agglomeration of chips does not occur, it spills freely through the holes of the matrix, the granule is not formed;
- in the range of moisture content of technological chips of 8–14%, the formation of granules occurs, but a large yield of agglomerated mass of chips is observed, thus, in the specified range of values of moisture content of chips, the granulation process is not effective;
- with an increase in the operating time of the granulator in the range of moisture content of technological chips of 8–14% leads to a significant increase in the temperature of the chips in the gap between the rollers and the matrix, which is accompanied by smoke and threatens to ignite the chips, on the surface of the granules, in this mode, a burnt is formed (Fig. 2a);
- the moisture content of technological chips in the range of 15–20% ensures optimal yield and quality of granules (Fig. 2b);
- an increase in humidity over 20% leads to a sharp deterioration in the quality of the granules, which is caused by an excess of moisture, which, after cooling the granule, leads to an unacceptable increase in internal stresses and destruction of the granule material (Fig. 2c).

Fig. 2 Reed fuel pellets: **a**—pellet with burnt (moisture content 8–14%); **b**—conditioned granule (moisture content 15–20%); **c**—granule with excess moisture (moisture over 20%). *Source* Developed and compiled by the authors

In the course of the research, it was noted that in the process of granulation, a process of significant change in the moisture content of the original technological chips is observed.

The initial moisture content of the chips, determined by the express method, was 19.4%. In the process of granulation, a significant amount of heat is released, which leads to heating of the chips and intensive evaporation of moisture, as a result of which its moisture content decreases. At the outlet of the granulator, the moisture content of the hot granule material is 12.1%. With further cooling of the granule material to the ambient temperature, the humidity decreases to 8.1%. Thus, in the process of granulation, the moisture content of the material decreases by almost 2.4 times from 19.4 to 8.1% (Fig. 3), this should be taken into account when developing an industrial-technological process.

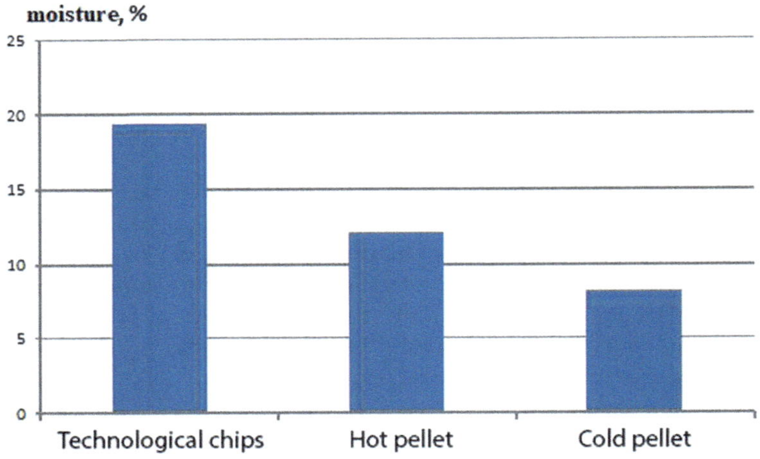

Fig. 3 Changes in the moisture content of the material during pelleting (process chips, hot granule, and cold granule). *Source* Developed and compiled by the authors

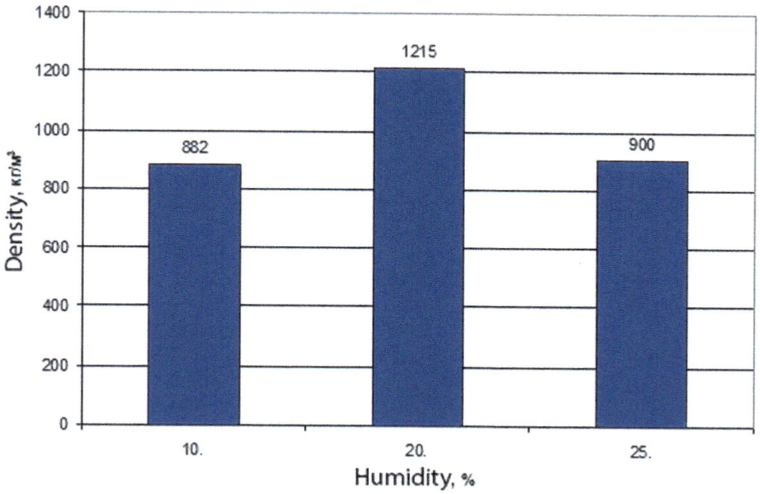

Fig. 4 Dependence of the pellet density on the process chips initial moisture. *Source* Developed and compiled by the authors

The influence of the initial moisture content of technological chips on the density of the granules has been established. For the experiment, we selected granules made from process chips of fine grinding with a moisture content of 10%, 20%, and 25%.

Analysis of the graph of the dependence of the density of the granules on the initial moisture content of the technological chips (Fig. 4) shows that when the moisture content of the chips deviates from the optimal value, the density of the granules, and, consequently, their quality, decreases.

4 Conclusions

In the course of laboratory studies, the influence of the moisture content of the original wood chips on the quality of fuel pellets was established: the optimal value of the moisture content of the initial technological chips of leafy cane biomass after the fine grinding operation is in the range of 15–20%, which significantly differs from the optimal initial moisture content of wood chips, which is 8 -12%.

References

Barmin, A. N., & Golub, V. B. (2000). An instructive lesson in the results of exploitation of reed thickets in the Volga delta. *Proceedings of the Samara Scientific Center of the Russian Academy of Sciences, 2*(2), 295–299.

Barz, M., Wichtmann, W., & Ahlhaus, T. (2006). Energetic utilization of common reed for combined heat and power generation. In *Proceedings of the 2nd International Baltic Bioenergy Conference*, Stralsund, 02–04. November 2006.

Information and Analytical Agency "INFOBIO", http://www.infobio.ru Accessed 11 Dec 2019

ISO 17225-6: (2014) Solid biomass fuels (Biofuels)—Specifications and fuel classes—Part 6: Graded Non-Wood Pellets

Kostin, V. E., Sokolova, N. A., Kablov, V. F., Kochetkov, V. G., & Khlobzheva, I. N. (2019). Features of the technology for the production of fuel pellets from cane, Interaction between enterprises and universities—Science, personnel, new technologies: A collection of articles. In *Report XV Interregional Scientific-Practical Conference, dedicated 65th anniversary of the city of Volgograd* (Volgograd, May 15–16, 2019). VPI (branch) of Volgograd State Technical University, Volzhsky, (pp. 96–98).

Kostin, V. E., Sokolova, N. A., & Parshev, S. S. (2012). Perspective technical and economic proposals for the production of fuel pellets. The interaction of enterprises and universities to improve the efficiency of production, management, and innovation: a collection of articles. In *Report VIII interregional Scientific-Practical Conference*, Volgograd, 17–18 Apr. VPI (branch) Volgograd State Technical University. Volgograd (pp. 148–149).

Kostin, V. E. Sokolova, N. A., & Alexandrova, V. O. (2012). Determination of the operational reserves of southern reed in the region and the ways of its rational use. In *11th Scientific-Practical Conference of the teaching staff of VPI (branch) of Volgograd State Technical University* (Volgograd, January 27–28, 2012). VPI (branch) Volgograd State Technical University. Volgograd, (pp. 162–164).

Kostin, V. E., Sokolova, N. A., & Chernonosova, L. Y. (2017a). Southern reed—For business use farmer. *Volga Region, 2*(55), 66–69.

Kostin, V. E., Sokolova, N. A., & Sukhov, A. A. (2017). Study of the energy potential of commercial cane tracts in the Leninsky district of the Volgograd Region. In *16th Scientific and Practical Conference of the teaching staff of VPI (branch) of Volgograd State Technical University* (Volgograd, January 23–27, 2017). At 2 o'clock, Part 2. Volzhsky Politech Institute (branch) of Volgograd State Technical University. Volgograd (pp. 14–17).

Safonov, A. O. (2011). Automation of control of a pellet press for the production of pellets from wood waste [Electronic resource]. Polythematic network electronic scientific journal of the Kuban State Agrarian University. *Scientific Journal of Kuban State Agrarian University, 74*(10), 1–8. Kuban State Agrarian University http://ej.kubagro.ru/2011/10/pdf/21.pdf (Date of Access 07 Feb 2014)

Zotova, E. V., Safonov, A. O., & Platonov, A. D. (2014). Analytical study of the parameters that determine the technology of wood pellet production. *Forestry Journal, 1.*

Conclusion

A System View of the Sustainable Development of the Geo-Economy of the Future

So, the sustainable development of the geo-economy of the future means, firstly, a balance in nature, in which society and the economy can develop without damage to the environment, that is, green growth. This means refusal from exhaustive and consumer practices of using the Earth's resources and accepting responsibility for the heritage of future generations and the future of nature. A particularly acute problem is the revealed environmental imbalance of economic growth in developed and developing countries. In the geo-economy of the future, environmental costs should be equally distributed among all participants in international relations.

Secondly—the mass character of the practices of responsible production and consumption with increased attention to the issues of environmental quality and ecological efficiency. This requires a revision of the existing concept of the quality of goods and services through the inclusion of environmental characteristics in the basic interpretation of quality. The separate assessment of technical and environmental quality characteristics should be replaced by an integrated assessment in the geo-economics of the future.

Third—the balanced development of urban and rural areas, as well as food security. The prevailing urbanistic philosophy, in which rural areas are viewed as a source of resources for the development of the urban economy, and agriculture is less priority than industry and services, contradicts the vision of the geo-economy of the future. It should pay much more attention to the development of rural areas and equal territorial progress, as well as sectorial diversification of the economy.

This first volume of the book showed that such areas of sustainable development of the geo-economy as green growth, environmental protection, and sustainable agriculture are closely interrelated and must be implemented together. Implementing unidirectional measures can not only increase the imbalance in the SDGs but also harm initiatives in other directions.

E. G. Popkova and B. S. Sergi (eds.), *Geo-Economy of the Future*,
https://doi.org/10.1007/978-3-030-92303-7

The results obtained provided scientific, methodological, and empirical support for the implementation of SDG 3 (in terms of the contribution of the green economy to health), SDG 13 (combating climate change in terms of environmental quality and environmental efficiency), 14 and 15 (biodiversity conservation in terms of green economy and environmental management), as well as SDG 2 (sustainable agriculture and rural development).

Nevertheless, such significant issues and relevant directions for the development of the geo-economy of the future, such as responsible environmental management, sustainable societies, and alternative energy, remained unaffected. The second volume of this book is dedicated to them.

Sustainable Development of Geo-Economy: Look into the Future

Sustainable economic development is possible and accessible not only in the future but also at present—as this book has shown. The new perspective on geo-economy presented in it—from the standpoint of land-use—made it possible, firstly, to clarify the scientific concepts (directions of sustainable development of geo-economy) of the "green" economy and environmental management, environmental quality and environmental efficiency, state regulatory and legal regulation and technologies (under corporate governance) for responsible environmental management, responsible industries, sustainable regions, cities and communities, sustainable agriculture and rural development, and sustainable and alternative energy.

Secondly—to offer highly detailed and ready-to-implement application guidelines and progressive frameworks for sustainable economic development. Thirdly, to form a systematic view (scientific vision), to determine prospects and to propose recommendations for the mass implementation in practice of systemic recommendations on the simultaneous and comprehensive implementation of all directions of sustainable development of geo-economy for its balance and consistency. New scientific results and benefits have determined the contribution of this book to the world literature, as well as its significance:

- The theoretical significance and scientific value lie in the rethinking of the theory of geo-economy from the standpoint of Earth and Planetary Sciences.
- Policy implications consist in the development of detailed applicable recommendations for improving public governance for sustainable development of geo-economy.
- Management implications consist in proposing universal and specific (for individual countries, regions, and industries) recommendations for improving corporate governance by sustainable development of geo-economy.
- The social significance lies in scientific, methodological, and empirical (with the help of practice-oriented recommendations and frameworks) support for the practical implementation of SDG 12 (responsible consumption and production),

SDG 3 (in terms of the contribution of the green economy to health), SDG 13 (combating climate change), 14 and 15 (biodiversity conservation), SDG 16 (rule of law, justice, and strong institutions), SDG 11 (development of responsible industries, sustainable regions, cities and communities), SDG 2 (sustainable agriculture and rural development, as well as SDG 7 (sustainable and alternative energy).

Nevertheless, as scientific knowledge increased, this book revealed and updated two new scientific and practical issues. The first one is that the experience and characteristics of developed and developing countries in the field of sustainable development of geo-economy need more, in-depth, and targeted scientific research to offer appropriate specific recommendations. The question of how serious and surmountable the differences in the sustainable development of the geo-economy of the future in the development and developing countries are remains open.

The second issue is that, as the experience of 2020–2021 has shown, the COVID-19 pandemic and crisis have become a serious challenge for the sustainable development of the geo-economy. The question of what its future challenges are, also remains open. There is a need for scenario analysis, forecasting, development, and adoption of global preventive measures that will provide an effective response to the future challenges of sustainable development of geo-economy.

It is proposed to devote further research of the geo-economy of the future in a new perspective to the scientific study and search for solutions to the noted problems, the foundations of which are laid in this book.

GPSR Compliance

The European Union's (EU) General Product Safety Regulation (GPSR) is a set of rules that requires consumer products to be safe and our obligations to ensure this.

If you have any concerns about our products, you can contact us on ProductSafety@springernature.com

In case Publisher is established outside the EU, the EU authorized representative is:

Springer Nature Customer Service Center GmbH
Europaplatz 3
69115 Heidelberg, Germany

Batch number: 10739612

Printed by Printforce, the Netherlands